Advanced Array Signal Processing for Target Imaging and Detection

Advanced Array Signal Processing for Target Imaging and Detection

Editors

Jiahua Zhu
Xiaotao Huang
Jianguo Liu
Xinbo Li
Gerardo Di Martino
Shengchun Piao
Junyuan Guo
Wei Guo

Basel • Beijing • Wuhan • Barcelona • Belgrade • Novi Sad • Cluj • Manchester

Editors

Jiahua Zhu
National University of
Defense Technology
Changsha
China

Xiaotao Huang
National University of
Defense Technology
Changsha
China

Jianguo Liu
Imperial College London
London
UK

Xinbo Li
Jilin University
Changchun
China

Gerardo Di Martino
University of Naples
Federico II
Napoli
Italy

Shengchun Piao
Harbin Engineering
University
Harbin
China

Junyuan Guo
Harbin Engineering
University
Harbin
China

Wei Guo
National University of
Defense Technology
Changsha
China

Editorial Office
MDPI
St. Alban-Anlage 66
4052 Basel, Switzerland

This is a reprint of articles from the Special Issue published online in the open access journal *Remote Sensing* (ISSN 2072-4292) (available at: https://www.mdpi.com/journal/remotesensing/special_issues/0T467A37IU).

For citation purposes, cite each article independently as indicated on the article page online and as indicated below:

Lastname, A.A.; Lastname, B.B. Article Title. *Journal Name* **Year**, *Volume Number*, Page Range.

ISBN 978-3-7258-1313-1 (Hbk)
ISBN 978-3-7258-1314-8 (PDF)
doi.org/10.3390/books978-3-7258-1314-8

Contents

About the Editors

Jiahua Zhu

Jiahua Zhu (Senior Member, IEEE) received his B.S. degree in electronic engineering and his Ph.D. degree in information and communication engineering from the National University of Defense Technology, Changsha, China, in 2012 and 2018, respectively. He is currently an Associate Professor and the Master's Supervisor at the College of Meteorology and Oceanography, National University of Defense Technology. From November 2015 to November 2017, he was a Visiting Ph.D. Student at the School of Engineering, RMIT University, and the Department of Electrical and Electronic Engineering, University of Melbourne, Australia. His current research interests include waveform design and target detection for radar and sonar. He received the Best Paper Award in The 9th Research Symposium for Chinese Ph.D. Students and Scholars in Australia, 2016; the Excellent Paper Award in the 2021 IEEE/OES China Ocean Acoustics Conference; and the outstanding Ph.D. degree thesis of the Chinese People's Liberation Army (PLA), 2020. He is a Program Co-Chair and Regional Chair of the IEEE International Conference on Signal and Image Processing (ICSIP), an Editorial Board Member of the IEEE Signal Processing Letters and IET Signal Processing, a Topical Advisory Panel Member and the Lead Guest Editor of Remote Sensing, and the Guest Editor of the IEEE Journal of Selected Topics in Applied Earth Observations and Remote Sensing.

Xiaotao Huang

Xiaotao Huang (Senior Member, IEEE) received his B.S. and Ph.D. degrees in information and communications engineering from the National University of Defense Technology, Changsha, China, in 1990 and 1999, respectively. He is currently a Professor at the National University of Defense Technology. His research interests include radar theory, signal processing, and radio frequency signal suppression.

Jianguo Liu

Jianguo Liu (Senior Member, IEEE) received his M.Sc. degree in remote sensing for geology from China University of Geosciences, Beijing, China, in 1982, and his Ph.D. degree in remote sensing and image processing from Imperial College London, U.K., in 1991. He was a Reader in Remote Sensing until 2019 at the Department of Earth Science and Engineering, Imperial College London, and is currently an Emeritus Reader in remote sensing to carry out research in the same department after retirement. He is the principal author of two books on image processing and GIS for remote sensing and he authored or coauthored more than 130 research articles. His current research interests include subpixel technology for precise image registration and feature matching, 3D and motion data generation, illumination-invariant change detection, vision-based UAV navigation, and super-resolution reconstruction; image processing techniques for data fusion, filtering, and InSAR; and GIS multidata modeling for geohazard studies.

Xinbo Li

Xinbo Li received his B.Eng. degree in automation and his M.Sc. and Ph.D. degrees in control theory and control engineering from Jilin University, Changchun, China, in 2002, 2005, and 2009, respectively. From October 2007 to October 2008, he was a Visiting Ph.D. Student at the School of Electrical and Electronic Engineering, Nanyang Technological University (NTU), Singapore. He is currently a Professor at Jilin University. His research interests include piezoelectric drive control technology, array signal processing, and time–frequency analysis.

Gerardo Di Martino

Gerardo Di Martino (Senior Member, IEEE) was born in Naples, Italy, in 1979. He received his Laurea degree (cum laude) in telecommunication engineering and his Ph.D. degree in electronic and telecommunication engineering from the University Federico II, Naples, in 2005 and 2009, respectively. From 2009 to 2016, he was with the University of Naples Federico II, focusing on projects regarding applied electromagnetics and remote sensing topics. From 2014 to 2015, he was with the Italian National Consortium for Telecommunications (CNIT), Naples. In 2016, he joined the Regional Center Information Communication Technology (CeRICT), Naples. He is currently an Associate Professor of electromagnetics at the Department of Electrical Engineering and Information Technology, University of Naples Federico II. His research interests include microwave remote sensing and electromagnetics, with a focus on electromagnetic scattering from natural surfaces and urban areas, synthetic aperture radar (SAR) signal processing and simulation, information retrieval from SAR data, and electromagnetic propagation in urban areas. Prof. Di Martino is the Lead Associate Editor of the IEEE Geoscience and Remote Sensing Society section within IEEE Access. He is also an Associate Editor of the IEEE Journal of Selected Topics on Applied Earth Observations and Remote Sensing, Remote Sensing (MDPI), and Electronics (MDPI).

Shengchun Piao

Shengchun Piao received his B.E. and M.E. degrees from the Harbin Institute of Shipbuilding Engineering, Harbin, China, in 1991 and 1994, respectively, and his Ph.D. degree from Harbin Engineering University, Harbin, China, in 1999, all in underwater acoustic engineering. He is currently a Professor at Harbin Engineering University. His research interests include underwater sound propagation modeling, ocean acoustics, and underwater acoustic signal processing.

Junyuan Guo

Junyuan Guo received her B.E. and Ph.D. degrees in underwater acoustic engineering from Harbin Engineering University, Harbin, China, in 2011 and 2018, respectively. She is currently an Associate Professor at Harbin Engineering University. Dr. Guo was nominated for the Excellent Doctoral Dissertation Award by the Acoustical Society of China, and mainly engages in ocean sound field analysis and acoustic vector signal processing, including very low-frequency acoustic scattering calibration, miniaturized aperture acoustic vector sensor array signal processing, dim target detection, etc.

Wei Guo

Wei Guo received her B.E., M.E., and Ph.D. degrees in underwater acoustic engineering from Harbin Engineering University, Harbin, China, in 2013, 2015, and 2020, respectively. In the same year, as a Research Associate, she continued her research on underwater acoustics at the National University of Defense Technology, Changsha, China. Her major study is underwater acoustic signal processing including target detection and identification.

Preface

In recent decades, considerable progress has been made in the theory and methodology of array signal processing for airborne, ground, marine, and underwater target detection and imaging. However, further development faces increasing challenges regarding improving target illumination performance due to the influences of clutter, interference, and noise. It will be valuable to attain a comprehensive understanding of current array signal processing theory and approaches for detecting various targets in the air, on the land, in the sea, and underwater, and thus to solve future problems that will become relevant from the new application requirements.

This Reprint Book selected 22 papers within the research field of radar/sonar array signal processing and analysis, which would interest readers working on waveform design, detection, and imaging, as well as target recognition, etc. We hope these current works can provide positive reference to the students and staff who read this book.

Jiahua Zhu, Xiaotao Huang, Jianguo Liu, Xinbo Li, Gerardo Di Martino, Shengchun Piao,
Junyuan Guo, and Wei Guo
Editors

 remote sensing

 MDPI

Article

Radar and Jammer Intelligent Game under Jamming Power Dynamic Allocation

Jie Geng [1], Bo Jiu [1,*], Kang Li [1], Yu Zhao [1], Hongwei Liu [1] and Hailin Li [2]

[1]   National Laboratory of Radar Signal Processing, Xidian University, Xi'an 710071, China
[2]   Beijing Institute of Tracking and Telecommunication Technology, Beijing 100094, China
*   Correspondence: bojiu@xidian.edu.cn

Abstract: In modern electronic warfare, the intelligence of the jammer greatly worsens the anti-jamming performance of traditional passive suppression methods. How to actively design anti-jamming strategies to deal with intelligent jammers is crucial to the radar system. In the existing research on radar anti-jamming strategies' design, the assumption of jammers is too ideal. To establish a model that is closer to real electronic warfare, this paper explores the intelligent game between a subpulse-level frequency-agile (FA) radar and a transmit/receive time-sharing jammer under jamming power dynamic allocation. Firstly, the discrete allocation model of jamming power is established, and the multiple-round sequential interaction between the radar and the jammer is described based on an extensive-form game. A detection probability calculation method based on the signal-to-interference-pulse-noise ratio (SINR) accumulation gain criterion (SAGC) is proposed to evaluate the game results. Secondly, considering that the competition between the radar and the jammer has the feature of imperfect information, we utilized neural fictitious self-play (NFSP), an end-to-end deep reinforcement learning (DRL) algorithm, to find the Nash equilibrium (NE) of the game. Finally, the simulation results showed that the game between the radar and the jammer can converge to an approximate NE under the established model. The approximate NE strategies are better than the elementary strategies from the perspective of detection probability. In addition, comparing NFSP and the deep Q-network (DQN) illustrates the effectiveness of NFSP in solving the NE of imperfect information games.

Keywords: electronic warfare; intelligent game; jamming power dynamic allocation; neural fictitious self-play; deep reinforcement learning; Nash equilibrium

Citation: Geng, J.; Jiu, B.; Li, K.; Zhao, Y.; Liu, H.; Li, H. Radar and Jammer Intelligent Game under Jamming Power Dynamic Allocation. *Remote Sens.* **2023**, *15*, 581. https://doi.org/10.3390/rs15030581

Academic Editor: Andrzej Stateczny

Received: 24 November 2022
Revised: 11 January 2023
Accepted: 11 January 2023
Published: 18 January 2023

1. Introduction

In modern electronic warfare, the radar faces great challenges from different advanced jamming types [1]. Among different jamming types, main lobe jamming is especially difficult to deal with because the jammer and the target are close enough and both in the main lobe of the radar antenna [2].

Radar anti-main lobe jamming technologies mainly include passive suppression and active antagonism. The passive suppression methods mean that, after the radar is jammed, it can filter out the jamming signal by finding the separable domain between the target echo and the jamming signal [3–6]. In contrast to the passive suppression methods, active antagonism requires the radar to take measures in advance to avoid being jammed [7]. Common active countermeasures include frequency agility, waveform agility, pulse repetition frequency (PRF) agility, and joint agility [8]. Since the frequency-agile (FA) radar can randomly change the carrier frequency in each transmit pulse, it is difficult for the jammer to intercept and jam the radar, which is considered to be an effective means of anti-main lobe jamming [9,10]. In [11], frequency agility combined with the PRF jittering method for the radar transmit waveform was proposed to resist deception jamming. In [12], the authors proposed a moving target detection algorithm under the background of deception jamming based on FA radar.

The key to FA radar anti-jamming is the frequency-hopping strategy. For the purposes of the electronic counter-countermeasures (ECCM) considered in this paper, the radar needs to take different frequency-agile strategies to deal with different jamming strategies. How to design frequency-agile strategies according to the jammer's actions is of vital importance. For an effective anti-jamming system, the information about the environment and the jammer must be known; otherwise, the judgment of the radar is not credible [13]. Therefore, some researchers have introduced reinforcement learning (RL) algorithms to design anti-jamming strategies for FA radar. In [14], the authors designed a novel frequency-hopping strategy for cognitive radar against the jammer, and the radar does not need to know the operating mode of the jammer. The signal-to-interference-pulse-noise ratio (SINR) as a reward function was used in [14], and the interaction between the radar and the jammer was achieved by two methods, Q-learning and the deep Q-network (DQN), to learn the attack strategy of the jammer to avoid the radar being jammed. In [15], the authors designed an anti-jamming strategy for FA radar against spot jamming based on the DQN approach. Unlike the SINR reward function adopted in [14], Reference [15] used the detection probability as a reward for the radar to learn the optimal anti-jamming strategy. In [16], a radar anti-jamming scheme with the joint agility of the carrier frequency and pulse width was proposed. Different from the anti-jamming strategy design for the pulse-level FA radar in [14,15], Reference [17] studied the anti-jamming strategy for the subpulse-level FA radar, where the carrier frequency of the transmit signal can be changed both within and between pulses. In addition, a policy-gradient-based RL algorithm known as proximal policy optimization (PPO) was adopted in [17] to further improve the anti-jamming performance of the radar.

Currently, most of the research assumes that the jamming strategy is static, which means that the jammer is a dumb jammer who adopts a fixed jamming strategy. However, the jammer can also adaptively learn jamming strategies according to the radar's actions [18,19]. How to model and study intelligent games between the radar and the jammer is of great significance to modern electronic warfare.

The game analysis framework can generally be used to model and deal with multi-agent RL problems [20]. It is feasible to apply game theory to model the relationship between the radar and the jammer. In [21], the competition between the radar with constant false alarm processing and the self-protection jammer was considered based on the static game, and the Nash equilibrium (NE) was studied for different jamming types. In [22], the competition was also modeled by the static game, and the NE strategies could be obtained. In [23,24], a co-located multiple-input multiple-output (MIMO) radar and a smart jammer were considered, and the competition was modeled based on the dynamic game. From the perspective of mutual information, the NE of the radar and the jammer were solved.

Although the jammer is considered as a player, which has the same intelligence level as the radar, the established model is too ideal in the above-mentioned work. For example, the work based on static games cannot characterize the sequence decision-making between the radar and the jammer, and the work based on dynamic games only considers a single-round interaction. In real electronic warfare, the competition between the radar and the jammer is a multiple-round interaction with imperfect information [25]. In addition, with the advancement of jamming technology, the jammer can transmit spot jamming, which aims at multiple frequencies simultaneously [26]. How to establish a more realistic electronic warfare model becomes a preliminary step for designing anti-jamming strategies for the radar.

Therefore, this paper considered a signal model of the jammer as transmitting spot jamming with its central frequency aiming at different frequencies simultaneously, and the jamming power of each frequency can be arbitrarily allocated under the constraint condition. Extensive-form games [27] are proposed to model the relationship of the multiple-round sequence decision-making between the radar and the jammer. Imperfect information was also considered through the characteristics of the partial observation of two-player games. Under this model, the NE strategies of the competition between the radar and the jammer with jamming power dynamic allocation can be investigated. The main contributions of this work are summarized as follows:

- A mathematical model of jamming power discrete allocation is established. Different action spaces of the jammer can be obtained for different quantization steps of power. The smaller the quantization step, the larger the action space of the jammer. When the number of available actions is more, the jammer could find the optimal jamming strategy, and the conclusion is proven by simulation.
- A detection probability calculation method based on the SINR accumulation gain criterion (SAGC) is proposed. After the radar receives a target echo, it judges whether each subpulse is retained or discarded through the SAGC. The specific calculation procedure is that the radar uses the subpulse and the subpulse with the same carrier frequency retained in the past to calculate the coherent integration. If the SINR is improved, the subpulse is retained; otherwise, the subpulse is discarded. At the end of one coherent processing interval (CPI), the coherent integration results obtained from the retained subpulses are used to calculate the detection probability based on the SINR-weighting-based detection (SWD) [17,28] algorithm.
- Extensive simulations were carried out to demonstrate the competition results. Specifically, the training curves of the detection probability of the radar and whether the game between the radar and the jammer can converge to an NE under different quantization steps of power were investigated. The simulation results showed that: (1) the proposed SAGC outperformed another criterion; (2) the game can achieve an approximate NE; if the jammer action space is larger, the game can achieve an NE because the jammer can explore the best action; (3) the approximate NE strategies are better than elementary strategies from the perspective of detection performance.

The remainder of this paper is organized as follows. In Section 2, the signal model of the radar and the jammer is introduced and the jamming power allocation model is proposed. In Section 3, the game elements for the radar and the jammer are designed in detail. In Section 4, the deep reinforcement learning (DRL) and NFSP algorithms are described and the overall confrontation process between the radar and the jammer is given. Section 5 shows the results of the competition between the radar and the jammer under the system model, and Section 6 summarizes the work of this paper.

2. System Model

Consider a game between a subpulse-level FA radar [29] and a jammer. Compared with the pulse-level FA radar, the subpulse-level FA radar can further improve the anti-jamming performance of the radar [17].

2.1. The Signal Model of the Radar

Assume that the radar transmits N pulses in one CPI, which contains K subpulses. The mathematical expression of the nth pulse is

$$
\begin{aligned}
s_{TX}(t,n) &= \sum_{k=0}^{K-1} u(t - nT_r - kT_c) \exp[j2\pi(f_0 + a_{n,k}\Delta f)t] \\
&= \sum_{k=0}^{K-1} \mathrm{rect}(t - nT_r - kT_c) \exp\left[j\pi\gamma(t - nT_r - kT_c)^2\right] \exp[j2\pi(f_0 + a_{n,k}\Delta f)t],
\end{aligned}
\tag{1}
$$

where $u(t) = \mathrm{rect}(t) \exp\left[j\pi\gamma t^2\right]$ is the complex envelope of the signal, T_r and T_c are the pulse repetition interval (PRI) and the subpulse width, respectively, f_0 denotes the initial carrier frequency, and Δf is the step size between two subcarriers. As for $a_{n,k}$, it represents the frequency hopping code; assume that the number of available frequencies for the radar is M, then $a_{n,k} \in \{0, 1, \cdots, M-1\}$. Here, $\mathrm{rect}(t)$ is a rectangular function and is described by

$$\mathrm{rect}(t) = \left\{ \begin{array}{l} 1, 0 < t < T_c \\ 0, \mathrm{otherwise} \end{array} \right. .$$
(2)

Take $K = 4$ and $M = 5$ as an example. The time–frequency diagram of the radar transmit waveform is illustrated in Figure 1.

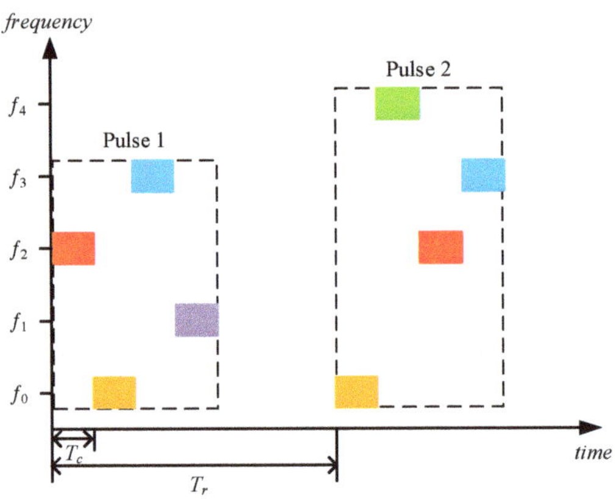

Figure 1. Time–frequency diagram of the subpulse FA radar waveform.

Assume that there is only one target. The nth target echo can be expressed as follows:

$$s_{RX}(t, n) = \sum_{k=0}^{K-1} \sigma_k u(t - nT_r - kT_c - \tau_0) \exp[j2\pi(f_0 + a_{n,k}\Delta f)(t - \tau_0)],$$
(3)

where σ_k is the subpulse echo amplitude corresponding to the carrier frequency of that subpulse and $\tau_0 = 2R/c$ denotes the time delay of the target echo.

2.2. The Signal Model of the Jammer

The jammer considered in this paper is a self-protection jammer that works in a transmit/receive time-sharing mode. Therefore, the jammer cannot receive and transmit signals at the same time. The jamming type is spot jamming. To accurately implement spot jamming, the jammer needs to intercept a portion of the radar waveform and measure its carrier frequency, which is called look-through [30]. After that, the jammer transmits a jamming signal based on the carrier frequency of the intercepted radar waveform. Therefore, the signal model of the jammer consists of two parts: interception and transmission.

For the convenience of analysis, it is assumed that the interception duration T_l of the jammer is an integer multiple of the radar subpulse duration T_c:

$$T_l = ZT_c \ (0 \leq Z \leq K).$$
(4)

Suppose that the delay of the nth pulse transmitted by the radar reaching the jammer is τ'; the interception action of the jammer with respect to this pulse can be expressed as

$$J_{RX}(t, n) = \mathrm{rect}\left[\frac{T_c(t - nT_r - \tau')}{T_l}\right] s_{TX}(t - \tau', n).$$
(5)

If the jammer transmits spot jamming that aims at G frequencies simultaneously, the expression of the jamming signal is

$$J_{TX}(t) = \sum_{j=0}^{G-1} \xi_j \text{rect}\left(\frac{t - nT_r - \tau' - T_l}{KT_c - T_l}T_c\right) \exp(j2\pi f_j t), \tag{6}$$

where f_j is the jth central frequency of the spot jamming and ξ_j is the Gaussian process with variable variance representing the jamming power allocated to the jth frequency.

The time–frequency diagram of the signal transmitted by the jammer is shown in Figure 2. The dashed box indicates that the period is an interception. Different colors mean that the jammer transmits the spot jamming with its central frequency aiming at different frequencies simultaneously in the remaining subpulses.

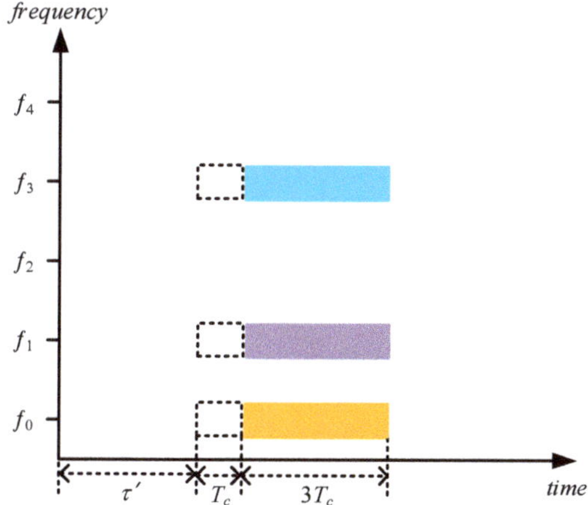

Figure 2. Time–frequency diagram of jammer's transmitted signal.

2.3. The Discrete Allocation Model of Jamming Power

As described in Section 2.2, the jammer transmits the jamming signal, which aims at multiple frequencies simultaneously and allocates the total power reasonably to these frequencies. To simplify the analysis, it was assumed that the total power of the jammer is normalized to 1. Besides, assume that the jammer cannot allocate its power to each frequency arbitrarily, which is restricted by $P_{\min}(0 < P_{\min} \leq 1)$. In other words, $P_{\min}$ is the smallest unit of power allocation. Therefore, the power allocated by the jammer for each frequency is an integer multiple of $P_{\min}$. The smallest unit $P_{\min}$ is defined as the "quantization step". According to the total power and "quantization step", the number of samples of jamming power is

$$N = \frac{1}{P_{\min}}. \tag{7}$$

The number of frequencies available for the radar is M; denote the number of power samples distributed by the jammer to these frequencies as $N_0, N_1, \cdots, N_{M-1}$, then the percentage of power allocated to each frequency is

$$P_i = N_i \times P_{\min}, i \in \{0, 1, \cdots, M-1\}. \tag{8}$$

The allocation model should satisfy the following constraints:

$$\text{s.t.} \begin{cases} 0 \leq N_i \leq N \\ \sum_{i=0}^{M-1} N_i = N \end{cases}. \tag{9}$$

The competition between the radar and the jammer is dynamic, which means they both optimize their strategies to maximize their performance. As for the jammer, its central frequency selection and power allocation strategy is not fixed and can be optimized by interacting with the radar.

3. Game Elements Design for Radar and Jammer

In a complex electronic warfare environment, the confrontation between the radar and the jammer is often multi-round and can be regarded as a sequential decision-making process. The interaction process between the radar and the jammer can be described as follows. The radar transmits the signal, and the jammer makes a decision based on the intercepted partial information of the radar. The radar analyses the behavior of the jammer or the possible jamming strategy based on the interfered echoes and improves the transmitting waveform in the next pulse to achieve the anti-jamming objective.

Each pulse transmitted by the radar corresponds to one competition between the radar and the jammer. At the end of one CPI, the radar will evaluate the anti-jamming performance of the entire process based on the information of all previous pulses. Extensive-form games are a model involving the sequential interaction of multiple agents [31], which can conveniently describe the relationship between the radar and the jammer. The essential elements of the game include actions, information states, and payoff functions.

Generally, the interaction process between the radar and the jammer can be modeled by game theory, in which the radar and the jammer are players in a game.

3.1. Radar Actions

The target of the subpulse-level FA radar is to adopt an appropriate frequency-hopping strategy to deal with interference, and each transmitted pulse is one competition, so the action of the radar is defined as the carrier frequency combination of subpulses. Given the number of subpulses K in one pulse and the number of available carrier frequencies M, the action of the radar can be expressed as $\mathbf{a}_t^r = \left[a_{t,1}^r, \cdots, a_{t,K}^r \right]$, which is a vector with size $1 \times K$. Each element $a_{t,i}^r \in \{0, \cdots, M-1\}$ represents the subcarrier of the ith subpulse of the tth pulse. For example, $a_{t,i}^r = 2$ indicates that the subcarrier is f_2. Based on the number of subpulses and the available frequencies, it can be known that the total number of actions of the radar is $\mathcal{A}_R = M^K$.

3.2. Jammer Actions

The action of the jammer consists of two parts: interception and transmission. To simplify the analysis, assume that the total duration of these two actions is equal to the duration of the radar pulse. According to the number of subpulses, the interception action of the jammer takes any value in set $\{0, 1, \cdots, K\}$, which denotes the number of look-through subpulses. If the value of the interception action is K, then the jammer does not transmit any jamming signal and only executes the look-through operation. The jammer transmits the jamming signal referring to the number of power samples allocated to different frequencies. Based on the number of available frequencies for the radar, $[N_0, \cdots, N_{M-1}]$ can represent this part of the action. The value of N_i is related to the quantization step of the jamming power $P_{\min}$ and should satisfy the allocation model in Section 2.3. Combining the two actions of interception and transmission, the complete action of the jammer is a vector with size $1 \times (M+1)$. It is worth noting that when the quantization step remains unchanged, unless the jammer intercepts all subpulses, the number of the jammer action in the second part is the same. Take $K = 2$, $M = 2$, $P_{\min} = 0.5$ as an example. According to the jamming power allocation model, it can be known that there are three allocation schemes, which is the number of the jammer actions in the second part, as shown in Table 1.

Table 1. Jamming power allocation schemes.

index	1	2	3
scheme	[0, 2]	[2, 0]	[1, 1]

The interception action can be 0, 1, and 2. Only when the interception code is 2, the second part of the jammer action is all 0. Under other codes, the transmission action can be any of the cases in Table 1. Therefore, the total number of jammer actions is $\mathcal{A}_J = 2 \times 3 + 1 = 7$. The complete actions of the jammer are shown in Table 2.

Table 2. The action set of the jammer.

action number	1	2	3	4	5	6	7
action vector	[1, 0, 2]	[1, 2, 0]	[1, 1, 1]	[2, 0, 0]	[0, 0, 2]	[0, 2, 0]	[0, 1, 1]

3.3. Information States

In the competition between the radar and the jammer, the radar decides the action at the next moment according to the behavior of the jammer, and so does the jammer. The information state is defined as the player's actions and partial observations of adversary actions at all historical times. Partial observation makes the player unable to fully obtain the opponent's actions, which reflects the imperfect information of the game. When calculating the information state of the jammer at time t, the radar has executed action $\mathbf{a}_t^r$. Since the action of the jammer always lags behind the radar in timing, the current radar action $\mathbf{a}_t^r$ is not available to the jammer. This also reflects the existence of imperfect information. The information states of the radar and the jammer are given as follows:

$$\mathbf{s}_t^r = \left[\mathbf{a}_0^r, \mathbf{o}_0^j, \cdots, \mathbf{a}_{t-1}^r, \mathbf{o}_{t-1}^j \right], \tag{10}$$

$$\mathbf{s}_t^j = \left[\mathbf{o}_0^r, \mathbf{a}_0^j, \cdots, \mathbf{o}_{t-1}^r, \mathbf{a}_{t-1}^j \right], \tag{11}$$

where $\mathbf{o}_{t-1}^j$ denotes the partial observation of the jammer action by the radar at time $t - 1$. $\mathbf{o}_{t-1}^r$ represents the partial observation of the radar action by the jammer at time $t - 1$.

3.4. Payoff Functions

The payoff function is used to evaluate the value of the agent's policy. After the agent makes an action according to the information state, it will obtain a feedback signal from the environment. The agent judges the value of that action according to the feedback information to guide subsequent learning. Therefore, the agent will formulate a payoff function as the feedback. Through the payoff function, it can achieve the expected objective. Detection probability is an important performance indicator of the radar, which can be used as the feedback for the anti-jamming strategies' design. However, in practical signal processing, the radar calculates the detection probability based on the information of all pulses after one CPI ends. The game between the radar and the jammer is based on a single pulse, so taking the detection probability as a payoff function will bring the problem of a sparse reward. For each echo received by the radar, the SINR of the echo can be calculated. The existence of jamming signals will reduce the SINR. Thus, it is feasible to use the SINR as a reward to guide anti-jamming strategies' learning for the radar and can avoid the sparse reward. The calculation formulas [32] for the signal power and jamming power of the kth subpulse echo are

$$P_{r_k} = \frac{P_T G_R^2 \lambda_k^2 \sigma_k}{(4\pi)^3 R^4}, \tag{12}$$

$$P_{j_k} = \frac{P_J G_R G_J \lambda_k^2}{(4\pi)^2 R^2}, \tag{13}$$

where P_T and G_R are the radar transmission power and antenna gain, respectively, R represents the distance between the radar and the target, λ_k and σ_k are the wavelength and radar cross section (RCS) corresponding to the kth subpulse carrier frequency, and P_J and G_J are the jammer transmission power and antenna gain. Therefore, the mathematical expression for calculating the SINR of the kth subpulse is

$$\mathrm{SINR}_k = \frac{P_{r_k}}{P_N + P_{j_k} \cdot \mathbf{1}(f_k = f_j)}, \tag{14}$$

where P_{r_k} and P_{j_k} are the signal power and jamming power of the kth subpulse echo, respectively; P_N is the system noise power of the radar receiver, and it can be estimated by

$$P_N = \bar{k} T_0 B_n, \tag{15}$$

where $\bar{k} = 1.38 \times 10^{-23}$ J/K is the Boltzmann constant, $T_0 = 290$ K is the effective noise temperature, and B_n is the bandwidth of a subpulse.

In (14), P_{j_k} is the jamming power entering the radar receiver, but it exists only when the central frequency f_j of the jamming signal is equal to the subpulse carrier frequency f_k. Otherwise, it is 0. Therefore, $\mathbf{1}(x)$ can be expressed by

$$\mathbf{1}(x) = \begin{cases} 1, \text{if } x \text{ is true} \\ 0, \text{elsewhere} \end{cases}. \tag{16}$$

Therefore, the payoff function of the radar can be expressed as follows:

$$R_t^r = \sum_{k=0}^{K-1} \mathrm{SINR}_k. \tag{17}$$

Due to the hostile relationship between the radar and the jammer, they can be regarded as a two-player zero-sum (TPZS) game, so the payoff function of the jammer is given as follows:

$$R_t^j = -\sum_{k=0}^{K-1} \mathrm{SINR}_k. \tag{18}$$

3.5. Detection Probability Calculation Method Based on SINR Accumulation Gain Criterion

In Section 3.4, the target echo power, jamming power, and noise power can be estimated. Based on this information, the coherent integration of each carrier frequency is obtained according to the SINR accumulation gain criterion (SAGC). Then, the detection probability is calculated by the SWD algorithm [17,28]. The calculation step of the SAGC is given below:

(1) Let SINR_k^n denote the coherent integration of f_k from n pulses. Here, we take two carrier frequencies f_1, f_2, two subpulses, and one CPI containing four pulses as an example. Therefore, the value of k is 1 and 2, and the value of n is 1 to 4. Let the initial thresholds of the SINR of these two frequencies be T_1 and T_2, respectively.

(2) After the radar receives the first pulse echo, if the carrier frequencies of the two subpulses are $[f_2, f_1]$, the signal power is $[P_{r_2}, P_{r_1}]$, and the noise power is $[P_N, P_N]$, the jamming power of each subpulse is determined as $[P_{j_2}, P_{j_1}]$ based on the central frequency and power allocation schemes of the jamming signal. According to the above information of the first pulse echo, the coherent integration of each frequency can be calculated (since there is only one pulse and the carrier frequency is different, the SINR is calculated directly).

$$\mathrm{SINR}_1^1 = \frac{P_{r_1}}{P_N + P_{j_1}}, \tag{19}$$

$$\mathrm{SINR}_2^1 = \frac{P_{r_2}}{P_N + P_{j_2}}. \tag{20}$$

Judgment: if $\text{SINR}_1^1 > T_1$, retain the subpulse whose carrier frequency is f_1 and update the value of T_1 with SINR_1^1. Otherwise, discard the subpulse whose carrier frequency is f_1, and still use the initial T_1 as the threshold. In the same way, it is determined whether the subpulse whose carrier frequency is f_2 is reserved or discarded. Assume that both subpulses are retained here, then $T_1 = \text{SINR}_1^1$, $T_2 = \text{SINR}_2^1$.

(3) After the radar receives the second echo, if the frequency is $[f_1, f_2]$, the signal power is $[P_{r_1}, P_{r_2}]$, and the noise power is $[P_N, P_N]$, the jamming power of each subpulse is determined as $[P_{j_1}, P_{j_2}]$ according to the jamming signal. Each subpulse is coherently integrated with the same carrier frequency as the subpulse reserved in the first pulse.

Firstly, add the first subpulse to calculate the coherent integration of f_1:

$$\text{SINR}_1^2 = \frac{\left(\sqrt{P_{r_1}} + \sqrt{P_{r_1}}\right)^2}{P_N + P_{j_1} + P_N + P_{j_1}}, \tag{21}$$

and if $\text{SINR}_1^2 > T_1$, reserve the subpulse with carrier frequency f_1 in the second echo and update the value of T_1 with SINR_1^2. Otherwise, discard the subpulse, and do not update the value of T_1.

Next, append the second subpulse to compute the coherent integration of f_2:

$$\text{SINR}_2^2 = \frac{\left(\sqrt{P_{r_2}} + \sqrt{P_{r_2}}\right)^2}{P_N + P_{j_2} + P_N + P_{j_2}}, \tag{22}$$

and if $\text{SINR}_2^2 > T_2$, retain the subpulse with carrier frequency f_2 in the second echo and update the value of T_2 with SINR_2^2. Otherwise, discard the subpulse, and the value of T_2 is not updated.

(4) After receiving the third echo, the radar takes the same operation: adding subpulses in turn to calculate the coherent integration of each frequency and comparing with the thresholds to determine whether to retain the subpulses and update the thresholds. Until the end of one CPI, the obtained SINR is used as the final coherent integration of each frequency.

It is important to note that, although the symbols of the jamming powers of different echoes are the same, their values are different and depend on the specific jamming situation.

SAGC focuses on the impact of a single subpulse on the overall effect, rather than just the subpulse itself. Another advantage of SAGC is that the coherent integration of all frequencies is immediately available as the last pulse is judged.

4. Approximate Nash Equilibrium Solution Based on Neural Fictitious Self-Play

4.1. Deep Reinforcement Learning

RL problems can be described by a Markov decision process (MDP) [33]. At time t, the agent observes the environment state s_t and selects action a_t according to the strategy $\pi(a_t|s_t)$. After acting, the agent will obtain a reward signal r_{t+1} indicating the quality of the action and make the environment enter a new state s_{t+1}. The objective of the agent is to maximize the cumulative reward through continuous interaction with the environment, which is given in (23):

$$\pi^* = \operatorname*{argmax}_{\pi} \mathbb{E}[R_t|\pi], \tag{23}$$

where $R_t = \sum_{k=0}^{\infty} \gamma^k r_{t+1+k}$ is a discounted long-term reward with $\gamma \in [0,1)$ denoting the discount factor.

Value-based and policy-gradient-based methods are two commonly used methods for solving RL problems. The value-based methods need to estimate the state–action value function and then obtain the optimal strategy through the value function. The policy-gradient-based methods calculate the gradient of the objective function on the policy parameters to solve the optimal strategy. The policy-gradient-based methods are usually used to deal with high-dimensional and continuous action space problems. Since the action space of the radar and jammer considered in this paper is discrete and not so large, the value-based method was used to solve the optimal strategy.

The long-term expected reward when starting in a specific state s following the policy π is called the state value function, which is defined as

$$V_\pi(s) = \mathbb{E}_\pi\left[\sum_{k=0}^{\infty} \gamma^k r_{t+1+k}|s_t = s\right]. \tag{24}$$

The state–action value function denotes the long-term expected return after executing action a in state s according to policy π, which is defined as

$$Q_\pi(s,a) = \mathbb{E}_\pi\left[\sum_{k=0}^{\infty} \gamma^k r_{t+1+k}|s_t = s, a_t = a\right]. \tag{25}$$

The relationship between the state value function and the state–action value function is

$$V_\pi(s) = \mathbb{E}_{a\sim\pi(a|s)}[Q_\pi(s,a)]. \tag{26}$$

$Q_\pi(s,a)$ can guide the agent's decision. If the agent adopts the greedy strategy, it chooses the action that maximizes $Q(s,a)$ at each moment. If the agent executes the $\varepsilon - \text{greedy}(Q)$ strategy, it selects the action that maximizes $Q(s,a)$ with probability $1 - \varepsilon$ and randomly chooses an action from the action space with probability ε. The agent follows the $\varepsilon - \text{greedy}(Q)$ policy to balance exploration and exploitation when it acts [33].

$$\varepsilon - \text{greedy}(Q) \leftarrow \begin{cases} \arg\max_{a'} Q(s,a'), \text{with probability } 1 - \varepsilon \\ \text{random}(\mathcal{A}), \quad \text{with probability } \varepsilon \end{cases} \tag{27}$$

Estimates for the optimal action values can be learned using Q-learning [34]. In standard Q-learning, the estimation accuracy is increased by visiting states during the exploration phase and replacing the value of each state–action pair using the Bellman optimality equation:

$$Q(s,a) \leftarrow Q(s,a) + \alpha\left[r + \gamma\max_{a'} Q(s',a') - Q(s,a)\right], \tag{28}$$

where $\alpha \in [0,1)$ is the learning rate.

Deep reinforcement learning (DRL) combines deep neural networks and RL, introduces an approximate representation of the value function, and solves the problem of instability in the learning process based on two key technologies of experience replay and target network [20]. DQN [35] is a typical value-based DRL algorithm, which means it needs to estimate the state–action value function from the samples. The loss function to update the parameters of the neural network of the DQN is given in (29):

$$\mathcal{L}\left(\theta^Q\right) = \mathbb{E}_{\{s,a,r,s'\}\in\mathcal{D}_{RL}}\left\{\left[r + \max_{a'} Q\left(s',a'\middle|\theta^{Q'}\right) - Q\left(s,a\middle|\theta^Q\right)\right]^2\right\}. \tag{29}$$

4.2. Neural Fictitious Self-Play

The confrontation between the radar and the jammer has the characteristics of multiple-round sequential decision-making, which allows us to model their interactions with extensive-form games. Moreover, due to the transmit/receive time-sharing working mode of the jammer, the game has imperfect information. NFSP is an end-to-end DRL algorithm for solving the approximate NE of extensive-form games with imperfect information and does not need any prior knowledge [36]. NFSP includes DRL and supervised learning (SL) when solving strategies, and both of them can only be applied to problems with a discrete action space. Combined with the model established in this paper, NFSP is feasible to solve NE.

NFSP agents learn directly from the experience of interacting with other agents in the game based on DRL. Each NFSP agent contains a memory buffer $\mathcal{D}_{RL}$ that stores the transition experience $\{s_t, a_t, r_{t+1}, s_{t+1}\}$ and a memory buffer $\mathcal{D}_{SL}$ that stores the best response $\{s_t, a_t\}$. NFSP treats these buffers as two separate datasets suitable for deep reinforcement learning and supervised classification, respectively. The agent trains the value network parameters θ^Q from the data in $\mathcal{D}_{RL}$ using an off-policy RL algorithm based on experience replay. The value network defines the agent's best response policy $\varepsilon - \text{greedy}(Q)$. The agent trains a separate neural network to imitate its own past best response behavior using supervised classification data in $\mathcal{D}_{SL}$. This network achieves the mapping of states to action probabilities. Define the action probability distribution of the network output as the agent's historical average policy Π. Based on the above two strategies, the NFSP agent chooses action a_t in state s_t from a mixture of its two policies, and it can be expressed as follows:

$$
\sigma \leftarrow \begin{cases} \varepsilon - \text{greedy}(Q), & \text{with probability } \eta \\ \Pi, & \text{with probability } 1 - \eta \end{cases} ,
\tag{30}
$$

where η is anticipatory parameter. Store $\{s_t, a_t\}$ in $\mathcal{D}_{SL}$ if and only if the agent chooses an action based on $\varepsilon - \text{greedy}(Q)$.

NFSP also utilizes two innovations to ensure that the resulting algorithm is stable and can be simultaneously self-play learning [36]. First, it uses reservoir sampling [37] to avoid the window effect caused by sampling in a finite memory buffer. Second, it uses anticipatory dynamics [38] to enable each agent to sample its own best response behavior and more effectively track changes in the opponent's behavior.

NFSP uses the value-based DQN algorithm to solve the best response strategy. The double-DQN method solves the overestimation problem by separating the selection of the target action and the calculation of the target Q value, and it can find better strategies [39]. The DRL algorithm combined with the dueling network architecture has a dramatic performance improvement [40]. Therefore, this paper adopted the double-DQN method combined with the dueling architecture to solve the best response. The loss functions for updating the parameters of the value network and the supervised network are given in (31) and (32), respectively [36,39].

$$
\mathcal{L}\left(\theta^Q\right) = \mathbb{E}_{\{s,a,r,s'\} \in \mathcal{D}_{RL}} \left\{ \left[r + Q\left(s', \arg\max_{a'} Q\left(s', a' \middle| \theta^Q \right) \middle| \theta^{Q'} \right) - Q\left(s, a \middle| \theta^Q \right) \right]^2 \right\}
\tag{31}
$$

$$
\mathcal{L}\left(\theta^\Pi\right) = \mathbb{E}_{\{s,a\} \in \mathcal{D}_{SL}} \left[-\log \Pi\left(s, a \middle| \theta^\Pi \right) \right]
\tag{32}
$$

4.3. The Complete Game Process between the Radar and Jammer

In the previous section, the confrontation relationship between the radar and the jammer was modeled through extensive-form games, and the NFSP method was given to solve the NE of the game. This subsection presents the complete competition process of the multiple-round interaction between the radar and the jammer, as shown in Algorithm 1.

Algorithm 1 The complete game process between the radar and jammer.

1: Determine the radar action space according to the number of carrier frequencies and the number of subpulses
2: Determine the jammer action space based on the quantization step of jamming power
3: **for** each CPI **do**
4:     Set the policy represented by (30) according to η
5:     The radar observes the initial information state
6:     **for** each pulse **do**
7:         The radar chooses the transmission waveform as an NFSP agent
8:         The jammer determines the number of look-through subpulses and the power allocation scheme as an NFSP agent
9:         The radar receives an echo containing jamming signals
10:         Calculate the SINR payoff according to (14)
11:         The radar and jammer update their respective information states based on observed adversary behavior
12:         Store transition experience $\{s_t, a_t, r_{t+1}, s_{t+1}\}$ in their respective $\mathcal{D}_{RL}$
13:         **if** radar or jammer action a_t is obtained by $\varepsilon - \text{greedy}(Q)$ **then**
14:             Store $\{s_t, a_t\}$ in their respective $\mathcal{D}_{SL}$
15:         **end if**
16:         **for** each subpulse **do**
17:             Judge whether to retain subpulses and update thresholds based on SAGC
18:         **end for**
19:     **end for**
20:     Calculate detection probability based on SWD
21:     Update network parameters θ^Q by (31)
22:     Update network parameters θ^Π by (32)
23:     Update target network parameters $\theta^{Q'}$ after a fixed number of iterations $\theta^{Q'} \leftarrow \theta^Q$
24: **end for**

---

5. Experiments

This section shows the competition results between the radar and the jammer under the jamming power dynamic allocation. The simulation experiments included detection probability training curves, a performance comparison between different quantization steps of jamming power, the verification of the approximate NE, the visualization of approximate NE strategies, etc. The basic simulation parameters are shown in Table 3.

Table 3. Basic simulation parameters.

Parameters	Value
radar transmission power: P_T	30 kW
radar antenna gain: G_R	32 dB
radar initial carrier frequency: f_0	3 GHz
the number of pulses in one CPI: N	8
the number of subpulses in one pulse: K	3
the number of frequencies for the radar: M	3
bandwidth of each subpulse: B_n	5 MHz
time width of each subpulse: T_c	10 μs
range between the radar and the jammer: R	100 km
false alarm rate	10^{-4}
jammer transmission power: P_J	1 W
jammer antenna gain: G_J	0 dB
quantization step of jamming power: $P_{\min}$	0.2
initial thresholds of SAGC	0.5

According to Table 3, $M = 3, K = 3$, so the total number of actions of the radar is $\mathcal{A}_R = 27$. To decorrelate the subpulse echoes of different carrier frequencies, let the frequency step size $\Delta f = 100$ MHz [17]. It was assumed that the RCS of the target does not fluctuate at the same frequency, but the RCS may be different at different frequencies [26]. Without loss of generality, the RCS corresponding to the three carrier frequencies was set to $[15, 3, 1]$ m^2. The number of samples of jamming power is five when $P_{min} = 0.2$. Based on P_{min} and M, it can be known that there are 21 allocation schemes. Combining with K, then the total number of jammer actions is $\mathcal{A}_J = 64$. The radar actions and jammer actions are given in Figure 3.

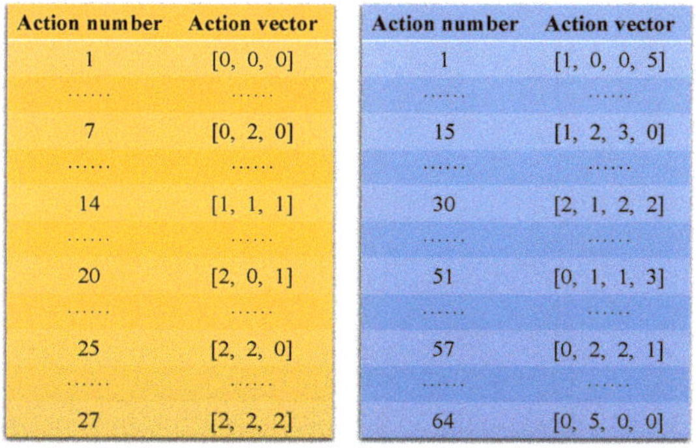

Action number	Action vector
1	[0, 0, 0]
......	......
7	[0, 2, 0]
......	......
14	[1, 1, 1]
......	......
20	[2, 0, 1]
......	......
25	[2, 2, 0]
......	......
27	[2, 2, 2]

(a) radar actions

Action number	Action vector
1	[1, 0, 0, 5]
......	......
15	[1, 2, 3, 0]
......	......
30	[2, 1, 2, 2]
......	......
51	[0, 1, 1, 3]
......	......
57	[0, 2, 2, 1]
......	......
64	[0, 5, 0, 0]

(b) jammer actions

Figure 3. The relationship between action number and action vector for the radar and the jammer.

As described in Section 4.2, this paper used the NFSP algorithm to train the radar and jammer. The NFSP algorithm contains a value network and a supervised network. Multilayer perceptron (MLP) [41] was used to parameterize these two networks in the experiments. The network information for DRL with the dueling architecture and SL is shown in Table 4 and Table 5, respectively.

Table 4. DRL network architecture.

Layer	Input	Output	Activation Function
MLP1	state size	256	LeakyReLU
MLP2	256	256	LeakyReLU
MLP3 of Branch 1	256	128	LeakyReLU
MLP4 of Branch 1	128	1	/
MLP3 of Branch 2	256	128	LeakyReLU
MLP4 of Branch 2	128	action number	/

Table 5. SL network architecture.

Layer	Input	Output	Activation Function
MLP1	state size	256	LeakyReLU
MLP2	256	256	LeakyReLU
MLP3	256	128	LeakyReLU
MLP4	128	action number	Softmax

The learning rates for DRL and SL were set to be 0.001 and 0.0001, respectively. The capacity for DRL memory $\mathcal{D}_{RL}$ and SL memory $\mathcal{D}_{SL}$ was 150,000 and 500,000. The update frequency of the target network parameters in the double-DQN was 4000. The anticipatory parameter η of the mixed strategy was 0.1. The exploration rate of $\varepsilon - \text{greedy}(Q)$ was 0.06 at the beginning and gradually decayed to 0 with the increase of the number of episodes.

5.1. The Training Curve of Detection Probability

Let the game between the radar and the jammer go on for 400,000 episodes. Perform 1000 Monte Carlo adversarial experiments on the resulting policy every 2000 episodes to estimate the detection probability of the radar. The training curve is shown in Figure 4.

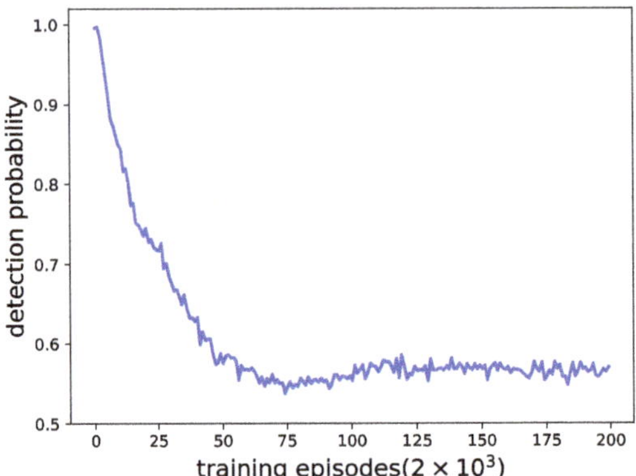

Figure 4. The detection probability curve.

It can be seen from Figure 4 that, as the number of training episodes increases, the detection probability gradually becomes stable and converges to 0.57.

In target detection theory, the detection probability is determined by the threshold and test statistic. If the statistical properties of the noise are known, the value of the threshold can be derived from the false alarm rate in constant false alarm rate (CFAR) detection. Then, the detection probability is determined by the test statistic. It can be known from the SWD algorithm that the SINR after coherent integration of each channel will affect the expression of the test statistic. Therefore, the results of the coherent integration directly affect the detection performance of the radar. Section 3.5 proposes to calculate the coherent integration of each frequency based on SAGC. It is clear from the calculation procedure of SAGC that the key to this criterion is the setting of the initial thresholds of the SINR. To illustrate the influence of the initial thresholds on the detection probability, five initial thresholds were set, as shown in Table 6. The radar and jammer strategies trained when $P_{\min} = 0.2$ were used to perform 1000 Monte Carlo experiments under different thresholds to obtain the variation of the detection probability with the thresholds. Figure 5 presents the result of this experiment.

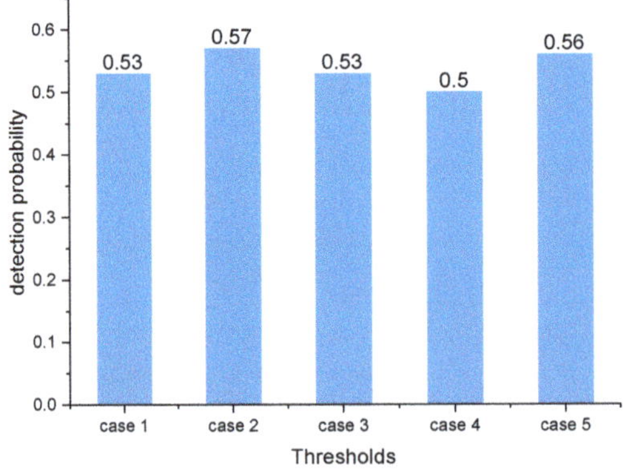

Figure 5. Detection probability of SAGC at different initial thresholds.

Table 6. Different initial thresholds.

Thresholds	T_0	T_1	T_2
Case 1	0	0	0
Case 2	0.5	0.5	0.5
Case 3	1	1	1
Case 4	1	0.5	0
Case 5	0	0.5	1

A coherent integration calculation method based on a fixed threshold criterion (FTC) was also adopted as a comparison. This method also needs to set thresholds. The calculation procedure is to retain the subpulse as long as the SINR is greater than the threshold. At the end of one CPI, the coherent integration for each frequency is calculated using the retained subpulses. Different from SAGC, the thresholds of FTC are unchanged in the whole training process, and the judgment of the current subpulse is only related to its SINR, not to the past retained subpulses. In contrast, the thresholds of SAGC are dynamic, and the judgment of the current subpulse needs to be combined with the past retained subpulses. Figure 6 shows the effect of different fixed thresholds (same as Table 6) on the detection probability under FTC. The experimental approach is to perform 1000 Monte Carlo with the radar and jammer strategies trained when $P_{min} = 0.2$.

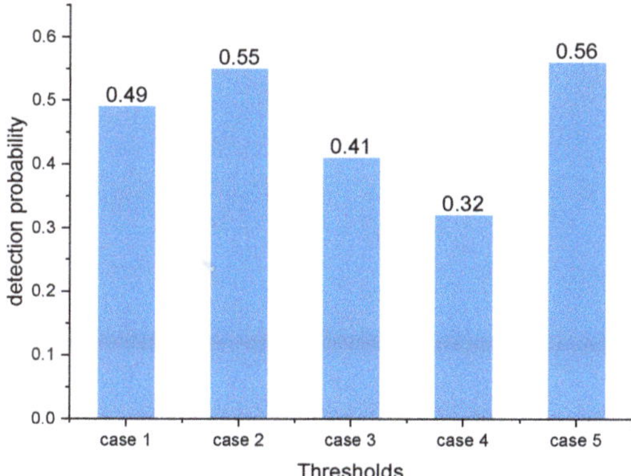

Figure 6. Detection probability of FTC in different fixed thresholds.

Conclusion: According to Figures 5 and 6, SAGC outperformed FTC. The reason for this result is whether to eliminate each subpulse depends not only on its SINR, but also on its contribution to coherent integration in SAGC. However, FTC only considers the subpulses themselves and does not care about the results of the coherent integration of all pulses.

5.2. Performance Comparison between Different Quantization Steps of Jamming Power

This subsection studies the performance comparison between different quantization steps of jamming power. Four quantization steps were set in the experiment: $P_{min} = 1$, $P_{min} = 0.5$, $P_{min} = 0.2$, and $P_{min} = 0.1$. The number of power samples in these four cases was 1, 2, 5, and 10, respectively. Therefore, the number of jammer action spaces corresponding to these situations was $\mathcal{A}_J^1 = 10$, $\mathcal{A}_J^{0.5} = 19$, $\mathcal{A}_J^{0.2} = 64$, and $\mathcal{A}_J^{0.1} = 199$. Figure 7 shows the jammer actions under different quantization steps.

(a) quantization step is 1

Action number	Action vector
1	[1, 0, 0, 1]
2	[1, 0, 1, 0]
3	[1, 1, 0, 0]
4	[2, 0, 0, 1]
5	[2, 0, 1, 0]
6	[2, 1, 0, 0]
7	[3, 0, 0, 0]
8	[0, 0, 0, 1]
9	[0, 0, 1, 0]
10	[0, 1, 0, 0]

(b) quantization step is 0.5

Action number	Action vector
1	[1, 0, 0, 2]
......	......
4	[1, 1, 0, 1]
......	......
9	[2, 0, 2, 0]
......	......
14	[0, 0, 0, 2]
......	......
17	[0, 1, 0, 1]
18	[0, 1, 1, 0]
19	[0, 2, 0, 0]

(c) quantization step is 0.1

Action number	Action vector
1	[1, 0, 0, 10]
......	......
60	[1, 7, 3, 0]
......	......
130	[2, 9, 0, 1]
......	......
165	[0, 3, 6, 1]
......	......
192	[0, 7, 2, 1]
......	......
199	[0, 10, 0, 0]

Figure 7. The relationship between action number and action vector for the jammer under different quantization steps.

The detection probability curves under different quantization steps are shown in Figure 8. From Figure 8, if the quantization step of jamming power is smaller, the detection performance of the radar is worse. However, the total number of jammer actions will increase accordingly, and the convergence speed will become slower. It can also be seen from Figure 8 that, when the quantization step is 0.1 and 0.2, the convergence results of the detection probability are consistent. This shows that the jamming effect of the jammer has performance boundaries.

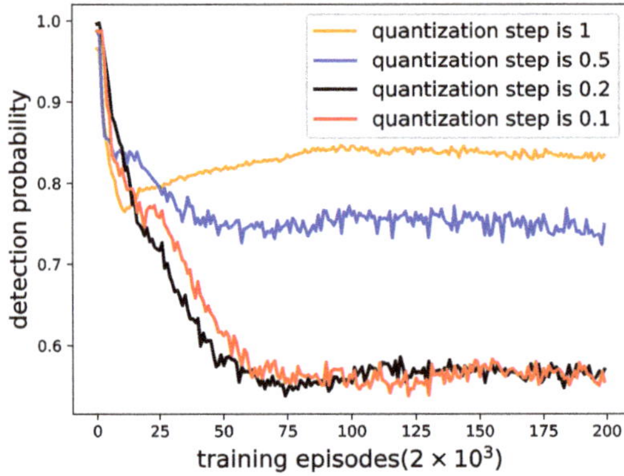

Figure 8. Detection probability curves of different quantization steps.

To verify whether the competition between the radar and the jammer can converge to an NE at the end of the training, the exploitability of the strategy profile needs to be evaluated. Exploitability is a metric that describes how close a strategy profile is to an NE [42–44]. A perfect NE is a strategy profile (σ_1, σ_2) that satisfies the following conditions:

$$\begin{cases} u_1(\sigma_1, \sigma_2) \geq \max u_1(\sigma_1', \sigma_2) \\ u_2(\sigma_1, \sigma_2) \geq \max u_2(\sigma_1, \sigma_2') \end{cases}. \tag{33}$$

An approximate NE or ϵ-NE is a strategy profile that satisfies the following conditions:

$$\begin{cases} u_1(\sigma_1, \sigma_2) + \epsilon \geq \max u_1(\sigma_1', \sigma_2) \\ u_2(\sigma_1, \sigma_2) + \epsilon \geq \max u_2(\sigma_1, \sigma_2') \end{cases}. \tag{34}$$

For a perfect NE, its exploitability is 0. The exploitability of ϵ-NE is ϵ. The closer the exploitability is to 0, the closer the strategy profile is to the NE. The exploitability curves under different quantization steps are shown in Figure 9.

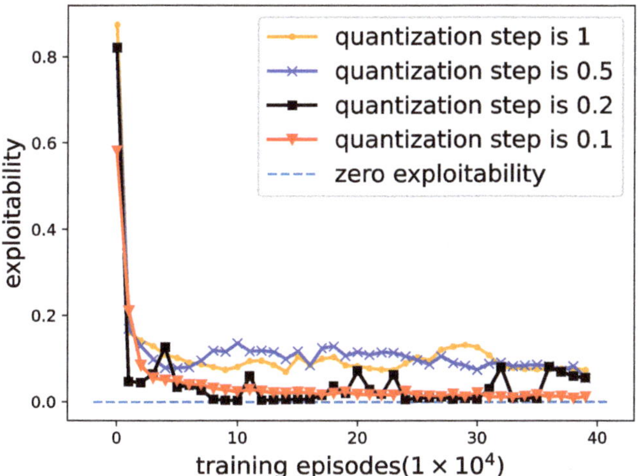

Figure 9. Exploitability curves of different quantization steps.

It can be seen from Figure 9 that, under different quantization steps, the exploitability curves gradually decrease and are close to 0. The exploitability when the quantization step is 0.1 and 0.2 can converge to 0. When the quantization step is 0.5 and 1, the exploitability converges to 0.05 and 0.07, respectively. This shows that the strategy profile of the radar and jammer can achieve an approximate NE under different quantization steps.

Conclusion: If the quantization step of jamming power is smaller, the total number of jammer actions will increase accordingly. Therefore, the jammer could explore the optimal jamming strategy so that the game between the radar and the jammer can achieve a real NE.

5.3. Visualization of Approximate Nash Equilibrium Strategies

Section 5.2 shows that the game between the radar and the jammer can converge to an approximate NE under different quantization steps of jamming power. Therefore, this subsection visualizes the approximate NE strategies. Through Figures 3 and 7, the corresponding relationship between the action number and action vector can be understood. The radar action vector is transformed into frequency, and the jammer action vector is transformed into power percentage for strategy research.

The strategies of the radar and jammer can be expressed in a three-dimensional coordinate system, in which the x-axis represents the action index, the y-axis represents the pulse index, and the z-axis represents the probability. Therefore, the meaning of the coordinates (x, y, z) of any point is that the probability of choosing action x at the yth pulse is z.

In Figures 10–13, (a) and (b) are the X-Y views of their strategies. The X-Y view shows the probability distribution of the radar or jammer's selection action on each pulse. (c) and (d) are the Y-Z views of their strategies. From the Y-Z view, it can be seen that the radar or jammer selects the action with the highest probability on each pulse.

In Figure 10, the radar prefers to select Actions 1 and 14, indicating that the carrier frequency combination of the transmitted signal is $[f_0, f_0, f_0]$ and $[f_1, f_1, f_1]$, respectively. The jammer tends to choose actions 165 and 192, representing that the power ratio allocated to f_0, f_1, and f_2 is $[0.3, 0.6, 0.1]$ and $[0.7, 0.2, 0.1]$. $\text{RCS}(f_0) > \text{RCS}(f_1) > \text{RCS}(f_2)$. The larger the RCS, the stronger the target echo power, so the jammer will allocate more power to reduce the SINR of the radar receiver. Jammer Action 192 allocates the most jamming power to f_0, while there is little difference in jamming power between f_1 and f_2. Thus, the radar should choose f_1 with a larger RCS, corresponding to Radar Action 14. Jammer Action 165 allocates the most jamming power to f_1, so the radar selects f_0 with the largest RCS, corresponding to Radar Action 1.

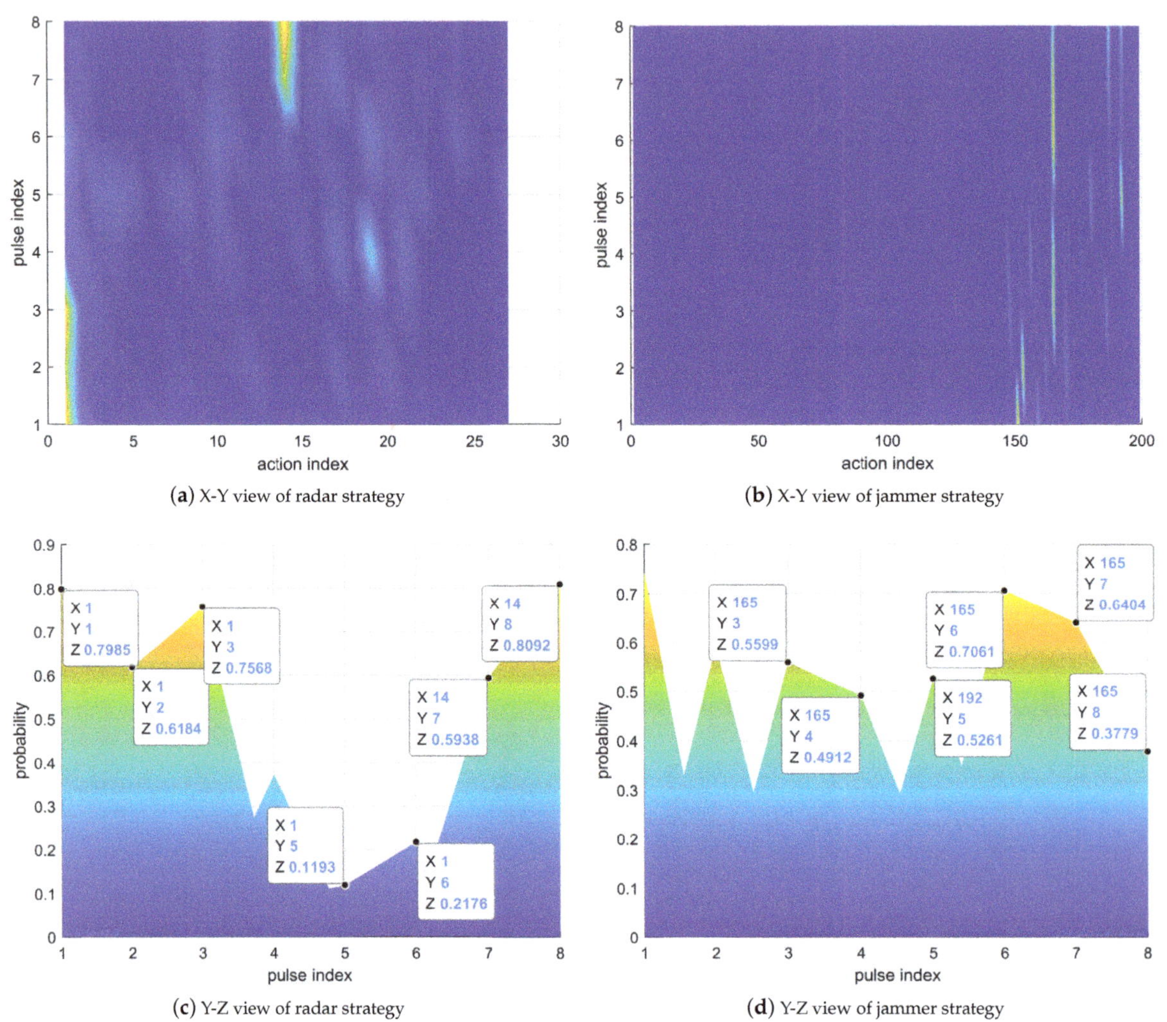

(**a**) X-Y view of radar strategy

(**b**) X-Y view of jammer strategy

(**c**) Y-Z view of radar strategy

(**d**) Y-Z view of jammer strategy

Figure 10. Approximate NE strategies with a quantization step of 0.1.

In Figure 11, the radar selects Action 1 with the highest probability. The jammer tends to select Action 57, indicating that the power allocated to the three frequencies is $[0.4, 0.4, 0.2]$. Although the power allocated by Jammer Action 57 to f_2 is the smallest, the RCS corresponding to f_2 is also the smallest, and the echo power is correspondingly the smallest. The jamming power of f_0 and f_1 is the same, but the RCS of f_0 is the largest. Therefore, the radar selects $[f_0, f_0, f_0]$, that is Action 1, which can ensure the maximum output SINR.

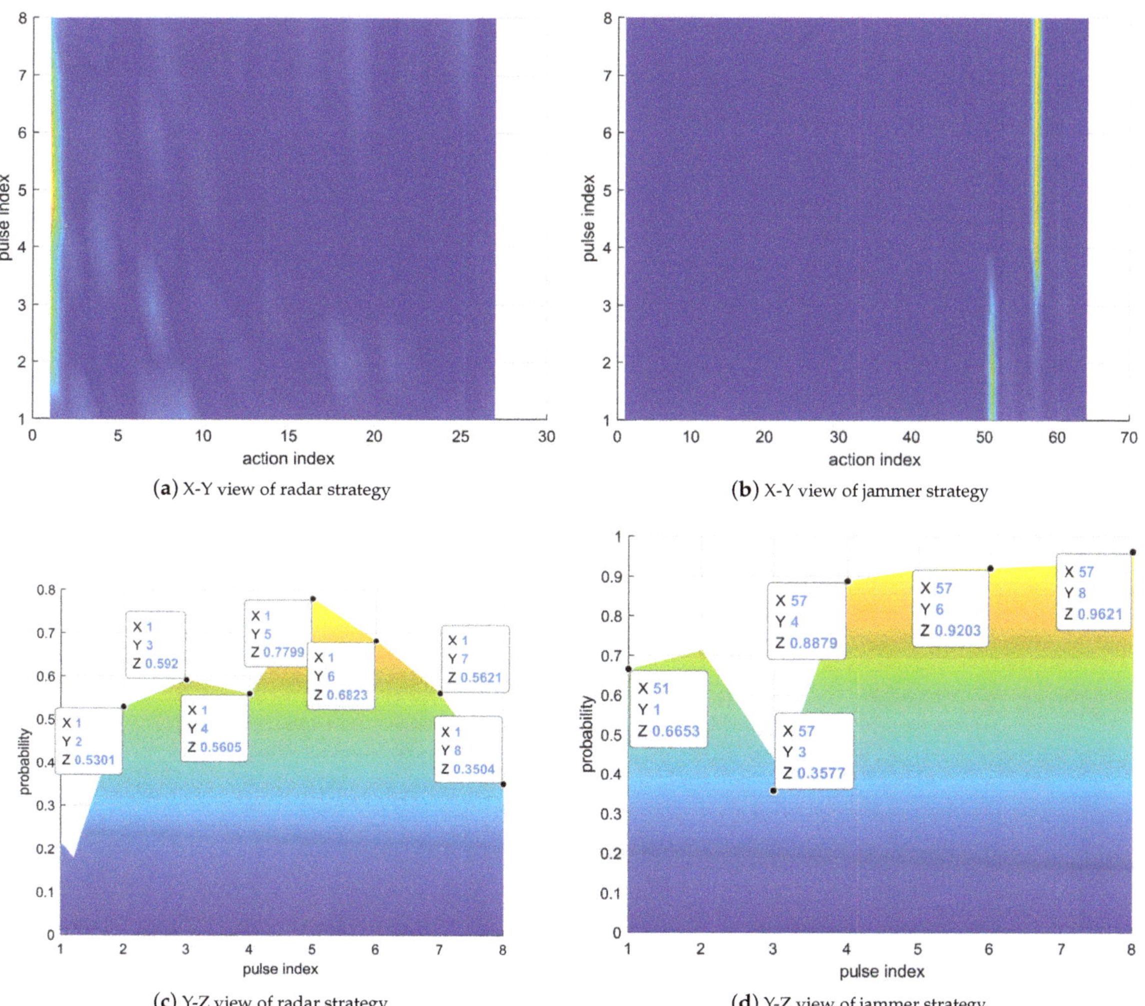

Figure 11. Approximate NE strategies with a quantization step of 0.2.

In Figure 12, the radar selects Action 27, meaning that the combination of the carrier frequency of the transmitted signal is $[f_2, f_2, f_2]$. The jammer selects Action 18, representing the power allocation scheme as $[0.5, 0.5, 0]$. The strategy of the jammer is to evenly distribute the power to the two frequencies with the first- and second-largest RCS. At this time, the radar selection Action 27 can ensure that all subpulses will not be jammed and the radar can obtain a larger SNR.

In Figure 13, the radar selects Action 14, which means the carrier frequency combination of the transmitted signal is $[f_1, f_1, f_1]$. The jammer selects Action 10, representing that the power allocation scheme is $[1, 0, 0]$, that is all the power is allocated to f_0 with the largest RCS. In this case, the quantization step of power is 1, so the jammer can only use all the jamming power to jam one frequency. At this time, the subcarrier of the subpulse of the radar is all f_1, which is the frequency of the second-largest RCS. In this way, it can not only avoid being jammed, but also ensure a large output SNR.

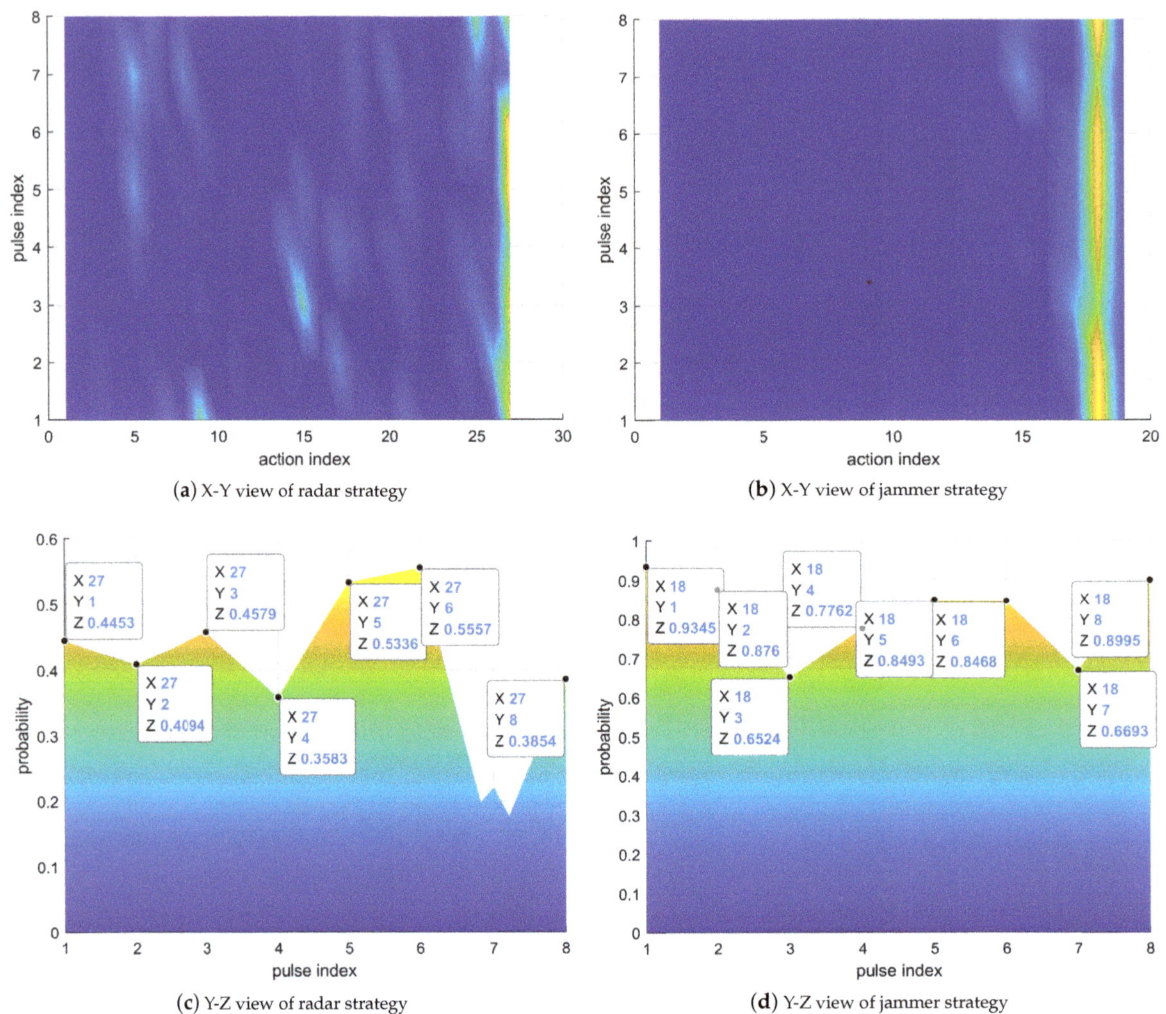

Figure 12. Approximate NE strategies with a quantization step of 0.5.

In these four different scenarios, when the game converges to the NE, the strategy of the jammer is that it does not perform the look-through operation. This shows that, when the jammer is regarded as an agent, it can learn the carrier frequency information of the radar through the interaction with the radar, so it only needs to optimize the power allocation strategy. In real electronic warfare, due to the limited confrontation time, the jammer cannot fully know the available frequencies of the radar, that is the jammer needs to intercept the subpulse of the radar most of the time, which indicates that the strategy of the jammer must deviate from the NE. Therefore, the radar can achieve better performance.

It can also be seen from Figures 10–13 that, no matter what the quantization step of jamming power is, the NE strategies of the radar and the jammer are mixed strategies. The radar and the jammer select actions from their respective action sets with a probability. This is the characteristic of imperfect information games.

Conclusion: Imperfect information games require stochastic strategies to achieve optimal performance [36].

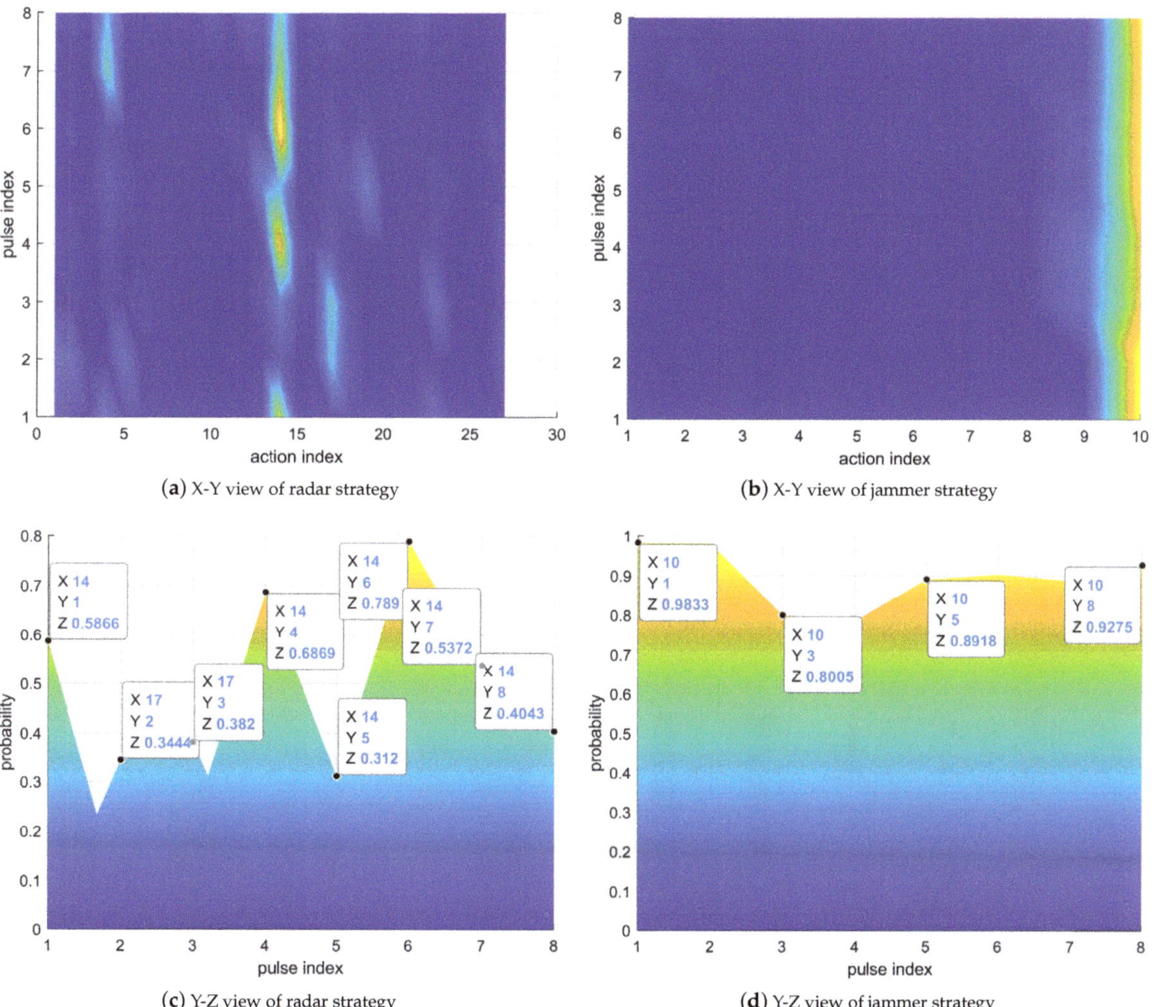

(**a**) X-Y view of radar strategy

(**b**) X-Y view of jammer strategy

(**c**) Y-Z view of radar strategy

(**d**) Y-Z view of jammer strategy

Figure 13. Approximate NE strategies with a quantization step of 1.

5.4. Comparison to Elementary Strategies

This subsection verifies the performance of the approximate NE strategies (ANESs) by comparing them with the elementary strategies.

Assume that the radar can choose two elementary strategies, which are the constant strategy (CS) and the stepped frequency strategy (SFS). The CS means that the carrier frequency of the radar is unchanged. Since the radar has three available frequencies, the CS includes three cases, denoted as CS0, CS1, and CS2. The SFS means that the carrier frequency of the radar increases or decreases step by step between pulses, and these two situations are recorded as SFS-up and SFS-down.

Two elementary strategies for the jammer were considered, which are the constant strategy (CS) and the swept strategy (SS). The CS means that the central frequency of the jamming signal remains unchanged. Similar to the CS of the radar, the CS of the jammer is also denoted as CS0, CS1, and CS2. The SS is similar to the SFS of the radar, and these two situations are recorded as SS-up and SS-down.

We made one side of the radar and jammer adopt the ANES, and the other side adopts the elementary strategies. In addition to the elementary strategies, the radar and the jammer also adopt the ANES as a comparison. The results of 1000 Monte Carlo experiments are shown in Figures 14–17.

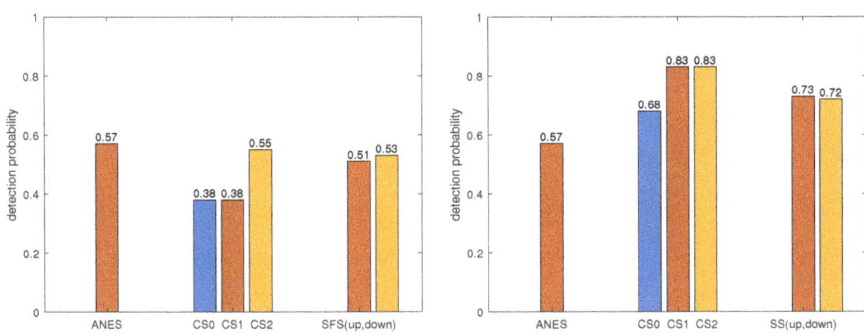

(**a**) The radar with elementary strategies and the jammer with ANES

(**b**) The radar with ANES and the jammer with elementary strategies

Figure 14. The quantization step of jamming power is 0.1.

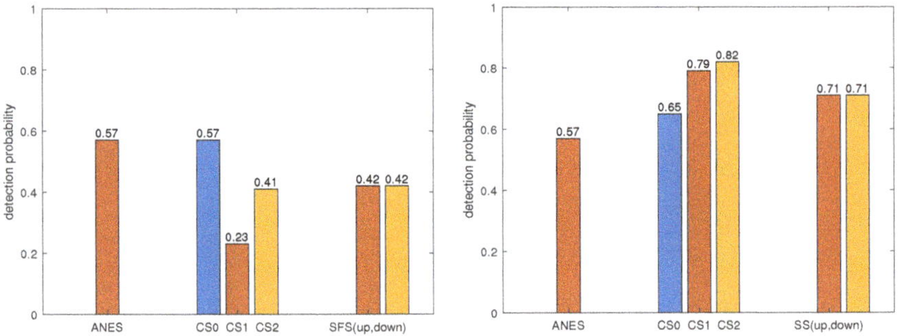

(**a**) The radar with elementary strategies and the jammer with ANES

(**b**) The radar with ANES and the jammer with elementary strategies

Figure 15. The quantization step of jamming power is 0.2.

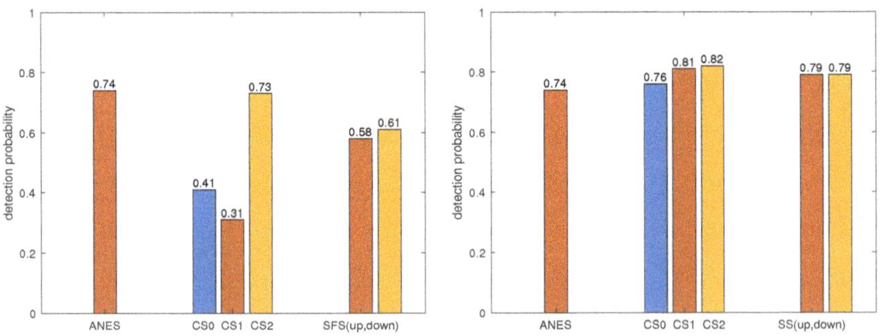

(**a**) The radar with elementary strategies and the jammer with ANES

(**b**) The radar with ANES and the jammer with elementary strategies

Figure 16. The quantization step of jamming power is 0.5.

In Figure 15, the detection probability of the radar adopting CS0 and the ANES is the same because these two strategies are similar in this jamming situation.

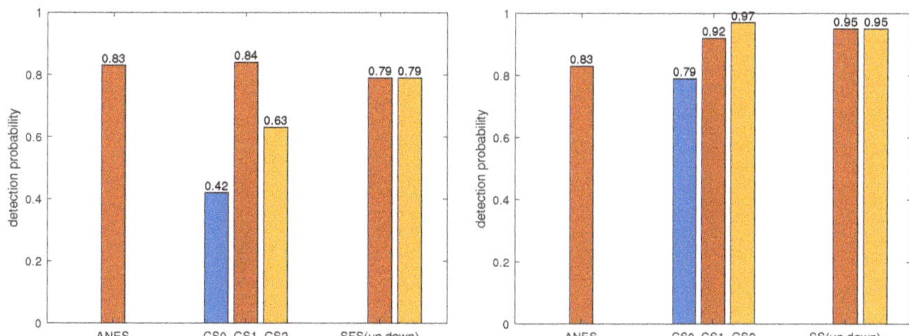

(**a**) The radar with elementary strategies and the jammer with ANES

(**b**) The radar with ANES and the jammer with elementary strategies

Figure 17. The quantization step of jamming power is 1.

Similarly, in Figure 16, since the CS2 and ANES of the radar are the same, there is little difference in their detection performance.

In Figure 17, the ANES of the radar is the same as CS1, and the ANES of the jammer is the same as CS0. Therefore, the performance of one side adopting the ANES and the other taking the elementary strategy is basically the same as that of both adopting the ANES.

From Figures 14–17, the practical implication of the NE can be known, that is, as long as one side deviates from the NE, its performance will decrease. For the jammer, performance degradation refers to an increase in the detection probability of the radar.

Conclusion: The approximate NE strategies obtained in this paper are better than the elementary strategies from the perspective of detection probability.

5.5. Comparison to DQN

This subsection discusses the performance of the DQN in multi-agent imperfect information games. Two forms of the DQN were considered: DQN greedy and DQN average. DQN greedy chooses the action that maximizes the Q value in each state, so it learns a deterministic policy. DQN average draws on the idea of NFSP and also trains the historical average strategy through the supervised learning model, but the average strategy does not affect the agent's decision. Therefore, the agent chooses an action only based on $\varepsilon - \text{greedy}(Q)$ at each moment, not based on a mixed policy. DQN average can be achieved by setting the anticipatory parameter $\eta = 1$ in the NFSP algorithm. Because the NFSP agent in this paper solves the best response by the dueling double-DQN, DQN greedy and DQN average also adopt this method.

In Figure 18, the detection probability and exploitability curves of DQN greedy fluctuate markedly. Its exploitability cannot converge to 0, indicating that DQN greedy cannot achieve NE. Although the training curve of the detection probability of DQN average can be stable, its policy is highly exploitable. DQN average cannot reach an NE either.

Conclusion: DQN greedy learns a deterministic policy. Such strategies are insufficient to behave optimally in multi-agent domains with imperfect information. DQN average learns the best responses to the historical experience generated by other agents, but the experiences are generated only based on $\varepsilon - \text{greedy}$. These experiences are both highly correlated over time and highly focused on a narrow distribution of states [36]. Thus, the DQN average performs worse than NFSP.

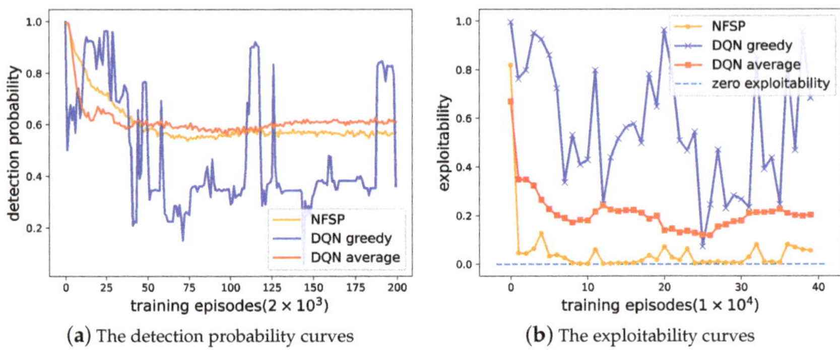

(a) The detection probability curves   (b) The exploitability curves

Figure 18. Comparison of three methods.

5.6. Performance Comparison with Existing Methods

To verify the effectiveness of the strategy obtained in this paper, a comparison between the proposed method and existing resource allocation methods was designed. The work in [17] is the strategy design problem based on RL, so the radar and the jammer interact with one of them as the agent and the other as the environment when applying this method to the established model of this paper. The strategy for the radar and jammer is solved independently rather than based on game theory. The work in [24] was based on the Stackelberg game and concluded that the jamming strategy is related to the target characteristic when the signal power is fixed. The method proposed in [25] was applied to the non-resource allocation scene, and the radar echo was processed by directly eliminating the jammed pulse. In addition to the above-mentioned methods, there is a common and without loss of generality method of allocating all power to the frequency with the second-largest RCS. This allocation strategy was proven by [25] to be feasible. This allocation method is denoted as a constant allocation strategy (CAS). The comparison result is given in Table 7.

Table 7. The comparison between the proposed method and other existing methods.

	This Paper	Method in [17]	Method in [24]	Method in [25]	CAS
detection probability	0.57	0.61	0.65	0.61	0.79
exploitability	0	0.2	0.08	0.07	0.22

In Table 7, in addition to the proposed method in this paper, the exploitability of the other existing allocation methods cannot reach 0. Therefore, only the strategy obtained in this paper is an NE.

6. Conclusions

In this paper, the intelligent game between the subpulse-level FA radar and the self-protection jammer under the jamming power dynamic allocation was investigated. Specifically, the discrete allocation model of jamming power was established and the corresponding relationship between the quantization step of power and the available actions of the jammer was obtained. Furthermore, an extensive-form game model was used to describe the multiple-round sequence decision-making characteristics between the radar and jammer. A detection probability calculation method based on SAGC was proposed to evaluate the competition results. Then, due to the feature of the imperfect information game between the radar and jammer, we utilized NFSP, an end-to-end DRL method, to solve the NE of the game. Finally, simulations verified that the game between the radar and the jammer can converge to the approximate NE under the established model, and the approximate NE strategies are better than the elementary strategies from the perspective of detection probability. The comparison of NFSP and the DQN demonstrated the advantages of NFSP in finding the NE of imperfect information games.

In the future, we should investigate the radar anti-jamming game with the continuous allocation of jamming power, in which the jammer has a continuous action space, and an algorithm to design the strategy for the radar and jammer should also be proposed.

Author Contributions: Conceptualization, J.G.; methodology, J.G. and K.L.; software, J.G.; validation, J.G., K.L. and Y.Z.; formal analysis, J.G.; investigation, B.J., H.L. (Hongwei Liu) and H.L. (Hailin Li); writing—original draft preparation, J.G.; writing—review and editing, B.J., K.L., Y.Z., H.L. (Hongwei Liu) and H.L. (Hailin Li); supervision, B.J.; funding acquisition, B.J., K.L. and H.L. (Hongwei Liu). All authors have read and agreed to the published version of the manuscript.

Funding: This work was supported in part by the National Natural Science Foundation of China under Grant 62201429 and 62192714, the Fund for Foreign Scholars in University Research and Teaching Programs (the 111 project) (No. B18039), the stabilization support of National Radar Signal Processing Laboratory under Grant KGJ202X0X, the Fundamental Research Funds for the Central Universities (QTZX22160).

Data Availability Statement: Not applicable.

Acknowledgments: The authors would like to thank all the Reviewers and Editors for their comments on this paper.

Conflicts of Interest: The authors declare no conflict of interest.

References

1. Ge, M.; Cui, G.; Yu, X.; Kong, L. Mainlobe jamming suppression with polarimetric multi-channel radar via independent component analysis. *Digit. Signal Process.* **2020**, *106*, 102806. [CrossRef]
2. Li, K.; Jiu, B.; Liu, H.; Pu, W. Robust antijamming strategy design for frequency-agile radar against main lobe jamming. *Remote Sens.* **2021**, *13*, 3043. [CrossRef]
3. Su, B.; Wang, Y.; Zhou, L. A mainlobe interference cancelling method. In Proceedings of the 2005 IEEE International Symposium on Microwave, Antenna, Propagation and EMC Technologies for Wireless Communications, Beijing, China, 8–12 August 2005; Volume 1, pp. 23–26.
4. Luo, Z.; Wang, H.; Lv, W.; Tian, H. Mainlobe anti-jamming via eigen-projection processing and covariance matrix reconstruction. *IEICE Trans. Fundam. Electron. Commun. Comput. Sci.* **2017**, *100*, 1055–1059. [CrossRef]
5. Ge, M.; Cui, G.; Yu, X.; Huang, D.; Kong, L. Mainlobe jamming suppression via blind source separation. In Proceedings of the 2018 IEEE Radar Conference (RadarConf18), Oklahoma City, OK, USA, 23–27 April 2018; pp. 0914–0918.
6. Ge, M.; Cui, G.; Yu, X.; Kong, L. Main lobe jamming suppression via blind source separation sparse signal recovery with subarray configuration. *IET Radar Sonar Navig.* **2020**, *14*, 431–438. [CrossRef]
7. Neri, F. *Introduction to Electronic Defense Systems*; SciTech Publishing: Raleigh, NC, USA, 2006.
8. De Martino, A. *Introduction to Modern EW Systems*; Artech House: London, UK, 2018.
9. Zhou, R.; Xia, G.; Zhao, Y.; Liu, H. Coherent signal processing method for frequency-agile radar. In Proceedings of the 2015 12th IEEE International Conference on Electronic Measurement & Instruments (ICEMI), Qingdao, China, 16–18 July 2015; Volume 1, pp. 431–434.
10. Axelsson, S.R. Analysis of random step frequency radar and comparison with experiments. *IEEE Trans. Geosci. Remote. Sens.* **2007**, *45*, 890–904. [CrossRef]
11. Quan, Y.; Wu, Y.; Li, Y.; Sun, G.; Xing, M. Range Doppler reconstruction for frequency agile and PRF-jittering radar. *IET Radar Sonar Navig.* **2018**, *12*, 348–352. [CrossRef]
12. Quan, Y.; Li, Y.; Wu, Y.; Ran, L.; Xing, M.; Liu, M. Moving target detection for frequency agility radar by sparse reconstruction. *Rev. Sci. Instruments* **2016**, *87*, 094703. [CrossRef] [PubMed]
13. Aziz, M.M.; Maud, A.R.M.; Habib, A. Reinforcement Learning Based Techniques for Radar Anti-Jamming. In Proceedings of the 2021 International Bhurban Conference on Applied Sciences and Technologies (IBCAST), Islamabad, Pakistan, 12–16 January 2021; pp. 1021–1025.
14. Li, K.; Jiu, B.; Liu, H.; Liang, S. Reinforcement learning based anti-jamming frequency hopping strategies design for cognitive radar. In Proceedings of the 2018 IEEE International Conference on Signal Processing, Communications and Computing (ICSPCC), Qingdao, China, 14–16 September 2018; pp. 1–5.
15. Li, K.; Jiu, B.; Liu, H. Deep q-network based anti-jamming strategy design for frequency agile radar. In Proceedings of the 2019 International Radar Conference (RADAR), Toulon, France, 23–27 September 2019; pp. 1–5.
16. Ailiya; Yi, W.; Yuan, Y. Reinforcement learning-based joint adaptive frequency hopping and pulse-width allocation for radar anti-jamming. In Proceedings of the 2020 IEEE Radar Conference (RadarConf20), Florence, Italy, 21–25 September 2020; pp. 1–6.

17. Li, K.; Jiu, B.; Wang, P.; Liu, H.; Shi, Y. Radar active antagonism through deep reinforcement learning: A Way to address the challenge of mainlobe jamming. *Signal Process.* **2021**, *186*, 108130. [CrossRef]
18. Wang, L.; Peng, J.; Xie, Z.; Zhang, Y. Optimal jamming frequency selection for cognitive jammer based on reinforcement learning. In Proceedings of the 2019 IEEE 2nd International Conference on Information Communication and Signal Processing (ICICSP), Weihai, China, 28–30 September 2019; pp. 39–43.
19. Liu, H.; Zhang, H.; He, Y.; Sun, Y. Jamming Strategy Optimization through Dual Q-Learning Model against Adaptive Radar. *Sensors* **2021**, *22*, 145. [CrossRef]
20. Dong, H.; Ding, Z.; Zhang, S. *Deep Reinforcement Learning*; Springer: Berlin, Germany, 2020.
21. Bachmann, D.J.; Evans, R.J.; Moran, B. Game theoretic analysis of adaptive radar jamming. *IEEE Trans. Aerosp. Electron. Syst.* **2011**, *47*, 1081–1100. [CrossRef]
22. Norouzi, T.; Norouzi, Y. Scheduling the usage of radar and jammer during peace and war time. *IET Radar Sonar Navig.* **2012**, *6*, 929–936. [CrossRef]
23. Song, X.; Willett, P.; Zhou, S.; Luh, P.B. The MIMO radar and jammer games. *IEEE Trans. Signal Process.* **2012**, *60*, 687–699. [CrossRef]
24. Lan, X.; Li, W.; Wang, X.; Yan, J.; Jiang, M. MIMO radar and target Stackelberg game in the presence of clutter. *IEEE Sens. J.* **2015**, *15*, 6912–6920. [CrossRef]
25. Li, K.; Jiu, B.; Pu, W.; Liu, H.; Peng, X. Neural Fictitious Self-Play for Radar Antijamming Dynamic Game With Imperfect Information. *IEEE Trans. Aerosp. Electron. Syst.* **2022**, *58*, 5533–5547. [CrossRef]
26. Chen, B. *Modern Radar System Analysis And Design*; Publishing House of Xidian University: Xi'an, China, 2012.
27. Myerson, R.B. *Game Theory: Analysis of Conflict*; Harvard University Press: Cambridge, MA, USA, 1997.
28. Liu, H.; Zhou, S.; Su, H.; Yu, Y. Detection performance of spatial-frequency diversity MIMO radar. *IEEE Trans. Aerosp. Electron. Syst.* **2014**, *50*, 3137–3155.
29. Bică, M.; Koivunen, V. Generalized multicarrier radar: Models and performance. *IEEE Trans. Signal Process.* **2016**, *64*, 4389–4402. [CrossRef]
30. Adamy, D. *EW 101: A First Course in Electronic Warfare*; Artech House: Norwood, MA, USA, 2001; Volume 101.
31. Heinrich, J.; Lanctot, M.; Silver, D. Fictitious self-play in extensive-form games. In Proceedings of the International Conference on Machine Learning, Lille, France, 7–9 July 2015; pp. 805–813.
32. Richards, M.A. *Fundamentals of Radar Signal Processing*; McGraw-Hill Education: New York, NY, USA, 2014.
33. Sutton, R.S.; Barto, A.G. *Reinforcement Learning: An Introduction*; MIT Press: Cambridge, MA, USA, 2018.
34. Watkins, C.J.; Dayan, P. Q-learning. *Mach. Learn.* **1992**, *8*, 279–292. [CrossRef]
35. Mnih, V.; Kavukcuoglu, K.; Silver, D.; Rusu, A.A.; Veness, J.; Bellemare, M.G.; Graves, A.; Riedmiller, M.; Fidjeland, A.K.; Ostrovski, G.; et al. Human-level control through deep reinforcement learning. *Nature* **2015**, *518*, 529–533. [CrossRef]
36. Heinrich, J.; Silver, D. Deep reinforcement learning from self-play in imperfect-information games. *arXiv* **2016**, arXiv:1603.01121.
37. Vitter, J.S. Random sampling with a reservoir. *ACM Trans. Math. Softw. (TOMS)* **1985**, *11*, 37–57. [CrossRef]
38. Shamma, J.; Arslan, G. Dynamic fictitious play, dynamic gradient play, and distributed convergence to Nash equilibria. *IEEE Trans. Autom. Control* **2005**, *50*, 312–327. [CrossRef]
39. Van Hasselt, H.; Guez, A.; Silver, D. Deep reinforcement learning with double q-learning. In Proceedings of the AAAI Conference on Artificial Intelligence, Phoenix, AZ, USA, 12–17 February 2016; Volume 30.
40. Wang, Z.; Schaul, T.; Hessel, M.; Hasselt, H.; Lanctot, M.; Freitas, N. Dueling network architectures for deep reinforcement learning. In Proceedings of the International Conference on Machine Learning, New York, NY, USA, 20–22 June 2016; pp. 1995–2003.
41. Goodfellow, I.; Bengio, Y.; Courville, A. *Deep Learning*; MIT Press: Cambridge, MA, USA, 2016.
42. Zinkevich, M.; Johanson, M.; Bowling, M.; Piccione, C. Regret minimization in games with incomplete information. In Proceedings of the Conference on Neural Information Processing Systems (NeurIPS), Vancouver, BC, Canada, 3–6 December 2007.
43. Johanson, M.; Waugh, K.; Bowling, M.; Zinkevich, M. Accelerating best response calculation in large extensive games. In Proceedings of the Twenty-second International Joint Conference on Artificial Intelligence, Barcelona, Spain, 16–22 July 2011.
44. Davis, T.; Burch, N.; Bowling, M. Using response functions to measure strategy strength. In Proceedings of the AAAI Conference on Artificial Intelligence, Ann Arbor, MI, USA, 1–4 June 2014; Volume 28.

Article

High-Resolution SAR Imaging with Azimuth Missing Data Based on Sub-Echo Segmentation and Reconstruction

Nan Jiang [1], Jiahua Zhu [2,*], Dong Feng [1], Zhuang Xie [1], Jian Wang [1] and Xiaotao Huang [1]

[1] College of Electronic Science and Technology, National University of Defense Technology, Changsha 410008, China; jiangnan@nudt.edu.cn (N.J.)
[2] College of Meteorology and Oceanography, National University of Defense Technology, Changsha 410008, China
* Correspondence: zhujiahua1019@hotmail.com

Abstract: Due to the substantial electromagnetic interference, radar interruptions, and other factors, the SAR system may fail to receive valid data in some azimuth areas. This phenomenon is known as Azimuth Missing Data (AMD). If classical SAR imaging algorithms are performed directly using AMD echo, the imaging results may be defocused or even display false targets, which seriously affects the accuracy of the image. Thus, we proposed a Sub-echo Segmentation and Reconstruction Azimuth Missing Data SAR Imaging Algorithm (SSR-AMDIA) to solve the problem of incomplete echo SAR imaging in this article. Instead of using the motion compensation step of the Polar Format algorithm (PFA) to recover the full echo from the AMD echo, the proposed SSR-AMDIA eliminates the effect of the planar approximation in PFA and expands the maximum depth of focus (DOF). The raw AMD echo was first subjected to range compression and Range Cell Migration Correction (RCMC), after which the AMD-RCMC echo was divided along the range direction. Then, we constructed a series of phase compensation functions based on the sub-segment AMD-RCMC echoes to guarantee the perfect recovery of the full RCMC echoes corresponding to the sub-scenes. Finally, we combined them to obtain the complete RCMC echo, and an excellent focused imaging result was then obtained via azimuth compression. Simulation and experimental data verified the effectiveness of the proposed algorithm. Furthermore, we derived the mathematical expressions for the two-dimensional maximum DOFs of the proposed algorithm. In contrast to the State-Of-the-Art (SOA) AMDIA, the SSR-AMDIA can obtain a superior imaging performance in a larger imaging scope under the conditions of most AMD cases.

Keywords: azimuth missing data; maximum depth of focus; SAR imaging scene size; segmentation SAR imaging

Citation: Jiang, N.; Zhu, J.; Feng, D.; Xie, Z.; Wang, J.; Huang, X. High-Resolution SAR Imaging with Azimuth Missing Data Based on Sub-Echo Segmentation and Reconstruction. *Remote Sens.* **2023**, *15*, 2428. https://doi.org/10.3390/rs15092428

Academic Editor: Domenico Velotto

Received: 22 March 2023
Revised: 24 April 2023
Accepted: 3 May 2023
Published: 5 May 2023

1. Introduction

Unavoidable interference between SAR systems and imaging scenes, interruptions in SAR systems for different purposes, or new SAR mission requirements will result in the azimuth missing data (AMD) [1,2]. If the conventional SAR imaging algorithm is used directly in the AMD echo case, false targets or severe defocus will be produced in the final imaging results [3].

In order to overcome the AMD-SAR imaging challenge, an auto-regressive linear prediction approach used initially in the discontinuous aperture SAR imaging [4]. However, it only improves image quality if the Azimuth Missing Ratio (AMR) is below 30%. The equal-gap AMD-SAR imaging problem was solved by P. Stoica and J. Li. They proposed the Gapped-data APES (GAPES) algorithm based on the Amplitude and Phase EStimation (APES) algorithm [5–7]. In order to enhance AMD-SAR imaging performance in random AMD conditions, they took advantage of the Expectation-Maximization algorithm and then further presented the Missing-data APES (MAPES) algorithm [8]. However, its reliability decreases rapidly when the AMR increases, and the computational complexity is relatively expensive. To address this problem under the high AMR conditions, a random Missing-data Iterative Adaptive algorithm (MIAA) was proposed in [9]. The maximum AMR threshold can achieve nearly 80%. Compared with the MAPES, the MIAA's recovery performance is greatly improved when the AMR is higher than 60%. However, due to the fact that the MIAA involves numerous matrix inversions and iterations, its computing cost will be insurmountable for large-scene AMD-SAR imaging.

The partial data SAR imaging problem has been addressed from a new perspective since the Compressed Sensing (CS) technology was proposed [10,11]. Various CS-based sparse SAR imaging algorithms and methods have been proposed and improved in recent years [12–14]. A segmented reconstruction algorithm for the large-scene sparse SAR imaging was proposed in [15]. The whole scene is split into a set of small sub-scenes. With the appropriate increase in the segment number, the reconstruction time and running memory can be greatly reduced. Additionally, ref. [16] proposed an improved method to speed up the sparse SAR imaging and reduce the memory requirement using the Non-Uniform Fast Fourier Transform (NUFFT).It applies interpolation coefficients instead of multiplication of observation matrices and vectors, leading to a smaller computational complexity and memory usage. Furthermore, since the strong scattering points are rebuilt directly, the imaging accuracy will be severely degraded under low echo signal-to-noise ratio (SNR).

For this problem, an Azimuth Missing Data Imaging Algorithm (AMDIA) was proposed in 2018 [17]. It estimates and recovers the full echo of sparse targets from the AMD echo. The CS methods cannot reconstruct the complete SAR echo in the time domain because it is not sparse. Hence, influenced by PFA's motion compensation approach [18,19], Literature [17] discovered that multiplying the dense SAR echo with a Phase Compensation Function (PCF) in the range-frequency domain can yield a sparser signal in the Doppler domain. Next, a phase-compensated complete echo can be recovered from the phase-compensated AMD echo using the CS method. Then, by multiplying the phase-compensated complete echo with the conjugate of the previous PCF, the complete echo can be estimated. Lastly, using the traditional SAR imaging algorithms, the final image can be focused via the estimated full echo. Compared with the sparse SAR imaging algorithms, the AMDIA can obtain an excellent-focused image even at low SNR due to the two-dimensional Matched Filtering process [3]. Its improved algorithms have developed rapidly in these years. K. Liu improved the imaging capabilities of the AMDIA by extending it into the spaceborne FMCW SAR system [20,21]. J. Wu suggested a sparsity adaptive StOMP algorithm for AMD-SAR imaging [22]. It exhibits excellent recovery performance when the prior sparsity is unknown. In 2022, we proposed a Moving Target AMD-SAR Imaging (MTIm-AMD) method based on the AMDIA [23]. Since the motion parameters are considered, the PCF is modified to be more efficient, and hence the moving target can be well-focused in the AMD case. Moreover, we proposed an Enhancement AMDIA (EnAMDIA) to improve the AMD-SAR imaging performance in [24]. The EnAMDIA recovers the RCMC echo instead of the time domain echo. Therefore, it demonstrates a more accurate recovery and a more moderate computational burden.

However, the PCFs of all the above-mentioned AMD-SAR imaging algorithms are designed based on one reference point, which is generally regarded as the scene centroid. Therefore, the phase compensation error for the non-centered targets will increase significantly with the expansion of the imaging scene. Once the imaging scene is larger than the limit of the focusable region, the PCF of the State-Of-the-Art (SOA) AMDIA will result in unsatisfactory sparsity of the phase-compensated signal. Therefore, the estimation accuracy of the complete echo will decrease. It indicates that the imaging performance of SOA-AMDIA is unsatisfactory when the imaging scene is relatively large.

Therefore, to enlarge the maximum focusable region under the AMD conditions, an improved Sub-echo Segmentation and Reconstruction AMDIA (SSR-AMDIA) is proposed in this paper. We consider enhancing the phase-compensated signal's sparsity and then enlarging the imaging scene limits in azimuth and range direction, respectively. First, we apply RCMC processing on the raw AMD echo and then design a new PCF for AMD-RCMC echo. Since the range migration is removed, a sparser signal can be obtained along the azimuth direction. Subsequently, the AMD-RCMC echo is split into a series of AMD-RCMC sub-echoes along the range direction. Each sub-scene's centroid is regarded as the reference point for phase compensation. Instead of designing a PCF for the whole scene, many PCFs are redesigned for each sub-scene. Therefore, each phase-compensated RCMC (PC-RCMC) sub-echo is sparser, which implies that the complete RCMC sub-echoes can be estimated more precisely. Finally, by combining the reconstructed RCMC sub-echoes, a reliable complete RCMC echo is obtained. A superior imaging result of the edge targets can be obtained via azimuth compression.

The main innovations of the article consist in:

1. We first rebuilt the full RCMC echo rather than the full raw echo. The SOA-AMDIA only focuses on reconstructing the full raw echo before range compression and RCMC processing, resulting in an inaccurate reconstruction of azimuth far-field targets. Thus, the proposed algorithm first eliminates the negative effect of range migration on echo recovery. It significantly reduces the azimuth far-field target's residual phase error, and expands the azimuth maximum Depth of Focus (DOF) of the sparse domain signal. Additionally, the computational cost can also be reduced if the range direction's targets are adequately sparse.

2. We first exploited range segmentation to improve the SOA-AMDIA. Instead of using one PCF for the whole imaging scene, we redesigned a series of PCFs for each sub-scene. It ensures the significant reduction of the range far-field target's residual phase error, and the imaging range limits can be eliminated with a reasonable segmentation strategy.

3. We also carried out the mathematical derivation for the two-dimensional maximum DOFs of the proposed algorithm. The advantage of the proposed SSR-AMDIA over SOA-AMDIA for the imaging scene scope is theoretically verified.

The rest of this article is organized as follows. In Section 2, the SAR echo models are introduced. In Section 3, the proposed SSR-AMDIA is derived in detail. In Section 4, the azimuth maximum DOF, range segmentation strategy and the computational complexity of the proposed algorithm are analyzed and mathematically derived. The findings of the simulation and measured experiment are shown and discussed in Section 5. Finally, Section 6 serves as our conclusion.

2. SAR Signal Models

Typically, the linear frequency modulation signal is used as the transmitted signal $s_t(t)$ in SAR systems to obtain a uniform signal bandwidth [25]. Due to the linear modulation of frequency, its phase is a quadratic function with respect to fast time t. The complex form of $s_t(t)$ can be expressed as

$$s_t(t) = \beta_t w_r(t) \exp(j2\pi f_c t) \exp\left(j\pi K_r t^2\right) \tag{1}$$

K_r is chirp rate, and f_c is the center frequency. β_t is the chirp signal's amplitude, while $w_r(t)$ is the range windowing function.

2.1. Complete SAR Echo Model

The de-carried echo of a stationary point scatterer $P(x, y, 0)$ is presented in

$$s_r(t, \eta) = \beta_r w_r\left(t - \frac{2R_P(\eta)}{c}\right) w_a(\eta) \exp\left\{\frac{-j4\pi f_c R_P(\eta)}{c}\right\}$$
$$\times \exp\left\{j\pi K_r\left(t - \frac{2R_P(\eta)}{c}\right)^2\right\} + n \tag{2}$$

where η, β_r, c, n, and $w_a(\eta)$ are the slow time dimension during the imaging interval, back-scattered coefficient, light speed, random noise, and azimuth windowing function, respectively. $R_P(\eta)$ is the instantaneous distance between the stationary target $P(x, y, 0)$ and the platform position (x_i, y_i, h), which can be demonstrated as

$$R_P(\eta) = \sqrt{(x_i - x)^2 + (y_i - y)^2 + h^2} \tag{3}$$

Assuming that the SAR system moves at a constant velocity v_a, thus the azimuth position $y_i = y_a + v_a\eta$, where y_a denotes radar initial azimuth position.

2.2. SAR Echo Model with Azimuth Missing Data

The AMD echo types include periodic and random AMD echo. The FMCW SAR system or the anti-jamming SAR system is the primary cause of the periodic AMD echo. In contrast, the occlusion or inevitable interference are the primary causes of the random AMD echo [17,21]. Figure 1 compares the complete, periodic missing, and random missing signals.

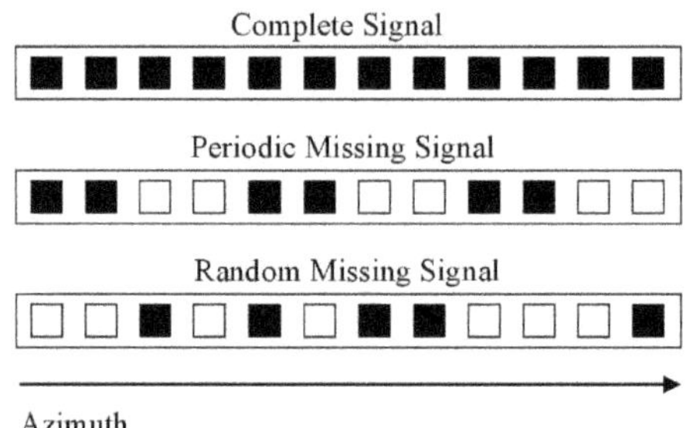

Figure 1. The comparison between the complete, periodical missing, and random missing signal.

White squares represent missing samples, while black squares represent valid samples.

Suppose that the azimuth and range sample numbers are denoted by N_A and N_R, respectively, and that the total echo size is $N_A \times N_R$. Then, assume that N_M ($N_M < N_A$) is the total number of missing azimuth samples, and G_m is their corresponding location set. In order to determine the AMD echo model, we define an azimuth missing matrix Λ_m, that is

$$\Lambda_m = diag\left[\lambda_{N_1}, \cdots, \lambda_{N_i}, \cdots, \lambda_{N_A}\right] \tag{4}$$

where

$$\begin{cases} \lambda_{N_i} = 0, & when \ N_i \in G_m \\ \lambda_{N_i} = 1, & else \end{cases} \tag{5}$$

Hence, the time domain AMD echo s_m is obtained by

$$s_m = \Lambda_m s_r \tag{6}$$

If all zero row vectors of s_m are removed, a small size echo s_y is acquired as

$$s_y = \widetilde{I}_y s_m \tag{7}$$

where $\widetilde{I}_y$ represents the deformed identity matrix, and it can be denoted as

$$\widetilde{I}_y \iff I(N_i, :)\big|_{N_i \in G_m} = \varnothing \tag{8}$$

where *I* and $\varnothing$ represent the identity matrix and empty set, respectively.

3. Sub-Echo Segmentation and Reconstruction Azimuth Missing Data SAR Imaging Algorithm

The detailed steps of the proposed SSR-AMDIA are demonstrated in Figure 2. First of all, the raw AMD echo is range compressed, and the range cells migration is corrected. Then, the entire AMD-RCMC echo is split into a series of AMD-RCMC sub-echoes along the range direction. Next, as the critical step, a set of PCFs are redesigned based on the RCMC sub-echoes. By multiplying each RCMC sub-echo with its corresponding PCF in the range-frequency domain, each complete PC-RCMC sub-echo can obtain a sparse representation in the Doppler domain. Subsequently, the accurate estimations of the complete PC-RCMC sub-echoes are recovered from the AMD-PC-RCMC sub-echoes using the CS method. In this article, we employ the Generalized Orthogonal Matching Pursuit (GOMP) algorithm [26] to reconstruct the full PC-RCMC sub-echoes in order to remove any potential error impacts of various CS techniques, as in the SOA-AMDIA [3,17,20,27]. Then, the accuracy estimations of the complete RCMC sub-echoes can be obtained by multiplying the complete PC-RCMC sub-echoes with the conjugation of the previously mentioned PCFs. Finally, by combining each reconstructed RCMC sub-echo, the reconstructed RCMC echo of the entire imaging scene can be acquired, and then after azimuth compression, a satisfied imaging result can be obtained.

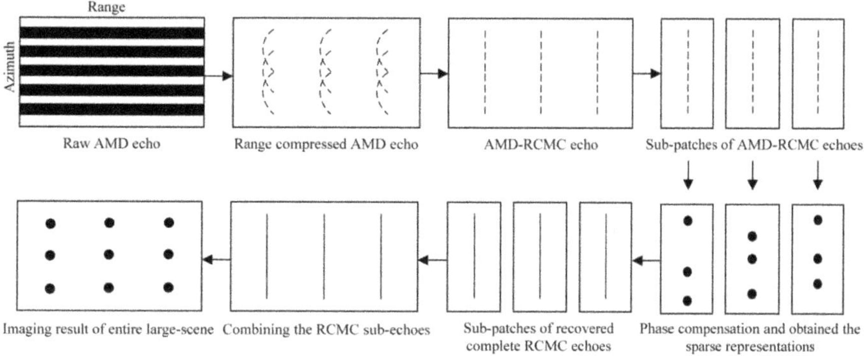

Figure 2. Flowchart of the proposed SSR-AMDIA.

Next, we describe the specific derivation steps of the proposed SSR-AMDIA in detail.

3.1. Range Compression and Range Cell Migration Correction

First, $s_m(t, \eta)$ should accomplish the range Fourier transform to obtain the range-frequency domain signal $S_m(f_r, \eta)$, which can be expressed as

$$S_m(f_r, \eta) = w_r(f_r)w_a(\eta)\exp\left(\frac{-j\pi f_r^2}{K_r}\right)\exp\left\{\frac{-j4\pi(f_c + f_r)R_P(\eta)}{c}\right\} \tag{9}$$

where f_r denotes the range frequency.

Then, the range compressed signal $s_{mrc}(t, \eta)$ can be obtained after range compression, and the range Doppler domain signal $S_{mrc}(t, f_\eta)$ can be presented as

$$S_{mrc}(t, f_\eta) = p_r\left(t - \frac{2R_{rd}(f_\eta)}{c}\right)w_a(f_\eta)\exp\left(\frac{-j4\pi f_c R_{rd}(f_\eta)}{c}\right) \tag{10}$$

where p_r stands for the sinc function and f_η is the azimuth frequency. The Doppler instantaneous distance $R_{rd}(f_\eta)$ can be written as

$$R_{rd}(f_\eta) \approx R_0 + \frac{v_a^2}{2R_0}\left(\frac{f_\eta}{K_a}\right)^2 \tag{11}$$

where R_0 is shortest distance between the platform and the target. K_a is Doppler chirp rate. The second term of (11) is the range cells migration term.

After RCMC and the azimuth Inverse Fourier Transform (AIFT), the AMD-RCMC signal $s_{mrcmc}(t,\eta)$ is demonstrated as

$$s_{mrcmc}(t,\eta) = p_r\left(t - \frac{2R_0}{c}\right)w_a(\eta)\exp\left(\frac{-j4\pi f_c R_P(\eta)}{c}\right) \tag{12}$$

3.2. Reconstructing the Sub-Echoes

To reduce the residual phase error of the range far-field targets, the AMD-RCMC echo is split into K sub-patches along the range direction. Since target range locations are determined after RCMC, the range segmentation will not distort the adjacent sub-scenes. To redesign the more effective PCFs, k-th AMD-RCMC sub-echo $s_{mrcmc}^{k_{th}}(t,\eta)$ transforms to range-frequency domain, which is described as

$$S_{mrcmc}^{k_{th}}(f_r,\eta) = w_r(f_r)w_a(\eta)\exp\left(\frac{-j4\pi f_r R_0}{c}\right)\exp\left(\frac{-j4\pi f_c R_P(\eta)}{c}\right) \tag{13}$$

Thus, k-th redesigned PCF $\theta_{\text{ref}}^{k_{th}}(f_r,\eta)$ is defined as

$$\theta_{\text{ref}}^{k_{th}}(f_r,\eta) = \exp\left(\frac{j4\pi f_r R_{\text{ref0}}^{k_{th}}}{c}\right)\exp\left(\frac{j4\pi f_c R_{\text{ref}}^{k_{th}}(\eta)}{c}\right) \tag{14}$$

where $R_{\text{ref0}}^{k_{th}}$ is the shortest slant range of k-th reference point $P_{\text{ref}}^{k_{th}}(x_{\text{ref}}^{k_{th}}, y_{\text{ref}}^{k_{th}}, 0)$. The $R_{\text{ref}}^{k_{th}}$ represents the slant range between $P_{\text{ref}}^{k_{th}}(x_{\text{ref}}^{k_{th}}, y_{\text{ref}}^{k_{th}}, 0)$ and the moving platform (x_i, y_i, h). It can be expressed as

$$R_{\text{ref}}^{k_{th}}(\eta) = \sqrt{(x_i - x_{\text{ref}}^{k_{th}})^2 + \left(y_i - y_{\text{ref}}^{k_{th}}\right)^2 + h^2} \tag{15}$$

Next, the azimuth missing redesigned PCF $\theta_{m\text{ref}}^{k_{th}}$ is acquired based on (6), that is

$$\theta_{m\text{ref}}^{k_{th}} = \Lambda_m \theta_{\text{ref}}^{k_{th}} \tag{16}$$

The PCF designation is the key step of the proposed SSR-AMDIA. A sparse PC-RCMC sub-echo $S_{\text{pc}}^{k_{th}}(t, f_\eta)$ that is likewise the waiting-recovering signal may be generated in the Doppler domain by multiplying $S_{\text{rcmc}}^{k_{th}}(f_r,\eta)$ by $\theta_{\text{ref}}^{k_{th}}(f_r,\eta)$, which is represented by

$$S_{\text{pc}}^{k_{th}}(t, f_\eta) = \text{FFT}_a\left[\text{IFFT}_r\left[S_{\text{rcmc}}^{k_{th}}(f_r,\eta)\theta_{\text{ref}}^{k_{th}}(f_r,\eta)\right]\right] \tag{17}$$

where $\text{FFT}_a[\cdot]$ and $\text{IFFT}_r[\cdot]$ are azimuth Fast Fourier Transform and range Inverse Fast Fourier Transform, respectively.

Since the main purpose of the proposed SSR-AMDIA is to reconstruct $S_{\text{pc}}(t, f_\eta)$, its sparsity is vital for the signal reconstruction.

To evaluate the focusing performance of the redesigned PCFs, $S_{\text{pc}}(t, f_\eta)$ results obtained by different methods are shown in Figure 3. There are nine targets in the imaging scene, and the SOA-AMDIA's $S_{\text{pc}}(t, f_\eta)$ is shown in Figure 3a. Obviously, only the center point is well-focused. The significant defocus can be easily found on the edge targets.

Therefore, the effectiveness of SOA-AMDIA's PCF for larger scenarios is limited. Moreover, Figure 3b is imaged using the proposed SSR-AMDIA before the sub-echo segmentation. Compared with the Figure 3a, the azimuth far-field targets at the center range are well-focused. However, it still cannot remove the residual phase error caused by the range differences. The range edge targets are still defocused. When a series of segmented $\theta_{\text{ref}}^{k_{th}}(f_r, \eta)$ are used, the most sparse $S_{\text{pc}}(t, f_\eta)$ can be obtained by observing Figure 3c. The two-dimensional residual phase errors of the borderline targets are significantly reduced. The focusing performance of the redesigned PCFs is validated.

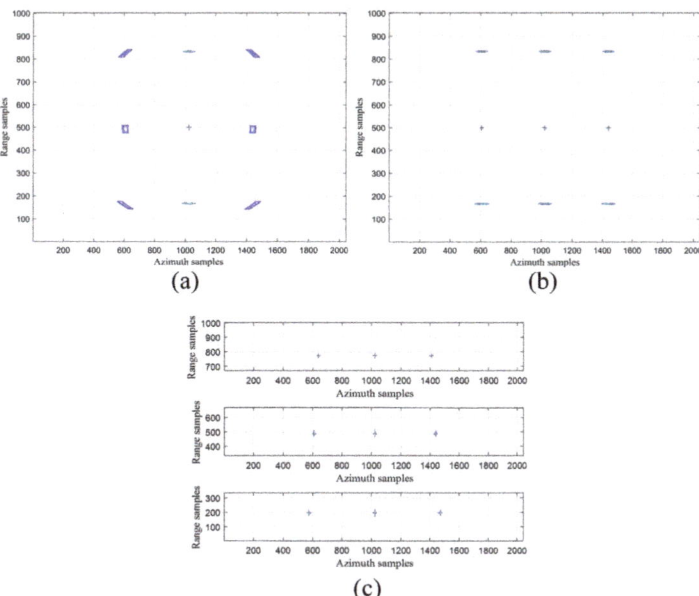

Figure 3. $S_{\text{pc}}(t, f_\eta)$ results obtained by (**a**) the SOA-AMDIA; (**b**) the proposed SSR-AMDIA without range segmentation; (**c**) the proposed SSR-AMDIA.

Next, the detailed reconstruction steps of the proposed SSR-AMDIA are introduced as follows. Firstly, the small size phase-compensated signal $s_{y\text{pc}}^{k_{th}}$ can be demonstrated as the same as (7)

$$s_{y\text{pc}}^{k_{th}} = \widetilde{\boldsymbol{I}}_y s_{m\text{pc}}^{k_{th}} \tag{18}$$

$s_{y\text{pc}}^{k_{th}}(t, \eta)$ must be segmented into several one-dimensional range signals in order to accommodate the one-dimensional signal recovery processing. The q-th ($1 \leq q \leq N_R$) range signal can be expressed as $s_{y\text{pc}}^{k_{th}}(t_q, \eta)$, where N_R denotes the number of entire range gates. In the proposed SSR-AMDIA, the q-th estimated range signal $S_{\text{pc}}^{k_{th}}(t_q, f_\eta)$ is regarded as the signal x in CS method, while $S_{y\text{pc}}^{k_{th}}(t_q, \eta)$ is considered as the compressed signal vector y. Accordingly, since $S_{\text{pc}}^{k_{th}}(t_q, f_\eta)$ is direct sparse, $\boldsymbol{\Phi}_{y\text{AIFT}}$ is understood as being the sensing matrix A. First, the complete AIFT matrix $\boldsymbol{\Phi}_{\text{AIFT}}$ is illustrated as

$$\boldsymbol{\Phi}_{\text{AIFT}} = \frac{1}{N_A}\begin{bmatrix} \exp\left(j\frac{2\pi\eta_1\eta_1}{\eta_{N_A}}\right) & \cdots & \exp\left(j\frac{2\pi\eta_1\eta_{N_A}}{\eta_{N_A}}\right) \\ \vdots & \ddots & \vdots \\ \exp\left(j\frac{2\pi\eta_{N_A}\eta_1}{\eta_{N_A}}\right) & \cdots & \exp\left(j\frac{2\pi\eta_{N_A}\eta_{N_A}}{\eta_{N_A}}\right) \end{bmatrix} \tag{19}$$

Similar to the Equations (6) and (16), the partial missing AIFT matrix $\mathbf{\Phi}_{m\,\text{AIFT}}$ is obtained by

$$\mathbf{\Phi}_{m\,\text{AIFT}} = \mathbf{\Lambda}_m \mathbf{\Phi}_{\text{AIFT}} \tag{20}$$

Similar to (7), the small size AIFT matrix $\mathbf{\Phi}_{y\,\text{AIFT}}$ is gained by

$$\mathbf{\Phi}_{y\,\text{AIFT}} = \tilde{\mathbf{I}}_y \mathbf{\Phi}_{m\,\text{AIFT}} \tag{21}$$

Its size is equal to $(N_A - N_M) \times N_A$. Consequently, the sub-echoes reconstructing process can be formulated as

$$\begin{aligned} &\min_{S_{\text{pc}}^{k_{th}}(t_q, f_\eta)} \quad ||S_{\text{pc}}^{k_{th}}(t_q, f_\eta)||_1, \\ &\text{s.t. } ||\mathbf{\Phi}_{y\,\text{AIFT}} S_{\text{pc}}^{k_{th}}(t_q, f_\eta) - s_{y\text{pc}}^{k_{th}}(t_q, \eta)||_2 \le \epsilon \end{aligned} \tag{22}$$

where ϵ denotes the threshold value.

To eliminate the possible error effects of different recovery methods, the estimated $\widehat{S}_{\text{pc}}^{k_{th}}(t_q, f_\eta)$ is reconstructed using the GOMP algorithm in this article. Table 1 or [26] contains the specific steps for the GOMP algorithm.

Table 1. The specific steps for the GOMP algorithm based on $s_{y\text{pc}}^{k_{th}}(t_q, \eta)$.

Step 1	Input the indices number of each selection P, the maximum number of iterations I_{max}, the threshold parameter ϵ and $\mathbf{\Phi}_{y\,\text{AIFT}}$;
Step 2	Initialize the iteration parameter $It=1$, let the residue signal $r^0 = s_{y\text{pc}}^{k_{th}}(t_q, \eta)$, and set a new sensing matrix $\mathbf{B}^0 = \varnothing$;
Step 3	Let $It = It + 1$;
Step 4	Calculate the largest P values in $\left|\left\langle r^{It-1}, \mathbf{\Phi}_{y\,\text{AIFT}} \right\rangle\right|$ from the largest to smallest and then the corresponding $\phi_{\max_p}$ are selected;
Step 5	Update matrix $\mathbf{B}^{It} = \mathbf{B}^{It-1} \cup \left[\phi_{\max_1}, \cdots, \phi_{\max_P}\right]$ and calculate the estimated value of complete signal vector by $\hat{\boldsymbol{\alpha}} = \left(\left(\mathbf{B}^{It}\right)^H \mathbf{B}^{It}\right)^{-1} \left(\mathbf{B}^{It}\right)^H s_{y\text{pc}}^{k_{th}}(t_q, \eta)$, where H represents the conjugate transpose operation;
Step 6	Update residue signal $r^{It} = s_{y\text{pc}}^{k_{th}}(t_q, \eta) - \mathbf{B}^{It}\hat{\boldsymbol{\alpha}}$;
Step 7	If $It = I_{max}$ or $||r^{It}||_2 \le \epsilon$, let $\widehat{S}_{\text{pc}}^{k_{th}}(t_q, f_a) = \hat{\boldsymbol{\alpha}}$. Else go to Step 3.

Once the reconstructed one-dimensional signals $\widehat{S}_{\text{pc}}^{k_{th}}(t_q, f_\eta)$ are combined, the k-th segment sub-echo $\widehat{S}_{\text{pc}}^{k_{th}}(t, f_\eta)$ can be acquired. It follows that the conjugation of $\theta_{\text{ref}}^{k_{th}}(f_r, \eta)$ must be compensated in order to obtain the k-th segment $\hat{s}_{\text{rcmc}}^{k_{th}}(t, \eta)$, which is represented by

$$\widehat{S}_{\text{rcmc}}^{k_{th}}(f_r, \eta) = \widehat{S}_{\text{pc}}^{k_{th}}(f_r, \eta) conj\left(\theta_{\text{ref}}^{k_{th}}(f_r, \eta)\right) \tag{23}$$

3.3. Combining the Sub-Echoes and Entire Scene Imaging

To obtain the reconstructed complete RCMC echo $\hat{s}_{\text{rcmc}}$ of entire imaging scene, a series of $\hat{s}_{\text{rcmc}}^{k_{th}}$ are combined in sequence, that is

$$\hat{s}_{\text{rcmc}} = \left\lfloor \hat{s}_{\text{rcmc}}^{1_{st}}, \cdots, \hat{s}_{\text{rcmc}}^{k_{th}}, \cdots, \hat{s}_{\text{rcmc}}^{K_{th}} \right\rfloor \tag{24}$$

where $\lfloor \cdot \rfloor$ denotes the combination operation.

The normalized recovery error results between $s_{\text{rcmc}}(t, \eta)$ and $\hat{s}_{\text{rcmc}}(t, \eta)$ are illustrated in Figure 4. Figure 4a is obtained using the SOA-AMDIA and Figure 4b is obtained using the proposed SSR-AMDIA. In order to quantitatively evaluate the reconstruction performance, the average normalized recovery error E_a is defined as

$$E_a = \mathrm{mean}\left(\sum^{t \in Q_R} \sum_{\eta=0}^{N_A-1} \left(\left| \frac{s_{\mathrm{rcmc}}(t,\eta)}{\max(s_{\mathrm{rcmc}}(t,\eta))} \right| - \left| \frac{\widehat{s}_{\mathrm{rcmc}}(t,\eta)}{\max(\widehat{s}_{\mathrm{rcmc}}(t,\eta))} \right| \right) \right) \tag{25}$$

where Q_R represents the range cells set corresponding to the presence of targets after RCMC, and $\mathrm{mean}(\cdot)$ denotes the average function. According to Figure 4, E_a of Figure 4a is equal to 0.087, which is almost half to that of Figure 4b, which is equal to 0.179. Therefore, the proposed SSR-AMDIA obviously has a better reconstruction performance.

Finally, since a satisfied complete RCMC echo of the entire scene is estimated, an excellent-focused image can be obtained via azimuth compression.

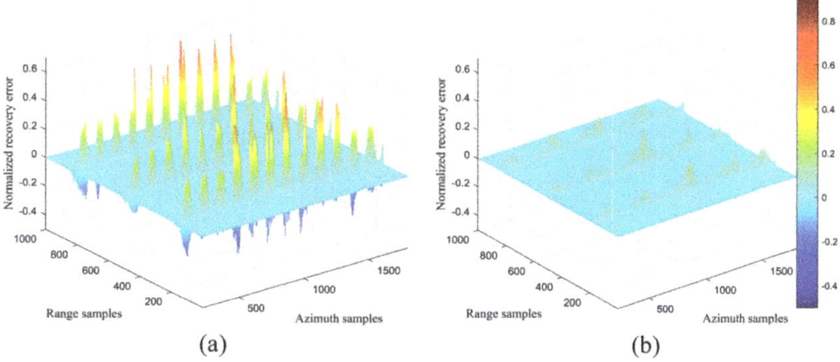

Figure 4. The normalized recovery error between $s_{\mathrm{rcmc}}(t,\eta)$ and $\widehat{s}_{\mathrm{rcmc}}(t,\eta)$ by using (**a**) the SOA-AMDIA; (**b**) the proposed SSR-AMDIA.

4. Parameter Analysis

As mentioned before, we consider extending the maximum DOFs of SOA-AMDIA in azimuth and range directions, respectively. Thus, the range segmentation is applied in the proposed SSR-AMDIA. However, the identical segmentation idea cannot be exploited to enlarge the azimuth imaging scope. Azimuth segmentation may result in too few azimuth samples available in some sub-apertures, especially in the case of random missing. Sub-apertures with a high AMR will be detrimental to complete sub-aperture reconstruction [3]. Hence, although the proposed SSR-AMDIA extends the azimuth maximum DOF, it has a limitation. Moreover, there is no limit to the maximum range imaging scope of the proposed SSR-AMDIA under a proper segmentation. We next analyze the azimuth maximum DOF and range segmentation strategy of the proposed algorithm. The computational complexity advantage is also investigated.

4.1. Azimuth Maximum Depth of Focus

In 2005, B. Rigling analyzed the imaging scene size limits for the PFA in the monostatic SAR system situation [28]. When the absolution value of residual Quadratic Phase Error after the PFA (PFA-QPE) $|\Phi_{\mathrm{QPE}}| < \pi/4$, the PFA's maximum well-focused radius r_{max} can be expressed as

$$r_{\max} < \rho_a \sqrt{\frac{2R_{\mathrm{ref0}}}{\lambda}} \tag{26}$$

where ρ_a and λ denote the azimuth resolution and wavelength, respectively. The prerequisite for applying (26) is that there are no unknown motion measurement errors during the flight. Otherwise, the phase errors will lead to an irreparable defocus to the image long before the far-field approximation (26) breaks down.

In 2016, L. Gorham and B. Riging further derived the imaging scene size limits for the PFA in the linear flight case [18]. The residual PFA-QPE Φ_{QPE} can be written as

$$\Phi_{\text{QPE}} = -\frac{L_a^2 \pi}{2\lambda}\left(\frac{1}{R_{P0}} - \frac{2}{R_{\text{ref0}}} - \frac{y^2}{R_{P0}^3} + \frac{R_{P0}}{R_{\text{ref0}}^2}\right) \tag{27}$$

where L_a is the length of synthetic aperture and R_{P0} can be calculated as

$$R_{P0} = R_P|_{y_i=0} = \sqrt{(x_i - x)^2 + (0 - y)^2 + h^2} \tag{28}$$

The value of Φ_{QPE} is related to the position of targets. Assume $f_c = 1$ GHz, $R_{\text{ref0}} = 3300$ m, Φ_{QPE} result in different positions is shown in Figure 5a. The maximum focus area is limited to a circle of which radius equals 148.3 m after the PFA imaging. However, since the SOA-AMDIA only utilizes the phase compensation process of PFA, the maximum DOF of $S_{\text{pc}}(t, f_\eta)$ decreases rapidly [29,30].

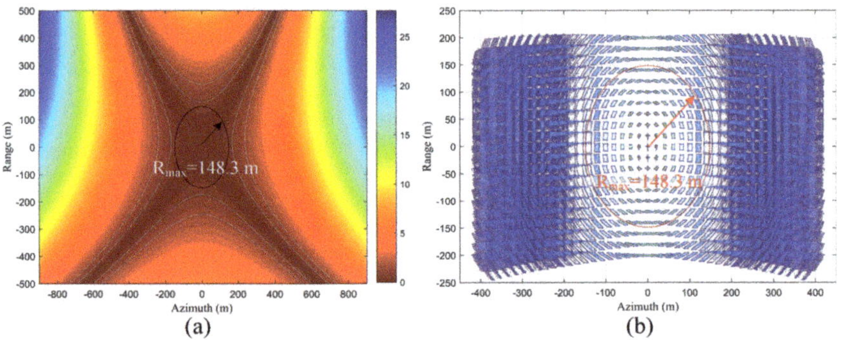

Figure 5. (**a**) Residual Quadratic Phase Error Φ_{QPE} after PFA. The inner white contour line represents an error of $\pi/4$, while the outer white contour line represents an error of $\pi/2$. The black contour circle denotes the maximum well-focused radius r_{max} deduced in [28], where the $r_{max} = 148.3$ m. (**b**) Simulated image of $S_{\text{pc}}(t, f_\eta)$ obtained by the SOA-AMDIA. The meaning of the red contour circle is the same as the black contour circle in (**a**).

To illustrate this phenomenon more clearly, we form $S_{\text{pc}}(t, f_\eta)$ using the SOA-AMDIA with a regular grid in actual coordinates in Figure 5b. Due to the impact of the range cells migration and the residual phase error, the azimuth edge targets of Figure 5b are distorted and defocused severely while only the centroid is excellent-focused. Although it is clearly sparser than the dense time domain signal, the maximum DOF of $S_{\text{pc}}(t, f_\eta)$ is only 25 m, much less than 148.3 m. The aforementioned conclusion has been verified in simulation. The low-quality $S_{\text{pc}}(t, f_\eta)$ will impair full echo reconstruction. The applicable imaging scope of the SOA-AMDIA will be significantly reduced.

In response to this problem, the range cells migration effect of the raw echo is removed by the proposed SSR-AMDIA. Then, a new PCF is redesigned to enhance the sparsity of $S_{\text{pc}}(t, f_\eta)$ and to extend the maximum azimuth DOF.

First, the residual phase error Φ_{E} between the scattering point and the reference point can be represented as

$$\Phi_{\text{E}} = -\frac{4\pi\Delta R(\eta)}{\lambda} \tag{29}$$

where $\Delta R = R_P - R_{\text{ref}}$ is the differential slant range, and the expressions of R_P and R_{ref} can be found in (3) and (15), respectively. The second-order Taylor series approximation of ΔR is performed as

$$\Delta R(\eta) \approx \Delta R|_{\eta=0} + \left.\frac{\partial \Delta R}{\partial \eta}\right|_{\eta=0}\eta + \left.\frac{\partial^2 \Delta R}{\partial^2 \eta}\right|_{\eta=0}\frac{\eta^2}{2}$$
$$\approx (R_{P0} - R_{\text{ref0}}) + \left(\left.\frac{\partial R_P}{\partial \eta}\right|_{\eta=0} - \left.\frac{\partial R_{\text{ref}}}{\partial \eta}\right|_{\eta=0}\right)\eta + \left(\left.\frac{\partial^2 R_P}{\partial \eta^2}\right|_{\eta=0} - \left.\frac{\partial^2 R_{\text{ref}}}{\partial \eta^2}\right|_{\eta=0}\right)\frac{\eta^2}{2} \tag{30}$$

Let the monostatic SAR system moves at a constant speed in the azimuth direction ($+y$ direction), the first and second derivatives of R_P and R_{ref} can be calculated as

$$\begin{cases} \dfrac{\partial R_P}{\partial \eta} = -\dfrac{(y - y_i)}{R_P}\dfrac{\partial y_i}{\partial \eta} \\[2mm] \dfrac{\partial R_{\text{ref}}}{\partial \eta} = \dfrac{y_i}{R_{\text{ref}}}\dfrac{\partial y_i}{\partial \eta} \\[2mm] \dfrac{\partial^2 R_P}{\partial \eta^2} = -\dfrac{1}{R_P}\left((y - y_i)\dfrac{\partial^2 y_i}{\partial \eta^2} - \left(\dfrac{\partial y_i}{\partial \eta}\right)^2 + \left(\dfrac{\partial R_P}{\partial \eta}\right)^2\right) \\[2mm] \dfrac{\partial^2 R_{\text{ref}}}{\partial \eta^2} = \dfrac{1}{R_{\text{ref}}}\left(y_i\dfrac{\partial^2 y_i}{\partial \eta^2} + \left(\dfrac{\partial y_i}{\partial \eta}\right)^2 - \left(\dfrac{\partial R_{\text{ref}}}{\partial \eta}\right)^2\right) \end{cases} \tag{31}$$

where $\partial y_i/\partial \eta = L_a/2$ and $\partial^2 y_i/\partial \eta^2 = 0$.

The residual QPE after RCMC and phase compensation (PC-RCMC-QPE) $\widetilde{\Phi}_{\text{QPE}}$ is thus computed as

$$\begin{aligned} \widetilde{\Phi}_{\text{QPE}} &= -\frac{2\pi}{\lambda}\left(\left.\frac{\partial^2 R_P}{\partial \eta^2}\right|_{\eta=0} - \left.\frac{\partial^2 R_{\text{ref}}}{\partial \eta^2}\right|_{\eta=0}\right) \\ &= \frac{L_a^2\pi\left(y^2 R_{\text{ref0}} + R_{P0}^3 - R_{P0}^2 R_{\text{ref0}}\right)}{2\lambda R_{P0}^3 R_{\text{ref0}}} \end{aligned}. \tag{32}$$

Suppose $f_c = 1$ GHz, $R_{\text{ref0}} = 3300$ m, PC-RCMC-QPE $\widetilde{\Phi}_{\text{QPE}}$ result in different positions is shown in Figure 6a and we form the calculated $S_{\text{pc}}(t, f_\eta)$ by (17) when $K = 1$ with a regular grid in actual coordinates in Figure 6b.

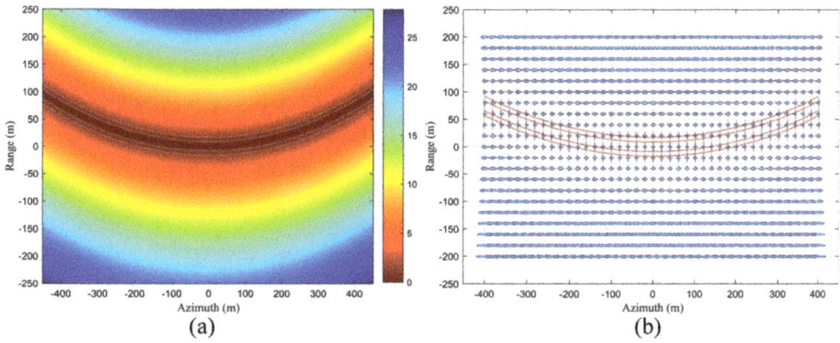

Figure 6. (**a**) Residual Quadratic Phase Error $\widetilde{\Phi}_{\text{QPE}}$ after the RCMC and phase compensation. The inner white contour line represents an error of $\pi/4$, while the outer white contour line represents an error of $\pi/2$. (**b**) Simulated image of $S_{\text{pc}}(t, f_\eta)$ obtained by (17) when $K = 1$. The meaning of the red contour lines is the same as the white contour lines in sub-figure (**a**).

Figure 6b is significantly sparser than Figure 5b. The azimuth maximum DOF is substantially enlarged. Let $\left|\tilde{\Phi}_{\mathrm{QPE}}\right| = \pi/2$ and $x = 0$, the azimuth coordinate y_{QPE} can be calculated as

$$y_{\mathrm{QPE}} = \pm\sqrt{\left(\frac{L_a^2 R_{\mathrm{ref0}}^3}{L_a^2 + \lambda R_{\mathrm{ref0}}}\right)^{2/3} - R_{\mathrm{ref0}}} \tag{33}$$

Substitute the above-mentioned simulation parameters into (33), then two farthest points, $P_{y_1}(0, -194.05, 0)$ and $P_{y_2}(0, 194.05, 0)$, that can achieve an excellent focus are obtained. Thus, the azimuth maximum DOF $\Delta y_{\mathrm{QPE}} = 194.05 - (-194.05) = 388.10$ m, which is much larger than that in Figure 5b.

A strong sparsity of $S_{\mathrm{pc}}(t, f_\eta)$ will facilitate the signal reconstruction. Hence, the azimuth maximum DOF of $S_{\mathrm{pc}}(t, f_\eta)$ given by (33) is conservative compared to that of the final image. We found that $P_y(0, \pm 400, 0)$ in the final image can still be accurately focused using the proposed algorithm with the above-mentioned simulation parameters. Conversely, the azimuth maximum DOF using SOA-AMDIA only reaches about ± 70 m under identical simulation conditions.

4.2. Range Segmentation Strategy

Figure 6b exhibits the limited range maximum DOF of $S_{\mathrm{pc}}(t, f_\eta)$ as well. Motivated by [15,19], the range segmentation is applied to expand the imaging scene scope in range direction. By observing Equation (32), the contour shape of $\tilde{\Phi}_{\mathrm{QPE}} = 0$ is a circle in the plane of $y = 0$, of which radius equals R_{ref0}. Since the imaging scene is limited in the plane of $z = 0$, $\tilde{\Phi}_{\mathrm{QPE}} = 0$ can only be obtained at two points $(0, 0, 0)$ and $(2R_{\mathrm{ref0}}, 0, 0)$. Similarly, the contour shapes of $\tilde{\Phi}_{\mathrm{QPE}} = -\pi/2$ and $\tilde{\Phi}_{\mathrm{QPE}} = \pi/2$ are two circles in the plane of $y = 0$ with different radius. It indicates that only four targets can ensure $\left|\tilde{\Phi}_{\mathrm{QPE}}\right| = \pi/2$ in the $y = z = 0$ case.

Let $\left|\tilde{\Phi}_{\mathrm{QPE}}\right| = \pi/2$ and $y = 0$, the range coordinate x_{QPE} can be calculated as

$$\begin{cases} x_{\mathrm{QPE}} = x_i \pm \sqrt{\left(\dfrac{L_a^2 R_{\mathrm{ref0}}}{L_a^2 - \lambda R_{\mathrm{ref0}}}\right)^2 - h^2}, & \tilde{\Phi}_{\mathrm{QPE}} = -\dfrac{\pi}{2} \\[3ex] x_{\mathrm{QPE}} = x_i \pm \sqrt{\left(\dfrac{L_a^2 R_{\mathrm{ref0}}}{L_a^2 + \lambda R_{\mathrm{ref0}}}\right)^2 - h^2}, & \tilde{\Phi}_{\mathrm{QPE}} = \dfrac{\pi}{2} \end{cases} \tag{34}$$

Hence, $P_{x_1}(-17.3, 0, 0)$, $P_{x_2}(17.1, 0, 0)$, $P_{x_3}(6582.9, 0, 0)$ and $P_{x_4}(6617.3, 0, 0)$ are obtained by substituting the above-applied simulation parameters into (34). Obviously, only $P_{x_1}(-17.3, 0, 0)$ and $P_{x_2}(17.1, 0, 0)$ are located in the imaging scene. The range maximum DOF is $\Delta x_{\mathrm{QPE}} = 17.1 - (-17.3) = 34.4$ m.

Obviously, when there is no target existing in the q-th range profile, the value of $S_{\mathrm{mpc}}(t_q, f_\eta)$ is equal to zero. The zero row vectors do not need to be reconstructed. We assume that the number of the range profiles that exist targets equals N_E. Therefore, the N_E range profiles of AMD-RCMC echo should be split into K sub-patches. The size of each sub-echo equals $(N_A - N_M) \times N_E/K$, where $N_E/K \leq \Delta x_{\mathrm{QPE}}/\Delta x$ and Δx denotes the interval of adjacent range cells.

Figure 7 demonstrates the simulated images of $S_{\mathrm{pc}}^{k_{th}}(t, f_\eta)$ based on the aforementioned range segmentation strategy. By adequately segmenting the imaging scene within the azimuth maximum DOF, all targets corresponding to $S_{\mathrm{pc}}^{k_{th}}(t, f_\eta)$ can be well-focused. The proposed SSR-AMDIA can guarantee the estimation accuracy of the complete echo in a larger imaging scene.

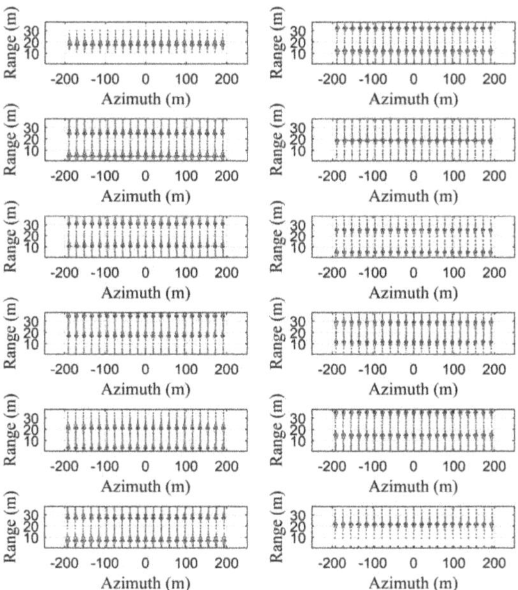

Figure 7. Simulated images of $S_{\text{pc}}^{k_{th}}(t, f_\eta)$ obtained by (17) when $K = 12$.

Moreover, since the proposed algorithm performs RCMC on the raw data before the range segmentation, the range position information of the sparse target is determined. Thus, it will not deteriorate the reconstruction error.

4.3. Computational Complexity

Assume that the number of entire range gates equals N_R and N_E sparse targets spread out along the range direction ($N_E \leq N_R$). Since the range position information of the target is hidden in all range profiles before range compression and RCMC, no matter how many sparse targets exist, N_R times reconstructions are required using the SOA-AMDIA. Contrarily, the proposed SSR-AMDIA only needs N_E times reconstructions to complete echo reconstruction. Suppose the computational complexity of one-dimensional GOMP algorithm equals $\mathcal{O}$, then the computational complexity of the SOA-AMDIA is $N_R\mathcal{O}$, while that of the proposed SSR-AMDIA is equal to $N_E\mathcal{O}$. It implies that the computational complexity of the proposed SSR-AMDIA is N_E/N_R times to that of SOA-AMDIA. Obviously, when fewer range profiles exist sparse targets, the computational complexity advantage of the proposed SSR-AMDIA is prominent.

5. Simulation and Real-Measured Experiment Validation

5.1. Simulation Verification of the Proposed SSR-AMDIA

An AMD-SAR imaging simulation is performed to evaluate the validation of the proposed SSR-AMDIA. Simulation parameters are shown in Table 2.

After allowing for 64 azimuth missing samples, we estimate that there are 64 available samples, making the AMR equals 50%. 600 range cells are segmented and reconstructed. Let the segment number $K = 12$. Thus the range cell number of each sub-patch equals 50.

Figure 8a depicts a grid of point targets spaced at 20 m intervals extending from -200 m to $+200$ m in both range and azimuth directions. The scenario geometry is chosen to accentuate the defocus effects. The imaging result obtained using the SOA-AMDIA is demonstrated in Figure 8b. In Figure 8b, only the targets near the center point can be excellent-focused, while the peripheral targets are defocused. Contrarily, all targets can be clearly imaged by using the proposed SSR-AMDIA in the AMD echo situation, as shown in Figure 8c.

To exhibit the imaging performance of the proposed SSR-AMDIA in more detail, two far-field targets $P_A(200, 200, 0)$ and $P_B(-200, -200, 0)$ are selected and marked with yellow squares in Figure 8b,c. These squares are zoomed for clearer exhibit in Figure 8d–g. In comparison to the Figure 8d,f, the false targets are eliminated using the proposed SSR-AMDIA, as shown in Figure 8e,g. While the azimuth resolution is maintained around 1 m, the azimuth Peak Side-lobe Ratio (PSLR) of P'_A and P'_B can reach -10 dB, which are much superior to that of P_A and P_B. It can be observed that the PSLR results of P'_A and P'_B are not as good as the ideal PSLR result. This is because the proposed SSR-AMDIA is an aperture estimation algorithm. Therefore, there are inevitably estimation errors in the estimated aperture signal, resulting in imperfect focus in the final imaging result. However, compared to SOA-AMDIA, its focusing performance has been significantly improved. Thus, the effectiveness of the proposed SSR-AMDIA is verified. The limits of the imaging scene size have been significantly expanded. The imaging quality of far-field targets has been improved. Moreover, the running times of the SOA-AMDIA and proposed SSR-AMDIA are calculated by the average of 50 times Monte Carlo simulations. The simulations are manipulated with the laptop that was configured with the Intel Core i5-1135G7 CPU, eight cores, and 16-GB RAM. The running time of the SOA-AMDIA equals 459.52 s while that of the proposed SSR-AMDIA equals 238.33 s, almost half of the former result. Therefore, the computational complexity advantage of the proposed SSR-AMDIA has been verified.

Table 2. Key parameters for simulation.

Parameters	Value
Central frequency / f_c	1 GHz
Shortest central slant range / R_{ref0}	3300 m
Signal frequency bandwidth / B	100 MHz
Range sampling rate / f_s	200 MHz
Pulse repetition frequency / PRF	197 Hz
Range samples / N_R	1002
Azimuth samples / N_A	2048
Azimuth missing ratio / AMR	50%

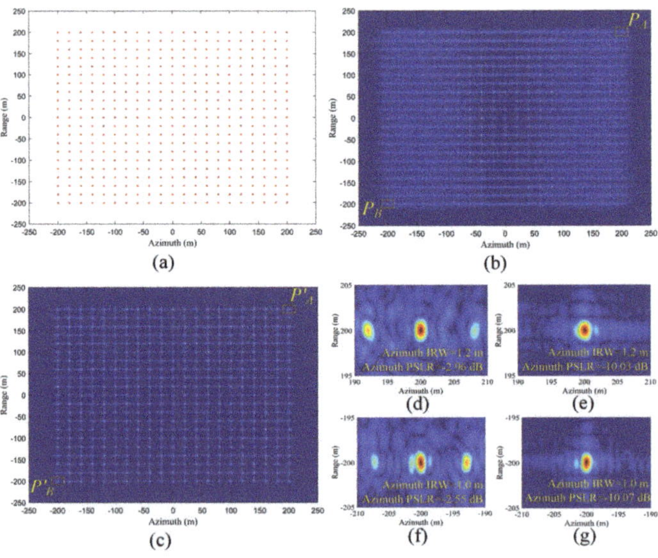

Figure 8. (**a**) Synthetic point targets grid. (**b**) Simulated image obtained by the SOA-AMDIA. (**c**) Simulated image obtained by the proposed SSR-AMDIA. (**d**) Zoomed P_A. (**e**) Zoomed P'_A. (**f**) Zoomed P_B. (**g**) Zoomed P'_B.

5.2. Measured Data Verification of the Proposed SSR-AMDIA

In order to further explore the theoretical analyses offered in this study, a real measured SAR experiment is designed and implemented based on a 77GHz millimeter-wave radar. As shown in Figure 9a, the radar is placed on an electric track 1.40 m above the ground and moved in the azimuth direction at a speed of 2.13 cm/s, forming a linear aperture with a length of 1.57 m. When the millimeter-wave radar stops, the SAR system collects a two-dimensional SAR echo with a size of 1024×1960.

First, to determine the azimuth and range maximum DOF, $\tilde{\Phi}_{QPE}$ is analyzed based on the measured SAR parameters. Substitute the related experiment parameters into (33) and (34), the two-dimensional maximum DOF can achieve 2.04 m and 0.32 m, respectively. We place five triangle reflectors in the scene based on the theoretical analysis. Target1 (11.10, −1.10), Target2 (11.10, 1.12), Target4 (9.24, −1.10), and Target5 (9.28, 1.02) are placed on the a square's four vertices. Additionally, Target3 (10.12, 0.00) is placed in the center, as shown in Figure 9b. Figure 9c displays the image result obtained using the Range Doppler algorithm with the real measured complete echo.

The AMD echo in the first experiment is produced by a periodic gap that occurs every 40 pulses. Hence, the AMR is equal to 50%. Figure 10a,b illustrate the final images focused by using SOA-AMDIA and the proposed SSR-AMDIA, respectively. Obviously, the SOA-AMDIA cannot effectively reconstruct a satisfied image. On the other hand, all five point targets are accurately focused through the proposed SSR-AMDIA, as shown in Figure 10b. It implies that SSR-AMDIA can significantly improve the imaging performance in a larger imaging scope. Therefore, the effectiveness of the proposed algorithm is successfully verified in the real SAR data. Image Entropy (IE) is also presented to assess the imaging performance of the above-mentioned two imaging algorithms. The lower the IE, the superior the focus of the imaging algorithm. The IE values corresponding to Figure 10a,b are 1.533 and 1.352, respectively. Compared with Figure 9c, the imaging result obtained using the proposed SSR-AMDIA is almost identical. Additionally, the IE result of Figure 10b reaches the equivalent level of Figure 9c. Thus, the proposed algorithm obviously obtains a superior focusing performance.

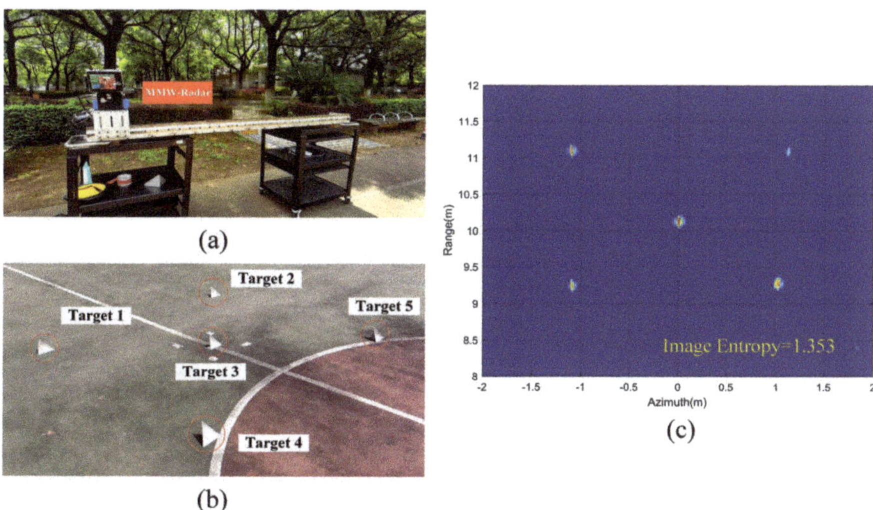

Figure 9. (**a**) The 77 GHz millimeter-wave SAR system for the real measured experiment. The electric track length equals 1.57 m and the radar height equals 1.40 m. (**b**) The large imaging scene consists of five triangle reflectors. They are Target1 (11.10, −1.10), Target2 (11.10, 1.12), Target3 (10.12, 0.00), Target4 (9.24, −1.10), and Target5 (9.28, 1.02). (**c**) The image result obtained by using the Range Doppler algorithm with the real measured complete echo.

Additionally, to comprehensively investigate the applicability of the proposed algorithm, a random AMD echo is set in the second experiment. Suppose that the SAR system is randomly subjected to 10 strong interferences during the data acquisition, and each interference causes 5% aperture data loss. In this case, AMR still equals 50% and the data size of random AMD echo is still 1024×980. Figure 11 demonstrates the imaging results comparison, which leads to the identical conclusion to that of Figure 10. The proposed algorithm can still accurately focus all targets while all targets obtained using the SOA-AMDIA are defocused. Compared with the IE = 1.525 obtained in Figure 11a, the IE can achieve 1.379 using the proposed SSR-AMDIA. Therefore, its effectiveness is fully verified on the measured SAR data once more.

Moreover, the running times of the SOA-AMDIA and proposed SSR-AMDIA are calculated by the average of 50 times Monte Carlo simulations. Table 3 shows the running time results. It can be found that the running time of the proposed SSR-AMDIA is much smaller than that of the SOA-AMDIA under both 50% periodic missing and 50% random missing conditions. Specifically, the proposed SSR-AMDIA needs 7.92 s to reconstruct the complete echo under periodic conditions, which is only about 1/7 of the existing method. Furthermore, in the random case, the running time of the proposed algorithm can be about 4 times faster than the SOA-AMDIA. A superior imaging result may be acquired more efficiently with the proposed SSR-AMDIA.

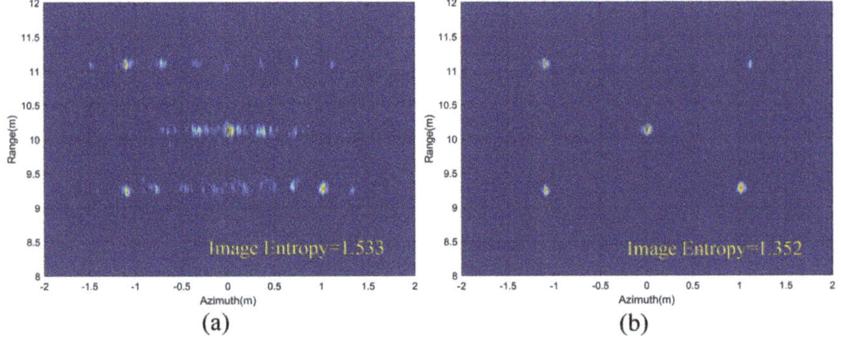

Figure 10. (**a**) Real measure data image targets obtained by the SOA-AMDIA with 50% periodic AMD echo. (Note that since the scene center point in this experiment is located at half of the maximum slant range, which is the (15, 0), all targets cannot be well-focused using the SOA-AMDIA.) (**b**) Real measure data image obtained by the proposed SSR-AMDIA with 50% periodic AMD echo.

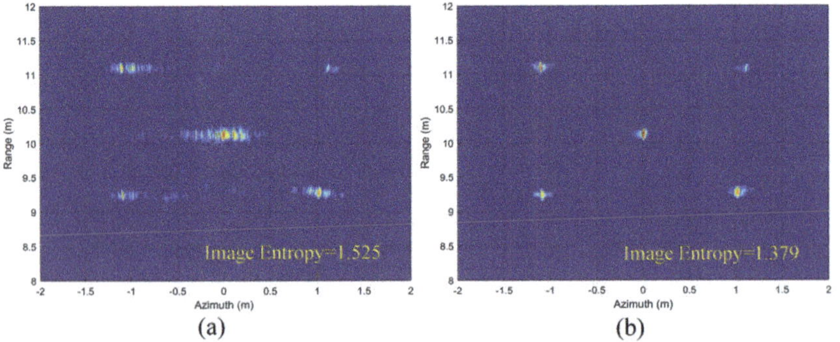

Figure 11. (**a**) Real measure data image obtained by the SOA-AMDIA with 50% random AMD echo. (Note that since the scene center point in this experiment is located at half of the maximum slant range, which is the (15, 0), all targets cannot be well-focused using the SOA-AMDIA.) (**b**) Real measure data image obtained by the proposed SSR-AMDIA with 50% random AMD echo.

Table 3. Running times of two AMD-SAR imaging algorithms under the real measured SAR data condition.

	SOA-AMDIA	SSR-AMDIA
50% Periodic Missing	53.25	7.92
50% Random Missing	29.39	7.04

5.3. Imaging Performance Effects on Different Azimuth Missing Ratios

Moreover, in order to evaluate the imaging performance effects of the proposed SSR-AMDIA on different AMRs, a series of simulations based on real measured SAR data are designed and implemented. We assume that the radar system will be subjected to 11–17 strong interferences during the motion, and each substantial interference will result in 5% azimuth data loss. Therefore, the AMR will gradually increase from 55% to 85%. The imaging results obtained by the proposed SSR-AMDIA are shown in Figure 12.

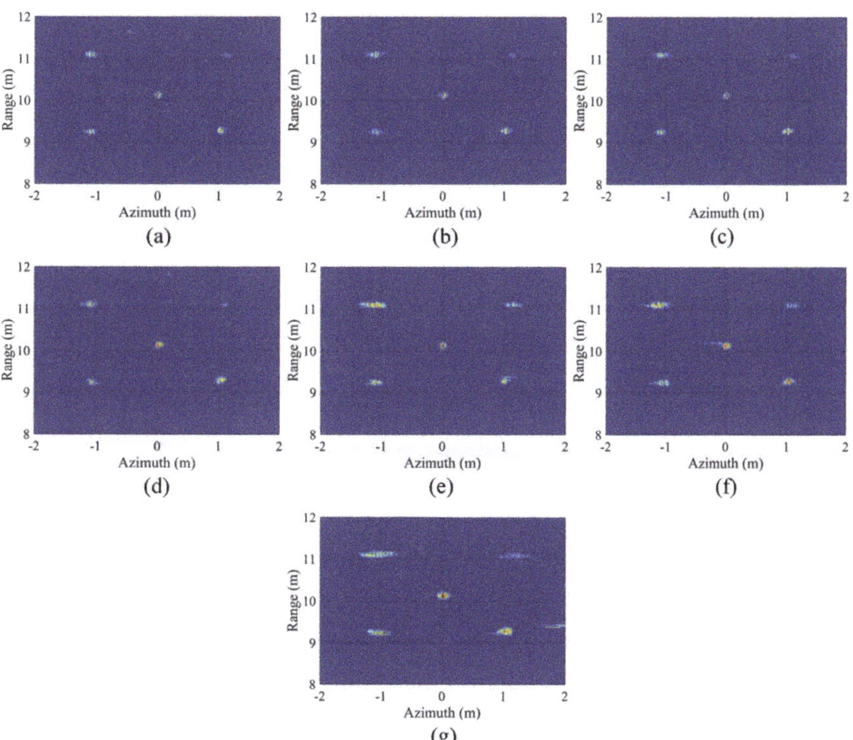

Figure 12. Image result obtained by the proposed SSR-AMDIA with the real measured SAR data when (**a**) AMR = 55%; (**b**) AMR = 60%; (**c**) AMR = 65%; (**d**) AMR = 70%; (**e**) AMR = 75%; (**f**) AMR = 80%; (**g**) AMR = 85%.

Obviously, when AMR $\leq$ 70%, the proposed SSR-AMDIA can obtain satisfactory imaging results. Targets in the imaging scene are well-focused. When AMR $>$ 70%, the imaging quality of SSR-AMDIA gradually decreases, and the imaging results have obvious side-lobes along the azimuth direction. This situation will further deteriorate as AMR rises. In order to measure the imaging quality of the proposed method in different AMR cases, we also introduce IE to quantitatively analyze the imaging results, as Figure 13a shows.

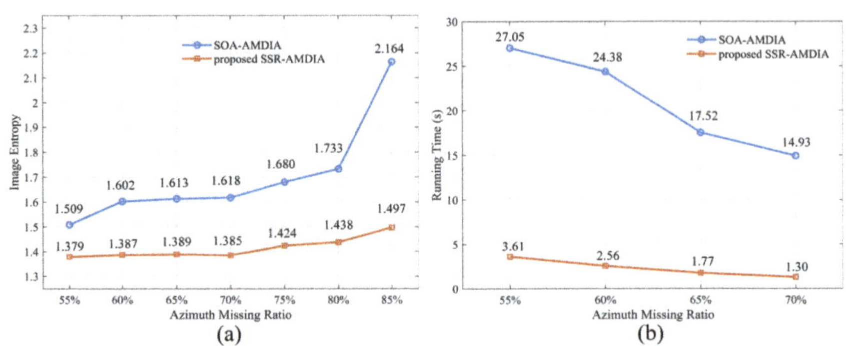

Figure 13. (**a**) Image Entropy results obtained by the SOA-AMDIA and the proposed SSR-AMDIA in different AMR cases. (**b**) Running times of the SOA-AMDIA and the proposed SSR-AMDIA in different AMR cases.

In all AMR cases, the image focusing performance of the proposed SSR-AMDIA is superior to that of the SOA-AMDIA. It can be found that the IE has a significant increase, reaching 1.424, when AMR = 75%. It is identical to the conclusion in Figure 12, indicating that when AMR > 70%, the imaging performance of the proposed SSR-AMDIA decreases obviously.

Additionally, we calculated and analyzed the running time of the two algorithms under different AMR conditions through 50 Monte Carlo experiments. The results are shown in Figure 13b. Same as the previous simulation and experimental results, the running time of SSR-AMDIA is much smaller than SOA-AMDIA. Concretely, the SSR-AMDIA takes only a few seconds to reconstruct the complete signal, which is an order of magnitude less than the SOA-AMDIA. So far, the imaging performance effects on different AMRs have been comprehensively analyzed. The proposed AMD-SAR imaging algorithm has obvious advantages.

6. Conclusions

In this paper, we propose SSR-AMDIA to solve the AMD-SAR imaging problem. The effectiveness of the proposed algorithm has been verified in both simulations and the measured SAR experiments. Additionally, we derive the two-dimensional maximum DOFs of the proposed algorithm and perform a rigorous theoretical analysis of the SSR-AMDIA's imaging scope. Compared with SOA-AMDIA, the proposed method can eliminate the limitation of the range maximum focusable scope, while the azimuth maximum focusable scope can be expanded by about 6 times. Our work found that the SOA-AMDIA has unacceptable focusable imaging size in the case of small shortest instantaneous distance or synthetic aperture length. The algorithm proposed in this paper can cope with this problem well and improve the applicability of AMD-SAR imaging. Furthermore, the imaging performance effects on different AMRs have been investigated. When AMR ≤ 70%, the proposed SSR-AMDIA can obtain satisfactory imaging results. It indicates that the proposed algorithm can handle most AMR cases. Furthermore, with the multi-dimensional development of radar signals, our next logical step is to deal with the AMD-SAR imaging problems under various robust waveforms [31,32].

Author Contributions: Conceptualization, N.J.; Software, J.W.; Formal analysis, Z.X.; Writing—review & editing, J.Z. and D.F.; Supervision, X.H. All authors have read and agreed to the published version of the manuscript.

Funding: This research was funded by National Natural Science Foundation of China grant number 62101562.

Data Availability Statement: Not applicable.

Conflicts of Interest: The authors declare no conflict of interest.

References

1. Pinheiro, M.; Rodriguez-Cassola, M.; Prats-Iraola, P.; Reigber, A.; Krieger, G.; Moreira, A. Reconstruction of coherent pairs of synthetic aperture radar data acquired in interrupted mode. *IEEE Trans. Geosci. Remote Sens.* **2014**, *53*, 1876–1893. [CrossRef]
2. Qian, Y.; Zhu, D. SAR Image Formation From Azimuth Periodically Gapped Raw Data Via Complex ISTA. In Proceedings of the 2019 6th Asia-Pacific Conference on Synthetic Aperture Radar (APSAR), Xiamen, China, 26–29 November 2019; IEEE: Piscataway, NJ, USA, 2019; pp. 1–5.
3. Jiang, N.; Feng, D.; Wang, J.; Huang, X. Missing data SAR imaging algorithm based on two dimensional frequency domain recovery. In Proceedings of the 2022 IEEE International Geoscience and Remote Sening Symposium, Kuala Lumpur, Malaysia, 17–22 July 2022; IEEE: Piscataway, NJ, USA, 2022.
4. Salzman, J.; Akamine, D.; Lefevre, R.; Kirk, J.C. Interrupted synthetic aperture radar (SAR). *IEEE Aerosp. Electron. Syst. Mag.* **2002**, *17*, 33–39. [CrossRef]
5. Li, J.; Stoica, P. An adaptive filtering approach to spectral estimation and SAR imaging. *IEEE Trans. Signal Process.* **1996**, *44*, 1469–1484. [CrossRef]
6. Stoica, P.; Larsson, E.G.; Li, J. Adaptive filter-bank approach to restoration and spectral analysis of gapped data. *Astron. J.* **2000**, *120*, 2163. [CrossRef]
7. Larsson, E.G.; Stoica, P.; Li, J. Amplitude spectrum estimation for two-dimensional gapped data. *IEEE Trans. Signal Process.* **2002**, *50*, 1343–1354. [CrossRef]
8. Wang, Y.; Stoica, P.; Li, J.; Marzetta, T.L. Nonparametric spectral analysis with missing data via the EM algorithm. *Digit. Signal Process.* **2005**, *15*, 191–206. [CrossRef]
9. Stoica, P.; Li, J.; Ling, J.; Cheng, Y. Missing data recovery via a nonparametric iterative adaptive approach. In Proceedings of the 2009 IEEE International Conference on Acoustics, Speech and Signal Processing, Taipei, Taiwan, 19–24 April 2009; IEEE: Piscataway, NJ, USA, 2009; pp. 3369–3372.
10. Donoho, D.L. Compressed sensing. *IEEE Trans. Inf. Theory* **2006**, *52*, 1289–1306. [CrossRef]
11. Candès, E.J.; Romberg, J.; Tao, T. Robust uncertainty principles: Exact signal reconstruction from highly incomplete frequency information. *IEEE Trans. Inf. Theory* **2006**, *52*, 489–509. [CrossRef]
12. Yang, J.; Thompson, J.; Huang, X.; Jin, T.; Zhou, Z. Random-frequency SAR imaging based on compressed sensing. *IEEE Trans. Geosci. Remote Sens.* **2012**, *51*, 983–994. [CrossRef]
13. Bi, H.; Zhu, D.; Bi, G.; Zhang, B.; Hong, W.; Wu, Y. FMCW SAR sparse imaging based on approximated observation: An overview on current technologies. *IEEE J. Sel. Top. Appl. Earth Obs. Remote Sens.* **2020**, *13*, 4825–4835. [CrossRef]
14. Zhou, K.; Li, D.; He, F.; Quan, S.; Su, Y. A Sparse Imaging Method for Frequency Agile SAR. *IEEE Trans. Geosci. Remote Sens.* **2022**, *60*, 1–16. [CrossRef]
15. Yang, J.; Thompson, J.; Huang, X.; Jin, T.; Zhou, Z. Segmented reconstruction for compressed sensing SAR imaging. *IEEE Trans. Geosci. Remote Sens.* **2013**, *51*, 4214–4225. [CrossRef]
16. Sun, S.; Zhu, G.; Jin, T. Novel methods to accelerate CS radar imaging by NUFFT. *IEEE Trans. Geosci. Remote Sens.* **2014**, *53*, 557–566.
17. Qian, Y.; Zhu, D. High-resolution SAR imaging from azimuth periodically gapped raw data via generalised orthogonal matching pursuit. *Electron. Lett.* **2018**, *54*, 1235–1237. [CrossRef]
18. Gorham, L.A.; Rigling, B.D. Scene size limits for polar format algorithm. *IEEE Trans. Aerosp. Electron. Syst.* **2016**, *52*, 73–84. [CrossRef]
19. Chen, J.; An, D.; Wang, W.; Luo, Y.; Chen, L.; Zhou, Z. Extended Polar Format Algorithm for Large-Scene High-Resolution WAS-SAR Imaging. *IEEE J. Sel. Top. Appl. Earth Obs. Remote Sens.* **2021**, *14*, 5326–5338. [CrossRef]
20. Liu, K.; Yu, W.; Lv, J. Azimuth interrupted FMCW SAR for high-resolution imaging. *IEEE Geosci. Remote Sens. Lett.* **2020**, *19*, 4001105. [CrossRef]
21. Liu, K.; Yu, W.; Lv, J.; Tang, Z. Parameter Design and Imaging Method of Spaceborne Azimuth Interrupted FMCW SAR. *IEEE Geosci. Remote Sens. Lett.* **2021**, *19*, 4015505. [CrossRef]
22. Wu, J.; Feng, D.; Wang, J.; Huang, X. SAR Imaging from Azimuth Missing Raw Data via Sparsity Adaptive StOMP. *IEEE Geosci. Remote Sens. Lett.* **2021**, *19*, 4501605. [CrossRef]
23. Jiang, N.; Wang, J.; Feng, D.; Kang, N.; Huang, X. SAR Imaging Method for Moving Target With Azimuth Missing Data. *IEEE J. Sel. Top. Appl. Earth Obs. Remote Sens.* **2022**, *15*, 7100–7113. [CrossRef]
24. Jiang, N.; Xin, Q.; Zhu, J.; Feng, D.; Wang, J.; Huang, X. Enhancement synthetic aperture imaging algorithm with azimuth missing data. *Electron. Lett.* **2023**, *59*, e12729. [CrossRef]
25. Cumming, I.G.; Wong, F.H. Digital processing of synthetic aperture radar data. *Artech House* **2005**, *1*, 108–110.
26. Wang, J.; Kwon, S.; Shim, B. Generalized orthogonal matching pursuit. *IEEE Trans. Signal Process.* **2012**, *60*, 6202–6216. [CrossRef]
27. Liu, K.; Yu, W. Interrupted FMCW SAR imaging via sparse reconstruction. In Proceedings of the IGARSS 2020—2020 IEEE International Geoscience and Remote Sensing Symposium, Waikoloa, HI, USA, 26 September–2 October 2020; IEEE: Piscataway, NJ, USA, 2020; pp. 1564–1567.

28. Rigling, B.D.; Moses, R.L. Taylor expansion of the differential range for monostatic SAR. *IEEE Trans. Aerosp. Electron. Syst.* **2005**, *41*, 60–64. [CrossRef]
29. Walker, J.L. Range-Doppler imaging of rotating objects. *IEEE Trans. Aerosp. Electron. Syst.* **1980**, *AES-16*, 23–52. [CrossRef]
30. Quegan, S. Spotlight Synthetic Aperture Radar: Signal Processing Algorithms. *J. Atmos. Sol.-Terr. Phys.* **1997**, *59*, 597–598. [CrossRef]
31. Xie, Z.; Fan, C.; Xu, Z.; Zhu, J.; Huang, X.; Senior Member, I. Robust joint code-filter design under uncertain target interpulse fluctuation. *Signal Process.* **2022**, *201*, 108687. [CrossRef]
32. Xie, Z.; Xu, Z.; Fan, C.; Han, S.; Huang, X. Robust Radar Waveform Optimization Under Target Interpulse Fluctuation and Practical Constraints Via Sequential Lagrange Dual Approximation. *IEEE Trans. Aerosp. Electron. Syst.* **2023**, *early access*.

*remote sensing*

MDPI

Communication

Enhanced Doppler Resolution and Sidelobe Suppression Performance for Golay Complementary Waveforms

Jiahua Zhu [1,2], Yongping Song [3,*], Nan Jiang [3], Zhuang Xie [3], Chongyi Fan [3] and Xiaotao Huang [3]

[1] College of Meteorology and Oceanography, National University of Defense Technology, Changsha 410073, China; zhujiahua1019@hotmail.com
[2] Hunan Key Laboratory for Marine Detection Technology, Changsha 410073, China
[3] College of Electronic Science and Technology, National University of Defense Technology, Changsha 410073, China; jiangnan@nudt.edu.cn (N.J.); xiezhuang18@nudt.edu.cn (Z.X.); chongyifan@nudt.edu.cn (C.F.); xthuang@nudt.edu.cn (X.H.)
* Correspondence: songyongping08@nudt.edu.cn

Abstract: An enhanced Doppler resolution and sidelobe suppression have long been practical issues for moving target detection using Golay complementary waveforms. In this paper, Golay complementary waveform radar returns are combined with a proposed processor, the pointwise thresholding processor (PTP). Compared to the pointwise minimization processor (PMP) illustrated in a previous work, which could only achieve a Doppler resolution comparable to existing methods, this approach essentially increases the Doppler resolution to a very high level in theory. This study also introduced a further filtering process for the delay-Doppler map of the PTP, and simulations verified that the method results in a delay-Doppler map virtually free of range sidelobes.

Keywords: complementary waveforms; pointwise thresholding processor; Doppler resolution; sidelobe suppression

Citation: Zhu, J.; Song, Y.; Jiang, N.; Xie, Z.; Fan, C.; Huang, X. Enhanced Doppler Resolution and Sidelobe Suppression Performance for Golay Complementary Waveforms. *Remote Sens.* **2023**, *15*, 2452. https://doi.org/10.3390/rs15092452

Academic Editor: Andrzej Stateczny

Received: 22 March 2023
Revised: 22 April 2023
Accepted: 28 April 2023
Published: 6 May 2023

1. Introduction

Due to complementarity, Golay complementary waveforms are effective at producing a satisfactory resolution range in a delay-Doppler map, as well as theoretical free-range side lobes at zero Doppler. Nevertheless, obvious range sidelobes are induced in nonzero Doppler intervals by their sensitivity to Doppler mismatch during matched filtering, and it is hard for conventional sidelobe suppression methods such as windowing to eliminate them.

A decade ago, Calderbank and Pezeshki et al. addressed the above problem by carefully designing Golay complementary waveforms in a specific transmitted order, named the Prouhet–Thue–Morse (PTM) design, which caused a satisfactory reduction in the range sidelobes in a narrow band around zero Doppler in the delay-Doppler map [1,2]. In a similar way, Suvorova et al. extended the idea of transmitted order design from the PTM sequence to the Reed–Müller codes, achieving a minimum range sidelobe level at a given Doppler bin in the delay-Doppler map [3]. This was further studied by Dang et al., who presented a binomial design (BD) algorithm that assigns weights to the matched filtering sequence of Golay complementary waveforms and significantly expands the sidelobe blanking area at the cost of an obvious decrease in the Doppler resolution [4]. Based on these works, Wu et al. employed semidefinite programming as a novel method to design complementary waveforms for improved sidelobe suppression as well as Doppler resolution [5,6].

However, the aforementioned methods only either preserve the Doppler resolution (none of them exceed the resolution of the conventional Golay pair) or enlarge the side-lobe blanking area, but they cannot achieve both at the same time. From another view, pointwise processing [7] (or cell-by-cell processing in some publications [8,9]) has been extensively researched in the processing of radar images to integrate the advantages of two figures and produce a further improvement in the signal-to-noise ratio (SNR). In our early work [10], a pointwise minimization processor (PMP) was proposed to combine the delay-Doppler maps of the BD algorithm with a weighted average Doppler (WD) algorithm under Golay waveforms, which maintained the Doppler resolution as well as a large side-lobe suppression area. However, as described before, the results of the PMP cannot exceed the original Doppler resolution of the waveform. Therefore, another designed pointwise thresholding processor (PTP) is proposed in this research to replace the PMP, which can further increase the Doppler resolution. A further filtering process is then applied based on the delay-Doppler map of the PTP, which almost eliminates the range sidelobes.

In the remainder of the manuscript, a brief introduction to Golay pairs and PMP is first given in Section 2, then the PTP is introduced. Section 3 presents the simulation results of the PMP and the PTP and compares the performance under fixed and randomized scenarios. The delay-Doppler map of a further filtering process after the PTP is also simulated to illustrate the improved sidelobe suppression performance. The conclusion and future directions are discussed in Section 4.

2. Golay Complementary Waveforms and Pointwise Processors

2.1. Golay Pairs

A Golay pair (complementary waveforms) consists of two length L sequences, $x(l)$ and $y(l)$, with several unimodular (±1) values/chips in each sequence [11]. The time width of each pair is LT_c (T_c for each chip). This waveform scheme is well known for its complementarity, i.e., the autocorrelation of the sequence pair is

$$C_x(k) + C_y(k) = 2L\delta(k), \quad k = -(L-1), ..., (L-1) \tag{1}$$

where $C_x(k)$ and $C_y(k)$ are the autocorrelation outputs of $x(l)$ and $y(l)$ at lag k, respectively, and $\delta(k)$ is the Kronecker delta function.

The sequence pair cannot be transmitted in the time domain before modulating a baseband pulse $\Omega(t)$ with unit energy on each chip, which means the transmitted sequences are given as

$$\begin{cases} x(t) = \sum\limits_{l=0}^{L-1} x(l)\Omega(t - lT_c) \\ y(t) = \sum\limits_{l=0}^{L-1} y(l)\Omega(t - lT_c) \end{cases} \tag{2}$$

where

$$\int_{-T_c/2}^{T_c/2} |\Omega(t)|^2 dt = 1. \tag{3}$$

Next, the transmission of either $x(t)$ or $y(t)$ is determined by a (P, Q) pulse train. Here, $P = \{p(n)\}_{n=0}^{N-1}$ is a binary sequence and the transmitted pulses are presented as

$$z_P(t) = \sum_{n=0}^{N-1} p(n)x(t - nT) + [1 - p(n)]y(t - nT) \tag{4}$$

where $P = \{0, 1, 0, 1 ...\}$ denotes the standard transmission order and T represents the pulse repetition interval (PRI). On the other hand, $Q = \{q(n)\}_{n=0}^{N-1}$ stands for the positive real number weights on the radar returns, where an all 1 sequence is the standard weighting. Next, the signal for matched filtering is written as

$$z_Q(t) = \sum_{n=0}^{N-1} q(n)\{p(n)x(t - nT) + [1 - p(n)]y(t - nT)\} \tag{5}$$

Specifically, the BD algorithm [4] designs Q as a binomial sequence, i.e., $Q = \{C_{N-1}^n\}_{n=0}^{N-1}$.

According to [12], we then calculate the delay-Doppler map of Golay complementary waveforms as follows:

$$\chi(t, F_D) = \int_{-\infty}^{+\infty} z_P(s) \exp(j2\pi F_D s) z_Q^*(t - s) \, ds \tag{6}$$

where "*" denotes the complex conjugation.

Based on the Equations (4) and (5), the delay-Doppler map of the Golay pair is further expanded as

$$\begin{aligned} \chi(t, F_D) = &\frac{1}{2} \sum_{k=-L+1}^{L-1} [C_x(k) + C_y(k)] \sum_{n=0}^{N-1} q(n) \exp(j2\pi F_D nT) C_\Omega(t - kT_c - nT) \\ &- \frac{1}{2} \sum_{k=-L+1}^{L-1} [C_x(k) - C_y(k)] \sum_{n=0}^{N-1} \left\{ \begin{array}{l} (-1)^{p(n)} q(n) \exp(j2\pi F_D nT) \\ C_\Omega(t - kT_c - nT) \end{array} \right\} \end{aligned} \tag{7}$$

The first item contains $[C_x(k) + C_y(k)]$, which is an impulse function due to the complementarity; thus, the sidelobe in the delay-Doppler map is only influenced by the second item.

For transmission order design methods such as standard order and PTM design, $q(n)$ is always 1; thus, the sub-item

$$\Theta = \sum_{n=0}^{N-1} (-1)^{p(n)} \exp(j2\pi F_D nT) \tag{8}$$

reaches 0 at $\theta = 2\pi n F_D T = \frac{2n\pi}{N}$, which means it is free of a sidelobe along the θ-Doppler axes and the Doppler resolution of the target is $\frac{2\pi}{N}$. However, a significant sidelobe can be observed at other Doppler axes other than θ. Moreover, for the PTM, it is easy to calculate that Θ is negligible if $2\pi F_D T$ is small.

For receiving weight design methods, i.e., the BD algorithm, $p(n)$ alternates between 1 and 0, while $q(n)$ is the coefficient of the binomial. Then, the sub-item Θ is further expressed as

$$\Theta = \sum_{n=0}^{N-1} (-1)^n C_{N-1}^n \exp(j2\pi F_D nT) = [1 - \exp(j2\pi F_D T)]^{N-1} \tag{9}$$

Obviously, the expression does not have zero points (which means the Doppler resolution is poor), while the value is an exponential function and its absolute value exponentially increases as $2\pi F_D T$ increases. This is the reason why a large sidelobe blanking area is generated.

The delay-Doppler maps of previous works are plotted in Figure 1 for a better understanding. As is demonstrated, the delay-Doppler maps of standard order and PTM design are divided into "grids" by the zero points, while the BD algorithm obtains a large blanking area with an exponential-like sidelobe on the side. The delay resolutions of the aforementioned approaches are all $2T_c$, which is the width of the impulse function.

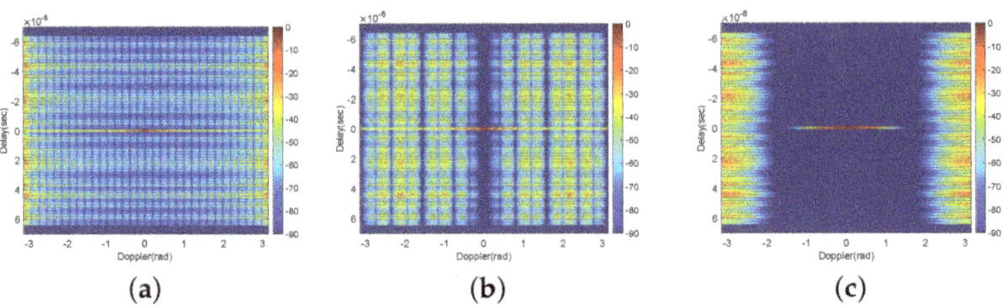

Figure 1. Delay-Doppler maps: (**a**) standard order; (**b**) PTM design; (**c**) BD algorithm (the unit of the colorbar is dB, $N = 32$).

2.2. Pointwise Minimization Procedure

Figure 2 describes the procedure presented in our early work [10], where $\chi_{\mathrm{BD}}(t, F_{\mathrm{D}})$, $\chi_{\mathrm{WD}}(t, F_{\mathrm{D}})$ and $\chi(t, F_{\mathrm{D}})$, respectively, stand for the delay-Doppler maps of the BD algorithm, the WD algorithm and this procedure. Specifically the "Pointwise Processor" in [10], represents the *pointwise minimization processor (PMP)*, whose result is denoted as $\chi_{\mathrm{PMP}}(t, F_{\mathrm{D}})$.

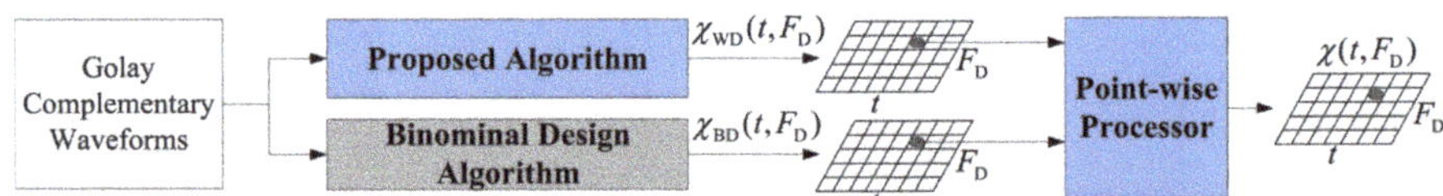

Figure 2. Demonstration of the procedure: pointwise processor PMP or PTP.

The WD algorithm is demonstrated based on [3] to select the the optimal transmission order of Golay complementary waveforms (a standard weighted Q is set for this algorithm), which minimizes the sidelobes near a known Doppler value. Here, we employ the mean target Doppler $\bar{f}_{\mathrm{d}}$ associated with their amplitudes to bring the sidelobe blanking area closer to the weak targets [13]:

$$\bar{f}_{\mathrm{d}} = \begin{cases} \dfrac{\sum_{h=1}^{H} \hat{f}_{\mathrm{d}_h}}{H} & \text{same } \hat{A}_h, \\[2ex] \dfrac{\sum_{h=1}^{H} (1-\hat{A}_h)\hat{f}_{\mathrm{d}_h}}{\sum_{h=1}^{H} (1-\hat{A}_h)} & \text{otherwise.} \end{cases} \tag{10}$$

where $\hat{A}_h$ and $\hat{f}_{\mathrm{d}_h}$ are the normalized amplitude and Doppler of the hth target, respectively, and H is the number of targets in the delay-Doppler map. A tracker is usually used to estimate the target magnitude and Doppler from the past detections [14].

Already existing methods, such as [1–4], etc., are able to suppress the range sidelobes and improve the signal-to-noise ratio (SNR) near the targets to different extents. However, they cannot reduce the overall sidelobe magnitude in the underlying surveillance window. Therefore, a PMP was proposed in our previous paper as a nonlinear pointwise processor, which achieves sidelobe suppression involving sidelobe power reduction and does not cause target loss, as verified by technical simulations.

$$\chi_{\mathrm{PMP}}(t, F_{\mathrm{D}}) = \min\{\chi_{\mathrm{BD}}(t, F_{\mathrm{D}}), \chi_{\mathrm{WD}}(t, F_{\mathrm{D}})\} \tag{11}$$

The advantage of the PMP is that it maintains the ideal large range sidelobe blanking region (in which the range sidelobes are less than -90 dB) provided by the BD algorithm and the acceptable Doppler resolution of the WD algorithm, based on the assumption that the targets are stable during the whole radar illumination. As described before, a drawback of the PMP is that it can only retain the improved Doppler resolution and the lower sidelobes of the two approaches. It still needs further enhancement when the Doppler resolution and sidelobe suppression performance fail to meet the requirements; thus, we propose the PTP in the following.

2.3. Pointwise Thresholding Procedure

The "Pointwise Processor" in Figure 2 is defined as the *pointwise thresholding processor* (PTP) and $\chi_{\text{PTP}}(t, F_D)$ is defined as the output of the PTP. Then, the processor is expressed as

$$\chi_{\text{PTP}}(t, F_D) = \begin{cases} \chi_{\text{WD}}(t, F_D) + \chi_{\text{BD}}(t, F_D), & |\chi_{\text{WD}}(t, F_D) - \chi_{\text{BD}}(t, F_D)| < thr(\text{dB}) \\ 0, & \text{otherwise} \end{cases} \tag{12}$$

where the threshold *thr* is artificially delimited considering the magnitude difference of targets and sidelobes. As the most important parameter in the processor, it directly influences the sidelobe level and the resolution (and also the performance of further filtering that will be discussed later). A small value will have no effect on the enhancement in Doppler resolution and sidelobe blanking performance, while a too large threshold may also blank the targets. Obviously $thr \geq 0$ dB, and we commonly consider it nonsensical if $thr > 10$ dB, since the radar return fluctuation of two illuminations caused by target micro-motion and other interferences normally cannot reach such a high level. In this paper, we choose $thr = 2$ dB as an example [15], but further research needs to be performed for better determination of the practical threshold. Under the same assumption, the PTP is expected to bring a further increase in the Doppler resolution and suppress the sidelobe magnitude compared to the PMP, which will be illustrated by the simulations in the next section.

3. Simulation and Further Discussion

The PTP is verified in simulations with the following global parameters when no other demonstration is presented. The targets in the simulations are set as Swerling II targets with 10% fluctuation in the radar cross-section (RCS).

f_c, carrier frequency: 1 GHz;
B, bandwidth: 50 MHz;
f_{ts}, time sampling rate: 2 B;
f_{ds}, Doppler sampling rate: 0.01 rad;
T, PRI: 50 μs;
N, pulse number: 32;
L, chip number of Golay pair: 64;
T_c, chip interval: 0.1 μs;
$E \sim \mathcal{CN}(0, 1)$, complex Gaussian zero-mean white noise: -10 dB (i.e., $SNR = 10$ dB).

3.1. Fixed Scenario

We first consider a fixed scenario with three targets (one weak and two strong targets), whose ground truth locations and magnitudes are listed in Table 1 and shown in Figure 3. Note that two strong targets can only be separated from the Doppler.

Table 1. Simulated target locations in the fixed scenario.

Target	Delay	Doppler	Magnitude
Target No. 1	$\tau_1 = 16.6$ μs	$f_{d_1} = -0.4$ rad	0 dB
Target No. 2	$\tau_2 = 16.6$ μs	$f_{d_2} = -0.9$ rad	0 dB
Target No. 3	$\tau_3 = 22$ μs	$f_{d_3} = 2.4$ rad	-20 dB

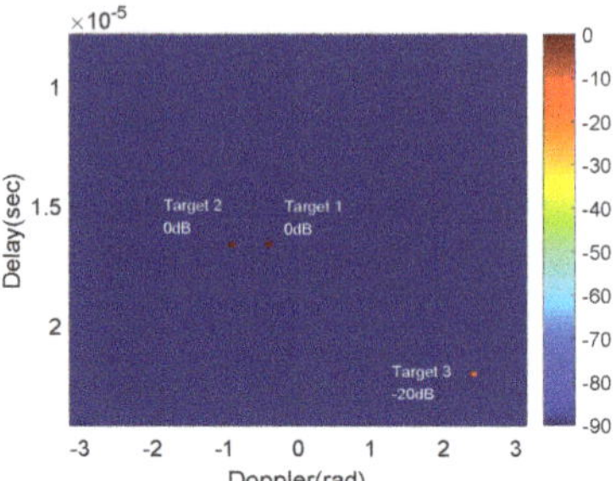

Figure 3. The ground truth locations and magnitudes of targets.

The delay-Doppler maps of the BD and WD algorithms and the outputs of the PMP and PTP with *thr* = 2 dB are given in Figure 4, and the comparison results of PMP and PTP in terms of the delay cross-section and Doppler cross-section, respectively, at all the targets location are then illustrated in Figures 5 and 6. Obviously, though PMP maintains the Doppler resolution of the WD algorithm, it performs much worse than the PTP. In addition, the overall SNR of the PTP in theory is also remarkably higher than the PMP (which may result in a higher performance during target detection). However, the processing times of the PMP and PTP will be twice those of separately using the BD or WD algorithm.

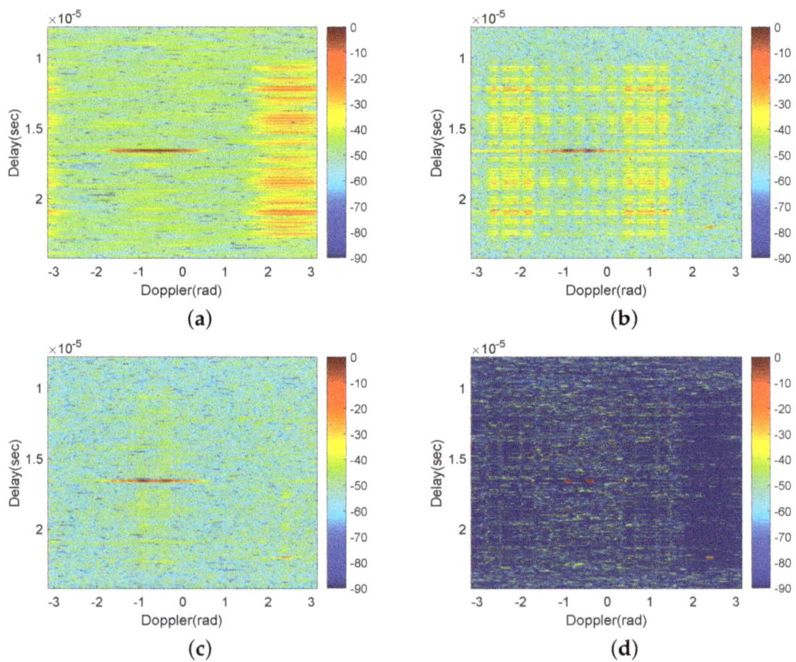

Figure 4. The results (in dB) of (**a**) the BD algorithm; (**b**) the WD algorithm; (**c**) the PMP; (**d**) the PTP.

The PTP results at different thresholds are also compared in Figure 7. When *thr* = 1 dB, the weak target is nearly blanked by the PTP, while two false targets near the strong targets may be detected if *thr* is increased to 8 dB. This further verified the previous illustration of the thresholding.

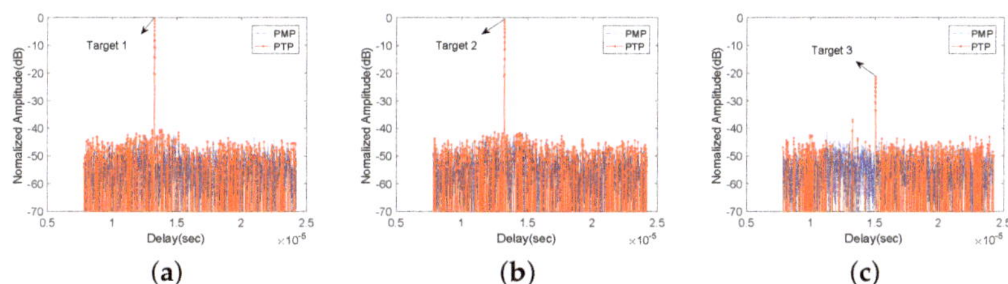

Figure 5. The delay cross-section of (**a**) target 1; (**b**) target 2; (**c**) target 3 using the PMP and PTP.

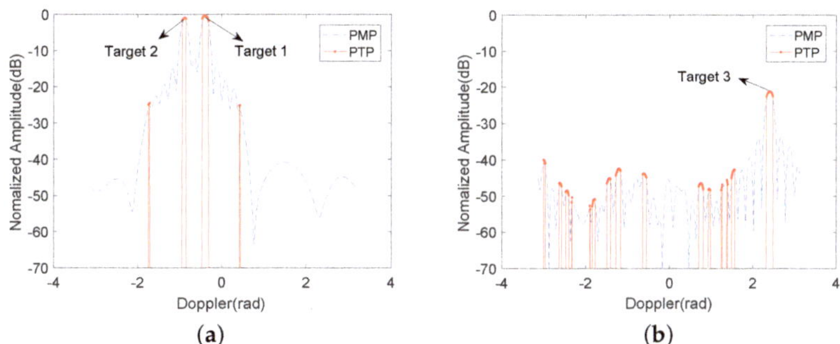

Figure 6. The Doppler cross-section of (**a**) target 1 and target 2 and (**b**) target 3 using the PMP and PTP.

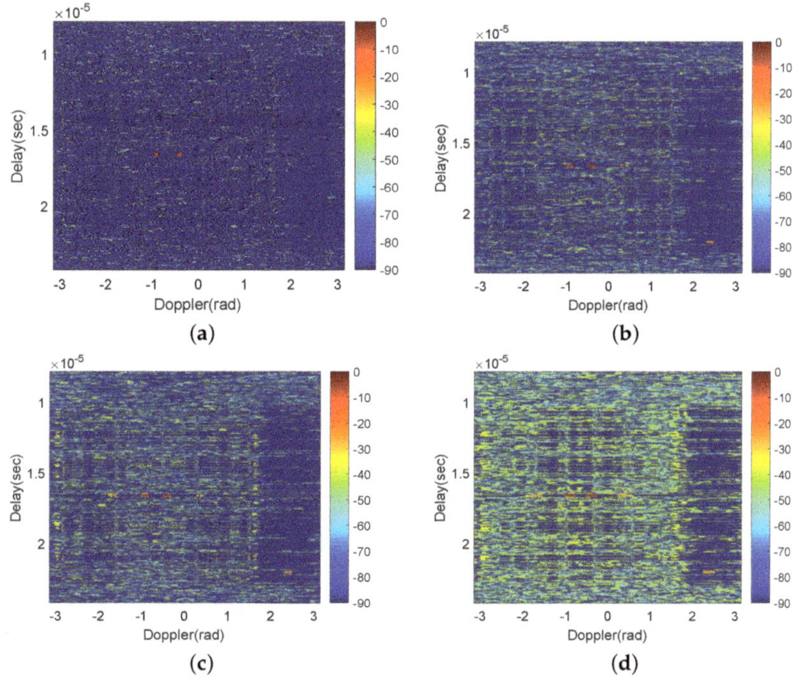

Figure 7. Delay-Doppler maps of the PTP when (**a**) *thr* = 1 dB; (**b**) *thr* = 2 dB; (**c**) *thr* = 4 dB; (**d**) *thr* = 8 dB (the unit of the colorbar is dB).

3.2. Further Filtering for the PTP

Even though we have discussed the better target detection ability of the PTP than PMP, the targets are still not easily visually detected. Therefore, a further filtering for PTP is proposed in this subsection, which picks the targets through its displayed characters in the map and almost eliminates the range sidelobes after filtering.

According to the delay-Doppler maps we obtain, it is found that the targets are displayed typically (approximate to a rectangle, whose size is related to the threshold *thr* as well as the pulse number N) in the results of the PTP, whereas the range sidelobes are usually irregular and cannot be described by the common shape. This character gives us a chance to further filter the targets and suppress the sidelobes in the images. A particular rectangle with a similar size to the target can be employed to search the delay-Doppler map after the PTP, and the target is considered to be found when all the values in the rectangle are higher than -90 dB (since the original effect of the PTP suppresses some of the range sidelobes lower than -90 dB). Note that the size of the target should be evaluated first before the above operation.

Based on the previous illustration in [13] and our subsequent analysis, we learned that the delay resolution of the target after PTP mainly depends on the threshold *thr*, while the Doppler resolution is primarily influenced by the pulse number N. On the other hand, the original delay and Doppler resolution of target can be analytically calculated as $2T_c$ and $\frac{2\pi}{N}$, as described in Section 2.1, which means they occupy 20 and 19 pixels in the delay and Doppler axes, respectively. Though it is still hard to analytically calculate the size of this particular rectangle, which is expressed as the number of pixels occupied in the row (delay axis, R_r) and column (Doppler axis, R_c) of the delay-Doppler map, a numerical fitting could be adopted to obtain an experimental Equation (13) with a certain calculable basis to illustrate the R_r and R_c of the rectangle. By observing Figures 4 and 7, we find that the delay and Doppler pixels of the target both shrink by about a half after the PTP, while the number of delay pixels needs to be further increased, which is nearly equal to the value of *thr*. Therefore, the experimental Equation (13) is written as follows, where the operator "round" means calculating to the nearest integer.

$$\begin{cases} R_r = \mathrm{round}(thr + T_c f_{ts}) \\ R_c = \mathrm{round}\left(\frac{\pi}{N f_{ds}}\right) \end{cases} \tag{13}$$

This experimental equation is only used for an explanation of the filtering process that is handled in this work. A more careful deduction of the calculation of this rectangle in practical studies will be the next step of our research.

The outputs of further filtering of the PTP at different thresholds and pulse numbers are demonstrated in Figure 8, which exhibit that further filtering makes the delay-Doppler map almost free of range sidelobes and the targets can be clearly visually recognized. Nevertheless, the lower threshold may lead to a higher probability of miss detection (Target 3 is lost under $thr = 1$ dB), and the increase in pulse number may generate more false targets (some false targets arise when $N = 64$).

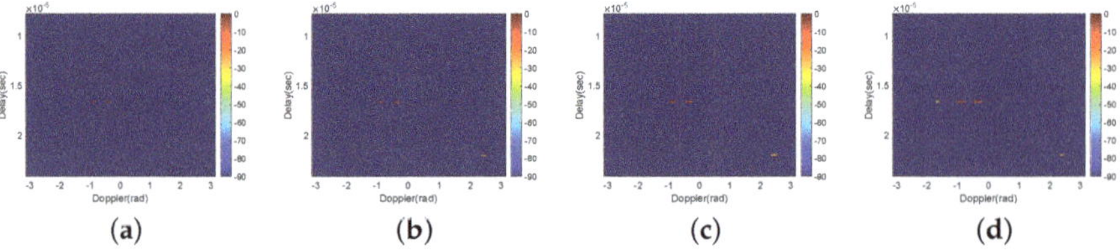

(a)   (b)   (c)   (d)

Figure 8. *Cont.*

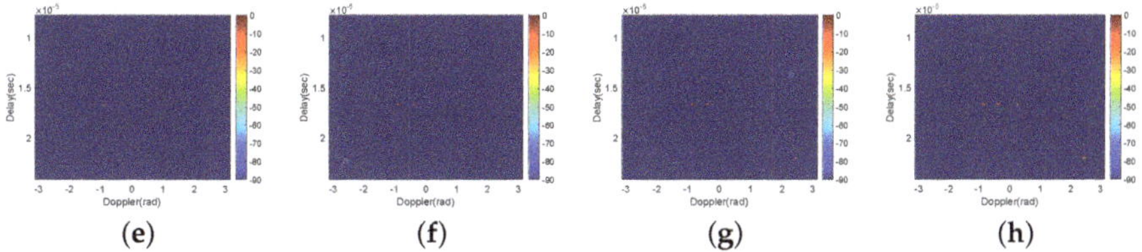

(e)   (f)   (g)   (h)

Figure 8. The outputs of further filtering of the PTP when (**a**) *thr* = 1 dB, N = 32; (**b**) *thr* = 2 dB, N = 32; (**c**) *thr* = 4 dB, N = 32; (**d**) *thr* = 8 dB, N = 32; (**e**) *thr* = 1 dB, N = 64; (**f**) *thr* = 2 dB, N = 64; (**g**) *thr* = 4 dB, N = 64; (**h**) *thr* = 8 dB, N = 64 (the unit of the colorbar is dB).

3.3. Randomized Scenario

In this scenario, several cases of Swerling II targets with different numbers are uniformly distributed in the delay-Doppler map. Based on the previous global parameters, we consider the following four cases:

(1)   Target number: 2 (one strong and one weak);
(2)   Target number: 3 (one strong and two weak);
(3)   Target number: 4 (two strong and two weak);
(4)   Target number: 5 (three strong and two weak).

thr and N are fixed as 2 dB and 32, respectively, in all cases.

For the sake of a clearer explanation, the target detection thresholds in these cases are set to the magnitude of the weakest target, which means that all the targets can be detected but false targets may also exist in the range sidelobes. Again, the proper setting of realistic detection thresholds requires further consideration.

A Monte Carlo simulation was operated for 1000 iterations for each case above, and the number of correct detections was calculated. A correct detection is counted when targets are all detected without any false targets. The correct detection occurrences of the PMP, the PTP and the PTP after further filtering are shown in Figure 9.

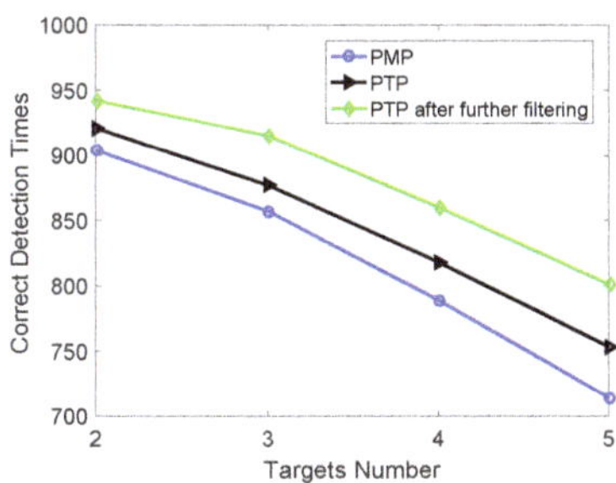

Figure 9. Correct detection occurrences of the PMP, the PTP and the PTP after further filtering.

As is observed, the PTP provided in this paper outperforms the previous proposed PMP due to a higher overall SNR, as discussed before. The output of the PTP after further filtering has even more correct detection occurrences than the others since the improved range sidelobe effect is achieved.

4. Conclusions

In this paper, a signal processing method involving a PTP is proposed for Golay complementary waveforms to achieve an essentially enhanced Doppler resolution compared to the previously proposed PMP, which can only maintain the original Doppler resolution of this waveform scheme. To solve the visual recognition problem of targets in a delay-Doppler map, further filtering of the PTP by extracting the targets more precisely through a particular rectangle is also employed for a significant improvement in range sidelobe suppression. The performance of the above methods are verified by simulation results. Our future research avenues may concern practical experiments of adaptive thresholding (such as constant false alarm rate, CFAR) and target detection under this waveform scheme, as well as some robust waveform optimization methods for complex target detection [16,17].

Author Contributions: Conceptualization and supervision, X.H.; methodology, J.Z. and Y.S.; software validation, J.Z., N.J. and Z.X.; writing—original draft preparation J.Z.; writing—review and editing, N.J., Z.X. and C.F.; funding acquisition, J.Z. All authors have read and agree to the published version of the manuscript.

Funding: This work was partly supported by the National Natural Science Foundation of China under grant 62101573 and the scientific research project of National University of Defense Technology under grant ZK20-35.

Data Availability Statement: Not applicable.

Acknowledgments: The authors would like to thank Bill Moran and Xuezhi Wang from the University of Melbourne and RMIT University for discussions on this work, and would also like to thank the editors and reviewers for their valuable suggestions for the improvement in this manuscript.

Conflicts of Interest: The authors declare no conflict of interest.

References

1. Pezeshki, A.; Calderbank, A.R.; Moran, W.; Howard, S.D. Doppler resilient Golay complementary waveforms. *IEEE Trans. Inform. Theory* **2008**, *54*, 4254–4266. [CrossRef]
2. Calderbank, R.; Howard, S.; Moran, B. Waveform diversity in radar signal processing. *IEEE Signal Process. Mag.* **2009**, *26*, 32–41. [CrossRef]
3. Suvorova, S.; Howard, S.; Moran, B.; Calderbank, R.; Pezeshki, A. Doppler resilience, Reed-Muller codes and complementary waveforms. In Proceedings of the 2007 Conference Record of the Forty-First Asilomar Conference on Signals, Systems and Computers, Pacific Grove, CA, USA, 4–7 November 2007; pp. 1839–1843.
4. Dang, W.; Pezeshki, A.; Howard, S.; Moran, W.; Calderbank, R. Coordinating complementary waveforms for sidelobe suppression. In Proceedings of the 2011 Conference Record of the Forty Fifth Asilomar Conference on Signals, Systems and Computers (ASILOMAR), Pacific Grove, CA, USA, 6–9 November 2011; pp. 2096–2100.
5. Wu, Z.; Wang, C.; Jiang, P.; Zhuo, Z. Range-Doppler sidelobe suppression for pulsed radar based on Golay complementary codes. *IEEE Signal Process. Lett.* **2020**, *27*, 1205–1209. [CrossRef]
6. Wu, Z.; Wang, C.; Jiang, P.; Zhuo, Z. Doppler resilient complementary waveform design for active sensing. *IEEE Sen. J.* **2020**, *20*, 9963–9976. [CrossRef]
7. Zhu, J.; Song, Y.; Fan, C.; Huang, X. Nonlinear processing for enhanced delay-Doppler resolution of multiple targets based on an improved radar waveform. *Signal Process.* **2017**, *130*, 355–364. [CrossRef]
8. Richards, M.A.; Scheer, J.A.; Holm, W.A. *Principles of Modern Radar Volume I-Basic Principles*; Scitech Publishing: Raleigh, NC, USA, 2010.
9. Ciuonzo, D. On time-reversal imaging by statistical testing. *IEEE Signal Process. Lett.* **2017**, *24*, 1024–1028. [CrossRef]
10. Zhu, J.; Wang, X.; Huang, X.; Suvorova, S.; Moran, B. Range sidelobe suppression for using Golay complementary waveforms in multiple moving target detection. *Signal Process.* **2017**, *141*, 28–31. [CrossRef]
11. Golay, M. Complementary series. *IRE Trans. Inform. Theory* **1961**, *7*, 82–87. [CrossRef]
12. Richards, M.A. *Fundamentals of Radar Signal Processing*; McGraw-Hill Education: New York, NY, USA, 2005.
13. Zhu, J.; Wang, X.; Huang, X.; Suvorova, S.; Moran, B. Golay Complementary Waveforms in Reed-Müller Sequences for Radar Detection of Nonzero Doppler Targets. *Sensors* **2018**, *18*, 192. [CrossRef] [PubMed]
14. Dang, W. Signal Design for Active Sensing. Ph.D. Dissertation, Colorado State University, Fort Collins, CO, USA, 2014.
15. Rasool, S.B.; Bell, M.R. Biologically inspired processing of radar waveforms for enhanced delay-Doppler resolution. *IEEE Trans. Signal Process.* **2011**, *59*, 2698–2709. [CrossRef]
16. Xie, Z.; Fan, C.; Xu, Z.; Zhu, J.; Huang, X. Robust joint code-filter design under uncertain target interpulse fluctuation. *Signal Process.* **2022**, *201*, 108687. [CrossRef]
17. Xie, Z.; Xu, Z.; Fan, C.; Han, S.; Huang, X. Robust radar waveform optimization under target interpulse fluctuation and practical constraints via sequential lagrange dual approximation. *IEEE Trans. Aero. Electron. Syst.* 2023, *accepted for publication*. [CrossRef]

remote sensing

Article

A Novel 3D ArcSAR Sensing System Applied to Unmanned Ground Vehicles

Yangsheng Hua [1,2], Jian Wang [1,*], Dong Feng [1] and Xiaotao Huang [1]

[1] College of Electronic Science and Technology, National University of Defense Technology, Changsha 410073, China; huayangsheng01@163.com (Y.H.); fengdong09@nudt.edu.cn (D.F.); xthuang@nudt.edu.cn (X.H.)

[2] Shanghai Electro-Mechanical Engineering Institute, Shanghai 201109, China

* Correspondence: jianwang_uwb@nudt.edu.cn; Tel.: +86-731-87003353

Abstract: Microwave radar has advantages in detection accuracy and robustness, and it is an area of active research in unmanned ground vehicles. However, the existing conventional automotive corner radar, which employs real-aperture antenna arrays, has limitations in terms of observable angle and azimuthal resolution. This paper proposes a novel 3D ArcSAR method to address this issue, which combines rotational synthetic aperture radar (SAR) and direction estimation algorithms. The method aims to reconstruct 3D images of 360° scenes and offers distinctive advantages in both azimuthal and altitudinal sensing. Nevertheless, due to the unique structural characteristics of vehicle SAR, it is limited to receiving only a single snapshot signal for 3D sensing. We propose a resolution algorithm based on ArcSAR and the iterative adaptive approach (IAA) to resolve the limitation. Furthermore, the errors in altitude angle estimation of the proposed algorithm and conventional algorithms are analyzed under various conditions, including different target spacing and signal-to-noise ratio (SNR). Finally, we design and implement a prototype of the 3D ArcSAR sensing system, which utilizes a millimeter-wave MIMO radar system and a rotating scanning mechanical system. The experimental results obtained from this prototype effectively validate the effectiveness of the proposed method.

Keywords: ArcSAR; direction estimation; unmanned ground vehicle; IAA; back projection algorithm (BPA); millimeter-wave radar; 3D sensing system

Citation: Hua, Y.; Wang, J.; Feng, D.; Huang, X. A Novel 3D ArcSAR Sensing System Applied to Unmanned Ground Vehicles. *Remote Sens.* **2023**, *15*, 4089. https://doi.org/10.3390/rs15164089

Academic Editor: Massimiliano Pieraccini

Received: 27 May 2023
Revised: 10 August 2023
Accepted: 11 August 2023
Published: 20 August 2023

1. Introduction

Unmanned ground vehicles, as a new type of transportation, represent the future trend in vehicle development. In recent years, many countries worldwide have recognized the significance of unmanned ground vehicles and implemented them as a national strategy supported by relevant policies. Related technologies have found initial applications in various fields, such as intelligent transportation, logistics, mines, and ports [1].

Among the key technologies in unmanned ground vehicles, real-time environmental sensing and understanding are prerequisites for platforms to accomplish their tasks autonomously [2,3]. Various environmental sensing technologies are advancing rapidly, each with its own advantages and disadvantages. Optical sensing technology excels at structured road and target recognition but falls short in detection accuracy. Multi-line LiDAR is too expensive for widespread implementation in ordinary vehicles. Infrared and ultrasonic sensing technologies have a short detection range [4–6]. In addition, the performance of these technologies degrades in challenging weather conditions such as darkness, rain, snow, and smoke, as well as in unstructured environments like vegetation cover [7].

As a novel means of environmental sensing for unmanned ground vehicles, microwave radar has inherent advantages in detection accuracy, cost, and environmental adaptability [8–10]. Consequently, it has garnered considerable attention in this research domain. Conventional automotive radar solutions, such as Continental's corner radar ARS408, employ real-aperture antenna arrays, which exhibit limitations in terms of observable angle and azimuthal resolution [11]. ARS408 radar has azimuthal resolutions of 12.3° and 3.2° within observable angles of $\pm60°$ and $\pm30°$, respectively [12]. Moreover, the azimuthal resolution decreases as the target deviates from the array's normal direction.

While the forward-looking direction is important for vehicle environmental sensing, the sense of rear and side information is equally vital for urban driving and detecting surrounding obstacles. Particularly in field environments, the mobility of unmanned ground vehicles relies on obtaining complete 360° environmental information. Unfortunately, the observable angle of conventional automotive corner radar is insufficient to meet the requirements of panoramic information. Consequently, existing solutions employ at least four or more corner radars to obtain panoramic information around the vehicle, thereby increasing sensor costs.

In response to the above situation, rotational SAR uses the rotating arm to detect outside the arc. By constructing an arc-shaped synthetic aperture, the effective observable angle can cover a 360° range. As a result, this novel SAR structure has been the subject of extensive research in recent years. In the field of ground deformation monitoring, this new all-time rotational SAR system is named ArcSAR [13,14]. Among them, the 2D ArcSAR has been verified using experiments [15,16]. In rotorcraft helicopters, ROSAR generates an arc-shaped synthetic aperture to observe the ground by rotating the helicopter rotors [17,18]. Here, we refer to the above SAR systems collectively as ArcSAR. While ArcSAR exhibits advantages in terms of azimuthal resolution and consistency, the current research efforts have mainly focused on 2D imaging, and the realization of 3D panoramic sensing based on microwave radar remains unaccomplished within the field of vehicle applications [19]. The absence of height information can result in distortions in the projected targets within 2D imaging results, leading to errors in target identification and distance estimation.

To address the problem of absent height information, this paper proposes a novel 3D ArcSAR method, which combines rotational SAR and direction estimation algorithms to reconstruct 3D images of 360° scenes. The 3D ArcSAR sensing system employs a millimeter-wave radar with multiple receivers installed at the end of the rotating arm to estimate the altitude angle, enabling 3D panoramic imaging during rotation. This ability enhances the accuracy of environmental understanding in road scenarios. Moreover, this system achieves panoramic 3D imaging with just one rotational scan by one radar.

As the 3D ArcSAR platform rotates during the scanning process, the location of the rotating arm is in constant change. Altitude angle estimation poses a challenge as only a single snapshot signal can be received per angle. Common resolution algorithms such as Music and APES require multiple snapshot signals to construct the sample covariance matrix, and the matrix would lose rank when only a snapshot signal is received [20–22]. Attempting to attain a full-rank sample covariance matrix by reducing the dimension of the matrix results in estimates with reduced resolution and robustness. The FFT algorithm does not require multiple snapshot signals, but its angle resolution is poor with a limited number of array antennas. Orthogonal Matching Pursuit (OMP) is a signal reconstruction algorithm that requires the number of targets to be known in advance [23,24]. IAA is a spectral estimation algorithm based on weighted least squares (WLS) estimation proposed by Li Jian at the University of Florida in 2010 [25–27], which has demonstrated good results in linear aperture experiments. In order to overcome the single snapshot limitation, we designed an IAA resolution algorithm based on circular apertures, which exhibits improved adaptability for unmanned ground vehicles.

In this paper, we propose the 3D ArcSAR method as a solution to the deficiencies in the sensing performance of conventional automotive radar in both azimuthal and altitudinal directions. To overcome the limitation that the vehicle SAR can only receive a single snapshot signal in 3D sensing, we designed a dedicated IAA resolution algorithm for this specific case. The main contributions of this paper are summarized as follows.

- The 3D ArcSAR method is proposed for reconstructing panoramic 3D images with rotational SAR and direction estimation techniques. This method effectively addresses the limitations of existing automotive corner radar systems, including shortcomings in altitude direction sensing, azimuthal resolution, and consistency. The 3D ArcSAR sensing system achieves panoramic 3D imaging with only one radar and one rotation scan. Therefore, it reduces the number and complexity of devices required and can better meet the sensing needs of unmanned ground vehicles.

- A resolution algorithm based on IAA was designed specifically for 3D ArcSAR. This algorithm overcomes the limitation of receiving only a single snapshot signal per angle. We analyze the errors in altitude angle estimation for both the proposed algorithm and conventional algorithms under varying conditions, such as target spacing and SNR. The proposed algorithm has superior resolution in the case of single snap, small antenna arrays, and an unknown number of targets compared to other existing methods.

- The 3D ArcSAR prototype is designed based on a millimeter-wave radar system and a rotating mechanical system. The radar system employs a low-cost, readily available commercial off-the-shelf (COTS) radar, facilitating easy deployment. The rotating mechanical system is designed to adapt to different scenes, with adjustable rotation speed and arm length. Additionally, it can be remotely controlled by a computer to facilitate experiments. The 3D ArcSAR prototype validates the superior resolution accuracy performance of the proposed algorithm and can be further utilized for experiments in various scenes in the future.

The remaining sections of this paper are organized as follows. Section 2 provides a detailed description of the model structure and the fundamental principles underlying rotational SAR imaging and direction estimation in 3D ArcSAR. In Section 3, simulations of 3D ArcSAR are conducted and the errors in altitude angle estimation for both the proposed algorithm and conventional algorithms under varying conditions are analysed. Section 4 presents the designed prototype of the 3D ArcSAR sensing system and provides the experimental results obtained with this prototype. Finally, conclusions are drawn in Section 5.

2. Three-Dimensional ArcSAR and Signal Processing

2.1. Signal Model and Geometry Model

2.1.1. 2D Structure of ArcSAR

Figure 1 shows a schematic diagram of ArcSAR's 2D structure. It consists of a rotating arm r extending from the central location of the rotating platform and a SAR radar system P for transmit and receive signals. The x and y axes indicate two mutually perpendicular directions parallel to the ground.

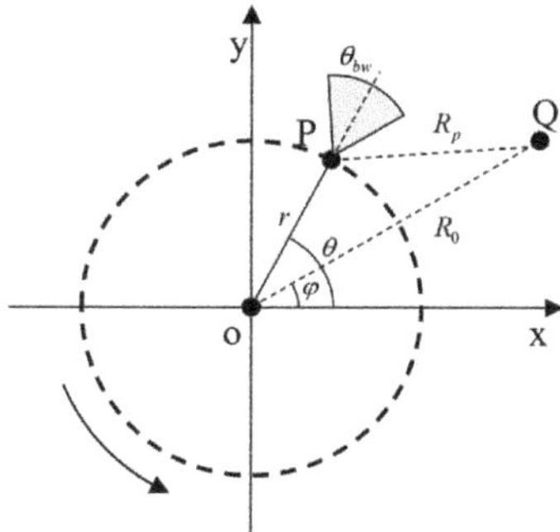

Figure 1. Schematic diagram of ArcSAR's 2D structure.

The rotating plane of the SAR radar system is chosen as the imaging plane. Q is the target (assume Q is on the imaging plane). θ is the rotating angle of the rotating arm. φ is the azimuth angle of the target. θ_{bw} is the antenna beam width. R_0 is the distance between Q and the center of the platform. R_P is the distance between Q and P, which can be expressed as

$$R_P = \sqrt{R_0{}^2 + r^2 - 2R_0 r \cos(\theta - \varphi)} \tag{1}$$

The time difference τ_i between the transmit signal and the receive signal can be expressed as

$$\tau_i = 2R_P/c \tag{2}$$

Take the linear frequency modulation continuous wave (LFMCW) as an example, and the transmit and receive signal models of LFMCW can be expressed as

$$S_{Tx}(t) = \text{rect}(t/T_c) \cdot \exp[j2\pi(f_0 t + \tfrac{1}{2}Kt^2)] \tag{3}$$

$$S_{Rx}(t) = \text{rect}[(t - \tau_i)/T_c] \cdot \delta_P \cdot \exp\{j2\pi[f_0(t - \tau_i) + \tfrac{1}{2}K(t - \tau_i)^2]\} \tag{4}$$

where t and T_c represent the time variation in the fast time dimension and the sweep time of each chirp signal. f_0 and K represent the carrier frequency and LFMCW slope. δ_P represents the scattering coefficient.

In addition, the beamwidth of the antenna is considered. After dechirp processing, the echo signal $S_{if}(t,\theta)$ can be expressed as

$$\begin{aligned} S_{if}(t,\theta) = {} & \text{rect}[(t - \tau_i)/T_c] \cdot \text{rect}((\theta - \varphi)/\theta_{bw}) \cdot \delta_P \cdot \\ & \exp[j2\pi(f_0 \tau_i + K\tau_i t - \tfrac{1}{2}K\tau_i{}^2)] \end{aligned} \tag{5}$$

2.1.2. Direction Estimation with Multiple Receivers

Figure 2 shows the geometric structure of the 3D ArcSAR sensing system, which acquires altitude information in the altitude direction by an antenna array. Combined with the antenna array and the azimuthal synthetic aperture produced by the antenna sweep, the system is able to generate panoramic 3D images. Here, the antenna array is configured as a uniform array. The z-axis indicates the direction perpendicular to the ground.

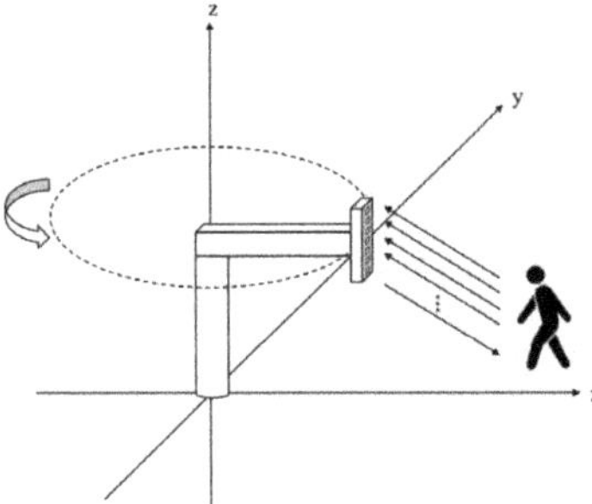

Figure 2. Geometric structure of the 3D ArcSAR sensing system.

Figure 3 shows the distance difference in receive signals between the different antennas. According to the fundamental principle in the direction estimation with multiple receivers, the signal travels $d\sin(\theta)$ further in the second antenna than the first.

The speed of electromagnetic waves is equal to the speed of light c. Then, the above distance difference is calculated to the time difference, assuming that the array configuration consists of M antennas, with the first antenna serving as the reference point. Therefore, the time difference of each antenna is

$$\Delta t = [0, \frac{d\sin(\theta)}{c}, \frac{2d\sin(\theta)}{c}, \cdots, \frac{(M-1)\sin(\theta)}{c}] \tag{6}$$

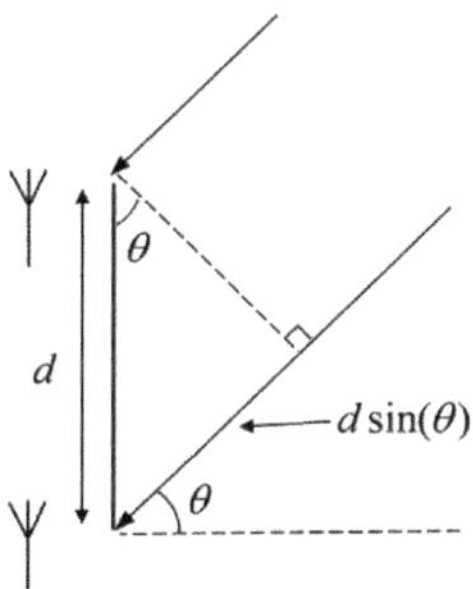

Figure 3. Schematic diagram of the distance difference in receive signals between the different antennas.

Relative to the first antenna, the phase difference of the signal arriving at each antenna is

$$\Delta\phi = [0, 2\pi f_0 \frac{d\sin(\theta)}{c}, 2\pi f_0 \frac{2d\sin(\theta)}{c}, \cdots, 2\pi f_0 \frac{(M-1)d\sin(\theta)}{c}]$$ (7)

The receive signal is defined as $s(N)$, where N is expressed as the number of snapshots of the receive signal. Then, the signal received by the multiple antenna array is

$$\mathbf{X}_{M\times N} = a(\theta)s(N) = [1, e^{-j2\pi f_0 \frac{d\sin(\theta)}{c}}, e^{-j2\pi f_0 \frac{2d\sin(\theta)}{c}}, \cdots, e^{-j2\pi f_0 \frac{(M-1)d\sin(\theta)}{c}}]^T s(N)$$ (8)

where $a(\theta)_{M\times 1}$ represents the steering vector, and it contains the angle information.

For a uniform linear array, the signal containing information about l directions can be expressed as

$$\mathbf{X}_{M\times N} = a(\theta_1)s_1(N) + a(\theta_2)s_2(N) + \cdots + a(\theta_l)s_l(N)$$ (9)

For a given antenna array, the mathematical form of the steering vector is known. As a result, FFT can be applied for altitude angle estimation according to Equation (8).

L directions are chosen equally within the angle range, and the steering matrix can be expressed as

$$\mathbf{A}_{M\times L} = [a(\theta_1), a(\theta_2), \cdots, a(\theta_L)]$$ (10)

The inner product of this steering matrix $\mathbf{A}$ and the receive signal with the target angle θ_0 can be expressed as

$$y = \mathbf{A}^H \mathbf{X} = \mathbf{A}^H a(\theta_0)s_0(N)$$ (11)

where the inner product result y is a scalar array. It goes through all the angles to find the maximum value, and the angle corresponding to the maximum value represents the estimated target angle.

In addition, for a uniform linear array with a known antenna spacing and number of antennas, the Rayleigh limit resolution of the altitude angle estimation is

$$\rho_s = \frac{\lambda}{M \times d} \times 0.886$$ (12)

where M represents the number of antennas. d represents the antenna spacing, and $\lambda/2$ is generally chosen.

2.2. Panoramic Image Formation by the Back Projection Algorithm

The Back Projection Algorithm (BPA) is a widely used SAR imaging algorithm. Unlike the existing ArcSAR frequency domain algorithms, which require a certain imaging angle, BPA utilizes azimuth accumulation imaging, enabling imaging at any azimuth position without restrictions on the size of the imaging angle. This feature makes BPA particularly advantageous for vehicle-mounted ArcSAR real-time perception, as it allows imaging results to be obtained at any desired azimuth position during platform rotation. Furthermore, BPA exhibits excellent parallelism, enabling real-time processing through multi-core parallel computing [28].

BPA can be described in four key steps:

1. Distance compression.

Distance compression is the matched filtering process. In the dechirp system, the received echo signal from each antenna can be directly sampled and FFT-processed to obtain the distance compression results. Without considering the beamwidth of the antennas, the result of the FFT of the signal (5) can be expressed as

$$S(f) = D\sin c[T_c(f - K\tau_i)]\exp(-j2\pi f\tau_i) \tag{13}$$

$$D = T_c\exp(j2\pi f_0\tau_i)\exp(j\pi K\tau_i^2) \tag{14}$$

where D represents the parameter unrelated to f, and τ_i represents the time difference between the transmit and receive signal at the target Q position. Equation (13) can observe that the target Q is compressed to the frequency $f = K\tau_i$, which gives preliminary distance information.

2. Data interpolation resampling.

In step 1, only the signal time difference between the radar system P and the target Q is considered for a particular moment. However, in order to cover the entire sensing range, it is essential to calculate the signal time differences for the complete range.

Interpolation resampling is employed subsequent to the distance compression process. The detection range is partitioned into a grid, with n distance gates set in each direction. The time difference between the transmit and receive signal at each distance gate position is represented by $\tau(n)$. Additionally, the frequency is represented by $K\tau(n)$. The signal after interpolation resampling can be expressed as

$$S(n) = D\sin c[T_cK(\tau(n) - \tau_i)]\exp(-j2\pi\tau_i \cdot K\tau(n)) \tag{15}$$

It is evident that the magnitude of $S(n)$ reaches its maximum when $\tau(n)$ is taken to τ_i.

3. Compensating for delayed phase.

The delayed phase is the additional phase of the echo signal relative to the transmit signal. In order to achieve phase-coherent accumulation of the echo signal from the same target at different moments, it is necessary to compensate for the delayed phase at each moment. The delayed phase compensation factor can be expressed as

$$G_{rcmc}(f_0) = \exp[j4\pi f_0\tau_i(n)] \tag{16}$$

The delayed phase compensation factor (16) and the data in step 2 are multiplied to obtain the echo signal amplitude $Y(n)$, which can be expressed as

$$Y(n) = S(n) \cdot G_{rcmc}(n) \tag{17}$$

4. Phase-coherent accumulation.

The position of the signal transmitted by the SAR system is constantly changing. In ArcSAR, the radar system P can receive the echo signal of the target Q on a segment of the circular trajectory. The angle size of this circular trajectory is equal to the beamwidth of the antenna.

Following that, a section of azimuthal data with an angle range equal to the antenna beamwidth are selected, and a window function with the same angle range is used to weight this section of data. The data values $Y_w(n)$ under this scene can be obtained as

$$Y_w(n) = Y(n) \cdot window(n) \tag{18}$$

where $window(n)$ represents the weighted window function.

The above four steps are systematically repeated for every position within the entire $360°$ scene. Finally, the resulting output is a 2D panoramic image.

It is known that the radial resolution is related to the signal bandwidth, and the azimuth resolution is related to the synthetic aperture length. However, in ArcSAR, the synthetic aperture length is determined by combining the arm length and the antenna beamwidth. As a result, the theoretical resolution in the radial and azimuth directions can be expressed as

$$\Delta r = \frac{0.886 \times c}{2B_r} \times k_r \tag{19}$$

$$\Delta \theta = \frac{0.886 \times \lambda}{4r \sin(\theta_{bw}/2)} \times k_\theta \tag{20}$$

where B_r and λ represents the bandwidth and central wavelength of the signal. The weighted window process widens the main lobe. Moreover, k represents the compensation for the weighted window function. For example, the compensation of the Hanning window is about 1.6269.

2.3. IAA-Based Angle Estimation in Altitude Direction

Due to the structure of the 3D ArcSAR sensing system, the antenna array obtains only one snapshot signal for each azimuth direction. As a result, common algorithms such as APES and Music are unsuitable for this scenario. These algorithms require multiple snapshots of the signal when constructing the sample covariance matrix since a full-rank matrix cannot be achieved from a single snapshot signal alone. Attempting to attain a full-rank sample covariance matrix by reducing the dimension of the matrix results in estimates with reduced resolution and robustness. The FFT algorithm, which employs the inner product of the echo signal and the steering vector for angle estimation, can be applied with a single snapshot. However, it is affected by sidelobe interference, resulting in a reduction in estimation resolution.

In contrast, IAA is an iterative algorithm that utilizes the relationship between the sample covariance matrix and the complex scattering coefficients. It initializes the iterations with the FFT results and constructs the signal covariance matrix with the results of the previous iteration. This algorithm can reduce the effect of sidelobes while maintaining the same number of receivers, resulting in enhanced resolution and improved 3D imaging capabilities.

Similar to the FFT, the signal model can be mathematically expressed as

$$X = \mathbf{A}S + w \tag{21}$$

where $X_{M \times 1}$ represents the echo signal received by M antennas on the 3D ArcSAR at a particular moment, commonly called the measurement signal. $\mathbf{A}_{M \times L}$ is the steering matrix and $S_{L \times 1}$ is the estimated signal. w is additive white Gaussian noise (AWGN).

The idea of the least square estimation is to find a linear unbiased estimate $\hat{S}$, such that $\left\| X - \mathbf{A}\hat{S} \right\|^2$ is minimized. Also, by introducing the weighting matrix $\mathbf{W}$, the optimization problem of WLS can be expressed as

$$\min_{\alpha}(X - \mathbf{A}\hat{S})^H \mathbf{W}(X - \mathbf{A}\hat{S}) \tag{22}$$

The solution of the above equation is as follows:

$$\hat{S} = (\mathbf{A}^H \mathbf{W} \mathbf{A})^{-1} \mathbf{A}^H \mathbf{W} X \tag{23}$$

where $(\cdot)^H$ represents the conjugate transpose. IAA is equivalent to the WLS problem when $\mathbf{W}$ is the interference-plus-noise covariance matrix of the signal X.

$\mathbf{P}$ is a diagonal matrix of $L \times L$, and the elements on the diagonal of $\mathbf{P}$ are the scattering coefficients in the L directions.

$$\mathbf{P} = E[SS^H] \tag{24}$$

where $E[\cdot]$ represents the mathematical expectation.

Assuming that $\mathbf{A}$ and $\mathbf{P}$ are known, the signal covariance matrix $\mathbf{R}$ can be expressed as

$$\mathbf{R}_{M \times M} = \mathbf{A} \mathbf{P} \mathbf{A}^H \tag{25}$$

In practice, only $\mathbf{A}$ is known about the echo signal, while $\mathbf{P}$ is unknown. $\mathbf{A}$ can be expressed as

$$\mathbf{A}_{M \times L} = [a(\theta_1), a(\theta_2), \cdots, a(\theta_L)] \tag{26}$$

Assuming that the scattering coefficient of each direction in $\mathbf{P}$ is expressed by P_l, the interference-plus-noise covariance matrix $\mathbf{U}$ is defined as

$$\mathbf{U} = \mathbf{R} - P_l a(\theta_l) a(\theta_l)^H \tag{27}$$

The scattering coefficient is replaced by the unbiased estimate of it, which is obtained by substituting Equation (23) into (24)

$$P_l = \left| \frac{a(\theta_l)^H \mathbf{U}^{-1} X}{a(\theta_l)^H \mathbf{U}^{-1} a(\theta_l)} \right|^2 \tag{28}$$

According to the principle of matrix inversion, Equation (28) can also be written as

$$P_l = \left| \frac{a(\theta_l)^H \mathbf{R}^{-1} X}{a(\theta_l)^H \mathbf{R}^{-1} a(\theta_l)} \right|^2 \tag{29}$$

From the above equations, P_l can be calculated from $\mathbf{R}$. $\mathbf{R}$ should be initialized as the unit matrix first. Since $\mathbf{P}$ is unknown, the initialized P_l can be expressed as

$$P_l = \frac{\left| a(\theta_l)^H X \right|^2}{\left(a(\theta_l)^H a(\theta_l) \right)^2} \tag{30}$$

When $\mathbf{R}$ is a unit matrix, the result of IAA is equivalent to FFT. IAA takes the result of FFT as the initial value of $\mathbf{P}$ and then iteratively updates $\mathbf{R}$ and $\mathbf{P}$ based on their interrelationships until the convergence condition is reached. In general, the estimation performance does not improve significantly after more than eight iterations. At this time, it is reasonable to consider the algorithm to have reached convergence.

In practice, when the condition number of $\mathbf{R}$ is large, the calculation of matrix inversion may be inaccurate or even non-invertible. To solve this problem, a diagonal loading approach can be taken to regularize $\mathbf{R}$. Thus, $\mathbf{R}$ can be expressed as

$$\mathbf{R} = \mathbf{A} \mathbf{P} \mathbf{A}^H + \Delta \tag{31}$$

where $\mathbf{\Delta}$ is the diagonal matrix. The elements on the diagonal represent the estimates of the noise, and these can be expressed as

$$\Delta_m = \left| \frac{e_m{}^H \mathbf{R}^{-1} X}{e_m{}^H \mathbf{R}^{-1} e_m} \right|^2, m = 1, 2, \cdots, M \tag{32}$$

where e_m is column m of the unit matrix.

The specific steps of the IAA based on 3D ArcSAR include the following:

1. Determine the spatial coordinates of the targets and extract the signals from multiple receivers corresponding to these specific locations.
2. Employ IAA to estimate the altitude angle at the designated target locations.
3. Generate M sets of SAR images of the 2D scenes through the application of BPA to process the received signal from each antenna. Obtain a complete 3D panoramic image combined with the height dimension information obtained by IAA.

First, initialize the scattering coefficient estimate $P_{l(0)}$ and the noise estimation matrix $\mathbf{\Delta}_{(0)}$.

$$P_{l(0)} = \frac{\left| a(\theta_l)^H X \right|^2}{\left(a(\theta_l)^H a(\theta_l) \right)^2}, l = 1, 2, \cdots, L \tag{33}$$

$$\mathbf{\Delta}_{(0)} = \mathbf{0}_{M \times M} \tag{34}$$

Perform iterations according to the interrelationship of Equations (35)–(39) until convergence.

$$\mathbf{R}_{(n)} = \mathbf{A} \mathbf{P}_{(n-1)} \mathbf{A}^H + \mathbf{\Delta}_{(n-1)} \tag{35}$$

$$\Delta_{m(n)} = \left| \frac{e_{m(n)}{}^H \mathbf{R}_{(n)}^{-1} X}{e_{m(n)}{}^H \mathbf{R}_{(n)}^{-1} e_{m(n)}} \right|^2, m = 1, 2, \cdots, M \tag{36}$$

$$\mathbf{\Delta}_{(n)} = diag(\Delta_{1(n)}, \Delta_{2(n)}, \cdots, \Delta_{M(n)}) \tag{37}$$

$$P_{l(n)} = \left| \frac{a(\theta_l)^H \mathbf{R}_{(n)}^{-1} X}{a(\theta_l)^H \mathbf{R}_{(n)}^{-1}(\theta_l)} \right|^2, l = 1, 2, \cdots, L \tag{38}$$

$$\mathbf{P}_{(n)} = diag(P_{1(n)}, P_{2(n)}, \cdots, P_{L(n)}) \tag{39}$$

Finally, the obtained scattering coefficient of the echoes can be expressed as

$$\sigma_n = \left| \sqrt{P_l} \right| \tag{40}$$

where the altitude angle estimation of the corresponding location is combined with the 2D SAR panoramic imaging results to obtain a 3D image of the scene.

3. Parametric Analysis in Simulations

The 3D ArcSAR method proposed in this paper was evaluated using simulation experiments. This section presents an analysis of the performance of ArcSAR 2D imaging and 3D imaging. Additionally, the errors in altitude angle estimation between the proposed algorithm and the common algorithms are analyzed, considering different target spacing and SNR.

3.1. Imaging Simulation

3.1.1. Imaging of 2D ArcSAR

In order to verify the performance of 2D ArcSAR imaging, we set up the simulation experiments. The main parameters are shown in Table 1.

Table 1. Experimental parameters of 2D ArcSAR imaging.

SNR	Initial Frequency	Signal Bandwidth	FM Slope	AD Frequency	Turntable Arm Length	Beamwidth
20dB	77.12 GHz	1.365 GHz	30 MHz/us	25.5 MHz	0.41 m	70°

According to Equations (19) and (20), the theoretical resolution can be obtained under these experimental parameters. In particular, Hanning windows are added to the distance processing, and cos windows are added to the azimuthal processing. The radial resolution is 0.158m, and the azimuthal resolution is 0.34°.

In this experiment, we set six point targets surrounding the system. These targets were placed at 12 m, 15 m, and 18 m, with azimuth angles of 90° and 150°. The imaging results of the target array are shown in Figure 4. The horizontal axis of the coordinate system indicates the azimuthal direction, and the vertical axis indicates the radial direction.

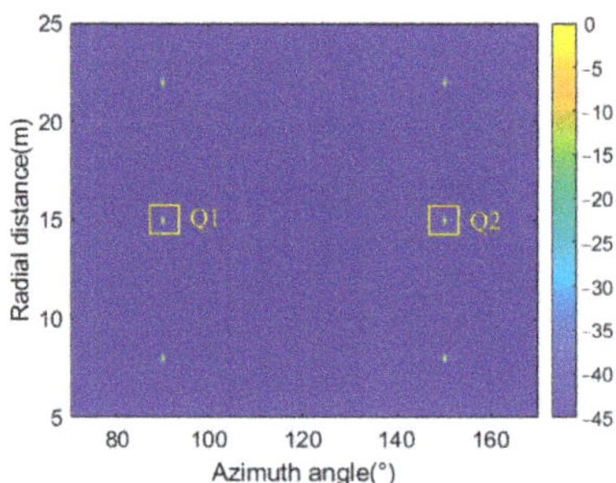

Figure 4. Simulation of imaging results for the point targets.

Subsequently, we selected Q1 and Q2 for analysis, marked in Figure 4. The imaging results of these two targets were sliced, and the corresponding impulse response results are presented in Figures 5 and 6. Their quality parameters are shown in Table 2.

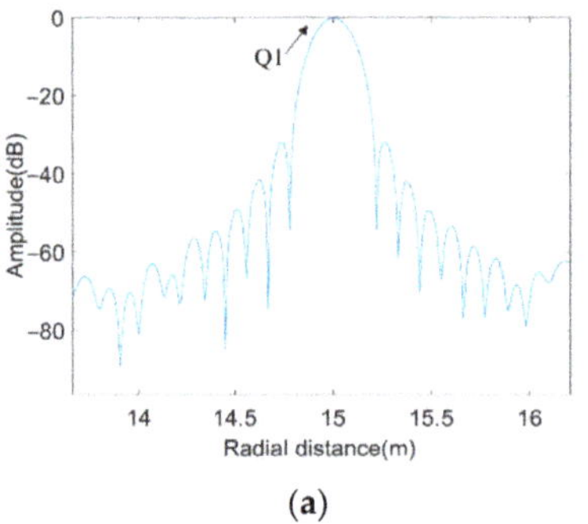
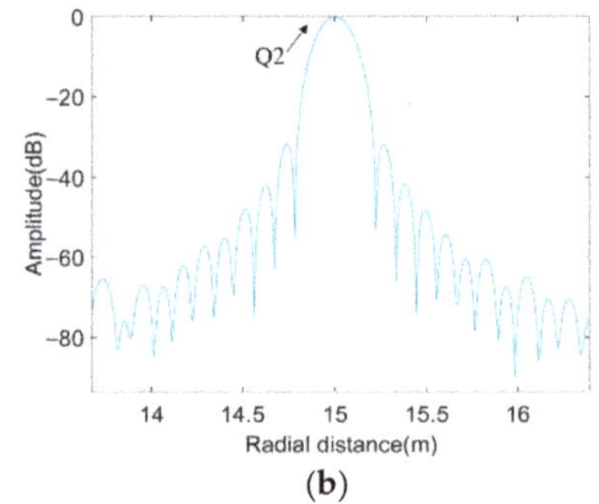

(a)   (b)

Figure 5. Radial impulse responses: (**a**) Q1; (**b**) Q2.

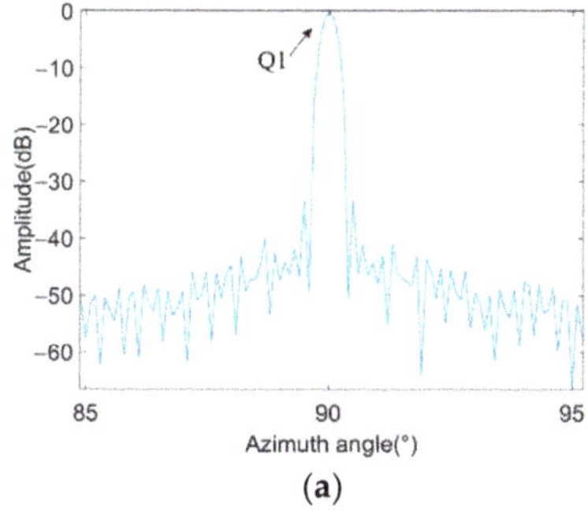 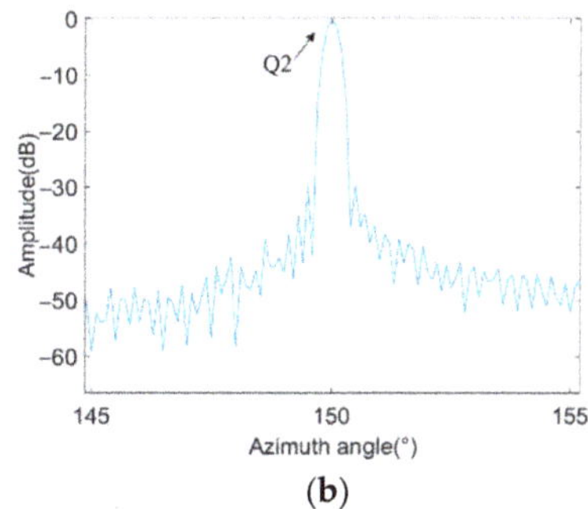

Figure 6. Azimuthal impulse responses: (**a**) Q1; (**b**) Q2.

Table 2. Simulation results quality parameters.

Quality Parameters	Point Target Q1	Point Target Q2
Radial resolution (m)	0.1588	0.1576
Radial PSLR (dB)	−31.6067	−31.5194
Radial ISLR (dB)	−30.7331	−30.6482
Azimuth resolution (°)	0.3433	0.3467
Azimuth PSLR (dB)	−31.4269	−28.2715
Azimuth ISLR (dB)	−21.5856	−18.3847

The inclusion of different window functions in the radial and azimuthal directions resulted in different main lobe spreading and sidelobe attenuation. Specifically, Hanning windows were applied for distance processing, while cos windows were applied for azimuthal processing. According to Table 2, we observed that the radial and azimuthal resolutions were in general agreement with the theoretical values. The results of Peak Sidelobe Ratio (PSLR) and Integrated Sidelobe Ratio (ISLR) were within the reasonable range.

3.1.2. Imaging of 3D ArcSAR

The proposed 3D ArcSAR method was analyzed via a simulation, with an SNR of 10 dB and an AD frequency of 45.5 MHz. The rest of the main parameters are shown in Table 1. The number of antenna array elements was set to 16, and the antenna array had a length of 14.4mm and a spacing of 0.96mm. Referring to Equation (12), the Rayleigh limit resolution of this antenna array was 6.35°.

The parameters of the set coordinate system are expressed as the distance between the target and the turntable center, the azimuthal angle, and the altitudinal angle, respectively. In this experiment, we placed seven targets in the 3D scene, which were located at (10 m, 90°, 0°), (15 m, 80°, 0°), (15 m, 100°, 0°), (15 m, 90°, 0°), (15 m, 90°, 6°), (20 m, 90°, 0°) and (20 m, 90°, 12°), and these target location relationships in the 3D scene are shown in Figure 7. Among them, the turntable center of the 3D ArcSAR sensing system was located at (0 m, 90°, 0°).

Following the rotational scan, the received signals from each antenna undergo 2D ArcSAR imaging processing. Figure 8 displays the eighth antenna's 2D imaging results. It is important to highlight that the 2D imaging results exhibited a stacked masking phenomenon at (15 m, 90°) and (20 m, 90°), where two targets with differing heights were located.

The above results were directly employed for 3D image processing, as shown in Figure 9.

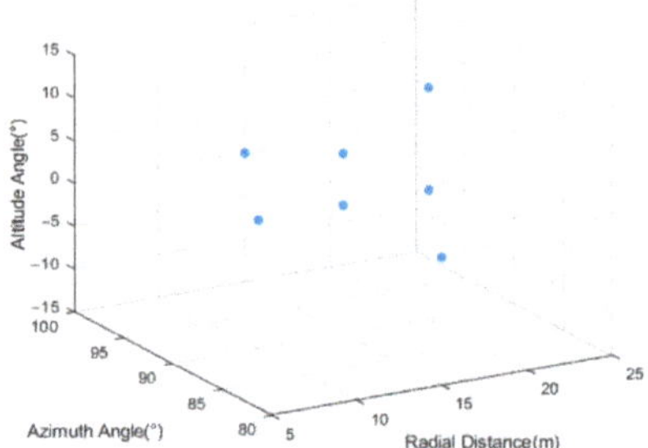

Figure 7. Target location relationships in 3D scenes.

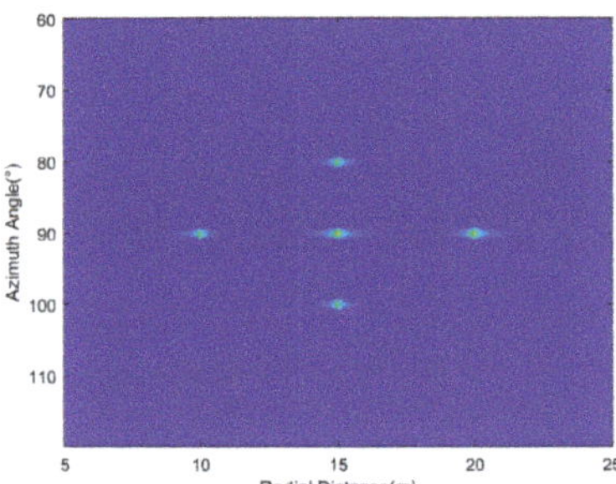

Figure 8. Two-dimensional imaging results of the eighth antenna.

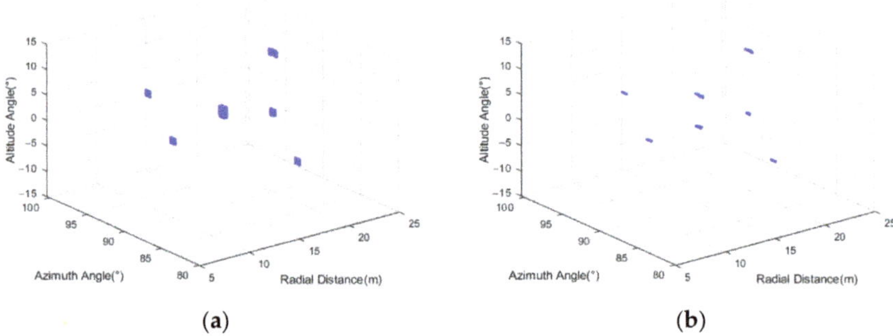

Figure 9. Imaging results of 3D ArcSAR: (**a**) FFT; (**b**) IAA.

The target spacing at (15 m, 90°) and (20 m, 90°) were 6° and 12°. The phase information of the echo signals at these two locations was extracted, and the angle estimation was performed by FFT, OMP, and IAA, respectively. The achieved angle estimation results are depicted in Figure 10. Due to the target spacing at (15 m, 90°) exceeding the Riley limit resolution, FFT could not distinguish the two targets. Conversely, both OMP and IAA exhibited resolutions that surpassed the Rayleigh limit resolution. Figure 10a shows that the angle estimation results of OMP were inaccurate. Consequently, the next subsection thoroughly analyzes the errors in altitude angle estimation of different algorithms, considering the influence of different target spacing and SNR.

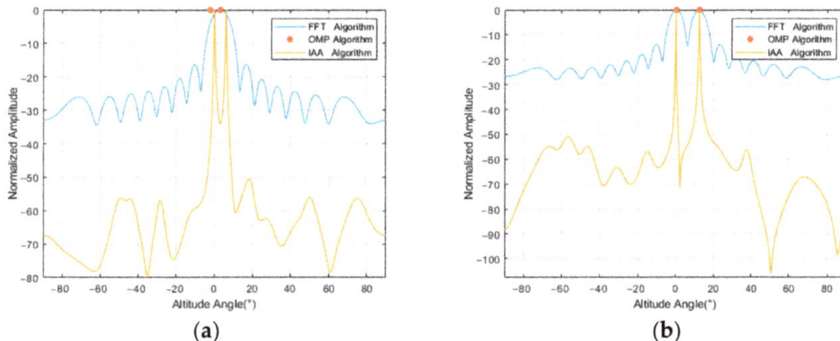

Figure 10. The altitude angle estimation results: (**a**) target spacing is 6°; (**b**) target spacing is 12°.

3.2. Performance Analysis of Different Algorithms Influenced by Different Factors

3.2.1. Target Spacing

Music can attain the full-rank sample covariance matrix by reducing the dimensionality of the matrix, enabling angle estimation under a single snap signal. Therefore, we added a contrastive analysis of Music in this section, where we processed a single snap signal of 16 antenna array elements into a 9-snap signal of 8 elements.

We analyzed the accuracy of angle estimation under different target spacing. Target 2 was set at a fixed altitude angle of 0°, while the altitude angle of Target 1 was variable. The angle spacing between them was linearly varied from 15° to 3.8° with steps of 0.05°. Assuming that the SNR was 35dB and there were 16 antenna array elements, the altitude direction of the targets was estimated by different resolution algorithms. The errors in altitude angle estimation were then determined through a comparison between the estimated and true values. The relative error analysis for Target 1 is shown in Figure 11.

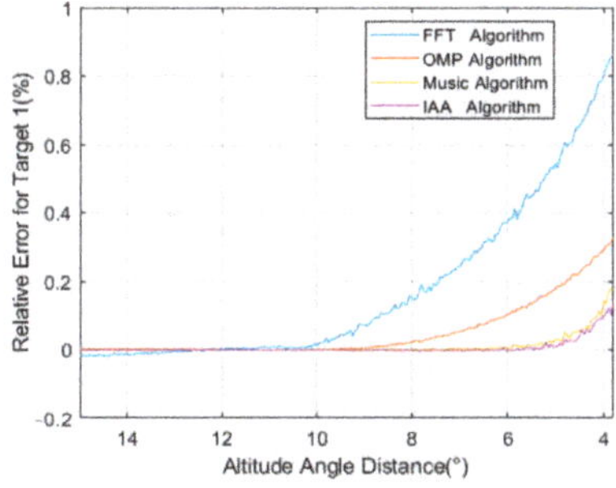

Figure 11. The relative error in the angle estimation for Target 1 under different target spacing.

The results indicate that the IAA algorithm outperformed the FFT, OMP, and Music algorithms in terms of angle resolution. The error in altitude angle estimation increased gradually from approximately 12° and 9° for FFT and OMP, respectively, while the values were about 5.5° for Music and IAA. However, Music exhibited a slightly higher angle estimation error compared to IAA for narrow spacing conditions.

3.2.2. SNR

We analyzed the angle estimation performance under different SNRs. In this simulation, the number of antenna elements was still fixed at 16. To ensure the optimal resolution for all algorithms at different SNRs, the target spacing was set to 12°. We analyzed where the SNR ranged from −15 dB to 35 dB in steps of 0.25 dB, repeated 10,000 times. The analysis of the relative error for Target 1 is shown in Figure 12.

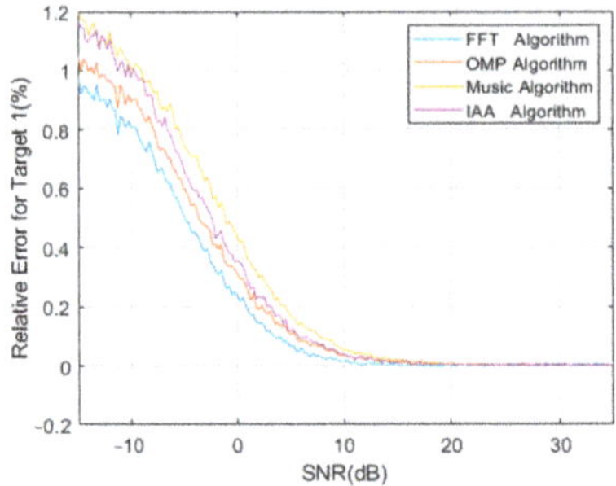

Figure 12. The relative error in the angle estimation for Target 1 under different SNRs.

The findings demonstrate that as the SNR decreased, the angle resolution also decreased. All these algorithms exhibited robustness to changes in the SNR. Specifically, FFT exhibited superior resolution performance under low-SNR conditions, while IAA excelled under high SNR conditions. Music with a reduced matrix dimension had no performance advantage.

However, similar to OMP, Music requires the target signal to be sparse in spatial location and necessitates prior knowledge of the number of targets. These requirements can present challenges for accurate angle estimation in practical, real-world scenarios. As a result, Music and OMP may not be suitable for applications when an unknown number of scattering targets are stacked in similar locations.

4. Imaging Analysis in Experiments

4.1. The Experimental System

There is little domestic and international research on the 3D SAR imaging of rotational structures. In order to validate the feasibility of the 3D ArcSAR sensing system and its associated imaging processing techniques, it was necessary to construct a 3D ArcSAR imaging platform.

We designed the 3D ArcSAR sensing system prototype to facilitate the acquisition of 3D panoramic imaging data. The prototype consisted of a millimeter-wave MIMO radar system and a rotating scanning mechanical system, as shown in Figure 13. The radar was mounted at the end of the rotating arm, while the MIMO antenna was set up in the altitude direction. By rotating the arm, the system can generate an azimuthal synthetic aperture and uses the MIMO antenna for altitude angle estimation. With the capability to be mounted on the roof of an unmanned ground vehicle, this system enables the acquisition of high-quality 3D imaging results in a 360° range surrounding the vehicle.

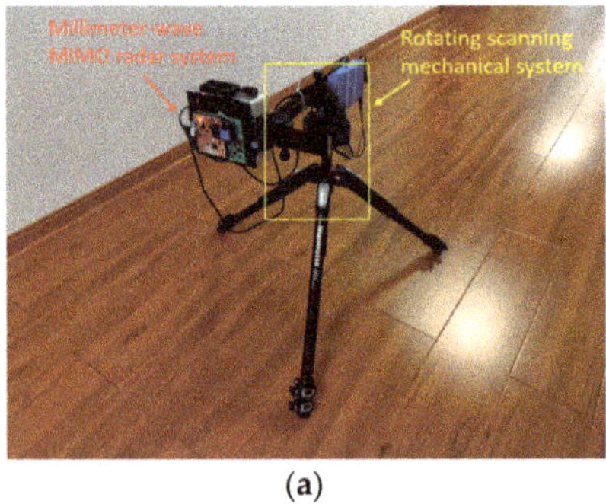

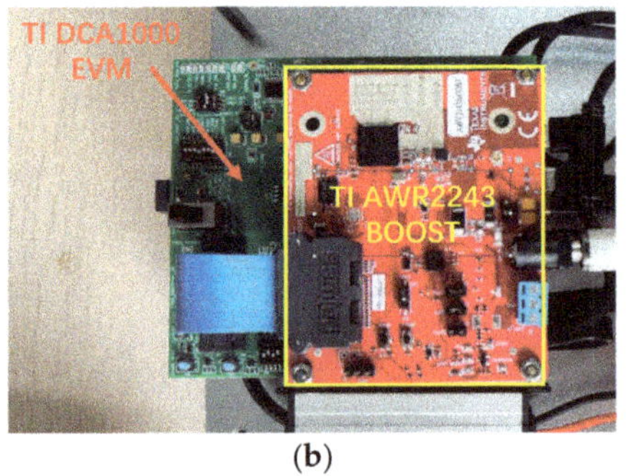

(a)   (b)

Figure 13. (**a**) The 3D ArcSAR sensing system. (**b**) The millimeter-wave MIMO radar system.

The millimeter-wave MIMO radar system consists of the TI AWR2243BOOST radar and the TI DCA1000EVM real-time data capture adapter. TI AWR2243BOOST is equipped with three transmit antennas and four receive antennas, where the separation between adjacent receive antennas is set as $\lambda/2$. However, these transmit antennas are not arranged in parallel. Therefore, only the first and third transmit antennas can be used to achieve the desired MIMO radar. The angle resolution of the millimeter-wave MIMO radar is about 12.69°. The rotating scanning mechanical system consists of a rotary base, a rotating arm, and a rotating motor. The rotating motor is electrically assisted and can be remotely controlled for speed and steering by a computer. The modular design of the mechanical system allows the flexible adjustment of the arm length to accommodate diverse experimental requirements.

4.2. Two-Dimensional Experimental Results of ArcSAR

In this experiment, metal corner reflectors are used to simulate point targets. The main parameters of the ArcSAR system based on millimeter-wave MIMO radar are shown in Table 3.

Table 3. Experimental parameters.

Initial Frequency	FM Slope	AD Frequency	Pulse Width	Turntable Arm Length	Beamwidth	Rotational Speed
77 GHz	30 MHz/us	22.5 MHz	45.5 us	0.41 m	70°	12°/s

We placed two metal corner reflectors, Q1 and Q2, spaced 90° apart. Both Q1 and Q2 were situated at a radial distance of about 8.4 m from the rotation center. Figure 14 shows the relationship diagram between the targets and the ArcSAR system's location.

Figure 14. The relationship diagram between the targets and the ArcSAR system's location.

To show the ArcSAR 2D imaging results visually, we converted the polar coordinate system of the imaging results into the rectangular coordinate system. The imaging results of the two metal corner reflectors Q1 and Q2 in the polar coordinate system and rectangular coordinate system are shown in Figure 15a,b.

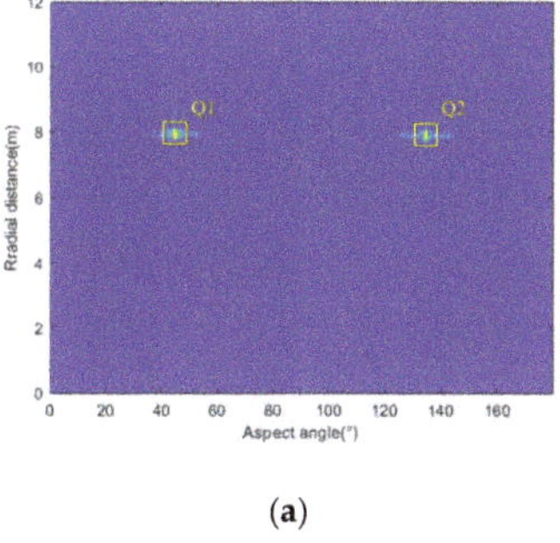

(**a**)

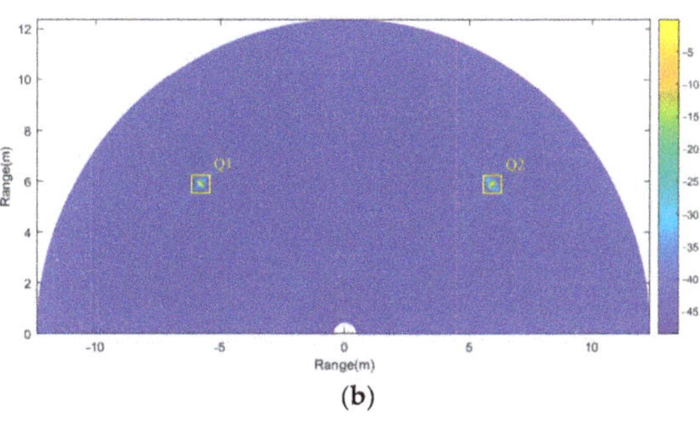

(**b**)

Figure 15. Two-dimensional imaging results of the targets: (**a**) polar coordinate system; (**b**) rectangular coordinate system.

Subsequently, the imaging results for Q1 and Q2 were analyzed. The imaging results were sliced, and the corresponding impulse response results in the azimuthal direction are shown in Figure 16. Their quality parameters are shown in Table 4.

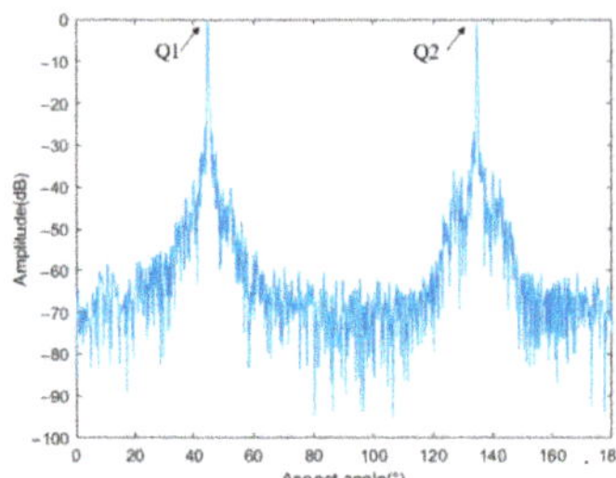

Figure 16. Azimuthal impulse responses of Q1 and Q2.

Table 4. Quality parameters of experimental results.

Quality Parameters	Point Target Q1	Point Target Q2
Azimuth resolution (°)	0.3833	0.3690
Azimuth PSLR (dB)	−25.7565	−26.0115
Azimuth ISLR (dB)	−11.3737	−11.1613

Based on the above results, it can be observed that the measured azimuthal resolution deviated by approximately 10% from the theoretical value. This discrepancy can be attributed to the comparatively lower SNR encountered in the real measurement environment. The results of PSLR and ISLR were within the reasonable range.

4.3. Three-Dimensional ArcSAR Experiments

In this experimental phase, the primary focus was on verifying the altitudinal resolution performance of the 3D ArcSAR system. The FM slope and turntable arm length were adjusted to 50 MHz/us and 0.395 m, respectively. The rest of the main parameters were the same as in Table 3, with the antenna array comprising eight elements.

Two reflectors were positioned at different altitudinal locations, and the altitudinal distance between them was 0.64 m. The millimeter-wave radar system's plane was situated at a height of 0.43 m, while the horizontal distance between the rotating scanning mechanical system's center and the targets measured 1.53 m. The true values of the targets' altitudinal angles were determined to be about −7.81° and 15.69°, respectively. Figure 17 displays the spatial relationship between the targets and the 3D ArcSAR system.

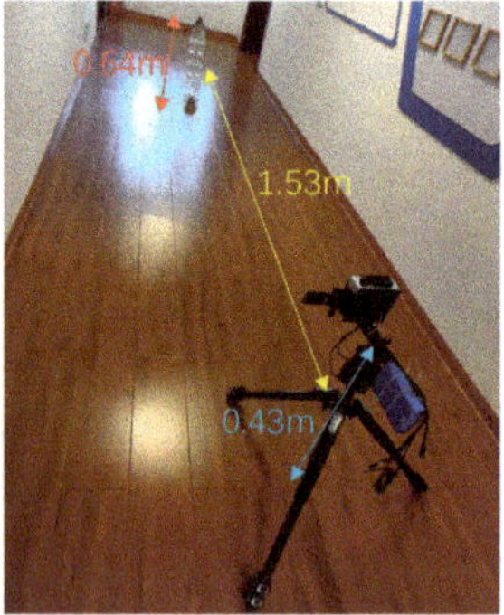

Figure 17. The spatial relationship between the targets and the 3D ArcSAR system.

Considering the proximity of these two targets in the 2D imaging, it was necessary to perform angle estimation in the altitudinal direction. The 2D and 3D imaging results acquired through the utilization of the proposed algorithm for stacked masking targets are shown in Figure 18. Observing Figure 18a suggests that the two targets do not seem to be entirely stacked. However, it is known after processing that these two targets still have stacked parts of each other.

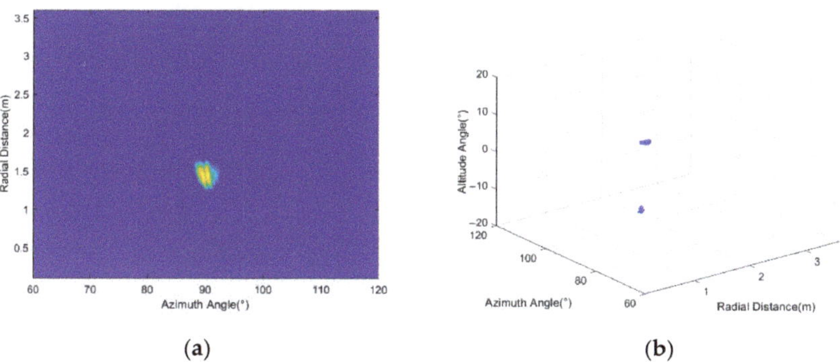

(**a**)   (**b**)

Figure 18. Imaging results under stacked masking targets: (**a**) 2D; (**b**) 3D.

The echo data from the stacked location of the two targets were analyzed, and the altitude angle estimation results after processing are shown in Figure 19. Both OMP and IAA could distinguish the two targets well, and IAA was able to determine that the angles of the two targets in the altitudinal direction were about $-6.6°$ and $13.3°$. This angle spacing was larger than the Rayleigh limit resolution of the AWR2243BOOST radar with two-transmitter and four-receiver antennas. Consequently, FFT could also distinguish the two targets. However, the angle estimation results of FFT were not too satisfactory due to factors such as a low SNR and sidelobe interference within the real measurement.

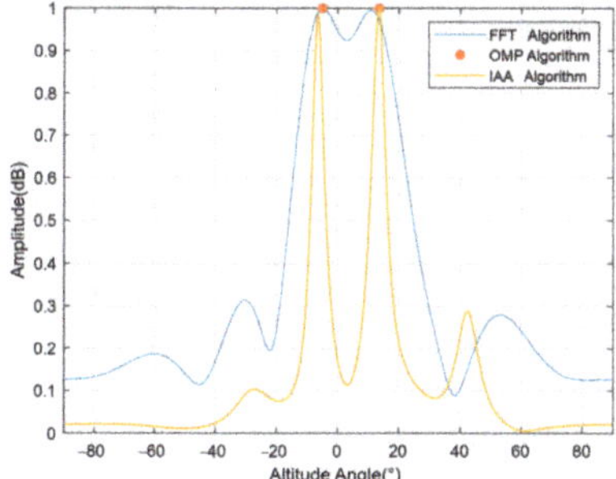

Figure 19. The altitude angle estimation results under stacked masking targets.

Furthermore, we conducted experiments involving a near-circular step structure, as shown in Figure 20, to evaluate the performance of the 3D ArcSAR imaging method in real-world scenarios. Due to the particular structure of the steps, each step can occupy a different location in the 2D plane and altitudinal direction. Therefore, this scene was well suited to verifying the real effect of the 3D ArcSAR imaging method, and this step structure comprised five levels. The 3D ArcSAR system was located at an approximate distance of 8 m from the steps, with the radar antenna elevated to a height of about 0.55 m over the ground.

In order to expand the detection distance and enhance the azimuth resolution of the 3D ArcSAR system, the pulse width was adjusted to 68.3 us, and the turntable arm length was increased to 0.5085 m here.

Figure 20. The relationship diagram between the steps and the 3D ArcSAR system's location.

The proposed algorithm was used to image the targets, and the 2D and 3D imaging results are shown in Figure 21.

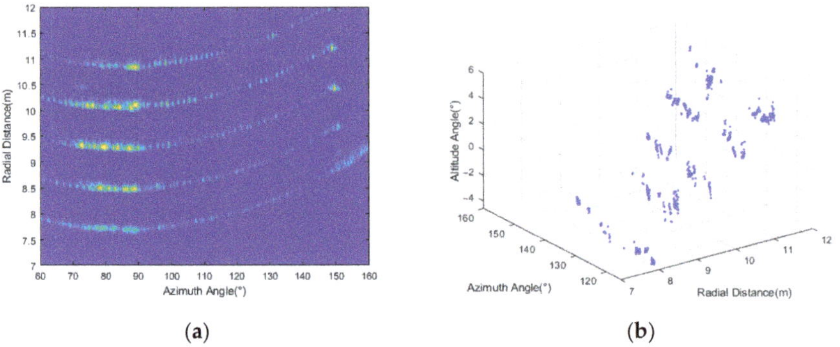

(a)   (b)

Figure 21. Imaging results of the steps: (**a**) 2D; (**b**) 3D.

Figure 21 demonstrates that the various levels of steps exhibited distinguishable differences in radial distance and altitudinal direction. The proposed method successfully facilitated the imaging and distinction of the steps. It should be noted that Figure 21b portrays altitude angles below $0°$ in the imaging results, which was due to the radar antenna's location above some of the steps. These experimental results can verify the rationality and accuracy of the designed 3D ArcSAR system.

5. Conclusions

The proposed 3D ArcSAR method enables the reconstruction of 3D panoramic images through panoramic rotational SAR and direction estimation. To address the limitations encountered in vehicle 3D sensing, particularly the reliance on a single snapshot signal, we designed an IAA-based resolution algorithm specifically for the 3D ArcSAR sensing system. At the same time, the errors in altitude angle estimation for both the proposed algorithm and the conventional algorithms were analyzed, considering various factors such as target spacing and SNR. It was demonstrated that the proposed algorithm has superior resolution in the case of single snap, small antenna arrays, and an unknown number of targets compared to other existing methods.

Additionally, we designed a prototype of 3D ArcSAR based on a 77GHz COTS radar and a rotating scanning mechanical system. This prototype serves the purpose of validating the proposed algorithms presented in this paper, and can be further utilized for experiments in various scenes in the future.

While our primary focus was on addressing the specific challenges of unmanned ground vehicle applications, we recognize that the proposed 3D ArcSAR sensing system holds significant potential for remote sensing applications. We propose the following potential "remote sensing" applications:

1. Environmental Monitoring: The 3D ArcSAR system's panoramic imaging capability can be leveraged for monitoring large-scale environmental areas, such as forests, coastal regions, and agricultural landscapes. It can provide valuable insights into vegetation growth, land use changes, and environmental dynamics.

2. Disaster Management: During natural disasters such as floods, earthquakes, and landslides, the 3D ArcSAR system can efficiently gather information on the affected areas and aid in disaster response and management efforts. The system's ability to acquire 3D images in real-time can be crucial for assessing the extent of damage and planning relief operations.

3. Climate Change Studies: The system's capacity to monitor vast areas with high accuracy enables data collection for climate change studies. It can be used to monitor ice caps, glaciers, and polar regions, providing critical data for understanding climate change patterns.

However, the current work is still at a preliminary stage. In the next stage, we plan to conduct further in-depth research on various aspects, including the experimental verification of complex targets, more efficient ArcSAR imaging algorithms, and the impact on the ArcSAR imaging results under the motion of unmanned ground vehicles.

Author Contributions: Conceptualization, Y.H., J.W. and X.H.; methodology, Y.H. and J.W.; software, J.W.; validation, Y.H., J.W. and D.F.; formal analysis, Y.H.; investigation, Y.H.; resources, J.W., X.H. and D.F.; data curation, Y.H. and J.W.; writing—original draft preparation, Y.H.; writing—review and editing, J.W. and X.H.; visualization, Y.H.; supervision, J.W.; project administration, J.W. and X.H.; funding acquisition, J.W. and D.F. All authors have read and agreed to the published version of the manuscript.

Funding: This research was funded by the National Natural Science Foundation of China, grant number 62101562.

Data Availability Statement: The experiment data are available on request.

Conflicts of Interest: The authors declare no conflict of interest.

References

1. Wang, S.; Jiang, X. Three-Dimensional Cooperative Positioning in Vehicular Ad-hoc Networks. *IEEE Trans. Intell. Transp. Syst.* **2021**, *22*, 937–950. [CrossRef]
2. Sun, S.; Petropulu, A.P.; Poor, H.V. MIMO Radar for Advanced Driver-Assistance Systems and Autonomous Driving: Advantages and Challenges. *IEEE Signal Process. Mag.* **2020**, *37*, 98–117. [CrossRef]
3. Engels, F.; Heidenreich, P.; Wintermantel, M.; Stacker, L.; Al Kadi, M.; Zoubir, A.M. Automotive Radar Signal Processing: Research Directions and Practical Challenges. *IEEE J. Sel. Top. Signal Process.* **2021**, *15*, 865–878. [CrossRef]
4. Bilik, I. Comparative Analysis of Radar and Lidar Technologies for Automotive Applications. *IEEE Intell. Transp. Syst. Mag.* **2023**, *15*, 244–269. [CrossRef]
5. Li, Y.; Ibanez-Guzman, J. Lidar for Autonomous Driving: The Principles, Challenges, and Trends for Automotive Lidar and Perception Systems. *IEEE Signal Process. Mag.* **2020**, *37*, 50–61.
6. Subedi, D.; Jha, A.; Tyapin, I.; Hovland, G. Camera-LiDAR Data Fusion for Autonomous Mooring Operation. In Proceedings of the 2020 15th IEEE Conference on Industrial Electronics and Applications (ICIEA), Kristiansand, Norway, 9 November 2020; pp. 1176–1181.
7. Gong, J.; Duan, Y.; Liu, K.; Chen, Y.; Xiong, G.; Chen, H. A robust multistrategy unmanned ground vehicle navigation method using laser radar. In Proceedings of the 2009 IEEE Intelligent Vehicles Symposium, Xi'an, China, 3 June 2009; pp. 417–424.
8. Li, Y.; Shang, X. Multipath Ghost Target Identification for Automotive MIMO Radar. In Proceedings of the 2022 IEEE 96th Vehicular Technology Conference (VTC2022-Fall), London, UK, 26 September 2022; pp. 1–5.
9. Gusland, D.; Torvik, B.; Finden, E.; Gulbrandsen, F.; Smestad, R. Imaging radar for navigation and surveillance on an autonomous unmanned ground vehicle capable of detecting obstacles obscured by vegetation. In Proceedings of the 2019 IEEE Radar Conference (RadarConf), Boston, MA, USA, 22 April 2019; pp. 1–6.
10. Zhu, J.; Song, Y.; Jiang, N.; Xie, Z.; Fan, C.; Huang, X. Enhanced Doppler Resolution and Sidelobe Suppression Performance for Golay Complementary Waveform. *Remote Sens.* **2023**, *15*, 2452. [CrossRef]
11. Infineon and Oculii Partner to Accelerate Imaging Radar Software Technology Solutions in Automotive Applications. Available online: https://www.businesswire.com/news/home/20191218005069/en/Infineon-Oculii-Partner-Accelerate-Imaging-Radar-Software (accessed on 18 December 2019).

12. Domhof, J.; Happee, R.; Jonker, P. Multi-sensor object tracking performance limits by the Cramer-Rao lower bound. In Proceedings of the 2017 20th International Conference on Information Fusion (Fusion), Xi'an, China, 10 July 2017; pp. 1–8.

13. Gao, Z.; Jia, Y.; Liu, S.; Zhang, X. A 2-D Frequency-Domain Imaging Algorithm for Ground-Based SFCW-ArcSAR. *IEEE Trans. Geosci. Remote Sens.* **2022**, *60*, 1–14. [CrossRef]

14. Zhang, Y.; Wang, Y.; Schmitt, M.; Yuan, C. Efficient ArcSAR Focusing in the Wavenumber Domain. *IEEE Trans. Geosci. Remote Sens.* **2022**, *60*, 1–10. [CrossRef]

15. Pieraccini, M.; Miccinesi, L. ArcSAR: Theory, Simulations, and Experimental Verification. *IEEE Trans. Microw. Theory Tech.* **2017**, *65*, 293–301. [CrossRef]

16. Lin, Y.; Liu, Y.; Wang, Y.; Ye, S.; Zhang, Y.; Li, Y.; Li, W.; Qu, H.; Hong, W. Frequency Domain Panoramic Imaging Algorithm for Ground-Based ArcSAR. *Sensors* **2020**, *20*, 7027. [CrossRef]

17. Li, W.; Liao, G.; Zhu, S.; Zeng, C.; Xu, J. A Novel Clutter Suppression Method Based on Time-Doppler Chirp Varying for Helicopter- Borne Single-Channel RoSAR System. *IEEE Geosci. Remote Sens. Lett.* **2022**, *19*, 1–5. [CrossRef]

18. Li, W.; Liao, G.; Zhu, S.; Xu, J. A Novel Helicopter-Borne RoSAR Imaging Algorithm Based on the Azimuth Chirp z transform. *IEEE Geosci. Remote Sens. Lett.* **2019**, *16*, 226–230. [CrossRef]

19. Nan, Y.; Huang, X.; Guo, Y.J. A Panoramic Synthetic Aperture Radar. *IEEE Trans. Geosci. Remote Sens.* **2022**, *60*, 1–13. [CrossRef]

20. Li, J.; Stoica, P. An adaptive filtering approach to spectral estimation and SAR imaging. *IEEE Trans. Signal Processing.* **1996**, *44*, 1469–1484. [CrossRef]

21. Schmidt, R.O. Multiple emitter location and signal parameter estimation. *IEEE Trans. Antennas Propagation.* **1986**, *34*, 276–280. [CrossRef]

22. Shang, X.; Liu, J.; Li, J. Multiple Object Localization and Vital Sign Monitoring Using IR-UWB MIMO Radar. *IEEE Trans. Aerosp. Electron. Systems.* **2020**, *56*, 4437–4450. [CrossRef]

23. Simoni, R.; Mateos-Nunez, D.; Gonzalez-Huici, M.A.; Correas-Serrano, A. Height estimation for automotive MIMO radar with group-sparse reconstruction. *Electr. Eng. Syst. Sci.* **2019**. [CrossRef]

24. Jafri, M.; Srivastava, S.; Anwer, S.; Jagannatham, A.K. Sparse Parameter Estimation and Imaging in mmWave MIMO Radar Systems With Multiple Stationary and Mobile Targets. *IEEE Access* **2022**, *10*, 132836–132852. [CrossRef]

25. Roberts, W.; Stoica, P.; Li, J.; Yardibi, T.; Sadjadi, F. Iterative Adaptive Approaches to MIMO Radar Imaging. *IEEE J. Sel. Top. Signal Process* **2010**, *4*, 5–20. [CrossRef]

26. Yardibi, T.; Li, J.; Stoica, P.; Xue, M.; Baggeroer, A.B. Source Localization and Sensing: A Nonparametric Iterative Adaptive Approach Based on Weighted Least Squares. *IEEE Trans. Aerosp. Electron. Syst.* **2010**, *46*, 425–443. [CrossRef]

27. Guo, Y.; Zhang, D.; Li, Q.; Zhang, P.; Liang, Y. FIAA-Based Super-Resolution Forward-Looking Radar Imaging Method for Maneuvering Platforms. In Proceedings of the 2022 3rd China International SAR Symposium (CISS), Shanghai, China, 2 November 2022; pp. 1–5.

28. Suren, C.; Evelina, A.; Anreas, K.; Alessandro, M. Reviewing GPU architectures to build efficient back projection for parallel geometries. *J. Real-Time Image Process.* **2020**, *17*, 1331–1373.

Article

A Space–Time–Range Joint Adaptive Focusing and Detection Method for Multiple Input Multiple Output Radar

Jian Guan [1], Xiaoqian Mu [1,*], Yong Huang [1], Baoxin Chen [2], Ningbo Liu [1] and Xiaolong Chen [1]

[1] Marine Target Detection Research Group, Naval Aviation University, Yantai 264001, China; guanjian96@tsinghua.org.cn (J.G.); huangjiaqi_2013@sohu.com (Y.H.); lnb198300@sohu.com (N.L.); cxlcxl1209@sohu.com (X.C.)
[2] 92337 Troop, PLA, Dalian 116000, China; chenbx19@sohu.com
* Correspondence: mxq1995@sohu.com

Abstract: The Multiple Input Multiple Output (MIMO) radar, as a new type of radar, emits orthogonal waveforms, which provide it with waveform diversity characteristics, leading to increased degrees of freedom and improved target detection performance. However, it also poses challenges such as difficulty in meeting higher data demand, separating waveforms, and suppressing the multidimensional sidelobes (range sidelobes, Doppler sidelobes, and angle sidelobes) of targets. Phase-coded signals are frequently employed as orthogonal transmission signals in the MIMO radar. However, these signals exhibit poor Doppler sensitivity, and the intra-pulse Doppler frequency shift can have an impact on the effectiveness of the matching filtering process. To address the aforementioned concerns, this paper presents a novel approach called the Space–Time–Range Joint Adaptive Focusing and Detection (STRJAFD) method. The proposed method utilizes the Mean Square Error (MSE) criterion and integrates spatial, temporal, and waveform dimensions to achieve efficient adaptive focusing and detection of targets. The experimental results demonstrate that the proposed method outperforms conventional cascaded adaptive methods in effectively addressing the matching mismatch issue caused by Doppler frequency shift, achieving super-resolution focusing, possessing better suppression effects on three-dimensional sidelobes and clutter, and exhibiting better detection performance in low signal-to-clutter ratio and low signal-to-noise ratio environments. Furthermore, STRJAFD is unaffected by coherent sources and demands less data.

Keywords: MIMO radar; Space–Time–Range; Joint Adaptive Focusing and Detection; Doppler frequency shift; MSE; three-dimensional sidelobe; clutter and noise suppression

Citation: Guan, J.; Mu, X.; Huang, Y.; Chen, B.; Liu, N.; Chen, X. A Space–Time–Range Joint Adaptive Focusing and Detection Method for Multiple Input Multiple Output Radar. *Remote Sens.* **2023**, *15*, 4509. https://doi.org/10.3390/rs15184509

Academic Editors: Gerardo Di Martino, Jiahua Zhu, Xinbo Li, Shengchun Piao, Junyuan Guo, Wei Guo, Xiaotao Huang and Jianguo Liu

Received: 30 July 2023
Revised: 8 September 2023
Accepted: 10 September 2023
Published: 13 September 2023

1. Introduction

Amidst the continuously evolving electromagnetic environment, where the cluttered background and target characteristics are becoming increasingly complex and diverse, radar systems face numerous challenges, and the difficulty of detecting targets is also on the rise [1]. The fixed transmission waveforms and operating modes utilized by conventional radar systems exhibit limited adaptability to the changing detection requirements in complex operational environments. Hence, it is crucial to investigate and advance novel radar systems and approaches to target detection [2,3].

The Multiple Input Multiple Output (MIMO) radar, as a new type of radar [4,5], employs orthogonal waveforms for achieving waveform diversity. Its versatile operational capability enables omnidirectional coverage of the airspace [6]. This not only enhances the degree of freedom and spatial resolution and expands the dimension of signal processing, but also benefits the performance of object detection [7].

The initial step in signal processing for the MIMO radar involves the separation of waveforms, which is facilitated by their orthogonality. This is followed by spatial and temporal processing, and ultimately leads to the detection of targets. The successful implementation of waveform separation in MIMO radar systems primarily depends on good orthogonal waveforms and the application of efficient filtering techniques. As the most frequently employed form of MIMO radar orthogonal waveform, the phase-coded waveform exhibits a distinct ambiguity function graph resembling a pushpin and does not suffer from range–Doppler coupling issues [8,9]. However, it is sensitive to Doppler frequency shifts, and the higher the Doppler frequency, the worse the matching filtering effect. Regarding the Doppler sensitivity issue of phase-coded signals, references [10,11] have proposed corresponding Doppler compensation methods, but both require target Doppler estimation, and the compensation effect in practical applications is not ideal. Currently, the commonly used waveform separation filtering methods include matched filtering and adaptive pulse compression, where the performance of matched filtering mainly depends on the performance of orthogonal waveforms [12]. Adaptive pulse compression, as an improved method of matched filtering, can adaptively estimate the filter weights of each distance unit and waveform, improving the separation level of waveforms [13]. However, its performance depends on the appropriate number of iterations, which requires consideration of the degree of contrast between strong and weak scatterers, thereby posing a challenge in the selection process.

Traditional spatial and temporal processing methods include beamforming, pulse accumulation, and Space–Time Adaptive Processing (STAP) [14,15], as well as the Iterative Adaptive Approach (IAA) [16,17]. STAP is an adaptive clutter suppression method that combines the spatial and temporal domains. This approach has been proven to yield significant improvements in clutter suppression and target separation capabilities. The STAP technique in the MIMO radar increases the degree of freedom, which is beneficial for improving the processing performance [18]. However, the estimation of the covariance matrix needs to meet the Reed I S, Mallett J D, Brennan L E (RMB) criterion [19], requiring a large number of independent and identically distributed samples, which can lead to problems such as large amount of computation and long operation time. Although methods such as those in [20,21] for dimensionality reduction and rank reduction have been continuously proposed, the fundamental resolution of this problem has not yet been achieved. For IAA, it can achieve effective angle Doppler imaging after matched filtering, thereby achieving Space–Time joint processing [22].

For MIMO radar systems, two commonly employed pre-detection methods are Matched Filtering–Beam Forming–Discrete Fourier Transform (MF-BF-DFT) [23] and Adaptive Pulse Compression–Beam Forming–Discrete Fourier Transform (APC-BF-DFT) [24]. The MF-BF-DFT method is the simplest and most convenient method for engineering applications, while the APC-BF-DFT method improves the effect of suppression on range sidelobes compared to the former method but increases computation complexity. However, the above two methods have poor suppression effects on three-dimensional (3D) sidelobes, and do not have clutter and noise suppression capabilities, so that the effect after cascaded processing does not meet good expectations. Adaptive Pulse Compression–Iterative Adaptive Approach–Space–Time Adaptive Processing (APC-IAA-STAP) [22,25] is a new adaptive iterative processing method that has been proposed in recent years. Compared with the previous two methods, it has better clutter and noise suppression capabilities and can achieve good suppression of 3D sidelobes. However, this method is a cascaded two-step iterative processing method, which makes it challenging to choose the appropriate number of iterations for each stage. Additionally, as it is still a cascaded method, it may not provide adequate performance for the overall suppression of 3D sidelobes. Therefore, further improvements are necessary to enhance its processing performance.

In light of the aforementioned scenario, this study presents a novel approach called the Space–Time–Range Joint Adaptive Focusing and Detection (STRJAFD) method. The proposed method utilizes the Mean Squared Error (MSE) criterion and integrates spatial, temporal, and waveform dimensions. The aforementioned approach has several advantages:

(1)	The technique effectively improves the matching mismatch problem caused by Doppler frequency shift;
(2)	The technique has good clutter- and noise-suppression effects;

(3) The technique has good sidelobe-suppression effects (range sidelobes, Doppler side-lobes, and angle sidelobes);

(4) The technique is capable of efficiently segregating waveforms;

(5) The technique is not affected by coherent source cancellation;

(6) The technique necessitates a minimal amount of data.

In this paper, a Space–Time–Range Joint Adaptive Focusing and Detection method is proposed to suppress 3D sidelobes and clutter, and improve detection performance with cluttered and noisy backgrounds. The content of this paper is organized as follows. In Section 2, the construction of the MIMO radar Space–Time–Range echo signal model is discussed; meanwhile, the principle and process of the Space–Time– Range Joint Adaptive Focusing and Detection method are introduced. In Section 3, we discuss the experimental simulations in different scenarios that were conducted to demonstrate STRJAFD's feasibility and effectiveness. Section 4 analyzes the reasons why the proposed method is superior to traditional cascaded adaptive methods. Section 5 summarizes the conclusions drawn from this study and outlines future prospects.

The main contribution of this study includes four aspects. Firstly, a Space–Time–Range Joint Adaptive Focusing and Detection algorithm is proposed, which is a three-dimensional joint adaptive processing method and can achieve high-resolution focusing. Secondly, the proposed method indirectly improves the matching mismatch issue caused by Doppler frequency shift through the design of expected signals, solving the problem that matching filtering and adaptive pulse compression are difficult to solve in the distance dimension, compared to conventional cascaded adaptive methods (MF-BF-DFT, APC-BF-DFT, and APC-IAA-STAP). Thirdly, through the effective design of the covariance matrix and the iterative filtering processing based on the MSE criterion, the proposed method can effectively suppress three-dimensional sidelobes in the distance, space, and Doppler dimensions, and its sidelobe suppression effect is far superior to the cascaded single dimension adaptive processing method (MF, APC, BF, and DFT). And, it can simultaneously suppress clutter and noise, which cannot be achieved by beamforming and Doppler processing, and the suppression effect of three-dimensional joint processing is also better than that of two-dimensional joint processing (STAP). Fourthly, a relatively complete Space–Time–Range three-dimensional echo signal model of the MIMO radar was established, taking into account the influence of Doppler frequency shift on phase-encoded signals.

2. Space–Time–Range Joint Adaptive Focusing and Detection Method

2.1. Space–Time–Range Model of MIMO Radar Echo Signal

Assuming that the MIMO radar is an equidistant linear array consisting of M transmitting elements and N receiving elements, with a carrier frequency of f_0, a wavelength of λ, a distance between receiving elements of $d_r = \lambda/2$, a distance between transmitting elements of $d_t = Nd_r$, a pulse repetition period of T, and a carrier speed of v_0, and assuming that the orthogonal encoded signals transmitted are all narrowband signals, then the transmitted signals of each transmitting element are $s_1(t)$, $s_2(t)$, $\cdots$, $s_M(t)$, denoted as $S(t) = [s_1(t), s_2(t), \cdots, s_M(t)]^{\mathrm{T}}$, and P is the encoding length of the signal within one pulse period.

Assuming that the target is non fluctuating and located in the r-th distance unit, with a direction of θ and a speed of v_t, then the echo signal model of the single pulse is as described in Equation (1).

$$X(t) = \xi(r, \theta, f_d)\boldsymbol{b}(\theta)\boldsymbol{a}^{\mathrm{T}}(\theta)S(t)e^{-j2\pi f_d t}, \tag{1}$$

where $\xi(r, \theta, f_d)$ is the true complex amplitude of the target with direction θ and Doppler frequency f_d in the distance unit r, and $\boldsymbol{a}(\theta)$, $\boldsymbol{b}(\theta)$, and f_d are the emission guidance vector, reception guidance vector, and Doppler frequency information of the target; their specific formulae are shown in Equations (2)–(4), respectively.

$$f_d = \frac{2(v_0 \sin\theta + v_t)}{\lambda},$$
(2)

$$\boldsymbol{a}(\theta) = \left[1,\ e^{-j\frac{2\pi}{\lambda}d_t \sin\theta},\ e^{-j\frac{2\pi}{\lambda}2d_t \sin\theta},\cdots,e^{-j\frac{2\pi}{\lambda}(M-1)d_t \sin\theta}\right]^{\mathrm{T}},$$
(3)

$$\boldsymbol{b}(\theta) = \left[1,\ e^{-j\frac{2\pi}{\lambda}d_r \sin\theta},\ e^{-j\frac{2\pi}{\lambda}2d_r \sin\theta},\cdots,e^{-j\frac{2\pi}{\lambda}(N-1)d_r \sin\theta}\right]^{\mathrm{T}},$$
(4)

After analog-to-digital conversion, a discrete target echo signal model is obtained, as seen in Equation (5).

$$\mathbf{X}(l) = \xi(r,\theta,f_d)\boldsymbol{b}(\theta)\boldsymbol{a}^{\mathrm{T}}(\theta)\mathbf{S}e^{-j2\pi(l-1)f_d T}\ \ l = 1,2,\cdots,L,$$
(5)

where $\mathbf{S} = [s_1 \cdot s_f,\ s_2 \cdot s_f,\ \cdots,s_M \cdot s_f]^{T} \in \mathbb{C}^{M\times P}$, $s_f = \left[e^{-j2\pi\frac{1}{p}f_d T},e^{-j2\pi\frac{2}{p}f_d T},\cdots,e^{-j2\pi f_d T}\right]^{\mathrm{T}}$, and $[s_1,\ s_2,\ \cdots,s_M]$ is the orthogonal phase-encoding waveform used in the paper.

The echo signals of the L pulses of the target located in the distance unit r are described in (6).

$$\begin{aligned}\mathcal{X}(r) &= \xi(r,\theta,f_d)\left[\boldsymbol{b}(\theta) \otimes \left(\boldsymbol{a}^{\mathrm{T}}(\theta)\mathbf{S}\right)\right] \circ \boldsymbol{g}(f_d)\\ &= \xi(r,\theta,f_d)\boldsymbol{b}(\theta) \circ \left(\mathbf{S}^{\mathrm{T}}\boldsymbol{a}(\theta)\right) \circ \boldsymbol{g}(f_d)\end{aligned},$$
(6)

where $\boldsymbol{g}(f_d) = [1,\ e^{-j2\pi f_d T},\ e^{-j2\pi 2 f_d T},\cdots,e^{-j2\pi(L-1)f_d T}]^{\mathrm{T}}$, $\otimes$ is the Kronecker product, and $\circ$ is the outer product of a tensor.

If there are K targets with different directions and Doppler frequencies in the distance unit r, then the above Equation (6) becomes Equation (7).

$$\mathcal{X}(r) = \sum_{j=1}^{K} \xi(r,\theta_j,f_{dj})\boldsymbol{b}(\theta_j) \circ \left(\mathbf{S}^{\mathrm{T}}\boldsymbol{a}(\theta_j)\right) \circ \boldsymbol{g}(f_{dj}),$$
(7)

where $\xi(r,\theta_j,f_{dj})$ is the true complex amplitude of the target with direction θ_j and Doppler frequency f_{dj} in the distance unit r, and $\boldsymbol{a}(\theta_j)$, $\boldsymbol{b}(\theta_j)$, and f_{dj} are the emission guidance vector, reception guidance vector, and Doppler frequency information of the j-th target, $j = 1, 2, \cdots, K$.

Assuming that each distance unit is affected by the clutter scattering points of different surrounding azimuth units, based on this, the spatial angle $(-\pi/2,\ \pi/2)$ is evenly divided into N_c parts, and the true complex amplitude of the clutter scattering bodies in each distance azimuth unit is recorded as $\xi(r,\theta_i)$. At the same time, it is assumed that it follows a complex Gaussian distribution with a mean of 0 and a variance of σ_c^2, and that they are statistically independent of each other [14]. At last, the clutter model is described in Equation (8).

$$\mathcal{C}(r) = \sum_{i=1}^{N_c} \xi(r,\theta_i)\boldsymbol{b}(\theta_i) \circ \left(\mathbf{S}^{\mathrm{T}}\boldsymbol{a}(\theta_i)\right) \circ \boldsymbol{g}(f_{di}),$$
(8)

where $f_{di} = \frac{2(v_0 \sin\theta_i + v_i)}{\lambda}$, and v_i represents the velocity of the scatterer with an azimuth of θ_i relative to the carrier, which is set to 0 in this paper. At this point, the clutter Doppler is a function of azimuth θ_i. As a result, the clutter Doppler is proportional to the normalized angle of $\frac{d_r \sin\theta_i}{\lambda}$ and has a constant proportional coefficient.

When there is a target in the r-th distance unit, the Space–Time–Waveform echo tensor $\mathcal{Y}(r) \in \mathbb{C}^{P\times N\times L}$ is as described in Equation (9).

$$\mathcal{Y}(r) = \mathcal{X}(r) + \mathcal{C}(r) + \mathcal{N}(r),$$
(9)

When there is no target in the r-th distance unit, the Space–Time–Waveform echo tensor $\mathcal{Y}(r) \in \mathbb{C}^{P \times N \times L}$ is as described in Equation (10).

$$\mathcal{Y}(r) = \mathcal{C}(r) + \mathcal{N}(r), \tag{10}$$

where $\mathcal{C}(r)$ is the clutter in the r-th distance unit and $\mathcal{N}(r)$ is the noise in the r-th distance unit.

In fact, the echo data of each distance unit come from the superposition of P waveform echo data, and the specific superposition mode is shown in Figure 1.

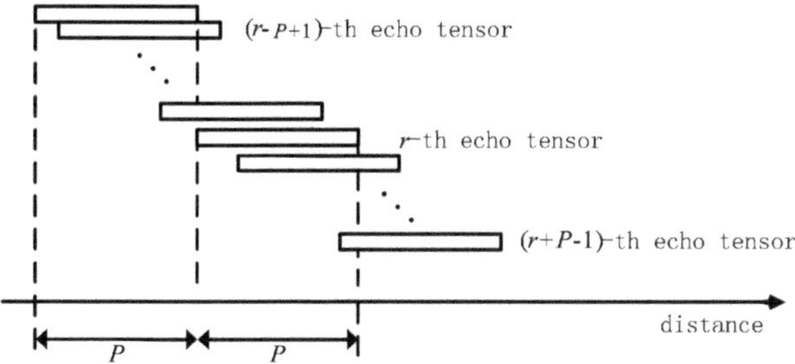

Figure 1. Superimposing waveform echo to obtain the final echo data.

Assuming that there are a total of $Q + 2P - 1$ Space–Time–Waveform echo tensors, the MIMO radar Space–Time–Range signal model $\boldsymbol{\mathcal{Y}}(r) \in \mathbb{C}^{P \times N \times L}$ before matched filtering can be described by Equation (11).

$$\boldsymbol{\mathcal{Y}}(r) = \sum_{p=-(P-1)}^{P-1} \mathcal{Y}(r + p) \; P \leq r \leq Q + P, \tag{11}$$

where $\mathcal{Y}(r + p)$ represents the $(r + p)$-th Space–Time–Waveform echo tensor. When there is a target in the $(r + p)$-th distance unit, the $\mathcal{Y}(r + p)$ is as described in Equation (12); when there is no target in the $(r + p)$-th distance unit, the $\mathcal{Y}(r + p)$ is as described in Equation (13).

$$\mathcal{Y}(r + p) = \mathcal{X}(r + p) + \mathcal{C}(r + p) + \mathcal{N}(r + p), \tag{12}$$

$$\mathcal{Y}(r + p) = \mathcal{C}(r + p) + \mathcal{N}(r + p), \tag{13}$$

where $\mathcal{X}(r + p)$ is as described in Equation (14) and $\mathcal{C}(r + p)$ is as described in Equation (15).

$$\mathcal{X}(r + p) = \sum_{j=1}^{K} \xi(r, \theta_j, f_{dj}) \boldsymbol{b}(\theta_j) \circ \left(\mathbf{J}^{\mathrm{T}}(p) \mathbf{S}^{\mathrm{T}} \boldsymbol{a}(\theta_j) \right) \circ \boldsymbol{g}(f_{dj}), \tag{14}$$

$$\mathcal{C}(r + p) = \sum_{i=1}^{N_c} (r, \theta_i) \boldsymbol{b}(\theta_i) \circ \left(\mathbf{J}^{\mathrm{T}}(p) \mathbf{S}^{\mathrm{T}} \boldsymbol{a}(\theta_i) \right) \circ \boldsymbol{g}(f_{di}), \tag{15}$$

where $\mathbf{J}(p)$ is a Stochastic matrix to realize waveform superposition, which is described in Equation (16).

$$\mathbf{J}_{i,j}(p) = \begin{cases} 1, & \text{if } i - j + p = 0 \\ 0, & \text{if } i - j + p \neq 0 \end{cases}' \tag{16}$$

Based on $\boldsymbol{\mathcal{Y}}(r) \in \mathbb{C}^{P \times N \times L}$, the Space–Time–Range echo data model $\boldsymbol{\mathcal{Y}} \in \mathbb{C}^{Q \times N \times L}$ received by the MIMO radar can be constructed, as shown in Figure 2.

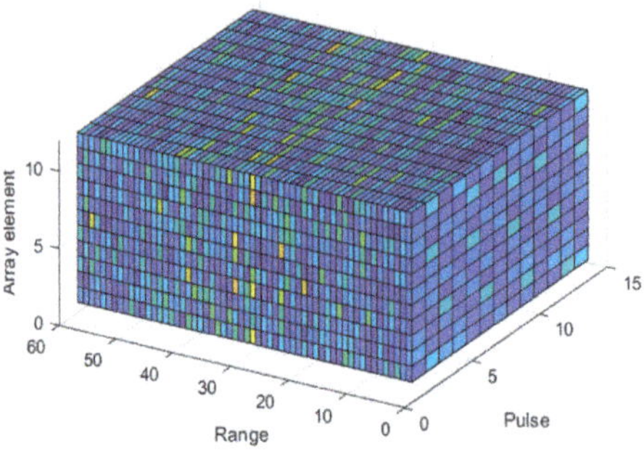

Figure 2. Space–Time–Range echo of MIMO radar.

2.2. The Principle of Space–Time–Range Joint Adaptive Focusing and Detection Algorithm

Among the existing array signal processing methods, single-dimensional adaptive processing methods, including beamforming in the spatial domain, coherent integration in the time domain, and matched filtering in the waveform dimension, essentially focus on accumulating energy in each dimension. However, for three-dimensional echo data of the MIMO radar in cluttered and noisy environments, cascaded single-dimensional adaptive processing is prone to forming sidelobes in each dimension. Moreover, there is no clutter and noise suppression function, resulting in the inadequate utilization of the structural information of the three-dimensional echo data, which can easily lead to a poor detection performance under a low signal-to-clutter ratio (SCR) and low signal-to-noise ratio (SNR).

To address this issue, this paper proposes a Space–Time–Range Joint Adaptive Focusing and Detection method based on the Mean Squared Error (MSE) criterion that combines spatial, temporal, and waveform dimensions. Firstly, the Space–Time–Range Joint Adaptive Focusing (STRJAF) is applied to the three-dimensional echo data to improve the matching mismatch caused by Doppler frequency shift and effectively suppress sidelobes. At the same time, it also helps to suppress clutter and noise, thereby enhancing the SCR and SNR. Finally, the test statistics are constructed based on the STRJAF results to achieve target detection.

The target energy in the radar echo data can be analyzed from the perspective of waveform dimension before matched filtering. In the MIMO radar echo, the echo energy of a target is distributed within P (encoding length of waveform) distance units that include its location. Therefore, when processing the received Space–Time–Range 3D echo data tensor $\mathcal{Y} \in \mathbb{C}^{Q \times N \times L}$, this paper divides it into $Q - P + 1$ Space–Time–Waveform scale $(P \times N \times L)$ 3D subtensors $\mathcal{Y}(r) \in \mathbb{C}^{P \times N \times L}$ $r = 1, 2, \cdots, Q - P + 1$, using P (encoding length of waveform) as a scale in the distance dimension. These subtensors can be regarded as $Q - P + 1$ new "three-dimensional distance units" to ensure that the target energy can be fully contained in a certain "three-dimensional distance unit". Due to the inability to directly process 3D data, this paper converts the "3D distance unit" $\mathcal{Y}(r) \in \mathbb{C}^{P \times N \times L}$ into a one-dimensional vector $x(r) \in \mathbb{C}^{PNL \times 1}$ $r = 1, 2, \cdots, Q - P + 1$.

The Mean Squared Error (MSE) criterion is a classical adaptive filter optimization criterion. Its core idea is to minimize the Mean Squared Error between the filtered output and the expected output, so as to achieve the purpose of filter optimization.

In actual situations, the target direction and Doppler frequency are unknown. Therefore, this paper uniformly divides the spatial angle $(-\pi/2, \pi/2)$ into N_s parts, and uniformly divides the normalized Doppler frequency $(-0.5, 0.5)$ into N_f parts, so that the target function based on the MSE criterion is as described in Equation (17).

$$
\begin{aligned}
J(r, \theta, f_d) &= E\left[|y(r) - d(r, \theta, f_d)|^2\right] = E\left[|w(r, \theta, f_d)x(r) - d(r, \theta, f_d)|^2\right] \\
&= w(r, \theta, f_d)^{\mathrm{H}} \mathbf{R}(r) w(r, \theta, f_d) + E\left[|d(r, \theta, f_d)|^2\right] - w(r, \theta, f_d)^{\mathrm{H}} r_{xd} - r_{xd}^{\mathrm{H}} w(r, \theta, f_d)
\end{aligned}
\tag{17}
$$

Among them, $x(r)$ is the echo signal of the r-th distance unit, $y(r)$ is the output after adaptive filtering, $d(r,\theta,f_d)$ is the expected output of the r-th distance unit when the direction is θ, the normalized Doppler frequency is f_d, $w(r,\theta,f_d)$ is the Space–Time–Range joint adaptive filter of the r-th distance unit when the direction is θ, the normalized Doppler frequency is f_d, r_{xd} is the correlation vector between the echo signal $x(r)$ and the expected output $d(r,\theta,f_d)$, and $\mathbf{R}(r)$ is the covariance matrix.

The formula obtained by solving the objective function is described in Equation (18).

$$w(r,\theta,f_d) = \mathbf{R}(r)^{-1}r_{xd},\tag{18}$$

where $r_{xd} = x(r)d(r,\theta,f_d)^*$.

The obtained solution reveals that the utilization of the MSE criterion necessitates the availability of a covariance matrix and a predetermined expected output. Consequently, the meticulous design of both the covariance matrix and initial expected output assumes paramount significance.

In this paper, first and foremost, we propose a method to design the initial expected output using the Space–Time–Waveform joint adaptive filter, which is described in Equation (19).

$$d(r,\theta,f_d) = \frac{h(r,\theta,f_d)^{\mathrm{H}}x(r)}{h(r,\theta,f_d)^{\mathrm{H}}h(r,\theta,f_d)},\tag{19}$$

where $h(r,\theta,f_d)$ is the Space–Time–Waveform joint adaptive filter, which is capable of achieving a satisfactory initial result by applying joint adaptive matched filtering in three dimensions (Space, Doppler frequency, and waveform) to the input data, $h(r,\theta,f_d) = b(\theta) \otimes g(f_d) \otimes \left(a^{\mathrm{T}}(\theta)\mathbf{S}\right)$.

To address the matching mismatch caused by intra-pulse Doppler frequency shift, this paper proposes a solution. Instead of directly compensating for it, we will set $h(r,\theta,f_d)$ as the product of the transmit steering vector and the intra-pulse waveform after the intra-pulse Doppler frequency shift, followed by the Kronecker product of the receive steering vector and the Doppler steering vector. This approach effectively eliminates the matching mismatch without direct compensation for the intra-pulse Doppler frequency shift. By designing the initial expected output in this way, we could achieve an indirect improvement in the matching mismatch.

Then, we designed the covariance matrix. The covariance matrix of each distance unit was designed to the sum of the covariance matrices of $2P - 1$ distance units (including this distance unit) in all Doppler frequency and spatial directions, which is described in Equation (20).

$$\mathbf{R}(r) = \sum_{p=-(P-1)}^{P-1} \sum_{\theta=-\pi/2}^{\pi/2} \sum_{f_d=-PRF/2}^{PRF/2} x(r+p,\theta,f_d)x(r+p,\theta,f_d)^{\mathrm{H}},\tag{20}$$

The energy output after filtering is described in Equation (21).

$$\begin{aligned}
z(r,\theta,f_d) &= y(r)^{\mathrm{H}}y(r)\\
&= \left(w(r,\theta,f_d)^{\mathrm{H}}x(r)\right)^{\mathrm{H}}\left(w(r,\theta,f_d)^{\mathrm{H}}x(r)\right),
\end{aligned}\tag{21}$$

Finally, iterative processing is performed on the filtered output. We used the filtered output $z(r,\theta,f_d)$ as the expected output $d(r,\theta,f_d)$ for the next iteration input, iterating until the desired filtered output had been obtained. And, the output at this point was the final Space–Time–Range Joint Adaptive Focusing output.

In this paper, the Space–Time–Range data used were Array Element–Pulse–Range data, which can be considered equivalent to Beam–Pulse–Range data, Array Element–Doppler–Range data, and Beam–Doppler–Range data. The difference lies in whether beamforming or Doppler transformation is performed. Therefore, the Space–Time–Range Joint Adaptive Focusing method proposed in this paper actually includes four sub methods: Array Element–Pulse–Range Joint Adaptive Focusing, Beam–Doppler–Range Joint Adaptive Focusing, Array Element–Doppler–Range Joint Adaptive Focusing, and Beam–Doppler–Range Joint Adaptive Focusing. These four sub methods are essentially equivalent, and so the remaining three sub methods will not be elaborated upon in the paper.

After achieving Space–Time–Range Joint Adaptive Focusing, the problem of MIMO radar target detection can be expressed as in Equation (22).

$$
\begin{aligned}
&H_0 : z(r) = z_{clutter}(r) + z_{noise}(r) \\
&H_1 : z(r) = z_{target}(r, \theta, f_d) + z_{clutter}(r) + z_{noise}(r)
\end{aligned}, \tag{22}
$$

Based on this, the test statistics are constructed in Equation (23).

$$
T_{\text{STRJAFD}} = z(r) \underset{H_0}{\overset{H_1}{\gtrless}} \Lambda_0, \tag{23}
$$

where Λ_0 is the detection threshold.

Figure 3 is the flow chart of the STRJAFD method. Based on the flow chart, we summarize and sort out the steps of the STRJAFD method.

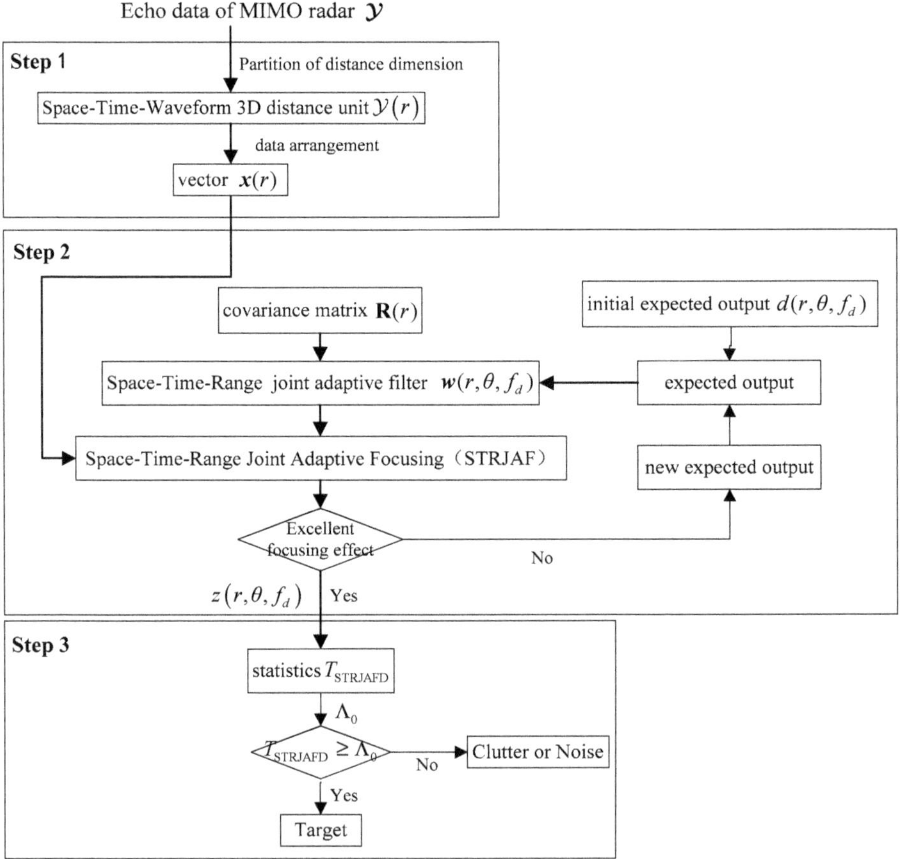

Figure 3. Flow chart of STRJAFD.

Step 1: MIMO radar echo data processing. ① Receive MIMO radar Space–Time–Range three-dimensional echo data $\mathcal{Y} \in \mathbb{C}^{Q \times N \times L}$; ② divide it into multiple Space–Time–Waveform "three-dimensional distance units" $\mathcal{Y}(r) \in \mathbb{C}^{P \times N \times L}$ $r = 1, 2, \cdots, Q - P + 1$; ③ convert the "three-dimensional distance units" into one-dimensional vector data $x(r) \in \mathbb{C}^{PNL \times 1}$.

Step 2: The Space–Time–Range Joint Adaptive Focusing based on the MSE criterion. ① Calculation of initial expected output $d(r, \theta, f_d)$ and covariance matrix $\mathbf{R}(r)$; ② calculation of Space–Time–Range joint adaptive filter $w(r, \theta, f_d) = \mathbf{R}(r)^{-1} r_{xd}$; ③ Space–Time–Range Joint Adaptive Focusing for echo data $x(r) \in \mathbb{C}^{PNL \times 1}$; ④ judgment on whether the output of the focus is satisfactory. If the focusing result is excellent, output it directly as the final result. If the focusing result is not particularly excellent, use this focusing result as the expected output for the next iteration and continue the iterative operation until an excellent focusing effect is achieved. (In fact, this step is an iterative process. After setting the number of iterations, it is assumed that the results before reaching the number of iterations are not satisfactory).

Step 3: Target detection. ① Build detection statistics based on focused output; ② judge whether to exceed the threshold; ③ output of detection results.

3. Results

In this section, we firstly investigate the focusing performance of STRJAF on multiple targets in a cluttered background (background for ground detection) and pure noise background (background for airspace detection), respectively. The purpose is to assess the method's ability to suppress sidelobe, clutter, and noise in both environments. Additionally, it compares the proposed method with the existing cascaded methods. Subsequently, the impact of intra-pulse Doppler frequency shift on the STRJAF method is examined. Finally, the detection effect under a cluttered background is experimentally verified and compared with existing cascaded methods.

3.1. Focusing Results with Cluttered and Noisy Backgrounds

Assuming that the number of transmitting Array Elements is 4, the number of receiving Array Elements is 8, and the number of pulses is 16, the orthogonal polyphase-encoding sequence is used as the intra-pulse encoding waveform, and the waveform encoding length is 8. The angle–Doppler–distance information of the simulation target was set as is shown in Table 1.

Table 1. Description of objectives.

Target Number	Angle–Doppler–Distance
1	$(0°, 0.3, 80)$
2	$(0°, -0.1, 80)$
3	$(0°, 0.1, 80)$
4	$(40°, 0.1, 80)$
5	$(-40°, 0.1, 80)$
6	$(0°, 0.1, 70)$
7	$(0°, 0.1, 90)$

In this paper, we compare the proposed method with three other approaches: MF-BF-DFT, APC-BF-DFT, and APC-IAA-STAP. The target's signal-to-noise ratio was set to 20 dB, and the clutter-to-noise ratio (CNR) was set to 0 dB. The number of iterations for APC and IAA were both set to 4. Furthermore, the number of angle units was set to 37, the number of Doppler units was set to 41, and the number of distance units was set to 140.

3.1.1. Focusing Results with Noisy Background

Firstly, under the background for airspace detection (noisy background), this paper selected Targets 1–7 to study the focusing imaging effects of various methods in the angle–normalized Doppler dimension and the angle–distance dimension. The experimental results are shown in Figures 4 and 5, respectively.

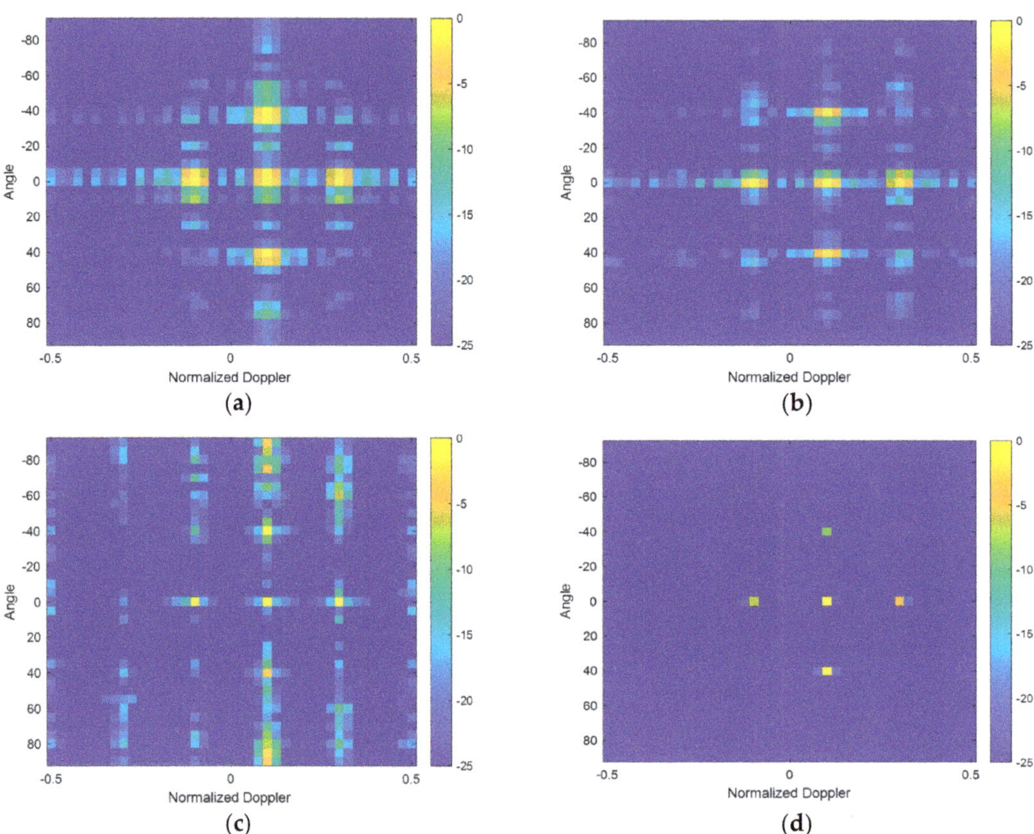

Figure 4. Angle–normalized Doppler frequency focusing results of four methods with noisy background. (**a**) MF-BF-DFT; (**b**) APC-BF-DFT; (**c**) APC-IAA-STAP; and (**d**) STRJAF.

The experimental results demonstrate that the proposed 3D joint focusing method STRJAF performs with excellent suppression capabilities for three-dimensional sidelobes (range sidelobe, Doppler sidelobe, and angle sidelobe) under a noisy background. Furthermore, it can achieve super-resolution and boasts exceptional noise reduction abilities. The focusing effect of the proposed method is slightly superior to that of the semi-cascaded and semi-joint method "APC-IAA-STAP", and significantly surpasses traditional cascading methods such as "MF-BF-DFT" and "APC-BF-DFT".

3.1.2. Focusing Results with Cluttered Background

Under the background for ground detection (cluttered background), this paper selected Targets 1–5 to study the focusing imaging effects of various methods in the angle–normalized Doppler dimension and the normalized Doppler–range dimension. The experimental results are shown in Figures 6 and 7, respectively.

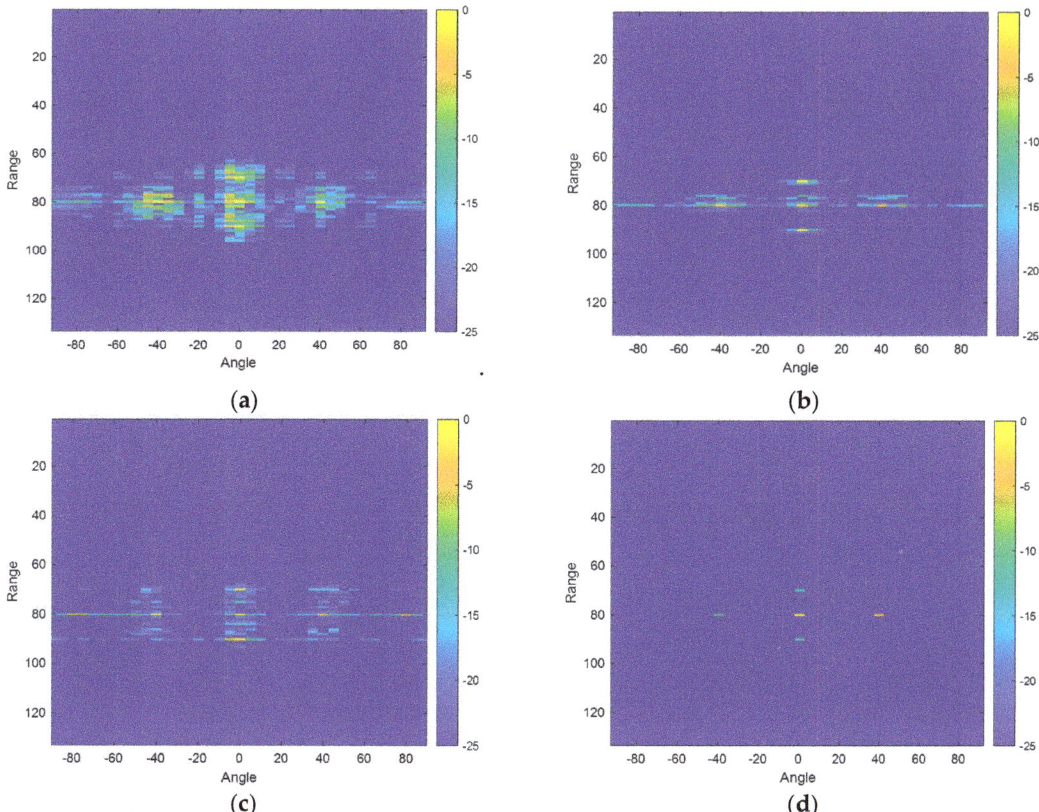

Figure 5. Angle–Range focusing results of four methods. (**a**) MF-BF-DFT; (**b**) APC-BF-DFT; (**c**) APC-IAA-STAP; and (**d**) STRJAF.

From the experimental results, it can be seen that under the cluttered background, there are clutter ridges in the diagonal position of the two-dimensional image in the Space–Time dimension. It is obvious that the broadening of the clutter ridges in the cascaded methods ("MF-BF-DFT" and "APC-BF-DFT") is more severe than that in the semi-cascaded and semi-combined method ("APC-IAA-STAP"). As the SCR decreases, this situation will become more apparent. Meanwhile, the proposed 3D joint focusing method STRJAF still has good clutter suppression and focusing effects in cluttered environments, being significantly superior to traditional cascaded methods.

In summary, the proposed STRJAF method can effectively suppress clutter and noise and effectively suppress three-dimensional sidelobes for focusing, and has super-resolution ability that is significantly superior to traditional cascading methods, regardless of whether it is a cluttered or noisy background.

3.2. The Influence of Intra-Pulse Doppler Frequency Shift

In order to study the impact of intra-pulse Doppler frequency shift on the proposed method, this paper selects two targets with an angle, normalized Doppler, and distance units of (0°, 0, 100) and (0°, 1, 140), representing two cases of no intra-pulse Doppler frequency shift and a maximum intra-pulse Doppler frequency shift. We set the number of transmitting Array Elements to 4, the number of receiving Array Elements to 4, the number of pulses to 8, and the SNR to 0 dB. The experimental results after processing with STRJAF and cascading methods under different code length conditions in a noisy background are shown in Figures 8 and 9, respectively.

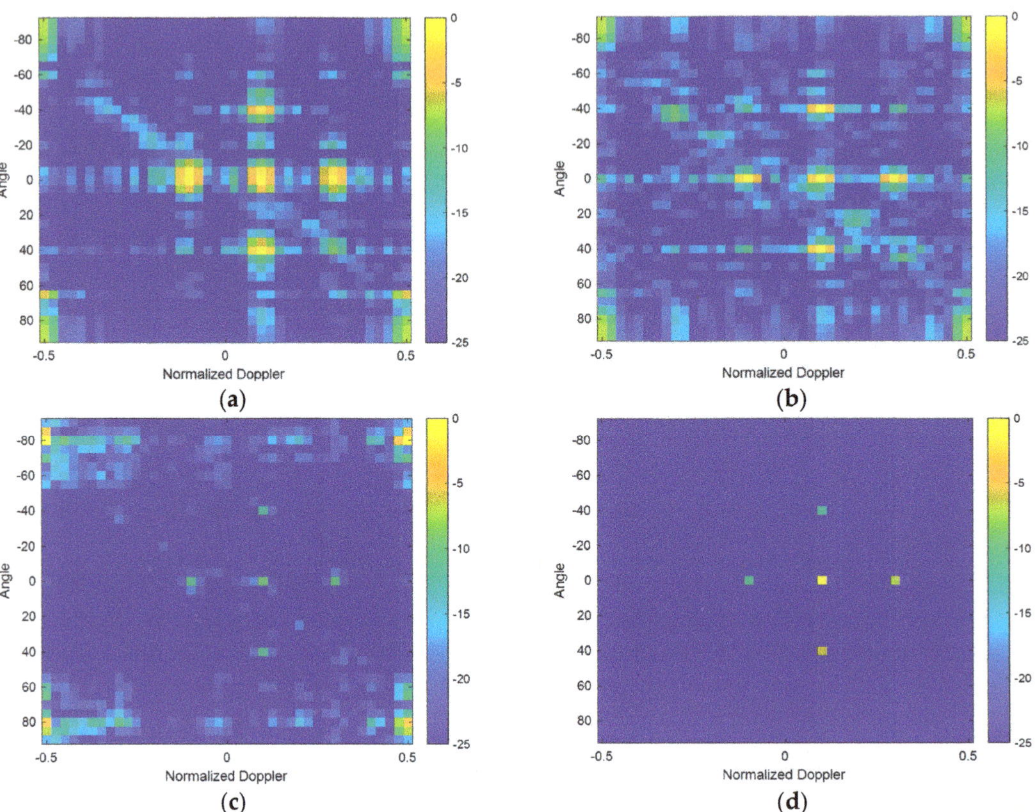

Figure 6. Angle–normalized Doppler frequency focusing results of four methods with cluttered background. (**a**) MF-BF-DFT; (**b**) APC-BF-DFT; (**c**) APC-IAA-STAP; and (**d**) STRJAF.

From the experimental results, it can be verified that when there is intra-pulse Doppler frequency shift, matched filtering and adaptive pulse compression methods are prone to being mismatched in the distance dimension, resulting in a decrease in the effectiveness of matched filtering and pulse compression, which is not conducive to the accumulation of target energy. Moreover, the larger the Doppler frequency shift and the longer the code length of the encoded signal, the more significant the mismatch. The proposed method STRJAF is not affected by the intra-pulse Doppler frequency shift, ensuring the effective accumulation of target energy, which proves the effectiveness of the proposed method in improving the intra-pulse Doppler frequency shift problem.

3.3. Detection Results with Noisy and Cluttered Backgrounds

In this section, this paper investigates the detection performance of the proposed method with noisy and cluttered backgrounds and compares it with traditional cascaded processing methods. Assuming that the number of transmitting Array Elements is 4, the number of receiving Array Elements is 4, and the number of pulses is 8, an orthogonal polyphase-encoding sequence is used as the intra-pulse encoding waveform, with a waveform encoding length of 8, and the false alarm rate is set to 10^{-2}. With a cluttered background, the detection probability at each signal-to-clutter ratio was obtained through

1000 simulations. The signal-to-clutter ratio here is defined as the signal-to-clutter ratio before pulse compression and coherent accumulation.

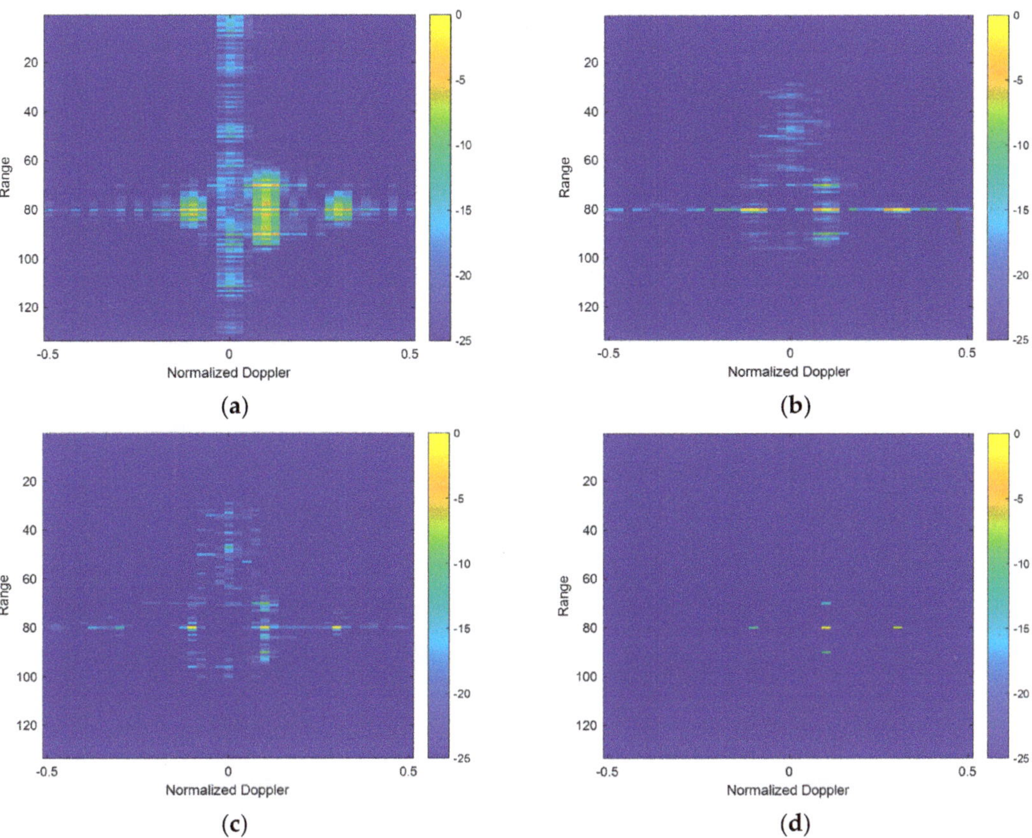

Figure 7. Range-normalized Doppler frequency focusing results of four methods. (**a**) MF-BF-DFT; (**b**) APC-BF-DFT; (**c**) APC-IAA-STAP; and (**d**) STRJAF.

3.3.1. The Effect of Intra-Pulse Doppler Frequency Shift on Detection Performance

Firstly, the impact of intra-pulse Doppler frequency shift on target detection performance was studied under two conditions of Doppler frequency shift and no Doppler frequency shift in the echo signal. At the same time, comparisons were made between three methods, "MF-BF-DFT", "APC-BF-DFT", and "APC-IAA-STAP", using the Constant False Alarm Rate (CFAR) method.

At first, the target detection performance without intra-pulse Doppler frequency shift was studied, and targets with an angle, normalized Doppler, and distance units of ($10°$, 0, 50) were selected for detection experiments and used as the reference standards. The experimental results are shown in Figures 10 and 11.

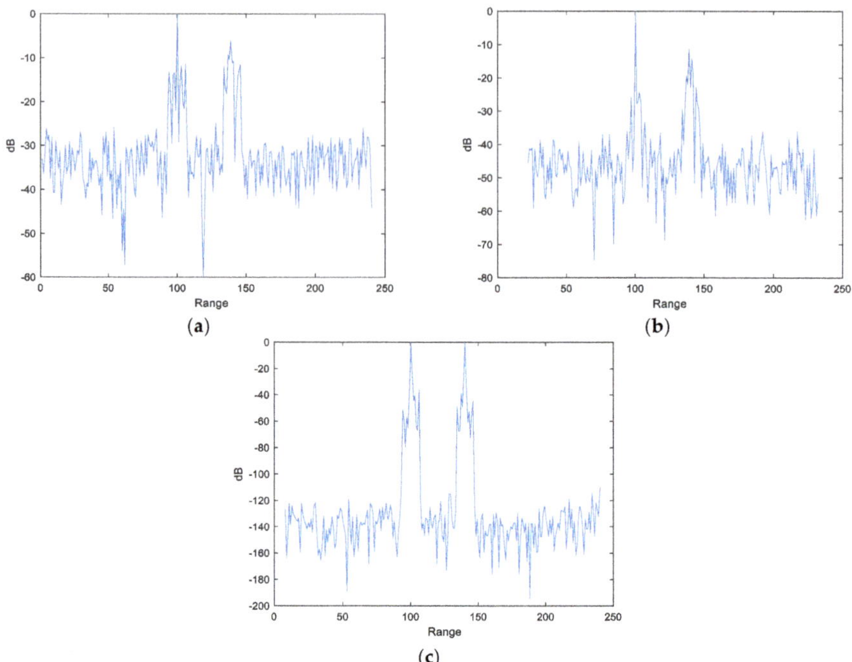

Figure 8. The influence of intra-pulse Doppler frequency shift on three methods when the code length is 8. (**a**) MF-BF-DFT; (**b**) APC-BF-DFT; and (**c**) STRJAF.

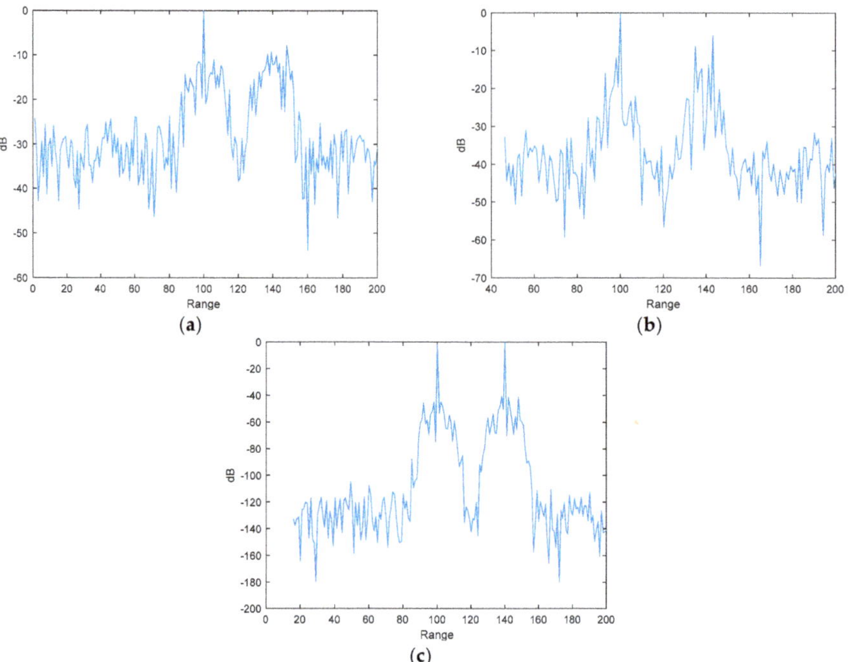

Figure 9. The influence of intra-pulse Doppler frequency shift on three methods when the code length is 16. (**a**) MF-BF-DFT; (**b**) APC-BF-DFT; and (**c**) STRJAF.

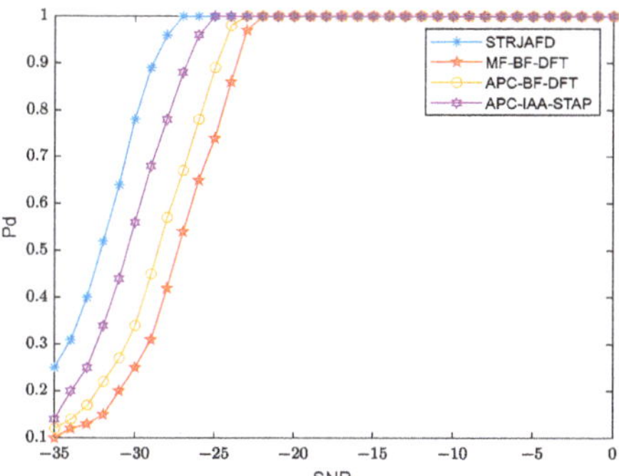

Figure 10. Comparison of test results with a noisy background.

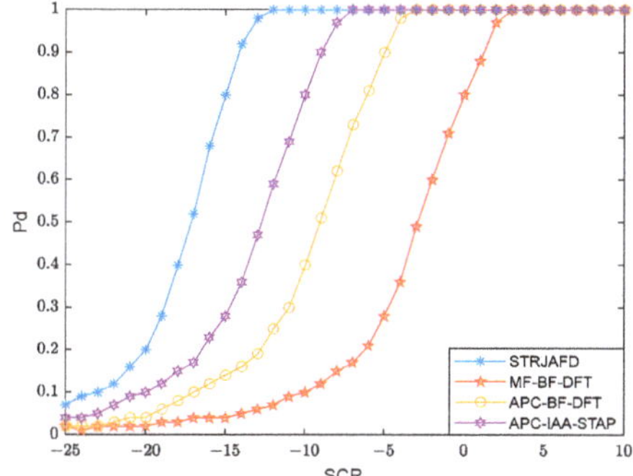

Figure 11. Comparison of test results with a cluttered background.

Then, the impact of intra-pulse Doppler frequency shift on detection performance with a cluttered background was studied, and targets with an angle, normalized Doppler, and distance units of $(10°, 0.5, 50)$ were selected as experimental targets. The experimental results are shown in Figure 12.

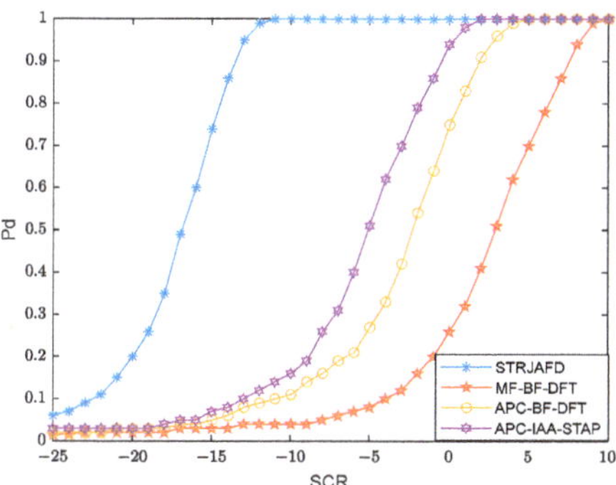

Figure 12. Detection results under large intra-pulse Doppler frequency shift.

The experimental results show that the detection results of various methods do not differ significantly under the condition of no intra-pulse Doppler frequency shift with a noisy background. However, with a cluttered background, the proposed detection method has good performance and robustness, with smaller fluctuations in the detection performance curve, regardless of whether there is intra-pulse Doppler frequency shift or not. In contrast, the traditional cascaded method shows a significant decline in detection performance when intra-pulse Doppler frequency shift is present. Therefore, compared to the traditional method, the proposed method demonstrates obvious superiority.

3.3.2. The impact of Interference Targets on Detection Performance

Then, the impact of interference targets on target detection performance was studied under the presence of interference targets and compared with three methods, "MF-BF-DFT", "APC-BF-DFT", and "APC-IAA-STAP", using the Small Of-Constant False Alarm Rate (SO-CFAR).

The target with an angle, normalized Doppler, and range units of (10°, 0, 50) was still selected as the detection target, while the target with (20°, 0, 46) was selected as the interference target, and we set the energy of the interference target to 0 dB. The target detection results with noisy and cluttered backgrounds are shown in Figures 13 and 14, respectively.

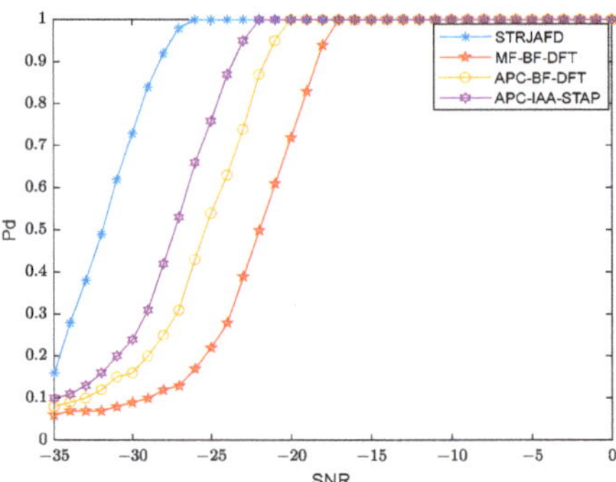

Figure 13. Comparison of detection results with a noisy background.

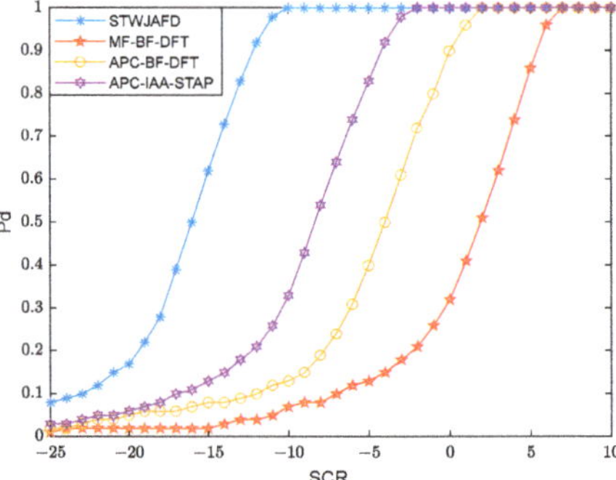

Figure 14. Comparison of detection results with a cluttered background.

The experimental results show that the interference target has little impact on the target detection performance of the proposed method, indicating that the proposed method can effectively suppress sidelobes, while the detection performance of the cascade method has a certain degree of decline, indicating that it can be significantly affected by the interference target sidelobes.

4. Discussion

Traditional processing and detection methods aim to enhance SCR and SNR by accumulating target energy from each dimension, thereby improving the detection performance. Beamforming, coherent integration, and matched filtering (pulse compression) essentially accumulate energy in spatial, temporal, and distance dimensions. However, with the advancement of radar systems, echo data dimensions have expanded from one-dimensional and two-dimensional to three-dimensional or even higher dimensions. The cascaded single-dimensional energy accumulation method cannot effectively handle multidimensional sidelobes in multidimensional data. To further enhance signal processing and target detection performance, joint processing and detection methods are necessary. Based on this idea, this paper has proposed a Space–Time–Range 3D Joint Adaptive Focusing and Detection method, which achieves the joint suppression of 3D sidelobes and accumulates target energy. At the same time, by suppressing clutter and noise, the SCR/SNR is further improved, and the performance of target detection is improved.

5. Conclusions

In this paper, we have proposed a novel Space–Time–Range Joint Adaptive Focusing and Detection method to address the challenges associated with the multidimensional sidelobes (range sidelobes, Doppler sidelobes, and angle sidelobes) of targets, clutter, and noise in the echo background, and Doppler frequency shift-induced mismatch in matched filtering within existing MIMO radar target detection processes. The experimental results show that the proposed method effectively suppresses sidelobe interference and exhibits good focusing capabilities, achieving superior clutter and noise suppression while improving SCR and SNR. Moreover, it demonstrates excellent detection performance in low-SCR and -SNR environments, significantly enhancing the target detection performance.

In this study, we utilized the phase-coded signal as the MIMO radar transmission signal for research. In fact, different transmission signals possess distinct characteristics and are suitable for different scenarios. In the future, we will conduct research on other types of waveforms and use them as signals for MIMO radar transmission to verify whether the proposed method still performs well under different waveform signals and different scenarios, and whether it still has significant advantages compared to other methods.

In addition, the computational cost of the proposed method is relatively high, and research on its fast implementation version will also be a part of future research.

Author Contributions: Conceptualization, J.G., X.M. and B.C.; methodology, X.M., Y.H. and B.C.; software, X.M., B.C. and Y.H.; validation, X.M., B.C. and N.L.; formal analysis, X.M.; investigation, X.C.; resources, N.L. and X.C.; data curation, Y.H.; writing—original draft preparation, X.M.; writing—review and editing, X.M.; visualization, X.M.; supervision, Y.H.; project administration, J.G.; funding acquisition, J.G. All authors have read and agreed to the published version of the manuscript.

Funding: This work was supported in part by the National Natural Science Foundation of China under grants 62222120, 62101583, and 61871392, Taishan Scholars Program (tsqn202211246), and in part by the Natural Science Foundation of Shandong Province under grant ZR2021YQ43.

Data Availability Statement: Not applicable.

Conflicts of Interest: The authors declare no conflict of interest.

References

1. Lin, B.; Yang, X.; Wang, J.; Wang, Y.; Wang, K.; Zhang, X. A Robust Space Target Detection Algorithm Based on Target Characteristics. *IEEE Geosci. Remote Sens. Lett.* **2022**, *19*, 8012405. [CrossRef]
2. Ye, S.; He, Q.; Wang, X. MIMO Radar Moving Target Detection in Clutter Using Supervised Learning. In Proceedings of the 2021 IEEE Radar Conference (RadarConf21), Atlanta, GA, USA, 7–14 May 2021; pp. 1–5.
3. Chen, X.; Chen, B.; Guan, J.; Huang, Y.; He, Y. Space-Range-Doppler Focus-Based Low-observable Moving Target Detection Using Frequency Diverse Array MIMO Radar. *IEEE Access* **2018**, *6*, 43892–43904. [CrossRef]
4. Li, J.; Stoica, P. MIMO Radar with Colocated Antennas. *IEEE Signal Process. Mag.* **2007**, *24*, 106–114. [CrossRef]
5. Fishler, E.; Haimovich, A.; Blum, R.; Chizhik, D.; Cimini, L.; Valenzuela, R. MIMO radar: An idea whose time has come. In Proceedings of the 2004 IEEE Radar Conference (IEEE Cat. No. 04CH37509), Philadelphia, PA, USA, 29 April 2004; pp. 71–78.
6. Zhou, J.; Li, H.; Cui, W. Low-Complexity Joint Transmit and Receive Beamforming for MIMO Radar with Multi-Targets. *IEEE Signal Process. Lett.* **2020**, *27*, 1410–1414. [CrossRef]
7. Wang, M.; Gao, F.; Jin, S.; Lin, H. An Overview of Enhanced Massive MIMO with Array Signal Processing Techniques. *IEEE J. Sel. Top. Signal Process.* **2019**, *13*, 886–901. [CrossRef]

8.   Zoltowski, M.; Shuman, M.; Rangaswamy, M. Virtual Waveform Diversity with Phase-Coded Radar Waveforms. In Proceedings of the 2021 55th Asilomar Conference on Signals, Systems, and Computers, Pacific Grove, CA, USA, 31 October–3 November 2021; pp. 1048–1052.
9.   Davis, R.M.; Fante, R.L.; Perry, R.P. Phase-coded waveforms for radar. *IEEE Trans. Aerosp. Electron. Syst.* **2007**, *43*, 401–408. [CrossRef]
10. Qureshi, T.R.; Zoltowski, M.D.; Calderbank, R. A novel approach to Doppler compensation and estimation for multiple targets in MIMO radar with unitary waveform matrix scheduling. In Proceedings of the 2012 IEEE International Conference on Acoustics, Speech and Signal Processing (ICASSP), Kyoto, Japan, 25–30 March 2012; pp. 2473–2476.
11. Wang, H.; Zhu, X. The parameter estimation of transmit diversity MIMO radar with iteratively adaptive pulse compression and Doppler compensation. In Proceedings of the 2015 International Conference on Wireless Communications & Signal Processing (WCSP), Nanjing, China, 15–17 October 2015; pp. 1–5.
12. Zhu, J.; Song, Y.; Jiang, N.; Xie, Z.; Fan, C.; Huang, X. Enhanced Doppler Resolution and Sidelobe Suppression Performance for Golay Complementary Waveforms. *Remote Sens.* **2023**, *15*, 2452. [CrossRef]
13. Guan, J.; Pei, J.; Huang, Y.; Chen, X.; Chen, B. Time-Range Adaptive Focusing Method Based on APC and Iterative Adaptive Radon-Fourier Transform. *Remote Sens.* **2022**, *14*, 6182. [CrossRef]
14. Cui, N.; Duan, K.; Xing, K.; Yu, Z. Beam-Space Reduced-Dimension 3D-STAP for Nonside-Looking Airborne Radar. *IEEE Geosci. Remote Sens. Lett.* **2022**, *19*, 3506505. [CrossRef]
15. Wen, C.; Huang, Y.; Peng, J.; Wu, J.; Zheng, G.; Zhang, Y. Slow-Time FDA-MIMO Technique with Application to STAP Radar. *IEEE Trans. Aerosp. Electron. Syst.* **2022**, *58*, 74–95. [CrossRef]
16. Yardibi, T.; Li, J.; Stoica, P.; Xue, M.; Baggeroer, A.B. Source Localization and Sensing: A Nonparametric Iterative Adaptive Approach Based on Weighted Least Squares. *IEEE Trans. Aerosp. Electron. Syst.* **2010**, *46*, 425–443. [CrossRef]
17. Tian, J.; Zhang, B.; Li, K.; Cui, W.; Wu, S. Low-Complexity Iterative Adaptive Approach Based on Range–Doppler Matched Filter Outputs. *IEEE Trans. Aerosp. Electron. Syst.* **2023**, *59*, 125–139. [CrossRef]
18. Chen, C.Y.; Vaidyanathan, P.P. MIMO Radar Space–Time Adaptive Processing Using Prolate Spheroidal Wave Functions. *IEEE Trans. Signal Process.* **2008**, *56*, 623–635. [CrossRef]
19. Reed, I.S.; Mallett, J.D.; Brennan, L.E. Rapid Convergence Rate in Adaptive Arrays. *IEEE Trans. Aerosp. Electron. Syst.* **1974**, *AES-10*, 853–863. [CrossRef]
20. Brigui, F.; Boizard, M.; Ginolhac, G.; Pascal, F. New Low-Rank Filters for MIMO-STAP Based on an Orthogonal Tensorial Decomposition. *IEEE Trans. Aerosp. Electron. Syst.* **2018**, *54*, 1208–1220. [CrossRef]
21. Feng, W.; Zhang, Y.; He, X. Clutter Rank Estimation for Reduce-Dimension Space-Time Adaptive Processing MIMO Radar. *IEEE Sens. J.* **2017**, *17*, 238–239. [CrossRef]
22. Feng, W.; Guo, Y.; He, X.; Liu, H.; Gong, J. Jointly Iterative Adaptive Approach Based Space Time Adaptive Processing Using MIMO Radar. *IEEE Access* **2018**, *6*, 26605–26616. [CrossRef]
23. Rytel-Andrianik, R. Efficient matched filtering and beamforming for coherent MIMO radar. In Proceedings of the 2016 IEEE International Symposium on Phased Array Systems and Technology (PAST), Waltham, MA, USA, 18–21 October 2016; pp. 1–6.
24. Malik, H.; Burki, J.; Mumtaz, M.Z. Adaptive Pulse Compression for Sidelobes Reduction in Stretch Processing Based MIMO Radars. *IEEE Access* **2022**, *10*, 93231–93244. [CrossRef]
25. Gao, Z.; Wang, X.; Huang, P.; Xu, W.; Zhang, Z. Arc Array Radar IAA-STAP Algorithm Based on Sparse Constraint. In Proceedings of the 2020 International Conference on Information Science, Parallel and Distributed Systems (ISPDS), Xi'an, China, 14–16 August 2020; pp. 277–282.

remote sensing

Article

Adaptive Beamforming with Sidelobe Level Control for Multiband Sparse Linear Array

Hongtao Li [1,†], Longyao Ran [1,†], Cheng He [2], Zhoupeng Ding [1] and Shengyao Chen [1,*]

1   School of Electronic and Optical Engineering, Nanjing University of Science and Technology, Nanjing 210094, China; liht@njust.edu.cn (H.L.); rly@njust.edu.cn (L.R.); dzp@njust.edu.cn (Z.D.)
2   Beijing Institute of Remote Sensing Equipment, Beijing 100005, China
*   Correspondence: chenshengyao@njust.edu.cn
†   These authors contributed equally to this work.

Abstract: Multiband antenna arrays have the capability of effectively working in multiple frequency bands and thus significantly simplify the antenna system. To further reduce the system overhead, this paper discusses the joint design of antenna selection and adaptive beamforming for multiband antenna arrays, where the sidelobe level is also controlled so as to alleviate the effect of unknown sporadic interference. Based on the maximum signal-to-interference-plus-noise ratio (SINR) criterion and sidelobe level constraints, the proposed multiband sparse array design is formulated into a nonconvex constrained nonlinear optimization problem with an $l_{0,2}$-mixed norm regularization. This problem ensures that the same antenna positions are selected at all operating frequencies while the beamformer weights of each frequency are optimized independently. By exploiting the semi-definite relaxation and the reweighted $l_{1,\infty}$-norm approximation, the problem is converted into a series of convex subproblems and is then effectively solved by the proposed iterative reweighted method. Numerical results show that the proposed multiband sparse array significantly reduces the sidelobe levels in all operating frequencies while maintaining the maximum SINR, so it provides superior performance of interference suppression.

Keywords: multiband antenna; sparse array; adaptive beamforming; sidelobe level control

Citation: Li, H.; Ran, L.; He, C.; Ding, Z.; Chen, S. Adaptive Beamforming with Sidelobe Level Control for Multiband Sparse Linear Array. *Remote Sens.* **2023**, *15*, 4929. https://doi.org/10.3390/rs15204929

Academic Editor: Dusan Gleich

Received: 30 August 2023
Revised: 10 October 2023
Accepted: 10 October 2023
Published: 12 October 2023

1. Introduction

A multiband antenna is a specialized type of antenna that is designed to effectively operate across multiple preset frequency bands simultaneously. This versatile technology substantially reduces the volume, cost, weight, and complexity associated with antenna systems. As a result, multiband antennas are increasingly being used in advanced communication and radar systems [1–3]. With the increasing requirement on the spatial resolution and capacity, several kinds of multiband arrays have been developed for the application of next-generation wireless communication [4]. However, in the utilization of medium- or large-scale multiband arrays, the cost, hardware complexity, and power consumption are high. Sparse arrays offer significant advantages in terms of reducing the system complexity and hardware overhead. Compared to conventional uniform linear arrays, sparse arrays use fewer antenna elements and radio-frequency channels while they have the same array aperture and suffer from only a little performance loss. Therefore, one promising direction in developing multiband antenna arrays is to design optimal sparse array configurations. Different from conventional sparse arrays working at a single frequency, the configuration of a multiband sparse array should possess the capability to deliver excellent performance across all operating frequencies, tailored to specific functions such as transmit beampattern synthesis or adaptive receive beamforming.

The design of narrowband sparse arrays, specifically focusing on single-frequency operation, has been widely explored in various tasks and performance metrics [5–22]. Depending on the application and the performance metrics, sparse array design can be divided into two categories: environment-independent or environment-dependent. In the environment-independent case, various structured sparse arrays, including minimum redundancy arrays [5], nested arrays [6], and co-prime arrays [7], have been developed to improve the direction-of-arrival (DOA) estimation performance, and then to provide good beamforming performance [8,9]. Furthermore, to obtain a sparse array with the smallest element number, sparsity-promoting algorithms for unstructured sparse arrays are used to synthesize the desired beampattern [10] or to improve the parameter estimation performance [11]. The representative algorithms include reweighted l_1-norm [10,12,13], mixed norm or norm combination [14], nonconvex l_p-norm ($0 < p < 1$) [15], soft-thresholding shrinkage [16], and Bayesian inference [17]. In the environment-dependent case, joint optimization of antenna position and receive beamformer has been utilized to maximize output signal-to-interference-plus-noise ratio (SINR) by exploiting environmental data. These methods have been implemented by using reweighted l_1-norm and semi-definite relaxation (SDR) [18], sequential convex approximation (SCA) [19], and the alternating direction method of multipliers (ADMM) [20], to name a few. Additional constraints have also been introduced to achieve sidelobe level (SLL) control [21]. To further minimize the number of required antennas, an l_0-norm concave approximation approach has been proposed in [22]. Since the unstructured sparse array designs are commonly coined as nonconvex constrained optimization problems, the main challenge is how to resolve these problems efficiently. Due to the powerful capability of deep neural networks (DNNs) in solving nonlinear problems and performing fast computations, a fully connected DNN has recently been applied to select antenna positions for adaptive beamforming [23,24].

Along with the continuous development of narrowband sparse arrays, wideband sparse array design has also been studied extensively in the past two decades [25–32]. Due to the significantly degraded performance of narrowband sparse arrays when the signal bandwidth increases and the narrowband hypothesis no longer holds, which results in a poor ability of interference suppression, it is necessary to consider wideband sparse array design. By utilizing a limited number of available antennas, the wideband sparse array design offers more degrees-of-freedom (DoFs) to control the beampattern over the frequencies of interest. In wideband beamforming, there are two commonly used implementation schemes: tapped delay line (TDL) filtering and discrete Fourier transform (DFT)-based sub-band processing. Concretely, TDL implements temporal filtering by using a TDL to capture the signal at different time instants, while DFT processes the signal in several narrow sub-bands via DFT [33].

Based on the TDL and DFT schemes, several different goals involving frequency-invariant (FI) beampattern synthesis, SLL control, and robust beampattern design [25–27] have been achieved by many wideband sparse array design methods. To be specific, FI beampattern synthesis is dedicated to generating a specific pattern regardless of the operation frequency, the SLL control aims to reduce the power of sidelobes around the mainlobe, and the robust beampattern design focuses on maintaining desired beampatterns that are not influenced by the array uncertainties or the changes of operating environment. Early methods such as simulated annealing [28,29] and genetic algorithms [30], which rely on heuristic methods, have been abandoned due to their high computational cost. Recently, the sparsity-promoting algorithm has emerged as a prevalent solution to optimizing the array design. For TDL implementation, FI beampatterns with a small number of antennas are synthesized by several effective algorithms, including reweighted l_1-norm [27], second-order cone programming (SOCP) [26], and the generalized matrix pencil method [34]. Although these algorithms demonstrate outstanding performance, they are still computationally expensive and thus not suitable for large-scale arrays. In contrast, DFT-based sub-band processing has become increasingly popular due to its remarkable computational efficiency [31,32]. In this approach, the wideband signal is divided into several narrow sub-bands via DFT. The beams in each sub-band are optimized by imposing group sparsity constraints through convex optimization techniques. This method has demonstrated commendable performance while demanding lower computational requirements than

TDL-based approaches. However, it requires storing blocked received signals and updating the weights block by block.

In this paper, we consider the multiband sparse array design for adaptive beamforming, which is partially distinct from existing narrowband and wideband sparse array design. Since the multiband array works simultaneously at multiple frequencies, it can be considered as a narrowband array at each frequency. That is to say, the design of a multiband sparse array is equivalent to the joint design of multiple narrowband sparse arrays with the same antenna positions. From another perspective, the multiband sparse array design can be considered as a special case of the DFT-based wideband sparse array design, in which only partial DFT bins exist. However, the number of DFT bins depends on the bandwidth of the multiband antenna. Hence, the DFT-based wideband schemes will be inefficient when the frequency spacing between adjacent operation frequency bands is large enough. For multiband sparse arrays, ref. [35] utilized the linear Cantor fractal array to construct a structured sparse multiband array and then offered a Kalman filtering-based adaptive beamformer. Ref. [31] considered the joint design of antenna selection and adaptive beamformer by using group sparse regularization. The array has the same antenna position in all frequencies, while the beamforming weights of each frequency are separately optimized. However, the SLL control of receive beampattern is not taken into account. Uncontrollable high sidelobes generated at some operating frequencies will reduce the interference suppression performance, especially when unknown sporadic interference appears.

Based on the above observations, this paper discusses the problem of multiband sparse array design for adaptive beamforming with SLL control. Concretely, we jointly design an antenna selection and adaptive beamformer under the maximum SINR criterion and the SLL constraints. Since it is essential for the antenna positions to be identical in all operating frequencies, we coin the proposed sparse array design as a nonconvex constrained nonlinear optimization problem with an $l_{0,2}$-mixed norm regularization. The proposed problem is intractable since the objective function and all constraints are nonconvex, and the beamforming weights of different frequencies are coupled in the objective function. By employing the reweighted norm transformation and SDR techniques, we construct an iterative reweighted method to solve this problem effectively. With the aid of the reweighted norm approximation technique, we first equivalently express the original problem as a series of $l_{1,\infty}$-norm regularized nonconvex constrained optimization subproblems. By using SDR and linear fractional SDR schemes, we then relax the $l_{1,\infty}$-norm regularized nonconvex subproblem to the corresponding convex subproblem, which is tractably resolved by off-the-shelf toolboxes. Numerical results demonstrate that the proposed method can effectively reduce the SLL across all operating frequencies, thereby enhancing its interference suppression performance.

The remainder of this paper is organized as follows. Section 2 introduces the signal model of adaptive beamforming for multiband arrays. Section 3 states the problem formulation of multiband sparse array design for maximizing the output SINR under SLL constraints and then provides an SDR-based iterative reweighted solution algorithm. Section 4 analyzes the computational complexity of the proposed algorithm. Numerical experiments are conducted in Section 5 to validate the superiority of the optimized multiband sparse array. Section 6 provides some discussions regarding the multiband sparse array design. Concluding remarks follow at the end.

Notations: Throughout this paper, lower-case bold characters and upper-case bold characters represent vectors and matrices, respectively. $(\cdot)^T$ indicates the transpose and $(\cdot)^H$ denotes the conjugate transpose. $|\cdot|$ is the modulus operator. $\mathbb{E}\{\cdot\}$ denotes the statistical expectation. $\text{Tr}(\cdot)$ and $\text{Rank}(\cdot)$ stand for the trace and the rank operations, respectively. $\mathbf{I}_N$ stands for an $N \times N$ identity matrix. $\mathbf{W} \succeq 0$ means that $\mathbf{W}$ is positive semi-definite. $\mathfrak{R}\{\cdot\}$ and $\mathfrak{I}\{\cdot\}$ represent the real and imaginary parts of the complex variables, respectively.

2. Signal Model

Assume that the multiband array, consisting of N uniformly spaced multiband antenna elements, has the capability of receiving narrowband signals belonging to M frequency bands centered at the frequency ω_i ($i = 1, \cdots, M$), respectively. Consider a desired source operating in the i-th band with the center frequency ω_i, while there exist P_i sources of interference. Both the desired source and interference signals impinge on the N-element multiband array. The baseband signal received by the multiband array at the frequency ω_i is given by

$$\mathbf{x}_{\omega_i} = \alpha_i \mathbf{a}(\theta_{s_i}, \omega_i) + \sum_{p_i=1}^{P_i} \beta_{p_i} \mathbf{a}(\theta_{p_i}, \omega_i) + \mathbf{v}_i, \tag{1}$$

where $\mathbf{v}_i \in \mathbb{C}^N$ is the additive, while Gaussian noises with variance $\sigma_{v_i}^2$, $\alpha_i, \beta_{p_i} \in \mathbb{C}$ are the complex amplitudes of the incident baseband source and the p_i-th interference source, respectively; $\mathbf{a}(\theta_{s_i}, \omega_i)$ and $\mathbf{a}(\theta_{p_i}, \omega_i)$ are the steering vectors at the frequency ω_i with respect to the desired source with the direction θ_{s_i} and the interference source with the direction θ_{p_i}, which are defined by

$$\mathbf{a}(\theta_{s_i}, \omega_i) = [1, \quad e^{j\frac{2\pi}{\lambda_{\omega_i}} d \cos \theta_{s_i}}, \ldots, e^{j\frac{2\pi}{\lambda_{\omega_i}} d(N-1) \cos \theta_{s_i}}]^T \tag{2}$$

where d is the element spacing and λ_{ω_i} is the wavelength at the frequency ω_i. To prevent spatial aliasing, we set $d = \frac{\lambda_{\omega_m}}{2}$, where ω_m is the highest frequency of $\{\omega_i\}_{i=1}^M$. Then the steering vector $\mathbf{a}(\theta_{s_i}, \omega_i)$ can be simplified as

$$\mathbf{a}(\theta_{s_i}, \omega_i) = [1, \quad e^{j\pi \frac{\omega_i}{\omega_m} \cos \theta_{s_i}}, \ldots, e^{j\pi \frac{\omega_i}{\omega_m} (N-1) \cos \theta_{s_i}}]^T \tag{3}$$

The received signal $\mathbf{x}_{\omega_i}$ is linearly combined by a beamformer at the receiver to maximize the output SINR. Denote $\mathbf{w}_i = [w_1, \ldots, w_N]^T \in \mathbb{C}^N$ as the beamformer weight vector. Then the output of the beamformer is

$$y_{w_i} = \mathbf{w}_i^H \mathbf{x}_{w_i}, \ i = 1, \ldots, M. \tag{4}$$

Let the adaptive beamformers be used at all frequencies $\{\omega_i\}_{i=1}^M$. Based on the maximum SINR (maxSINR) criterion, the optimal beamformers of all frequencies are determined by the following optimization problem:

$$\min_{\{\mathbf{w}_i\}_{i=1}^M} \quad \sum_{i=1}^M \mathbf{w}_i^H \mathbf{R}_{in_i} \mathbf{w}_i \tag{5}$$

$$\text{s.t.} \quad \mathbf{w}_i^H \mathbf{R}_{s_i} \mathbf{w}_i = 1, \quad \forall i \in 1, \ldots, M$$

where $\mathbf{R}_{s_i} = \sigma_i^2 \mathbf{a}(\theta_{s_i}, \omega_i) \mathbf{a}^H(\theta_{s_i}, \omega_i)$ is the covariance matrix of the desired signal, and $\sigma_i^2 = \mathbb{E}\{\alpha_i \alpha_i^H\}$ is the average power of the source at the i-th frequency. Similarly, $\mathbf{R}_{in_i} = \sum_{p_i=1}^{P_i} \left(\sigma_{p_i}^2 \mathbf{a}(\theta_{p_i}, \omega_i) \mathbf{a}^H(\theta_{p_i}, \omega_i) \right) + \sigma_{v_i}^2 \mathbf{I}_N$ is the interference-plus-noise covariance matrix (INCM), where $\sigma_{p_i}^2 = \mathbb{E}\{\beta_{p_i} \beta_{p_i}^H\}$ is the average power of the p_i-th interference source at the i-th frequency.

As for multiband uniform linear arrays, problem (5) can be decomposed into M independent subproblems. The optimal beamformer at the frequency ω_i is obtained by $\mathbf{w}_{opt_i} = \mathcal{P}\{\mathbf{R}_{in_i}^{-1} \mathbf{R}_{s_i}\}$ according to the principle of minimum variance distortionless response (MVDR), where the operator $\mathcal{P}\{\cdot\}$ extracts the principal eigenvector of the input matrix. We then obtain the optimal output SINR operating at the frequency ω_i as [36]

$$\text{SINR}_{opt_i} = \frac{\mathbf{w}_{opt_i}^H \mathbf{R}_{s_i} \mathbf{w}_{opt_i}}{\mathbf{w}_{opt_i}^H \mathbf{R}_{in_i} \mathbf{w}_{opt_i}} = \lambda_{\max}\{\mathbf{R}_{in_i}^{-1} \mathbf{R}_{s_i}\}, \ i = 1, \ldots, M, \tag{6}$$

where $\lambda_{\max}\{\cdot\}$ represents the principal eigenvalue of the matrix.

3. Proposed Multiband Sparse Array Design

To reduce the cost and system complexity of multiband arrays, this section addresses the issue of multiband sparse array design. In the narrowband case, sparse array design is equivalent to finding the beamforming weight $\mathbf{w}_i$, having only K non-zero entries at the frequency ω_i. As for the multiband sparse array design, the non-zero entries of $\mathbf{w}_i$ at all frequencies, $\{\omega_i\}_{i=1}^M$, should occupy the same antenna positions. In other word, the design of sparse beamforming weights, $\{\mathbf{w}_i\}_{i=1}^M$, are mutually coupled and thus cannot be resolved separately, which is different from that of the multiband uniform array in (6). On the other hand, the multiband sparse array often results in uncontrollable high sidelobe levels in some frequencies since all $\mathbf{w}_i$ have to locate at the same antenna positions, leading to the DoFs of antenna selection being considerably reduced. The designed beamformer will be sensitive to unknown sporadic interference in the high SLL region, which degrades the performance of interference suppression. Therefore, it is necessary to incorporate the SLL constraints into the multiband sparse array design. Based on these considerations, this section formulates the problem of multiband sparse array design under the MaxSINR criterion and SLL constraints and then provides an effective solution algorithm.

3.1. Problem Formulation

To proceed, we define the normalized array power response at the direction θ and the frequency ω_i as

$$B(\theta, \theta_{i,0}, \omega_i) \triangleq \frac{\left|\mathbf{w}^H \mathbf{a}(\theta, \omega_i)\right|^2}{\left|\mathbf{w}^H \mathbf{a}(\theta_{i,0}, \omega_i)\right|^2}, \tag{7}$$

where $\theta_{i,0}$ is the desired source direction at the frequency ω_i; that is, the angle pointing to the mainlobe. Denote the corresponding sidelobe region as Ω_i and discretize Ω_i to obtain a set of angles as $\{\theta_{i,l}\}, l = 1, \cdots, L_i$. The sidelobe steering vector is then $\mathbf{a}(\theta_{i,l}, \omega_i)$, and the normalized array power response at the direction $\theta_{i,l}$ is [21]

$$B(\theta_{i,l}, \theta_{i,0}, \omega_i) \triangleq \frac{\mathbf{w}^H \mathbf{a}(\theta_{i,l}, \omega_i) \mathbf{a}^H(\theta_{i,l}, \omega_i) \mathbf{w}}{\mathbf{w}^H \mathbf{a}(\theta_{i,0}, \omega_i) \mathbf{a}^H(\theta_{i,0}, \omega_i) \mathbf{w}}. \tag{8}$$

Therefore, SLL constraints at all frequencies, $\{\omega_i\}_{i=1}^M$, can be expressed as

$$B(\theta_{i,l}, \theta_{i,0}, \omega_i) \leq \delta_i, \ \forall i, \ \forall l, \tag{9}$$

where δ_i is the desired SLL at the frequency ω_i.

Note that the received multiband signal consists of M sub-bands. The multiband array correspondingly yields M beamformer weight vectors: $\mathbf{w}_1, \mathbf{w}_2, \ldots, \mathbf{w}_M$. Define the vector $\overline{\mathbf{w}}_n = [w_1(n), \cdots, w_i(n), \cdots, w_M(n)]^T \in \mathbb{C}^M$, where $w_i(n)$ is the n-th component of $\mathbf{w}_i$. That is to say, $\overline{\mathbf{w}}_n$ represents the beamforming weights of all M frequencies at the n-th antenna position. If we avoid the n-th antenna receiving the signal, the vector $\overline{\mathbf{w}}_n$ must be set to $\mathbf{0}_M$. This means that for all M sub-bands, the n-th entry of each $\mathbf{w}_i$ must be set to 0 at the same time. To effectively express the selection of K elements from N multiband antennas, we generate the concatenated vector $\hat{\mathbf{w}} \triangleq [\mathbf{w}_1^T, \mathbf{w}_2^T, \ldots, \mathbf{w}_M^T]^T \in \mathbb{C}^{NM}$ and define its $l_{0,2}$-mixed norm as $\|\hat{\mathbf{w}}\|_{0,2} \triangleq |\{n : \|\overline{\mathbf{w}}_n\|_2 \neq 0\}|$ [37]. The requirement on antenna selection is then expressed as

$$\|\hat{\mathbf{w}}\|_{0,2} \leq K. \tag{10}$$

Based on the MaxSINR criterion, the proposed multiband sparse array design under SLL constraints is then formulated into

$$\min_{\{\mathbf{w}_i\}_{i=1}^{M}} \sum_{i=1}^{M} \mathbf{w}_i^H \mathbf{R}_{in_i} \mathbf{w}_i$$

$$\text{s.t.} \quad \mathbf{w}_i^H \mathbf{R}_{s_i} \mathbf{w}_i \geq 1, \; \forall i,$$

$$B(\theta_{i,l}, \theta_{i,0}, \omega_i) \leq \delta_i, \; \forall i, \; \forall l, \tag{11}$$

$$\|\hat{\mathbf{w}}\|_{0,2} \leq K$$

In lieu of the sparsity constraint, the mixed $l_{0,2}$-norm can be used as a penalty term in the objective function to promote sparsity. Therefore, problem (11) is translated into the following optimization problem:

$$\min_{\{\mathbf{w}_i\}_{i=1}^{M}} \sum_{i=1}^{M} \mathbf{w}_i^H \mathbf{R}_{in_i} \mathbf{w}_i + \mu \|\hat{\mathbf{w}}\|_{0,2}$$

$$\text{s.t.} \quad \mathbf{w}_i^H \mathbf{R}_{s_i} \mathbf{w}_i \geq 1, \; \forall i, \tag{12}$$

$$\frac{\mathbf{w}_i^H \mathbf{A}(\theta_{i,l}, \omega_i) \mathbf{w}_i}{\mathbf{w}_i^H \mathbf{A}(\theta_{i,0}, \omega_i) \mathbf{w}_i} \leq \delta_i, \; \forall i, \; \forall l,$$

where μ is a regularized factor that controls the sparsity of the solution [37] and $\mathbf{A}(\theta_{i,l}, \omega_i) \triangleq \mathbf{a}(\theta_{i,l}, \omega_i)\mathbf{a}^H(\theta_{i,l}, \omega_i)$ for $l = 0, 1, \cdots, L_i$.

Unfortunately, solving problem (12) requires exhaustively searching all possible sparse combinations of $\hat{\mathbf{w}}$ due to the mixed $l_{0,2}$-norm. Therefore, (12) is a combinational optimization problem and cannot be solved in polynomial time [38]. Moreover, the two kinds of constraints are both nonconvex and thus increase the difficulty of problem solving. To this end, the following section will provide an SDR-based iterative reweighted method to solve problem (12) effectively.

3.2. Proposed SDR-Based Iterative Reweighted Algorithm

For the convenience of solving the group-sparse regularized problem, it is usual to replace the nonconvex $l_{0,2}$-norm by a convex $l_{1,\infty}$-norm as the group sparsity-inducing regularization [37], where the $l_{1,\infty}$-norm is defined as $\|\hat{\mathbf{w}}\|_{1,\infty} \triangleq \sum_{n=1}^{N} \|\overline{\mathbf{w}}_n\|_{\infty}$. Furthermore, we introduce the reweighted vector $\mathbf{u} = [u(1), u(2), \ldots, u(N)]^T$ to enhance the group sparsity [39], where $u(1), u(2), \ldots, u(N)$ are all positive numbers. Moreover, the square of $l_{1,\infty}$-norm does not change its original sparsity. Given all that, we adopt the squared reweighted $l_{1,\infty}$-norm $(\sum_{n=1}^{N} u(n)\|\overline{\mathbf{w}}_n\|_{\infty})^2$ in place of $\|\hat{\mathbf{w}}\|_{0,2}$, and therefore relax problem (12) to

$$\min_{\{\mathbf{w}_i\}_{i=1}^{M}} \sum_{i=1}^{M} \mathbf{w}_i^H \mathbf{R}_{in_i} \mathbf{w}_i + \mu\left(\sum_{n=1}^{N} u(n)\|\overline{\mathbf{w}}_n\|_{\infty}\right)^2$$

$$\text{s.t.} \quad \mathbf{w}_i^H \mathbf{R}_{s_i} \mathbf{w}_i \geq 1, \quad \forall i, \tag{13}$$

$$\frac{\mathbf{w}_i^H \mathbf{A}(\theta_l, \omega_i) \mathbf{w}_i}{\mathbf{w}_i^H \mathbf{A}(\theta_0, \omega_i) \mathbf{w}_i} \leq \delta_i, \quad \forall i, \; \forall l.$$

It can be noticed that the introduction of reweighted vector $\mathbf{u}$ to enhance the group sparsity stems from the original iterative reweighting scheme. As we known, l_0-norm is the natural representation of sparse antenna selection, but the minimization of l_0-norm is NP-hard and it is often relaxed as a l_1-norm. According to the iterative reweighting principle [39], the reweighted l_1-norm can well approximate to the l_0-norm, and thus has better sparsity than l_1-norm. With the help of reweighting, the contribution of nonzero entries with large amplitudes is gradually weakened, and the nonzero entries with small amplitudes therefore can be successfully found. As for problem (13), the reweighted vector $\mathbf{u}$ has the similar ability to improve the group sparsity of $l_{1,\infty}$-norm.

Denote $\tilde{\mathbf{w}}_i \triangleq [\Re\{\mathbf{w}_i^T\}, \Im\{\mathbf{w}_i^T\}]^T$ and define the matrices $\tilde{\mathbf{A}}(\theta_{i,l}, \omega_i)$ and $\tilde{\mathbf{R}}_{in_i}$ ($\tilde{\mathbf{R}}_{s_i}$) as

$$\tilde{\mathbf{A}}(\theta_{i,l}, \omega_i) \triangleq \begin{bmatrix} \Re\{\mathbf{A}(\theta_{i,l}, \omega_i)\} & -\Im\{\mathbf{A}(\theta_{i,l}, \omega_i)\} \\ \Im\{\mathbf{A}(\theta_{i,l}, \omega_i)\} & \Re\{\mathbf{A}(\theta_{i,l}, \omega_i)\} \end{bmatrix} \tag{14}$$

and

$$\tilde{\mathbf{R}}_{in_i} \triangleq \begin{bmatrix} \Re\{\mathbf{R}_{in_i}\} & -\Im\{\mathbf{R}_{in_i}\} \\ \Im\{\mathbf{R}_{in_i}\} & \Re\{\mathbf{R}_{in_i}\} \end{bmatrix}. \tag{15}$$

Problem (13) can then be rewritten as the following real number form:

$$\begin{aligned} \min_{\{\tilde{\mathbf{w}}_i\}_{i=1}^M} \quad & \sum_{i=1}^M \tilde{\mathbf{w}}_i^H \tilde{\mathbf{R}}_{in_i} \tilde{\mathbf{w}}_i + \mu \Big(\sum_{n=1}^N u(n) \|\overline{\mathbf{w}}_n\|_\infty \Big)^2 \\ \text{s.t.} \quad & \tilde{\mathbf{w}}_i^H \tilde{\mathbf{R}}_{s_i} \tilde{\mathbf{w}}_i \geq 1, \quad \forall i, \\ & \frac{\tilde{\mathbf{w}}_i^H \tilde{\mathbf{A}}(\theta_{i,l}, \omega_i) \tilde{\mathbf{w}}_i}{\tilde{\mathbf{w}}_i^H \tilde{\mathbf{A}}(\theta_{i,0}, \omega_i) \tilde{\mathbf{w}}_i} \leq \delta_i, \quad \forall i, \forall l. \end{aligned} \tag{16}$$

Due to the existence of non-continuous objective function and nonconvex quadratic or fractional constraints, it is still difficult to solve problem (16) directly. Therefore, we further relax (16) by using SDR and linear fractional SDR [21], simultaneously. To this end, we rewrite the quadratic objective function in (16) as

$$\tilde{\mathbf{w}}_i^H \tilde{\mathbf{R}}_{in_i} \tilde{\mathbf{w}}_i = \mathrm{Tr}(\tilde{\mathbf{w}}_i^H \tilde{\mathbf{R}}_{in_i} \tilde{\mathbf{w}}_i) = \mathrm{Tr}(\tilde{\mathbf{R}}_{in_i} \tilde{\mathbf{W}}_i), \tag{17}$$

where $\tilde{\mathbf{W}}_i = \tilde{\mathbf{w}}_i \tilde{\mathbf{w}}_i^H \in \mathbb{R}^{2N \times 2N}$. Similarly, we relax the linear fractional constraint in (16) to

$$\mathrm{Tr}((\tilde{\mathbf{A}}(\theta_{i,l}, \omega_i) - \delta_i \tilde{\mathbf{A}}(\theta_{i,0}, \omega_i)) \tilde{\mathbf{W}}_i) \leq 0. \tag{18}$$

Furthermore, we relax the squared reweighted $l_{1,\infty}$-norm by using convex SDP. Denote $\mathbf{U} \triangleq \mathbf{u}\mathbf{u}^T \in \mathbb{R}^{N \times N}$, $\mathbf{W}_i = \mathbf{w}_i \mathbf{w}_i^H \in \mathbb{C}^{N \times N}$, and $\hat{\mathbf{W}} \triangleq \max_{i=1,\dots,M} |\mathbf{W}_i| \in \mathbb{R}^{N \times N}$. By invoking the properties of rank relaxation, we can rewrite the squared reweighted $l_{1,\infty}$-norm as [37]

$$\begin{aligned} \Big(\sum_{n=1}^N u(n) \|\overline{\mathbf{w}}_n\|_\infty \Big)^2 &= \sum_{n_1=1}^N \sum_{n_2=1}^N \Big((\max_k u(n_1)|w_k(n_1)|)(\max_k u(n_2)|w_k(n_2)|) \Big) \\ &= \sum_{n_1=1}^N \sum_{n_2=1}^N u(n_1)u(n_2) \max_{i \in \{1,\dots,M\}} |\mathbf{W}_i(n_1, n_2)| \\ &= \sum_{n_1=1}^N \sum_{n_2=1}^N \mathbf{U}(n_1, n_2) \hat{\mathbf{W}}(n_1, n_2) \\ &= \mathrm{Tr}(\mathbf{U}\hat{\mathbf{W}}). \end{aligned} \tag{19}$$

Since $\hat{\mathbf{W}}$ is a real matrix, we can deduce that matrices $\hat{\mathbf{W}}$ and $\mathbf{W}_i$ satisfy the element-wise inequality as

$$|\mathbf{W}_i| \leq \hat{\mathbf{W}}, \; i = 1, \dots, M, \tag{20}$$

which is specifically expressed as

$$\|\sqrt{-1}(-\tilde{\mathbf{W}}_i(p, N+q) + \tilde{\mathbf{W}}_i(N+p, q)) + \tilde{\mathbf{W}}_i(p, q) + \tilde{\mathbf{W}}_i(N+p, N+q)\|_2 \leq \hat{\mathbf{W}}(p, q) \tag{21}$$

After the above relaxation process, problem (16) is converted into

$$\min_{\{\tilde{\mathbf{W}}_i\}_{i=1}^{M},\hat{\mathbf{W}}} \sum_{i=1}^{M} \mathrm{Tr}(\tilde{\mathbf{R}}_{in_i}\tilde{\mathbf{W}}_i) + \mu\, \mathrm{Tr}(\mathbf{U}^r\hat{\mathbf{W}})$$

$$\begin{aligned}
\text{s.t.}\quad & \mathrm{Tr}(\tilde{\mathbf{R}}_{s_i}\tilde{\mathbf{W}}_i) \geq 1,\ \forall i, \\
& \mathrm{Tr}((\tilde{\mathbf{A}}(\theta_{i,l},w_i) - \delta_i\tilde{\mathbf{A}}(\theta_{i,0},w_i))\tilde{\mathbf{W}}_i) \leq 0,\ \forall i,\ \forall l, \\
& \tilde{\mathbf{W}}_i \succeq 0,\ \forall i, \\
& \|\sqrt{-1}(-\tilde{\mathbf{W}}_i(p,N+q) + \tilde{\mathbf{W}}_i(N+p,q)) + \\
& \tilde{\mathbf{W}}_i(p,q) + \tilde{\mathbf{W}}_i(N+p,N+q)\|_2 \leq \hat{\mathbf{W}}(p,q), \\
& \forall p,q \in 1,\ldots,N,\ \forall i,
\end{aligned} \tag{22}$$

where the nonconvex constraints $\mathrm{Rank}(\tilde{\mathbf{W}}_i) = 1$ are discarded in the process of convex relaxation [40]. The superscript r of $\mathbf{U}$ represents the r-th reweighted iteration, and the iterative update formula of $\mathbf{U}$ is [39]

$$\mathbf{U}^r(p,q) = \frac{1}{|\hat{\mathbf{W}}^{r-1}(p,q)| + \varepsilon} \tag{23}$$

where ε is a small positive number.

By iteratively solving problem (22), we finally obtain the desired weight matrices $\tilde{\mathbf{W}}_i \in \mathbb{R}^{2N \times 2N}, i \in 1,\ldots,M$. The principal eigenvector $\tilde{\mathbf{w}}_i$ is then extracted from $\tilde{\mathbf{W}}_i$, i.e., $\tilde{\mathbf{w}}_i = \mathcal{P}\{\tilde{\mathbf{W}}_i\}$. Ultimately, we restore the multiband beamforming vectors by

$$\mathbf{w}_i = [\mathbf{I}_N \quad j\mathbf{I}_N]\tilde{\mathbf{w}}_i, \quad i = 1,\ldots,M. \tag{24}$$

For clarity, we summarize the proposed multiband sparse array design method in Algorithm 1.

Algorithm 1 Multiband Sparse Array Design with Sidelobe Level Control

---

Input: $N, K, \delta_i, \varepsilon, \mu_{min}, \mu_{max}$.
Initialization: Set $r = 0$, $\mathbf{U}^0$ is an $N \times N$ all-one matrix.
  1: **while** $\|\hat{\mathbf{w}}\|_{0,2} \neq K$ **do**
  2:    Obtain $\tilde{\mathbf{W}}_1^{r+1}\cdots\tilde{\mathbf{W}}_i^{r+1}\cdots\tilde{\mathbf{W}}_M^{r+1}, \hat{\mathbf{W}}^{r+1}$ using (22);
  3:    Obtain $\mathbf{w}_1^{r+1}\cdots\mathbf{w}_i^{r+1}\cdots\mathbf{w}_M^{r+1}$ using (24);
  4:    Obtain $\mathbf{U}^{r+1}$ using (23);
  5:    Update the value of μ by the binary search approach;
  6:    $r = r + 1$;
  7: **end while**
Output: Multiband beamforming weights $\mathbf{w}_1, \mathbf{w}_2, \cdots, \mathbf{w}_M$.

---

4. Analysis of Computational Complexity

This section analyses the computational complexity of the proposed algorithm. It is obvious that the computational complexity is primarily determined by solving the problem (22). For the problem (22), we use the off-the-shelf toolboxes, such as CVX, to effectively find the optimal solution, where the interior point method is invoked. Following [37], the worst-case complexity order of the problem (22) remains the same as the problem without antenna selection, which is only solving the variables $\{\tilde{\mathbf{W}}_i\}_{i=1}^{M}$. Therefore, the problem without the antenna selection has M matrix variables of size $2N \times 2N$, and $(M + ML)$ linear constraints. The interior point method will take $\mathcal{O}(\sqrt{MN}\log(1/\epsilon))$ iterations, where ϵ stands for the accuracy of the solution at the algorithm's termination, and each iteration requiring at most $\mathcal{O}(M^3N^6 + LM^2N^2 + M^2N^2)$ arithmetic operations [41]. Therefore, the overall worst-case complexity of the proposed algorithm is $\mathcal{O}(M^{3.5}N^{6.5} + LM^{2.5}N^{2.5} + M^{2.5}N^{2.5})\log(1/\epsilon)$.

5. Numerical Results

In this section, we evaluate the effectiveness of the proposed method for multiband sparse array design by several numerical experiments. We compare it with other typical algorithms in [21,31]. Specifically, Zheng considered the design of narrowband sparse arrays working at a single frequency under SLL constraint in [21], while Hamza designed a multiband sparse array without SLL control in [31]. It is worth pointing out that Zheng's and Hamza's methods cannot tackle the proposed problem (12) since Zheng's method is only applicable to the single frequency sparse array design while Hamza's method has no capability of controlling SLL. We only design several different single frequency sparse arrays by using Zheng's method and a multiband sparse array without SLL control by using Hamza's method as a benchmark. In fact, the proposed problem has less DoFs than Zheng's and Hamza's problems since it is limited by more constraints. In comparison with the proposed problem, Zheng's problem does not impose restrictions on the sparse weights of all single frequency arrays locating at the same antenna location, while Hamza's problem has no constraint on the SLLs. Therefore, from the perspective of system DoFs, the performance of the proposed problem would naturally not exceed those of Zheng's and Hamza's problems. However, thanks to the adopted solving scheme, the performance of the proposed method may be close to or even better than that of Zheng's method or Hamza's method, which is displayed in the following experiments.

In the experiments, the multiband array has the capability of effectively working at $M = 4$ frequencies, $\omega_1 = \omega_M$, $\omega_2 = 0.972\omega_M$, $\omega_3 = 0.944\omega_M$, and $\omega_4 = 0.931\omega_M$, respectively, where the maximum frequency is $\omega_M = 3.6$ GHz, which is commonly used in 5G communications and emerging integrated sensing and communication systems. We select $K = 20$ antennas from a uniform linear array with $N = 26$ locations. For the proposed algorithm, we set $\mu = 0.01$, $\varepsilon = 5 \times 10^{-4}$, and $\delta_i = -20$ dB for all four frequencies. We assume the desired source is located at the direction $80°$ and three interference sources are located at the directions $10°$, $120°$, and $140°$, respectively. The SNR of the desired source is 0dB and the INR of each interference source is 40dB.

5.1. Beamforming with Multiple Interferences at the Same Desired DOA

Since Zheng's method can only design a narrowband sparse array working at a single frequency, four optimal narrowband sparse arrays are independently designed at different frequencies, which are provided in Figure 1a–d. On the contrary, Hamza's method provides a multiband sparse array directly, and its optimal sparse array is shown in Figure 1e. The multiband sparse array obtained by the proposed method is illustrated in Figure 1f. As seen in Figure 2, all three methods form deep nulls at the directions of three interference sources, and thus effectively suppress the interference. Due to the lack of consideration for the sidelobe suppression in Hamza's method, its SLL is significantly higher than that in Zheng's method and the proposed one. The proposed method has almost the same SLLs as Zheng's method. That is to say, compared with Hamza's method, the proposed method and Zheng's method will be less sensitive to unknown sporadic interference and thus have superior capability of interference suppression. The null depths of all three methods at each frequency are shown in Table 1. In general, we find that the proposed method has a weakly shallower null than Zheng's and Hamza's methods, but it is still deep enough to effectively suppress the interferences. From Table 2, we observe that the proposed method has better output SINRs performance than Hamza's method and is slightly inferior to Zheng's method. It is worth pointing out that Zheng's method is actually the performance upper bound of the multiband sparse array design because it is not constrained by the consistency of antenna locations at each frequency and thus owns more DoFs of antenna selection.

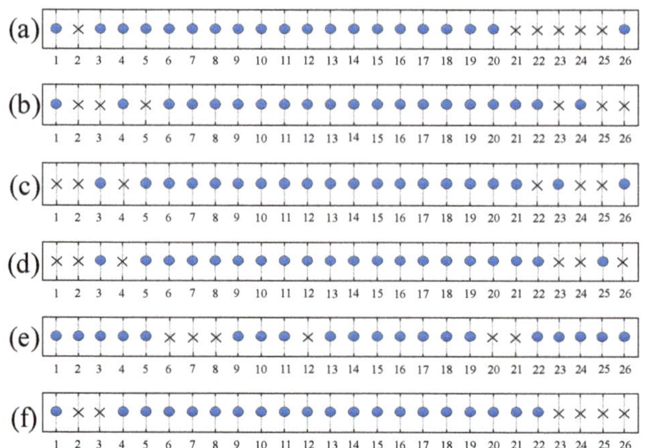

Figure 1. Sparse array configurations for the experiment 5.1, $N = 26, K = 20$. (**a**) Zheng's method for ω_1. (**b**) Zheng's method for ω_2. (**c**) Zheng's method for ω_3. (**d**) Zheng's method for ω_4. (**e**) Hamza's method. (**f**) Proposed method. (Dots mean selected antennas while crosses mean discarded antennas).

Table 1. Null depths (dB) of the three methods at each frequency for the experiment 5.1.

Hamza's method				
Interference \ Frequency	ω_1	ω_2	ω_3	ω_4
10°	−83.60	−83.60	−83.22	−77.56
120°	−74.35	−74.35	−75.84	−74.58
140°	−87.38	−87.38	−88.64	−82.39

Zheng's method				
Interference \ Frequency	ω_1	ω_2	ω_3	ω_4
10°	−80.88	−75.81	−74.66	−78.79
120°	−79.05	−78.02	−81.36	−84.77
140°	−82.02	−82.50	−77.95	−82.41

Proposed method				
Interference \ Frequency	ω_1	ω_2	ω_3	ω_4
10°	−79.33	−62.62	−84.10	−72.68
120°	−76.67	−61.02	−79.84	−68.79
140°	−75.59	−59.63	−78.29	−66.52

Table 2. Output SINR (dB) of the three methods at each frequency for the experiment 5.1.

Frequency	ω_1	ω_2	ω_3	ω_4
Hamza's method	11.22	11.74	11.55	11.20
Zheng's method	11.93	13.07	12.59	12.17
Proposed	11.78	10.34	12.07	11.77

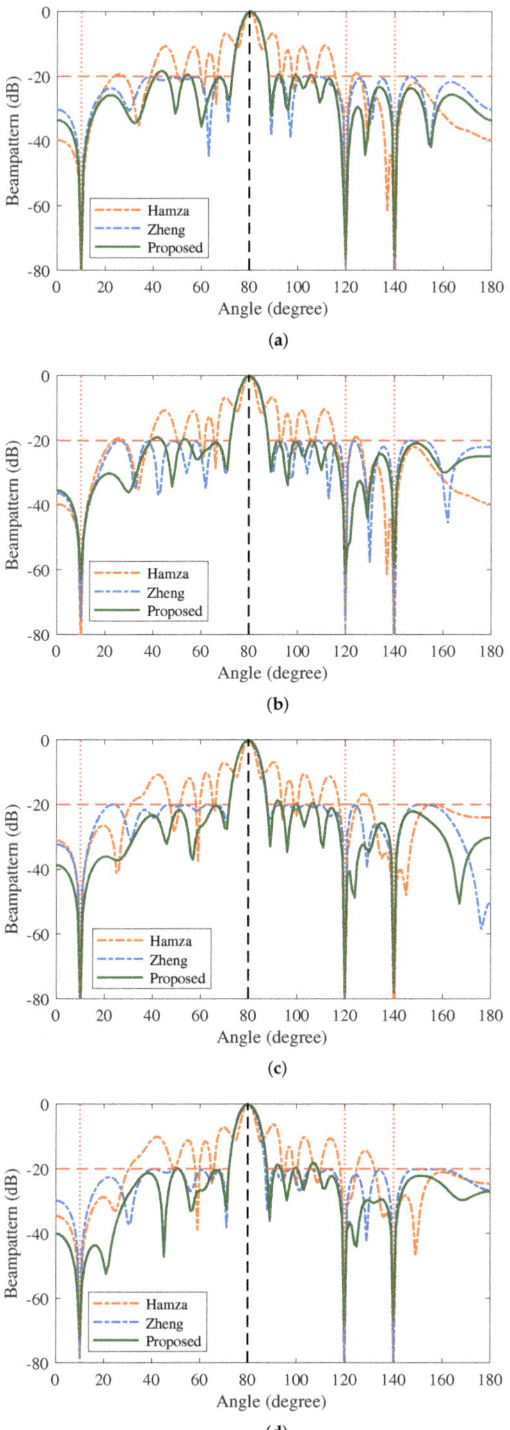

Figure 2. Normalized receive beampatterns of sparse arrays in Figure 1a–f at four different frequencies. (**a**) ω_1. (**b**) ω_2. (**c**) ω_3. (**d**) ω_4.

5.2. Beamforming with Multiple Interferences at the Distinct Desired DOAs

In multi-functional communication or radar systems, we also need to receive the desired signals in each frequency band with different DOAs, such as multi-user communications and multiband radars. In this experiment, we therefore set the mainlobe of each frequency with distinct DOAs. From ω_1 to ω_4, the desired directions are set sequentially as 70°, 75°, 80°, and 85°, while the sidelobes are correspondingly set as $\Theta_{SL,1} = [0°, 62°] \cup [78°, 180°]$, $\Theta_{SL,2} = [0°, 67°] \cup [83°, 180°]$, $\Theta_{SL,3} = [0°, 72°] \cup [88°, 180°]$, and $\Theta_{SL,4} = [0°, 77°] \cup [93°, 180°]$. The other parameters are set to be the same as in Section 5.1. The antenna selection results are shown in Figure 3, the normalized receive beampatterns of sparse arrays are displayed in Figure 4, and the null depths at each frequency are shown in Table 3. It can be observed from Figure 4 that the mainlobe always points to the desired directions for all three methods, and all three methods form deep null in the preset directions of interference sources. From Table 3, we observe that the proposed method yields almost the same null depths as Zheng's and Hamza's methods at each frequency. From Table 4, we can see that the proposed method achieves higher SINRs than Hamza's method, even though Hamza's method did not consider the SLL control. The output SINRs of the proposed method are still close to that offered by Zheng's method. These results reveal that the proposed method is an efficient method for multiband sparse array design.

Table 3. Null depths (dB) of the three methods at each frequency for the experiment 5.2.

Hamza's method				
Interference / Frequency	ω_1	ω_2	ω_3	ω_4
10°	−74.14	−84.16	−77.44	−77.22
120°	−79.17	−90.14	−87.49	−89.83
140°	−77.55	−77.62	−79.92	−80.79

Zheng's method				
Interference / Frequency	ω_1	ω_2	ω_3	ω_4
10°	−76.50	−80.15	−83.78	−85.28
120°	−74.86	−87.93	−82.38	−77.31
140°	−77.27	−76.28	−76.92	−90.73

Proposed method				
Interference / Frequency	ω_1	ω_2	ω_3	ω_4
10°	−81.45	−71.98	−77.27	−75.81
120°	−86.18	−74.81	−78.34	−93.93
140°	−82.68	−80.69	−79.41	−78.29

Table 4. Output SINR (dB) of the three methods at each frequency for the experiment 5.2.

Frequency	ω_1	ω_2	ω_3	ω_4
Hamza's method	11.21	11.28	11.01	11.15
Zheng's method	12.54	12.33	12.53	11.97
Proposed	12.09	12.11	12.22	11.59

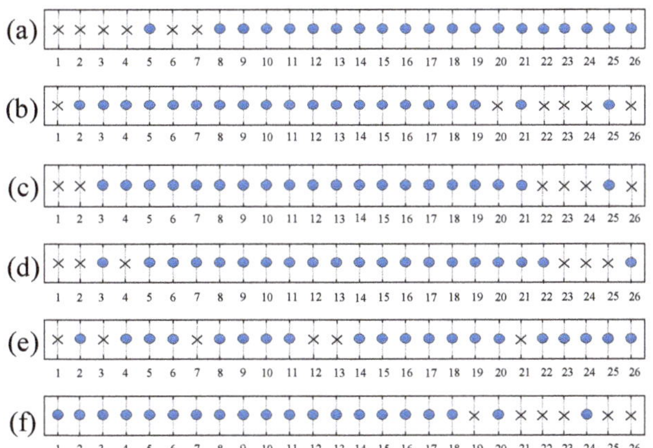

Figure 3. Sparse array configurations for the experiment 5.2, $N = 26, K = 20$. (**a**) Zheng's method for ω_1. (**b**) Zheng's method for ω_2. (**c**) Zheng's method for ω_3. (**d**) Zheng's method for ω_4. (**e**) Hamza's method. (**f**) Proposed method. (Dots mean selected antennas while crosses mean discarded antennas).

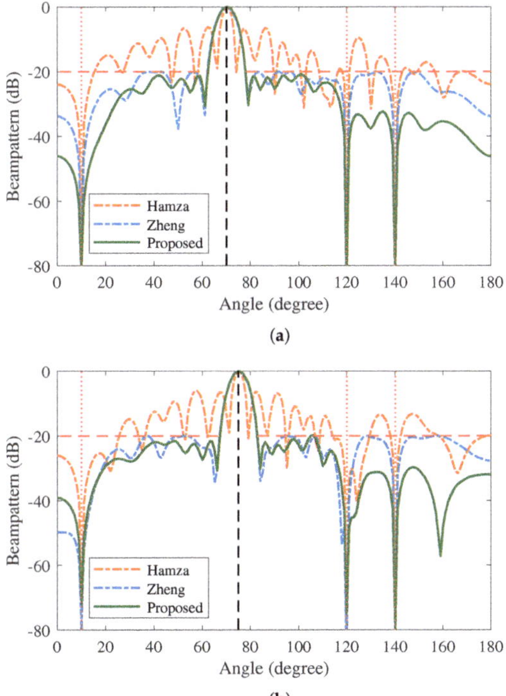

(**a**)

(**b**)

Figure 4. *Cont.*

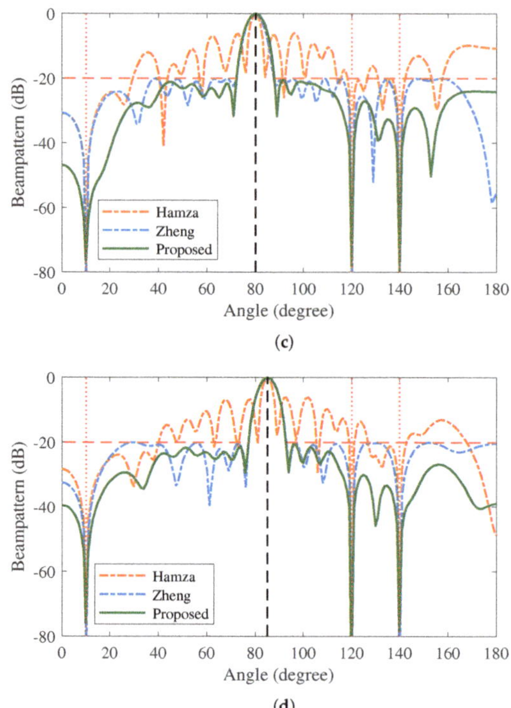

Figure 4. Normalized receive beampatterns of sparse arrays in Figure 3a–f at four different frequencies. (**a**) ω_1. (**b**) ω_2. (**c**) ω_3. (**d**) ω_4.

5.3. Nulling Forming at the Same Desired DOA

In real applications, it is often required to generate a deep null region to enhance anti-interference performance at preset directions. In this experiment, we replace the interference source at $120°$ by the null region in $[120°, 126°]$ with a null depth of -40 dB. As Hamza's method cannot work in this case, we only demonstrate the results of Zheng's method and the proposed method. The optimum sparse array configurations are presented in Figure 5. Based on these arrays, Figure 6a–d shows the beampatterns of null forming at the frequencies ω_1, ω_2, ω_3, and ω_4, separately. Table 5 provides the null depths of Zheng's and the proposed method at each frequency. It can be seen that both the proposed and Zheng's methods can form a deep null within the preset region $[120°, 126°]$ and the interference directions $10°$ and $140°$. Surprisingly, the proposed method generally has lower SLLs than Zheng's method in the whole sidelobe region. As seen from Table 6, the output SINRs of the proposed method are also close to that of Zheng's method at all frequencies, even though the constraints become stringent, which further validates the efficiency of the proposed method.

5.4. Nulling Anti-Interference Performance

In order to verify the interference suppression performance of the deep null in the proposed method and Zheng's method, we add an interference source at the angle direction $122°$ or $124°$ in the null region, respectively. With the same parameters as in Section 5.3, the INR of this interference source varies from 0 dB to 40 dB, and the variation of the output SINRs are respectively shown in Figure 7a,b. It can be observed from Figure 7 that after adding an interference source into the null region, the SINR does not change greatly as a whole compared with the null region without an interference source. We can conclude that the increasing INR of the interference source has a weak effect on the output SINR, which means that both the proposed method and Zheng's method have the capability of suppressing interference effectively in the null region.

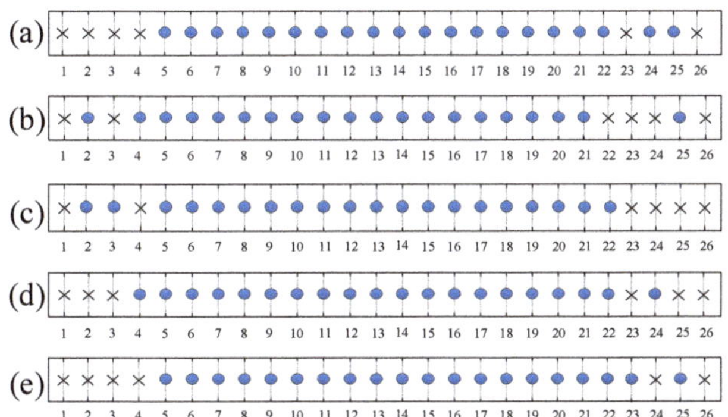

Figure 5. Sparse array configurations for the experiment 5.3, $N = 26, K = 20$. (**a**) Zheng's method for ω_1. (**b**) Zheng's method for ω_2. (**c**) Zheng's method for ω_3. (**d**) Zheng's method for ω_4. (**e**) Proposed method. (Dots mean selected antennas while crosses mean discarded antennas).

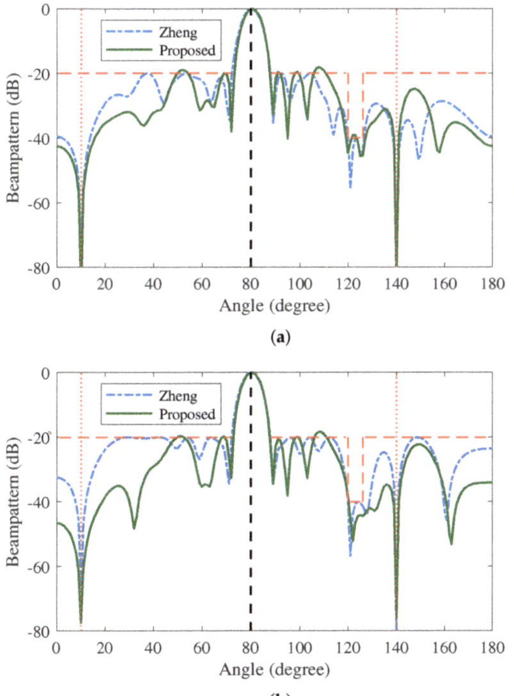

Figure 6. *Cont.*

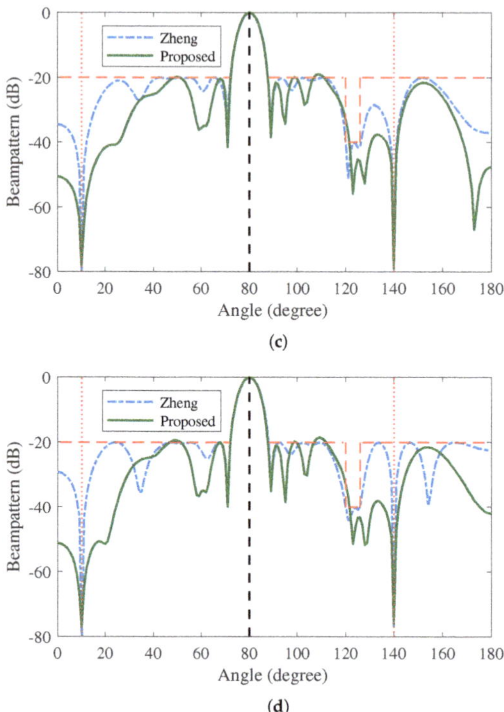

Figure 6. Normalized receive beampatterns of sparse arrays in Figure 5a–e at four different frequencies. (**a**) ω_1. (**b**) ω_2. (**c**) ω_3. (**d**) ω_4.

Table 5. Null depths (dB) of Zheng's and proposed methods at each frequency for the experiment 5.3.

	Zheng's method			
Interference \ Frequency	ω_1	ω_2	ω_3	ω_4
10°	−78.00	−70.18	−79.79	−83.28
140°	−85.39	−80.60	−70.88	−76.77
	Proposed method			
Interference \ Frequency	ω_1	ω_2	ω_3	ω_4
10°	−89.98	−77.42	−78.29	−77.00
140°	−85.42	−75.73	−78.98	−75.90

Table 6. Output SINR (dB) of Zheng's and proposed methods at each frequency for the experiment 5.3.

Frequency	ω_1	ω_2	ω_3	ω_4
Zheng's method	12.43	12.35	11.97	12.11
Proposed	12.31	11.78	11.71	12.07

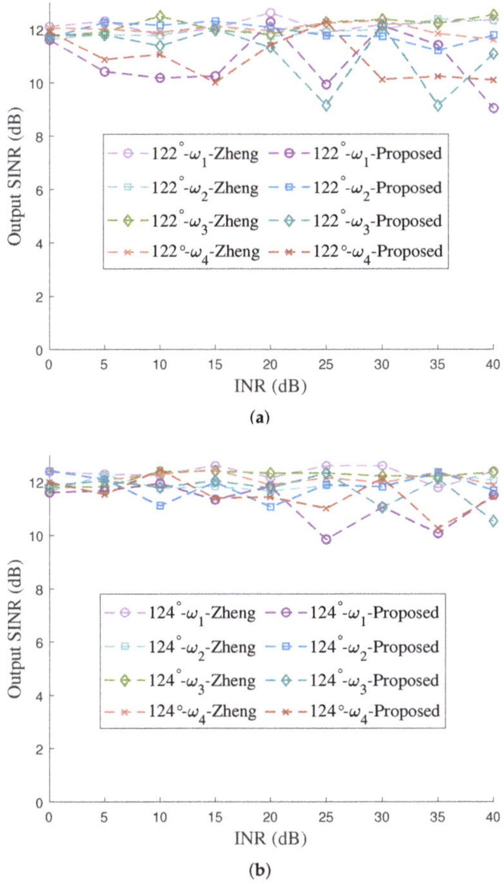

Figure 7. The output SINR with a varied interference in the nulling. (**a**) an interference source at 122°. (**b**) an interference source at 124°.

6. Discussion

Traditional sparse array designs for adaptive beamforming are usually discussed in the narrowband case or the wideband case. The multiband sparse array design is an emerging topic due to the rapid development of multi-functional communication and radar systems. Different from narrowband and wideband sparse arrays, the multiband sparse array has some unique characteristics, such as a large frequency gap between two adjacent operating frequencies and different desired source and/or interference directions at each operating frequency. Due to the group sparse regularization of beamforming weights at all operating frequencies, there exists a strong mutual coupling among all beamforming weights, and the objective function or constraints are necessarily nonconvex. On the other hand, the maximum SINR criterion often yields nonconvex quadratic equality constraints to fix the gain of desired directions at each operating frequency. Moreover, if we consider the SLL control, the SLL constraints are also nonconvex because they are the fractional quadratic functions of the beamforming weights. Therefore, the multiband sparse array design is commonly formulated into a complicated nonconvex constrained optimization problem, and its essence is how to effectively solving this problem.

This paper mainly employs several different kinds of convex relaxation techniques to tackle the problem (12), even though it uses the iterative reweighting scheme to promote the group sparse performance. From the perspective of optimization, the relaxation of constraints means that the constraints become more strict and thus the feasible set correspondingly becomes smaller. Therefore, the optimized sparse array of (22) is not necessarily the optimal sparse array of problem (12). Actually, SDR is a somewhat overly strict relaxation technique. To improve the performance of multiband sparse array design, we should utilize other loose relaxation techniques, such as SCA, the convex–concave procedure, and majorization–minimization, or we should handle problem (12) directly by using prevalent nonconvex optimization approaches, involving ADMM, quadratically constrained quadratic programming, and proximal operator algorithms.

7. Conclusions

This paper provided a multiband spare array design method for adaptive receive beamforming with SLL control. With the maximum SINR criterion and SLL constraints, we formulated the proposed joint design of antenna selection and adaptive beamformer as a group sparsity-regularized nonconvex constrained optimization problem. To deal with this intractable problem, we first translated the $l_{0,2}$-mixed norm regularization into a series of reweighted $l_{1,\infty}$-norm regularizations by employing the iterative reweighting technique. We then converted the $l_{1,\infty}$-norm regularized nonconvex optimization problem into the corresponding convex problem by using SDR and linear fractional SDR schemes. With the assistance of the iterative reweighting and SDR, we established the proposed SDR-based iterative reweighted algorithm. We also analyzed the computational complexity of the proposed algorithm. The numerical results verified that the proposed sparse array substantially reduces the SLL in all operating frequencies while maintaining the maximum output SINR performance at the same time, and its performance is approximate to the optimal sparse array designed separately at each frequency.

Author Contributions: Conceptualization , H.L.; methodology, H.L. and L.R.; software, H.L. and L.R.; supervision, H.L. and S.C.; validation, H.L., L.R. and C.H.; visualization, L.R., C.H. and Z.D.; writing— original draft, H.L. and L.R.; writing—review and editing, S.C. and Z.D.; funding acquisition, S.C. All authors have read and agreed to the published version of the manuscript.

Funding: This research was funded in part by the National Natural Science Foundation of China under Grant 62171224 and in part by the Natural Science Foundation of Jiangsu Province under Grant BK20221486.

Data Availability Statement: Not applicable.

Conflicts of Interest: The authors declare no conflict of interest.

Abbreviations

The following abbreviations are used in this manuscript:

SINR	Signal-to-interference-and-noise ratio
DOA	Direction-of-arrival
DoF	Degree-of-freedom
SDR	Semi-definite relaxation
SCA	Sequential convex approximation
ADMM	Alternating direction method of multipliers
DNN	Deep neural network
TDL	Tapped delay line
DFT	Discrete Fourier transform
FI	Frequency-invariant
SOCP	Second-order cone programming
SLL	Sidelobe level

References

1. He, X.Y.; Chang, L.; Chen, L.L. A multifunction broad-beam antenna with dual bands and dual circular-polarizations. In Proceedings of the 2016 IEEE MTT-S International Wireless Symposium (IWS), Shanghai, China, 14–16 March 2016; pp. 1–4. [CrossRef]
2. Sanad, M.; Antennas, A.; Hassan, N. A Multi-band Antenna Configuration for MIMO WiMax in Multi-standard Multifunction Handsets. In Proceedings of the 2009 IEEE Mobile WiMAX Symposium, Napa Valley, CA, USA, 9–10 July 2009; pp. 195–200. [CrossRef]
3. Haider, N.; Caratelli, D.; Yarovoy, A.G. Recent Developments in Reconfigurable and Multiband Antenna Technology. *Int. J. Antennas Propag.* **2013**, *2013*, 869170. [CrossRef]
4. Li, Y.; Sim, C.Y.D.; Luo, Y.; Yang, G. Multiband 10-Antenna Array for Sub-6 GHz MIMO Applications in 5-G Smartphones. *IEEE Access* **2018**, *6*, 28041–28053. [CrossRef]
5. Moffet, A. Minimum-redundancy linear arrays. *IEEE Trans. Antennas Propag.* **1968**, *16*, 172–175. [CrossRef]
6. Pal, P.; Vaidyanathan, P.P. Nested Arrays: A Novel Approach to Array Processing with Enhanced Degrees of Freedom. *IEEE Trans. Signal Process.* **2010**, *58*, 4167–4181. [CrossRef]
7. Vaidyanathan, P.P.; Pal, P. Sparse sensing with co-prime samplers and arrays. *IEEE Trans. Signal Process.* **2010**, *59*, 573–586. [CrossRef]
8. Zhou, C.; Gu, Y.; He, S.; Shi, Z. A Robust and Efficient Algorithm for Coprime Array Adaptive Beamforming. *IEEE Trans. Veh. Technol.* **2018**, *67*, 1099–1112. [CrossRef]
9. Zheng, Z.; Yang, T.; Wang, W.Q.; Zhang, S. Robust adaptive beamforming via coprime coarray interpolation. *Signal Process.* **2020**, *169*, 107382. [CrossRef]
10. Nai, S.E.; Ser, W.; Yu, Z.L.; Chen, H. Beampattern Synthesis for Linear and Planar Arrays with Antenna Selection by Convex Optimization. *IEEE Trans. Antennas Propag.* **2010**, *58*, 3923–3930. [CrossRef]
11. Wang, X.; Zhai, W.; Zhang, X.; Wang, X.; Amin, M.G. Enhanced Automotive Sensing Assisted by Joint Communication and Cognitive Sparse MIMO Radar. *IEEE Trans. Aerosp. Electron. Syst.* **2023**, *59*, 4782–4799. [CrossRef]
12. Fuchs, B. Synthesis of Sparse Arrays with Focused or Shaped Beampattern via Sequential Convex Optimizations. *IEEE Trans. Antennas Propag.* **2012**, *60*, 3499–3503. [CrossRef]
13. Nongpiur, R.C.; Shpak, D.J. Synthesis of Linear and Planar Arrays with Minimum Element Selection. *IEEE Trans. Signal Process.* **2014**, *62*, 5398–5410. [CrossRef]
14. Fuchs, B.; Rondineau, S. Array Pattern Synthesis with Excitation Control via Norm Minimization. *IEEE Trans. Antennas Propag.* **2016**, *64*, 4228–4234. [CrossRef]
15. Liang, J.; Zhang, X.; So, H.C.; Zhou, D. Sparse Array Beampattern Synthesis via Alternating Direction Method of Multipliers. *IEEE Trans. Antennas Propag.* **2018**, *66*, 2333–2345. [CrossRef]
16. Wang, X.; Aboutanios, E.; Amin, M.G. Thinned Array Beampattern Synthesis by Iterative Soft-Thresholding-Based Optimization Algorithms. *IEEE Trans. Antennas Propag.* **2014**, *62*, 6102–6113. [CrossRef]
17. Oliveri, G.; Massa, A. Bayesian Compressive Sampling for Pattern Synthesis with Maximally Sparse Non-Uniform Linear Arrays. *IEEE Trans. Antennas Propag.* **2011**, *59*, 467–481. [CrossRef]
18. Hamza, S.A.; Amin, M.G. Hybrid Sparse Array Beamforming Design for General Rank Signal Models. *IEEE Trans. Signal Process.* **2019**, *67*, 6215–6226. [CrossRef]
19. Zheng, Z.; Fu, Y.; Wang, W.Q. Sparse Array Beamforming Design for Coherently Distributed Sources. *IEEE Trans. Antennas Propag.* **2021**, *69*, 2628–2636. [CrossRef]
20. Huang, H.; So, H.C.; Zoubir, A.M. Sparse Array Beamformer Design via ADMM. In Proceedings of the 2022 IEEE 12th Sensor Array and Multichannel Signal Processing Workshop (SAM), Trondheim, Norway, 20–23 June 2022; pp. 336–340. [CrossRef]
21. Zheng, Z.; Fu, Y.; Wang, W.Q.; So, H.C. Sparse Array Design for Adaptive Beamforming via Semidefinite Relaxation. *IEEE Signal Process. Lett.* **2020**, *27*, 925–929. [CrossRef]
22. Wang, X.; Greco, M.S.; Gini, F. Adaptive Sparse Array Beamformer Design by Regularized Complementary Antenna Switching. *IEEE Trans. Signal Process.* **2021**, *69*, 2302–2315. [CrossRef]
23. Hamza, S.A.; Amin, M.G. Learning Sparse Array Capon Beamformer Design Using Deep Learning Approach. In Proceedings of the 2020 IEEE Radar Conference (RadarConf20), Florence, Italy, 21–25 September 2020. [CrossRef]
24. Hamza, S.A.; Amin, M.G.; Chalise, B.K. Phase-Only Reconfigurable Sparse Array Beamforming Using Deep Learning. In Proceedings of the ICASSP 2022—2022 IEEE International Conference on Acoustics, Speech and Signal Processing (ICASSP), Singapore, 23–27 May 2022. [CrossRef]
25. Crocco, M.; Trucco, A. Stochastic and Analytic Optimization of Sparse Aperiodic Arrays and Broadband Beamformers with Robust Superdirective Patterns. *IEEE Trans. Audio Speech Lang. Process.* **2012**, *20*, 2433–2447. [CrossRef]
26. Liu, Y.; Zhang, L.; Ye, L.; Nie, Z.; Liu, Q.H. Synthesis of Sparse Arrays with Frequency-Invariant-Focused Beam Patterns Under Accurate Sidelobe Control by Iterative Second-Order Cone Programming. *IEEE Trans. Antennas Propag.* **2015**, *63*, 5826–5832. [CrossRef]
27. Hawes, M.B.; Liu, W. Sparse array design for wideband beamforming with reduced complexity in tapped delay-lines. *IEEE/ACM Trans. Audio Speech Lang. Process.* **2014**, *22*, 1236–1247. [CrossRef]

28. Trucco, A. Synthesizing wide-band sparse arrays by simulated annealing. In Proceedings of the MTS/IEEE Oceans 2001. An Ocean Odyssey. Conference Proceedings (IEEE Cat. No.01CH37295), Honolulu, HI, USA, 5–8 November 2001; Volume 2, pp. 989–994. [CrossRef]
29. Doblinger, G. Optimized design of interpolated array and sparse array wideband beamformers. In Proceedings of the 2008 16th European Signal Processing Conference, Lausanne, Switzeerland, 25–29 August 2008; pp. 1–5.
30. Hawes, M.B.; Liu, W. Location Optimization of Robust Sparse Antenna Arrays with Physical Size Constraint. *IEEE Antennas Wirel. Propag. Lett.* **2012**, *11*, 1303–1306. [CrossRef]
31. Hamza, S.A.; Amin, M.G. Sparse Array Beamforming Design for Wideband Signal Models. *IEEE Trans. Aerosp. Electron. Syst.* **2021**, *57*, 1211–1226. [CrossRef]
32. Hamza, S.A.; Amin, M.G. Sparse Array DFT Beamformers for Wideband Sources. In Proceedings of the 2019 IEEE Radar Conference (RadarConf), Boston, MA, USA, 22–26 April 2019; pp. 1–5. [CrossRef]
33. Wei Liu, S.W. *Wideband Beamforming: Concepts and Techniques*; Wiley: Hoboken, NJ, USA , 2010.
34. Liu, Y.; Zhang, L.; Zhu, C.; Liu, Q.H. Synthesis of Nonuniformly Spaced Linear Arrays with Frequency-Invariant Patterns by the Generalized Matrix Pencil Methods. *IEEE Trans. Antennas Propag.* **2015**, *63*, 1614–1625. [CrossRef]
35. El-Khamy, S.E.; El-Sayed, H.F.; Eltrass, A.S. A new adaptive beamforming of multiband fractal antenna array in strong-jamming environment. *Wirel. Pers. Commun.* **2022**, *126*, 285–304. [CrossRef]
36. Monzingo, R.A.; Miller, T.W. *Introduction to Adaptive Arrays*; Scitech Publishing: Tamil Nadu, India, 2004.
37. Mehanna, O.; Sidiropoulos, N.D.; Giannakis, G.B. Joint Multicast Beamforming and Antenna Selection. *IEEE Trans. Signal Process.* **2013**, *61*, 2660–2674. [CrossRef]
38. Joshi, S.; Boyd, S. Sensor selection via convex optimization. *IEEE Trans. Signal Process.* **2008**, *57*, 451–462. [CrossRef]
39. Candes, E.J.; Wakin, M.B.; Boyd, S.P. Enhancing sparsity by reweighted l_1 minimization. *J. Fourier Anal. Appl.* **2008**, *14*, 877–905. [CrossRef]
40. Luo, Z.Q.; Ma, W.K.; So, A.M.C.; Ye, Y.; Zhang, S. Semidefinite Relaxation of Quadratic Optimization Problems. *IEEE Signal Process. Mag.* **2010**, *27*, 20–34. [CrossRef]
41. Karipidis, E.; Sidiropoulos, N.D.; Luo, Z.Q. Quality of Service and Max-Min Fair Transmit Beamforming to Multiple Cochannel Multicast Groups. *IEEE Trans. Signal Process.* **2008**, *56*, 1268–1279. [CrossRef]

 *remote sensing*

Article

Pentagram Arrays: A New Paradigm for DOA Estimation of Wideband Sources Based on Triangular Geometry

Mohammed Khalafalla, Kaili Jiang *, Kailun Tian, Hancong Feng, Ying Xiong and Bin Tang

School of Information and Communication Engineering, University of Electronic Science and Technology of China, Chengdu 611731, China; 201914010106@std.uestc.edu.cn (M.K.); 202311012422@std.uestc.edu.cn (K.T.); 202211012303@std.uestc.edu.cn (H.F.); xiongy@uestc.edu.cn (Y.X.); bint@uestc.edu.cn (B.T.)
* Correspondence: jiangkelly@uestc.edu.cn

Abstract: Antenna arrays are used for signal processing in sonar and radar direction of arrival (DOA) estimation. The well-known array geometries used in DOA estimation are uniform linear array (ULA), uniform circular array (UCA), and rectangular grid array (RGA). In these geometries, the neighboring elements are separated by a fixed distance $\lambda/2$ (λ is the wavelength), which does not perform well for d greater than $\lambda/2$. Uniform rectangular arrays introduce grating lobes, which cause poor DOA estimation performance, especially for wideband sources. Random sampling arrays are sometimes practically not realizable. Periodic geometries require numerous sensors. Based on the minimization of the number of sensors, this paper developed a novel pentagram array to address the problem of DOA estimation of wideband sources. The array has a fixed number of elements with variable element spacing and is abbreviated as (FNEVES), which offers a new idea for array design. In this study, the geometric structure is designed and mathematically analyzed. Also, a DOA signal model is designed based on a spherical radar coordinate system to derive its steering manifold matrix. The DOA estimation performance comparison with ULA and UCA geometries under the multiple signal classification (MUSIC) algorithm using different wideband scenarios is presented. For further investigation, more simulations are realized using the minimum variance distortionless (MVDR) technique (CAPON) and the subtracting signal subspace (SSS) algorithm. Simulation results demonstrate the effectiveness of the proposed geometry compared to its counterparts. In addition, the SSS, through the simulations, provided better results than the MUSIC and CAPON methods.

Keywords: direction of arrival estimation (DOA); pentagram geometry; steering manifold matrix; wideband signals; spherical radar coordinate system

Citation: Khalafalla, M.; Jiang, K.; Tian, K.; Feng, H.; Xiong, Y.; Tang, B. Pentagram Arrays: A New Paradigm for DOA Estimation of Wideband Sources Based on Triangular Geometry. *Remote Sens.* **2024**, *16*, 535. https://doi.org/10.3390/rs16030535

Academic Editor: Dusan Gleich

Received: 31 October 2023
Revised: 9 January 2024
Accepted: 25 January 2024
Published: 31 January 2024

1. Introduction

Antenna arrays have been commonly applied to address the problem of direction of arrival (DOA) estimation in many applications, such as wireless communication, sonar, and radar. The DOA estimation of sources is considered a critical problem for target determination in military radar applications. Spatial variety in communications system is achieved using DOA information [1].

Antenna array configurations are utilized to estimate the DOA of every signal in the case of multisource signals. By processing the signals received from the sensors in parallel, a higher signal-to-noise ratio (SNR) can be achieved by increasing the number of sensors in the array, and the implementation cost can be reduced [2]. DOA estimation of narrowband sources has been well studied theoretically in the literature [3]. The array model of narrowband sources is extremely simplified based on their properties [4]. However, to the best of the authors' knowledge, DOA estimation of wideband sources has not been sufficiently presented in the literature.

The first DOA estimation method for wideband sources is the incoherent signalsubspace method (ISM), which uses the discrete Fourier transform (DFT) along the temporal domain to decompose the incoherent wideband signal into many narrowband signals. It is assumed that in every frequency bin the frequency is time-invariant, ignoring the whole wideband information. Therefore, it is applied to2D problems using only a uniform linear array, has a large computational complexity related to the time domain representation of wideband signals, and is null for coherent signals [5,6]. The second method is the coherent signalsubspace (CSM) method, in which the wideband signal is transformed into a certain reference frequency using focusing matrices, and then the average of the covariance matrix is taken for de-correlation. It was implemented using a linear array and provided high accuracy for wideband DOA estimation with relatively low computational complexity; however, the focusing matrices require a priori knowledge [7], which is almost not available.

The least mean square (LMS) and sample matrix inversion algorithms are DOA estimation model-based techniques with higher computational complexity [8]. Eigen-analysis-based techniques exploit the phase differences of signals impinging on array sensors. The well-known MUSIC (multiple signal classification) algorithm uses the noise subspace after its separation, depending on eigen-analysis techniques [9].

The subtracting signal subspace (SSS) method estimates the DOAs using the signal subspace (SS) and array manifold vector (AMV) by exploiting the complementarity of the signal subspace (SS) and noise subspace (NS) after applying the eigen-analysis approach. Using signal subspace (SS) leads to much less computational burden compared with noise subspace (NS) [10]. CAPON [11] is a classical DOA technique that uses the minimum variance distortionless (MVDR) technique to estimate the power spectrum of the signal and find the peaks in the spatial power spectrum of the steering beams that correspond to the angles of the DOA.

To the best of our knowledge, researchers have been studying the problem of array geometry related to DOA estimation algorithms and wideband signals.

Many constraints in DOA estimation, such as angle (azimuth/elevation) estimation, angle resolution, and angle accuracy, are imposed by array geometries. The general 1D array geometry used in DOA estimation is uniform linear array (ULA), and 2D array geometries are rectangular grid array (RGA) [12] and uniform circular array (UCA) [13]. ULA has perfect directivity in a certain direction, forming a narrower main lobe, but its performance is affected by the variation of the azimuth direction. Its drawback is that it can only estimate the azimuth or elevation angle in one application. RGAs are uniform arrays that suffer from grating lobes because an extra main lobe with the same density appears on its opposite side. This represents their main disadvantage and makes them inappropriate for the application of wideband DOA algorithms.

UCAs are symmetrical arrays and are used to replace the ULA in the estimation of both azimuth and elevation angles [8], although they are high-side lobe-levels geometries and occupy large spaces for implementation. Reducing side lobe levels can be achieved by reducing the element spacing, but this increases the effect of the mutual coupling. Multi-rings are used as another choice to reduce these high side lobes, which has more advantages.

Smart antenna systems [14,15] were developed based on UCA geometries using optimization algorithms to study the mutual coupling between the array elements and their influence. To improve the performance, they utilize multi-ring arrays and a larger number of elements, which is more expensive. Sensor configurations in the aforementioned array geometries are determined by the fixed spacing distance d = $\lambda/2$ (λ is the wavelength of the carrier signal), which separates the neighboring elements on each axis. They suffer from performance limitations when the spacing distance d exceeds the value of $\lambda/2$ as the grating lobes start their appearances.

Random sampling of antenna apertures is used in [16], which is suitable for minimizing (compressing) the number of array sensors with main practical limitations. In [17], a geometry was obtained via random sampling of the antenna aperture and used by Ender to address the radar signal problems. This geometry has a measurement matrix with reduced mutual coherence between vectors and was applied only in some cases to compressive sensing algorithms such as pulse compression, radar DOA estimation, and imaging. Its implementation requires pre-processing techniques such as suppressing filters. In some ways, such geometries are practically impossible because some of the neighboring elements are located near each other.

Virtual sensors were innovated and used for DOA estimation for a larger number of signals than sensors by the authors in [18] via improving a subspace augmentation technique depending on the Khatri–Rao product. It was applied only to quasi-stationary signals based on some assumption, such as that element arrays are ULAs. To increase the number of virtual elements, researchers in [19] employed the nested arrays and estimated a greater number of signals than elements. However, it was obtained using ULAs, and its degrees of freedom depend on the total number of elements. Also, it suffers from the inherent drawbacks of the Khatri–Rao product technique.

A spatial smoothing technique based on removing the repeated rows from the signal vector was applied to the nested arrays to estimate more incoherent wideband sources than sensors [20]. It is unlike the augmentation technique and was presented as an alternative method for underdetermined DOA estimation by applying subspace-based methods directly without requiring any assumptions. It worked only for ULAs and was applied to wideband cases.

Co-arrays are used to increase the degrees of freedom for generating the covariance matrix from multi-time-domain snapshots of data measurements when applying the MU-SIC algorithm as an advanced technique. Researchers in [21] used two linear sub-arrays to create an array geometry based on co-arrays to enhance the DOA estimation. Co-prime arrays are studied on two bases: the first is based on reducing the spacing of the inter-sensors of the array by compressing one of the arrays, which is known as a co-prime array by compressed inter-element spacing (CACIS). The second is known as a co-prime array with displaced sub-arrays (CADiS), which effectively improves the degrees of freedom in the formation of the covariance matrix unless restricted to the linear arrays. They depend on uniform linear sub-arrays with more elements.

The aperiodic array in [22] was designed based on fractal geometries and used to effectively solve the practical implementation obstacle of random aperture sampling. Two frameworks of classical array processing, including Sierpinski carpet planar arrays and Cantor linear arrays, were used, but the results were not applied to the wideband sources.

A genetic algorithm (GA) based on an optimization scheme and a little perturbation in the inflation method was utilized by the authors in [23] to create antenna array styles with aperiodic tiling, avoiding grating lobes and low side-lobe levels through wide bandwidths. This method uses a larger number of sensors and is appropriate for conventional algorithms of array signal processing; however, it is not applied to the DOA estimation of wideband sources.

Another optimization scheme that is inspired by quasicrystals in materials physics was developed in [24] to generate an antenna-array mode over a disturbance of initial conditions for aperiodic geometry that can be applicable for both narrowband and wideband signals. The structure of the array output for wideband signals was used in [6], and wideband signals were modeled as a rational transfer function driven by white noise; then delays on every mode were estimated using modal decomposition. This technique is suitable and can be applied to the DOA estimation of wideband sources using compressive sensing algorithms.

Scholars in [25] showed that, for a flat power spectral density of the signal, an array manifold vector for wideband signals can be used as an alternative to the conventional array manifold for narrowband signals. It depends on the covariance matrix, which is generated by the array geometries and employs only their spatial information.

For the MUSIC algorithm [9], the spatial spectrum is computed depending on the orthogonality between the noise subspace (NS) and the array manifold vector (AMV); for the SSS algorithm [10], it is calculated utilizing the signal subspace (SS) and the array manifold vector (AMV); and for the CAPON algorithm [11], it is computed using the inversion of the covariance matrix (CM) and the array manifold vector (AMV). These DOA estimation techniques exploit the fact that the DOAs define the signal subspace. The reason for choosing these algorithms is that their performance essentially depends on the covariance matrix that is generated by our proposed geometry according to its selected element spacing.

In the eigenvalue decomposition of the covariance matrix for wideband sources, the mixing of different frequency components increases the number of significant eigenvalues to be greater than the number of sources. In other words, the separation of signal subspaces and noise subspaces from the covariance matrix will become more difficult with increasing bandwidth [26]. This means that reinvention and designing array geometries to generate covariance matrices so that their subspaces can be easily separated are promising field studies.

The main goal of this work is to derive an array manifold (vector) matrix related to the new array geometry to generate a developed covariance matrix that allows the aforementioned DOA estimation techniques to be applied directly to wideband signals, taking the whole wideband information with good accuracy. This can be achieved with a perfect array configuration based on the number of elements and their spacing settings. The proposed array geometry exploits its fixed number of elements with variable spacing to offer many configurations that can be suitable for wideband signals. This paper offers an important contribution to the field of array signal processing, whereas the proposed geometry presents a worthy way to remove the challenge of solving the DOA estimation problems of wideband sources and simplify their computation complexity. In a specific manner, the contributions and innovative points of this paper are listed as follows:

- Developing a new paradigm of pentagram arrays based on triangular geometry.
- Explanation of the theoretical principle of the superposition techniques.
- Clarification of the advantages of limiting the number of sensors in the array.
- The significance of designing an array with variable element spacing.
- Ability to maximize and minimize the array apertures.
- Addressing the issue of the DOA manifold matrix ambiguity problem.
- Application for DOA estimation algorithms of both azimuth and elevation angles.
- Conducting a large number of simulation experiments and analyses to validate the effectiveness of the geometry under different algorithms.

In this work, a novel pentagram antenna array geometry is proposed. This new geometry integrates the characteristics of linear and circular geometries. It tacitly includes five symmetrical ULAs and two concentric UCAs. Its diagram is designed, and its mathematical demonstration is proved. The steering manifold matrix of the new antenna array is derived based on the DOA signal model using a spherical polar radar coordinate system. The performance analysis is investigated for incoherent wideband DOA estimation using different scenarios of wideband signals. Also, for computation simplicity for wideband signals, we exploited the complementary nature of the time-domain and frequency-domain of the signal by using the array output signal in frequency-domain $X(f)$ instead of its representation in time-domain $X(t)$ for computation of the array covariance matrix and eigenvalue decomposition.

Since the objective of this paper is the study of the 1-DOA estimation of incoherent wideband sources, different cases will be considered in this study depending on the frequency, bandwidth, and number of sources of wideband signals. The scenarios are the following:

- A single wideband signal (pulse signal) comes from a single azimuth DOA.
- Two wideband signals with different frequencies and zero bandwidths (pulse signals) come from two closely related azimuth DOAs.
- Two wideband signals with different frequencies and zero bandwidth (pulse signals) come from two far azimuth DOAs.
- Two wideband signals with different frequencies and zero bandwidths (pulse signals) come from two far azimuth DOAs using different SNR values.
- Two wideband signals with different frequencies and different bandwidths (without overlapping) come from two far azimuth DOAs.
- Two wideband signals with the same frequencies and different bandwidths (with overlap) come from two far azimuth DOAs.

-   Four wideband signals with different frequencies and zero bandwidth (pulse signals) come from four far azimuth DOAs.
-   Four wideband signals with the same frequencies and different bandwidths (with overlap) come from four far azimuth DOAs.

The rest of the paper is organized as follows: In Section 2, we study the structure and mathematical computation of the proposed geometry. In Section 3, as a main contribution in the area, we discuss the DOA signal model for our proposed geometry and derive its steering manifold matrix. The simulation setup and the performance results of the proposed array geometry are included in Section 4. Section 5 contains some discussions. Finally, in Section 6, we present conclusions and future work.

Notations: The scalar quantities are represented by lowercase letters; vectors are denoted by boldface lowercase; and matrixes are denoted by boldface uppercase letters; $\measuredangle$ denotes the angle; the factor $(\bar{a})$ is a unit vector.

Superscripts: $(\cdot)^T$ means transpose; $(\cdot)^H$ denotes conjugate transpose; and $E[\cdot]$ refers to the expected value. The I_M is the $M \times M$ identity matrix and the $\|\cdot\|$ represents the matrix norm.

2. The Proposed Array Geometry

2.1. Pentagram and Triangular Geometry

Our proposed antenna array geometry is inspired by the five-pointed star (pentagram) and based on triangular geometry. It is implemented using the specific superposition method of three triangles. The three triangles are copies of one type of triangle (the isosceles triangle) [27]. The isosceles triangle is depicted in Figure 1.

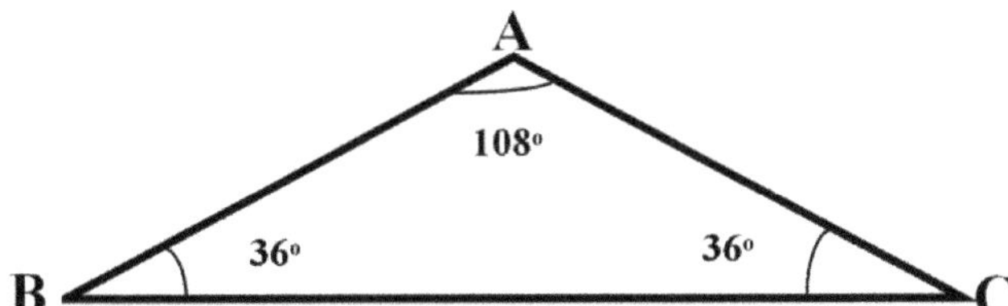

Figure 1. Isosceles triangle (base of the proposed array geometry).

All three angles of the triangle have known values. The two of them that face the two equal sides of the isosceles are equals ($\measuredangle$ ABC = ACB = 36°), and the third one that faces the longest side is bigger than them ($\measuredangle$ BAC = 108°). These three triangles are arranged in a superposed manner to form five-star (pentagram) geometry, as shown in Figure 2.

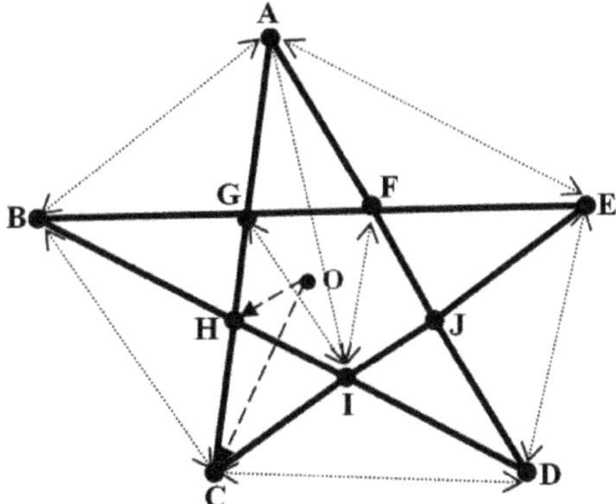

Figure 2. Three-superposed triangles (five-star pentagram) formation.

2.2. Pentagram Geometry Analysis Based on Triangular Geometry

According to the geometry of the isosceles triangle [27], we can make a geometry analysis for our proposed pentagram array geometry in Figure 2 as follows:

- The five outer vertices of the superposed triangles (pentagram) are named A, B, C, D, and E, and the five inner vertices are named F, G, H, I, and J.
- All five angles of the five outer vertices ($\angle$ FAG $=\angle$ GBH $=\angle$ HCI $=\angle$ IDJ $=\angle$ JEA $= 36°$) have equal values of 36°.
- All five inner angles of the five inner vertices ($\angle$ JFG $=\angle$ FGH $=\angle$ GHI $=\angle$ HIJ $=\angle$ IJF $= 108°$) have equal values of 108°.
- The lengths of the triangle sides between any two points (AB, BC, CD, DE, and EA) of the five outer vertices are equal; we named them L, and we have five Ls in total.
- The lengths of the triangle sides between the five outer vertices and any point of the five inner vertices (AF, AG, BG, BH, CH, CI, DI, DJ, EJ, and EF) are equal; we named them X, and we have ten Xs in total.
- The lengths of the triangle sides between two points of the five inner vertices (FG, GH, HI, IG, and JF) are equal (interconnection distances not included); we named them Y, and we have five Ys in total.
- The two straight lengths (interconnection) of triangle sides between any one point and the two points (opposite sides) of five interspersions (IF, IG, HJ, HF, and GJ) are equal; we named it P, and we have five Ps in total.
- The five lengths of triangle sides between the five headsails and the opposite (one) point of the five inner vertices (AI, BJ, CF, DG, and EH) are equal; we named them Z, and we have five Zs in total.

All these lengths of triangle sides in this geometry can be calculated by applying the trigonometry laws depending on the above-mentioned values of angles. Here, we use the law of sine [27] in Equation (1) below to calculate the lengths of the sides of a plane triangle when the angels are known, as shown in Figure 3.

$$\frac{a}{\sin \alpha} = \frac{b}{\sin \beta} = \frac{c}{\sin \gamma} \tag{1}$$

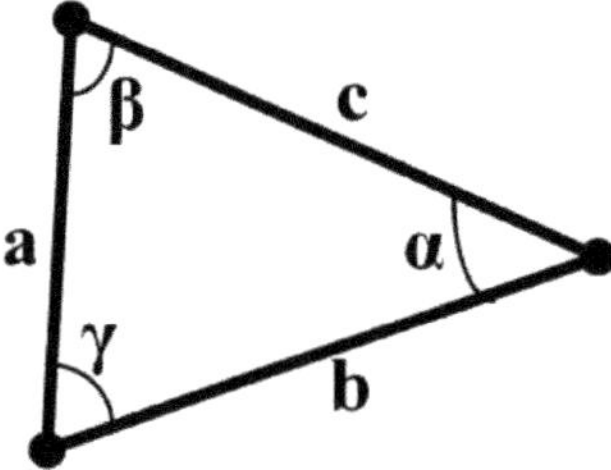

Figure 3. Law sine of a plane triangle.

After applying the law of sine to our proposed array geometry, we obtained the relations between the mentioned lengths of the triangle sides using these equations:

$$\frac{Y}{\sin 36°} = \frac{X}{\sin 72°} \tag{2}$$

So

$$Y = \frac{\sin 36°}{\sin 72°}X = 0.618X \tag{3}$$

Or

$$X = 1.6182Y \tag{4}$$

Also

$$\frac{L}{\sin 36°} = \frac{2X + Y}{\sin 72°} \tag{5}$$

So

$$L = \frac{\sin 36°}{\sin 72°}(2X + Y) = 0.618(2X + Y) = 1.236X + 0.618Y = 1.618X \tag{6}$$

Also

$$\frac{P}{\sin 36°} = \frac{X + Y}{\sin 72°} \tag{7}$$

So

$$P = \frac{\sin 36°}{\sin 72°}(X + Y) = 0.618(X + Y) = 0.618X + 0.618Y \tag{8}$$

Also

$$\frac{Z}{\sin 108°} = \frac{X + Y}{\sin 54°} = \frac{Y}{\sin 18°} \tag{9}$$

So

$$Z = \frac{\sin 108°}{\sin 54°}(X + Y) = \frac{\sin 108°}{\sin 18°}(Y) = 1.17557(X + Y) = 3.07768Y = 1.902X \tag{10}$$

Referring to Figure 2, it is assumed that the point O is located at the center of the pentagram geometry. We can see that the pentagram geometry appears as two circles. The small inner circle has a small radius that indicates the distance OH, where the five inner vertices of the pentagram lie on its circumference, and the outer circle has a large radius, which indicates the distance OC, where the five outer vertices of the pentagram lie on its circumference. These two radiuses are given by the related equations below:

The outer radius of the pentagram geometry (OC)

$$OC = \frac{\sin 126°}{\sin 36°}(X) = 1.376X = 2.616(OH) \tag{11}$$

The inner radius of the pentagram geometry (OH)

$$OH = \frac{\sin 18°}{\sin 36°}(X) = 0.526X = 0.382(OC) \tag{12}$$

It is worth noting from the above equations that if we specify any value for the lengths of the triangle sides X or Y, we can know the other lengths of the triangle sides (L, P, and Z). Also, we can note that these sides have different length values from each other.

2.3. Pentagram Geometry Structure

From the previous Sections 2.1 and 2.2, we obtained the first inspired idea of our proposed novel antenna array geometry, exploiting the different lengths of the pentagram. So, we propose to study this (pentagram) geometry as a new antenna array geometry by considering the lengths of the triangle sides as actual distances between the array (sensor) elements. In this way, we can locate ten antenna elements (sensors) at the outer and inner vertices (ten) of the triangles. So, the form of the aperture we now have and the number of sensors are usually fixed. The size of the aperture can be maximized or minimized depending on the distances between the array elements.

According to this property, we can define a new type of antenna array geometry that has a fixed number of elements with variable element spacing (FENVES). The second inspired idea that we obtained was the sparse array configuration of our proposed antenna array according to the inherent sparse property of the pentagram geometry as based on triangular geometry using a fewer number of elements. Also, the proposed new geometry can be described as non-uniform, non-random array geometry due to the differences in the spacing elements (differences in lengths of the triangle sides) and known values of the element spacing.

Now, the goal is to choose the appropriate array configuration for our new antenna array geometry, which makes it applicable for DOA estimation for wideband signals (sources). As we mentioned before, the number of elements (sensors) of our new antenna array (as a factor affecting DOA estimation) is fixed at ten elements (M = 10 sensors). The only variable parameter in our array configuration is the distance between the array elements (usually named d in the literature), which determines the sensor configuration. According to the inherently sparse geometry of the pentagram, the distances between the array elements are related to their positions. Every element has a differentiated property of distance from its adjacent neighbors.

For every sensor from the five sensors that are located at the outer vertices of the pentagram (A, B, C, D, and E), its differentiated distances with its adjacent element neighbors are (X, X, L, L, and Z), and for every sensor from the five sensors that are located at the inner vertices of the pentagram (F, G, H, I, and J), its differentiated distances with its (adjacent elements) neighbors are (X, X, Y, Y, P, P, and Z). This differentiated property in the distances of the pentagram geometry, of course, has an effect on the sparsity degree of the antenna array. One of our goals is to exploit this property to achieve the desired sensor configuration for our new array geometry. It is clear that, from Equations (2)–(12), these distances depend on each other. If we set the distances X or Y for our sensor configuration of array geometry to some value related to the (λ) wavelength of wideband signals (X = $\lambda/2$ or Y = 2λ), then all these distances will automatically relate to the λ. The distances X or Y can be set to the suitable value related to the λ in the case of wideband signals, as below:

$$X = \frac{\lambda}{2} \text{ or } X = \frac{\lambda_{min}}{2} \text{ or } X = \frac{\lambda_{max}}{2}, \text{ where } \lambda = \frac{c}{f_\circ}, \lambda_{min} = \frac{c}{f_{max}}, \lambda_{max} = \frac{c}{f_{min}}, f_{min} = f_\circ - \frac{BW}{2} \text{ and } f_{max} = f_\circ + \frac{BW}{2} \tag{13}$$

Or

$$Y = \frac{\lambda}{2} \text{ or } Y = \frac{\lambda_{max}}{2} \text{ or } Y = \frac{\lambda_{min}}{2} \text{ where } \lambda = \frac{c}{f_\circ}, \lambda_{min} = \frac{c}{f_{max}}, \lambda_{max} = \frac{c}{f_{min}}, f_{min} = f_\circ - \frac{BW}{2} \text{ and } f_{max} = f_\circ + \frac{BW}{2} \tag{14}$$

where $f_\circ$ is the center frequency and BW is the bandwidth of the signal.

So, we can arrive at the suitable array configuration for our new array geometry by setting the separation between adjacent elements.

To evaluate the performance of our new (novel) array geometry with the desired sensor configuration, we need other array geometries for comparison. Here we use uniform linear array (ULA) and uniform circular array (UCA) geometries, two of the well-known array geometries found in the literature.

3. DOA Array Signal Model

The direction of arrival (DOA) is the angle between the array normal and the direction vector of the plane wave. Consider the direction of arrival estimation system, which contains Q wideband signals (sources) arriving from different directions impinging on our proposed new antenna array, which is inspired by pentagram geometry with 10sensors (M = 10 array elements) named as A, B, C, D, E, F, G, H, I, and J, as depicted in Figure 4. The depicted diagram describes the radar coordinate system (spherical polar) of the DOA signal model using our new antenna array. The elevation angle (measured clockwise from the Z-axis) and azimuth angle (measured counterclockwise from the X-axis) of the default signal source (target) K are θ_K and ϕ_K, respectively.

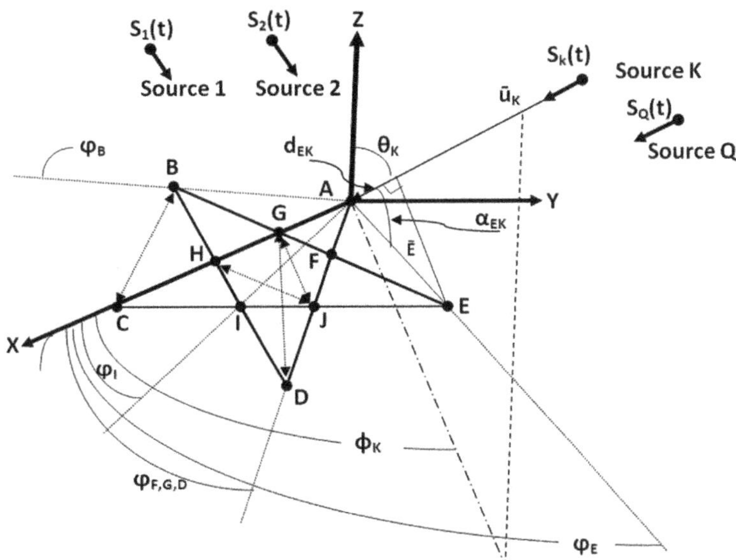

Figure 4. The coordinate system of pentagram antenna array receiving Q wideband signals from different directions.

For convenience, we chose sensor A, which is located at the origin of the coordinate system (the earth's surface lies in the X-Y plane), as the reference node. Assume the Q targets in the far-field sensing field emit wideband signals consisting of N time-domain snapshots (number of samples). At ith sensor (any sensor except A) of this array, there is a time delay $\tau_{q,i}$ for target q to arrive at that sensor. The time delay depends on the straight distance and the projection angle αAi between the A reference element and the ith element. The output of the ith sensor in the time-domain can be described as

$$x_i(t) = \sum_{q=1}^{Q} s_q\big(t - \tau_{q,i}\big) + n(t) \tag{15}$$

where $s_q(t) = e^{j2\pi ft}$ denotes the qth wideband signal, $\tau_{q,i}$ is the time delay of qth signal at ith sensor, f is the signal frequency, t is the time interval, i indicates the sensors of A, B, C, D, E, F, G, H, I, and J, and q = 1,2,...,Q. Then, the total received signal in the frequency-domain, $X(f)$, that includes directions both of elevation angle θ_K and azimuth angle ϕ_K corrupted by noise is given by the relation described below:

$$X(f) = [x_1(f), x_2(f), \ldots, x_M(f)]^T = \sum_{q=1}^{Q} a\big(\theta_q, \phi_q, f_q\big) s(f) + n(f) = A(\theta, \phi, f) S(f) + N(f) \tag{16}$$

where

$$S(f) = [s_1(f_1), s_2(f_2), \ldots, s_Q(f_Q)]^T \tag{17}$$

is the $(Q \times N)$ incident signals in the frequency-domain; N is the number of snapshots.

$$N(f) = [n_1(f_1), n_2(f_2), \ldots, n_M(f_M)] \tag{18}$$

is the $(M \times N)$ array that contains noise vectors with complex normal distribution (zero mean and variance σ) $CN(0, \sigma^2 I)$ and $A(\theta, \phi, f)$ is the steering matrix for Q vectors (Q columns); it is an $(M \times Q)$ array matrix and is defined as follows:

$$A(\theta, \phi, f) = [a(\theta_1, \phi_1, f_1), a(\theta_2, \phi_2, f_2), \ldots, a(\theta_Q, \phi_Q, f_Q)] \tag{19}$$

Referring to Figure 4, the sensors A, G, H, and C are located on the X-axis (sensor A is the reference), and the phase angles φ between the reference element A and the other elements can be calculated as follows:

$$\varphi_G = \varphi_H = \varphi_C = 0° \tag{20}$$

$$\varphi_I = \frac{36°}{2} = 18° \tag{21}$$

$$\varphi_F = \varphi_J = \varphi_D = 36° \tag{22}$$

$$\varphi_E = 72° \tag{23}$$

$$\varphi_B = 360° - 72° = 288° \tag{24}$$

The projection angle α_{EK} between the reference element vector $\bar{u}_K$ of the incident signal $S_K(t)$ and the element E as shown in the above model due to the incident plane waves can be calculated by finding the dot product between unit vectors $\bar{u}_K$ and $\bar{E}$ ($\bar{E}$ is the vector length between the reference element A and the element E) as follows:

$$\alpha_{EK} = \cos^{-1}\left[\frac{\bar{u}_K \cdot \bar{E}}{\|\bar{u}_K\| \cdot \|\bar{E}\|}\right] \tag{25}$$

Then, the unit vector $\bar{E}$ can be expressed as:

$$\bar{E} = E\cos\varphi_E \bar{a}_x + E\sin\varphi_E \bar{a}_y \tag{26}$$

where E and φ_E are the distance and phase angles between the reference element A and element E, respectively. Also, the unit vector $\bar{u}_K$, which contains the directions of θ_K and ϕ_K for any arrival signal source, can be given by:

$$\bar{u}_K = \cos\phi_K \sin\theta_K \bar{a}_x + \sin\phi_K \sin\theta_K \bar{a}_y + \cos\theta_K \bar{a}_z \tag{27}$$

where $\bar{a}_x$ $\bar{a}_y$, and $\bar{a}_z$ are unit vectors for Cartesian coordinates. Then, by substituting these two Equations (26) and (27) in the Equation (25), we obtain:

$$\alpha_{EK} = \cos^{-1}\left[\frac{(\cos\phi_K \sin\theta_K \bar{a}_x + \sin\phi_K \sin\theta_K \bar{a}_y + \cos\theta_K \bar{a}_z) \cdot (E\cos\varphi_E \bar{a}_x + E\sin\varphi_E \bar{a}_y)}{\|(\cos\phi_K \sin\theta_K \bar{a}_x + \sin\phi_K \sin\theta_K \bar{a}_y + \cos\theta_K \bar{a}_z)\| \cdot \|(E\cos\varphi_E \bar{a}_x + E\sin\varphi_E \bar{a}_y)\|}\right] \tag{28}$$

So now, after algebraic manipulation, we obtain:

$$\alpha_{EK} = \cos^{-1}[\sin\theta_K \cos(\phi_K - \varphi_E)] \tag{29}$$

By obtaining this projection angle, it can be used to achieve the additional distance named α_{EK} in Figure 4 that needs to be traveled by the Kth signal to arrive at element E as follows:

$$d_{EK} = E\cos\alpha_{EK} = E\cos\left[\cos^{-1}[\sin\theta_K \cos(\phi_K - \varphi_E)]\right] = E\sin\theta_K \cos(\phi_K - \varphi_E) \tag{30}$$

where E and φ_E are the distance and phase angles between the reference element A and element E, respectively, and θ_K and ϕ_K are the elevation angle and azimuth angle, respectively. From Section 2.2, we assumed that the distance between the reference element A and element E is equal to L and we found that $\varphi_E = 72°$. Then,

$$d_{EK} = L\sin\theta_K \cos(\phi_K - 72°) \tag{31}$$

For wideband signals, we can define c as the speed of light and f as the signal frequency. According to Equation (31), the corresponding phase difference ψ_{EK} for the time delay τ_{EK} depending on the distance d_{EK} can be expressed by this equation:

$$\psi_{EK} = 2\pi f_K \frac{d_{EK}}{c} = \frac{2\pi f_K}{c} L sin\theta_K cos(\phi_K - 72°) \tag{32}$$

Now we conduct the same procedures for the residual elements of our pentagram array for the same incident signal $S_K(t)$ as below:

For sensor B, we find that:

$$d_{BK} = Bcos\alpha_{BK} = Bcos[cos^{-1}[sin\theta_K cos(\phi_K - \varphi_B)]] = Bsin\theta_K cos(\phi_K - \varphi_B) \tag{33}$$

where the distance between the reference element A and element B is equal to L, and we found that $\varphi_B = 288°$. So,

$$d_{BK} = L\, sin\theta_K cos(\phi_K - 288°) \tag{34}$$

So, the corresponding phase difference ψ_{BK} for τ_{BK} and d_{BK} can be expressed by this equation:

$$\psi_{BK} = 2\pi f_K \frac{d_{BK}}{c} = \frac{2\pi f_K}{c} L sin\theta_K cos(\phi_K - 288°) \tag{35}$$

For sensor I, we can write:

$$d_{IK} = Icos\alpha_{IK} = Icos[cos^{-1}[sin\theta_K cos(\phi_K - \varphi_I)]] = Isin\theta_K cos(\phi_K - \varphi_I) \tag{36}$$

where the distance between the reference element A and element I is known as Z, and we found that $\varphi_I = 18°$. So,

$$d_{IK} = Z\, sin\theta_K cos(\phi_K - 18°) \tag{37}$$

Then, the corresponding phase difference ψ_{IK} for τ_{IK} and d_{IK} can be given by:

$$\psi_{IK} = 2\pi f_K \frac{d_{IK}}{c} = \frac{2\pi f_K}{c} Z sin\theta_K cos(\phi_K - 18°) \tag{38}$$

For sensor F, we can state:

$$d_{FK} = Fcos\alpha_{FK} = Fcos[cos^{-1}[sin\theta_K cos(\phi_K - \varphi_F)]] = Fsin\theta_K cos(\phi_K - \varphi_F) \tag{39}$$

where the distance between the reference element A and element F is known as X, and we found that $\varphi_F = 36°$. So,

$$d_{FK} = Xsin\theta_K cos(\phi_K - 36°) \tag{40}$$

and the corresponding phase difference ψ_{FK} for τ_{FK} and d_{FK} can be given by:

$$\psi_{FK} = 2\pi f_K \frac{d_{FK}}{c} = \frac{2\pi f_K}{c} Xsin\theta_K cos(\phi_K - 36°) \tag{41}$$

For the sensors J and D, which have the same phase angle with the sensor F, we can obtain their time difference of arrival and corresponding phase difference by tacking only the distances between them and the reference A, so:

For sensor J:

$$d_{JK} = Jcos\alpha_{JK} = Jcos[cos^{-1}[sin\theta_K cos(\phi_K - \varphi_J)]] = Jsin\theta_K cos(\phi_K - \varphi_J) \tag{42}$$

where the distance between the reference element A and element J is known as $(X + Y)$, and we found that $\varphi_J = 36°$. So,

$$d_{JK} = (X + Y)\, sin\theta_K cos(\phi_K - 36°) \tag{43}$$

and the corresponding phase difference ψ_{JK} for τ_{JK} and d_{JK} can be given by:

$$\psi_{JK} = 2\pi f_K \frac{d_{JK}}{c} = \frac{2\pi f_K}{c}(X + Y)sin\theta_K cos(\phi_K - 36°) \tag{44}$$

For sensor D:

$$d_{DK} = Dcos\alpha_{DK} = Dcos[cos^{-1}[sin\theta_K cos(\phi_K - \varphi_D)]] = Dsin\theta_K cos(\phi_K - \varphi_D) \tag{45}$$

where the distance between the reference element A and element D is known as $(2X + Y)$, and we found that $\varphi_D = 36°$. So,

$$d_{DK} = (2X + Y) \, sin\theta_K cos(\phi_K - 36°) \tag{46}$$

and the corresponding phase difference ψ_{DK} for τ_{DK} and d_{DK} can be given by:

$$\psi_{DK} = 2\pi f_K \frac{d_{DK}}{c} = \frac{2\pi f_K}{c}(2X + Y)sin\theta_K cos(\phi_K - 36°) \tag{47}$$

For the sensors G, H, and C, which are located at the same axis as the reference element A, their phase angle will be equal to $0°$, as mentioned before. Then, we can obtain their time difference of arrival and corresponding phase difference as follows:
For sensor G:

$$d_{GK} = Gcos\alpha_{GK} = Gcos[cos^{-1}[sin\theta_K cos(\phi_K - \varphi_G)]] = Gsin\theta_K cos(\phi_K - \varphi_G) \tag{48}$$

where the distance between the reference element A and element G is known as X, and we found that $\varphi_G = 0°$. So,

$$d_{GK} = X \, sin\theta_K cos\phi_K \tag{49}$$

and the corresponding phase difference ψ_{GK} for τ_{GK} and d_{GK} can be given by:

$$\psi_{GK} = 2\pi f_K \frac{d_{GK}}{c} = \frac{2\pi f_K}{c}Xsin\theta_K cos\phi_K \tag{50}$$

For sensor H:

$$d_{HK} = Hcos\alpha_{HK} = Hcos[cos^{-1}[sin\theta_K cos(\phi_K - \varphi_H)]] = Hsin\theta_K cos(\phi_K - \varphi_H) \tag{51}$$

where the distance between the reference element A and element H is known as $(X + Y)$, and we found that $\varphi_H = 0°$. So,

$$d_{HK} = (X + Y) \, sin\theta_K cos\phi_K \tag{52}$$

and the corresponding phase difference ψ_{HK} for τ_{HK} and d_{HK} can be given by:

$$\psi_{HK} = 2\pi f_K \frac{d_{HK}}{c} = \frac{2\pi f_K}{c}(X + Y)sin\theta_K cos(\phi_K) \tag{53}$$

For sensor C:

$$d_{CK} = Ccos\alpha_{CK} = Ccos[cos^{-1}[sin\theta_K cos(\phi_K - \varphi_C)]] = Csin\theta_K cos(\phi_K - \varphi_C) \tag{54}$$

where the distance between the reference element A and element C is known as $(2X + Y)$, and we found that $\varphi_C = 0°$. So,

$$d_{CK} = (2X + Y)sin\theta_K cos\phi_K \tag{55}$$

and the corresponding phase difference ψ_{CK} for τ_{CK} and d_{CK} can be given by:

$$\psi_{CK} = 2\pi f_K \frac{d_{CK}}{c} = \frac{2\pi f_K}{c}(2X + Y)sin\theta_K cos(\phi_K) \tag{56}$$

Finally, we can obtain the steering vector (column) of any Kth signal for our novel array geometry (pentagram) by using the formula below:

$$a(\theta_K, \phi_K, f_K) = \left[e^{-j\psi_{AK}}, e^{-j\psi_{BK}}, e^{-j\psi_{CK}}, e^{-j\psi_{DK}}, e^{-j\psi_{EK}}, e^{-j\psi_{FK}}, e^{-j\psi_{GK}}, e^{-j\psi_{HK}}, e^{-j\psi_{IK}}, e^{-j\psi_{JK}} \right]^T \tag{57}$$

The array manifold matrix, or steering matrix $A(\theta, \phi, f)$, is an $(M \times Q)$ frequency-dependent array matrix. For our novel array geometry (pentagram), M = 10 elements (A, B, C, D, E, F, G, H, I, and J sensors), and the array manifold matrix that describes the response of our proposed antenna array for Q signals arriving from different directions is formed by stacking Q steering vectors or Q columns (which are functions of directions θ, ϕ, and frequency f) $a(\theta_K, \phi_K, f_K)$. Referring to Equation (19), it can be described as:

$$A(\theta, \phi, f) = \begin{bmatrix} e^{-j\psi_{A1}} & e^{-j\psi_{B1}} & e^{-j\psi_{C1}} & e^{-j\psi_{D1}} & e^{-j\psi_{E1}} & e^{-j\psi_{F1}} & e^{-j\psi_{G1}} & e^{-j\psi_{H1}} & e^{-j\psi_{I1}} & e^{-j\psi_{J1}} \\ \vdots & \vdots & \vdots & \vdots & \vdots & \vdots & \vdots & \vdots & \vdots & \vdots \\ e^{-j\psi_{AK}} & e^{-j\psi_{BK}} & e^{-j\psi_{CK}} & e^{-j\psi_{DK}} & e^{-j\psi_{EK}} & e^{-j\psi_{FK}} & e^{-j\psi_{GK}} & e^{-j\psi_{HK}} & e^{-j\psi_{IK}} & e^{-j\psi_{JK}} \\ \vdots & \vdots & \vdots & \vdots & \vdots & \vdots & \vdots & \vdots & \vdots & \vdots \\ e^{-j\psi_{AQ}} & e^{-j\psi_{BQ}} & e^{-j\psi_{CQ}} & e^{-j\psi_{DQ}} & e^{-j\psi_{EQ}} & e^{-j\psi_{FQ}} & e^{-j\psi_{GQ}} & e^{-j\psi_{HQ}} & e^{-j\psi_{IQ}} & e^{-j\psi_{JQ}} \end{bmatrix}^T \tag{58}$$

where A, B, C, D, E, F, G, H, I, and J are the array sensors (elements), and 1, K, and Q are the signal sources (targets). According to the Q numbers of incident signals with a single snapshot and referring to Equation (17), the signal $S(f)$ is the $(Q \times 1)$ vector, which can be described as follows:

$$S(f) = \left[s_1(f_1), \ldots, s_K(f_K), \ldots, s_Q(f_Q) \right]^T \tag{59}$$

where 1, K, and Q are the incident signals. In the case of a multiple number of snapshots, N, then the signal $S(f)$ is the $(Q \times N)$ matrix and can be as follows:

$$S(f) = \begin{bmatrix} s_{11}(f_1) & s_{21}(f_2) & \cdots & s_{Q1}(f_Q) \\ \vdots & \vdots & \vdots & \vdots \\ s_{1K}(f_1) & s_{2K}(f_2) & \cdots & s_{QK}(f_Q) \\ \vdots & \vdots & \vdots & \vdots \\ s_{1N}(f_1) & s_{2N}(f_2) & \cdots & s_{QN}(f_Q) \end{bmatrix}^T \tag{60}$$

where 1, K, and Q are the signals and 1, 2...N are the snapshots in the signal.

In order to obtain the DOAs of the incident signals, the above array manifold matrix $A(\theta, \phi, f)$ of Equation (58) must satisfy the restricted isometry propriety (RIP) condition according to the locations of the array sensors (elements) of the selected sensor array configuration. In Ref. [28], the RIP condition is described as: any two columns selected from array manifold matrix $A(\theta, \phi, f)$ must be linearly independent of each other. In other words, the correlation coefficient between each two columns in the array manifold matrix $A(\theta, \phi, f)$ affects the performance of the DOAs. The number of columns of the array manifold matrix $A(\theta, \phi, f)$ is determined by the number of incident signals (sources). We can define the correlation coefficient μ_{ij} by this equation:

$$\mu_{ij} = \frac{\left| a^H_i a_j \right|}{\|a_i\|_2 \|a_j\|_2} \tag{61}$$

where a_i and a_j are the ith and jth columns of $A(\theta, \phi, f)$, respectively. The correlation coefficient μ_{ij} depends on the array configuration as it relates to the number of elements (sensors) and element spacing (d). In our new array geometry, the number of elements (M = 10) is fixed, and element spacing (d) is a variable depending on the selected setting. The positions of the array elements are already known (the outer and inner vertices of the triangles) and are determined according to the underlying formation of the array geometry. This means the array elements can be practically arranged and designed using suitable element spacing (d).

By making this important note, we avoid the practically realizable problem that happens when some of the elements lie closer to each other, as in the case of random geometry. As we mentioned before in Section 2.3, the spacing (d) can be set to some values related to the λ, λ being the wavelength of the carrier signal. Many values of the correlation coefficient μ_{ij} with a fixed number of elements and different element spacings can be obtained. In other words, this offers some benefits related to the degrees of freedom for creating our proposed array covariance matrix. Although the fixed number of elements binds these values, they nevertheless offer some advantages related to simplifying the computations of the correlation coefficient μ_{ij}.

Experiments in the literature [29] showed that the correlation coefficient μ_{ij} reduces with the increasing number of elements at fixed element spacing (d = $\lambda/2$) and increases ($\mu_{ij} = 1$) when increasing the element spacing (d = 2λ), indicating that the array manifold matrix $A(\theta, \phi, f)$ violates the RIP condition and its steering vectors become ambiguous (this is known as the manifold ambiguity problem), leading to DOA estimation failure. The fixed small number of elements (M = 10) as in our new array geometry leads to a degree of free distribution depending only on the element spacing (d) setting, so the elements can be close to or far away from each other. In other words, we can set the element spacing that leads to minimizing or maximizing the antenna aperture, which has some bearing on the output of the DOA estimation algorithms.

Here, our study becomes related to aperture sampling and the antenna aperture itself. We will take the minimization ability of the antenna aperture in this proposed antenna array as a new compressive idea for compressing the size of the antenna aperture. The only design parameter for our array geometry is the element spacing (d). The motivation behind setting the spacing elements is to minimize the mutual coherence (small value of correlation coefficient μ_{ij}) of the columns of the array manifold matrix in order to make it well suited for DOA estimation algorithms for wideband sources.

Referring to Equation (18), the complex normal distribution noise matrix $N(f)$ for incident signals with a single snapshot is the $(M \times 1)$ vector and can be expressed as:

$$N(f) = [n_A(f_A), n_B(f_B), n_C(f_C), n_D(f_D), n_E(f_E), n_F(f_F), n_G(f_G), n_H(f_H), n_I(f_I), n_J(f_J)]^T \quad (62)$$

where $A, B, C, D, E, F, G, H, I,$ and J are the array elements.

In the case of a multiple number of snapshots N, then the noise $N(f)$ is the $(M \times N)$ matrix and can be as follows:

$$N(f) = \begin{bmatrix} n_{A1}(f_A) & n_{B1}(f_B) & n_{C1}(f_C) & n_{D1}(f_D) & n_{E1}(f_E) & n_{F1}(f_F) & n_{G1}(f_G) & n_{H1}(f_H) & n_{I1}(f_I) & n_{J1}(f_J) \\ n_{A2}(f_A) & n_{B2}(f_B) & n_{C2}(f_C) & n_{D2}(f_D) & n_{E2}(f_E) & n_{F2}(f_F) & n_{G2}(f_G) & n_{H2}(f_H) & n_{I2}(f_I) & n_{J2}(f_J) \\ \vdots & \vdots & \vdots & \vdots & \vdots & \vdots & \vdots & \vdots & \vdots & \vdots \\ n_{AN}(f_A) & n_{BN}(f_B) & n_{CN}(f_C) & n_{DN}(f_D) & n_{EN}(f_E) & n_{FN}(f_F) & n_{GN}(f_G) & n_{HN}(f_H) & n_{IN}(f_I) & n_{JN}(f_J) \end{bmatrix}^T \quad (63)$$

where $A, B, C, D, E, F, G, H, I,$ and J are the array elements, and $1, 2 \ldots N$ are the snapshots of the signal.

4. Results

4.1. Simulation Setup

As the objective of this paper focuses on the DOA estimation of wideband sources, many scenarios of wideband signals and DOA azimuth, as mentioned in Section 1, have been studied using computer software simulations. Each scenario was performed under a specific simulation parameter setting.

To evaluate the performance of our geometry configuration, we need other array geometries for comparison. When we look at the formation appearance of the proposed geometry, we can see that it contains five groups of linear arrays with four elements in each group. Also, it looks like two concentric circular arrays; each array has five elements. The five elements that are located on the inner vertices of the pentagram form a uniform circular array with a small radius far from the center of the pentagram, and those five elements that are located on the outer vertices shape a large uniform circular array with a large radius far from the same center. The relationship between these two radiuses was explained in Section 2.2. So according to these descriptions, we use the well-known ULA (uniform linear array) geometry and UCA (uniform circular array) for DOA estimation and performance comparison. We assume a UCA with ten (M = 10) elements and a 36-degree fixed center angular separation between them. The radius (R) of the circle in every simulation scenario will be set to a value that is larger than the inner and smaller than the outer radiuses of the pentagram geometry. Also, we consider a ULA with ten horizontally stacked elements and element spacing (d = 0.5λ, λ is the wavelength that is used for the simulation) for all simulations. The reason for choosing such ULA and UCA geometries is that our proposed geometry formation only depends on the element spacing and aperture size due to its fixed element number. ULA and UCA have fixed element spacing (d) and a variable numbers of elements.

The experimental simulations were performed using MATLAB/R2018a computer software. It is a technical computing environment for matrix computation, signal processing, and graphics visualization.

4.1.1. Simulation Setup for Proposed Geometry Configuration

In this section, we select suitable parameters to form the sensor configuration of our new array geometry. The parameters that were used for geometry formation and performance comparison are listed in Table 1. For simplicity, in this DOA estimation performance comparison, we use one source that emits a wideband signal with zero bandwidth (a pulse signal). The element spacing (d) of the three geometries is set to some value related to the wavelength (λ) of the incident signal. For UCA geometry, its radius is set to a value that is greater than the inner and less than the outer radius of the pentagram geometry mentioned above. The inner radius of the pentagram array in this case is equal to 1.622.λ and the outer radius is equal to 4.244.λ. Because the antenna aperture does have some bearing on the output of the DOA algorithm, in this section we will investigate the DOA estimation performance of the three geometries under the MUSIC algorithm.

Table 1. Simulation parameters for DOA estimation based on proposed, ULA, and UCA geometries.

Parameter	Symbol	ULA/UCA/Proposed	Notes
Number of sensors	M	10	Fixed number in the proposed geometry
Sampling frequency(GHz)	f_s	16	For simulation
Center frequency (GHz)	f_0	7.1	One signal is considered
Bandwidth (GHz)	BW	0	Pulse signal
Wavelength of the incident signal (m)	λ	0.042	$\lambda = c/f_0$
Source elevation DOA (degrees)	θ_K	90	Elevation kept as fixed
Source azimuth DOA (degrees)	ϕ_K	0	One source is considered
Snapshots	N	100	Number of samples
Speed of propagation (light) m/s	c	3×10^8	-
Signal-to-noise ratio (in dB)	SNR	10	-
Element spacing (in wavelength)	d	$d = 0.5\lambda/R = 2.5\lambda/d_X = 3.084\lambda$	Element spacing for ULA/radius of UCA/base element spacing for proposed geometry
Element angle position (rad)	Phi	$-/(2\pi/M)/-$	Element distribution of the geometry

4.1.2. Simulation Setup for DOA Angular Accuracy and Resolution Performance of the Proposed Geometry

In this area, we adjust the simulation parameters to provide the angular accuracy and resolution of DOA estimation for the proposed geometry compared with ULA and UCA geometries. We select two wideband sources that lie close to each other and emit two wideband signals with different frequencies and zero bandwidths (pulse signals). The two sources are separated by six azimuth degrees from each other. As the frequencies and wavelengths of the signal sources are changed from the previous Section 4.1.1, the element spacing of the geometries will change. This change in the array aperture will affect the performance of the DOA estimation algorithm. Since we use two different wideband signals in this scenario, we have two values of wavelengths (λ), which are λ_1 for the first signal and λ_2 for the second signal. For setting the element spacing (d) of the three geometries to some value related to the wavelength (λ), we set the wavelength (λ) for simulation to a value that lies in the range between the two wavelength values λ_1 and λ_2 as λ greater than λ_2 and less than λ_1. The inner radius of the pentagram array in this case is equal to 1.324λ, and the outer radius is equal to 3.468λ. The parameters that are used for DOA angular resolution and accuracy performance comparison are fixed as the same values as those listed in Table 1 for the three geometries, except some parameters, which will take the new values as listed in Table 2.

Table 2. Simulation parameters for DOA estimation angular accuracy and resolution under MUSIC algorithm based on proposed, ULA, and UCA geometries.

Parameter	Symbol	ULA/UCA/Proposed	Notes
Center frequency (GHz)	f_0	7.1 and 13.7	Two signals (different frequencies) are considered
Bandwidth (GHz)	BW	0 and 0	Pulse signals
Wavelength of the simulation (m)	λ	0.0375	$\lambda_2 < \lambda < \lambda_1$
Wavelength for signal 7.1 (m)	λ_1	0.0423	$\lambda_1 = c/f_0$
Wavelength for signal 13.4 (m)	λ_2	0.0224	$\lambda_2 = c/f_0$
Source azimuth DOA (degrees)	ϕ_K	15 and 21	Two sources are considered
Element spacing (in wavelength)	d	$d = 0.5\lambda/R = 2.0\lambda/d_X = 2.520\lambda$	Element spacing for ULA/radius of UCA/base element spacing for proposed geometry

4.1.3. Simulation Setup for DOA Estimation Comparison of the Proposed Geometry with UCA and ULA Geometries

In this part, we will set the parameters of the DOA estimation simulation comparison between the proposed geometry and UCA and ULA geometries under the MUSIC algorithm. Here, we use the same two incident wideband signals that were used in the previous Section 4.1.2 for the two sources, only changing their azimuth DOAs. As the actual azimuth DOAs of the incident signal sources are changed, the proposed geometry needs to adjust its element spacing. In this case, the outer radius of the pentagram is equal to 3.472λ and the inner radius is equal to 1.327λ. The radius of the UCA geometry is set to a value that lies between these two values. The parameters that are used for DOA estimation comparison are fixed at the same values as those listed in Tables 1 and 2 for the three geometries, except for the parameters that are revalued as listed in Table 3.

Table 3. Simulation parameters for DOA estimation comparison of the proposed geometry with UCA and ULA geometries under MUSIC algorithm.

Parameter	Symbol	ULA/UCA/Proposed	Notes
Source azimuth DOA (degrees)	ϕ_K	18 and 54	Two sources are considered
Element spacing (in wavelength)	d	$d = 0.5\lambda / R = 2.0\lambda / d_X = 2.523\lambda$	Element spacing for ULA/radius of UCA/base element spacing for proposed geometry

4.1.4. Simulation Setup for Proposed Geometry Performance Analysis

In this part, the performance of the proposed geometry is analyzed using different SNR values and the MUSIC DOA algorithm. The selected three values range from low to high SNR, which are -25 dB, 0 dB, and 25 dB, respectively. The parameters that are used for performance analysis that are only related to the proposed geometry are fixed at the same values as those listed in Tables 1–3, except for the parameters that shall take the new values listed in Table 4. The outer radius and the inner radius of the pentagram have the same values as in the previous Section 4.1.3.

Table 4. Simulation parameters for DOA estimation performance analysis with different SNRs.

Parameter	Symbol	Proposed	Notes
Signal-to-noise ratio (in dB)	SNR	-25, 0, and 25	The values range from low to high

4.1.5. Simulation Setup for DOA Estimation Comparison Based on Different Frequencies with Different Bandwidths without Overlapping

The bandwidth of the wideband signal affects the performance of the DOA estimation algorithm based on the array geometry. Here, we change the bandwidths (different center frequencies and different bandwidths without overlapping) of the incident wideband signals for the same two sources that were used in Sections 4.1.2 and 4.1.3. The outer radius of the pentagram here is equal to 3.095λ and the inner radius is equal to 1.182λ. So, the radius of the UCA geometry is again set to a value that lies between these two values. The parameters that are used for DOA estimation comparison are fixed at the same values as those listed in Table 1, Table 2, and Table 3 for the three geometries, except for the parameters that are revalued as listed in Table 5.

Table 5. Simulation parameters for DOA estimation performance comparison based on different bandwidths without overlapping.

Parameter	Symbol	ULA/UCA/Proposed	Notes
Bandwidth (GHz)	BW	1.5 and 2.3	1.5 bandwidth of the signal 7.1 2.3 bandwidth of the signal 13.4
Element spacing (in wavelength)	d	$d = 0.5\lambda / R = 2.0\lambda / d_X = 2.520\lambda$	Element spacing for ULA/radius of UCA/base element spacing for proposed geometry

4.1.6. Simulation Setup for DOA Estimation Comparison Based on Same Frequencies with Different Bandwidths with Overlapping

In this section, we set the two wideband signals with the same center frequencies and different bandwidths with frequency overlapping. The same two bandwidth values in Section 4.1.5 remain unchanged. This simulation studies the influence of frequency overlapping on the performance of the DOA estimation algorithm based on the array geometry. Now, we change the center frequencies (same center frequencies and different bandwidths with overlapping) of the incident wideband signals for the same two sources that were used in Sections 4.1.1, 4.1.3, and 4.1.5. The outer radius of the pentagram here is equal to 3.451λ and the inner radius is equal to 1.318λ. Then, the radius of the UCA geometry is set to a value that lies between these two values. The parameters that are used for DOA estimation comparison are fixed as the same values as those listed in Tables 1, 3 and 5 for three geometries, except for the parameters that are revalued as listed in Table 6.

Table 6. Simulation parameters for DOA estimation performance comparison based on bandwidths with overlapping.

Parameter	Symbol	ULA/UCA/Proposed	Notes
Center frequency (GHz)	f_0	8.0 and 8.0	Two signals (same frequencies)
Wavelength of the simulation (m)	λ	0.0375	$\lambda = \lambda_1 = \lambda_2 = c/f_0$
Element spacing (in wavelength)	d	$d = 0.5\lambda / R = 2.0\lambda / d_X = 2.508\lambda$	Element spacing for ULA/radius of UCA/base element spacing for proposed geometry

4.1.7. Simulation Setup for DOA Estimation Comparison Based on a Larger Number of Sources with Different Frequencies and Zero Bandwidths without Overlapping

Here, we used four sources which emit four wideband signals with different center frequencies and zero bandwidths (pulse signals scenario). In this case, we assume that the first two different signals are coming from the opposite directions of the incoming directions of the other two signals. Regarding these different frequencies in this scenario, we have four different values of wavelengths (λ), which are λ_1, λ_2, λ_3, and λ_4 for every signal such that $\lambda_1 > \lambda_2 > \lambda_3 > \lambda_4$. For setting the element spacing (d) of the three geometries to some value related to the wavelength (λ), we set the wavelength (λ) for simulation to a value that lies in the range between λ_2 and λ_3 as λ greater than λ_3 and less than λ_2, which is the same value as in Sections 4.1.2, 4.1.3, 4.1.4, 4.1.5, and 4.1.6, respectively. The outer and inner radius of the pentagram and the radius of the UCA geometry remain the same as in Section 4.1.3. The simulation parameters selected here are the same as in Table 1, except for the parameter named element spacing (in wavelength), which takes the same values as in Table 3 for the three geometries. The other revalued parameters are listed in Table 7.

Table 7. Simulation parameters for DOA estimation performance comparison based on greater number of sources with different frequencies and zero bandwidths.

Parameter	Symbol	ULA/UCA/Proposed	Notes
Center frequency (GHz)	f_0	3.6, 7.1, 9.2, and 13.4	Four signals (different frequencies) are considered
Wavelength of the simulation (m)	λ	0.0375	$\lambda_4 < \lambda_3 < \lambda < \lambda_2 < \lambda_1$
Wavelength for signal 1 (m)	λ_1	0.0833	$\lambda_1 = c/f_0$
Wavelength for signal 2 (m)	λ_2	0.0423	$\lambda_2 = c/f_0$
Wavelength for signal 3 (m)	λ_3	0.0326	$\lambda_3 = c/f_0$
Wavelength for signal 4 (m)	λ_4	0.0224	$\lambda_4 = c/f_0$
Source azimuth DOA (°)	ϕ_K	$-54, -18, 18,$ and 54	Four sources are considered

4.1.8. Simulation Setup for DOA Estimation Comparison Based on Greater Number of Sources with Same Frequencies and Different Bandwidths with Overlapping

This section is the same as the previous Section 4.1.7, as we used the same four sources, all coming from their same actual azimuth DOAs. In this case, we only changed their center frequencies to be one for all of them and their bandwidths to be different from each other. The simulation examines the impact of bandwidth and frequency overlapping on the DOA algorithm performance of wideband sources. In this case, we have the same center frequency for all four signals, so we have the same four values of wavelengths (λ), which are $\lambda_1 = \lambda_2 = \lambda_3 = \lambda_4$. For setting the element spacing (d) of the three geometries to some value related to the wavelength (λ), we set the wavelength (λ) for simulation to any value of these values. The outer radius of the pentagram here is equal to 1.512λ and the inner radius is equal to 0.578λ. So, the radius of the UCA geometry is again set to a value that lies between these two values. The simulation parameters that are used here are the same as in Tables 1 and 7 for three geometries, except for some parameters that take the new values in Table 8.

Table 8. Simulation parameters for DOA estimation performance comparison based on a larger number of sources with same frequencies and different bandwidths.

Parameter	Symbol	ULA/UCA/Proposed	Notes
Center frequency (GHz)	f_0	8.0, 8.0, 8.0, and 8.0	Four signals (same frequency)
Wavelength of the simulation (m)	λ	0.0375	$\lambda = \lambda_1 = \lambda_2 = \lambda_3 = \lambda_4 = c/f_0$
Bandwidth (GHz)	BW	1.5, 2.3, 3.5, and 4.3	1.5 bandwidth of the signal 1 2.3 bandwidth of the signal 2 3.5 bandwidth of the signal 3 4.3 bandwidth of the signal 4
Element spacing (in wavelength)	d	$d = 0.5\lambda/R = 1.0\lambda/d_X = 1.099\lambda$	Element spacing for ULA/radius of UCA/base element spacing for proposed geometry

4.1.9. Simulation Setup for Different DOA Algorithm Performances Comparison Based on Proposed Geometry

Here, we selected another two different DOA algorithms to investigate the effect of the array geometry configuration and the antenna aperture on the DOA algorithms' performance. The DOA estimation performance of the two algorithms will be compared with that of the MUSIC algorithm. These two DOA algorithms are CAPON and SSS.

The settings of the simulation parameters used here for the proposed geometry are the same as those related to the proposed geometry in Sections 4.1.6, 4.1.7, and 4.1.8, respectively. Then the three algorithms (MUSIC, CAPON, and SSS) are applied for DOA estimation based on the proposed geometry under these settings. The SSS algorithm has an additional parameter named scalar value, which will take its value as listed in Table 9.

Table 9. Simulation parameters for different DOA algorithms performance comparison.

Parameter	Symbol	MUSIC/CAPON/SSS	Notes
Scalar value (in wavelength)	ε	$-/-/\lambda$	Small scalar value added to avoid possible singularities (only for SSS)

4.2. Simulation Results

Various software simulations were performed under different wideband signals and DOA azimuth scenarios to investigate the theoretical assumptions of the proposed geometry. Each simulation was implemented according to its related settings, which were specified in Section 4.1.

4.2.1. The 1-DOA Estimation of Proposed Geometry vs. ULA and UCA Geometries under the MUSIC Algorithm

The performance of the MUSIC algorithm for the proposed, UCA and ULA geometries was evaluated using the simulation parameters given in Table 1. The DOA estimation performance results are shown in Figure 5. The red circle represents the true DOA from $0°$ azimuth DOA. The blue line, purple line, and green line represent the DOAs estimated by the MUSIC algorithm based on ULA geometry, UCA geometry, and proposed geometry, respectively.

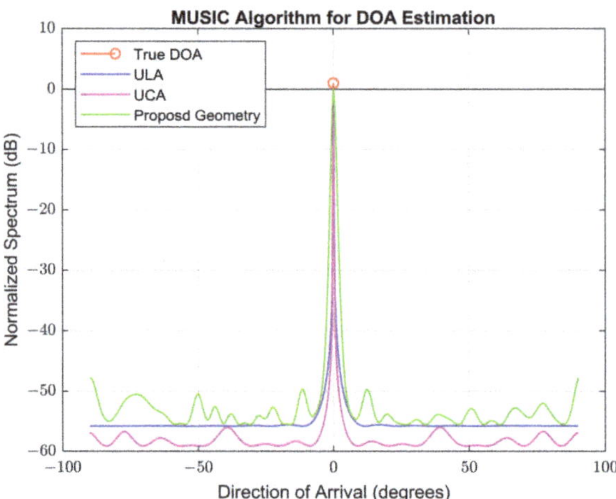

Figure 5. The DOA estimation performance of the proposed geometry vs. ULA and UCA under the MUSIC algorithm.

For one wideband signal source with zero bandwidth (pulse signal) given in Table 1, i.e., 0° DOA azimuth, with elevation kept at 90°, the DOAs estimated by the MUSIC algorithm based on the proposed geometry and the other two geometries (ULA and UCA) were located at the correct DOAs appearing at 0° DOA azimuth, as shown in Figure 5. In other words, the MUSIC algorithm based on the proposed geometry detected the wideband source effectively, the same as it did based on the ULA and UCA geometries. On the other hand, the proposed geometry obeys the DOA MUSIC algorithm and is valid for achieving accurate DOA estimation for wideband sources. The reason for this accurate DOA estimation with the proposed geometry is the perfect setting of the element spacing that is used to form the pentagram array to generate a manifold matrix that has signal and noise subspaces that can be easily separated.

The DOA performance of the MUSIC algorithm for the proposed geometry is much more convincing compared to the ULA and UCA geometries for the simulation parameters given in Table 1, provided that our new proposed geometry with the same number of sensors has a significantly better configuration based on the chosen element spacing, which benefits the conformability of the RIP condition as a separation of signal and noise subspaces. Also, the simulation shows that the MUSIC algorithm achieved a higher and narrower peak-to-floor ratio (PFR) of the normalized spatial power for the UCA geometry than the ULA and proposed geometries, although the ULA has a flatter normalized spatial power at the floor than the other two geometries. In this scenario, the smallest interelement spacing (Y) between the five sensors at the inner vertices (F, G, H, I, and J) of the pentagram array as defined in Section 2.3 equals to 0.080 m (Y = 0.618X = 0.618 × 3.084 × 0.042 = 0.080 m), and the largest one is the spacing between the sensor C and the reference sensor A, which equals to 0.339 m (d_C = 2X + Y = 2 × 3.084 × 0.042 + 0.080 = 0.339 m). This largest spacing determines the size of the aperture of the proposed geometry array, which can be acceptable in logical and practical considerations.

4.2.2. The 1-DOA Angular Accuracy and Resolution Performance of the Proposed Geometry under the MUSIC Algorithm

The 1-DOA estimation angular accuracy and resolution performance of the proposed, ULA and UCA geometries under the MUSIC algorithm were investigated using the simulation parameters given in Tables 1 and 2. The results of the 1-DOA estimation comparison are shown in Figure 6. The red circles again represent the true DOAs from 15° and 21° azimuth DOA, respectively. The blue line, purple line, and green line represent the DOAs estimated by the MUSIC algorithm based on the ULA, UCA, and proposed geometry, respectively.

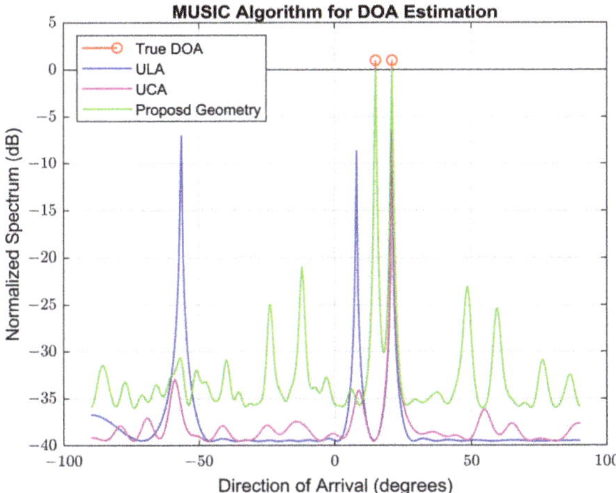

Figure 6. The 1-DOA angular accuracy and resolution performance for the proposed geometry, ULA, and UCA under the MUSIC algorithm.

For the simulation parameters given in Tables 1 and 2, and the two wideband signal sources with zero bandwidths (pulse signals), i.e., 15° and 21° DOA azimuth, with elevation kept at 90°, the DOAs estimated by the MUSIC algorithm based on the proposed geometry are located at the correct DOAs appearing at 15° and 21° DOA azimuth, respectively. At the same time, only one DOA peak is estimated by the MUSIC algorithm based on the ULA and UCA geometries, which is located at the correct DOA of the second source and appears at 21° DOA azimuth, and it failed to estimate the DOA peak of the first source, which should be located at 15° DOA azimuth, as shown in Figure 6. For ULA geometry, another two DOA peaks are estimated by the MUSIC algorithm, approximately appearing at −58° and 8° DOAs azimuth, respectively. This makes us think that even the first DOA peak that is located at the correct DOA azimuth of the second source and appears at 21° DOA azimuth is most likely a coincidence due to these two false DOA peaks. In contrast, for the UCA geometry, only one DOA peak is estimated by the MUSIC algorithm that is located at the correct DOA azimuth of the second source and appears at 21° DOA azimuth. Thus, again, it failed to estimate the DOA peak for the first source that lies at 15° DOA azimuth.

It is worth noting from Figure 6 that the MUSIC algorithm, based on the proposed geometry, estimated the two wideband sources effectively even though they are close to each other. Also, the simulation shows that the MUSIC algorithm estimated the DOAs of the two sources with a high peak-to-floor ratio (PFR) of the normalized spatial power with notable decreasing double-side peaks using the proposed geometry.

In this scenario, the smallest interelement spacing (Y) equals 0.058 m (Y = 0.618X = 0.618 × 2.520 × 0.0375 = 0.058 m) and the largest spacing equals 0.247 m (d_C = 2X + Y = 2 × 2.520 × 0.0375 + 0.058 = 0.247 m), which are still reasonable for design considerations. This simulation scenario shows that the proposed geometry has better angular accuracy and resolution performance under the MUSIC algorithm for wideband sources compared to the ULA and UCA geometries.

4.2.3. The 1-DOA Estimation Performance Comparison of the Proposed Geometry with UCA and ULA Geometries

This simulation scenario is the same as that in Section 4.2.2, as we only changed the actual azimuth DOA of the two wideband sources to be separated by some degrees from each other while leaving their signal characteristics unchanged to see how the proposed geometry deals with the variation in the actual DOA to keep its DOA estimation accuracy. This DOA estimation performance comparison is performed using the simulation parameters given in Tables 1–3. The simulation results of the DOA estimation performance comparison are shown in Figure 7. Again, the red circles represent the true DOAs from 18° and 54° azimuth DOA, respectively. The blue line, purple line, and green line represent the DOAs estimated by the MUSIC algorithm based on the ULA, UCA, and proposed geometry, respectively. For the simulation parameters given in Tables 1–3, and the two wideband signal sources, i.e., 18° and 54° DOA azimuth, with elevation kept at 90°, the DOAs estimated by the MUSIC algorithm based on the proposed geometry are located at the correct DOAs appearing at 18° and 54° DOA azimuth, respectively. As in the previous Section 4.2.2, only one DOA is estimated by the MUSIC algorithm based on the ULA and UCA geometries, which is located at the correct DOA of the second source and appears at 54° DOA azimuth, and it failed to estimate the first source, which should be located at 18° DOA azimuth, as shown in Figure 7. The MUSIC algorithm based on ULA geometry estimated another two DOA peaks, which approximately appear at −25° and 9° DOAs, respectively. This again supports the probability of the first DOA peak being coincidentally located at the correct DOA azimuth of the second source and appearing at 54° DOA azimuth. In comparison, the MUSIC algorithm based on UCA geometry estimated only one DOA peak that was located at the correct DOA azimuth of the second source, appearing at 54° DOA azimuth, and failed to estimate the DOA peak for the first source, which lies at 18° DOA azimuth.

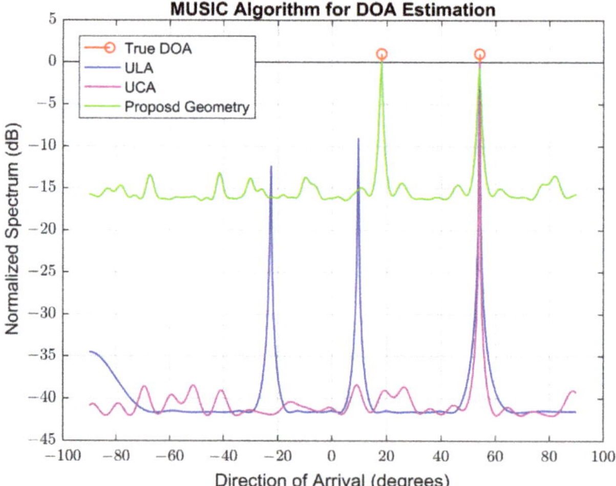

Figure 7. The DOA estimation performance comparison of the proposed geometry with UCA and ULA geometries under the MUSIC algorithm.

According to the simulation results in Figure 7, the MUSIC algorithm, based on the proposed geometry, detected and estimated the two wideband sources effectively even though they are separated by some degrees from each other. Also, the simulation shows that the MUSIC algorithm estimated the DOAs of the two sources with a low peak-to-floor ratio (PFR) of the normalized spatial power using the proposed geometry compared to the UCA and ULA geometries.

In this simulation, we found that the proposed geometry provided high DOA accuracy with high resolution by insignificantly maximizing its aperture by increasing the element spacing to a larger value than those in Section 4.2.2 to adapt the variation in the DOA of the sources. The element spacing between the five elements (A, B, C, D, and E) at the outer vertices of the pentagram array is increased to 2.523λ instead of 2.520λ in Section 4.2.2. The smallest interelement spacing (Y) and the largest spacing can be the same as their values in

Section 4.2.2 due to these little increases in design considerations and measurement errors. In this scenario, we show that the proposed geometry can adapt to the variation in the DOA of the wideband sources to provide high-accuracy DOA estimation with high resolution performance by resetting its element spacing as its number of elements remains fixed.

4.2.4. The 1-DOA Estimation Performance Analysis of the Proposed Geometry under the MUSIC Algorithm

The 1-DOA estimation performance analysis of the proposed geometry under the MUSIC algorithm was investigated using the same simulation parameters given in Tables 1–3, except those parameters which take their values as specified in Table 4. The results of the DOA estimation performance analysis are shown in Figure 8. The red circles again represent the true DOAs from 18° and 54° azimuth DOA. The green line, purple line, and blue line represent the DOAs estimated by the MUSIC algorithm based on the proposed geometry with SNR equal to −25 dB, 0 dB, and 25 dB, respectively.

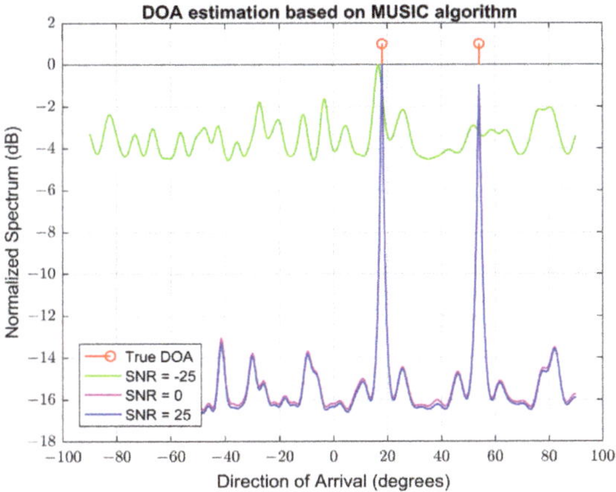

Figure 8. The proposed geometry DOA estimation performance analysis under the MUSIC algorithm.

For the simulation parameters given in Tables 1–4 and two wideband signal sources with zero bandwidths (pulse signals), i.e., 18° and 54° DOA azimuth, with elevation kept at 90°, the DOAs estimated by the MUSIC algorithm based on the proposed geometry are located at the correct DOAs appearing at 18° and 54° DOA azimuth, respectively, for three different SNR values, as shown in Figure 8. It is worth noting from Figure 8 that the MUSIC algorithm, based on the proposed geometry, estimated the two wideband sources effectively even at low SNR levels. Also, the simulation shows that the MUSIC algorithm estimated DOAs with a high peak-to-floor ratio (PFR) of the normalized spatial power at high SNR levels using the proposed geometry. This means that the increase in SNR improved the DOA estimation performance. The explanation is that for the proposed geometry, there is no ability for two or more sensors to be located very near each other at one or many locations. For this reason, the rows of the array manifold matrix would be differentiated, generating a sufficient rank covariance matrix. Again, these results stated that our new proposed geometry with a perfect sensor configuration has significantly accurate DOA estimation performance regardless of the variation in SNR. In this simulation, the smallest interelement spacing (Y) and the largest spacing (d_C) have the same values as in Section 4.2.3.

4.2.5. The 1-DOA Estimation Performance Comparison of the Proposed Geometry with UCA and ULA Geometries Based on Different Frequencies with Different Bandwidths without Overlapping

For simplicity purposes, in the wideband DOA estimation comparison in previous Sections 4.2.2–4.2.4, two wideband signals with zero bandwidth (pulse signals) were considered. Here, we assumed that the same two wideband signals are not pulse signals, and they have different bandwidths without overlapping between them. Thus, the two different frequencies of the two signals remain unchanged.

This scenario studies the effect of the bandwidth of the incident wideband signals on the array geometry configuration, the antenna aperture, and the DOA algorithm's performance, as the signal frequency components are spreading along these bandwidths. The DOA estimation performance comparison is provided using the simulation parameters given in Tables 1–3 and 5. The simulation results of the DOA estimation performance comparison in this case are shown in Figure 9.

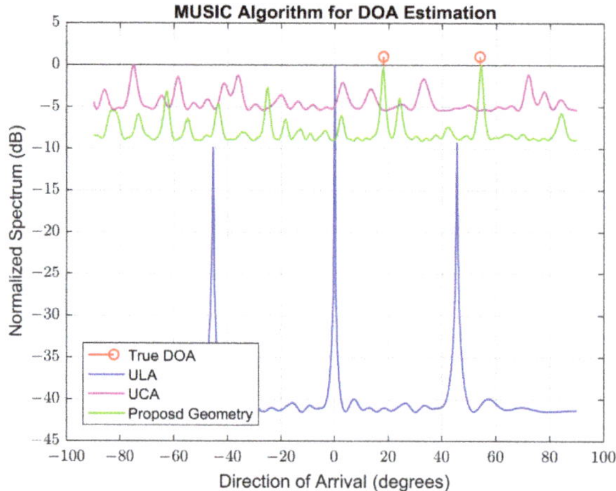

Figure 9. The DOA estimation performance comparison based on the proposed geometry, UCA, and ULA geometries under the MUSIC algorithm using different frequencies with different bandwidths without overlapping.

As usual, the red circles represent the true DOAs from 18° and 54° azimuth DOA, respectively. The blue line, purple line, and green line represent the DOAs estimated by the MUSIC algorithm based on the ULA, UCA, and proposed geometry, respectively.

For the simulation parameters given in Tables 1–3 and 5 and the two wideband signal sources with different bandwidths without overlapping, i.e., 18° and 54° DOA azimuth, with elevation kept at 90°, the DOAs estimated by the MUSIC algorithm based on the proposed geometry are located at the correct DOAs appearing at 18° and 54° DOA azimuth, respectively, as shown in Figure 9. This means that the MUSIC algorithm, based on the proposed geometry, estimated the DOAs of the two wideband sources accurately. From Figure 9, it is clear that the MUSIC algorithm based on UCA geometry failed to estimate the correct DOA of the two wideband sources by providing messy results for DOA estimation, as many low and wide peaks appear in the normalized power spectrum. Based on ULA geometry, the MUSIC algorithm also failed to estimate the correct DOA of the two wideband sources and estimated three DOA peaks appearing at −48°, 0°, and 48° DOAs, respectively. As we can see in this case and due to the bandwidths of the two signals, the MUSIC algorithm based on the ULA and UCA geometries failed to estimate the DOAs of the two or even one wideband source.

In this case, the proposed geometry provided good DOA results by minimizing its aperture by decreasing the element spacing to a smaller value than those values in Sections 4.2.2 and 4.2.3 to adapt the variation in the bandwidths of the wideband signals. The element spacing between the five elements (A, B, C, D, and E) at the outer vertices of the pentagram array is decreased to 2.249λ instead of 2.520λ in Section 4.2.2 and 2.523λ in Section 4.2.3. The smallest interelement spacing (Y) in this case is equal to 0.052 m ($Y = 0.618X = 0.618 \times 2.249 \times 0.0375 = 0.052$ m) and the largest spacing equals 0.221 m

(d_C = 2X + Y = 2 × 2.249 × 0.0375 + 0.052 = 0.221 m), which are acceptable in practical implementations. Again, in this scenario, we show that the proposed geometry can be used for the DOA of the wideband sources to provide good results of DOA estimation by adjusting its element spacing in case of changing the bandwidths of the incoming signals.

This simulation result in Figure 9 demonstrates the outgeneralization of the proposed geometry over its counterparts (ULA and UCA), as it can be used for DOA estimation of wideband sources even at different signal bandwidths.

4.2.6. The 1-DOA Estimation Performance Comparison of the Proposed Geometry with UCA and ULA Geometries Based on the Same Frequencies with Different Bandwidths with Overlapping

In the previous Section 4.2.5, we assumed that the two wideband signals have different frequencies and different bandwidths without frequency overlapping. In this part, we considered the frequency overlapping between these signals and assumed that they had the same frequencies and different bandwidths. This scenario discusses the influence of the frequency overlapping of the wideband signals due to their bandwidth on the performance of the DOA estimation as the signals share some frequency components that are already present in these bandwidths. In the literature, frequency overlapping represents the main reason that most of the DOA algorithms are not applicable to DOA estimation of wideband sources, especially since they use subspace methods, as the separation of these subspaces of the covariance matrix becomes difficult. The DOA estimation performance comparison is provided using the simulation parameters given in Tables 1, 3, 5, and 6. The simulation results of the DOA estimation performance comparison for this scenario are shown in Figure 10. In Figure 10, the red circles represent the true DOAs from 18° and 54° azimuth DOA, respectively. The blue line, purple line, and green line represent the DOAs estimated by the MUSIC algorithm based on the ULA, UCA, and proposed geometry, respectively. For the simulation parameters given in Tables 1, 3, 5, and 6, and the two wideband signal sources with the same frequencies and different bandwidths with overlapping, i.e., 18° and 54° DOA azimuth, with elevation kept at 90°, the DOAs estimated by the MUSIC algorithm based on the proposed geometry are located at the correct DOAs appearing at 18° and 54° DOA azimuth, respectively, as shown in Figure 10. The result is that the MUSIC algorithm, based on the proposed geometry, estimated the DOAs of the two wideband sources effectively. From Figure 10, it can be seen that the MUSIC algorithm based on UCA geometry failed to estimate the correct DOAs of the two wideband sources, providing only one DOA peak at the left end of the normalized spectrum. Also, it is clear that in the same figure, the MUSIC algorithm based on ULA geometry failed to estimate the correct DOAs of the two wideband sources and estimated three DOA peaks appearing at −45°, 0°, and 65° DOAs, respectively. It is clear from Figure 10 that, because of the frequency overlapping of the two wideband signals, the MUSIC algorithm based on the ULA and UCA geometries failed to estimate the DOAs of the two wideband sources. In this case, the element spacing between the five elements (A, B, C, D, and E) at the outer vertices of the pentagram array is increased to 2.508λ instead of 2.249λ in Section 4.2.5, 2.520λ in Section 4.2.2, and 2.523λ in Section 4.2.3. The smallest interelement spacing (Y) in this case is equal to 0.058 m (Y = 0.618X = 0.618 × 2.508 × 0.0375 = 0.058 m) and the largest spacing is equal to 0.246 m (d_C = 2X + Y = 2 × 2.508 × 0.0375 + 0.058 = 0.246 m), which are also convincing for practical considerations.

Again, this scenario shows that the proposed geometry proved its validity for a DOA of the wideband sources, providing good results of DOA estimation depending on changing its element spacing to generate a covariance matrix from which its signal and noise subspaces can be easily separated in the case of frequency overlapping of the incoming wideband signals. The simulation results in Figure 10 show the outperformance of the proposed geometry over its counterparts (ULA and UCA). The simulation shows that the MUSIC algorithm, based on the proposed geometry, estimated the DOAs of the two sources with one redundant peak with less spectrum amplitude, which can be minimized using other DOA estimation algorithms, as will be demonstrated later in Section 4.2.9.

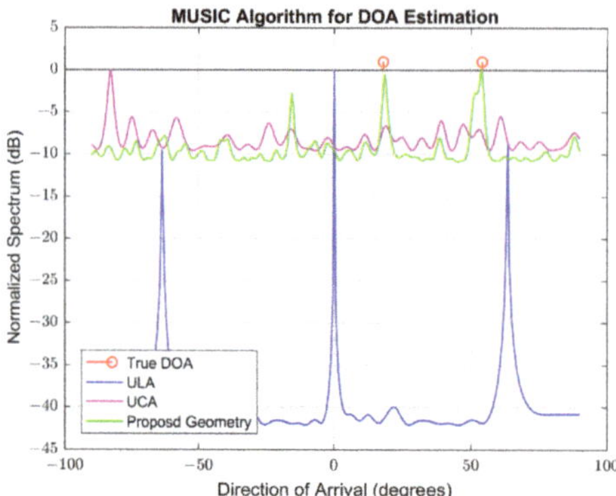

Figure 10. The DOA estimation performance comparison based on the proposed geometry, UCA, and ULA geometries under the MUSIC algorithm using the same frequencies with different bandwidths with overlapping.

4.2.7. The 1-DOA Estimation Performance Comparison of the Proposed Geometry with UCA and ULA Geometries Based on a Larger Number of Wideband Sources

In the previous Sections 4.2.1–4.2.6, we used one and two wideband sources with different frequencies and bandwidth scenarios. In the literature, using a larger number of array elements provides high DOA estimation accuracy. According to the proposed geometry, we only have ten array elements. Also, the MUSIC algorithm can only estimate a smaller number of sources than the number of array elements. Due to these notes, we used four wideband sources with different frequencies and zero bandwidths (pulse signals) in this simulation.

The DOA estimation performance comparison for four wideband sources is presented using the simulation parameters given in Tables 1, 3, and 7. The DOA estimation performance comparison simulation results of this scenario are shown in Figure 11. Constantly, the red circles represent the true DOAs from $-54°$, $-18°$, $18°$, and $54°$ azimuth DOA, respectively. The blue line, purple line, and green line represent the DOAs estimated by the MUSIC algorithm based on the ULA, UCA, and proposed geometry, respectively. Using the simulation parameters given in Tables 1, 3, and 7, and the four wideband signal sources with different frequencies and zero bandwidths (pulse signals), i.e., $-54°$, $-18°$, $18°$, and $54°$ DOA azimuth, with elevation kept at $90°$, the DOAs estimated by the MUSIC algorithm based on the proposed geometry are located at the correct DOAs appearing at $-54°$, $-18°$, $18°$, and $54°$ DOA azimuth, respectively, as shown in Figure 11. This indicates that the MUSIC algorithm, based on the proposed geometry, estimated the DOAs of the four wideband sources effectively.

From the same Figure 11, we see that the MUSIC algorithm based on UCA geometry estimated only one DOA peak for the four wideband sources at the correct DOA azimuth of the fourth source, appearing at $54°$ DOA azimuth, and failed to estimate the correct DOAs of the other three wideband sources. Also, based on ULA geometry, the MUSIC algorithm failed to estimate the correct DOAs of the four wideband sources by estimating seven DOA peaks, and only one of these peaks is estimated at the correct DOA azimuth of the fourth source, appearing at $54°$ DOA azimuth. The other six DOA peaks for the three wideband sources were estimated at random DOA azimuth, approximately appearing at $-75°$, $-20°$, $-15°$, $-11°$, $15°$, and $75°$ DOA azimuth, respectively.

Referring to Figure 11, we can note that the simulation results explain the power of the MUSIC algorithm based on the proposed geometry to estimate the DOAs of four wideband signals, with each of them coming from opposite directions, whereas this is unobtainable when using ULA and UCA geometries. In this part, the element spacing, the smallest interelement, and the largest spacing have the same values as in Section 4.2.3. This simulation shows the strength of the proposed geometry.

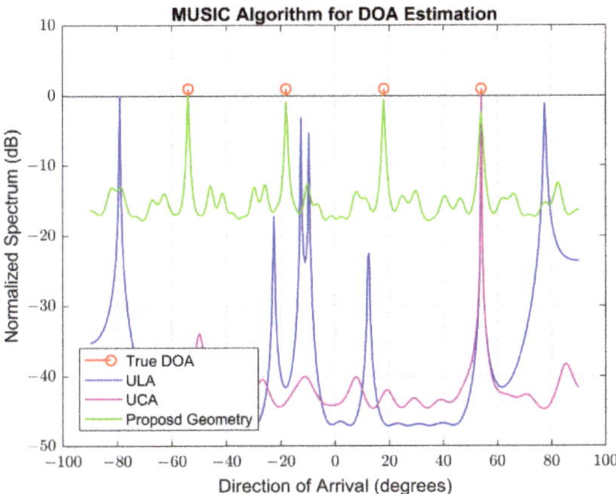

Figure 11. The DOA estimation performance comparison based on the proposed geometry, UCA, and ULA geometries under the MUSIC algorithm using four wideband sources without frequency overlapping.

4.2.8. The 1-DOA Estimation Performance Comparison of the Proposed Geometry with UCA and ULA Geometries Based on a Larger Number of Wideband Sources with Frequency Overlapping

In this section, we use the same four wideband sources with their actual azimuth DOAs that were used in Section 4.2.7. Here, we only changed their corresponding wideband signals, such that we used frequency overlapping between them with the same center frequency and different bandwidths. The DOA estimation performance comparison is examined using the simulation parameters given in Tables 1, 7, and 8. The DOA estimation performance comparison simulation results of this part are shown in Figure 12.

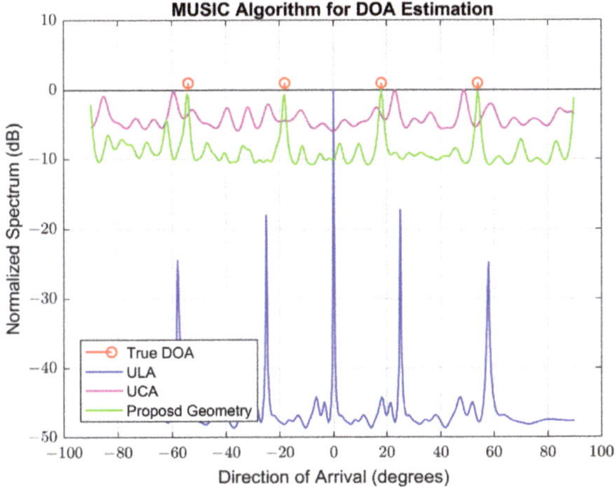

Figure 12. The DOA estimation performance comparison based on the proposed geometry, UCA, and ULA geometries under the MUSIC algorithm using four wideband sources with frequency overlapping.

Indelibly, the red circles represent the true DOAs from $-54°$, $-18°$, $18°$, and $54°$ azimuth DOA, respectively. The blue line, purple line, and green line represent the DOAs estimated by the MUSIC algorithm based on the ULA, UCA, and proposed geometry, respectively. According to the simulation parameters given in Tables 1, 7, and 8, and the four frequency overlapping wideband sources with the same frequencies and different bandwidths, i.e., $-54°$, $-18°$, $18°$, and $54°$ DOA azimuth, with elevation kept at $90°$, the MUSIC algorithm based on the proposed geometry estimated the DOAs that are located at the correct DOAs of the four wideband sources appearing at $-54°$, $-18°$, $18°$, and $54°$ DOA azimuth, respectively, as shown in Figure 12. This implies that the MUSIC algorithm, based on the proposed geometry, estimated the DOAs of the four wideband sources.

From the same Figure 12, we see that the MUSIC algorithm based on UCA geometry failed to estimate the correct DOAs of the four wideband sources by estimating four false wide and short DOA azimuth peaks approximately appearing at $-85°$, $-60°$, $23°$, and $50°$ DOA azimuth, respectively. Likewise, the MUSIC algorithm based on ULA geometry failed to estimate the correct DOAs of the four wideband sources by estimating false five DOA azimuth peaks approximately appearing at $-56°$, $-24°$, $0°$, $25°$, and $60°$ DOA azimuth, respectively.

As we can see from Figure 12, the MUSIC algorithm, based on the proposed geometry, fully estimated the correct DOAs of the four wideband sources under frequency overlapping and contrary direction conditions of the wideband signals using a suitable sensor configuration. In this case, the element spacing between the five elements (A, B, C, D, and E) at the outer vertices of the pentagram array is decreased to 1.099λ instead of 2.508λ in Section 4.2.6, 2.249λ in Section 4.2.5, 2.520λ in Section 4.2.2, and 2.523λ in Sections 4.2.3 and 4.2.7. The smallest interelement spacing (Y) in this case equals 0.025 m ($Y = 0.618X = 0.618 \times 1.099 \times 0.0375 = 0.025$ m) and the largest spacing equals 0.107 m ($d_C = 2X + Y = 2 \times 1.099 \times 0.0375 + 0.025 = 0.107$ m), which can be realizable in design considerations.

The simulation shows that the MUSIC algorithm, based on the proposed geometry, estimated the DOAs of the four wideband sources with two redundant peaks with less amplitude at the two ends of the normalized spectrum, which can be minimized using other DOA estimation algorithms as discussed in Section 4.2.9.

This simulation also reflects the validity and outstanding nature of the proposed geometry of a larger number of wideband sources with frequency-overlapping signals.

4.2.9. The 1-DOA Estimation Performance Comparison of Different DOA Algorithms Based on the Proposed Geometry

In this section, the performance of the proposed geometry under different DOA algorithms is introduced. The performance is analyzed by applying the two DOA algorithms (CAPON and SSS) to Sections 4.2.6, 4.2.7, and 4.2.8, respectively. Then their performance is compared with that of the MUSIC algorithm in these same sections.

The simulation parameters are the same as those used in these three sections for the three DOA algorithms. The SSS algorithm has another parameter named scalar value, whose value is listed in Table 9.

The results of the 1-DOA estimation performance comparison of the CAPON and SSS algorithms with the MUSIC algorithm as in Section 4.2.6 are shown in Figure 13a–c. The red circles represent the true DOAs from $18°$ and $54°$ azimuth DOA, with elevation kept at $90°$. The black line, green line, and blue line represent the DOAs estimated by the MUSIC, CAPON, and SSS algorithms, respectively, based on the proposed geometry.

In this DOA performance comparison, for the two wideband sources with frequency overlapping and other setting considerations in Section 4.2.6, i.e., $18°$ and $54°$ DOA azimuth, with elevation kept at $90°$, the DOAs estimated by all three different DOA algorithms based on the proposed geometry are located at the correct DOAs appearing at $18°$ and $54°$ DOA azimuth, respectively, as shown in Figure 13a–c.

Also, the results of the 1-DOA estimation performance comparison of the CAPON and SSS algorithms with the MUSIC algorithm as in Section 4.2.7 are shown in Figure 14a–c. Usually, the red circles represent the true DOAs from $-54°$, $-18°$, $18°$, and $54°$ azimuth DOA, with elevation kept at $90°$. Again, the black line, green line, and blue line represent the DOAs estimated by the MUSIC, CAPON, and SSS algorithms, respectively, based on the proposed geometry.

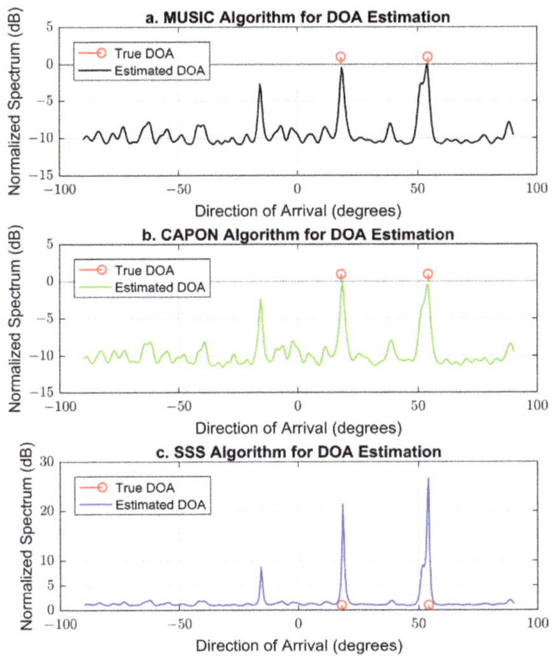

Figure 13. (**a–c**) The performance comparison of different DOA algorithms based on the proposed geometry using two wideband sources with different bandwidths and frequency overlapping.

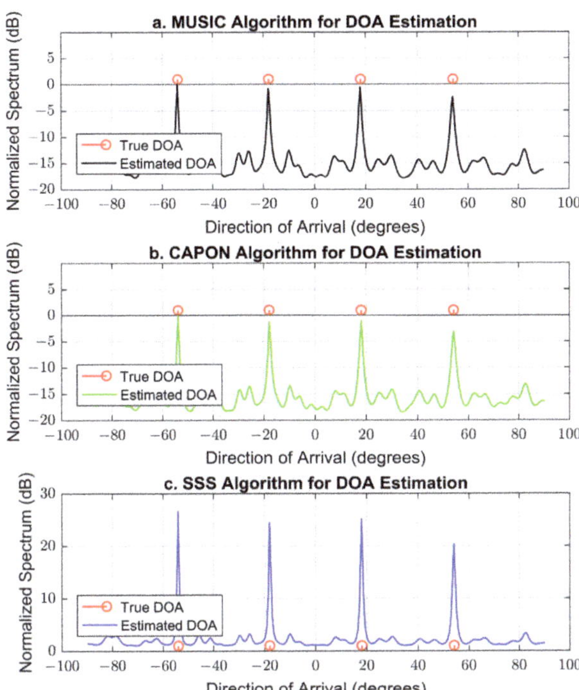

Figure 14. (**a–c**) The performance comparison of different DOA algorithms based on the proposed geometry using four wideband sources with different frequencies without overlapping.

In this DOA performance comparison, for the four wideband sources with different frequencies without overlapping and other setting considerations applied in Section 4.2.7, i.e., $-54°$, $-18°$, $18°$, and $54°$ DOA azimuth, with elevation kept at $90°$, the DOAs estimated by all three different DOA algorithms based on the proposed geometry are located at the correct DOAs appearing at $-54°$, $-18°$, $18°$, and $54°$ DOA azimuth, respectively, as shown in Figure 14a–c.

Finally, the outcomes of the 1-DOA estimation performance comparison of the CAPON and SSS algorithms with the MUSIC algorithm as in Section 4.2.8 are shown in Figure 15a–c. Generally, the red circles represent the true DOAs from $-54°$, $-18°$, $18°$, and $54°$ azimuth DOA, with elevation kept at $90°$. Commonly, the black line, green line, and blue line represent the DOAs estimated by the MUSIC, CAPON, and SSS algorithms, respectively, based on the proposed geometry.

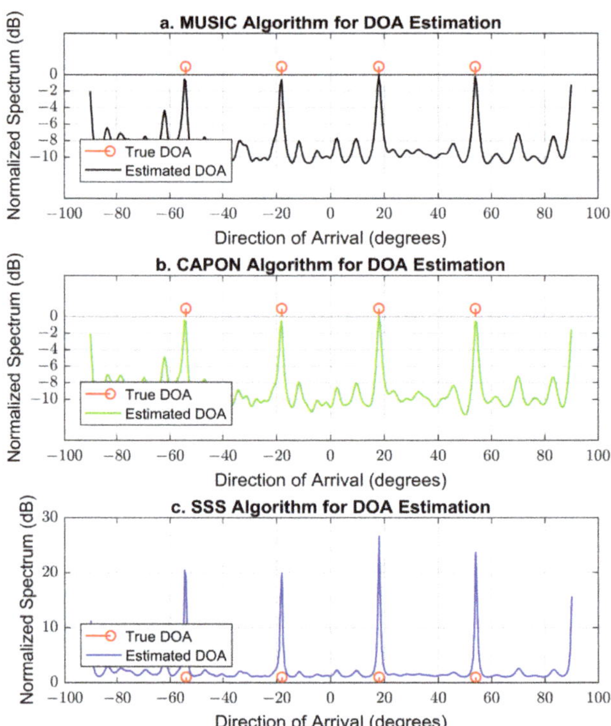

Figure 15. (**a–c**) The performance comparison of different DOA algorithms based on the proposed geometry using four wideband sources with same frequency with overlapping.

In this DOA performance comparison scenario, for the four wideband sources with the same frequencies with overlapping and other setting considerations used in Section 4.2.8, i.e., $-54°$, $-18°$, $18°$, and $54°$ DOA azimuth, with elevation kept at $90°$, the DOAs estimated by all three different DOA algorithms based on the proposed geometry are located at the correct DOAs appearing at $-54°$, $-18°$, $18°$, and $54°$ DOA azimuth, respectively, as shown in Figure 15a–c.

As we can see from Figures 13a–c–15a–c, all three DOA algorithms fairly estimated the correct DOAs under different sensor configurations of the proposed geometry. Thus, we state that these sensor configurations, which were created by different perfect element spacing for the proposed geometry, generated manifold matrices in which their signal and noise subspaces have easy separation.

As we mentioned before, the motivation behind choosing appropriate element spacing for the proposed geometry is to achieve an array configuration that works well for DOA estimation algorithms for wideband sources. These simulation results show that the proposed geometry generates a developed covariance matrix, which allows the aforementioned DOA estimation techniques to be applied directly to wideband signals.

Also, it is clear from Figures 13c and 15c that the DOA estimation performance of the proposed geometry under the SSS algorithm is better than that of the other two algorithms by minimizing the redundant peaks in cases of increasing the bandwidth or frequency overlapping of the incident wideband signals. This is because SSS uses signal subspace to construct its spatial spectrum instead of noise subspace and minimizes the side-lobe levels by subtracting its normal pseudo-spectrum from unity and using a small scalar value to avoid possible singularities.

The proposed geometry exploits its differentiated property for sensor configurations to formulate an array manifold matrix that has sufficient rank covariance to obey the RIP condition and make the separation of signal and noise subspaces easy even for different wideband signal scenarios. In other words, the proposed geometry solved the problems of manifold matrix ambiguity and avoiding grating lobes when the element spacing exceeds 0.5λ using gathering or superposition techniques on triangular geometries.

It is worth noting from these simulation results that the proposed geometry is valid for achieving accurate results of DOA estimation using different DOA algorithms and wideband signals, avoiding additional or preprocessing requirements for taking the whole wideband frequency information.

All previous investigated simulation scenarios showed that our proposed pentagram geometry has perfect performance with a small number of sensors and various sensor configurations.

5. Discussion

After we carried out and completed the study of the investigated simulations in Section 4.2, and for practical considerations, we had many findings that are summarized in the below discussion:

From Section 4.2.1, the proposed geometry used a large interelement spacing ($d_X = 3.084\lambda$) to form element configuration and maximized antenna aperture for dealing with a single wideband pulse signal coming from a single direction.

Referring to Sections 4.2.2, 4.2.3, and 4.2.7, when the number of wideband sources increased without changing their frequencies, the proposed geometry decreased its interelement spacing ($d_X = 2.520\lambda$ and $d_X = 2.523\lambda$) to form element configuration and minimized antenna aperture for dealing with a large number of wideband sources regardless of their incoming directions or separations.

According to Sections 4.2.5 and 4.2.6, when the bandwidths of wideband signals increased regardless of their overlapping frequencies, the proposed geometry decreased its interelement spacing ($d_X = 2.249\lambda$ and $d_X = 2.503\lambda$) to form element configuration and minimized antenna aperture for dealing with the increasing bandwidths of the wideband signals.

From Sections 4.2.7 and 4.2.8, when the bandwidths of wideband signals increased with overlapping frequencies, the proposed geometry decreased its interelement spacing ($d_X = 1.099\lambda$) to form element configuration and minimized antenna aperture for dealing with the increasing bandwidths of the wideband signals and overlapping frequencies existing in these bandwidths.

According to these results, we make some comments and guidelines on the DOA performance of the proposed geometry as follow:

- The proposed geometry used a fixed number of elements and variable element spacing to form various element configurations.

- These element configurations are selected and determined in accordance with the characteristics of the incident wideband signals and sources (from previous results, when bandwidth increases, element spacing decreases).
- Every element configuration is related to a specific antenna aperture and generates a particular manifold matrix.
- This particular manifold matrix has independent columns and satisfies the RIP condition.
- The covariance matrix for the incident wideband sources is obtained using this manifold matrix.
- Finally, the subspaces of this covariance matrix or its inversion are used to obtain unambiguous DOAs of the incident wideband sources by directly applying different DOA algorithms.

6. Conclusions and Future Works

In this study, a novel pentagram antenna array geometry was proposed based on triangular geometry as its main contribution. The geometry was constructed using the gathering and superposition of three isosceles triangles. Its geometry and mathematical model were analyzed. It has a fixed number of elements (10 sensors) with variable element spacing and could be classified as a FNEVES array geometry. The aperture was designed and can be minimized or maximized according to the setting of the element spacing, which offers a creative approach to array design using a fixed number of sensors to build an array with many various apertures. The main motivation of this geometry is the setting of element spacing to obtain an array manifold matrix that satisfies the RIP condition to avoid the manifold matrix ambiguity problem. This motivation offers benefits related to the degrees of freedom for creating a perfect array covariance matrix, which allows DOA estimation algorithms to be applied directly to wideband signals without needing additional signal preprocessing or separation techniques. The array geometry can be applied to both narrowband and wideband sources, as well as to both elevation and azimuth DOA estimation.

In this paper, we only investigated the performance of the proposed geometry for the DOA of wideband signals to estimate the azimuth angle (ϕ_K) using the MUSIC algorithm. A large number of simulation experiments and analyses were conducted under different wideband signal scenarios based on frequencies and bandwidths. Its performance showed better results when compared with those of the UCA and ULA geometries. For further verification, its performance was also examined using the SSS and CAPON algorithms, which provided considerable results. In addition, the SSS method had better results compared to the MUSIC and CAPON methods.

The simulation presented DOA results with good accuracy and showed that the novel geometry is wellsuited for wideband sources and DOA algorithms (the classic scenario). As a consequence, the geometry effectively solved the DOA manifold ambiguity problem for wideband sources by avoiding the grating lobes using gathering and superposition techniques for triangular geometries. It is important to mention here that the duality between the time-domain and frequency-domain was exploited by using the array output signal in the frequency-domain X(f) instead of its representation in the time-domain X(t) in the computation of the array covariance matrix and eigenvalue decomposition.

From a cost perspective, our geometry exploits the properties of a fixed (few) number of elements and variable element spacing to outperform its counterparts (UCA and ULA).

Our future work will be extended to study the performance of the array geometry in the following cases: DOA estimation of both azimuth angle (ϕ_K) and elevation angle (θ_K), DOA estimation of narrowband signals (sources), DOA estimation of wideband coherent signals (sources), DOA estimation of (narrowband and wideband) compressive sensing scenarios, real demonstration and experimental setup, and application of the other DOA estimation state-of-the-art algorithms.

Author Contributions: Conceptualization, M.K. and B.T.; methodology, M.K.; software, K.T., H.F. and Y.X.; validation, M.K., B.T. and K.J.; formal analysis, M.K., K.T. and H.F.; investigation, M.K., K.J., K.T., H.F., Y.X. and B.T.; resources, B.T.; data curation, M.K.; writing—original draft preparation, M.K.; writing—review and editing, M.K. and K.J.; visualization, M.K. and Y.X.; supervision, B.T.; project administration, M.K., B.T. and K.J.; funding acquisition, B.T. and K.J. All authors have read and agreed to the published version of the manuscript.

Funding: This research was funded in part by the Key Project of the National Defense Science and Technology Foundation Strengthening Plan 173 of funder grant number 2022-JCJQ-ZD-010-12 and in part by the National Natural Science Foundation of China under grant 62301119.

Data Availability Statement: Data are contained within the article.

Acknowledgments: The authors would like to thank all editors and reviewers for their valuable comments and suggestions.

Conflicts of Interest: The authors declare no conflicts of interest.

References

1. Chandran, S. *Advances in Direction-of-Arrival Estimation*; Artech House: Norwood, MA, USA, 2006.
2. Fallahi, R.; Roshandel, M. Effect of mutual coupling and configuration of concentric circular array antenna on the signal-to-interference performance in CDMA systems. *Prog. Electromagn. Res.* **2007**, *76*, 427–447. [CrossRef]
3. Krim, H.; Viberg, M. Two decades of array signal processing research: The parametric approach. *IEEE Signal Process. Mag.* **1996**, *13*, 67–94. [CrossRef]
4. Stoica, P.; Moses, R.L. *Introduction to Spectral Analysis*; Prentice Hall: Upper Saddle River, NJ, USA, 1997.
5. Wax, M.; Shan, T.J.; Kailath, T. Spatio-temporal spectral analysis by eigenstructure methods. *IEEE Trans. Acoust. Speech Signal Process.* **1984**, *32*, 817–827. [CrossRef]
6. Su, G.; Morf, M. The signal subspace approach for multiple wide-band emitter location. *IEEE Trans. Acoust. Speech Signal Process.* **1983**, *31*, 1502–1522.
7. Wang, H.; Kaveh, M. Coherent signal-subspace processing for the detection and estimation of angles of arrival of multiple wide-band sources. *IEEE Trans. Acoust. Speech Signal Process.* **1985**, *33*, 823–831. [CrossRef]
8. Naidu, P.S. *Sensor Array Signal Processing*; CRC Press: New York, NY, USA, 2001.
9. Schmidt, R.O. Multiple emitter location and signal parameter estimation. *IEEE Trans. Antennas Propag.* **1986**, *34*, 276–280. [CrossRef]
10. Al-Sadoon, M.A.G.; Abduljabbar, N.A.; Ali, N.T.; Asif, R.; Zweid, A.; Alhassan, H.; Noras, J.M.; Abd-Alhameed, R.A. A More Efficient AOA Method for 2D and 3D Direction Estimation with Arbitrary Antenna Array Geometry. In *BROADNETS 2018, LNICST*; Sucasas, V., Mantas, G., Althunibat, S., Eds.; Springer: Berlin/Heidelberg, Germany, 2019; Volume 263, pp. 419–430.
11. Capon, J. High-resolution frequency-wavenumber spectrum analysis. *Proc. IEEE* **1969**, *57*, 1408–1418. [CrossRef]
12. Trees, H.L.V. *Optimum Array Processing, Part IV of Detection, Estimation and Modulation Theory*; Wiley Interscience: New York, NY, USA, 2002.
13. Wang, T.; Yang, L.S.; Lei, J.M.; Yang, S.Z. A modified MUSIC to estimate DOA of the coherent narrowband sources based on UCA. In Proceedings of the International Conference on Communication Technology (ICCT'06), Guilin, China, 27–30 November 2006.
14. Mahmoud, K.R.; El-Adawy, M.; Ibrahem, S.M.M.; Bansal, R.; Zainud-Deen, S.H. A comparison between circular and hexagonal array geometries for smart antenna systems using particle swarm optimization algorithm. *Prog. Electromagn. Res.* **2007**, *72*, 75–90. [CrossRef]
15. Gozasht, F.; Dadashzadeh, G.R.; Nikmehr, S. A comprehensive performance study of circular and hexagonal array geometries in the lms algorithm for smart antenna applications. *Prog. Electromagn. Res.* **2007**, *68*, 281–296. [CrossRef]
16. Cand'es, E.; Romberg, J. Sparsity and incoherence in compressive sampling. *Inverse Prob.* **2007**, *23*, 969. [CrossRef]
17. Ender, J.H.G. On compressive sensing applied to radar. *Signal Process.* **2010**, *90*, 1402–1414. [CrossRef]
18. Ma, W.K.; Hsieh, T.H.; Chi, C.Y. DOA Estimation of quasi-stationary signals with less sensors than sources and unknown spatial noise covariance a khatri-rao subspace approach. *IEEE Trans. Signal Process.* **2010**, *58*, 2168–2180. [CrossRef]
19. Pal, P.; Vaidyanathan, P.P. Nested arrays: A novel approach to array processing with enhanced degrees of freedom. *IEEE Trans. Signal Process.* **2010**, *58*, 4167–4181. [CrossRef]
20. Han, K.; Nehorai, A. Wideband Direction of Arrival Estimation Using Nested Arrays. In Proceedings of the International Workshop on Computational Advances in Multi-Sensor Adaptive Processing (CAMSAP05), St. Martin, France, 15–18 December 2013; IEEE: New York, NY, USA, 2013; pp. 188–191.
21. Qin, S.; Zhang, Y.D.; Amin, M.G. Generalized co-prime array configurations for direction-of-arrival estimation. *IEEE Trans. Signal Process.* **2015**, *63*, 1377–1390. [CrossRef]
22. Werner, D.H.; Haupt, R.L.; Werner, P.L. Fractal antenna engineering: The theory and design of fractal antenna arrays. *IEEE Antennas Propag. Mag.* **1999**, *41*, 37–59. [CrossRef]

23. Spence, T.G.; Werner, D.H. Design of broadband planar arrays based on the optimization of aperiodic tilings. *IEEE Trans. Antennas Propag.* **2008**, *56*, 76–86. [CrossRef]
24. Asghar, Z.S.; Ng, B.P. Aperiodic geometry design for DOA estimation of broadband sources using compressive sensing. *Signal Process.* **2019**, *155*, 96–107. [CrossRef]
25. Agrawal, M.; Prasad, S. Broadband DOA Estimation Using Spatial-Only Modeling of Array Data. *IEEE Trans. Signal Process.* **2000**, *48*, 663–670. [CrossRef]
26. Zatman, M. How Narrow Is Narrowband? *IEE Proc. Radar Sonar Navig.* **1998**, *145*, 85–91. [CrossRef]
27. Whitney, L.E.; FSA; MAAA. *Math Handbook of Formulas, Processes and Tricks, Geometry Version 2.9*; Earl Whitney: Reno, NV, USA, 2015.
28. Candes, E.; Romberg, J.; Tao, T. Robust uncertainty principles: Exact signal reconstruction from highly incomplete frequency information. *IEEE Trans. Inf. Theory* **2004**, *52*, 489–509. [CrossRef]
29. Cui, A.; Xu, T.; Yu, W.; He, P.; Xu, Z. An Array Interpolation Based Compressive Sensing DOA Method for Sparse Array. In Proceedings of the 3rd International Conference on Imaging, Signal Processing and Communication, Vancouver, BC, Canada, 26–28 August 2019; IEEE: New York, NY, USA, 2019; pp. 24–27.

 *remote sensing*

Communication

A Fast Power Spectrum Sensing Solution for Generalized Coprime Sampling

Kaili Jiang *, Dechang Wang, Kailun Tian, Yuxin Zhao, Hancong Feng and Bin Tang

School of Information and Communication Engineering, University of Electronic Science and Technology of China, Chengdu 611731, China; 202221011414@std.uestc.edu.cn (D.W.); 202311012422@std.uestc.edu.cn (K.T.); 202111012012@std.uestc.edu.cn (Y.Z.); 202211012303@std.uestc.edu.cn (H.F.); bint@uestc.edu.cn (B.T.)
* Correspondence: jiangkelly@uestc.edu.cn

Abstract: With the growing scarcity of spectrum resources, wideband spectrum sensing is necessary to process a large volume of data at a high sampling rate. For some applications, only second-order statistics are required for spectrum estimation. In this case, a fast power spectrum sensing solution is proposed based on the generalized coprime sampling. The solution involves the inherent structure of the sensing vector to reconstruct the autocorrelation sequence of inputs from sub-Nyquist samples, which requires only parallel Fourier transform and simple multiplication operations. Thus, it takes less time than the state-of-the-art methods while maintaining the same performance, and it achieves higher performance than the existing methods within the same execution time without the need to pre-estimate the number of inputs. Furthermore, the influence of the model mismatch has only a minor impact on the estimation performance, allowing for more efficient use of the spectrum resource in a distributed swarm scenario. Simulation results demonstrate the low complexity in sampling and computation, thus making it a more practical solution for real-time and distributed wideband spectrum sensing applications.

Keywords: genralized coprime sampling; power spectrum sensing; non-sparsity; blind sensing; cyclostationary

Citation: Jiang, K.; Wang, D.; Tian, K.; Zhao, Y.; Feng, H.; Tang, B. A Fast Power Spectrum Sensing Solution for Generalized Coprime Sampling. *Remote Sens.* **2024**, *16*, 811. https://doi.org/10.3390/rs16050811

Academic Editors: Jiahua Zhu, Gerardo Di Martino, Xiaotao Huang, Jianguo Liu, Xinbo Li, Shengchun Piao, Junyuan Guo and Wei Guo

Received: 12 January 2024
Revised: 23 February 2024
Accepted: 23 February 2024
Published: 26 February 2024

1. Introduction

The demand for spectrum resources is increasing due to the rapid development of low-orbit satellite constellation systems (e.g., SpaceX, OneWeb), 5G/6G networks, and the Internet of Things (IoT) [1,2]. These applications are driving an unprecedented increase in demand for wideband spectrum sensing. Correspondingly, direct sampling requires a high-speed analog-to-digital converter (ADC) [3] based on the Shannon–Nyquist sampling theorem, increasing data volume and energy consumption.

Currently, the most widely used methods are sweep-tune sampling and filter band sampling. Both of these methods fall under the category of low-speed sampling. However, the scanning scheme has a detection latency and may miss short-lived signals [4]. In addition, the filter band scheme has a complicated structure and is prone to serious channel crosstalk [5]. Consequently, there has been a trend towards using wideband spectrum sensing as a guide.

The recent compressive sensing (CS) theory provides a sub-sampling scheme that offers low-speed and large instantaneous bandwidth by utilizing the sparsity in the frequency domain [6,7]. The typical CS schemes include the analog-to-information converter (AIC), multi-coset sampling (MCS), and multi-rate sampling (MRS). The typical AIC architecture is the modulated wideband converter (MWC), which presents significant challenges for the implementation of the Nyquist-rate pre-randomizing [8]. The MCS architecture is currently a preferred scheme for the ADC with high-speed and high-precision, which cannot obtain a high significant bit because of the mismatching between multiple channels [9]. The MRS architecture is currently a preferred scheme for the sparse array signal acquisition, whose performance is limited by time synchronization accuracy [10]. Obviously, the engineering implementation of these schemes presents major difficulty, which restricts their applications. Furthermore, the CS-based methods [3,11] for sparse signal recovery have high computational complexity and are extremely sensitive to model matching and noise, which mainly include greedy methods, convex methods, and others. Therefore, reducing the dependence of signal processing algorithms on model mismatches between the digital and the analog world is an interesting research idea.

To overcome these difficulties, the approach to spectrum estimation has shifted from processing the original signal to analyzing its second-order statistics. In this context, the compressive covariance sensing (CCS) theory provides a wideband spectrum sensing scheme that operates at a low-speed and has a large instantaneous bandwidth. It is reliable even in environments with a low signal-to-noise ratio (SNR) and in non-sparse conditions. Generally, according to the methods of computation, the CCS-based wideband power spectrum sensing can be divided into the time-domain approach [12] and the frequency-domain approach [13]. The time-domain approach establishes a relationship between the original inputs and the output samples through the selection matrix with zero and one elements under the equivalent Nyquist-rate sampling. As well as this, the frequency-domain approach builds the relationship between both frequency representations of them. However, these studies mainly focus on the MWC and MCS schemes.

For the MRS scheme, the existing CCS-based methods include entropy function minimization [14], matrix norm minimization [15], and Toeplitz matrix completion [16], among others. However, all these methods are based on the reconstruction of the covariance matrix and the use of multiple signal classification (MUSIC) algorithm, which have high computational complexity and require pre-estimation of the number of signals. Additionally, they are also sensitive to model matching.

Furthermore, a computationally efficient method is developed, which is based on the relationship between the autocorrelation sequence and sub-sampling samples of the MCS scheme [17]. Building on this, a fast solution for generalized coprime sampling is introduced, which utilizes only parallel FFT and multiplication operations. As a result, it achieves a reduced time and low estimation error, presenting a trade-off between system performance and the number of degrees of freedom (DOFs). Moreover, model mismatch has minimal effect on performance, making a more practical solution for real-time and distributed wideband spectrum sensing applications.

The rest of this paper is organized as follows: Section 2 describes the signal model and the proposed fast power spectrum sensing solution. Section 3 conducts an analysis and validation of the proposed solution through simulation. The discussion is presented in Section 4.

Notations: The bold characters denote vectors. The notations $\mathbb{R}$, $\mathbb{N}$, and $\mathbb{N}^+$ represent the set of real numbers, nonnegative integers, and positive integers, respectively. The superscripts $(\cdot)^T$ and $(\cdot)^H$ indicate the transpose and conjugate transpose of a vector or a matrix, respectively. The operator $\circ$ signifies the Hadamard product, $|\cdot|^2$ signifies the element-wise square modulus of a vector, and ceil$(\cdot)$ signifies round up to an integer. The symbols F_a and F_a^{-1} mean the a-point fast Fourier transofrm (FFT) and the inverse fast Fourier transofrm (IFFT), respectively.

2. Materials and Methods

2.1. Signal Model

The generalized coprime sampling architecture comprises two uniform sub-Nyquist sampling channels, whose sampling periods are coprime multiples of the Nyquist sampling period. The introduction of two additional operations, the multiple coprime unit factor $p \in \mathbb{N}^+$ and the non-overlapping factor $q \in \mathbb{N}^+$, enhances the number of DOFs and improves the estimation accuracy. Consequently, the coprime sampling scheme is presented with sampling intervals $r_0 T_s$ and $r_1 T_s$, as depicted in Figure 1. Without loss of generality, it is assumed that $r_0 < r_1$ with $r_0, r_1 \in \mathbb{N}^+$, where r_0 and r_1 are coprime. Additionally, the sampling interval T_s corresponds to the Nyquist sampling rate f_s.

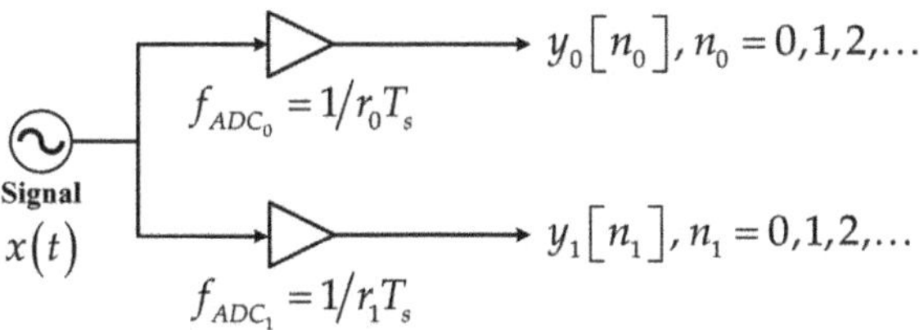

Figure 1. Coprime sampling scheme.

For a wide-sense stationary or cyclostationary process $X(t), t \in \mathbb{R}$, it consists of a number of frequencies that are confined within the bandwidth $B_s \leq f_s/2$. The outputs of two uniform sub-Nyquist sampling channels can be expressed as

$$
\begin{aligned}
y_0[n_0] = x[r_0 n_0] = X(r_0 n_0 T_s), \ n_0 \in \mathbb{N} \\
y_1[n_1] = x[r_1 n_1] = X(r_1 n_1 T_s), \ n_1 \in \mathbb{N}
\end{aligned}
\tag{1}
$$

where $x[n], n \in \mathbb{N}$ denotes the Nyquist sampling samples, and the highest sampling rate of the coprime sampling system is given by $1/(r_0 T_s) = f_s/r_0$.

Accordingly, the elements of the sensing vector corresponding to the two coprime samplers can be denoted as

$$
a_0[i] = \begin{cases} 1, & i = r_0 l_0 + k r_0 r_1 \\ 0, & \text{elsewhere} \end{cases}
\tag{2}
$$

and

$$
a_1[i] = \begin{cases} 1, & i = r_1 l_1 + (k+q) r_0 r_1 \\ 0, & \text{elsewhere} \end{cases}
\tag{3}
$$

where $l_0 = 0, 1, \ldots, r_1 - 1, l_1 = 0, 1, \ldots, r_0 - 1$, and $k = 0, 1, \ldots, p - 1$.

From a data acquisition perspective, the output samples obtained from the generalized coprime sampling scheme are a subset of the Nyquist samples, positioned at

$$
\mathbb{P} = \{r_0 l_0 + k r_0 r_1\} \cup \{r_1 l_1 + (k+q) r_0 r_1\}
\tag{4}
$$

Based on the sensing vectors, the relationship between the elements of the generalized coprime sampling vector and the Nyquist sampling vector can be expressed as

$$
y[n] = \begin{cases} x[n], & n \in \mathbb{P} \\ 0, & \text{elsewhere} \end{cases}
\tag{5}
$$

and the elements of the associated sensing vector are defined as

$$
a[n] = \begin{cases} 1, & n \in \mathbb{P} \\ 0, & \text{elsewhere} \end{cases}
\tag{6}
$$

As a result, there is

$$
\mathbf{y}[n] = \mathbf{a}[n] \circ \mathbf{x}[n]
\tag{7}
$$

where $\mathbf{x}[n] = [x[0], x[1], \ldots, x[N-1]]^T$, $\mathbf{a}[n] = [a[0], a[1], \ldots, a[N-1]]^T$, and $\mathbf{y}[n] = [y[0], y[1], \ldots, y[N-1]]^T$ are all vectors of size $N \times 1$.

2.2. Proposed Fast Solution

Considering the widely-used unbiased estimation of the autocorrelation sequence for the output of generalized coprime sampling, the elements $r_y[m]$ can be expressed as

$$r_y[m] = \frac{1}{N} \cdot \mathbf{y}[n]\mathbf{y}^H[n-m] \tag{8}$$

Substituting Equation (7) into Equation (8) results in

$$
\begin{aligned}
r_y[m] &= \frac{1}{N} \cdot (\mathbf{a}[n] \circ \mathbf{x}[n])(\mathbf{a}^H[n-m] \circ \mathbf{x}^H[n-m]) \\
&= \frac{1}{N} \cdot ((\mathbf{a}[n]\mathbf{a}^H[n-m]) \circ (\mathbf{x}[n]\mathbf{x}^H[n-m])) \\
&= r_a[m] \circ r_x[m]
\end{aligned}
\tag{9}
$$

where $|m| \leq N-1$. Thus, the power spectrum can be obtained by performing FFT on the autocorrelation sequence $\{r_x[m]\}$, which is derived from the autocorrelation sequence $\{r_y[m]\}$ and $\{r_a[m]\}$. Therefore, a computationally efficient practical solution is employed to obtain the estimation of the autocorrelation sequences. The steps are as follows:

Step 1: Pad vectors $\mathbf{a}[n]$ and $\mathbf{y}[n]$ with an additional N zeros.

$$a_{2N}[n] = \begin{cases} a_N[n], & 0 \leq n \leq N-1 \\ 0, & N \leq n \leq 2N-1 \end{cases} \tag{10}$$

and

$$y_{2N}[n] = \begin{cases} y_N[n], & 0 \leq n \leq N-1 \\ 0, & N \leq n \leq 2N-1 \end{cases} \tag{11}$$

Step 2: Calculate the autocorrelation sequence based on the power spectrum estimation of vector $\mathbf{a}_{2N}[n] = [a[0], a[1], \ldots, a[2N-1]]^T$ and $\mathbf{y}_{2N}[n] = [y[0], y[1], \ldots, y[2N-1]]^T$ by involving FFT and IFFT.

$$\hat{\mathbf{r}}_a'[k] = F_{2N}^{-1}|F_{2N}\mathbf{a}_{2N}|^2/N \tag{12}$$

and

$$\hat{\mathbf{r}}_y'[k] = F_{2N}^{-1}|F_{2N}\mathbf{y}_{2N}|^2/N \tag{13}$$

where $k = 0, 1, \ldots, 2N-1$.

Step 3: Truncate the autocorrelation sequence of interest according to the frequency resolution of the system Δf.

$$\hat{r}_y[m] = \begin{cases} \hat{r}_y'[m], & 0 \leq m \leq M-1 \\ \hat{r}_y'[m+2N], & -M+1 \leq m \leq -1 \end{cases} \tag{14}$$

and

$$\hat{r}_a[m] = \begin{cases} \hat{r}_a'[m], & 0 \leq m \leq M-1 \\ \hat{r}_a'[m+2N], & -M+1 \leq m \leq -1 \end{cases} \tag{15}$$

where $M = ceil(fs/2/\Delta f) + 1$.

Step 4: Compute the autocorrelation sequence of the inputs using the obtained sequences $\{\hat{r}_y[m]\}$ and $\{\hat{r}_a[m]\}$.

$$\hat{r}_x[m] = \hat{r}_y[m]/\hat{r}_a[m] \tag{16}$$

where $m = -M+1, \ldots, -1, 0, 1, \ldots, M-1$.

Step 5: Obtain the power spectrum estimation by taking the FFT of the vector $\hat{\mathbf{r}}_x[m] = [\hat{r}_x[-M+1], \ldots, \hat{r}_x[1], \hat{r}_x[0], \hat{r}_x[1], \ldots, \hat{r}_x[M-1]]^T$.

$$\hat{\mathbf{S}}_x(\omega) = |F_{2M-1}\hat{\mathbf{r}}_x[m]| \tag{17}$$

The block diagram that illustrates the fast power spectrum sensing solution for generalized coprime sampling is depicted in Figure 2. The proposed solution is efficient because it only involves FFT/IFFT operations and some basic multiplication operations. Moreover, as shown in the red section of Figure 2, the autocorrelation sequence $\{\hat{r}_a[m]\}$ of the sensing vector can be pre-calculated offline. This calculation solely depends on the generalized coprime sampling scheme.

Consequently, the computational complexity of the proposed solution involves performing the FFT on a $(2N)$-point sequence twice, resulting in $(2N)\log(2N)$ floating-point operations according to (13). In addition, the FFT is performed on a $(2M-1)$-point sequence once, leading to $(2M-1)\log(2M-1)$ floating-point operations according to (17). After incorporating $2M-1$ multiplication calculations, the total computational complexity requires $(4N)\log(2N) + (2M-1)\log(2M-1) + (2M-1)$ floating-point operations. This leads to a lower computational complexity compared to the state-of-the-art methods. Meanwhile, it is feasible to effectively compute the FFT operations in parallel, making it a more suitable practical solution for real-time wideband power spectrum sensing applications.

As researched in the state-of-art, there is an example of wideband spectrum sensing based on the MWC architecture with a 1 GHz bandwidth, which requires a spectrum resolution of 10 kHz. The time-domain approach involves at least 10^{14} floating-point operations in total, assuming that the number of sampling branches is set to 8, the downsampling factor sets to 25, then the number of output samples need to be 4000, and the number of samples used to calculate the correlation matrix is set to 100. Moreover, the more efficient time-domain approach has the same computational complexity as the frequency-domain approach, which involves more than 10^7 floating-point operations in total under the same assumption. Additionally, regarding the MCS schemes, 1.08×10^7 floating-point operations are needed. In comparison, the proposed method involves 8.805×10^6 floating-point operations in total.

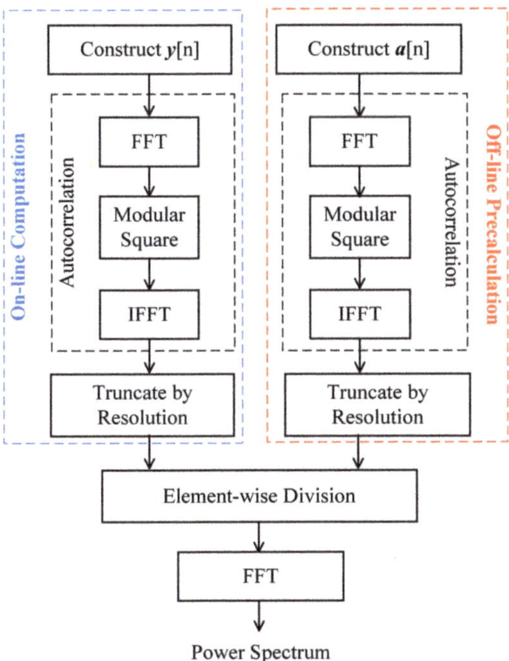

Figure 2. Block diagram of the proposed fast power spectrum sensing solution.

3. Results

In the experiments, it is assumed that there are I inputs with identical powers, which are distributed in the frequency band $[2, 18]$ GHz. Subsequently, the coprime integers $r_0 = 3$, $r_1 = 4$ and the Nyquist sampling rate $f_s = 32$ GHz are set. Furthermore, the relative root mean square error (RMSE) is adopted to evaluate the performance of the proposed fast power spectrum sensing method, which is defined as follows:

$$\text{Relative RMSE}(f_i) = \frac{1}{f_s} \sqrt{\frac{1}{500I} \sum_{j=1}^{500} \sum_{i=1}^{I} (\hat{f}_i(j) - f_i)^2} \tag{18}$$

where $\hat{f}_i(j)$ is the estimation of f_i from the jth Monte Carlo trial, and five hundred Monte Carlo trials are conducted.

3.1. Estimated Power Spectrum Performance

Herein, the estimated power spectrum results are initially displayed, with $p = 3000$ and an input SNR of 15 dB. As shown in Figure 3, there are $I = 50$ mono-frequency pulse (MP) signals, which are randomly distributed. Figure 4 depicts $I = 20$ binary phase shift keying (BPSK) signals for 1 M symbols per second with random frequency and code. Figure 5 presents $I = 2$ linear frequency modulation (LFM) signals with 10 GHz bandwidth under ± 6 GHz initial frequencies. Figure 6 shows $I = 21$ mixture signals of three types. As can be observed, all frequencies are estimated accurately with the proposed method.

Furthermore, there are multiple LFM signals that share the same carrier frequency but have different quadratic modulation coefficients. The frequency and bandwidth of these signals are randomly selected. Compared with Figure 7, the power spectrum estimation shows an increase in the number of pseudo-spectra as more signals are aliased, and as the number of aliased signals increases, as shown in Figure 8 and Figure 9, respectively.

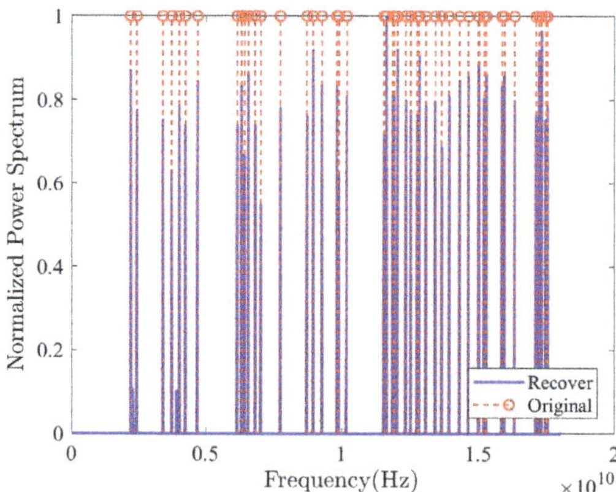

Figure 3. Estimated power spectrum of MP signals ($I = 50$, SNR = 15 dB).

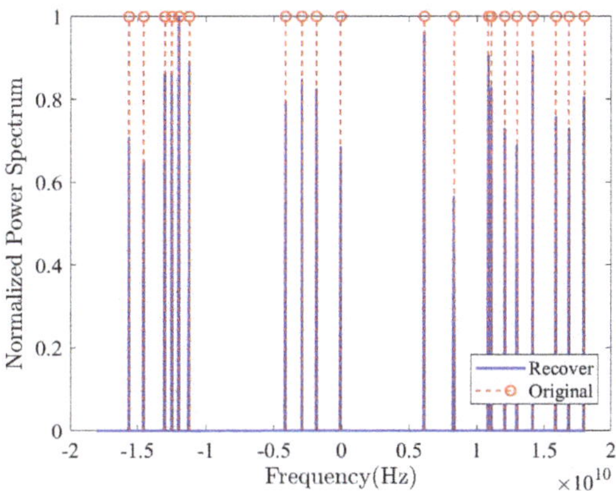

Figure 4. Estimated power spectrum of BPSK signals ($I = 20$, SNR $= 15$ dB).

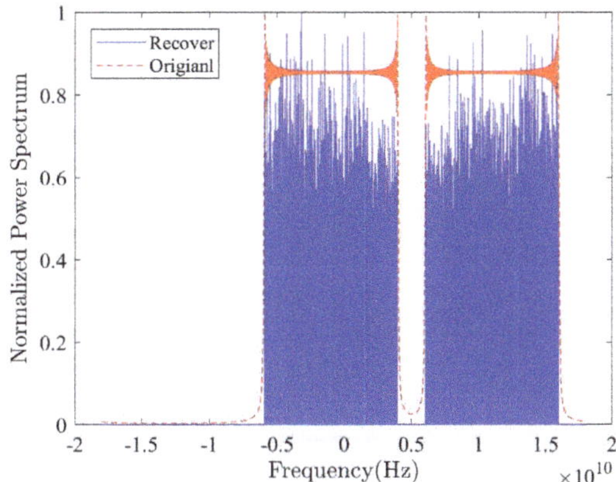

Figure 5. Estimated power spectrum of LFM signals ($I = 2$, SNR $= 15$ dB).

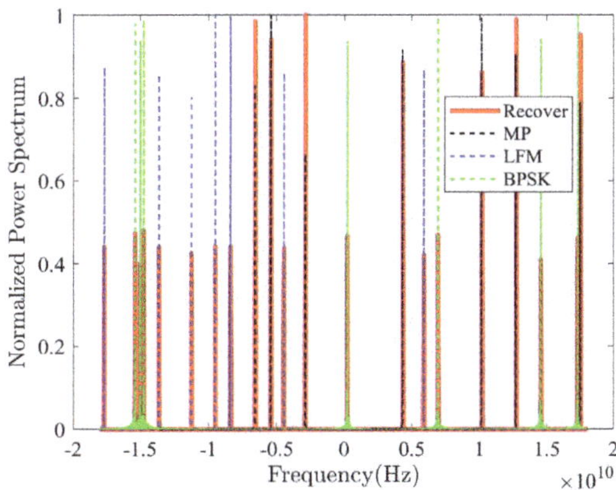

Figure 6. Estimated power spectrum of mixed signals ($I = 21$, SNR $= 15$ dB).

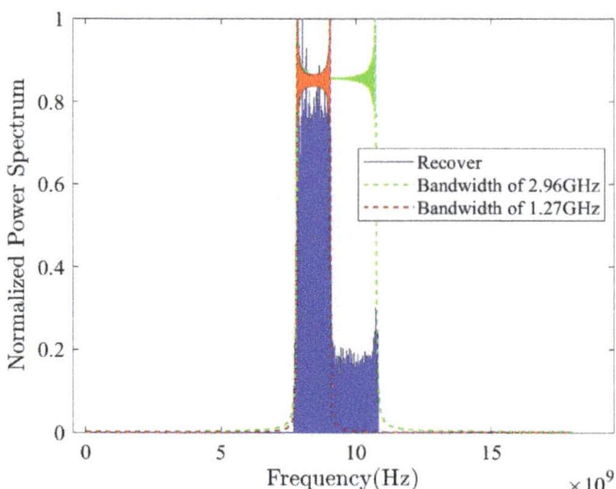

Figure 7. Estimated power spectrum of LFM signals with the same carrier frequency ($I = 2$, SNR = 15 dB).

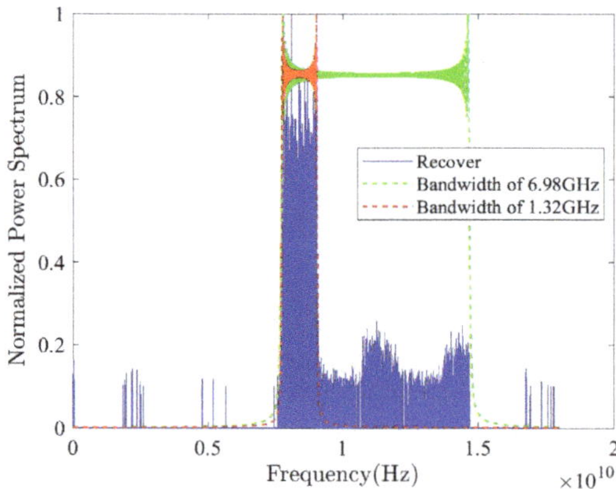

Figure 8. Estimated power spectrum of LFM signals with the same carrier frequency ($I = 2$, SNR = 15 dB).

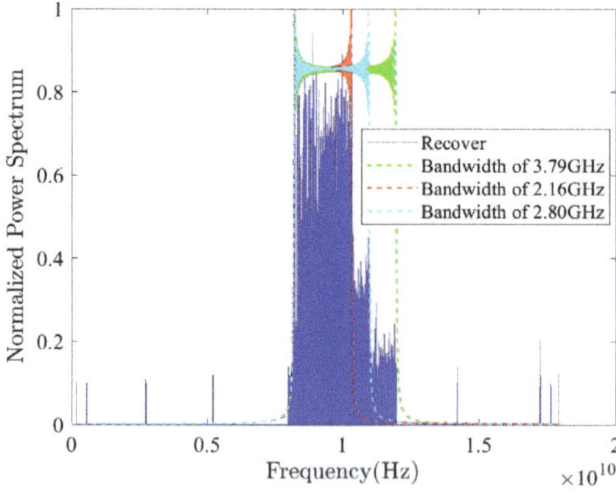

Figure 9. Estimated power spectrum of LFM signals with the same carrier frequency ($I = 3$, SNR = 15 dB).

3.2. Relative RMSE Performance

As depicted in Figure 10, the RMSE results are compared as a function of the input SNR, where $I = 18$ MP signals are utilized and the frequency is randomly selected. It is observed that the RMSE tends to stabilize when SNR is greater than -2 dB for $p = 300$. Under the same conditions, the original method exhibits better performance at a low SNR by utilizing the Toeplitz matrix completion. Meanwhile, the estimation performance improves as p increases, due to the fact that the DOF increases with p, leading to improved resolution. Clearly, the performance of the MCS scheme is the worst with the same number of samples as $p = 3000$, which is due to the probability of signal loss during the process of power spectrum estimation. As expected, Figure 11 presents the same result. However, the selection of coprime sampling rate makes less of a difference to performance when p is greater than 1000. This is because the system redundancy under simulation is sufficient.

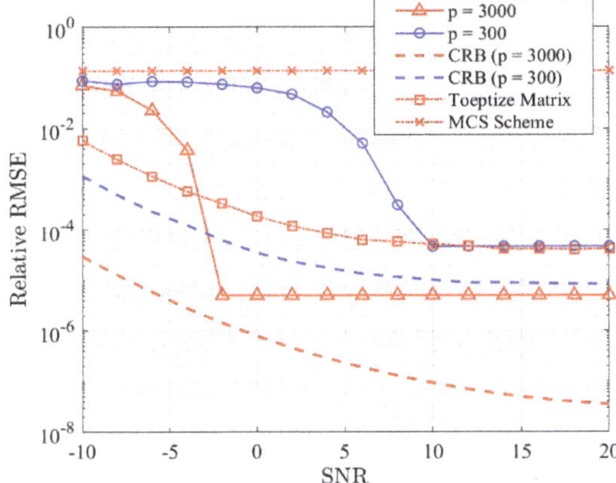

Figure 10. Relative RMSE versus SNR ($I = 18$).

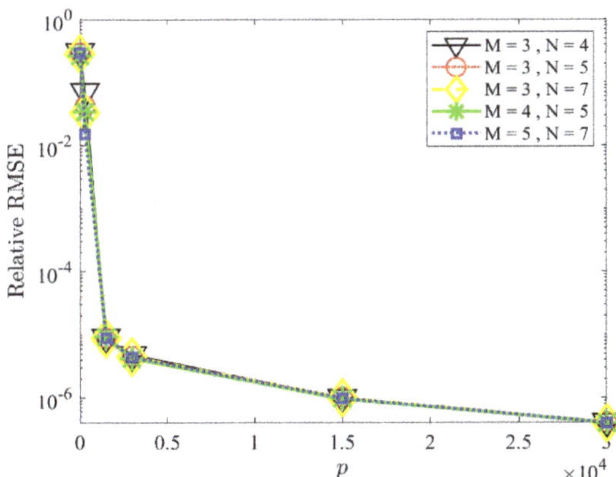

Figure 11. Relative RMSE versus p ($I = 18$, SNR $= 0$ dB).

3.3. Execution Time Performance

Furthermore, the multiple coprime unit factor p not only affects the resolution, but also determines the execution time of algorithms. Consequently, the execution time results, as a function of p, are compared as illustrated in Figure 12, where $I = 10$ MP signals are utilized and the frequency is randomly selected at 0dB SNR. The computing environment is based on Windows 11, equipped with an AMD Ryzen 5 3500U processor, Radeon Vega Mobile Gfx at 2.10 GHz, and 20.0 GB of RAM from Lenovo in Beijing, China. It is evident that the execution time of the proposed method has significant advantages over the matrix completion method. Analogous to Figure 13, the proposed method surpasses the matrix completion method under identical execution time. Additionally, the proposed method requires less time than the matrix completion method under the same performance. However, both methods depend on a larger number of samples. Therefore, a tradeoff exists between execution time and system performance.

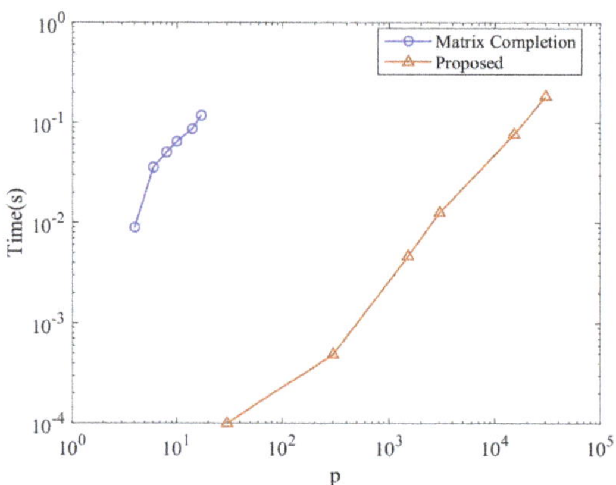

Figure 12. Execution time versus p ($I = 10$, SNR $= 0$ dB).

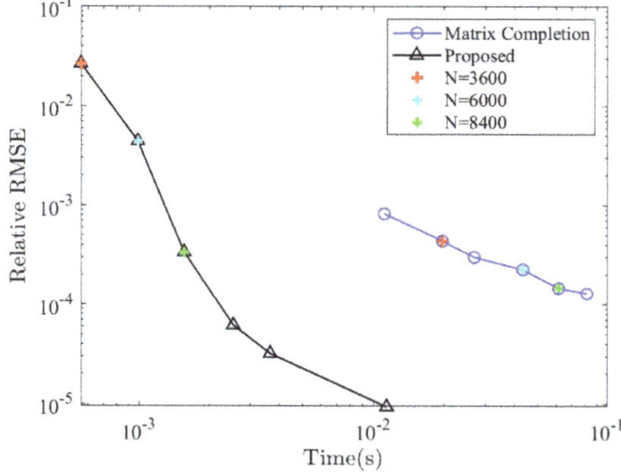

Figure 13. Execution time versus relative RMSE under the same number of samples ($I = 18$, SNR $= 5$ dB).

3.4. Model Mismatching Performance

Finally, the influence of the model matching degree between the sensing vector and measurements on the performance is discussed in Figure 14 with the 5 dB SNR. Here, $I = 18$ frequencies are randomly selected for MP signals, and different time delays are used, which are unknown to the sensing vector. As a result, the influence of the model mismatch causes minimal fluctuation in RMSE within 200 µs. This is interesting, as it suggests that the fact potentially enables more efficient utilization of the spectrum resource in a distributed swarm scenario.

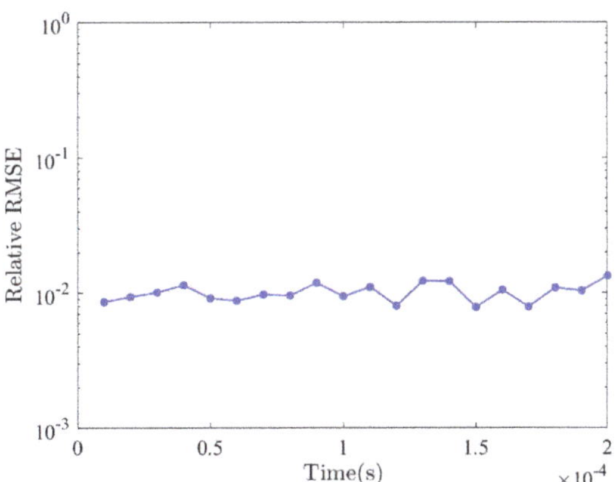

Figure 14. Relative RMSE versus model matching ($I = 18$, SNR $= 5$ dB).

4. Discussion

A fast power spectrum sensing solution for generalized coprime sampling is proposed that only uses the parallel FFT and simple multiplication operations. It has obvious advantages over existing methods in terms of spectrum estimation performance and execution time. Moreover, there is no need to pre-estimate the number of inputs. Furthermore, the influence of the model mismatch has minimal impact on the spectrum estimation performance. That makes it more suitable for further discussion on its application in the distributed swarm scenario.

Author Contributions: Conceptualization, K.J.; methodology, K.J. and D.W.; software, D.W.; validation, K.T., Y.Z. and H.F.; formal analysis, D.W.; investigation, K.J.; resources, B.T.; data curation, K.T.; writing—original draft preparation, K.J.; writing—review and editing, B.T.; visualization, K.J. and D.W.; supervision, B.T.; project administration, B.T.; funding acquisition, B.T. and K.J. All authors have read and agreed to the published version of the manuscript.

Funding: This research was funded in part by the Key Project of the National Defense Science and Technology Foundation Strengthening Plan 173 of Funder grand number 2022-JCJQ-ZD-010-12, and in part by the National Natural Science Foundation of China under Grant 62301119.

Data Availability Statement: Data are contained within the article.

Acknowledgments: The authors would like to thank all editors and reviewers for their valuable comments and suggestions.

Conflicts of Interest: The authors declare no conflicts of interest.

References

1. Chae, K.; Park, J.; Kim, Y. Rethinking autocorrelation for deep spectrum sensing in cognitive radio networks. *IEEE Internet Things J.* **2023**, *10*, 31–41. [CrossRef]
2. Zhou, B.; Ma, X.; Kuang, T.; Li, J. New paradigm of electromagnetic spectrum space situation cognition: Spectrum semantic and spectrum behavior. *J. Data Acquis. Process.* **2022**, *37*, 1198–1207.
3. Manz, B. Technology survey: A sampling of analog-to-digital converter (adc) boards. *J. Electromagn. Domin.* **2021**, *44*, 25–30.
4. Fang, J.; Wang, B.; Li, H.; Liang, Y.-C. Recent advances on sub-nyquist sampling-based wideband spectrum sensing. *IEEE Wirel. Commun.* **2021**, *28*, 115–121. [CrossRef]
5. Pace, P.E. *Developing Digital RF Memories and Transceiver Technologies for Electromagnetic Warfare*, 1st ed.; Artech House: Boston, MA, USA, 2022.
6. Mishra, A.K.; Verster, R.S. *Compressive Sensing Based Algorithms for Electronic Defence*, 1st ed.; Springer International Publishing: Cham, Switzerland, 2018.
7. Wu, Y.; Rosca, M.; Lillicrap, T. Deep compressed sensing. *ICML* **2019**, *97*, 6850–6860.
8. Byambadorj, Z.; Asami, K.; Yamaguchi, T.J.; Higo, A.; Fujita, M.; Iizuka, T. Theoretical analysis of noise figure for modulated wideband converter. *IEEE Trans. Circuits Syst. I* **2020**, *67*, 298–308. [CrossRef]

9. Liu, S.; Zhao, L.; Li, S. A novel all-digital calibration method for timing mismatch in time-interleaved adc based on modulation matrix. *IEEE Trans. Circuits Syst. I* **2022**, *69*, 2955–2967. [CrossRef]

10. Huang, X.; Zhao, X.; Ma, J. Joint carrier and DOA estimation for multi-band sources based on sub-nyquist sampling coprime array with large time lags. *Signal Process.* **2022**, *195*, 108466. [CrossRef]

11. Tropp, J.A.; Wright, S.J. Computational methods for sparse solution of linear inverse problems. *Proc. IEEE* **2010**, *98*, 948–958. [CrossRef]

12. Xu, W.; Wang, S.; Yan, S.; He, J. An efficient wideband spectrum sensing algorithm for unmanned aerial vehicle communication networks. *IEEE Internet Things J.* **2019**, *6*, 1768–1780. [CrossRef]

13. Cohen, D.; Eldar, Y.C. Sub-nyquist sampling for power spectrum sensing in cognitive radios: A unified approach. *IEEE Trans. Signal Process.* **2014**, *62*, 3897–3910. [CrossRef]

14. Huang, S.; Tran, T.D. Sparse signal recovery via generalized entropy functions minimization. *IEEE Trans. Signal Process.* **2019**, *67*, 1322–1337. [CrossRef]

15. Jiang, K.; Xiong, Y.; Tang, B. Wideband spectrum sensing via derived correlation matrix completion based on generalized coprime sampling. *IEEE Access* **2019**, *7*, 117403–117410. [CrossRef]

16. Qin, S.; Zhang, Y.D.; Amin, M.G.; Zoubir, A.M. Generalized coprime sampling of toeplitz matrices for spectrum estimation. *IEEE Trans. Signal Process.* **2017**, *65*, 81–94. [CrossRef]

17. Yang, L.; Fang, J.; Duan, H.; Li, H. Fast compressed power spectrum estimation: Toward a practical solution for wideband spectrum sensing. *IEEE Trans. Wireless Commun.* **2020**, *19*, 520–532. [CrossRef]

Communication

Lightweight Deep Neural Network with Data Redundancy Removal and Regression for DOA Estimation in Sensor Array

Aifei Liu [1], Jiapeng Guo [1], Yauhen Arnatovich [1,*] and Zhiling Liu [2]

[1] The School of Software, Northwestern Polytechnical University, Xi'an 710072, China; liuaifei@nwpu.edu.cn (A.L.)

[2] Nanjing Electronic Equipment Institute, Nanjing 210007, China

* Correspondence: yauhen@nwpu.edu.cn

Abstract: In this paper, a lightweight deep neural network (DNN) for direction of arrival (DOA) estimation is proposed, of which the input vector is designed to remove data redundancy as well as remaining DOA information. By exploring the Vandermonde property of the steering vector of a uniform linear array (ULA), the size of the newly designed input vector is greatly reduced. Furthermore, the DOA estimation is designed as a regression problem instead of a classification problem; that is, the lightweight DNN designs the output vector as the estimated DOAs of sources, of which the size is much shorter than that of the spatial spectrum used as the output vector in the conventional DNN. The reductions in the sizes of input and output vectors lead to a reduction in the sizes of hidden layers, achieving lightweightness of the neural network. The analysis illustrates that when the number of sensors is 22, the number of parameters in the lightweight DNN is three orders of magnitude less than that in the conventional DNN. The simulation results demonstrate the lightweight DNN can provide high DOA estimation accuracy with the shortest testing time. It performs better than the conventional DNN. Furthermore, it is superior to traditional solutions such as the multiple signal classification (MUSIC) method and conventional beamforming (CBF) method in harsh conditions like low signal-to-noise ratios (SNRs), closely spaced sources, and few snapshots.

Keywords: DOA estimation; lightweight deep neural network; data redundancy; deep learning; regression

Citation: Liu, A.; Guo, J.; Arnatovich, Y.; Liu, Z. Lightweight Deep Neural Network with Data Redundancy Removal and Regression for DOA Estimation in Sensor Array. *Remote Sens.* **2024**, *16*, 1423. https://doi.org/10.3390/rs16081423

Academic Editors: Jiahua Zhu, Xiaotao Huang, Jianguo Liu, Xinbo Li, Gerardo Di Martino, Shengchun Piao, Junyuan Guo and Wei Guo

Received: 21 March 2024
Revised: 14 April 2024
Accepted: 15 April 2024
Published: 17 April 2024

1. Introduction

Direction of arrival (DOA) estimation is a widely studied topic in the signal processing area, which performs a key role in wireless communications, astronomical observation, and radar applications [1–5]. The conventional beamforming (CBF) method is a classical solution for DOA estimation. However, it suffers from Rayleigh limit. Subsequently, many traditional methods were proposed to meet the accuracy requirement and high resolution of DOA estimation, such as the minimum variance distortionless response (MVDR) beamformer (also referred to as the Capon beamformer) [6], multiple signal classification method (MUSIC) algorithm [7], estimation of signal parameters using rotational invariance techniques (ESPRIT) algorithm [8] and their variants [9–13]. However, the above-mentioned traditional methods require operations such as singular value decomposition and/or the inversion on the array covariance matrix of the received signal and/or spatial spectrum searching. As a result, their computational complexity is high, which makes it difficult for them to meet real-time requirements. Moreover, most of them have large estimation errors under harsh scenarios such as when the DOAs of source signals have small angular intervals or the signal-noise ratio (SNR) is low. To overcome the drawbacks of the traditional solutions, many studies use machine learning methods to solve the problem of DOA estimation, these methods first establish a training dataset with DOA labels, and then utilize existing machine learning techniques such as radial basis function (RBF) [14] and support vector regression (SVR) [15] to apply the derived mapping to the test data for DOA estimation. These methods require significant effort to learn the mapping during the training stage. However, once the mapping is learned and fixed after the training stage, they directly apply the mapping to process the testing data without labels to obtain DOA estimates. It is noted that the mapping only involves calculations of additions and multiplications, which avoids matrix inverse, decomposition, and spectrum searching. Thus, in the testing stage, they acquire higher computational efficiency compared to traditional methods [16], but they heavily rely on the generalization characteristics of machine learning technology. That is, only when the training data and test data have almost the same distribution, satisfactory test results can be obtained.

In recent years, DOA estimation based on deep learning methods has gained great attention due to its high accuracy and high computational efficiency during the testing phase. In 2015, a single-layer neural network model based on classification was designed to implement DOA estimation [17]. Since then, more and more improved neural networks aiming at solving DOA estimation have been proposed. In 2018, a deep neural network (DNN) was proposed, which contains a multitask auto-encoder and a set of parallel multi-layer classifiers, with the covariance vector of the array output as an input to the DNN, the auto-encoder decomposes the input vectors into sub-regions of space, then the classifiers output the spatial spectrum for DOA estimation [18]. In 2019, a deep convolutional neural network (CNN) was developed for DOA estimation by mapping the initial sparse spatial spectrum obtained from the covariance matrix to the true sparse spatial spectrum [19]. In 2020, a DeepMUSIC method was proposed for DOA estimation, by using multiple CNNs each of which is dedicated to learning the MUltiple SIgnal Classification (MUSIC) spectra of an angular sub-region [20]. In 2021, a CNN with 2D filters was developed for DOA prediction in the low SNR [21], by mapping the 2-D covariance matrix to the spatial spectrum labeled according to the true DOAs of source signals. In 2023, a DNN framework for DOA estimation in a uniform circular array was proposed, using transfer learning and multi-task techniques [22]. The existing results show that deep learning frameworks provide better performance than traditional methods in harsh conditions such as low SNRs and small angle intervals between the DOAs of two source signals.

It is noted that all of the above-mentioned DNN-based DOA estimation methods choose to use the whole array covariance matrix of the received signal or its upper triangular elements or their transformation as the input of the network, which contains lots of redundant information when the array is uniformly linear. In addition, most of them try to match DOA estimation with the classification problem and thus use the spatial spectrum (labeled by the true DOAs of source signals or given by the existing traditional MUSIC method) as their output vector. Therefore, in the existing DNN-based DOA estimation approaches, the data redundancy in the input vector and the large size of the output vector lead to large sizes of hidden layers and make the DNN models complex overall, resulting in low computational efficiency.

There are a few works [23–25] that use neural networks with regression for DOA estimation. In [23], the neural network and a particle swarm optimization (PSO) were combined for DOA estimation, which might be trapped into a minimum solution. In [24], a DNN with regression was developed to estimate the DOA of a single source signal, without considering the situation of multiple source signals. In [25], a DNN with regression was designed for DOA estimation of multiple source signals. However, it does not consider the data redundancy in a uniform linear array (ULA).

In this paper, we consider a ULA, which is the most generally adopted array geometry for DOA estimation due to its regular structure and well-developed techniques according to the Nyquist sampling theorem [26]. By exploring the property of the ULA, a lightweight DNN is proposed by designing an input vector with data redundancy removal and using the regression fashion for DOA estimation. The lightweight DNN significantly reduces the sizes of the input vector, hidden layers, and output vector, which leads to a reduction in the number of trainable parameters of the neural network and computational load. Meanwhile, the proposed lightweight DNN can preserve DOA estimation accuracy and performs better than the method in [25]. It is noted that by considering that the array signal is different from the image signal and DOA information is hidden in each element of the input vector obtained from the covariance matrix of the array signal, we utilize a fully connected deep neuron network to obtain the mapping from the input vector to the DOAs of source signals.

Throughout this paper, *, T, H, and E represent the conjugate, transpose, conjugate transpose, and expectation operations, respectively.

2. Background

Assume that K-independent far-field source signals $\{s_k(t)\}_{k=1}^{K}$ with a wavelength λ and DOAs of $\{\theta_k\}_{k=1}^{K}$ impinge on an M-element uniform linear array (ULA) with an inter-element spacing d. Moreover, it is assumed that the source signals and the array sensors are on the same plane. The received data of the array can be expressed as

$$\mathbf{r}(t) = \mathbf{A}\mathbf{s}(t) + \mathbf{n}(t), \tag{1}$$

where $\mathbf{n}(t)$ is an additive and zero-mean white Gaussian noise vector, $\mathbf{A} = [\mathbf{a}(\theta_1), ..., \mathbf{a}(\theta_K)]$, $\mathbf{s}(t) = [s_1(t), ..., s_K(t)]^{\mathrm{T}}$; In particular, $\mathbf{a}(\theta_k)$ is an M-dimensional steering vector, which is defined as

$$\mathbf{a}(\theta_k) = [1, e^{-j2\pi \frac{d\sin\theta_k}{\lambda}}, ..., e^{-j2\pi \frac{d\sin\theta_k}{\lambda}(M-1)}]^{\mathrm{T}}. \tag{2}$$

The array covariance matrix $\mathbf{R}$ can be expressed as

$$\mathbf{R} = \mathrm{E}[\mathbf{r}(t)\mathbf{r}^{\mathrm{H}}(t)] = \mathbf{A}\mathbf{R}_s\mathbf{A}^{\mathrm{H}} + \sigma_n^2\mathbf{I}_M, \tag{3}$$

where $\mathbf{R}_s = \mathrm{E}[\mathbf{s}(t)\mathbf{s}^{\mathrm{H}}(t)]$, σ_n^2 is the noise power, and $\mathbf{I}_M$ is an identity matrix with a size of $M \times M$. In practice, due to the finite snapshots, the covariance matrix $\mathbf{R}$ can be estimated as

$$\widehat{\mathbf{R}} = \frac{1}{N}\sum_{t=1}^{N}\mathbf{r}(t)\mathbf{r}^{\mathrm{H}}(t), \tag{4}$$

where N is the number of snapshots, and $\widehat{\bullet}$ means the approximation of the quantity above which it appears.

Equation (4) illustrates that $\widehat{\mathbf{R}}$ is a conjugate symmetric matrix. Utilizing this feature, many real-valued deep learning methods use the upper triangular elements as their input vectors [18,20]. Define the vector composed of the off-diagonal upper triangular elements of $\widehat{\mathbf{R}}$ by $\mathbf{z}$, that is

$$\mathbf{z} = [\widehat{R}(1,2), \cdots, \widehat{R}(1,M), \widehat{R}(2,3), \cdots, \widehat{R}(2,M), \cdots, \widehat{R}(M-1,M)]^{\mathrm{T}}. \tag{5}$$

It is noted that for a real-valued DNN network, the input vector needs to be real-valued. Therefore, by concatenating the real and imaginary parts of $\mathbf{z}$, we obtain $\tilde{\mathbf{z}}$ below.

$$\tilde{\mathbf{z}} = [Real(\mathbf{z}^T), Imag(\mathbf{z}^T)]^T / \|\mathbf{z}\|_2, \tag{6}$$

where $\| \cdots \|_2$ defines L_2 norm. $Real\{\bullet\}$ and $Imag\{\bullet\}$ represent the real and imaginary parts of a complex value, respectively.

In [18], a fully connected DNN method with classification was developed for DOA estimation, and it utilizes the vector $\tilde{\mathbf{z}}$ as its input, named as the conventional DNN in this paper. Note that the input vector $\tilde{\mathbf{z}}$ contains data redundancy and costs the computational load without performance improvement. Moreover, since the conventional DNN is based on classification fashion, its output is equal to $\lceil \frac{\theta_{max}-\theta_{min}}{\eta} \rceil$, where $[\theta_{min}, \theta_{max})$ is the angle-searching range of the sources, and η is the grid; with $\lceil x \rceil$ is equal to the smallest integer not smaller than x. Therefore, the size of its output vector is much larger than the number of DOAs of sources, which further increases the computational load.

In the following, we analyze the data redundancy in the ULA and design a new input vector that removes data redundancy and retains DOA information. In a sequence, the lightweight DNN is proposed by using the newly designed input vector and employing the regression fashion for DOA estimation.

3. Data Redundancy Removal

3.1. Development of Data Redundancy Removal

In this section, we first prove that the conventional input vector $\tilde{\mathbf{z}}$ in Equation (6) contains data redundancy. Afterwards, we propose a new input vector that removes data redundancy and retains DOA information.

According to Equations (2) and (3), the array covariance matrix $\mathbf{R}$ can be expanded as

$$\mathbf{R} = \sum_{k=1}^{K} \mathbf{a}(\theta_k)\mathbf{a}^{\mathrm{H}}(\theta_k)\sigma_{s_k}^2 + \sigma_n^2\mathbf{I}_M, \tag{7}$$

where $\sigma_{s_k}^2$ is the power of the $k-$th source signal.

Define the matrix $\mathbf{B}_k = \mathbf{a}(\theta_k)\mathbf{a}^H(\theta_k)$ and its element at m-th row and l-th column as $B_k(m, l)$. According to Equation (2), we obtain that

$$B_k(m, l) = e^{\varphi_k(m-l)}\sigma_{s_k}^2, \tag{8}$$

where $\varphi_k = -j2\pi\frac{d\sin\theta_k}{\lambda}$. Therefore, by substituting Equation (8) into Equation (7), we have

$$R(m, l) = \sum_{k=1}^{K} B_k(m, l) + sgn(m, l)\sigma_n^2 \quad = \sum_{k=1}^{K} e^{\varphi_k(m-l)}\sigma_{s_k}^2 + sgn(m, l)\sigma_n^2. \tag{9}$$

where $sgn(m, l) = \begin{cases} 1 & if \quad m = l \\ 0 & if \quad m \neq l \end{cases}$. As a result, from Equation (9), we observe Lemma 1 below.

Lemma 1. *When the array is ULA, all the elements along the sub-diagonal, super-diagonal and diagonal lines of the covariance matrix $\mathbf{R}$ are equal.*

Lemma 1 can be illustrated in Equation (10) below.

$$\mathbf{R} = \begin{bmatrix} \rho & \beta & v & \cdots & \epsilon \\ \beta^* & \rho & \beta & \ddots & \vdots \\ v^* & \beta^* & \ddots & \ddots & v \\ \vdots & \ddots & \ddots & \rho & \beta \\ \epsilon^* & \cdots & v^* & \beta^* & \rho \end{bmatrix} \tag{10}$$

where $\rho, \beta, v,$ and ϵ are elements of the covariance matrix $\mathbf{R}$.

As shown in Equation (5), the conventional input vector uses all the upper triangular elements of the covariance matrix $\mathbf{R}$, which contains duplicate information and leads to data redundancy according to *Lemma 1*.

On the other hand, from Equation (9), it is observed that the elements along the diagonal lines are affected by noise power and source signal power. However, they do not contain information about DOAs of sources. Thus, they shall not be involved in the input vector of the DNN model. In addition, by observing Equation (9), we define

$$\mathbf{z}_1 = [\widehat{R}(1,2), \ \widehat{R}(1,3), \quad \widehat{R}(1,4), \cdots, \ \widehat{R}(1,M)]^{\mathrm{T}}. \tag{11}$$

By considering the above-mentioned observation, *Lemma 1*, and the conjugate symmetric feature of the covariance matrix $\mathbf{R}$, we obtain that indeed for a real-valued DNN model and the expected covariance matrix $\mathbf{R}$, $\mathbf{z}_1$ contains all the useful elements relevant to the DOAs of sources and discards duplicate data in the vector $\mathbf{z}$ in Equation (5), leading to the removal of data redundancy.

It is worth noticing that in practice, due to the limit of the number of snapshots, the elements along any off-diagonal line of the estimated covariance matrix $\widehat{\mathbf{R}}$ are not exactly equal. On the other hand, as shown in Equation (9), the elements along each super-diagonal line contain the same information about DOAs. Thus, we propose to take the average of all the elements along each super-diagonal line of the estimated covariance matrix $\widehat{\mathbf{R}}$ to obtain a new vector without data redundancy, denoted as $\mathbf{z}_{sum}$. Define the i-th element of $\mathbf{z}_{sum}$ as $z_{sum}(i)$, we have

$$z_{sum}(i) = \frac{1}{M-i} \sum_{m=1}^{M-i} \widehat{R}(m, m+i), i = 1, 2, ..., M-1. \tag{12}$$

Therefore, according to Equations (11) and (12), we can construct two vectors (that is, $\tilde{\mathbf{z}}_1$ and $\tilde{\mathbf{z}}_{sum}$) as shown in Equations (13) and (14).

$$\tilde{\mathbf{z}}_1 = [Real(\mathbf{z}_1^{\mathrm{T}}), Imag(\mathbf{z}_1^{\mathrm{T}})]^{\mathrm{T}} / \|\mathbf{z}_1\|_2, \tag{13}$$

$$\tilde{\mathbf{z}}_{sum} = [Real(\mathbf{z}_{sum}^{\mathrm{T}}), Imag(\mathbf{z}_{sum}^{\mathrm{T}})]^{\mathrm{T}} / \|\mathbf{z}_{sum}\|_2. \tag{14}$$

It is noted that both $\tilde{\mathbf{z}}_1$ and $\tilde{\mathbf{z}}_{sum}$ remove data redundancy and can be used as the input vector of the DNN network theoretically. However, due to the limit of the number of snapshots in practice, the lightweight deep neural network (DNN) proposed in the following does not converge when the vector $\tilde{\mathbf{z}}_1$ is used as the input vector of the DNN. Therefore, we choose the vector $\tilde{\mathbf{z}}_{sum}$ as the input of the proposed DNN in the following, which ensures convergence. On the other hand, it is noted that the conventional input vector $\tilde{\mathbf{z}}$ using upper triangular elements as shown in Equation (5) has a dimension of $M(M-1)$. In contrast, the new input vector $\tilde{\mathbf{z}}_{sum}$ has a dimension of $2(M-1)$. Therefore, the new input vector reduces the dimension to $M/2$ times that of the conventional input vector. This implies that the nodes in the following hidden layers can be correspondingly reduced, which contributes to forming a lightweight DNN.

As a sequence, the data redundancy removal developed for the ULA above can be applied to the matrix $\widehat{\mathbf{R}}$ to obtain the input vector without data redundancy (i.e., $\tilde{\mathbf{z}}_{sum}$).

3.2. Analysis of Data Redundancy Removal

According to Equations (3) and (4), $\widehat{\mathbf{R}}$ is the maximum-likelihood estimate of the expected $\mathbf{R}$, and thus the estimation error $\Delta\mathbf{R}$ always exists [27]; that is,

$$\Delta\mathbf{R} = \widehat{\mathbf{R}} - \mathbf{R}. \tag{15}$$

In addition, the proposed lightweight DNN is based on the $\tilde{\mathbf{z}}_{sum}$ in Equation (14). In contrast, the method in [25] uses the vector composed of the off-diagonal upper triangular elements; that is, $\tilde{\mathbf{z}}$ in Equation (6). Both $\tilde{\mathbf{z}}_{sum}$ and $\tilde{\mathbf{z}}$ are based on the estimated covariance matrix $\widehat{\mathbf{R}}$. Consequently, these elements are also subject to estimation inaccuracies, which subsequently precipitate errors in DOA estimation. It is expected that a larger estimation error of $\tilde{\mathbf{z}}_{sum}$ or $\tilde{\mathbf{z}}$ leads to a higher DOA estimation error. We define the estimation error of $\tilde{\mathbf{z}}_{sum}$ by $\Delta\tilde{\mathbf{z}}_{sum}$, which is given as

$$\Delta\tilde{\mathbf{z}}_{sum} = \tilde{\mathbf{z}}_{sum} - \tilde{\mathbf{z}}_{sum}^{exp}, \tag{16}$$

where the elements of $\tilde{\mathbf{z}}_{sum}^{exp}$ are obtained by replacing the estimated covariance matrix $\widehat{\mathbf{R}}$ with the expected one $\mathbf{R}$ in Equation (12).

Similarly, we define the estimation error of $\tilde{\mathbf{z}}$ by $\Delta\tilde{\mathbf{z}}$, which is written as

$$\Delta\tilde{\mathbf{z}} = \tilde{\mathbf{z}} - \tilde{\mathbf{z}}^{exp}, \tag{17}$$

where the elements of $\tilde{\mathbf{z}}^{exp}$ are obtained by replacing the estimated covariance matrix $\hat{\mathbf{R}}$ with the expected one $\mathbf{R}$ in Equation (5). Define the 2-norm of the vectors $\Delta\tilde{\mathbf{z}}_{sum}$ and $\Delta\tilde{\mathbf{z}}$ as $\|\Delta\tilde{\mathbf{z}}_{sum}\|_2$ and $\|\Delta\tilde{\mathbf{z}}\|_2$, respectively.

In the following, a comparative analysis of the numerical outcomes for $\|\Delta\tilde{\mathbf{z}}_{sum}\|_2$ and $\|\Delta\tilde{\mathbf{z}}\|_2$ is presented. Assuming that the ULA consists of 22 elements with an inter-element spacing equal to $\frac{\lambda}{2}$. Supposing that there are two source signals with the same SNR impinge onto the array with DOAs of $\theta_1 = -40.55°$ and $\theta_2 = -36.3°$, respectively. The number of snapshots equals 400. The number of trials is 200.

When SNR = -10 dB, we obtain $\|\Delta\tilde{\mathbf{z}}_{sum}\|_2 = 0.19$, and $\|\Delta\tilde{\mathbf{z}}\|_2 = 0.37$. When SNR = 5 dB, we obtain $\|\Delta\tilde{\mathbf{z}}_{sum}\|_2 = 0.06$, and $\|\Delta\tilde{\mathbf{z}}\|_2 = 0.09$. Overall, $\|\Delta\tilde{\mathbf{z}}_{sum}\|_2 < \|\Delta\tilde{\mathbf{z}}\|_2$. This fact leads to better performance of the proposed lightweight DNN with its input as $\tilde{\mathbf{z}}_{sum}$, as comparisons of the method in [25] with its input as $\tilde{\mathbf{z}}$. This fact matches with numerical results in Section 5.

Furthermore, from the analysis above, we obtain that $\|\Delta\tilde{\mathbf{z}}_{sum}\|_2$ decreases as the SNR increases, which implies that the performance of the lightweight DNN with $\tilde{\mathbf{z}}_{sum}$ gets better as the SNR increases.

4. Lightweight DNN for DOA Estimation

In this section, we propose a lightweight DNN for DOA estimation, which is illustrated in Figure 1. As shown in Figure 1, the proposed lightweight DNN model utilizes the newly developed input vector $\tilde{\mathbf{z}}_{sum}$ as its input vector. Furthermore, different from the conventional DNN model with classification [18–21], the new DNN model is a regression model and has an output vector with a dimension equal to the number of sources, which approaches to the vector of true DOAs of sources in a regression fashion. It is noted that by considering the DOAs of sources are continuous values, the DNN model with regression can match the task of DOA estimation naturally. It is noted that in practice, prior to DOA estimation, the estimation of the number of sources can be accomplished by the classical methods such as the Minimum Description Length (MDL) and the Akaike Information Criterion (AIC) methods [28]. In addition, by considering that the array signal is different from the image signal and DOA information is hidden in each element of the input vector which is obtained from the covariance matrix of the array signal, we select a fully connected deep neuron network to extract the mapping from the input vector to the DOAs of source signals.

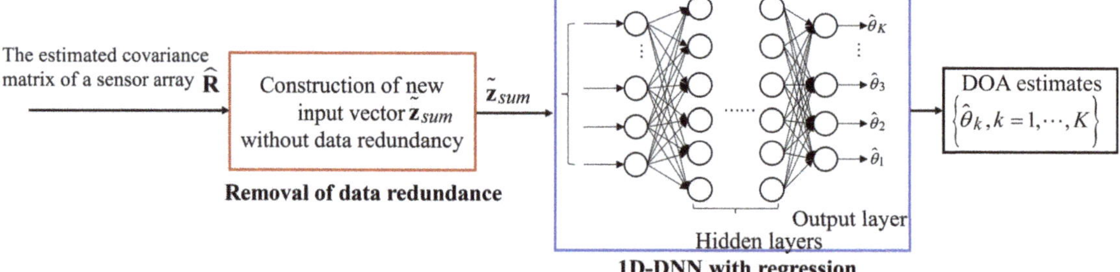

Figure 1. Proposed lightweight DNN for DOA estimation in a ULA array.

As shown in Figure 1, the proposed lightweight DNN is a fully connected network with regression and contains an input layer, several hidden layers with activation functions, and an output layer. Furthermore, each node of each layer in the network is connected to each node of the adjacent forward layer. The input data flows into the input layer, passes through the hidden layers, and turns into the output of the network, which gives DOA estimates. The detailed structure of the proposed lightweight DNN model and its construction of the training data set are given as follows. It is noted that both the proposed lightweight DNN method and the method in [25] use regression for DOA estimation. The difference between the proposed lightweight DNN method and the one in [25] is that the proposed lightweight DNN method removes the data redundancy and significantly reduces the trainable parameters, by analyzing the property of the covariance matrix of a ULA and the parameters of the network.

4.1. Detailed Structure of Lightweight DNN

The computations of hidden layers are feedforward as

$$\mathbf{h}_l = g_l(\mathbf{W}_{l,l-1}\mathbf{h}_{l-1} + \mathbf{b}_l), l = 1, 2, ..., L - 1, \tag{18}$$

where L is the total number of the layers except for the input layer; $\mathbf{h}_l$ represent the output vector of the *l-th* layer; $\mathbf{W}_{l,l-1}$ is the weight matrix between the *(l-1)-th* layer and *l-th* layer; $\mathbf{b}_l$ is the bias vector of the *l-th* layer; g_l is the activation function of the *l-th* layer. The activation function is set as $g_l(\bullet) = \tanh(\bullet)$, which is expressed as

$$tanh(\alpha) = \frac{e^\alpha - e^{-\alpha}}{e^\alpha + e^{-\alpha}}, \tag{19}$$

where α is a real value. The output vector of the output layer is given as

$$\mathbf{h}_L = \mathbf{W}_{L,L-1}\mathbf{h}_{L-1} + \mathbf{b}_L. \tag{20}$$

In the training phase, the proposed DNN is performed in a supervised manner with the training data-label set, and the parameters of the DNN are adjusted to make the output vector $\mathbf{h}_L$ approach to the label, which is composed of the DOAs of source signals. We define the number of input vectors by I. Then, the training data set can be expressed as $\Gamma = \{\mathbf{x}_{(1)}, ..., \mathbf{x}_{(I)}\}$ with its label set $\Psi = \{\bar{\boldsymbol{\theta}}_{(1)}, ..., \bar{\boldsymbol{\theta}}_{(I)}\}$, $\mathbf{x}_{(i)}$ and $\bar{\boldsymbol{\theta}}_{(i)}$ are the *i*-th input vector and its label, respectively. $\mathbf{x}_{(i)}$ is equal to $\tilde{\mathbf{z}}_{sum}$ generated in the *i*-th numerical experiment. $\bar{\boldsymbol{\theta}}_{(i)}$ is a K-dimensional vector composed of the true DOAs of sources in the *i*-th numerical experiment.

The set of all the trainable parameters in the lightweight DNN model can be collectively referred to as Ω. The update of Ω follows back-propagation towards minimizing the Mean Square Error(MSE) loss function as follows.

$$\widehat{\Omega} = argmin_\Omega \frac{1}{IK} \sum_{i=1}^{I} \|\mathbf{h}_{L,(i)} - \bar{\boldsymbol{\theta}}_{(i)}\|_2^2, \tag{21}$$

where $\|\bullet\|_2$ represents 2-norm, which measures the distance between the output vector of the network and the corresponding label, $\mathbf{h}_{L,(i)}$ represents the output vector of the network corresponding to the *i*-th input vector. In the testing phase, the output vector of the output layer gives the estimated values of the DOAs of source signals explicitly.

For the lightweight DNN model, we define the size of the input vector by $\tilde{J} = 2(M - 1)$. Note that with more layers and larger sizes of layers, the expressivity power of the network is increased during the training stage. However, the network tends to overfit the training data. As a result, in the testing stage, the performance is obviously degraded due to the lack of generalization. Furthermore, referring to [18], for the balance between the expressivity power with deeper network and aggravation with more network parameters, we set the number of hidden layers to be 2 and their sizes are equal to $\lfloor \frac{2}{3}\tilde{J} \rfloor$ and $\lfloor \frac{4}{9}\tilde{J} \rfloor$, respectively, where $\lfloor x \rfloor$ is equal to the largest integer not larger than x.

4.2. Construction of Training Data Set

Assuming that the searching angle range of the source signals is from θ_{min} to θ_{max}, the angular interval between two source signals in this range is defined as $\triangle$, which is sampled from a set of $[\triangle_{min}, \triangle_{min} + \triangle_d, \triangle_{min} + 2\triangle_d, ..., \triangle_{max}]$, where $\triangle_{min}$, $\triangle_{max}$, and $\triangle_d$ are the minimum angle interval between the DOAs of two source signals, the maximum angle interval, and an angle increment, respectively. In this way, any two source signals in this range that are spatially close to each other and those with large spacing can all be included in the training data set. Since the elements used as the input vector from the covariance matrix are not affected by the order of DOAs of source signals, with the DOA of the first source signal is sampled with a grid η from θ_{min} to $\theta_{max} - (K-1)\triangle$, and the DOA of the k-th source signal is $\theta_1 + (k-1)\triangle, k = 2, ..., K$. Furthermore, in order to adapt to the performance fluctuations in low SNRs, input vectors with multiple SNRs lower than 0dB are trained at the same time, making the lightweight DNN better adapted to unknown low and high SNRs during the testing phase.

4.3. Analysis of Number of Trainable Parameters

In this section, we present a comparative analysis of the proposed lightweight DNN, against the method in [25], the conventional DNN in [18], deep convolution network (DCN) in [19], and DeepMUSIC in [20], focusing on the number of trainable parameters. For the conventional DNN model in [18], we follow the setting in [18]. That is, for the autoencoder, we denote the size of each of the input and output layers as $J = M(M-1)$, define the number of each encoder and decoder has one hidden layer with a size of $\lfloor \frac{J}{2} \rfloor$, and define the number of spatial subregion as p. As a sequence, we obtain that for each of the multilayer classifiers after the autoencoder, the sizes of two hidden layers are equal to $\lfloor \frac{2}{3}J \rfloor$ and $\lfloor \frac{4}{9}J \rfloor$, respectively. In addition, the size of output layer (denoted as γ) for each multilayer classifier is equal to

$$\gamma = \lceil \frac{\theta_{max} - \theta_{min}}{\eta p} \rceil. \tag{22}$$

Correspondingly, according to the analysis in Section 3.1 for the proposed lightweight DNN, we have $\tilde{J} = \frac{2}{M}J$. By following the above-mentioned definitions and the structure of the lightweight DNN, conventional DNN, and method in [25], we can obtain the number of parameters in the three fully connected DNN models, as shown in Table 1.

Table 1. Analysis of number of trainable parameters in fully-connected DNN methods.

Number of Parameters	Autoencoder	Hidden Layer 1	Hidden Layer 2	Output Layer
Lightweight DNN	N.A.	$(\tilde{J}+1) \times \lfloor \frac{2}{3}\tilde{J} \rfloor$	$(\lfloor \frac{2}{3}\tilde{J} \rfloor + 1) \times \lfloor \frac{4}{9}\tilde{J} \rfloor$	$(\lfloor \frac{4}{9}\tilde{J} \rfloor + 1) \times K$
Method in [25]	N.A.	$(J+1) \times \lfloor \frac{2}{3}J \rfloor$	$(\lfloor \frac{2}{3}J \rfloor + 1) \times \lfloor \frac{4}{9}J \rfloor$	$(\lfloor \frac{4}{9}J \rfloor + 1) \times K$
Conventional DNN	$(J+1) \times \lfloor \frac{J}{2} \rfloor + (\lfloor \frac{J}{2} \rfloor + 1) \times J \times p$	$(J+1) \times \lfloor \frac{2}{3}J \rfloor \times p$	$(\lfloor \frac{2}{3}J \rfloor + 1) \times \lfloor \frac{4}{9}J \rfloor \times p$	$(\lfloor \frac{4}{9}J \rfloor + 1) \times \gamma \times p$

Table 2 shows the number of trainable parameters in DeepMUSIC and DCN by following the parameter settings in [19,20], which are mainly from the convolution layers and dense layers. For DeepMUSIC, C_{in1} represents the number of input channels, K_{s1} is the kernel size of the first two convolution layers, and K_{s2} is the kernel size of convolution layer 3 and convolution layer 4. N_f is the number of filters. C_{out1} and C_{out2} represent the sizes of the first and second dense layers, respectively. For DCN, K_{s3}, K_{s4}, K_{s5} and K_{s6} represent the kernel size of the first till fourth convolution layers, of which the number of filters are N_{f1}, N_{f2}, N_{f3}, and N_{f4}, respectively. There is no dense layer.

Table 2. Analysis of number of trainable parameters in CNN-based methods.

Number of Parameters	Convolution Layer 1	Convolution Layer 2	Convolution Layer 3	Convolution Layer 4	Dense Layer 1	Dense Layer 2
DeepMUSIC	$K_{s1}^2 \times C_{in1} \times N_f$	$K_{s1}^2 \times N_f \times N_f$	$K_{s2}^2 \times N_f \times N_f$	$K_{s2}^2 \times N_f \times N_f$	$M^2 \times N_f \times C_{out1}$	$C_{out1} \times C_{out2}$
DCN	$K_{s3} \times C_{in2} \times N_{f1}$	$K_{s4} \times N_{f1} \times N_{f2}$	$K_{s5} \times N_{f2} \times N_{f3}$	$K_{s6} \times N_{f3} \times N_{f4}$	N.A.	N.A.

When $\theta_{min} = -60°$, $\theta_{max} = 60°$, $p = 6$, $\eta = 1°$, $K = 2$, $C_{in1} = 3$, $C_{in2} = 2$, $K_{s1} = 5$, $K_{s2} = 3$, $K_{s3} = 25$, $K_{s4} = 15$, $K_{s5} = 5$, $K_{s6} = 3$, $N_f = 256$, $N_{f1} = 12$, $N_{f2}=6$, $N_{f3} = 3$, $N_{f4} = 1$, $C_{out1} = 1024$, $C_{out2} = 120$, the total parameters of the above-mentioned five deep learning methods versus the number of sensors are shown in Figure 2. From Figure 2, we can see that the number of trainable parameters in the lightweight DNN is significantly reduced compared to those of the conventional DNN, method in [25], and DeepMUSIC. In particular, when the number of sensors is 22, the number of parameters in the lightweight DNN is three orders, two orders, and five orders of magnitude less than that in the conventional DNN, the method in [25], and DeepMUSIC, respectively. This fact contributes to fitting the DNN-based DOA estimation into the embedded system. In addition, the lightweight DNN method has fewer parameters than the DCN method when the number of sensors is less than 22. The DCN method remains constant regardless of the number of sensors. This is because the input of the DCN method is the spatial spectrum proxy, which has a fixed length equal to $\lceil \frac{\theta_{max} - \theta_{min}}{\eta} \rceil$. On the other hand, the inputs of other methods are all explicitly relevant to the dimension of the array covariance matrix. Thus, their parameters are related to the number of sensors.

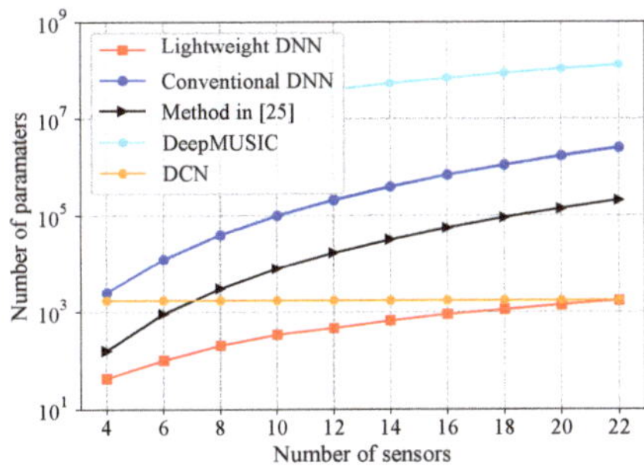

Figure 2. Trainable parameters in the DNN models versus number of sensors.

4.4. Analysis of Computational Complexity

Analogous to the approach detailed in [11], we quantify the primary computational complexity through the calculation of real-valued multiplications, as given in Table 3. In this table, L pertaining to the DCN denotes the length of the input vector, which is set as 120. In addition, we define

$$\tilde{\gamma} = \lceil \frac{\theta_{max} - \theta_{min}}{\eta} \rceil. \tag{23}$$

Note that when $\eta = 0.1$ and $M = 22$, we have $\tilde{\gamma} \gg J > \tilde{J} > M > K$ [11] and $(M-1)(M-K) \approx (M+1)M \approx J$. According to the settings in Section 4.3, it is found from Table 3 that the computational complexity of the CBF, MUSIC, DeepMUSIC, and DCN methods is significantly higher than that of the fully-connected DNN-based methods, which corresponds to the testing time in Table 4 below.

Table 3. Analysis of primary computational complexity.

Algorithms	Primary Computational Complexity
Lightweight DNN	$\mathcal{O}[\tilde{J} \times \lfloor \frac{2}{3}\tilde{J} \rfloor + \lfloor \frac{2}{3}\tilde{J} \rfloor \times \lfloor \frac{4}{9}\tilde{J} \rfloor + \lfloor \frac{4}{9}\tilde{J} \rfloor \times K]$
Method in [25]	$\mathcal{O}[J \times \lfloor \frac{2}{3}J \rfloor + \lfloor \frac{2}{3}J \rfloor \times \lfloor \frac{4}{9}J \rfloor + \lfloor \frac{4}{9}J \rfloor \times K]$
Conventional DNN	$\mathcal{O}[J \times \lfloor \frac{J}{2} \rfloor + \lfloor \frac{J}{2} \rfloor \times J \times p + J \times \lfloor \frac{2}{3}J \rfloor \times p + \lfloor \frac{2}{3}J \rfloor \times \lfloor \frac{4}{9}J \rfloor \times p + \lfloor \frac{4}{9}J \rfloor \times \gamma \times p]$
MUSIC	$4 \times \mathcal{O}[(M+1)(M-K)\tilde{\gamma} + M^2K]$
CBF	$4 \times \mathcal{O}[(M+1)M\tilde{\gamma}]$
DeepMUSIC	$\mathcal{O}[K_{s1}^2 \times C_{in1} \times N_f \times M^2 + K_{s1}^2 \times N_f \times N_f \times M^2 + K_{s2}^2 \times N_f \times N_f \times M^2 \times 2 + M^2 \times N_f \times C_{out1} + C_{out1} \times C_{out2}]$
DCN	$\mathcal{O}[K_{s3} \times C_{in2} \times N_{f1} \times L + K_{s4} \times N_{f1} \times N_{f2} \times L + K_{s5} \times N_{f2} \times N_{f3} \times L + K_{s6} \times N_{f3} \times N_{f4} \times L]$

Table 4. Averaged testing time for one trial.

Method	Light Weight DNN	Method in [25]	Conventional DNN	MUSIC with Gird $1°$	CBF with Grid $1°$	MUSIC with Gird $0.1°$	CBF with Grid $0.1°$	Deep MUSIC	DCN
Testing time/ms	0.9	1.1	3.3	3.7	2.3	22.9	17.9	26.3	20.4

5. Results

In this section, by conducting simulation experiments, the proposed lightweight DNN is compared with the conventional DNN [18], the method in [25], DeepMUSIC in [20] and DCN in [19] in terms of testing time and the root-mean-square-error (RMSE) of DOA estimation. In addition, the traditional spectrum-based methods such as MUSIC and CBF are also included for comparisons. Furthermore, the Cramér–Rao Bound (CRB) of DOA estimation [2] is given as a lower bound. The DNN models are implemented using TensorFlow as the backend. In the testing stage, for a fair comparison of testing time, all the above-mentioned methods are executed on the Intel(R) Core(TM) i7-8750H CPU at 2.20 GH.

5.1. Simulation Settings

Assuming that the ULA consists of 22 elements with an inter-element spacing equal to $\frac{\lambda}{2}$. Supposing that there are two source signals impinging onto the array, of which the DOA range is from $\theta_{min} = -60°$ to $\theta_{max} = 60°$. The angular interval between the DOAs of two source signals is from $\triangle_{min} = 2°$ to $\triangle_{max} = 40°$, with $\triangle_d = 2°$ and $\eta = 1°$. The SNRs for different source signals are equal and SNR_k is defined as the power ratio of the k-th source signal to noise in dB, which is given below.

$$SNR_k = 10log_{10} \frac{\sigma_{s_k}^2}{\sigma_{\tilde{n}}^2}. \tag{24}$$

For the DNN models, in the training phase, the snapshots are set as 400. In addition, using input vectors from multiple SNRs of $\{-13 \text{ dB}, -10 \text{ dB}, -5 \text{ dB}, 0 \text{ dB}\}$ to train the network simultaneously. Moreover, 10 groups of covariance vectors are collected for each direction setting with random noise. Therefore, $(118 + 116 + ... + 80) \times 4 \times 10 = 79,200$ input vectors are collected in the training dataset in total. The learning rate is $\mu = 0.001$ and the mini-batch size is 32, the order of training data is shuffled in each epoch.

5.2. MSE Loss during Training and Validation

In this section, as given in Figure 3, we provide the training and validation MSE loss of the proposed lightweight DNN versus the number of epochs by randomly dividing the training data into 80% for training and 20% for validation. From Figure 3, we observe that the training loss and validation loss gradually reduce when the number of epochs increases and converges at about 400 epochs. Furthermore, they are close to each other. Therefore, we conclude that the proposed lightweight DNN with the input vector after the removal of data redundancy can accomplish the task of DOA estimation well. The detailed performance of DOA estimation in the testing stage is given as follows.

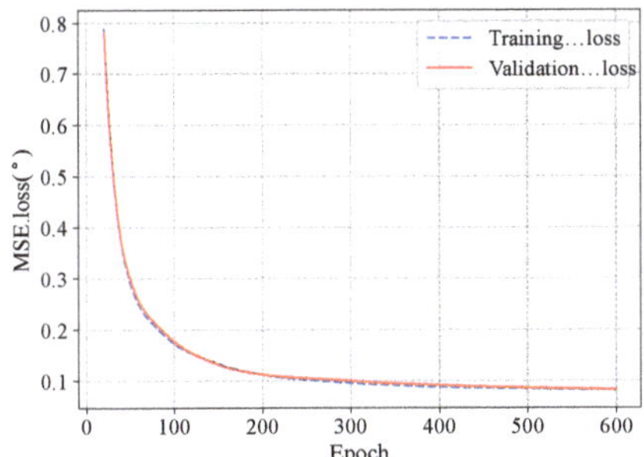

Figure 3. MSE loss versus epoch given by the proposed lightweight DNN.

5.3. RMSE versus SNRs and Testing Time

In the testing phase, in order to verify the generalization of the DNN models, the noises and source signals are different from those that appeared in the training phase. In addition, the DOAs of sources are set to be non-integer (that is, off-grid), which does not appear in the training stage. The DOAs of source signals are $(\theta_1, \theta_2) = (-40.55°, -36.3°)$. The RMSE is used to measure the testing performance of different methods, defined as

$$RMSE = \sqrt{\frac{1}{GK} \sum_{g=1}^{G} \sum_{k=1}^{K} |\widehat{\theta}_{k,g} - \theta_k|^2},$$ (25)

where G is the number of Monte Carlo simulation experiments, which is set as 200. $K = 2$. $\widehat{\theta}_{k,g}$ represents the DOA estimation value of the k-th source signal in the g-th experiment. In this part, SNR is taken from -16 dB to 10 dB with an interval of 2 dB and the number of snapshots is 400. The RMSE of DOAs estimated by the above-mentioned methods under different SNRs is given in Figure 4. Table 4 shows the averaged testing time for one trial. Figure 4 illustrates that the proposed lightweight DNN performs better than the method in [25], conventional DNN and the MUSIC and CBF methods with a grid of 1°. Its superiority is obvious when the SNR is lower than -8 dB. Moreover, the time spent by the lightweight DNN is about four times less than that spent by the MUSIC method with a grid of 1°. On the other hand, the proposed lightweight DNN has estimation accuracy lower than the MUSIC method with a grid of 0.1°. This is because the DNN-based approach yields biased estimators [20]. In contrast, the MUSIC method provides unbiased estimation when the source signals are uncorrelated and the number of arrays and snapshots is large [29,30]. It is noted that in Figure 4, the CBF method always fails because it suffers from the Rayleigh limit. In addition, the MUSIC method with a grid of 0.1° performs closely to the CRB when the SNR is larger than -8 dB. On the other hand, its performance deviates from the CRB when the SNR is larger than 5 dB. This phenomenon is caused by the limit of the searching grid in the MUSIC Method. Furthermore, as illustrated in Figure 4, the performance of the DeepMUSIC method is similar to that of the MUSIC method with a grid of 1°. This is because the label of the DeepMUSIC is the spatial spectrum of the MUSIC method and the grid in the DeepMUSIC method is equal to 1° to be consistent with the grid for other DNN methods. In addition, the DCN method performs better than the other methods except the lightweight DNN method, in most cases. In terms of testing time as given in Table 4, both DeepMUSIC and DCN methods cost much more than the lightweight DNN method.

It is noted that the higher estimation accuracy of the MUSIC method with a grid of $0.1°$ costs more spectrum searching load and the time it takes is about 25 times more than that by the lightweight DNN, as shown in Table 4. In addition, it is observed the time spent by the lightweight DNN is about 3 times less than that by the conventional DNN.

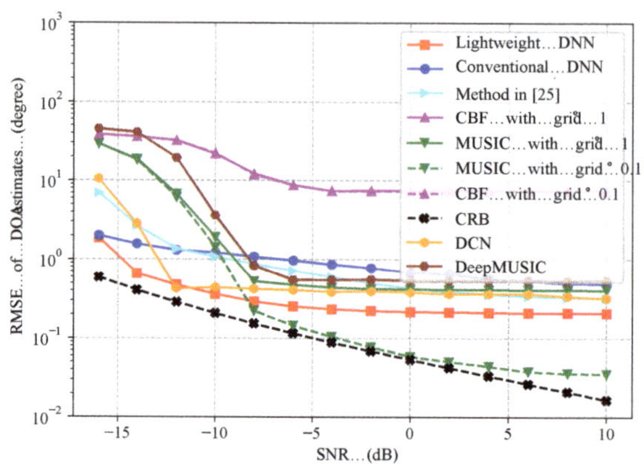

Figure 4. RMSE versus SNR when $(\theta_1, \theta_2) = (-40.55°, -36.3°)$.

5.4. RMSE versus DOA Separations

In this part, the RMSE of different methods is shown with the variation of intervals between the DOAs of two source signals. The DOA of the first source signal is set to be $-40.55°$ and the DOA of the second source signal is equal to $-40.55° + \widetilde{\Delta}$, where $\widetilde{\Delta}$ is taken from the set of $\{2.25°, 4.25°, 8.25°, \cdots, 32.25°, 36.25°\}$ in sequence. The number of snapshots is 400. When SNR is -2 dB, the RMSE of DOA estimated by the above-mentioned methods under different DOA separations is shown in Figure 5. From Figure 5, it can be seen that the lightweight DNN performs better than the MUSIC method and CBF method when their searching grid is set to be $1°$. Furthermore, it is always superior to the method in [25], conventional DNN method, DeepMUSIC method, and DCN method. Similar to the Figure 4, the MUSIC method with a grid of $0.1°$ approaches the CRB in most cases. However, it is noted that in a very small DOA separation such as $2.25°$, even the MUSIC method with a grid of $0.1°$ fails. In contrast, the lightweight DNN performs well. In addition, it is shown that the CBF method with a grid of $0.1°$ gradually approaches the CRB when the DOA separation increases.

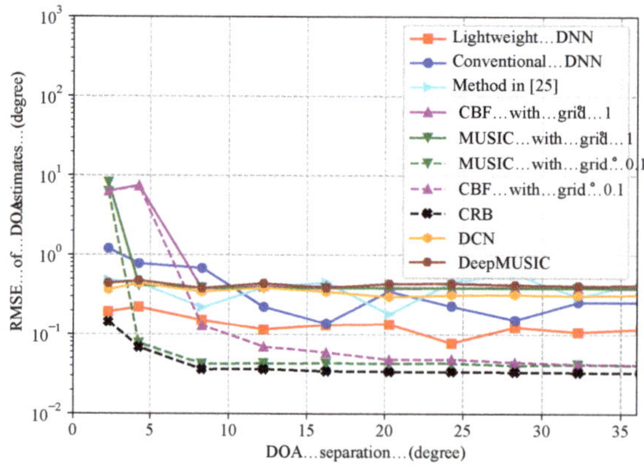

Figure 5. RMSE versus DOA separation when SNR = -2 dB.

5.5. RMSE versus Number of Snapshots

In this part, the number of snapshots is selected from the set of {30, 50, 100, 200, 300, 400, 500, 600, 700, 900, 1200, 1500, 1800, 2000}, the DOAs of the two source signals are $(\theta_1, \theta_2) = (-40.55°, -36.3°)$. Figure 6 shows the RMSE of all methods against the number of snapshots when the SNR is equal to -2 dB. From Figure 6, it is found that except for the CBF method, the other methods perform better when the number of snapshots increases. In addition, it is observed that the DNN models trained in the scenario of 400 snapshots are applicable to the scenarios of more snapshots and fewer snapshots. Furthermore, the lightweight DNN behaves significantly better than the conventional DNN when the number of snapshots is less than 900. As the number of snapshots increases, the estimation accuracy of lightweight DNN is still slightly higher than that of the conventional DNN, method in [25], DeepMUSIC method, and DCN method. It is worth noting that the lightweight DNN is always superior to the MUSIC method with a grid of 1° and it performs better than the MUSIC method with 0.1° when the number of snapshots is less than 100.

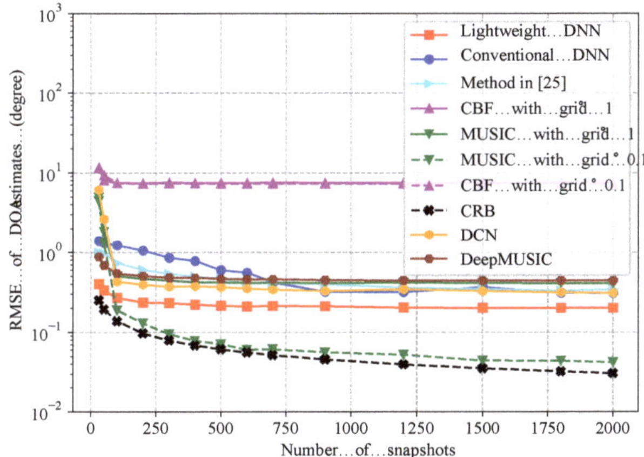

Figure 6. RMSE versus number of snapshots when SNR = -2 dB.

5.6. RMSE versus Power Ratio of Two Source Signals

The DOAs of the two closely spaced source signals are $(\theta_1, \theta_2) = (-40.55°, -38.3°)$. The number of snapshots for both source signals is 400. The SNR for the first source signal is fixed as -2 dB. Figure 7 demonstrates the RMSE versus the power ratio of the second source signal to the first source signal. From Figure 7, we observe that the RMSE of the lightweight DNN method increases from 0.2° to about 1.5° when the power ratio of the second source signal to the first one increases from 1 to 14. Similarly, the RMSE of the conventional DNN, the method in [25], DeepMUSIC, and DCN methods increase slightly with the increment of the power ratio. As shown in Figure 7, the MUSIC and CBF methods always fail because the DOAs of source signals are very close. It is noted that the CRB reduces a bit when the power ratio increases. This is because the power of the second source signal is increased with the increment of the power ratio. However, the CRB is limited by the closely spaced source signals.

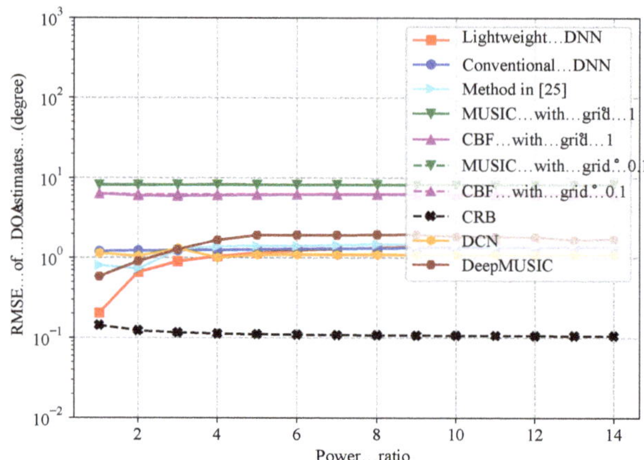

Figure 7. RMSE versus power ratio of second source signal to the first one when $(\theta_1, \theta_2) = (-40.55°, -38.3°)$.

6. Discussion

From the analysis above, it can be seen that the number of total parameters in the lightweight DNN model is significantly reduced compared to those DNN models that use the upper triangular elements of the covariance matrix as input. In particular, when the number of sensors is 22, it is 2 and 3 orders of magnitude less than that in the conventional DNN model and the method in [25], respectively. This fact makes the proposed lightweight DNN suitable for real-time embedded applications. Furthermore, it is noted that the lightweight DNN can preserve high accuracy of DOA estimation and perform better than the conventional DNN and method in [25]. In addition, it provides higher estimation accuracy and costs less trainable parameters and computational load than CNN-based methods such as DeepMUSIC and DCN. Also, it is illustrated that the lightweight DNN performs better than the spatial spectrum-based methods such as MUSIC and CBF method under harsh conditions such as low SNRs and/or closely spaced source signals and/or few snapshots. Moreover, its testing time is obviously shorter than that of the MUSIC and CBF method, due to the avoidance of spectrum searching and matrix decomposition. On the other hand, under good conditions such as high SNRs, the estimation accuracy of the lightweight DNN is lower than the MUSIC method with a grid of 0.1°. This is because the DNN-based methods are biased estimators while the MUSIC method can provide unbiased DOA estimation. It is noted that the MUSIC method with a grid of 0.1° provides higher estimation accuracy with a cost of a testing time of about 25 times more than that of the lightweight DNN. On the other hand, as shown in simulation results, the lightweight DNN can achieve high estimation accuracy such as 0.2° when the SNR is not extremely low (not lower than -6 dB) and the number of snapshots is not very small (not smaller than 100).

7. Conclusions

In order to make the DNN-based DOA estimation approaches real-time and less costly, we proposed a lightweight DNN model for a ULA. Compared to the conventional DNN model, the proposed lightweight DNN model has two improvements. Firstly, its input vector is designed by using the knowledge of ULA (that is, the steering vector of the ULA has the property of Vandermonde) to implement the removal of data redundancy as well as retain the DOA information. Therefore, the input vector is $\frac{M}{2}$ times less than the conventional DNN model, which contributes to reducing the sizes of the following hidden layers. Secondly, the output vector of the lightweight DNN model is designed in a regression fashion instead of classification, which has a size equal to the number of sources. Overall, the number of total parameters in the lightweight DNN model is significantly reduced compared to that in the existing DNN models. Furthermore, the lightweight DNN model performs better than the existing DNN models because the lightweight DNN model explores the characteristics of the signal received by the ULA for designing its input.

Author Contributions: Conceptualization and methodology, A.L., J.G. and Y.A.; software, validation, and visualization, J.G. and A.L.; formal analysis and investigation, J.G.; resources and data curation, A.L. and Z.L.; writing—original draft preparation and writing—review and editing, A.L., J.G. and Y.A.; supervision, A.L. and Z.L.; project administration, A.L.; funding acquisition, Y.A. All authors have read and agreed to the published version of the manuscript.

Funding: This work was supported in part by Natural Science Basic Research Program of Shaanxi Province, China (Program No. 2023-JC-YB-573), and in part by Basic Research Program of Taicang City, China, in 2023 (Program No. TC2023JC05), and in part by the Fundamental Research Funds for the Central Universities under Grants D5000210641 and G2022WD01022.

Data Availability Statement: Data are contained within the article.

Conflicts of Interest: The authors declare no conflicts of interest.

References

1. Godara, L.C. Application of antenna arrays to mobile communications. II. Beam-forming and direction-of-arrival considerations. *Proc. IEEE* **1997**, *85*, 1195–1245. [CrossRef] [CrossRef]
2. Van Trees, H.L. *Optimum Array Processing: Part IV of Detection, Estimation, and Modulation Theory*; John Wiley & Sons: Hoboken, NJ, USA, 2002.
3. Pesavento, M.; Trinh-Hoang, M.; Viberg, M. Three more decades in array signal processing research: An optimization and structure exploitation perspective. *IEEE Signal Process. Mag.* **2023**, *40*, 92–106. [CrossRef] [CrossRef]
4. Xiao, G.; Quan, Y.; Sun, Z.; Wen, B.; Liao, G. Design of a Digital Array Signal Processing System with Full Array Element. *Remote Sens.* **2023**, *15*, 4043. [CrossRef] [CrossRef]
5. Wen, F.; Shi, J.; Wang, X.; Wang, L. Angle estimation for MIMO radar in the presence of gain-phase errors with one instrumental Tx/Rx sensor: A theoretical and numerical study. *Remote Sens.* **2021**, *13*, 2964. [CrossRef] [CrossRef]
6. Capon, J. High-resolution frequency-wavenumber spectrum analysis. *Proc. IEEE* **1969**, *57*, 1408–1418. [CrossRef] [CrossRef]
7. Schmidt, R. Multiple emitter location and signal parameter estimation. *IEEE Trans. Antennas Propag.* **1986**, *34*, 276–280. [CrossRef] [CrossRef]
8. Roy, R.; Kailath, T. ESPRIT-estimation of signal parameters via rotational invariance techniques. *IEEE Trans. Acoust. Speech Signal Process.* **1986**, *37*, 984–995. [CrossRef] [CrossRef]
9. Haardt, M.; Nossek, J.A. Unitary ESPRIT: How to obtain increased estimation accuracy with a reduced computational burden. *IEEE Trans. Signal Process.* **1995**, *43*, 1232–1242. [CrossRef] [CrossRef]
10. Viberg, M.; Ottersten, B. Sensor array processing based on subspace fitting. *IEEE Trans. Signal Process.* **1991**, *39*, 1110–1121. [CrossRef] [CrossRef]
11. Yan, F.G.; Jin, M.; Liu, S.; Qiao, X.L. Real-valued MUSIC for efficient direction estimation with arbitrary array geometries. *IEEE Trans. Signal Process.* **2014**, *62*, 1548–1560. [CrossRef] [CrossRef]
12. Liu, A.; Liao, G. An eigenvector based method for estimating DOA and sensor gain-phase errors. *Digit. Signal Process.* **2018**, *79*, 116–124. [CrossRef] [CrossRef]
13. Tayem, N.; Kwon, H.M. Conjugate esprit (c-sprit). *IEEE Trans. Trans. Antennas Propag.* **2004**, *52*, 2618–2624. [CrossRef] [CrossRef]
14. Pastorino, M.; Randazzo, A. A smart antenna system for direction of arrival estimation based on a support vector regression. *IEEE Trans. Antennas Propag.* **2005**, *53*, 2161–2168. [CrossRef] [CrossRef]
15. El Zooghby, A.H.; Christodoulou, C.G.; Georgiopoulos, M. A neural network-based smart antenna for multiple source tracking. *IEEE Trans. Antennas Propag.* **2000**, *48*, 768–776. [CrossRef] [CrossRef]
16. Randazzo, A.; Abou-Khousa, M.A.; Pastorino, M.; Zoughi, R. Direction of arrival estimation based on support vector regression: Experimental validation and comparison with MUSIC. *IEEE Antennas Wirel. Propag. Lett.* **2007**, *6*, 379–382. [CrossRef] [CrossRef]
17. Xiao, X.; Zhao, S.; Zhong, X.; Jones, D.L.; Chng, E.S.; Li, H. A learning-based approach to direction of arrival estimation in noisy and reverberant environments. In Proceedings of the 2015 IEEE International Conference on Acoustics, Speech and Signal Processing (ICASSP), South Brisbane, Australia, 6 August 2015; pp. 2814–2818.
18. Liu, Z.M.; Zhang, C.; Philip, S.Y. Direction-of-arrival estimation based on deep neural networks with robustness to array imperfections. *IEEE Trans. Antennas Propag.* **2016**, *66*, 7315–7327. [CrossRef] [CrossRef]
19. Wu, L.; Liu, Z.M.; Huang, Z.T. Deep convolution network for direction of arrival estimation with sparse prior. *IEEE Signal Process. Lett.* **2019**, *26*, 1688–1692. [CrossRef] [CrossRef]
20. Elbir, A.M. DeepMUSIC: Multiple signal classification via deep learning. *IEEE Sens. Lett.* **2020**, *4*, 1–4. [CrossRef] [CrossRef]
21. Papageorgiou, G.K.; Sellathurai, M.; Eldar, Y.C. Deep networks for direction-of-arrival estimation in low SNR. *IEEE Trans. Signal Process.* **2021**, *69*, 3714–3729. [CrossRef] [CrossRef]

22. Labbaf, N.; Oskouei, H.D.; Abedi, M.R. Robust DoA estimation in a uniform circular array antenna with errors and unknown parameters using Deep Learning. *IEEE Trans. Green Commun. Netw.* **2023**, *7*, 2143–2152. [CrossRef] [CrossRef]
23. Ping, Z. DOA estimation method based on neural network. In Proceedings of the 2015 10th International Conference on P2P, Parallel, Grid, Cloud and Internet Computing (3PGCIC), Krakow, Poland, 4–6 November 2015; pp. 828–831.
24. Agatonovic, M.; Stankovic, Z.; Milovanovic, I.; Doncov, N.; Sit, L.; Zwick, T.; Milovanovic, B. Efficient neural network approach for 2D DOA estimation based on antenna array measurements. *Prog. Electromagn. Res.* **2013**, *137*, 741–758. [CrossRef] [CrossRef]
25. Xiong, Y.; Liu, A.; Gao, X.; Yauhen, A. DOA Estimation Using Deep Neural Network with Regression. In Proceedings of the 2022 5th International Conference on Information Communication and Signal Processing (ICICSP), Shenzhen, China, 26–28 November 2022; pp. 1–5.
26. Zhou, C.; Gu, Y.; Fan, X.; Shi, Z.; Mao, G.; Zhang, Y.D. Direction-of-arrival estimation for coprime array via virtual array interpolation. *IEEE Trans. Signal Process.* **2018**, *66*, 5956–5971. [CrossRef] [CrossRef]
27. Reed, I.S.; Mallett, J.D.; Brennan, L.E. Rapid convergence rate in adaptive arrays. *IEEE Trans. Aerosp. Electron. Syst.* **1974**, *10*, 853–863. [CrossRef] [CrossRef]
28. Wax, M.; Kailath, T. Detection of signals by information theoretic criteria. *IEEE Trans. Acoust. Speech Signal Process.* **1985**, *33*, 387–392. [CrossRef] [CrossRef]
29. Stoica, P.; Nehorai, A. MUSIC, maximum likelihood, and Cramer-Rao bound. *IEEE Trans. Acoust. Speech Signal Process.* **1989**, *37*, 720–741. [CrossRef] [CrossRef]
30. Li, F.; Liu, H.; Vaccaro, R.J. Performance analysis for DOA estimation algorithms: Unification, simplification, and observations. *IEEE Trans. Aerosp. Electron. Syst.* **1993**, *29*, 1170–1184. [CrossRef] [CrossRef]

Technical Note

Information Extraction and Three-Dimensional Contour Reconstruction of Vehicle Target Based on Multiple Different Pitch-Angle Observation Circular Synthetic Aperture Radar Data

Jian Zhang [1], Hongtu Xie [1,*], Lin Zhang [2] and Zheng Lu [3]

[1] School of Electronics and Communication Engineering, Shenzhen Campus of Sun Yat-Sen University, Shenzhen 518107, China; zhangj765@mail2.sysu.edu.cn
[2] Department of Early Warning Technology, Air Force Early Warning Academy, Wuhan 430019, China
[3] Institute of Remote Sensing Satellite, China Academy of Space Technology, Beijing 100094, China
* Correspondence: xiehongtu@mail.sysu.edu.cn

Citation: Zhang, J.; Xie, H.; Zhang, L.; Lu, Z. Information Extraction and Three-Dimensional Contour Reconstruction of Vehicle Target Based on Multiple Different Pitch-Angle Observation Circular Synthetic Aperture Radar Data. *Remote Sens.* **2024**, *16*, 401. https://doi.org/10.3390/rs16020401

Academic Editors: Gerardo Di Martino, Xiaotao Huang, Jianguo Liu, Xinbo Li, Shengchun Piao, Junyuan Guo, Wei Guo and Jiahua Zhu

Received: 20 December 2023
Revised: 15 January 2024
Accepted: 16 January 2024
Published: 20 January 2024

Abstract: The circular synthetic aperture radar (CSAR) has the ability of all-round continuous observation and high-resolution imaging detection, and can obtain all-round scattering information and higher-resolution images of the observation scene, so as to realize the target information extraction and three-dimensional (3D) contour reconstruction of the observation targets. However, the existing methods are not accurate enough to extract the information of vehicle targets. Through the analysis of the vehicle target scattering model and CSAR image characteristics, this paper proposes a vehicle target information extraction and 3D contour reconstruction method based on multiple different pitch-angle observation CSAR data. The proposed method creatively utilizes the projection relationship of the vehicle in 2D CSAR imaging to reconstruct the 3D contour of the vehicle, without prior information. Firstly, the CSAR data obtained from multiple different pitch-angle observations are fully utilized, and the scattering points of odd-bounce reflection and even-bounce reflection echoes are extracted from the two-dimensional (2D) coherent CSAR images of the vehicle target. Secondly, the basic contour of the vehicle body is extracted from the scattering points of the even-bounce reflected echoes. Then, the geometric projection relationship of the "top–bottom shifting" effect of odd-bounce reflection is used to calculate the height and position information of the scattering points of odd-bounce reflection, so as to extract the multi-layer 3D contour of the vehicle target. Finally, the basic contour and the multi-layer 3D contour of the vehicle are fused to realize high-precision 3D contour reconstruction of the vehicle target. The correctness and effectiveness of the proposed method are verified by using the CVDomes simulation dataset of the American Air Force Research Laboratory (AFRL), and the experimental results show that the proposed method can achieve high-precision information extraction and realize distinct 3D contour reconstruction of the vehicle target.

Keywords: circular synthetic aperture radar; odd-bounce reflection; even-bounce reflection; vehicle target; information extraction; 3D contour reconstruction; "top–bottom shifting" effect

1. Introduction

Synthetic aperture radars (SAR) obtain high resolution in the range direction by transmitting large-bandwidth signals, at the same time, the platform moves to observe the target at a large angle to obtain high resolution in the azimuth direction [1]. As one of the most productive sensors in the field of microwave remote sensing and advanced array signal processing, SAR has received rapid development and widespread attention. SAR imaging can obtain more electromagnetic scattering information of the observed target through the reconstruction of the target scattering function, which is helpful for the analysis, classification and identification of target characteristics [2]. SAR imaging offers significant advantages in remote sensing observation, due to its ability to operate regardless of external environmental conditions such as weather and light. It can provide all-weather and all-day reconnaissance capabilities, making it highly versatile [3]. As a result, SAR imaging has found wide-ranging applications in various fields. For instance, in the agricultural sector, SAR imaging enables the monitoring of crop growth, soil moisture levels, and the occurrence of pests and diseases. This accurate and timely data supports crop assessment and agricultural management [4]. In the domain of disaster forecasting, SAR imaging plays a vital role. Its capability to detect surface deformations allows for early warning of natural disasters such as earthquakes, volcanic eruptions, and floods, this information is crucial for rescue efforts and emergency response coordination [5].

In addition, SAR imaging has proven to be of great value in civilian applications such as marine surveying and mapping [6]. It provides detailed and precise data that aids in understanding oceanographic features, coastal erosion, and bathymetry, this information is essential for maritime industries, environmental monitoring, and coastal zone management [7]. Additionally, in military operations, SAR imaging facilitates battlefield reconnaissance. Its high-resolution imagery and ability to track enemy targets accurately contribute to situational awareness and support strategic decision-making. Furthermore, SAR technology aids in strategic early warning systems, enhancing national security and defense preparedness [8,9].

Circular SAR (CSAR) refers to a 360-degree circular movement of the radar platform around the target scene, and its antenna emission beam is always directed at the target scene [10]. Compared to traditional SAR with a straight flight trajectory, CSAR offers 360-degree observation, resulting in targets with more complete contours and better suppression of background [11]. The omnidirectional scattering characteristics of the target obtained by CSAR can effectively improve target detection performance. The biggest advantage of CSAR imaging is its ability to observe targets in all directions, allowing the resulting image to reflect the backscattering information of the target in all azimuths. This leads to higher image resolution and enables certain three-dimensional (3D) imaging capabilities [12]. The comprehensive coverage of CSAR imaging enables a more detailed and accurate representation of the target's backscattering characteristics, resulting in improved image resolution and the potential for 3D imaging [13].

In recent years, the reconstruction of 3D images of CSAR data has become a research hotspot [14–16]. At present, the main method for 3D image reconstruction of observation scene targets using CSAR is to use multi-baseline CSAR imaging technology [17,18]. However, it requires the acceptance and processing of multi-baseline CSAR data, which can be time-consuming and expensive in terms of hardware requirements. Additionally, this technology is associated with complex imaging algorithms and often exhibits low processing efficiency.

E. Dungan, from the American Air Force Research Laboratory (AFRL), has proposed a 3D image reconstruction method of vehicle contour based on single-baseline fully polarimetric CSAR data [19]. Compared to HoloSAR imaging [20], this method has the advantage of requiring only single-baseline fully polarimetric CSAR data. This significantly reduces the cost of acquiring 3D images and simplifies the imaging algorithm, leading to improved processing efficiency. The use of single-baseline fully polarimetric CSAR data allows for the acquisition of meaningful information with fewer data points, reducing the hardware and computational costs associated with data collection and processing. Additionally, the simplified imaging algorithm used in this method reduces the complexity of the reconstruction process, resulting in faster processing times and decreased computational resources required. However, this method rectangles the vehicle contour to form a multi-dimensional variable search process, which greatly increases the amount of calculation, reduces the efficiency of the algorithm. Also, this method weakens the contour features of the target vehicle, and it is not conducive to the subsequent vehicle classification and recognition. Although this method does not require multi-baseline CSAR data, it requires a variety of polarized CSAR data, which leads to an increase in the amount of computation and a high cost of 3D image reconstruction.

L. Chen, from the National University of Defense Technology, has proposed a 3D image reconstruction method of vehicle contours based on single baseline single polarization CSAR data [21]. This method starts from the incoherent imaging processing, the basic contour information is extracted based on the two-dimensional (2D) image of the vehicle, and then the height information of the scattered points of the attribute is deduced by using the "top–bottom shifting" effect, and then the 3D contour image of the target vehicle is reconstructed. Also, this method obtains high-precision vehicle size estimation results and has the unique advantages of high efficiency and low cost. However, the method introduces the ratio of the underbody profile to the roof profile as a priori information, and does not extract the multi-layer profiles of the vehicle.

In order to solve the above problems, this paper proposes a new method to extract information and 3D contour reconstruction of vehicle target, based on multiple different pitch-angle observation CSAR data. The research content of this paper is as follows: Section 2 researches the distribution characteristics of the electromagnetic reflection model of the vehicle target, and the 2D CSAR imaging characteristics of the vehicle target at different pitch-angle observations. Section 3 provides a detailed description of the proposed method for 3D reconstruction of vehicle target contours. In Section 4, the civilian vehicle dome (CVDomes) simulation dataset is used to verify the correctness of the theoretical analysis and the effectiveness of the proposed method. Finally, the research work of this paper is summarized, and the next research work is prospected in Section 5.

2. Vehicle Scattering Model

This section begins by analyzing the scattering characteristic models of radar electromagnetic wave odd-bounce reflection and even-bounce reflection of the vehicle targets under far-field conditions. Then, an analysis of the 2D CSAR imaging features of vehicle targets at multiple different pitch-angle observations is presented.

According to the electromagnetic theory, the high-frequency echo response of a complex target can be regarded as the sum of the attribute scattering centers of multiple standard scatterers [22]. The attribute scattering center contains relevant information such as the position, amplitude, and polarization of the target, which can better describe the scattering characteristics of the target on the SAR data.

2.1. Odd-Bounce Reflection

Figure 1 shows the "top–bottom shifting" model of the vehicle target and the backscattering of electromagnetic waves, and the electromagnetic echoes reflected by the vehicle target are mainly divided into odd-bounce reflection (blue lines in Figure 1) and even-bounce reflection (orange lines in Figure 1). Among them, odd-bounce reflection refers to the electromagnetic echo that returns to the antenna after an odd number of bounce reflection, mainly provided by the edges and corners of the vehicle target (such as the roof ridge, etc.). The edges and corners of the vehicle target are the height dimensional information with the vehicle target, which can form the 3D contour of the vehicle.

As shown in Figure 1a, when the ground plane $x - y$ is used as the imaging plane, the distance from the point A to the radar platform is R, the height from the point A to the imaging plane is h, and the pitch angle θ_e represents the observation angle from the radar to the point A. During imaging, the point A is projected onto the imaging plane A_P. This shift in the position of the projection, which is caused by the height of focus, is called the "layover effect" [23]. The "top–bottom shifting" distance is l, the slope distance from the antenna phase center to the target is R, if l is much smaller than R, that is $l \ll R$, this corresponds to the far-field condition. Then, the height of the point A to the ground plane $x - y$ is:

$$h = l\cot(\theta_e) \tag{1}$$

Therefore, starting from the CSAR 2D image of the vehicle, the edge position of the image of the vehicle ridge is obtained, which is the position of the odd-bounce reflected bright line. The shifting distance of the top and bottom of the ridge can be obtained, and then the height of the ridge h and its position coordinates are calculated.

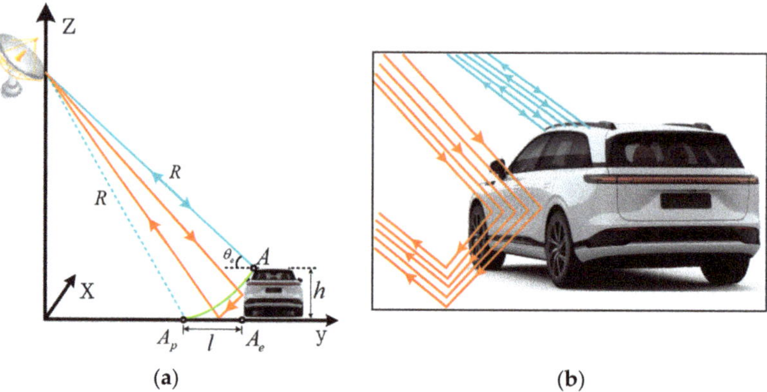

Figure 1. The "top–bottom shifting" model of a vehicle target and the backscattering of electromagnetic waves: (**a**) Schematic diagram of the calculation of the "top–bottom shifting" of odd-bounce reflection; (**b**) schematic diagram of odd-bounce reflection (orange lines) and even-bounce reflection (blue lines) backscattering.

2.2. Even-Bounce Reflection

Figure 2 shows the even-bounce reflection geometry path. Even-bounce reflection refers to the secondary bounce reflection of electromagnetic waves emitted by radar antennas and finally return to the receiving antenna [24]. The even-bounce reflection path can be assumed to consist of three parts, including the outbound R_1, the first bounce reflection R_2, and the second bounce reflection R_3. θ_e is the pitch angle from the radar to the point A. When the incident wave passes through the outbound journey R_1 and reaches a certain scattering center with a height h of the vehicle, there is a scattering angle $\Delta\theta$ between the first bounce reflection R_2 and the specular reflection angle θ_e. R_2 generates a second bounce reflection R_3 through specular reflection with the ground, and R_3 subsequently returns to the receiving antenna. It is worth noting that R_2 and R_3 are not the only path, vary within the orange shaded area, as shown in Figure 2.

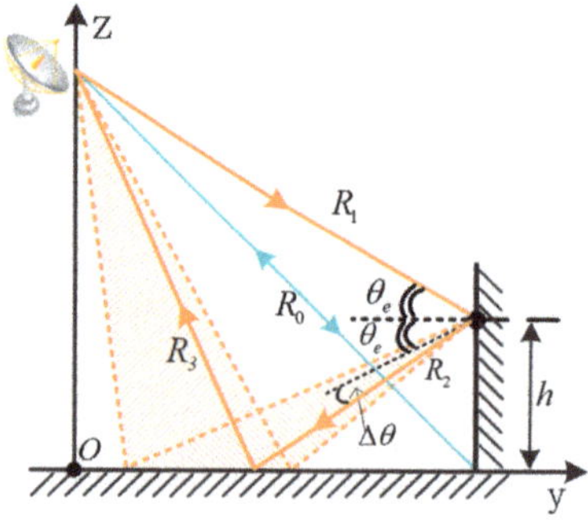

Figure 2. Geometry of the even-bounce reflection.

For a vehicle target parked on the flat ground, the side of the vehicle forms a dihedral reflection with the ground, the dihedral reflection is directional, and its reflected energy is concentrated in the vertical direction. The projection of the even-bounce reflected energy can be thought of as a distribution along the junction between the ground and the side of the vehicle, i.e., the distribution of the basic contour of the underside of the vehicle. Although the even-bounce reflection has undergone the secondary bounce reflection of different medium planes, the energy of the reflected echo is relatively large, due to the large reflection surface of the vehicle target. And the even-bounce reflection appears as a relatively bright and thick closed rectangular frame on the 2D CSAR image [25].

Figure 3 shows the vehicle CSAR imaging results at different pitch-angle observations. As shown in Figure 3d,e, there are scattered centers of discrete properties formed by even-bounce reflection in the vicinity of the vehicle contour. Due to the "top–bottom shifting" effect, the odd-bounce reflection forms an outer contour outside the vehicle's baseline, and as the pitch angle of the radar increases, the image of vehicle edge expands outward more.

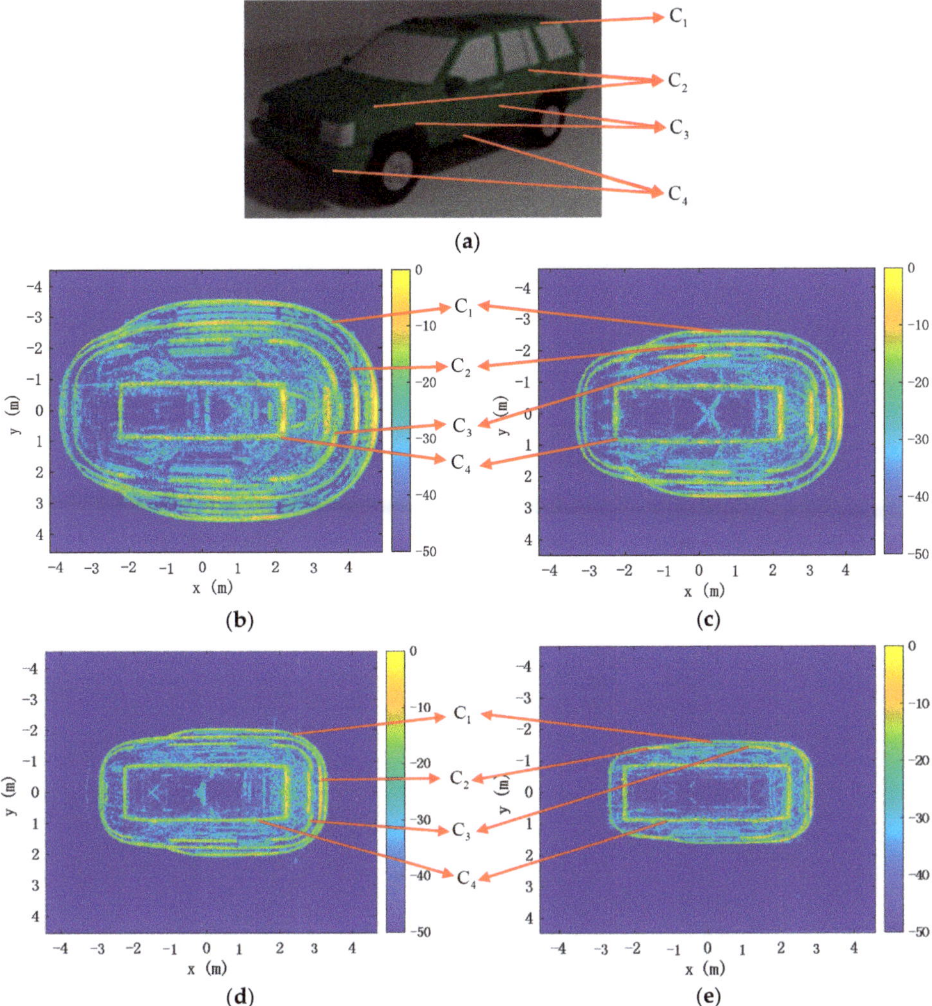

Figure 3. Simulated vehicle model and 2D CSAR images of the vehicle at different pitch-angle observations: (**a**) Photographs of simulated vehicle Jeep93; (**b**) 2D CSAR images of vehicles with the pitch angle of 60 degree; (**c**) 2D CSAR images of vehicles with the pitch angle of 50 degree; (**d**) 2D CSAR images of vehicles with the pitch angle of 40 degree; (**e**) 2D CSAR image of a vehicle with the pitch angle of 30 degree (the unit of the colorbar is dB).

2.3. CSAR Image Characteristics of the Vehicle Target at Different Pitch-Angle Observations

In the CSAR data of HH polarization, the reflected energy of the dihedral angle is larger and the contour is clearer, so the HH polarization is used in this method [26–29]. Due to the lower side lobes of coherent imaging, which is more conducive to extracting the edge of the vehicle, the proposed method chooses to process the coherent imaging. This method extracts the inner even-bounce reflection contour and the outer odd-bounce reflection contour of the vehicle from the CSAR image at the same time. Subsequently, the basic contour of the vehicle bottom is obtained from the even-bounce reflection, and the 3D contour of the vehicle with height information is obtained from the odd-bounce reflection. The proposed method makes full use of the data at 30°, 40°, 50° and 60° observation pitch angles, and the angle of view is more comprehensive, the extracted vehicle target information is richer, and the error of information extraction is smaller.

As can be seen from Figure 3, the body and the edge of the vehicle will form four relatively distinct curves. From the inside to the outside, the first curve C_1 is the top contour curve of the vehicle, which is composed of a circle of roof covers, and it is the highest, at about 1.63 m. The second curve, C_2, is the second layer of the vehicle formed by the door handle and the front cover, and its ridge curve is the second highest, at about 1.01–1.16 m. The third curve, C_3, is the curve formed at the top of the wheel, the defect position is the position of the wheel, and its height is the third highest, at about 0.64–0.68 m. The fourth closed curve C_4 is the imaging result of the dihedral angle formed between the body and the ground, which is the imaging result of the even-bounce reflection of electromagnetic waves. As a result, curve C_4 closely approximates a rectangular frame, capturing the basic contours of the vehicle body. The imaging position of curve C_4 is almost unchanged under different pitch-angle observations.

When the ridge is higher, the observed pitch-angle is larger, which will lead to a greater distance of the top-bottom shifting. Therefore, the image of the target vehicle in pitch-angle of 60 expends more than in pitch angle of 50, as shown in Figure 3b,c. When the pitch angle of observation is 30 degrees, the top–bottom shifting distance of the ridge is relatively small, so the first curve, C_1, and the second curve, C_2, coincide multiple times.

3. Information Extraction and 3D Contour Reconstruction

Based on the "top–bottom shifting" effect, a novel approach that exploits the projection imaging relationship between different pitch-angle observations under the same polarization is proposed to extract the 3D contour of the vehicle target.

3.1. Overall Framework

Firstly, coherent processing is used to generate CSAR vehicle images. The complete 360° total aperture data is divided into every 1° sub-aperture aperture without overlap. To achieve high-resolution and low side lobes in CSAR vehicle 2D images, the back projection algorithm (BPA) is employed, which is capable of adapting to the CSAR geometry [30,31]. Additionally, GPU acceleration is utilized for improved efficiency. Due to its lower sidelobes and ability to provide more detailed target imaging, coherent accumulation is performed to obtain a CSAR coherent image of the vehicle under consideration.

Secondly, the CSAR 2D image is converted to the polar coordinates. Since the CSAR image is a 360-degree surround image, the image of the vehicle's body and edges will be closed curves. In order to facilitate the extraction of contour curves, four sets of 2D CSAR images with multiple different pitch angles are transformed to polar coordinates, with the center of the image as the pole. The vehicle contour is changed from planar curves to a one-dimensional curve, so that the contour extraction and subsequent curve processing are easier.

Finally, the scattering points are obtained. Using the peak detection method, the odd-bounce reflection and even-bounce reflection scattering points of the four profile curves are extracted. As shown in Figure 4, the CSAR 2D image at the 60-degree pitch observation is transformed to polar coordinates, and the results (red dots in Figure 4) are extracted from odd-bounce and even-bounce reflection scattering points using the peak detection method. In Figure 4a, the leftmost red dots are the even-bounce reflection scattering points, and the other red dots are the odd-bounce reflection scattering points. And in Figure 4b, the scattering points that approximately make up the rectangular box inside are the even-bounce reflection scattering points; the three curves on the outside are the odd-bounce reflection scattering points.

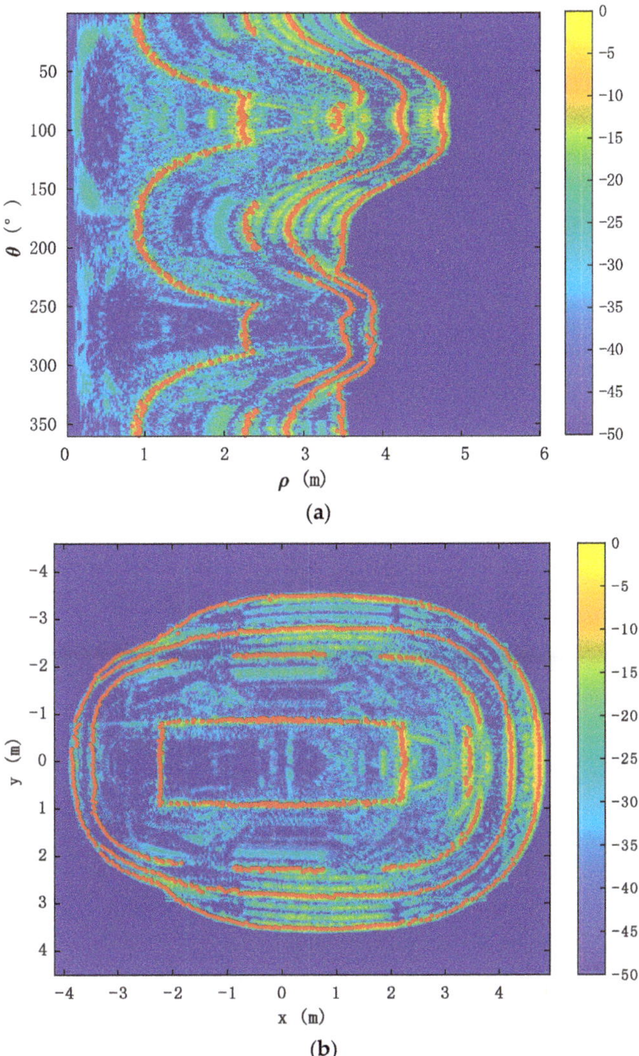

Figure 4. Extraction results of scattering points at the 60-degree pitch-angle observation (vehicle Jeep93): (**a**) The extraction result of the scattering points in polar coordinates; (**b**) the extraction result of the scattering points in Cartesian coordinates (the unit of the colorbar is dB).

3.2. Contour Extraction of Even-Bounce Reflection

After converting the CSAR 2D image to the polar coordinates, the basic contour of the vehicle consisting of the even-bounce reflection scattering points is extracted in the following two steps.

Step 1: Smooth filtering of even-bounce scatter points. The side of the vehicle forms a vertical dihedral angle with the ground. The reflected echo of this vertical dihedral angle remains relatively unchanged as the pitch observation angles vary. As a result, the imaging position of the even-bounce scattered bright line shows minimal variation at top view observation angles of 30°, 40°, 50°, and 60°. As shown in Figure 3b–e, the position of the rectangle in the center of the four images, i.e., the even-bounce scattering light line, is basically unchanged.

In our method, the even-bounce scattering points obtained from four pitch-angle observations are used for mean filtering and smoothing. In Figure 5a, the extracted results are obtained by applying mean filtering and smoothing to the even-bounce reflection scattering points. These results form the 2D projection of the even-bounce reflection contour, specifically representing the 2D projection of the basic contour of the vehicle body, from which the geometric feature parameters such as orientation, length, and width of the vehicle can be extracted.

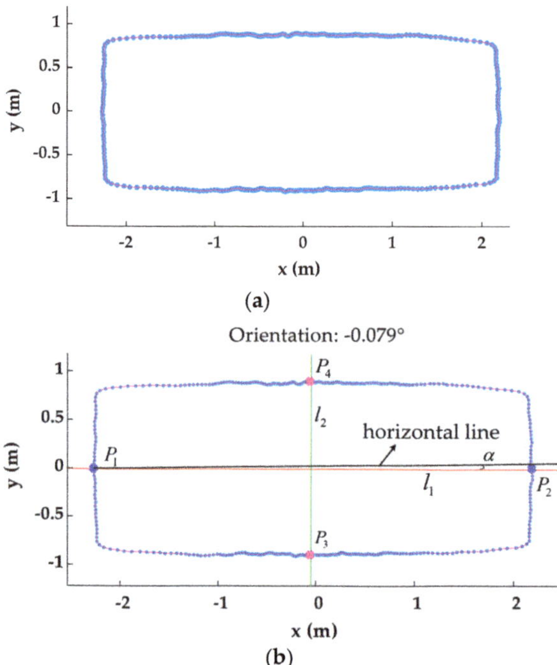

Figure 5. The results of the processing of the vehicle even-bounce reflection contour and the calculation of the length and width of the vehicle body (vehicle Jeep93): (**a**) The results of the even-bounce reflection scattering points extraction after the mean filtering smoothing process; (**b**) the results of extracting geometric feature parameters of the vehicle.

Step 2: Extract and calculate the geometric parameters of the vehicle body. According to the projection of the basic contour of the vehicle body obtained in step 1 on the 2D CSAR imaging plane, the principal axis direction is estimated by calculating the covariance matrix of the projection matrix. The principal axis direction represents the direction of the closed surface, and refers to the major axis direction of the basic outline of the vehicle body projection. By determining the intersection angle between the major axis and the X-axis of the ellipse with the same standard second-order central moment as the basic outline of the vehicle body projection, the orientation angle α is obtained. α represents the angle between the orientation of the vehicle and the horizontal line.

Then, a straight line l_1 is calculated, which passes through the projection of the basic outline of the vehicle body and has an inclination angle α. Next, the intersection points P_1 and P_2 of the straight line l_1 with the contour of the vehicle are obtained. Subsequently, the perpendicular bisector l_2 connecting P_1 and P_2 is found. Later, the intersection points P_3 and P_4 of l_2, and the projection of the basic contour of the vehicle body, are obtained.

The distance between P_1 and P_2 is the length of the vehicle body, and the distance between P_3 and P_4 is the width of the vehicle body. As shown in Figure 5b, the results are extracted for the geometric feature parameters such as orientation, length, and width of the vehicle. As shown in Figure 5b, the orientation of the vehicle is -0.079 degree with the horizontal line; the width and the length of the Jeep93 are 4.462 m and 1.796 m, respectively.

3.3. Contour Extraction of Odd-Bounce Reflection

Under different pitch-angle observations, the projection position shift will be different for imaging vehicle targets at a certain height, due to the "top–bottom shifting" effect. According to the odd-bounce reflection model in Section 2.1, there are:

$$htan(\theta_1) = l_1 = L_1 - L_0 \tag{2}$$

$$htan(\theta_2) = l_2 = L_2 - L_0 \tag{3}$$

$$htan(\theta_3) = l_3 = L_3 - L_0 \tag{4}$$

$$htan(\theta_4) = l_4 = L_4 - L_0 \tag{5}$$

where, h is the height of the vehicle ridge, which can be the different parts of the vehicle, including the front, rear, and sides. θ_1, θ_2, θ_3 and θ_4 correspond to θ_e values of 30, 40, 50, and 60 degrees, meaning the pitch-angle of radar observation are 30, 40, 50, and 60 degrees, respectively. $l_1 \sim l_4$ is the propagation distance of the "top–bottom shifting" in the 2D CSAR image at the 30~60 degree pitch-angle observations. $L_1 \sim L_4$ are the coordinate position of the vehicle ridge contour in the 2D CSAR image at the 30~60 degree pitch observation angles. And L_0 is the actual coordinate position of the simulated vehicle model.

By simultaneously solving Equations (2)–(5), we obtain

$$h = \frac{l_m - l_n}{\tan(\theta_m) - \tan(\theta_n)} \tag{6}$$

The value range of m is [2, 4] and the value range of n is [1, 3]. There are six combinations of the above equation and the average of the results of these six calculations can be used to obtain the height h of the vehicle edge. Returning to Equations (2)–(5), the coordinate position of the vehicle's rib contour can be obtained.

Therefore, after converting the CSAR 2D image to the polar coordinates, the curves composed of three odd-bounce reflection scattering points need to be extracted, namely C_1, C_2 and C_3 in Figure 3. Then, according to the different vehicle contour extensions in the two-dimensional CSAR image of the radar at the 30–60 degree pitch angle, the corresponding vehicle ridge height and its position coordinates of each contour curve are calculated by the joint relationship Equations (2)–(6).

The proposed method makes full use of different pitch-angle observations and uses multiple formulas to jointly calculate several times. It takes the average value to reduce the error and improves the accuracy of information extraction.

3.4. 3D Reconstruction of Vehicle Contour

From the above analysis, it can be seen that the roof contour containing height information can be extracted from the odd-bounce reflection. The position coordinates of the basic vehicle body contour consisting of the even-bounce reflection is obtained by the method shown in Section 3.2. Using the method in Section 3.3, the height of the vehicle ridge and its position coordinates composed of odd-bounce reflection can be obtained.

In the processing of fusing the basic vehicle body contour, the height of the vehicle ridge and its position coordinates, the base is the basic vehicle body contour, whose Z coordinate is zero in the xyz coordinate system. And the position coordinates of the vehicle ridge is in the same xyz coordinate system, whose Z coordinate is its height. Therefore, the whole contour of the vehicle can be reconstructed.

Based on the above analysis, the flow chart of the 3D contour reconstruction method of the vehicle target is shown in Figure 6. Firstly, the GPU-accelerated BPA is used to process the echo data for imaging, and the 2D CSAR images of the vehicle target to be reconstructed under four sets of different pitch-angle observations are obtained. Then, the polar coordinate transformation of the CSAR 2D image of the vehicle target is performed to extract the odd-bounce reflection and even-bounce reflection scattering points of the vehicle target. On the one hand, the scattering points formed by the even-bounce reflection are used to extract the basic contour of the vehicle body, and the geometric parameters such as the length and width of the vehicle. On the other hand, according to the "top–bottom shifting" model, the scattering points formed by the odd-bounce reflection are used to extract the height and position of the ridge. Finally, the basic contour of the vehicle body, as well as the height and position of the ridge, are fused to realize the reconstruction of the 3D contour of the vehicle.

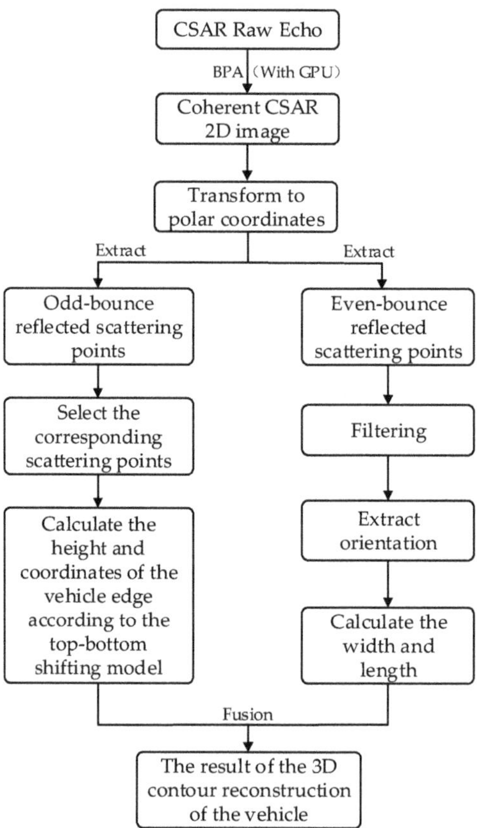

Figure 6. Flow diagram of the 3D contour reconstruction of the vehicle target.

4. Experiment and Analysis

Civilian vehicle domes (CVDomes)—a dataset of simulated X-band signatures of civilian vehicles is generated for $360°$ in azimuth direction [32]. The CVDomes dataset consists of simulation data with different pitch angles ranging from $30°\sim60°$. The main parameters of the CVDomes dataset are shown in Table 1.

Table 1. The main parameters of CVDomes dataset.

Parameter	Parameter Value
Carrier frequency	9.6 GHz
Signal bandwidth	5.35 GHz
Maximum no-blur distance	15 m
Pitch angle	$30°, 40°, 50°, 60°$

In order to prove the effectiveness of the proposed 3D contour reconstruction and information extraction method based on CSAR data for the vehicle target, the echo data processing of vehicles in CVDomes dataset is used to verify it. The vehicles Jeep93 and Jeep99 are processed using the proposed method; the results are shown in Figures 7 and 8, and Tables 2 and 3.

As shown in Figures 7a and 8a, h_{11} is the height of the side edge of the first vehicle prism curve (contours of the vehicle roof). h_{12} is the height of the rear edge of the first vehicle prism curve. h_{21} is the height of the front cover of the second vehicle prism curve. h_{22} is the height of the rear part of the second vehicle prism curve. h_{31} is the height of the front of the vehicle for the third prism curve (i.e., the curve at the top of the wheel). h_{32} is the height of the rear end of the third rim.

The method we propose can extract the multi-layered contours of the vehicle. Figures 7b and 8b display points in four different colors. Points of the same color are the result of contour reconstruction of the same layer. The purple points represent the layer where h_{11} and h_{12} are located, which is the highest layer. The green points represent the layer where h_{21} and h_{22} are located. The blue points represent the layer where h_{31} and h_{32} are located. The red points represent the basic contours of the vehicle, which includes information such as the length and width of the vehicle.

As shown in Figures 7 and 8, the proposed method effectively reconstructs the 3D contour of vehicles. The electromagnetic simulation dataset CVDomes used in this study had minimal external interference, and the signal bandwidth was extremely wide with high resolution. As a result, the reconstructed contours were found to be clear, smooth, and highly similar to the original model. The contour of the front of the vehicle is more rounded, closer to the real contour of the simulated vehicle model.

Tables 2 and 3 show a comparison of the geometric dimensions of the vehicle models Jeep93 and Jeep99, with the estimation results of the proposed method, as well as the comparison with paper [21]. Where, l and w are the length and width of the vehicle body, respectively. $u_{|\cdot|}$ is the sum of the error between the true value and the estimated value.

From Tables 2 and 3, comparing the length, width, and height of the extracted vehicle between the simulation model and the proposed method, it can be observed that the parameters are remarkably close, indicating the accuracy of the proposed method. Overall, compared with the results in paper [21], the proposed method extracts the clearer multi-layer contour of the vehicle target, and exhibits higher accuracy.

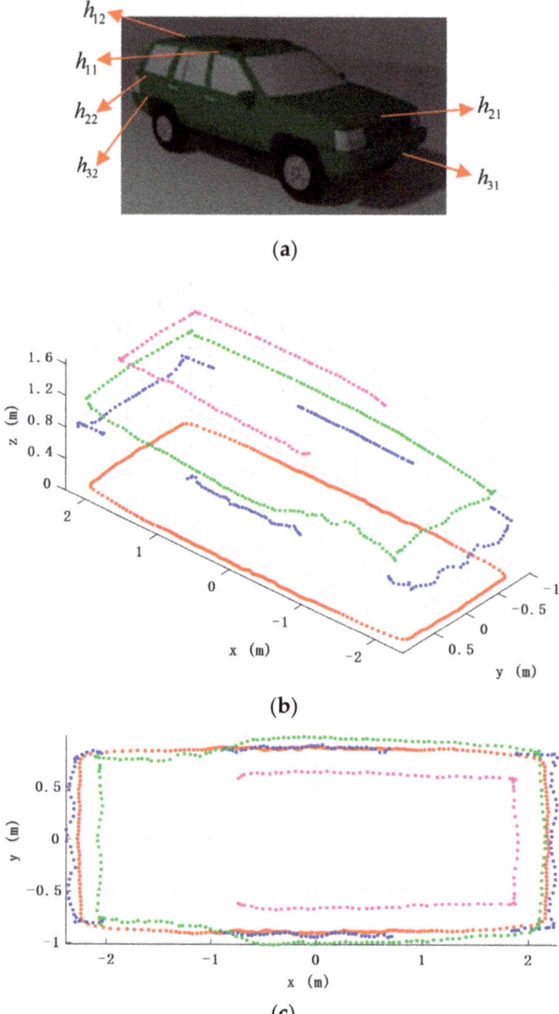

Figure 7. 3D contour reconstruction results of vehicle target Jeep93: (**a**) Photographs of simulated vehicle Jeep93; (**b**) 3D contour reconstruction results in 3D vision, the scattering points of the same color are the curves extracted from the same layer; (**c**) top view of the result of the 3D contour reconstruction of the vehicle.

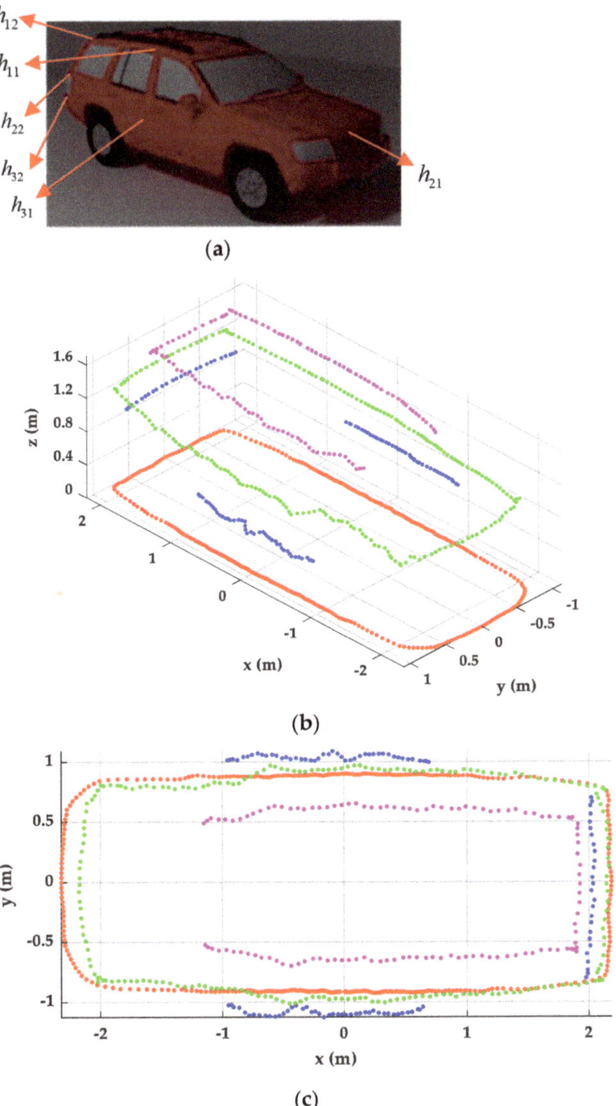

Figure 8. 3D contour reconstruction results of vehicle target Jeep99: (**a**) Photographs of simulated vehicle Jeep99; (**b**) 3D contour reconstruction results in 3D vision. The scattering points of the same color are the curves extracted from the same layer; (**c**) top view of the result of the 3D contour reconstruction of the vehicle.

Table 2. Comparison of the geometric dimensions of the vehicle model Jeep93 with the estimation results of the proposed method, and comparison with paper [21] (unit: mm).

Parameters	l	w	h_{11}	h_{12}	h_{21}	h_{22}	h_{31}	h_{32}
True value	4409	1755	1636	1634	1014	1158	653	675
Paper [21]	4351	1800	1722	1724	1046	1178	672	686
Our method	4462	1796	1648	1618	1032	1173	642	689

$$u_{|\Delta l|} = 53 \quad u^*_{|\Delta l|} = 58 \quad u_{|\Delta w|} = 41 \quad u^*_{|\Delta w|} = 45 \quad u_{|\Delta h|} = 86 \quad u^*_{|\Delta h|} = 258$$

"*" indicates the estimation results from paper [21].

Table 3. Comparison of the geometric dimensions of the vehicle model Jeep99 with the estimation results of the proposed method, and comparison with paper [21] (unit: mm).

Parameters	l	w	h_{11}	h_{12}	h_{21}	h_{22}	h_{31}	h_{32}
True value	4610	1826	1763	1616	1030	1131	655	1020
Paper [21]	4474	1825	1730	1594	990	1101	675	1046
Our method	4509	1813	1713	1628	1011	1116	645	1007

$$u_{|\Delta l|} = 101 \quad u^*_{|\Delta l|} = 136 \quad u_{|\Delta w|} = 13 \quad u^*_{|\Delta w|} = 1 \quad u_{|\Delta h|} = 119 \quad u^*_{|\Delta h|} = 171$$

"*" indicates the estimation results from paper [21].

5. Conclusions

This paper focuses on the problem of vehicle target information extraction and 3D contour reconstruction under CSAR 2D imaging, and then proposes a method by analyzing the CSAR image characteristics of the target vehicle under multiple different radar pitch-angle observations. The proposed method creatively utilizes the projection relationship of a vehicle in 2D CSAR imaging to reconstruct the 3D contour of the vehicle, without prior information. The basic contour of the vehicle body is extracted from the even-bounce reflection scattering points, and the height and position information of the vehicle ridge is extracted from the odd-bounce reflection points. Finally, the 3D contour of the vehicle can be reconstructed by fusing the vehicle body and the height and position of the vehicle ridge. The proposed method is applied to the reconstruction of CSAR simulation dataset, the high-quality 3D contour of the vehicle and high-precision vehicle size estimation results are obtained, which proves the correctness and effectiveness of this method.

The research results in this paper lay a foundation for future research on the target classification and recognition technology based on CSAR images, which have important practical value. By utilizing the reconstructed 3D contour, more accurate vehicle dimension information can be obtained, which is useful in areas such as vehicle industrial design, traffic planning and road safety. Additionally, the reconstructed 3D contour can be used for vehicle target detection and tracking, which is beneficial for fields like autonomous driving, traffic safety monitoring and smart transportation. The follow-up research is the high-precision imaging of bistatic CSAR, and the targets' classification and recognition after the 3D reconstruction of targets.

Author Contributions: Conceptualization, J.Z. and H.X.; methodology, J.Z. and H.X.; software, J.Z. and H.X.; validation, J.Z. and H.X.; formal analysis, J.Z. and H.X.; investigation, J.Z. and H.X.; resources, H.X. and Z.L.; data curation, H.X. and L.Z.; writing—original draft preparation, J.Z. and H.X.; writing—review and editing, J.Z., H.X., L.Z. and Z.L.; visualization, J.Z. and H.X.; supervision, H.X. and L.Z.; project administration, H.X. and L.Z.; funding acquisition, H.X., L.Z. and Z.L. All authors have read and agreed to the published version of the manuscript.

Funding: This research was co-supported by the Guangdong Basic and Applied Basic Research Foundation (Grants No. 2023A1515011588 and No. 2021A1515010768), the Shenzhen Science and Technology Program (Grant No. 202206193000001, 20220815171723002), the Beijing Nova Program (Grant No. Z201100006820103), the National Natural Science Foundation of China (Grant No. 62001523, No. 62203465, No. 62201614 and No. 6210593) and the Fundamental Research Funds for the Central Universities, Sun Yat-sen University (Grant No. 23lgpy45). Hongtu Xie is the corresponding author.

Data Availability Statement: The data presented in this study is available in: https://www.sdms.afrl.af.mil/content/public-data/s3_scripts/index.php?file=Civilian-Vehicle-SampleSet.zip, (accessed on 19 December 2023).

Acknowledgments: The authors would like to thank the editors and reviewers for their very competent comments and helpful suggestions to improve this paper. Moreover, the authors would like to thank the American Air Force Research Laboratory (AFRL) for providing the CSAR data.

Conflicts of Interest: The authors declare no conflicts of interest.

References

1. Xie, H.; An, D.; Huang, X.; Zhou, Z. Efficient Raw Signal Generation Based on Equivalent Scatterer and Subaperture Processing for One-Stationary Bistatic SAR Including Motion Errors. *IEEE Trans. Geosci. Remote Sens.* **2016**, *54*, 3360–3377. [CrossRef]
2. Zhu, J.; Chu, N.; Song, Y.; Yi, S.; Wang, X.; Huang, X.; Moran, B. Alternative signal processing of complementary waveform returns for range sidelobe suppression. *Signal Process.* **2019**, *159*, 187–192. [CrossRef]
3. Hu, X.; Xie, H.; Zhang, L.; Hu, J.; He, J.; Yi, S.; Jiang, H.; Xie, K. Fast Factorized Backprojection Algorithm in Orthogonal Elliptical Coordinate System for Ocean Scenes Imaging Using Geosynchronous Spaceborne-Airborne VHF UWB Bistatic SAR. *Remote Sens.* **2023**, *15*, 2215. [CrossRef]
4. Zhu, J.; Peng, C.; Zhang, B.; Jia, W.; Xu, G.; Wu, Y.; Hu, Z.; Zhu, M. An Improved Background Normalization Algorithm for Noise Resilience in Low Frequency. *J. Mar. Sci. Eng.* **2021**, *9*, 803. [CrossRef]
5. Dungan, K.E. Vehicle detection in wide-angle SAR. In *Algorithms for Synthetic Aperture Radar Imagery XXIII*; SPIE: Bellingham, WA, USA, 2016.
6. Gianelli, C.D.; Xu, L. Focusing, imaging, and ATR for the Gotcha 2008 wide angle SAR collection. In *Algorithms for Synthetic Aperture Radar Imagery XX*; SPIE: Bellingham, WA, USA, 2013; Volume 8746, pp. 174–181.
7. Dungan, K.E.; Nehrbass, J.W. Wide-area wide-angle SAR focusing. *IEEE Aerosp. Electron. Syst. Mag.* **2014**, *29*, 21–28. [CrossRef]
8. Li, D.; Wei, G.; Sun, B.; Wang, X. Recursive Sidelobe Minimization Algorithm for Back-Projection Imaging of Impulse-Based Circular Synthetic Aperture Radar. *IEEE Geosci. Remote Sens. Lett.* **2020**, *17*, 1732–1736. [CrossRef]
9. Saville, M.A.; Jackson, J.A.; Fuller, D.F. Rethinking vehicle classification with wide-angle polarimetric SAR. *IEEE Aerosp. Electron. Syst. Mag.* **2014**, *29*, 41–49. [CrossRef]
10. Feng, D.; An, D.; Wang, J.; Chen, L.; Huang, X. A Focusing Method of Buildings for Airborne Circular SAR. *Remote Sens.* **2024**, *16*, 253. [CrossRef]
11. Chen, L.; An, D.; Huang, X. Resolution Analysis of Circular Synthetic Aperture Radar Noncoherent Imaging. *IEEE Trans. Instrum. Meas.* **2020**, *69*, 231–240. [CrossRef]
12. Palm, S.; Oriot, H.M.; Cantalloube, H.M. Radargrammetric DEM Extraction Over Urban Area Using Circular SAR Imagery. *IEEE Trans. Geosci. Remote Sens.* **2012**, *50*, 4720–4725. [CrossRef]
13. Jia, G.; Buchroithner, M.F.; Chang, W.; Liu, Z. Fourier-based 2-D imaging algorithm for circular synthetic aperture radar: Analysis and application. *IEEE J. Sel. Top. Appl. Earth Obs. Remote Sens.* **2016**, *9*, 475–489. [CrossRef]
14. Nan, Y.; Huang, X.; Guo, Y.J. An Universal Circular Synthetic Aperture Radar. *IEEE Trans. Geosci. Remote Sens.* **2022**, *60*, 5601920. [CrossRef]
15. Austin, C.D.; Ertin, E.; Moses, R.L. Sparse multipass 3D SAR imaging: Applications to the Gotcha data set. In *Algorithms for Synthetic Aperture Radar Imagery XVI*; SPIE: Bellingham, WA, USA, 2009; Volume 7337, p. 733703.
16. Ferrara, M.; Jackson, J.A.; Austin, C. Enhancement of multi-pass 3D circular SAR images using sparse reconstruction techniques. In *Algorithms for Synthetic Aperture Radar Imagery XVI*; SPIE: Bellingham, WA, USA, 2009; Volume 7337, p. 733702.
17. Jiang, Y.; Deng, B.; Wang, H.; Zhuang, Z.; Wang, Z. Raw Signal Simulation for Multi-Circular Synthetic Aperture Imaging at Terahertz Frequencies. *IEEE Geosci. Remote Sens. Lett.* **2020**, *17*, 377–380. [CrossRef]
18. Luo, Y.; Chen, S.W.; Wang, X.S. Manmade-Target Three-Dimensional Reconstruction Using Multi-View Radar Images. In Proceedings of the IGARSS 2022—2022 IEEE International Geoscience and Remote Sensing Symposium, Kuala Lumpur, Malaysia, 17–22 July 2022; pp. 3452–3455.
19. Dungan, K.E.; Potter, L.C. 3-D imaging of vehicles using wide aperture radar. *IEEE Trans. Aerosp. Electron. Syst.* **2011**, *47*, 187–200. [CrossRef]
20. Feng, D.; An, D.; Chen, L.; Huang, X. Holographic SAR Tomography 3-D Reconstruction Based on Iterative Adaptive Approach and Generalized Likelihood Ratio Test. *IEEE Trans. Geosci. Remote Sens.* **2021**, *59*, 305–315. [CrossRef]

21. Chen, L.; An, D.; Huang, X.; Zhou, Z. A 3D Reconstruction Strategy of Vehicle Outline Based on Single-Pass Single-Polarization CSAR Data. *IEEE Trans. Image Process.* **2017**, *26*, 5545–5554. [CrossRef] [PubMed]
22. Potter, L.C.; Moses, R.L. Attributed scattering centers for SAR ATR. *IEEE Trans. Image Process.* **1997**, *6*, 79–91. [CrossRef] [PubMed]
23. Jakowatz, C.V.; Wahl, D.E.; Eichel, P.H.; Ghiglia, D.C.; Thompson, P.A. *Spotlight-Mode Synthetic Aperture Radar: A Signal Processing Approach*; Springer Science & Business Media: Berlin/Heidelberg, Germany, 1996.
24. Skolnik, M.I. *Radar Handbook*, 3rd ed.; McGraw-Hill: New York, NY, USA, 2008; pp. 16.7–16.10.
25. Fung, A.K. Theory of cross polarized power returned from a random surface. *Apply Sci. Res.* **1967**, *18*, 50–60. [CrossRef]
26. Peng, X.; Tan, W.; Hong, W.; Jiang, C.; Bao, Q.; Wang, Y. Airborne DLSLA 3-D SAR Image Reconstruction by Combination of Polar Formatting and L_1 Regularization. *IEEE Trans. Geosci. Remote Sens.* **2016**, *54*, 213–226. [CrossRef]
27. Ponce, O.; Prats, P.I.; Pinheiro, M.; Rodriguez, M.; Scheiber, R.; Reigber, A.; Moreira, A. Fully Polarimetric High-Resolution 3-D Imaging with Circular SAR at L-Band. *IEEE Trans. Geosci. Remote Sens.* **2014**, *52*, 3074–3090. [CrossRef]
28. Zyl, J.V.; Kim, Y. *Synthetic Aperture Radar Polarimetry*; John Wiley & Sons: Hoboken, NJ, USA, 2011.
29. Ponce, O.; Prats, P.; Scheiber, R.; Reigber, A.; Hajnsek, I.; Moreira, A. Polarimetric 3-D imaging with airborne holographic SAR tomography over glaciers. In Proceedings of the 2015 IEEE International Geoscience and Remote Sensing Symposium (IGARSS), Milan, Italy, 26–31 July 2015; pp. 5280–5283.
30. Chen, L.; An, D.; Huang, X. A Backprojection Based Imaging for Circular Synthetic Aperture Radar. *IEEE J. Sel. Top. Appl. Earth Obs. Remote Sens.* **2017**, *10*, 3547–3555. [CrossRef]
31. Xie, H.; Shi, S.; An, D.; Wang, G.; Wang, G.; Xiao, H.; Huang, X.; Zhou, Z.; Xie, C.; Wang, F.; et al. Fast Factorized Backprojection Algorithm for One-Stationary Bistatic Spotlight Circular SAR Image Formation. *IEEE J. Sel. Top. Appl. Earth Obs. Remote Sens.* **2017**, *10*, 1494–1510. [CrossRef]
32. Dungan, K.E.; Austin, C.; Nehrbass, J.; Potter, L. Civilian vehicle radar data domes. In *Algorithms for Synthetic Aperture Radar Imagery XVII*; SPIE: Bellingham, WA, USA, 2010; Volume 7699, p. 76990P.

remote sensing

MDPI

Article

Influence of Range-Dependent Sound Speed Profile on Position of Convergence Zones

Ziyang Li [1,2,3], Shengchun Piao [1,2,3], Minghui Zhang [1,2,3,*] and Lijia Gong [1,2,3]

[1] Acoustic Science and Technology Laboratory, Harbin Engineering University, Harbin 150001, China
[2] Key Laboratory of Marine Information Acquisition and Security, Harbin Engineering University, Ministry of Industry and Information Technology, Harbin 150001, China
[3] College of Underwater Acoustic Engineering, Harbin Engineering University, Harbin 150001, China
* Correspondence: zhangminghui@hrbeu.edu.cn

Abstract: Based on the Wentze–Kramers–Brillouin approximation, we derive formulae to calculate the position of convergence zones in a range-dependent environment with sound speed profiles varying in linear and ellipsoidal Gaussian eddy cases. Simulation results are provided for the linear and ellipsoidal Gaussian eddy cases. Experiment data are used for calculations considering linearly varying sound speed, and the findings suitably agree with the simulation results. According to the evaluated environment, the influence of the range-dependent sound speed profile on the position of the upper and lower convergence zones for different source depths is analyzed through simulations. The corresponding results show that the influence of the sound speed profile on the position of the upper convergence zone is greater for a shallower source. In contrast, the position of the lower convergence zone for large-depth reception is less affected.

Keywords: convergence zone; transmission loss; Wentze–Kramers–Brillouin approximation; coupled normal mode

Citation: Li, Z.; Piao, S.; Zhang, M.; Gong, L. Influence of Range-Dependent Sound Speed Profile on Position of Convergence Zones. *Remote Sens.* **2022**, *14*, 6314. https://doi.org/10.3390/rs14246314

Academic Editor: Andrzej Stateczny

Received: 16 November 2022
Accepted: 10 December 2022
Published: 13 December 2022

Publisher's Note: MDPI stays neutral with regard to jurisdictional claims in published maps and institutional affiliations.

1. Introduction

The marine hydrological environment, with its local sound speed profile structure, considerably influences long-range sound propagation and alters time and space correlations of underwater pressure fields. The sound speed profile in the deep sea causes rays to refract or reflect and gather in a certain area to form a spatial periodic high-intensity sound area called the convergence zone [1]. This zone is important for characterizing long-range sound propagation, and research on its characteristics in the deep sea has become a research hotspot.

Since the 1940s, several theoretical studies have been conducted on the sound propagation characteristics of the convergence zone, mainly focusing on the sound transmission loss and impact of environmental factors on the position of the convergence zone. Woezel [2] and Brekhovskikh [3] discovered sound fixing and ranging channels in the deep sea. Urick [4] found that, with increasing sound source depth, the convergence zone splits, with the inner zone moving inward and the outer zone moving outward. Schulkin [5] analyzed the influence of bubbles and surface waves on sound propagation in the convergence zone based on experimental data. Bongiovanni et al. [6] proposed a method to locate the convergence zone based on deep-sea surface temperature data and evaluated the prediction accuracy of this method using the parabolic equation method. Zhang [7] used the normal mode theory to approximate the pressure field for ideal shallow water mixed-layer conditions. Wang and Wang [8] analyzed the impact of strength, thickness, and depth of the main thermocline on the convergence zone through numerical simulations and found that the strength of the main thermocline has the greatest impact on the position of the convergence zone. Chen et al. [9] found that the Subtropical Mode Water in the Northwestern Pacific Ocean will help form a sonic duct which makes the convergence zone gain greater. Yang et al. [10] divided the sound speed profile in the North Atlantic into six types and analyzed the influence of different types on the position and gain of the convergence zone. Bi and Peng [11] found that, because of the earth's curvature, the location of the convergence zone moves toward the sound source, and its movement can reach 10 km at 1000 km in distance.

With increasing knowledge of the deep sea, research on the convergence zone is no longer limited to a range-independent environment but instead more focused on the impact of mesoscale spatial variability on sound propagation in the convergence zone. Henrick [12] analyzed the sound propagation characteristics of the convergence zone in a range-dependent environment with an eddy by establishing an ideal eddy parameter model. According to ray theory simulations, when the sound source is located at the center of a cold eddy, the convergence zone moves closer to the sound source, and when the sound source is far from the eddy center, the convergence zone movement reduces. Rudnick and Munk [13] found that the influence of internal waves on the sound speed gradient of a mixed layer causes the convergence zone to move towards the sound source. Colosi and Rudnick [14] analyzed experimental data and found that a change in the thickness of the mixed layer with range leads to energy leakage in the mixed layer, thus making the convergence zone close to the sound source and increasing the width. Li et al. [15] found that the convergence zone moves backwards and the width increases with an anticyclone eddy, whereas the cyclone eddy has the opposite effect. Colosi and Zincola-Lapin [16] analyzed the influence of mesoscale phenomena on mode coupling in the lower surface sound channel and found that the convergence zone moves toward the sound source, while the width increases under strong high-order coupling. White et al. [17] studied the impact of internal waves and tides on sound propagation and found that the second convergence zone moves farther from the sound source than from the first convergence zone for a 250 Hz sound source. Zhang et al. [18] analyzed the impact of mesoscale vortices on the convergence zone through the parabolic equation method, determining that warm vortices increase the distance of the convergence zone, while cold vortices make it smaller. Piao et al. [19] found that there is a convergence zone in the deep sea at large receiving depths and analyzed the position of the convergence zone using ray mode theory. Wu et al. [20] found that, under the influence of the Indian Ocean Dipole, the thermocline fluctuation at the location of the second convergence zone has an important influence on the formation and location of the third convergence zone. It is found that the location of the third convergence zone shifts 2–3 km toward the sound source in the experiment. Chandrayadula et al. [21] analyzed the influence of internal tides on the modes intensity and compared with a new hybrid broadband transport theory.

In long-range sound propagation, the influence of the range-dependent sound speed profile on the position of the convergence zone should be considered. Research on sound propagation in the deep sea mainly focuses on the characteristics of long-range sound propagation in a range-independent environment and the characteristics of the convergence zone near the sea surface, while scarce studies are available on the influence of range-dependent sound speed on the position of the convergence zone, especially the convergence zone at large receiving depths. Ocean acoustic tomography is very important. Studying the sensitivity of the convergence zone to the sound speed variation can provide theoretical support for ocean acoustic tomography. In this paper, the formulae to calculate the position of the convergence zone in a range dependent environment are derived based on the Wentze–Kramers–Brillouin (WKB) approximation for the position of the convergence zone in a range-independent environment. Two cases are considered. One is linearly varying sound speed, and the other is ellipsoidal Gaussian eddy. The formulae derived from the two cases are simulated and analyzed. Through the simulations, the correctness of the formulae is verified. The formula for the linear case is further verified with experimental data.

The remainder of this paper is organized as follows. In Section 2, the position of the convergence zone in a range-dependent environment is derived for sound speed profiles with linear variation and ellipsoidal Gaussian eddy case. In Section 3, we report simulation results using the derived formulae. In Section 4, the experimental setup is introduced, and experimental data are processed and analyzed. In Section 5, we discuss the experimental results. In Section 6, the conclusions are made.

2. Methods

2.1. Position of Convergence Zones

According to the normal mode theory, the pressure field of a single-frequency point source for range r and depth z can be expressed as (suppressing the harmonic time dependence, $e^{-i\omega t}$)

$$p(r,z) = A\sum_n \phi_n(z_s)\phi_n(z)H_0^{(1)}(k_n r),\tag{1}$$

where z is the receiver depth, z_s is the source depth, r is the horizontal range between the source and receiver, A is a normalization constant, $\phi_n(z)$ is the normal mode depth function, $H_0^{(1)}(k_n r)$ is the Hankel function of the first kind that satisfies the radiation condition over a large range, and k_n is the horizontal wavenumber.

The normal mode depth functions satisfy the following differential equation:

$$\frac{d^2\phi_n(z)}{dz^2} + \left[k_0^2 q^2(z) - k_n^2\right]\phi_n(z) = 0.\tag{2}$$

In addition, they satisfy orthonormality:

$$\int_0^H \phi_n(z)\phi_m(z)dz = \delta_{nm},\tag{3}$$

where $k_0 = \omega_0/c_0$, ω_0 is the angular frequency, c_0 is the reference sound speed, $q(z)$ is the depth-dependent refraction index, $c_0/c(z)$, and H is the sea depth. δ_{nm} is a Dirac function whose value is equal to 1 only when n is equal to m.

When the refraction index is a slowly varying function of depth, or when the frequency is high, the following WKB approximation is obtained:

$$\frac{\lambda_n(z)}{2\pi}\left(\left|\frac{d\kappa_n(z)}{dz}\right|/\kappa_n(z)\right) \ll 1,\tag{4}$$

where $\kappa_n^2(z) = k_0^2 q^2(z) - k_n^2$. $\lambda_n(z)$ is the wavelength. Using the WKB approximation and asymptotic approximation of the Hankel function in Equation (2), the phase function of the pressure field can be obtained as follows:

$$\theta = k_n r \pm \int_{z_s}^z \kappa_n(z')dz'.\tag{5}$$

Tindle and Guthrie [22] and Beilis [23] unveiled constructive interference between neighboring modes, resulting in a change of phase $\Delta\theta$ over a group of modes Δn being a multiple of 2π, that is, $\Delta\theta/\Delta n = 2\pi m$ and

$$\frac{\Delta\theta}{\Delta n} = 2\pi m = \left(\frac{\Delta k_n}{\Delta n}\right)r \mp \int_{z_s}^z \frac{k_n(\Delta k_n/\Delta n)}{\kappa_n(z')}dz'.\tag{6}$$

From Equation (6), the position of the convergence zone in a range-independent environment can be obtained as follows:

$$r = \pm\int_{z_s}^z \frac{k_n}{\kappa_n(z')}dz' + mD_n,\tag{7}$$

where $D_n = 2\pi/(\Delta k_n/\Delta n)$ is the modal skip distance.

When the sound speed varies with range, the peak positions of the convergence zone shift compared with those described by Equation (7). We consider an adiabatic approximation that discards any transfer of energy to other modes. Simultaneously, the refraction index is a slowly varying function of depth satisfying the condition of Equation (4). Hence, the variation in sound speed, $\delta c(z, r)$, is introduced to derive the formula for the position of the convergence zone in a range-dependent environment. The sound speed is defined as

$$c(z, r) = c(z) + \delta c(z, r), \tag{8}$$

where $c(z)$ is the sound speed at range 0 km. The corresponding phase function of the pressure field is given by

$$\theta = \int_0^r k_n(r')dr' \pm \int_{z_s}^z \kappa_n(z', r)dz', \tag{9}$$

where $\kappa_n(z, r) = \left[k_0^2 q^2(z, r) - k_n^2(r)\right]^{1/2}$ and $q(z, r) = c_0/c(z, r)$. Calculating $\Delta\theta/\Delta n = 2\pi m$ yields

$$2\pi m = \int_0^r \frac{\Delta k_n(r')}{\Delta n}dr' \mp k_n(r)\frac{\Delta k_n(r)}{\Delta n}\int_{z_s}^z \frac{1}{\kappa_n(z', r)}dz'. \tag{10}$$

All the other quantities of interest are defined in terms of $\delta c(z, r)$. The approximate wavenumber is given by

$$k_n(r) = k_n + \delta k_n(r), \tag{11}$$

where

$$\delta k_n(r) = \frac{-k_0^2}{k_n}\int_0^H \phi_n^2(z')q^3(z')\left(\frac{\delta c(z, r)}{c_0}\right)dz'$$

Hence,

$$\frac{\Delta\delta k_n(r)}{\Delta n} = \frac{-\Delta k_n}{\Delta n}\frac{\delta k_n(r)}{k_n}, \tag{12}$$

and

$$k_n(r)\frac{\Delta k_n(r)}{\Delta n} = k_n\frac{\Delta k_n}{\Delta n} + O\left\{[\delta k_n(r)]^2\right\}. \tag{13}$$

Further,

$$\kappa_n(z, r) = \kappa_n(z) + \delta\kappa_n(z, r), \tag{14}$$

where

$$\delta\kappa_n(z, r) = -\frac{\left[k_0^2 q^3(z)\delta c(z, r)/c_0 + k_n\delta k_n(r)\right]}{\kappa_n(z)}.$$

Consider the second integral part in Equation (10):

$$\begin{aligned}
I_n(z, r) &= \frac{1}{\kappa_n(z, r)} = \frac{1}{\kappa_n(z) + \delta\kappa_n(z, r)} \\
&= \frac{\kappa_n(z) - \delta\kappa_n(z, r)}{\kappa_n^2(z)} = I_n(z) - \frac{\delta\kappa_n(z, r)}{\kappa_n^2(z)}
\end{aligned} \tag{15}$$

Equation (15) shows that the integral term in the sound speed under varying range environment is composed of an integral term in a range-independent environment and a change term caused by variations in sound speed. We derive specific formulae for sound speed profiles with linear variation and ellipsoidal Gaussian eddy case.

2.2. Case of Linearly Varying Sound Speed

Consider the following specific form for the range variation of sound speed:

$$\delta c(z, r) = \sigma r c(z) w(z), \tag{16}$$

where σ is the intensity parameter of the range variation, which is small, $|\sigma|/k_0 \ll 1$, and $w(z)$ is a dimensionless function used to limit the range variation in a limited region of the environment, $0 < w(z) \leq 1$. In this form of $\delta c(z, r)$, Equation (15) can be rewritten as

$$I_n(z, r) = I_n(z) + \sigma r \hat{p}_n(z),\tag{17}$$

where

$$\hat{p}_n(z) = \frac{k_0^2 q^2(z) w(z) - k_n^2 a_n}{\kappa_n^3(z)},$$

and modal constant a_n is given by

$$a_n = \frac{k_0^2}{k_n^2} \int_0^H \phi_n^2(z') q^2(z') w(z') dz'.\tag{18}$$

Substituting the above partial expressions into Equation (10), we obtain

$$r(1 \mp \sigma L_n) + \frac{\sigma a_n}{2} r^2 = R_z + m D_n,\tag{19}$$

where $L_n = L_n(z) = k_n \int_{z_s}^z \hat{p}_n(z') dz'$ and $R_z = \pm k_n \int_{z_s}^z dz' I_n(z')$, being only related to the depth of the sound source and receiver.

When $\sigma = 0$, Equation (19) reduces to Equation (7). Compared with Equation (7), the change in the first term on the left-hand side of Equation (19) adds a linear range variation of sound speed, and the second term is a quadratic term related to range r in the phase integral term owing to the linear variation in sound speed. Solving the quadratic problem in r provides two roots. One root is $\frac{-(1 \mp \sigma L_n) - \sqrt{(1 \mp \sigma L_n)^2 + 2\sigma a_n(R_z + m D_n)}}{\sigma a_n} < 0$, which should be neglected. The other root is

$$r = \frac{-(1 \mp \sigma L_n) + \sqrt{(1 \mp \sigma L_n)^2 + 2\sigma a_n(R_z + m D_n)}}{\sigma a_n},\tag{20}$$

which indicates the position of the convergence zone in a range-dependent environment with a linear variation in sound speed.

2.3. Case of Ellipsoidal Gaussian Eddy

The sound speed variation for the ellipsoidal Gaussian eddy can be expressed as

$$\delta c(z, r) = A_0 \exp\left(-\left(\frac{r - r_0}{dr}\right)^2 - \left(\frac{z - z_0}{dz}\right)^2\right),\tag{21}$$

where A_0 is the amplitude of sound speed variation caused by the eddy, r_0 and z_0 determine the center position of the eddy, and dr and dz are characteristic parameters of the semi-major and semi-minor axes, respectively.

Equation (15) is thus rewritten as

$$I_n(z, r) = I_n(z) + \zeta(r) \hat{p}_n(z),\tag{22}$$

where

$$\zeta(r) = \frac{A_0}{c_0} \exp\left(-\left(\frac{r - r_0}{dr}\right)^2\right),$$

$$\hat{p}_n(z) = \frac{k_0^2 q^3(z) \exp\left(-\left(\frac{z - z_0}{dz}\right)^2\right) - k_n^2 a_n}{\kappa_n^3(z)},$$

$$a_n = \frac{k_0^2}{k_n^2} \int_0^H \phi_n^2(z')q^3(z') \exp\left(-\left(\frac{z'-z_0}{dz}\right)^2\right)dz'.$$

By substituting the above partial expressions into Equation (10), we obtain

$$r + \frac{A_0 a_n}{c_0} \frac{\sqrt{\pi}dr}{2} erf\left(\frac{r-r_0}{dr}\right) \mp \zeta(r)L_n(z) = R_z - R_{r_0} + mD_n, \tag{23}$$

where $erf(x)$ is a Gaussian error function, $L_n = L_n(z) = k_n \int_{z_s}^z \hat{p}_n(z')dz'$ and $R_z = \pm k_n \int_{z_s}^z dz' I_n(z')$ are only related to the depth of the sound source and receiver, and $R_{r_0} = \frac{A_0 a_n}{c_0} \frac{\sqrt{\pi}dr}{2} erf\left(\frac{r_0}{dr}\right)$ is related only to the eddy center position and characteristic width.

Compared with Equation (7), the left-hand side of Equation (23) adds two terms. Term $\frac{A_0 a_n}{c_0} \frac{\sqrt{\pi}dr}{2} erf\left(\frac{r-r_0}{dr}\right)$ is the result obtained by integrating the integral term with respect to r in the phase function, owing to the Gaussian eddy environment. Term $\zeta(r)L_n(z)$ is the product of the horizontal parameters and integral terms of vertical eddy parameters. Simultaneously, the right-hand side of Equation (23) adds R_{r_0} compared with the range-independent environment. R_{r_0} appears by the existence of eddy and is only related to the eddy center position and characteristic width. The position of the convergence zone in the ellipsoidal Gaussian eddy environment can be obtained by numerically calculating the root of Equation (23).

3. Simulations

We verified the derived formulae through simulations considering sound speed profiles with linear variation and ellipsoidal Gaussian eddy case.

3.1. Case of Linearly Varying Sound Speed

For the simulation, we considered a sea depth of 4500 m. The sound speed profile at a range of 0 km satisfied the Munk sound speed profile:

$$c(z) = c_0\{1 + \varepsilon[e^{-\eta} - (1-\eta)]\}, \tag{24}$$

where $\eta = 2(z-z_0)/B$, $B = 1000$ m, $z_0 = 1000$ m, $c_0 = 1500$ m/s, and $\varepsilon = 5.7 \times 10^{-3}$. The sedimentary layer was considered to be a fluid with a constant sound speed of 1650 m/s and density of 1.6 g/cm^3. The acoustic absorption coefficient of the sedimentary layer was 0.5 dB/λ.

Figures 1 and 2 show the sound speed profile and depth variation function $w(z)$ for the first 400 m. In this range-dependent environment with linearly varying sound speed, σ was set to -1.74×10^{-8}, such that the maximum sound speed variation in the range of 450 km was -12 m/s. The source depth was 200 m, and the receiver depth was 3500 m. Thus, the receiver depth was below the conjugate depth of the sound source.

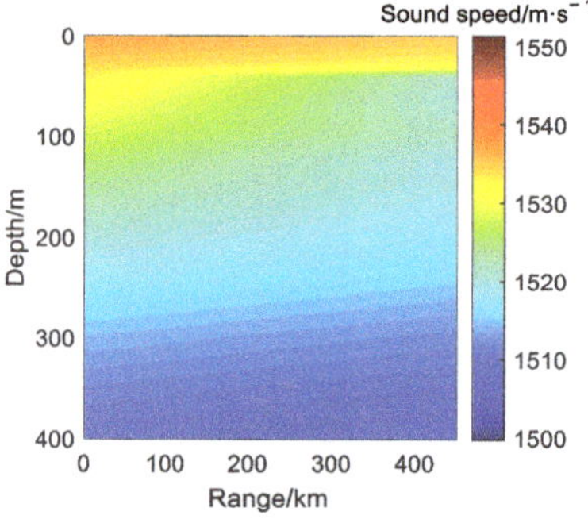

Figure 1. Sound speed profile with linear variation.

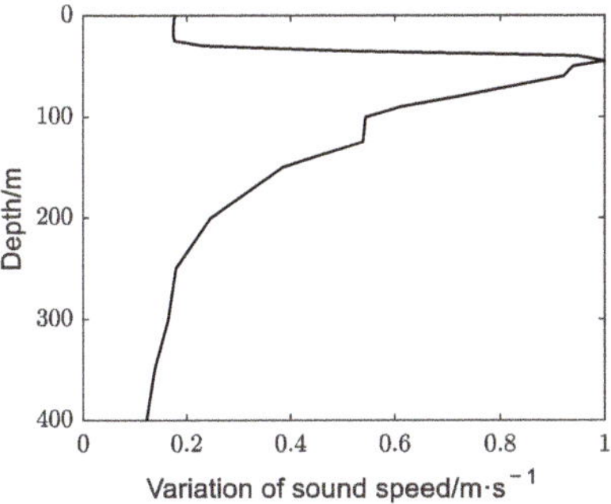

Figure 2. Depth variable function in linear variation case.

The transmission loss in the range-independent and range-dependent environments was calculated using the normal and coupled normal modes, respectively. For the range-dependent environment, adiabatic approximation, which assumes that no other energy transfers to other modes, was used. We focused on the position of the convergence zone. Thus, the modes with phase speeds less than the seabed sound speed, namely, the refracted and surface-reflected modes, were selected for the simulations to calculate the transmission loss. A comparison of the obtained transmission losses is shown in Figure 3.

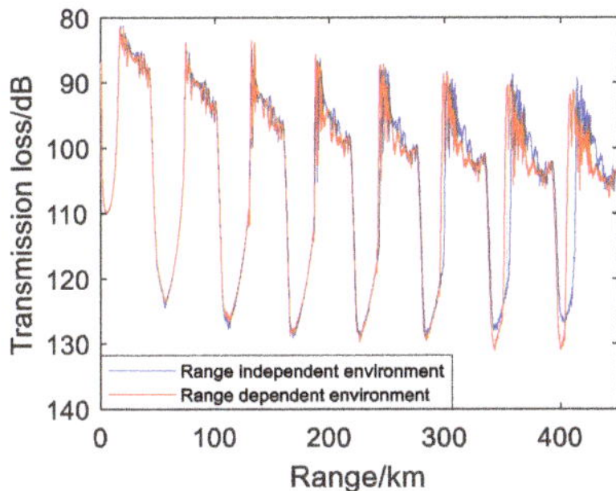

Figure 3. Transmission losses in range-independent and range-dependent environments for linearly varying sound speed.

To calculate the position of the convergence zone in a range-dependent environment with linear variation in sound speed according to Equation (20), the modal strength of the normal modes of each order must be calculated. The corresponding results are shown in Figure 4. The modal strength of the lower mode is 0, and only the part with the larger amplitude is shown in the figure.

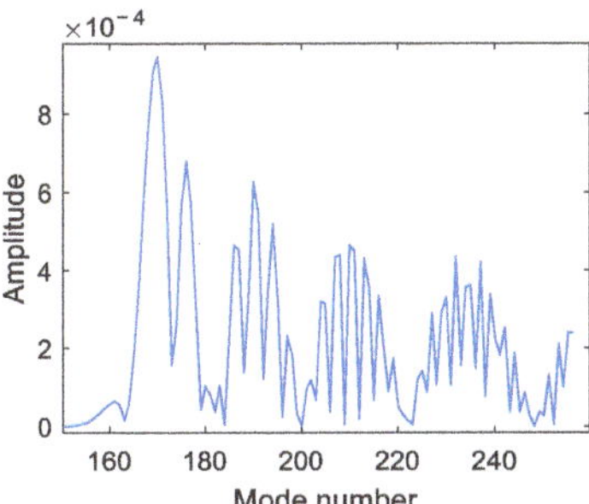

Figure 4. Mode amplitudes for refracted and surface-reflected modes considering linearly varying sound speed.

Figure 4 shows that mode 170 yields the strongest modal intensity at the receiver depth. Computationally, the spread in mode Δn is defined by two successive minima in the figure (i.e., $\Delta n = 10$). Table 1 lists the positions of the convergence zones calculated using Equation (20) and the coupled normal mode.

Table 1. Positions of convergence zones obtained using proposed formulation and coupled normal mode.

Number of Convergence Zones, m	r_{ri}/km	r_{rd}/km	r_{eq}/km
1	76.6	76.6	76.6
2	134.5	134.4	134.4
3	192.0	191.1	190.0
4	248.5	246.6	245.6
5	305.0	301.9	301.3
6	361.5	356.3	356.8
7	417.9	411.1	412.4

In Table 1, column r_{ri} lists the position of the convergence zone in the range-independent environment based on the normal mode. Column r_{rd} lists the position of the convergence zone in the range-dependent environment based on the coupled normal mode. Column r_{eq} lists the position of the convergence zone in the range-dependent environment obtained from our derived Equation (20).

For the first two convergence zones, Equation (20) better agrees with the coupled normal mode theory. With more convergence zones, there is an error between our formulation and the coupled normal mode, but the error at the seventh convergence zone is only 1.3 km. The error may be explained as follows. First, by calculating the modal strength at different ranges in sections, the maximum order of modal strength in the entire range is the 170th order of the normal mode, but two successive minima appear near the highest intensity mode change with increasing range. When calculating for 400 km, Δn should be 14. With this change, the position of the seventh convergence zone is 411.7 km, and the error with respect to the coupled normal mode result is smaller. Hence, the previous error is partially due to the error of D_n in Equation (20) because of the selection of Δn after increasing the range. Part of the error may also be caused by the fact that, although Δn orders of normal modes yield the strongest modal intensity in the convergence zone, other higher-order modes also have a certain impact. Therefore, the position of the convergence zone, according to Equation (20), shows some error with respect to the results of all the refracted and sea surface-reflected modes selected in the coupled normal mode. However, the error is negligible, indicating that Equation (20) can be used to accurately calculate the position of the convergence zone in a range-dependent environment with linearly varying sound speed.

3.2. Case of Ellipsoidal Gaussian Eddy

For the simulation, we considered a sea depth of 4500 m. The sound speed profile at a range of 0 km satisfied the Munk sound speed profile given by Equation (24). The Gaussian eddy parameters were set as follows: intensity of cyclonic eddy of -30 m/s, semi-major axis of 40 km, semi-minor axis of 400 m, eddy center $r_0 = 110$ km, and $z_0 = 0$ m. The sedimentary layer was considered as a fluid with a constant sound speed of 1650 m/s and density of 1.6 g/cm^3. The acoustic absorption coefficient of the sedimentary layer was 0.5 dB/λ.

Figure 5 shows the sound speed profile for the first 400 m. The source depth was 200 m, and the receiver depth was 3500 m. Thus, the receiver depth was below the conjugate depth of the sound source.

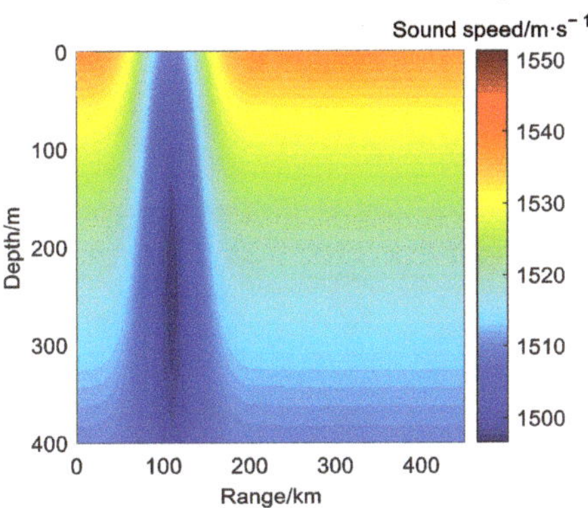

Figure 5. Sound speed profile of Gaussian eddy.

As for the linear variation in sound speed, the transmission loss in the range-independent and range-dependent environments was calculated using the normal and coupled normal modes, respectively. For the range-dependent environment, adiabatic approximation was used. The refracted and surface-reflected modes were selected in the simulation to calculate the transmission loss. The obtained transmission losses are shown in Figure 6.

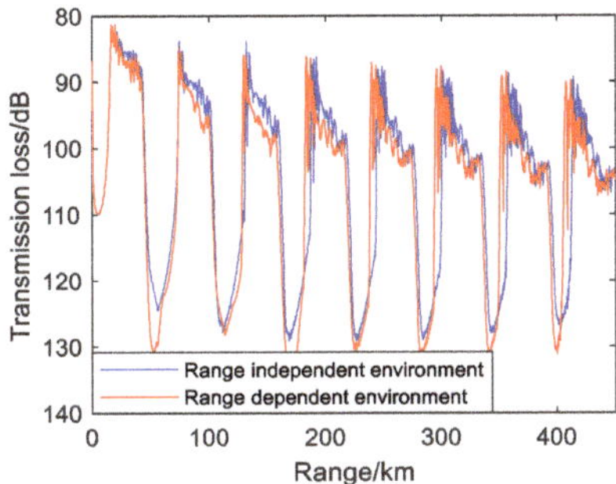

Figure 6. Transmission losses in range-independent and range-dependent environments for ellipsoidal Gaussian eddy case.

To calculate the position of the convergence zone in a range-dependent environment with a Gaussian eddy using Equation (23), the modal strength of the normal modes of each order must be calculated. The corresponding results are shown in Figure 7. The modal strength of the lower mode is 0, and only the part with the larger amplitude is shown in the figure.

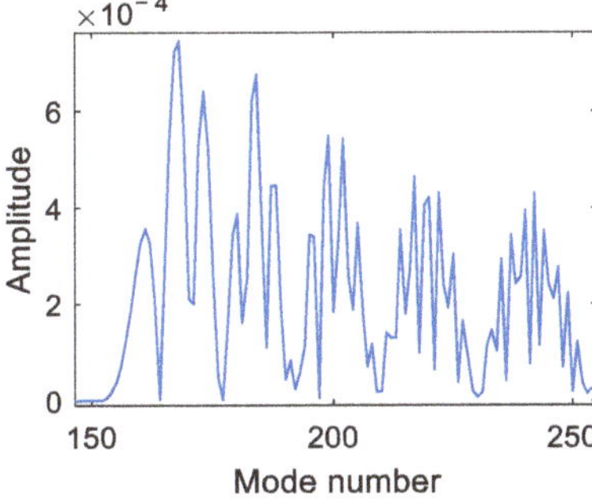

Figure 7. Mode amplitudes for refracted and surface-reflected modes considering ellipsoidal Gaussian eddy case.

Figure 7 shows that the strongest modal intensity at receiver depth is achieved at mode 168. Computationally, the spread in mode Δn is defined by two successive minima in the figure (i.e., $\Delta n = 10$). Table 2 lists the positions of the convergence zones calculated according to Equation (23) and the coupled normal mode.

Table 2. Positions of convergence zones obtained using proposed formulation and coupled normal mode.

m	r_{ri}/km	r_{rd}/km	r_{eq}/km
1	76.6	76.4	76.4
2	134.5	131.4	131.1
3	192.0	187.2	186.7
4	248.5	243.4	242.5
5	305.0	299.7	298.3
6	361.5	355.9	354.1
7	417.9	412.0	409.9

The columns of Table 2 have the same meanings as those of Table 1 but considering Equation (23) for column r_{eq}.

For the first convergence zone, Equation (23) agrees with the coupled normal mode theory. For other zones, there is an error between the two results, mainly because the skip distance is smaller than the numerical calculation. When considering the range of 400 km, an appropriate Δn value is 14 for the position of the convergence zone to be 410.5 km. However, the error is only about 0.5% between the result of Equation (23) and the coupled normal mode in the seventh convergence zone, indicating that Equation (23) can be used to accurately calculate the position of the convergence zone in a range-dependent environment with a Gaussian eddy.

4. Experiments

4.1. Experimental Setup and Data Processing

In the summer of 2014, a deep-sea long-range sound propagation experiment was conducted in the South China Sea. An experimental diagram is shown in Figure 8. A hydrophone was placed at a depth of 3146 m, and the explosions at a depth of 200 m was selected as the sound source.

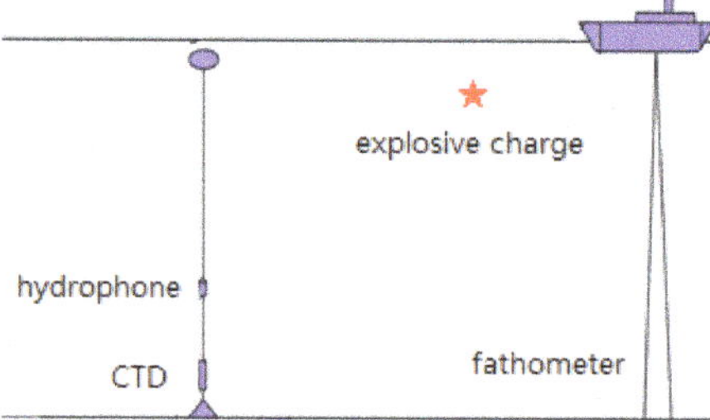

Figure 8. Diagram of experimental setup for data collection. The sound source is represented as a star mark.

Figure 9 shows the sea depth measured in the experiment, which was approximately 4300 m, within 450 km of the entire experimental range. Figure 10 shows the sound speed profile at the hydrophone position measured by a conductivity, temperature, depth (CTD) sensor. The CTD is placed at the bottom of the submersible buoy. With the deployment and recovery of the submersible buoy, the sound speed profile of the whole depth can be obtained. The axis of the sound channel was approximately 1000 m. Figure 9 shows that the seabed in the first 300 km is flat, and there are three seamounts in 300–450 km. To ignore the impact of sea depth changes on the convergence zone, the experimental data for the first 300 km were selected for analysis.

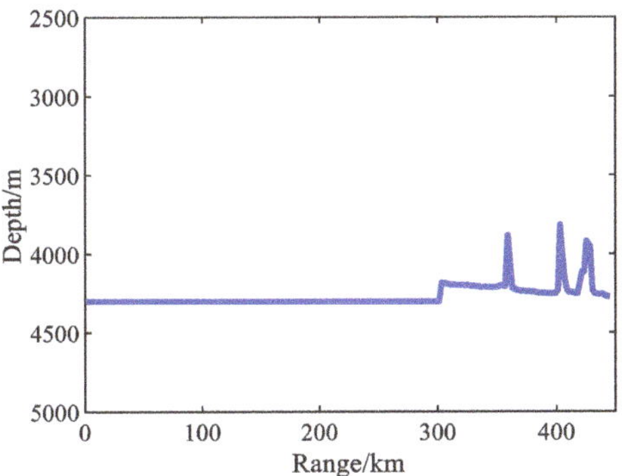

Figure 9. Bathymetry along experimental path.

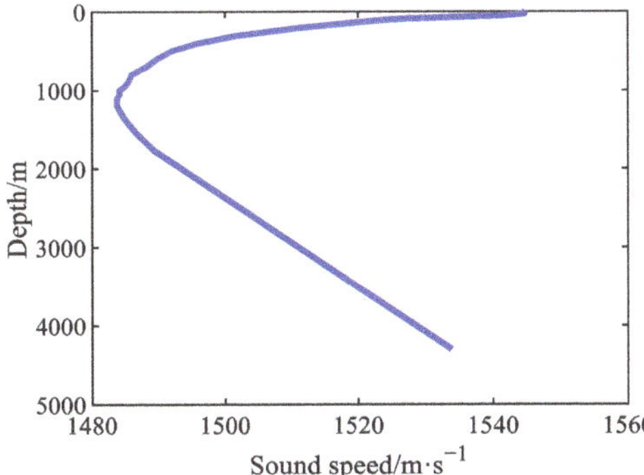

Figure 10. Sound speed profile at hydrophone position.

To study the influence of sound speed variation, data represented by the hybrid coordinate ocean model (HYCOM) were used to obtain the sound speed profile in different ranges. The HYCOM data included temperature, salinity, density, and other parameters. Shaji et al. [24] showed that HYCOM data can be used to simulate a real environment. The precision of the HYCOM data selected for this study was $1/12°$. By obtaining the temperature and salinity data from the experimental area and using Wood's empirical formula for sound speed, we obtain the following variation in sound speed:

$$c = 1450 + 4.206 \times T - 0.036 \times T^2 + 1.137 \times (S - 35) + 0.0175 \times D, \tag{25}$$

where c is the sound speed, T is the temperature, S is the salinity, and D is the depth.

The sound speed profile over the first 400 m of depth obtained from the HYCOM data is shown in Figure 11, where the solid black line represents the depth of the maximum sound speed in the mixed layer. The variation in sound speed is mainly concentrated at a depth above 200 m, and the depth of the maximum sound speed in the mixed layer changes slightly with range. To further analyze the characteristics of the surface sound speed in the experimental area, the corresponding parameters were calculated for the environment within 200 m.

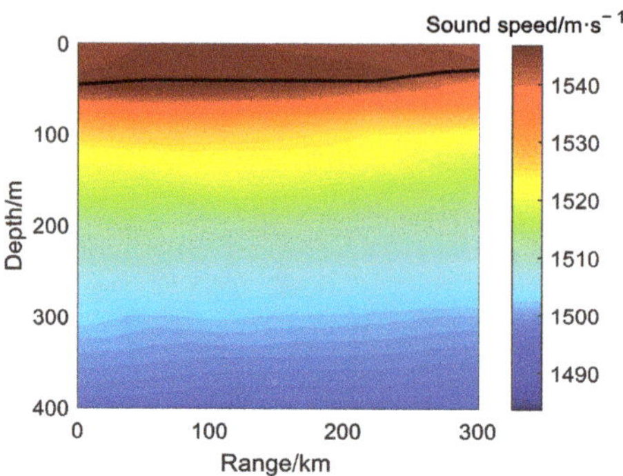

Figure 11. Sound speed profile obtained from HYCOM data.

Figure 12 shows the average sound speed, salinity, temperature, refractive index of sound speed, sound speed gradient, and root-mean-square error (RMSE) of the sound speed for depths above 200 m. In each graph, the circle indicates the depth of the maximum sound speed in the mixed layer, and the asterisk represents the depth of the maximum sound speed gradient. The salinity, temperature, sound speed gradient, and RMSE of the sound speed show negligible changes above the depth of the maximum sound speed, and all the parameters have obvious changes from the depth of the maximum sound speed. Through Figure 12a–c, it is found that temperature is the main influencing factor of sound speed. At the depth of maximum sound speed gradient, the RMSE of the sound speed also reaches its maximum. Below this depth, the RMSE of sound speed decreases again. Figure 12 shows that the depth variation function can be described by the RMSE of sound speed, namely, $w(z) = c_{rmse}/\max(c_{rmse})$ in Equation (16), where c_{rmse} is the RMSE of sound speed.

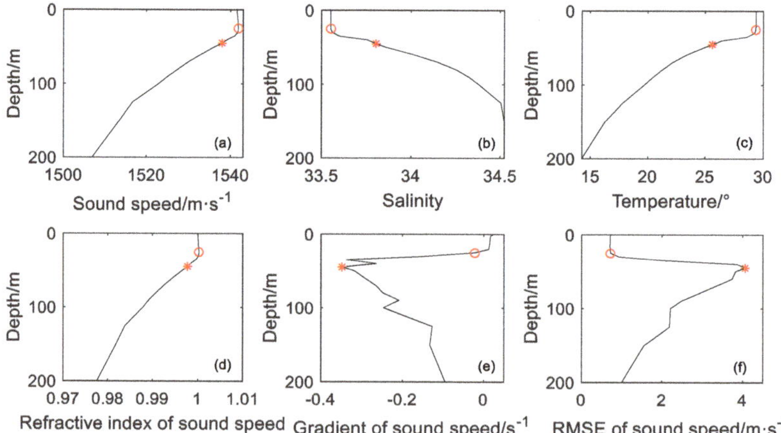

Figure 12. Parameters obtained for depths above 200 m. (**a**) Average sound speed, (**b**) salinity, (**c**) temperature, (**d**) refractive index of sound speed, (**e**) sound speed gradient, and (**f**) RMSE of sound speed. The circle indicates the depth of the maximum sound speed in the mixed layer, and the asterisk represents the depth of the maximum sound speed gradient.

4.2. Experimental Results and Analysis

The sound source used in the experiment was an explosion located at a depth of 200 m. When calculating the transmission loss curve of the sound pressure field, the broadband explosion source is usually filtered in a one-third octave with a certain center frequency, and then the filtered signal is processed. We filtered the received signal with a one-third octave of the central frequency of 200 Hz, and the experimental transmission loss was calculated as

$$\text{TL}(f_0|(r,z)) = \text{SL}(f_0) - (10\lg[E(f_0)] - M_v - \beta_m),\tag{26}$$

where $\text{SL}(f_0)$ is the source level of the explosion sound (199.8 dB in this experiment), $E(f_0)$ is the average energy within the signal bandwidth, M_v is the sensitivity of the hydrophone (-185 dB), and β_m is the amplification factor of the receiver (30 dB).

Figure 13 shows the transmission loss calculated from the experimental data and using normal mode theory. The asterisks indicate the results from experimental data, while the blue curve indicates the results of the range-independent environment, that is, the result calculated from the sound speed profile in Figure 10, and the red curve indicates the results of the range-dependent environment, that is, the result calculated from the sound speed profile in Figure 11. The change in sound speed profile is small within the first 150 km, and the position offset of the convergence zone caused by the sound speed variation is also small. After 150 km, the change in sound speed increases, and the position offset of the convergence zone also increases. Figure 13b–f show the transmission loss curves of the five convergence zones. The simulated transmission loss curve is more consistent with the experimental data when considering the sound speed variation.

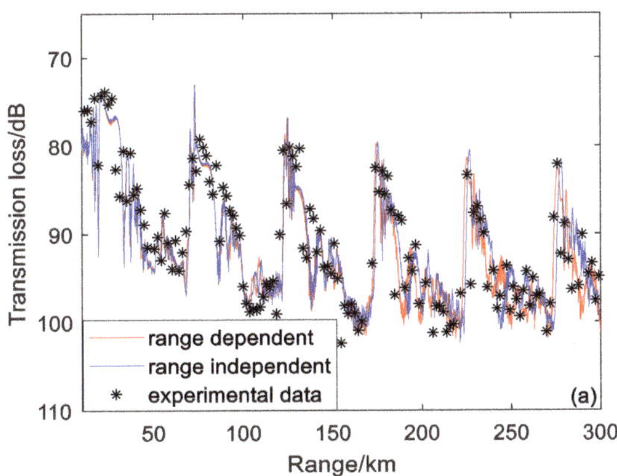

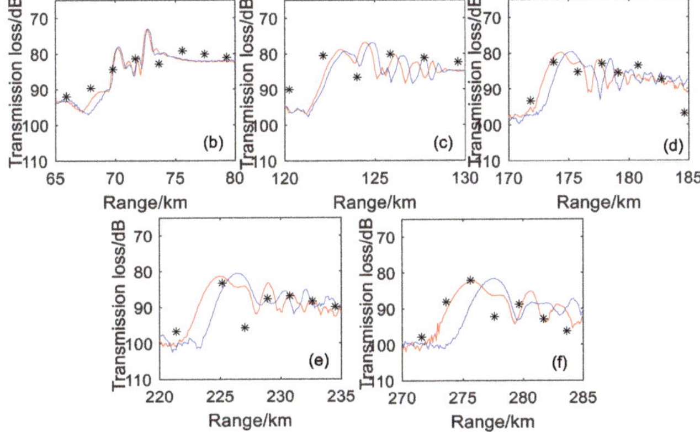

Figure 13. Transmission loss obtained from experiment and simulation. Results from (**a**) complete path and from (**b**–**f**) first to fifth convergence zones. In each graph, red line is the result of range dependent case. Blue line is the result of range independent case. The asterisk is the result of experimental data.

We used Equation (20) to calculate the position of the convergence zone under sound speed variations. The sound source was located at 200 m, and the receiver depth was 3146 m, being below the reciprocal depth of the sound source. The modal strength of the normal modes, for which the phase speed is smaller than the bottom sound speed, is shown in Figure 14. The modal strength of the lower mode is 0, and only the part with larger amplitude is shown in the figure.

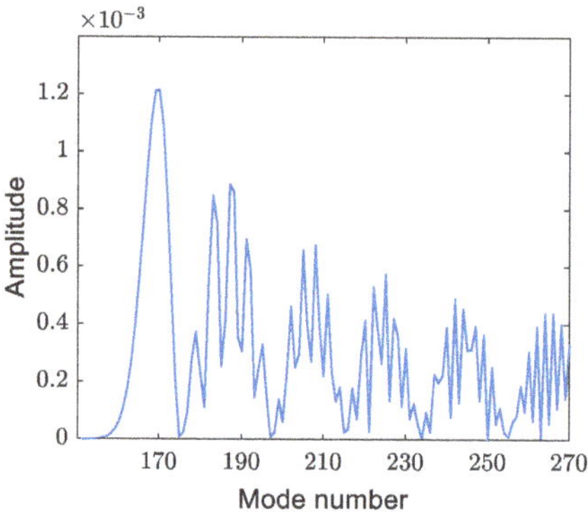

Figure 14. Mode amplitudes for refracted and surface-reflected modes obtained from experimental data.

Figure 14 shows that mode 170 yields the strongest modal intensity at the receiver depth. Computationally, the spread in mode Δn is defined by two successive minima in the figure (i.e., $\Delta n = 17$), and the WKB approximation (Equation (4)) is considered. From the coupled normal mode, the horizontal wavenumber of mode 170 is 0.9297. Substituting this value into expression $\kappa_n(z)$ and using Equation (4), we obtain 7.8×10^{-4}, being much less than 1 as required. Therefore, the refraction index is a slowly varying function of depth at this frequency. Equation (20) can be used to calculate the position of the convergence zone. The sound speed variation is calculated in segments with varying intensity parameter σ, such that $\delta c(z, r)$ is closer to the real sound speed profile. By substituting all the parameters into Equation (20), the position of the convergence zone is obtained. Table 3 lists the results obtained from our derived formula and coupled normal mode. The position of the convergence zone obtained in the experiment is listed in the fourth column.

Table 3. Positions of convergence zones obtained using proposed formulation and coupled normal mode on experimental data.

m	r_{ri}/km	r_{rd}/km	r_{ex}/km	r_{eq}/km
1	72.7	72.6	73.6	72.6
2	124.9	124.4	122.1	124.3
3	175.2	174.3	173.7	173.8
4	226.4	224.9	225.2	224.3
5	277.6	275.8	275.6	274.9

Equation (20) suitably provides the position of the convergence zone for a range-dependent sound speed profile. After the five convergence zones, there is a small difference of 0.9 km between the coupled normal mode and proposed formulation. The positions of the convergence zones obtained from the range-dependent and range-independent environments listed in Table 3 show that more convergence zones cause an increasing position offset owing to the sound speed variation. Because the variation in sound speed after 150 km accelerates, the absolute value of σ increases. Thus, according to Equation (20), the position offset of the convergence zone increases.

Figure 15 shows the transmission losses in range-dependent and range-independent environments for sound source and receiver depths of 200 m. The transmission loss curve is basically the same in both cases, and the position and gain of the convergence zone are equal because the sound speed variation does not change considerably before 150 km. After 150 km, owing to the increasing sound speed variation, the transmission loss curve begins to exhibit differences, and the position and gain of the convergence zone change between the environments. In Figure 15, the fifth convergence zone is enlarged locally. Through the transmission loss curve, a position offset of 2.4 km occurs at the fifth convergence zone owing to the sound speed variation. Compared with the transmission loss curve in Figure 13 and enlarged view, the convergence zone is shifted by approximately 1.8 km owing to the sound speed variation at a receiver depth of 3146 m. For the sound source below the mixed layer, the influences of the sound speed variation with range on the convergence zone near the sea surface and the convergence zone for large-depth reception are similar, and the influence on the convergence zone for large-depth reception is smaller.

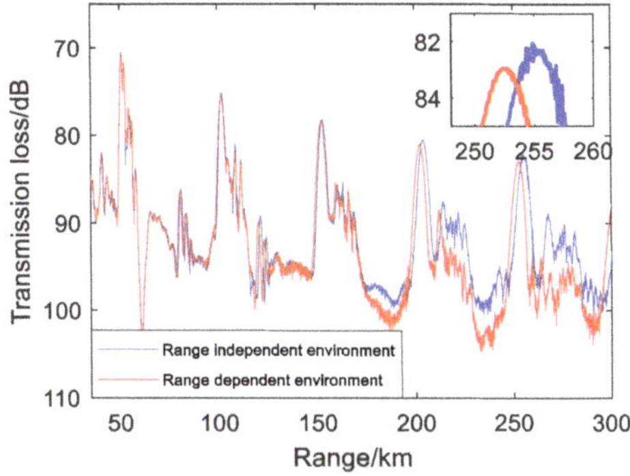

Figure 15. Transmission loss obtained from range-independent and range-dependent sound speed profiles with source and receiver depths of 200 m.

Figure 16 shows the results for a source depth of 100 m in the mixed layer. The receiver depths are 100 m and 3800 m, which is the reciprocal of 100 m. Before 150 km, given the small sound speed variation, the position and gain of the two types of convergence zones show negligible changes. When the sound speed variation changes notably after 150 km, it influences the position of the convergence zone. As the sound speed gradually decreases with range, the convergence zone shifts to the sound source, and the convergence zone near the sea surface is more affected than the convergence zone for large-depth reception. In this environment, the sound speed variation causes the convergence zone near the sea surface at the fifth convergence zone to shift by approximately 5 km, while the convergence zone for large depth reception only shifts by approximately 1 km.

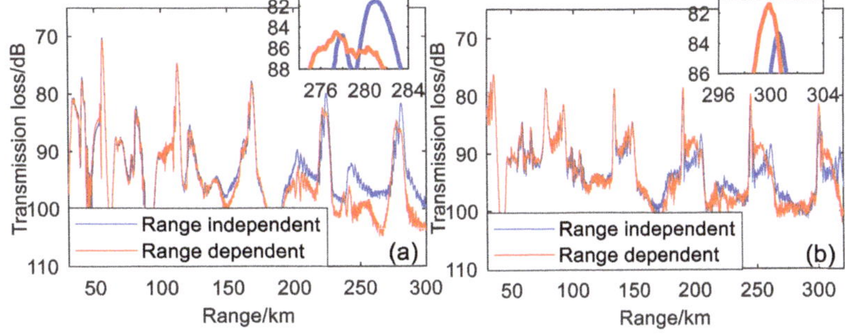

Figure 16. Transmission loss obtained from range-independent and range-dependent sound speed profiles at source depth of 100 m and receiver depths of (**a**) 100 and (**b**) 3800 m.

Comparing Figures 13, 15 and 16, the variation in sound speed has a greater influence on the convergence zone when the source is at the mixed layer than when the source is below it. According to Equation (20), the upper and lower limits of the integral of L_n in the molecule are related to the depth of the sound source, and the integral term contains depth function $w(z)$. When the sound source is located at the mixed layer, $w(z)$ is large, and the offset of the convergence zone distance increases. Comparing Figures 13 and 16b, as the receiver depth increases, the influence of the sound speed variation on the position of the convergence zone for large-depth reception decreases.

5. Discussion

The results of the experimental data show that the formula derived in this paper can accurately predict the position of the convergence zone when the sound speed varys linearly. As for the ellipsoidal Gaussian eddy case, the simulation results verify the correctness of the formula. The position of the convergence zone in other mesoscale phenomena cases may also be derived as long as the specific $\delta c(z, r)$ can be expressed. However, further research is still needed for other cases of stochastic sound speed variation. In addition, for real world problems, the roughness of the sea surface needs to be incorporated into the stochastic part of the problem. Simulation and experimental data reveal that the convergence zone for large-depth reception has a good convergence gain, and it is less affected by the sound speed variation. Accurate prediction of the position of the convergence zone will play a key role in ocean acoustic tomography.

6. Conclusions

Based on the WKB approximation, we derived the position formulae of the convergence zone in a range-dependent environment with sound speed profiles varying in linear and ellipsoidal Gaussian eddy cases. The formulae obtained under the two profiles were simulated. By comparing the calculation results obtained from the coupled normal mode and derived formula, we verified the accuracy of the formula of the convergence zone position under a range-dependent sound speed profile. In addition, the derived formula under linear variation was verified using experimental data. The experimental data and convergence zone position simulated by the coupled normal mode were compared to further verify the correctness of the derived formula.

A simulation was carried out based on experimental data, and the influence of the sound speed variation on the position of the convergence zone was analyzed when the sound source was located at and below the mixed layer. We can draw the following conclusions: (1) When the sound source is located below the mixed layer, the sound speed variation reduces the offset in the position of the convergence zone, but the offset near the sea surface is larger than that of the convergence zone for large-depth reception. In the fifth convergence zone, the offset for the zone near the sea surface is approximately 2.4 km, and the position offset for large-depth reception is approximately 1.8 km. (2) When the sound source is located at the mixed layer, the influence of sound speed variation on the position of the convergence zone near the sea surface is substantially increased. However, the influence on the position of the convergence zone for large-depth reception decreases. In the fifth convergence zone, the position offset near the sea surface is approximately 5 km, and the offset for large-depth reception is approximately 1 km. Through simulation and experimental data, the convergence zone for large-depth reception also has a good convergence gain, increasing the applicability of this type of convergence zone. When considering long-range target detection, the convergence zone for large-depth reception is more appropriate than that near the sea surface because the former is less affected by sound speed variations.

Author Contributions: Conceptualization, Z.L. and S.P.; methodology, Z.L.; software, Z.L.; validation, Z.L., S.P., M.Z. and L.G.; formal analysis, Z.L.; investigation, Z.L.; resources, Z.L.; data curation, Z.L.; writing—original draft preparation, Z.L.; writing—review and editing, M.Z. and L.G.; visualization, Z.L.; supervision, S.P.; project administration, Z.L.; funding acquisition, S.P. All authors have read and agreed to the published version of the manuscript.

Funding: This research was funded by the National Natural Science Foundation of China under grant number 12174048.

Institutional Review Board Statement: Not applicable.

Data Availability Statement: The data presented in this paper are available upon reasonable request to the corresponding author.

Acknowledgments: The authors would like to thank the anonymous reviewers for their careful assessment of our work.

Conflicts of Interest: The authors declare no conflict of interest.

References

1. Urick, R.J. *Principle of Underwater Sound*; McGraw-Hill: New York, NY, USA, 1983; pp. 79–162.
2. Woezel, J.L.; Ewing, M.; Pekeris, C.L. Explosion Sounds in Shallow Water. *Geol. Soc. Am. Mem.* **1948**, *27*, 1–62. [CrossRef]
3. Brekhovskikh, L.M. Propagation of acoustic and infrared waves in natural waveguides over long distances. *Sov. Phys. Usp.* **1960**, *3*, 159–166. [CrossRef]
4. Urick, R.J. Caustics and Convergence Zones in Deep-Water Sound Transmission. *J. Acoust. Soc. Am.* **1965**, *38*, 348–358. [CrossRef]
5. Schulkin, M. Surface-Coupled Losses in Surface Sound Channels. *J. Acoust. Soc. Am.* **1968**, *44*, 1152–1154. [CrossRef]
6. Bongiovanni, K.P.; Siegmann, W.L.; Ko, D.S. Convergence zone feature dependence on ocean temperature structure. *J. Acoust. Soc. Am.* **1996**, *100*, 3033–3041. [CrossRef]
7. Zhang, R.H. Normal mode sound field in shallow sea surface channel. *Acta Phys. Sin.* **1975**, *24*, 200–209. [CrossRef]
8. Wang, L.; Wang, K.P. Varieties of Sound Propagating in Convergence Zone Caused by Sound Spring Layer. *Mar. Geod. Cartogr.* **2014**, *34*, 40–42. [CrossRef]
9. Chen, C.; Yan, F.G.; Jin, T.; Zhou, Z.Q. Investigating acoustic propagation in the Sonic Duct related with subtropical mode water in Northwestern Pacific Ocean. *Appl. Acoust.* **2020**, *169*, 107478. [CrossRef]
10. Yang, F.; Wang, H.; Gao, W.D.; Meng, X.S. Zoning of sound speed profile types and characteristics of convergence zone in the deep North Atlantic Ocean. *Mar. Forecast.* **2021**, *38*, 103–110. [CrossRef]
11. Bi, S.Z.; Peng, Z.H. Effect of earth curvature on long range sound propagation. *Acta Phys. Sin.* **2021**, *70*, 114303. [CrossRef]
12. Henrick, R.F.; Siegmann, W.L. General analysis of ocean eddy effects for sound transmission applications. *J. Acoust. Soc. Am.* **1977**, *62*, 860–870. [CrossRef]
13. Rudnick, D.L.; Munk, W. Scattering from the mixed layer base into the sound shadow. *J. Acoust. Soc. Am.* **2006**, *120*, 2580–2594. [CrossRef]
14. Colosi, J.A.; Rudnick, D.L. Observations of upper ocean sound-speed structures in the North Pacific and their effects on long-range acoustic propagation at low and mid-frequencies. *J. Acoust. Soc. Am.* **2020**, *148*, 2040–2060. [CrossRef] [PubMed]
15. Li, J.X.; Zhang, R. Ocean mesoscale eddy modeling and its application in studying the effect on underwater acoustic propagation. *Mar. Sci. Bull.* **2011**, *30*, 37–46. [CrossRef]
16. Colosi, J.A.; William, Z.L. Sensitivity of mixed layer duct propagation to deterministic ocean features. *J. Acoust. Soc. Am.* **2021**, *149*, 1969–1978. [CrossRef]
17. White, A.W.; Henyey, F.S. Internal tides and deep diel fades in acoustic intensity. *J. Acoust. Soc. Am.* **2016**, *140*, 3952–3962. [CrossRef]
18. Zhang, L.; Liu, D. Deep-sea acoustic field effect under mesoscale eddy conditions. *Mar. Sci.* **2020**, *44*, 66–73. [CrossRef]
19. Piao, S.C.; Li, Z.Y. Lower turning point convergence zone in deep water with an incomplete channel. *Acta Phys. Sin.* **2021**, *70*, 024301. [CrossRef]
20. Wu, S.L.; Li, Z.L.; Qin, J.X.; Wang, M.Y. Influence of tropical dipole in the East Indian Ocean on acoustic convergence region. *Acta Phys. Sin.* **2022**, *71*, 134301. [CrossRef]
21. Chandrayadula, T.K.; Periyasamy, S.; Colosi, J.A. Observations of low-frequency, long-range acoustic propagation in the Philippine Sea and comparisons with mode transport theory. *J. Acoust. Soc. Am.* **2020**, *147*, 877–897. [CrossRef]
22. Tindle, C.T.; Guthrie, K.M. Rays as interfering modes in underwater acoustics. *J. Sound Vib.* **1974**, *34*, 291–295. [CrossRef]
23. Beilis, A. Convergence zone positions via ray mode theory. *J. Acoust. Soc. Am.* **1983**, *74*, 171–180. [CrossRef]
24. Shaji, C.; Wang, C.; Halliwell, G.R. Simulation of tropical Pacific and Atlantic Oceans using a HYbrid Coordinate Ocean Model. *Ocean Model.* **2005**, *9*, 253–282. [CrossRef]

remote sensing

MDPI

Article

Research on High Robustness Underwater Target Estimation Method Based on Variational Sparse Bayesian Inference

Libin Du, Huming Li, Lei Wang, Xu Lin and Zhichao Lv *

College of Ocean Science and Engineering, Shandong University of Science and Technology, Qingdao 266590, China; dulibin@sdust.edu.cn (L.D.); 2017051211@hrbeu.edu.cn (H.L.); skd996159@sdust.edu.cn (L.W.); linxu@sdust.edu.cn (X.L.)
* Correspondence: lvzhichao@hrbeu.edu.cn

Abstract: Pulse noise (such as glacier fracturing and offshore pile driving), commonly seen in the marine environment, seriously affects the performance of Direction-of-Arrival (DOA) estimation methods in sonar systems. To address this issue, this paper proposes a high robustness underwater target estimation method based on variational sparse Bayesian inference by studying and analyzing the sparse prior assumption characteristics of signals. This method models pulse noise to build an observation signal, completes the derivation of the conditional distribution of the observed variables and the prior distribution of the sparse signals, and combines Variational Bayes (VB) theory to approximate the posterior distribution, thereby obtaining the recovered signal of the sparse signals and reducing the impact of pulse noise on the estimation system. Our simulation results showed that the proposed method achieved higher estimation accuracy than traditional methods in both single and multiple snapshot scenarios and has practical potential.

Keywords: array signal processing; sparse Bayesian learning; direction estimation; pulse noise

1. Introduction

As an underwater sensor, a hydrophone can realize the real-time monitoring of various opportunistic sound sources and environmental noise in the ocean. However, estimating the distance and direction of the sound source solely based on a single hydrophone is difficult, and a single hydrophone has a low signal-to-noise ratio (SNR) and limited detection range. On the contrary, an array of hydrophones exhibits strong capabilities in estimating the direction and distance of the sound source and has a significantly higher SNR than a single hydrophone. Research on underwater acoustic arrays in terms of SNR improvement, distance, and direction estimation has become a significant topic.

Array signal processing is a technology that uses a group of sensor arrays to spatially sample signals and then uses corresponding signal processing algorithms to enhance and estimate the parameters of the received data. Compared to methods that use a single sensor to collect and process signals, array signal processing technology can achieve spatial gains by utilizing the spatial characteristics of the signal, thereby improving the accuracy of parameter estimation [1].

DOA estimation of underwater wave propagation is a significant research subject in array signal processing and has substantial theoretical and practical implications for underwater target detection and tracking. In the field of underwater acoustic signal processing, commonly used DOA estimation methods include conventional beamforming and the Multiple Signal Classification (MUSIC) algorithm. However, these methods have limited ability to estimate the direction of adjacent signal sources and may even fail in the case of impulsive noise [2]. Therefore, conducting research on the DOA estimation of underwater targets has significant theoretical and practical implications.

Citation: Du, L.; Li, H.; Wang, L.; Lin, X.; Lv, Z. Research on High Robustness Underwater Target Estimation Method Based on Variational Sparse Bayesian Inference. *Remote Sens.* **2023**, *15*, 3222. https://doi.org/10.3390/rs15133222

Academic Editor: Jaroslaw Tegowski

Received: 9 June 2023
Revised: 16 June 2023
Accepted: 16 June 2023
Published: 21 June 2023

Beamforming technology, based on array technology, is the main approach for high-precision target detection. In their research, Kawachi et al. demonstrated the design and testing of an echo-PIV system that efficiently mapped the interior and fluid flow of a submerged vessel using a single divergent signal wave and delay-and-sum processing. However, the DW-DAS echo-PIV method is not useful for sensing leakage points and underwater debris at relatively short distances and over a narrow field of view [3]. Meanwhile, Shostak et al. proposed a new method for estimating the distance to any underwater object or physical phenomenon by analyzing the curvature of the wavefront and its impact on the measuring sonar system of correlated noise. They also provided results that substantiated this method [4]. Li et al. demonstrated a method for estimating seabed parameters that used the spatial characteristics of the ocean's ambient noise without relying on matched-field processing [5]. Zhou et al. reported improvements over conventional PGC methods, and the hydroacoustic sensor system has great potential in large-scale multiplexing [6]. Li et al. investigated the effectiveness of array processing for the passive monitoring of gas seeps and proposed using beamforming methods to enhance the SNR and improve the productivity of passive acoustic systems [7]. Verdon et al. presented a case study showcasing the use of an "L"-shaped downhole fiber-optic array for monitoring microseismic activity [8]. Other influential work includes Schinault et al., 2019. The array, in its current state of development, is a low-cost alternative to obtain quality acoustic data from a towed array system. Their study demonstrated that this array could be used for observing whales and ship tonals at ranges up to 5 km, receiving acoustic signals from targets of interest with enhanced SNR and directional sensing capabilities. Marine mammal vocalizations have been captured by this prototype array, and whale species have been identified through visual observation [9]. Xie et al. proposed a robust wideband beamforming algorithm based on subspace spectrum separation for addressing the issue of manifold deviations that may occur in sensor arrays. In this algorithm, a sensor array manifold calibration method based on subspace partitioning was proposed, and an improved interference–noise covariance matrix reconstruction method based on spectral separation was derived. Firstly, the space was divided into several subspaces using the Capon spatial spectrum, and the array manifold deviation was calibrated. Then, noise and interference information was accurately extracted through spectral separation, and finally, the optimal beamformer was designed based on the extracted information. Simulation results showed that this algorithm had good performance for different types and ranges of array manifold deviations. Furthermore, the wideband interference will be further considered in future work [10]. Zou et al. developed a hybrid analytical–numerical method that combines the analytical technique with the acoustic superposition approach to predict the sound radiation of a spherical double-shell within the ocean's acoustic environment. Green's function was utilized to simultaneously analyze the coupled vibration of fluid–structure, near-field, and far-field sound radiation. To reduce the computational complexity, the near-field was simulated using the image source method, while the far-field was simulated using the normal mode method. This method was used to calculate the sound radiation field of a spherical double-shell with positive and negative gradient sound velocity profiles in a shallow ocean acoustic environment. However, there was no obvious interfering phenomenon in the contour of the sound pressure distribution when the spherical shell was at a certain submerging depth. This requires further study of the related mechanism [11]. The numerical results were compared with finite element calculation results, and the efficiency was improved without compromising the calculation accuracy. A deconvolution method for conventional beamforming (CBF) was proposed in reference [12], which showed theoretically higher array gain (AG) than CBF and provided the possibility of detecting weak signals using the SNR [12]. However, simulation data processing showed that effective AG decreased with a decreasing SNR. The method of output signal subspace deconvolution for CBF was used to recover most of the AG loss and track the azimuth and time of weak signals. Frequency difference beamforming (FDB) provided a robust estimate of the wave propagation direction by shifting the signal processing to lower frequencies.

Xie et al. proposed a deconvolution frequency difference beamforming (Dv-FDB) method to improve array performance, which produced narrower beams and lower sidelobes while maintaining robustness. Based on this, the R-L algorithm was used for deconvolution to make Dv-FDB's spatial spectrum clearer. Simulation and experimental results showed that Dv-FDB was superior to FDB in higher resolution and lower sidelobes while maintaining robustness. Existing R-L methods are limited to arrays with offset-invariant beam patterns [13]. Byun et al. (2020) proposed a multi-constraint method for matching field processing (MFP) to address the uncertainty of the array tilt, and the experimental results verified the robustness of MFP. In summary, beamforming-based target direction estimation algorithms for underwater acoustic arrays can improve the performance of weak signal detection and underwater noise suppression, but the computational complexity of these algorithms needs to be considered. An additional important source of mismatch is the array tilt, which has not received much attention, in spite of its significant impact, especially for a large array tilt observed in shallow environments [14]. Other influential works in this field included Zhu et al. and Zhang et al. [15,16].

Another representative class of estimation algorithms is the MUSIC algorithm. MUSIC is a high-resolution DOA estimation algorithm that was first introduced by Schmidt in 1986. It is a non-parametric algorithm that does not require any prior knowledge of signal statistics, and it is widely used in many fields, including radar, sonar, and wireless communications. The main idea behind the MUSIC algorithm is to transform the received signal into the frequency-domain and estimate the DOAs of the incoming signals based on the eigenvalues and eigenvectors of the covariance matrix of the received signal. Specifically, the MUSIC algorithm first divides the entire space into two subspaces: the signal subspace and the noise subspace. The signal subspace contains the eigenvectors corresponding to the signal, while the noise subspace contains the eigenvectors corresponding to the noise. The DOAs of the incoming signals are then estimated by calculating the peaks of the spectrum of the noise subspace. Compared to other DOA estimation algorithms, such as beamforming and the Estimation Signal Parameter via Rotational Invariance Techniques (ESPRIT), MUSIC has several advantages, including a high-resolution, robustness to noise, and the ability to handle both coherent and incoherent signals. However, it also has some limitations, such as sensitivity to array geometry, the need for an accurate estimation of the noise subspace, and computational complexity. Overall, MUSIC is a powerful and widely used DOA estimation algorithm that has applications in many fields, including signal processing, wireless communications, radar, and sonar.

Yi et al. utilized passive array sonar systems to track a changing number of underwater targets, also known as acoustic emitters [17]. However, the authors did not consider information fusion among multiple passive sonar's systems. Huang et al. addressed the problem of DOA estimation with one-bit quantized array measurements. Otherwise, the approximation error becomes relatively large at a high SNR, which deserves further Investigation [18]. Cheng et al. proposed a marine environment noise suppression method for multiple-input multiple-output (MIMO) applied to the DOA estimation of multiple targets. In future work, it is worth exploring further optimization of the noise suppression algorithm model to reduce the impact of pre-estimation results on the DOA estimation accuracy [19]. As the underwater detection platform has a limited size, the traditional bulky linear array is not feasible. To address this issue, Li et al. investigated the joint processing–MUSIC (JMUSIC) algorithm for estimating the DOA in shallow sea multi-path environments using a non-uniform line array of acoustic vector sensors. It is a pity that the authors only conducted research in an ideal situation and did not take into account complex situations [20]. Zhu et al. proposed a method for obtaining the optimal waveform estimation of source signals in a spatial scanning orientation through the estimation of the maximum posterior probability criterion and the iterative convergence process of the constraint equation. The experiment yielded excellent results in the case of single snapshots, but it is also worth paying attention to how fast the shots were [21]. Ahmed et al. conducted a comparative study of deterministic and heuristic algorithms for viable

DOA estimation for different dynamic objects in underwater environments. To achieve the precise positioning of underwater targets at a close range [22], Ahmed et al. utilized the Cuckoo Search Algorithm (CSA) and swarm intelligence to optimize DOA estimation with a Uniform Linear Array (ULA) in various underwater scenarios [23]. An et al. proposed a combination of a linear array composed of multiple mutually perpendicular sub-arrays, overcoming the ambiguity of a single linear array's port and starboard orientation [24]. Under normal circumstances, both Ahmed and An had achieved research results, but in unconventional situations, such as pulse environments, it is worth exploring the advanced nature of the algorithms.

In recent years, there has been significant development in DOA estimation algorithms based on SBL. SBL is a statistical inference technique that is used to estimate sparse signals from noisy and incomplete data. It is a type of Bayesian regularization method that aims to find the most probable solution to an inverse problem by incorporating prior knowledge and assumptions about the underlying signal. In the context of DOA estimation, SBL is used to estimate the sparse signal of the DOA parameters from the array measurements. The key idea of SBL is to formulate the DOA estimation problem as a Bayesian inference problem, where the unknown DOA parameters are modeled as random variables, and the prior distribution of the DOA parameters is assumed to be sparse. By incorporating the prior information about the sparsity of the DOA parameters, SBL can effectively suppress the noise and interference in the array measurements and accurately estimate the DOA parameters, even in the presence of a limited number of snapshots. SBL algorithms typically involve iterative optimization procedures that update the estimates of the unknown parameters based on the observed data and the prior distribution. These algorithms can be computationally intensive, but they have been shown to be effective in a wide range of DOA estimation applications, including radar, sonar, and wireless communications.

Wang et al. aimed at the problem of interactions among the hydrophone array elements of the actual sonar array, which causes estimation performance dropping of the array's DOA, and a DOA estimation method under uncertain interactions of the array elements was proposed. However, the author did not note the relevant signals [25]. In order to achieve the high-precision Direction-of-Arrival (DOA) estimation of array signals in complex underwater acoustic environments, the root off-grid sparse Bayesian learning (ROGSBL) algorithm was applied to an underwater acoustics field [26]. In 2022, Haodong Bai studied the efficient DOA processing algorithm under multi-snapshots by aiming at the problem that the DOA estimation method, based on sparse Bayesian learning under single snapshots, has a low estimation accuracy and a large number of operations for increasing the number of snapshots [27]. He et al. proposed the SS-OGSBI algorithm to solve the problem of off-grid DOA estimation under coherent sources [28]. Guo et al. applied sparse Bayesian learning to the DOA estimation of underdetermined broadband signals with mutual arrays in unknown noise fields [29], while Shen et al. proposed an off-grid DOA estimation method based on subspace fitting and block-SBL to address the poor performance of traditional SBL-based DOA estimation algorithms under low SNR conditions [30]. Other influential works in this field included Yu et al., Ma et al., Zhu et al., Jimenez-Martinez M and Zhang et al. [31–35].

Although researchers have provided answers to the questions raised and made significant contributions to the field of Direction-of-Arrival (DOA) estimation using beamforming in underwater acoustics, the algorithms themselves have limitations. The researchers conducted their studies under the background of Gaussian noise, and further investigation is necessary to determine the robustness of the algorithms in highly impulsive noise environments.

In this article, a high robustness underwater target estimation technique based on variational sparse Bayesian is put forward by studying and analyzing the sparse prior assumption characteristics of the signal. The method models the observed signal by modeling the pulse noise, completes the derivation of the conditional distribution of the observed variables and the prior distribution of the sparse signal, and then combines the approximate posterior distribution obtained by the VB method to obtain the recovered sparse signal, thereby reducing the impact of pulse noise on the estimation system. Finally, the performance of the aforementioned method was validated through simulation experiments.

2. Materials and Methods

2.1. Uniform Linear Array Signal Model

The linear array model is a fundamental mathematical framework for addressing the problem of sound source direction estimation. The model posits the existence of a linear array composed of multiple small sound sources, each of which continuously emits the same sound wave signal. These sound waves propagate through distinct paths to reach the receiving array, where the signal measured by each receiving element is expressed as a weighted sum of the signals stemming from each emitting source. More specifically, the linear array model comprises a transmit array and a receive array. Each sound-emitting source within the transmit array emits identical sound wave signals, which subsequently arrive at different receiving elements in the receive array via various propagation paths. The signal measured by each receiving element in the receive array is then computed as a weighted sum of the signals originating from the sound-emitting sources. These weighting coefficients reflect the path delay and attenuation factors experienced by the sound wave signal as it travels from the emitting source to the receiving element. Through the processing of signals within the linear array model, the direction of the sound source can be estimated. This involves calculating key parameters, such as the time delay and phase difference between individual receiving elements within the receiving array. Hence, the linear array model finds extensive applications in fields such as sound source direction estimation, sound beamforming, and signal source separation.

Consider an M-element ULA, the observation vector of the array can be defined as:

$$\mathbf{y}(t) = \mathbf{A}(\theta)\mathbf{S}(t) + \mathbf{n}(t), \quad t = 1, 2, \ldots, T \tag{1}$$

Here, $\mathbf{n}(t)$ represents independent identically distributed Gaussian white noise. The array manifold matrix is denoted by $\mathbf{A}(\theta)$ and denoted as $\mathbf{A}(\theta) = [\mathbf{a}(\theta_1), \mathbf{a}(\theta_2), \cdots, \mathbf{a}(\theta_n)]$. The matrix A of size $M \times N$ represents the phase information of the array, where N is the number of signal sources, and M is the number of sensors.

The covariance matrix for the array output is defined as follows:

$$\mathbf{R} = \widetilde{\mathbf{A}}(\theta)\mathbf{A}^{\mathbf{H}}(\theta) + \delta^2\mathbf{I} = \sum_{n=1}^{N} \mathbf{P}_n \mathbf{a}(\theta_n)\mathbf{a}^{\mathbf{H}}(\theta_n) + \delta^2\mathbf{I} \tag{2}$$

In this equation, $\mathbf{P}_n = \mathbf{E} = \left[\left|\widetilde{\mathbf{S}}_n(t)\right|^2\right]$.

In practical applications, the correlation matrix is commonly used to estimate the output covariance matrix of the array. The correlation matrix is represented as follows:

$$\widetilde{\mathbf{R}} = \frac{1}{T}\sum_{t=1}^{T} \mathbf{x}(t)\mathbf{x}^{\mathbf{H}}(t) \tag{3}$$

In order to explain the principle of ULA more clearly, it can be explained in more detail in Figure 1.

2.2. Pulse Noise Distribution Model—Student-t Distribution

In this section, we will present an exposition on the Student-t distribution from three perspectives: origin, definition, and frequency spectrum. Regarding the parameter settings of the noise model in the frequency spectrum section, we will adopt the parameters used in the experiment described in this article as the standard. The primary objective is to provide a more intuitive illustration of the advantages of replacing pulse noise with the Student-t distribution model.

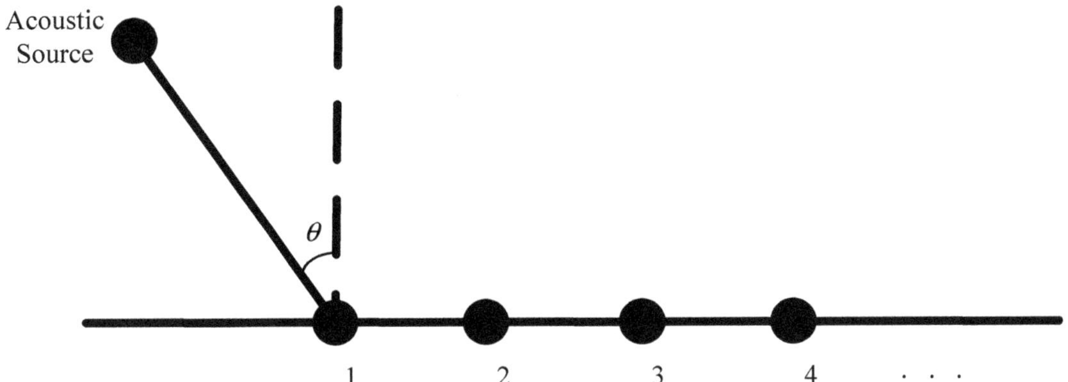

Figure 1. Uniform Linear Array signal model.

The main objective of array signal processing is to effectively remove noise from noisy observation data, thereby enabling the accurate recovery of the original signal and extraction of desired information. In many areas of array signal processing, narrowband signal models are commonly assumed, and noise is modeled as following a Gaussian distribution due to the fact that the Gaussian distribution satisfies the central limit theorem and has finite second-order and higher-order statistics. Additionally, signal characteristics can be represented by the mean and variance at any time. However, in real-world experimental environments, many types of noise do not adhere to a Gaussian distribution model. Such noises typically exhibit instantaneous pulse characteristics and more frequent abnormal data compared to Gaussian noise. Therefore, using a Gaussian distribution model to replace the noise model is not a realistic approach. For instance, if pulse noise is present in the DOA estimation environment, the noise distribution would have a heavy tail and a distribution with a heavier tail would be required to replace the Gaussian distribution.

Pulse noise models can be classified into two categories based on their generation mechanism: real physical statistical models and theoretical analytical models. Compared to physical statistical models, theoretical analytical models have relatively fixed mathematical expressions, which makes them more convenient for theoretical analysis. In the field of DOA estimation in the pulse environment, three models have been widely used, including the mixed Gaussian distribution model, the Alpha stable distribution model, and the Student-*t* distribution model. This paper primarily models pulse noise using the Student-*t* distribution, and the fundamental concepts of the Student-t distribution will be elaborated in detail below.

Gosset was a quality control officer at a brewery in 1908 when he discovered and proposed the Student-*t* distribution. At that time, he needed to study the variability of beer brewing in small sample sizes. However, since the data samples that he studied were very small, he could not use a traditional normal distribution for statistical analysis. To remedy this issue, Gosset examined the distribution of the population mean given the sample mean and sample standard deviation. He discovered that if the sample came from a normal distribution, the difference between the sample mean and the population mean could be described by a new distribution, which was later named the Student-t distribution. Gosset initially dubbed this distribution the "distribution of errors" because it was used to describe the error between the sample mean and the population mean. Later, the Student-t distribution became widely used in statistics, and it was named after Gosset's pen name, "Student". The Student-t distribution is a probability distribution that is commonly utilized to model data with heavy tails, i.e., tail probabilities that are significantly higher than those of a normal distribution.

The model of the Student-t distribution can be defined as follows:

$$n(t) \sim S(v|u, \mathbf{\Lambda}, \varsigma) \tag{4}$$

where u is the average of the M-dimensional vector, v, $\mathbf{\Lambda} = \mathrm{diag}(\Lambda_1, \Lambda_2, \cdots, \Lambda_M)$ denotes the precision matrix, and ς denotes the degree of freedom (DOF) parameter. The decay becomes slower as the DOF decreases. When the degrees of freedom decrease, the shape of the Student-t distribution changes, with the peak of the probability density function becoming lower and the tails becoming thicker. This makes it better suited for describing pulse noise. In order to explain the student's t-distribution more clearly, we did a simple simulation experiment and obtained the results shown in Figure 2.

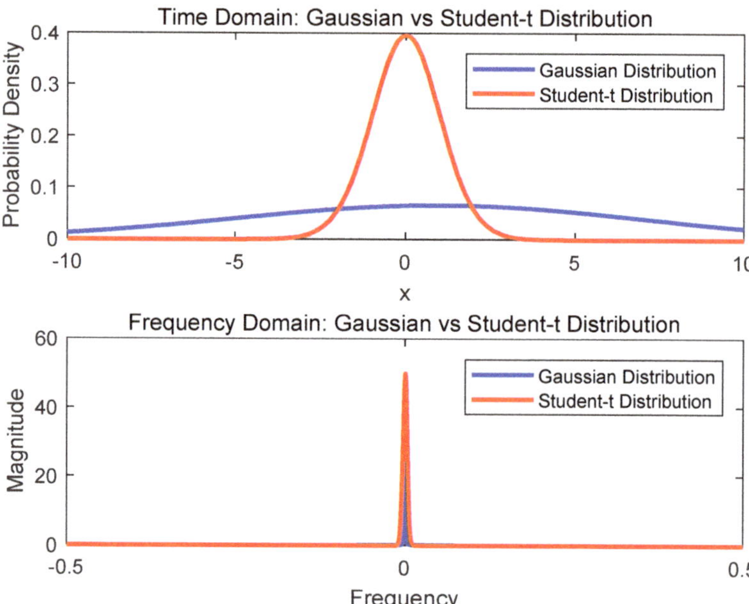

Figure 2. Student-t distribution in the time-domain spectrum.

This experiment was mainly used to plot the time-domain and frequency-domain graphs of the Gaussian distribution and Student-t distribution. We first set three parameters: mean = 1, standard deviation = 6, and degrees of freedom = 30. Then, it generated an x-axis vector containing 1000 points using the "linspace" function, used to represent the continuous variable, x, in the time-domain. Next, the probability density functions of the Gaussian distribution and Student-t distribution were calculated to generate the time-domain graphs of the two distributions. The program used the "plot" function to plot the time-domain graphs of the two distributions.

The next part of the experiment was used to plot the frequency-domain graphs of the Gaussian distribution and Student-t distribution. The experiment first used the "fft" function to calculate the Fourier transform of the time-domain graph and used the "fft-shift" function to center the result. Then, we used the "linspace" function to generate the continuous variable "freq" in the frequency domain and used the "plot" function to plot the frequency-domain graphs of the two distributions and added axis labels, legends, and titles.

When plotting the graphs, the program used the "hold on" function to make both graphs of the distributions plotted on the same figure. This is performed to better compare the differences between the two distributions.

In the present illustration, it can be observed that the time-domain spectrum of the Student-*t* distribution exhibits a shape similar to that of the Gaussian distribution. However, compared to the Gaussian distribution, under the parameters set in this experiment, the Student-*t* distribution is better able to model noise with local outliers, such as impulse noise. Additionally, in the frequency-domain spectrum, the frequency response of the Student-*t* distribution is smoother than that of the Gaussian distribution. That is, its amplitude changes more slowly with frequency, which also helps to reduce the impact of impulse noise in the high-frequency range. Therefore, in this paper, the Student-*t* distribution is adopted as the model for impulse noise.

2.3. Graphical Models

The interaction between entities involved in a probabilistic system is represented by a graphical model, where nodes represent random variables, and arrows depict dependencies between variables [36]. A directed arrow from node A to node B indicates that the value of random variable B depends on the value of random variable A. Graphical models can be categorized into directed graph models and undirected graph models [15,36]. This paper focuses on directed graph models, also known as Bayesian network graphical models [37].

The definition of a directed graph model is as follows:

Given the conditional probability distribution of each node in the graphical model, the formula for calculating the joint distribution over all variable sets is $p(x)$ [38].

$$p(x) = \prod_s p\left(x_s \middle| x_{\pi(s)}\right) \tag{5}$$

Figure 3 shows an example of a directed graph model. In this model, a, b, and d represent random variables, and each node in the graphical model represents a conditional probability density. If the probability density of the node is unknown, it can be parameterized by a set of parameters. The joint distribution of the probability density is then expressed as follows:

$$p(a,b,d) = p(a;\theta_1)p(b;\theta_2)p(d|a,b;\theta_3) \tag{6}$$

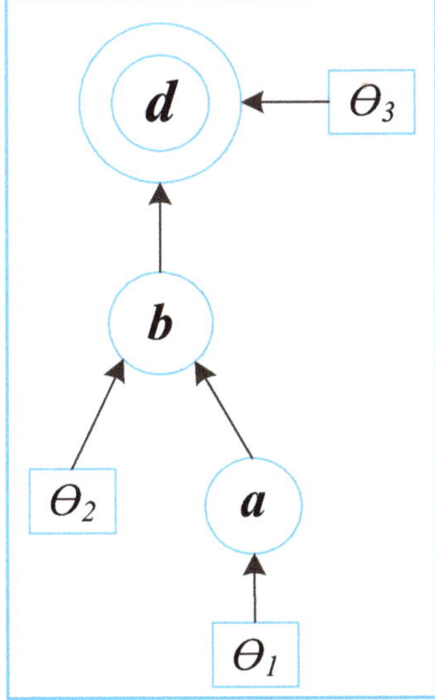

Figure 3. Example of a directed graph model.

The above expression can be simplified by considering the independence implied by the structure of the graphical model. Generally speaking, in the graphical model, each node is independent of its higher-level nodes. Therefore, expression (6) can be simplified as follows:

$$p(a,b,d) = p(a;\theta_1)p(b;\theta_2)p(d|a;\theta_3) \tag{7}$$

Another function of the graphical model is to arbitrarily distinguish random variables into those with directly observed results and those with hidden random variables without directly observed results [39]. In addition, the graphical model can be divided into parameterized graphical models and non-parameterized graphical models. If it is a parameterized graphical model, the parameters will appear in the conditional probability distribution of some nodes; that is, the probability models of these distributions are parameterized probability models.

3. Theoretical Model

In this paper, a high robustness underwater target estimation method based on variational sparse Bayesian inference is proposed by studying and analyzing the sparse prior assumption characteristics of the signal. The method models the observed signal by modeling the pulse noise, completes the derivation of the conditional distribution of the observed variables and the prior distribution of the sparse signal, and then combines the VB method to obtain the approximate posterior distribution, thereby obtaining the recovered signal of the sparse signal.

Firstly, it is assumed that there exist N narrowband signals impinging upon an M-element linear array.

$$\mathbf{Y} = \mathbf{\Phi X} + \mathbf{N} \tag{8}$$

In the equation, the observation matrix is represented by $\mathbf{Y} \in \mathbb{C}^{M \times L}$, $\mathbf{X} \in \mathbb{C}^{N \times L}$ represents the original signal, $\mathbf{N} \in \mathbb{C}^{M \times L}$ represents the noise matrix, and $\mathbf{\Phi} \in \mathbb{C}^{M \times N}$ represents the measurement matrix.

In the above Bayesian model, the joint distribution of all the observed variables and unknown variables is required, which usually includes the conditional distribution of the observed variables and the prior distribution of the sparse signal. The conditional distribution of the observed variables and the prior distribution of the sparse signal are derived, and the approximate posterior distribution is obtained through the VB method, thereby obtaining the recovered signal of the sparse signal.

3.1. Derivation of Conditional Distribution of Observation Variables

Modeling the pulse noise using the Student-*t* distribution, the probability density function is given as follows:

$$S(v|u,\Lambda,\varsigma) = \frac{\Gamma\left(\frac{M+\varsigma}{2}\right)}{\Gamma\left(\frac{\varsigma}{2}\right)(\varsigma\pi)^{\frac{M}{2}}}|\Lambda|^{\frac{1}{2}}\left[1 + \frac{(v-u)^{\mathrm{T}}\Lambda(v-u)}{\varsigma}\right]^{-\frac{M+\varsigma}{2}} \tag{9}$$

where $\Gamma(\cdot)$ denotes the Gamma function, assuming that all the columns of the noise matrix are independent and follow a zero-mean Student-*t* distribution [40].

By introducing the latent variable, λ, the Student-*t* distribution is an infinite mixture of Gaussian distributions with variances extended by gamma distributions [41].

$$S(\mathbf{v}|\mathbf{u},\mathbf{\Lambda},\zeta) = \int_0^\infty N\left(\mathbf{v}\Big|\mathbf{u},(\lambda\mathbf{\Lambda})^{-1}\right)G(\lambda|\zeta/2,\zeta/2)d\lambda \tag{10}$$

Here, $N(\cdot)$ And $G(\cdot)$ are the Gaussian distribution and Gamma distribution, respectively.

Therefore, the conditional distribution of the observed variables can be written as follows:

$$\left\{ \begin{array}{c} p(\mathbf{Y}|\mathbf{X},\Lambda,\zeta) = \prod\limits_{l=1}^{L} N\left(\mathbf{y}_l \middle| \Phi_l(\lambda\Lambda)^{-1}\right) \\ p(\lambda|\zeta) \neq G(\lambda|\zeta/2,\zeta/2) \end{array} \right. \tag{11}$$

By placing the Gamma distributions on each diagonal element of ζ and Λ, these equations can be obtained:

$$p(\zeta) = G(\zeta|c,d) \tag{12}$$

$$p(\Lambda) = \prod_{l=1}^{L} G(\Lambda_m|a_m,b_m) \tag{13}$$

In the equation, a_m, b_m, c, and d are hyperparameters of the Gamma distribution.

3.2. Derivation of Sparse Signal Prior Distribution

Assuming that all rows of the matrix, $\mathbf{X}$, are independent and follow a Gaussian distribution, the prior distribution of the sparse signal, s, can be obtained [41].

$$p(\mathbf{X}|\gamma) = \prod_{n=1}^{N} N\left(\mathbf{x}_{n,\cdot} \middle| 0, \gamma_n^{-1}\mathbf{I}_{L\times L}\right) \tag{14}$$

Here, $\gamma = [\gamma_1, \cdots, \gamma_n]$ represents the precision vector of the sparse signal, $\mathbf{X}$, and a Gamma distribution with hyperparameters ∂_n and β_n are used for each precision vector, γ_n, as follows [41,42]:

$$p(\gamma) = \prod_{n=1}^{N} G(\gamma_n|\partial_n,\beta) \tag{15}$$

According to Equations (12)–(15), the joint distribution of all the observed variables and unknown variables can be decomposed as follows:

$$\begin{aligned} p(\mathbf{Y},\mathbf{X},\gamma,\Lambda,\zeta,\lambda) \; &= p(\mathbf{Y}|\mathbf{X},,\Lambda,\lambda)p(\mathbf{X}|\gamma)p(\Lambda)p(\lambda|\zeta)p(\zeta) \\ &= \left(\prod_{l=1}^{L} p(\mathbf{y}_l|\mathbf{x}_l,\Lambda,\lambda)\right)\left(\prod_{n=1}^{N} p(\mathbf{x}_n|\gamma_n)p(\gamma_n)\right) \times \\ &\quad \left(\prod_{m=1}^{M} p(\Lambda_m)\right)p(\lambda|\zeta)p(\zeta) \end{aligned} \tag{16}$$

In Figure 4, more detailed dependencies can be obtained about variables and unknown variables.

3.3. Variational Bayes

Bayesian inference is based on the posterior distribution, $p(\Omega|\mathbf{Y}) = p(\mathbf{Y},\Omega)/p(\mathbf{Y})$, where Ω represents the set of all unknown variables. However, because the marginal distribution, $p(\mathbf{Y})$, can be difficult to handle, Bayesian inference often requires approximation. In the VB method, an approximation of $p(\Omega|\mathbf{Y})$, is made using a coefficient distribution, $q(\Omega) = q(\mathbf{X})q(\gamma)q(\Lambda)q(\zeta)q(\lambda)$, where $\Omega = \{\mathbf{X},\Lambda,\gamma,\lambda,\zeta\}$, and each approximate distribution in $q(\Omega)$ is obtained by computing the logarithmic expectation of (16) with respect to other distributions.

$$\ln q(\mathbf{X}) \propto \sum_{l=1}^{L} \left\{ \begin{array}{c} -\frac{1}{2}\mathbf{x}_l^T\left(E[\lambda]E[\Lambda] + \Phi^T \text{diag}E[\gamma]\Phi\right)\mathbf{x}_l + \\ \frac{1}{2}\left(\Phi^T \text{diag}E[\gamma]\mathbf{y}_l\right)^T + \frac{1}{2}\mathbf{x}_l^T\left(\Phi^T \text{diag}E[\gamma]\mathbf{y}_l\right) \end{array} \right\} \tag{17}$$

Here, $E[\cdot]$ denotes the expectation operator. By Equation (17), $q(\mathbf{X})$ can be calculated as follows:

$$q(\mathbf{X}) = \prod_{l=1}^{L} N(\mu_l,\Sigma) \tag{18}$$

where μ_l represents the expectation and Σ is the variance [43].

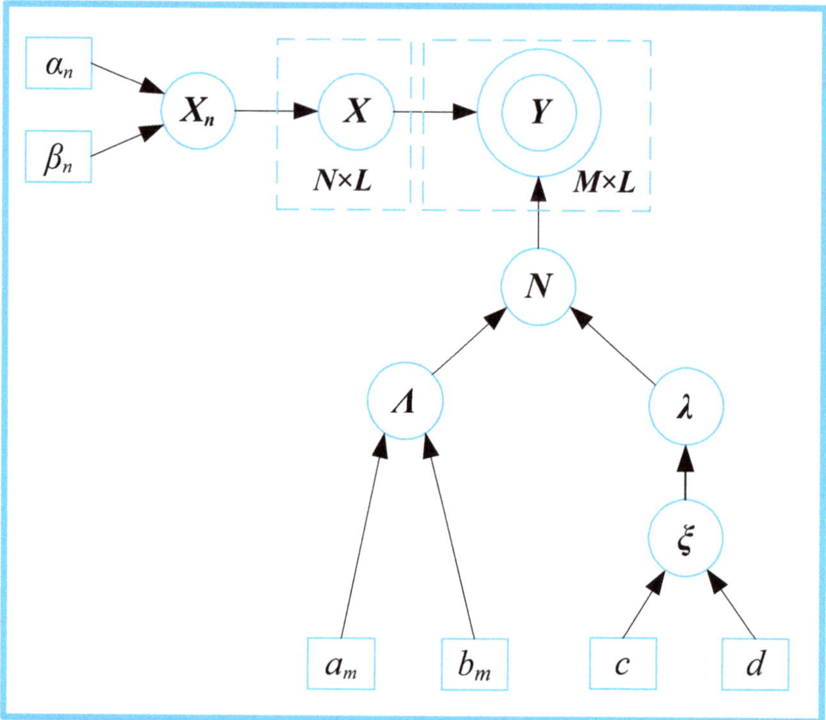

Figure 4. Graphical model.

$$\begin{cases} \boldsymbol{\Sigma} = \left(\mathbf{A} + \boldsymbol{\Phi}^T \mathbf{B} \boldsymbol{\Phi}\right)^{-1} \\ \boldsymbol{\mu}_l = \boldsymbol{\Sigma} \boldsymbol{\Phi}^T \mathbf{B} \mathbf{y}_l \ , \qquad l = 1, 2, \cdots, L \end{cases} \tag{19}$$

Similarly, $q(\gamma)$, $q(\boldsymbol{\Lambda})$, $q(\zeta)$, and $q(\lambda)$ is obtained one by one and expressed in the following form:

$$\begin{aligned} q(\zeta) &= \mathbf{G}(\zeta | c^*, d^*) \\ q(\lambda) &= \mathbf{G}(\lambda | \zeta_1^*, \zeta_2^*) \\ q(\gamma) &= \prod_{n=1}^{N} \mathbf{G}(\gamma_n | \alpha_n^*, \beta_n^*) \\ q(\boldsymbol{\Lambda}) &= \prod_{m=1}^{M} \mathbf{G}(\Lambda_m | a_m^*, b_m^*) \end{aligned} \tag{20}$$

The approximated hyperparameters are defined as follows [44,45]:

$$\begin{cases} c^* = c + \frac{1}{2} \\ d^* = d + \frac{1}{2}(\mathrm{E}[\lambda] - \mathrm{E}[\ln \lambda] - 1) \end{cases} \tag{21}$$

$$\begin{cases} \zeta_1^* = \frac{1}{2}(\mathrm{E}[\zeta] + M) \\ \zeta_2^* = \frac{1}{2} \sum_{m=1}^{M} \left\{ \mathrm{E}[\Lambda_m] \sum_{l=1}^{L} \mathrm{E}\left[(\mathbf{y}_l - \phi_m \mathbf{x}_l)^2\right] \right\} + \frac{1}{2}\mathrm{E}[\zeta] \end{cases} \tag{22}$$

$$\begin{cases} a_m^* = a_m + \frac{1}{2} \\ b_m^* = b_m + \frac{1}{2}\mathrm{E}[\lambda] \sum_{l=1}^{L} \mathrm{E}\left[(\mathbf{y}_l - \phi_m \mathbf{x}_l)^2\right] \end{cases} \tag{23}$$

$$\begin{cases} \alpha_n^* = \alpha_n + \frac{1}{2} \\ \beta_n^* = \beta_n + \frac{1}{2}\mathrm{E}\left[\mathbf{x}_n \mathbf{x}_n^T\right] \end{cases} \tag{24}$$

Based on the approximate distribution, (17), and Equation (20), the expectations of the approximated hyperparameters are expressed as follows [45]:

$$\mathrm{E}[\mathbf{x}_l] = \boldsymbol{\mu}_l$$
$$\mathrm{E}\left[\mathbf{x}_n\mathbf{x}_n^{\mathrm{T}}\right] = \left(\sum_{l=1}^{L}\left(\boldsymbol{\mu}_l\boldsymbol{\mu}_l^{\mathrm{T}} + \boldsymbol{\Sigma}\right)\right)_{nn} \tag{25}$$

$$\mathrm{E}[\lambda] = \zeta_1^*/\zeta_2^*$$
$$\mathrm{E}[\ln \lambda] = \varphi(\zeta_1^*) - \ln \zeta_2^* \tag{26}$$

$$\mathrm{E}[\Lambda_m] = a_m^*/b_m^*$$
$$\mathrm{E}[\gamma_n] = \alpha_n^*/\beta_n^*$$
$$\mathrm{E}[\zeta] = c^*/d^* \tag{27}$$

$$\mathrm{E}\left[(\mathbf{y}_l - \phi_m\mathbf{x}_l)^2\right] = \mathbf{y}_l^2 - 2\mathbf{y}_l\phi_m\mathrm{E}[\mathbf{x}_l] + \phi_m\left(\boldsymbol{\mu}_l\boldsymbol{\mu}_l^{\mathrm{T}} + \boldsymbol{\Sigma}\right)\phi_m^{\mathrm{T}} \tag{28}$$

Here, $\varphi(\cdot)$ represents the digamma function, and $\boldsymbol{\mu}_l$ represents the nth element on the main diagonal of the matrix.

Moreover, from Equation (25), the mean $\boldsymbol{\mu}_l$ gives an estimate of $\mathbf{x}_l$, and thus the recovery result of the sparse signal can be obtained by the following equation:

$$\hat{\mathbf{X}} = [\boldsymbol{\mu}_1, \cdots, \boldsymbol{\mu}_l] \tag{29}$$

where the expectation can be obtained by coupling the hyperparameters in (21)–(24).

Therefore, the solution can be obtained by iteratively computing (19) and (21)–(24) until convergence, leading to the optimal recovery result.

To summarize, the proposed algorithm is listed in Table 1.

Table 1. Flow of target azimuth estimation algorithm based on variational sparse Bayes.

Input	Observed Variables, s, and Measurement Matrix, X
1	Initialize hyperparameters $\{\alpha_n, \beta_n\}_{m=1}^{M}$, $\{\alpha_n, \beta_n\}_{n=1}^{N}$, c, and d, set stopping threshold, ε, and maximum iteration number, $I_{\max}$.
2	Calculate the mean and variance using Equation (19).
3	Update hyperparameters $\{a_m, b_m\}_{m=1}^{M}$, $\{\alpha_n^*, \beta_n^*\}_{n=1}^{N}$, c, and d separately using Equations (21)–(24).
4	If the maximum change of hyperparameters is less than the stopping threshold, stop iterating and go to step 5. Otherwise, go back to step 2 to continue iterating.
5	Output the recovery result.

4. Simulation Results

Consider a ULA with M = 30 elements spaced at half-wavelength. Three incoherent signal sources are assumed to be located in the far-field of the receiving array, an incident from the directions of $45°$, $60°$, and $90°$, with an SNR of 5 dB. The iteration number is set to 500, and the number of grid points is set to N = 181. The measurement noise is generated using $f = (1 - p)N(0, \delta^2) + pN(0, k\delta^2)$ [46], where $N(0, \delta^2)$ represents the background noise, $N(0, k\delta^2)$ represents the pulse noise, p represents the percentage of pulse noise, k represents the intensity of the pulse noise, and δ^2 represents the variance of the background noise. The parameters are set to $p = 0.3$, $k = 30$, and $\delta^2 = 1$. The initial parameters are set to 10^{-6} [47,48].

This section may be divided by subheadings. It should provide a concise and precise description of the experimental results, their interpretation, and the experimental conclusions that can be drawn.

4.1. Example 1: Single Snapshot Case

Figures 5 and 6 show the underwater DOA estimation results based on the improved SBL algorithm in both non-impulsive and impulsive noise environments. As can be seen from the figures, under the single snapshot condition, SBL can recover the signal well, regardless of whether it is in a non-impulsive or impulsive noise environment. The recovery result in the non-impulsive environment is better than that in the impulsive noise environment. From Figures 5d and 6d, it can be observed that the error between the two is not very large. Therefore, under the single snapshot condition, the algorithm has good estimation performance in the impulsive noise environment.

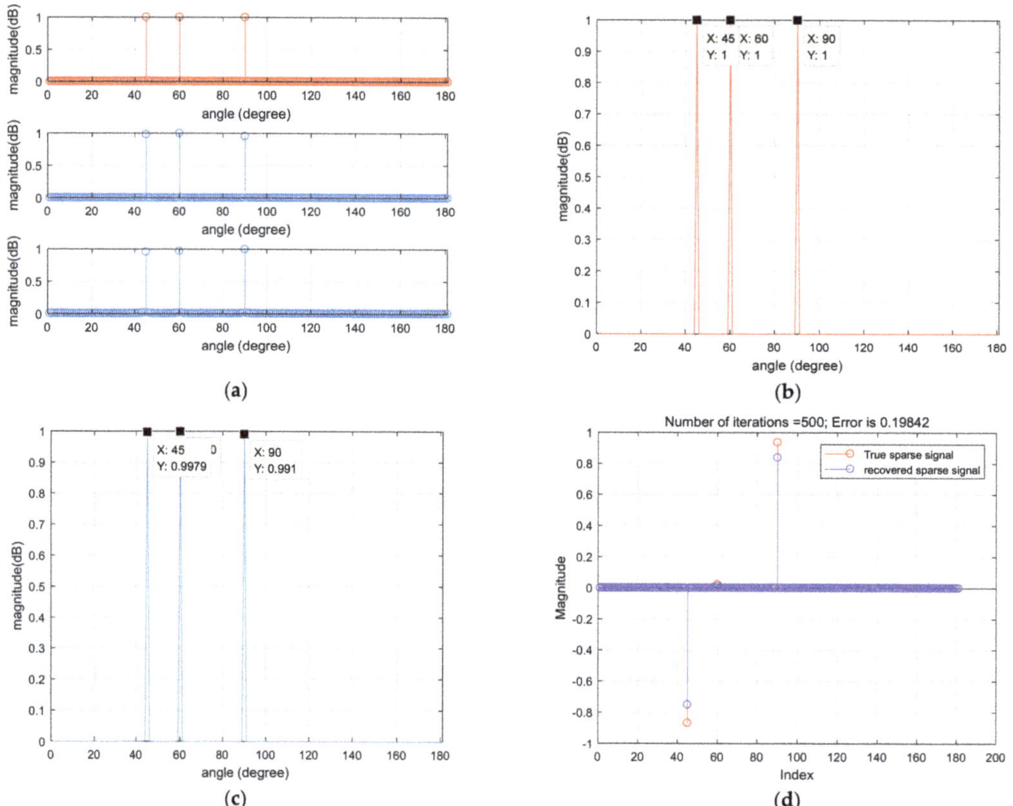

Figure 5. DOA estimation results of single snapshot in a non-pulsed environment. (**a**) Comparison of azimuth between the original signal and recovered signal; (**b**) bearing information of the original signal; (**c**) restoring the orientation information of the signal; (**d**) difference between estimated DOA and true DOA.

4.2. Example 2: Multiple Snapshot Case

In this section, the DOA estimation problem based on the SBL algorithm in the pulse noise environment under the multiple snapshot case is considered. The estimation results under Gaussian white noise are used as a reference to explore the estimation performance of the algorithm. The number of snapshots is set to 500, the frequency is $f = 1000$, and the sampling frequency is $f_s = 10 \times f$.

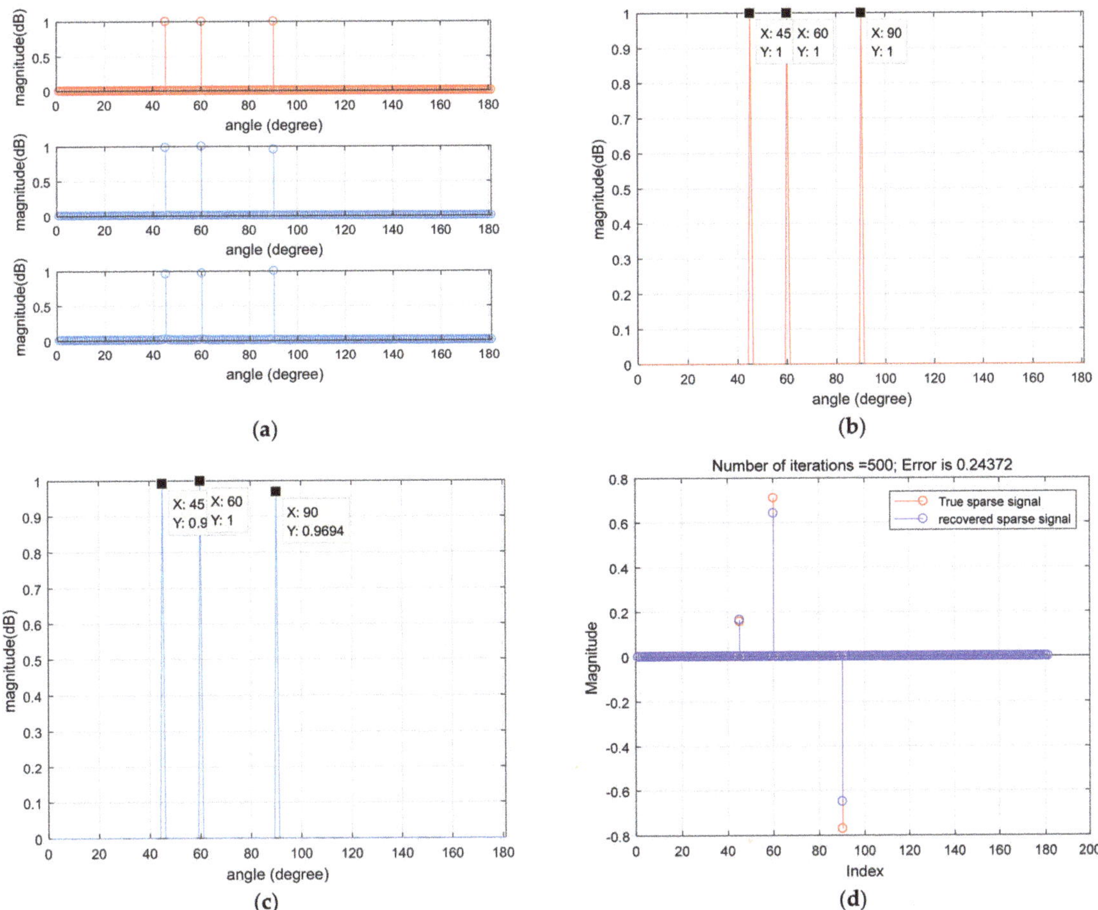

Figure 6. DOA estimation results of single snapshot under pulse environment. (**a**) Comparison of azimuth between the original signal and recovered signal; (**b**) bearing information of the original signal; (**c**) restoring the orientation information of the signal; (**d**) difference between estimated DOA and true DOA.

Figures 7 and 8 show the underwater DOA estimation results based on the improved variational Bayesian algorithm in both non-impulsive and impulsive noise environments under the multi-snapshot condition. As can be seen from the figures, the recovery results in the non-impulsive environment are better than those in the impulsive noise environment, but the errors between the recovered signal and the original signal are not large in either environment, which is consistent with theoretical expectations. On the other hand, from Figures 7d and 8d, as well as Figures 6d and 7d, it can be observed that the error between the recovered signal and the original signal under the single snapshot condition is larger than that under the multi-snapshot condition.

4.3. Example 3: The Comparison of Algorithm Performance

In this section, the root mean square error (RMSE) of the performance evaluation metric is introduced, and we analyzed the classic algorithms, CBF, MUSIC, and the proposed method. The basic experimental conditions were consistent with the first two experiments, with a snapshot number of 600, Monte Carlo iterations of 500, and an SNR ranging from −10 to 20. The results are shown in Figure 9.

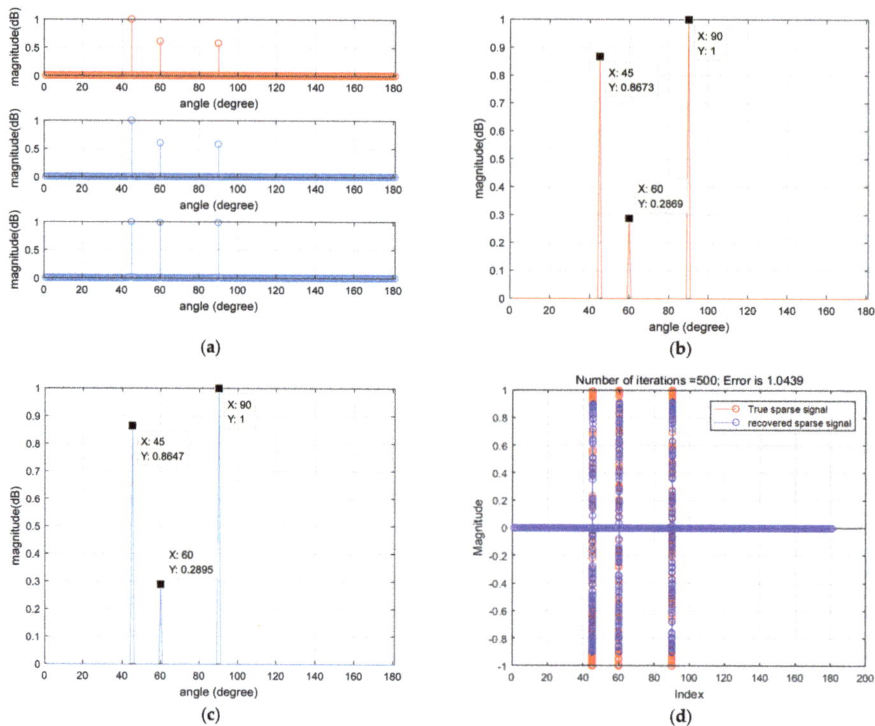

(a)

(b)

(c)

(d)

Figure 7. DOA estimation results of multi-snapshot in a non-pulsed environment. (**a**) Comparison of azimuth between the original signal and recovered signal; (**b**) bearing information of the original signal; (**c**) restoring the orientation information of the signal; (**d**) difference between estimated DOA and true DOA.

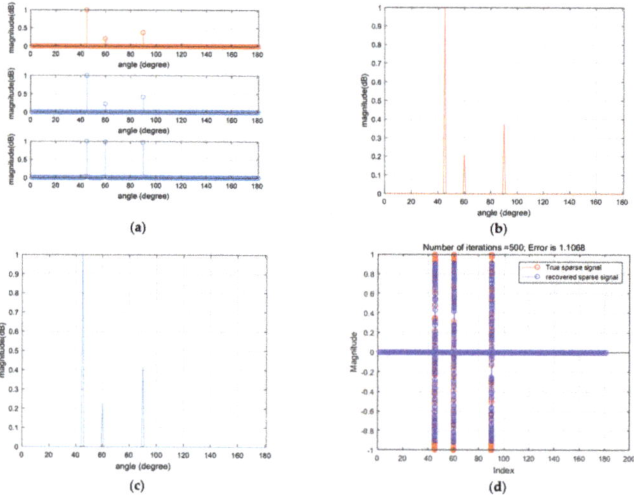

(a)

(b)

(c)

(d)

Figure 8. DOA estimation results of multi-fast in pulsed environment. (**a**) Comparison of azimuth between the original signal and recovered signal; (**b**) bearing information of the original signal; (**c**) restoring the orientation information of the signal; (**d**) difference between estimated DOA and true DOA.

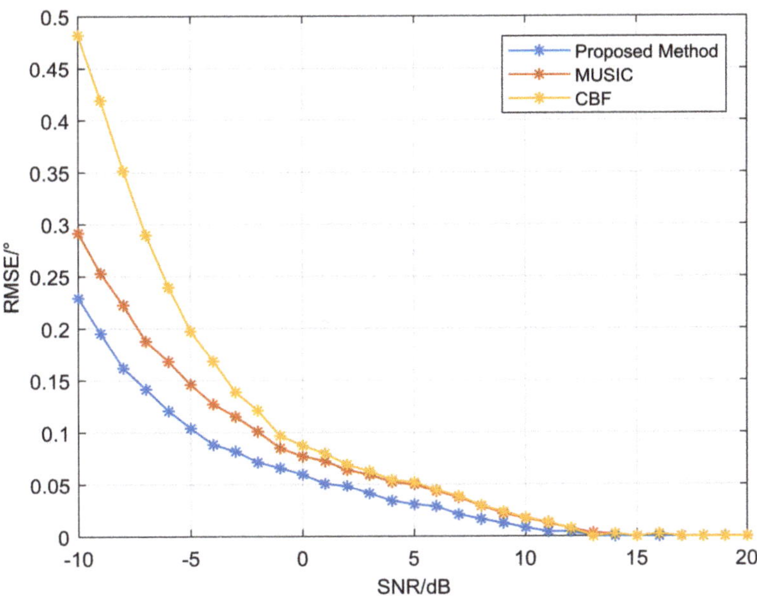

Figure 9. The comparison of algorithm performance.

The RMSE is a metric used to measure the difference between predicted and actual values of a variable. It is computed by taking the square root of the average of the squared differences between the predicted and actual values. The RMSE is commonly employed to evaluate the accuracy of predictive models or estimation methods, where a lower value indicates better performance. The formula for calculating RMSE is as follows:

$$\text{RMSE} = \sqrt{\frac{1}{PK}\sum_{p=1}^{P}\sum_{k=1}^{K}\left(\hat{\theta}_{p,n} - \theta_k\right)^2} \tag{30}$$

where K represents the number of signal sources, P represents the number of Monte Carlo experiments, and θ_k represents the estimated value of the Kth target angle in the Pth Monte Carlo experiment.

Figure 9 illustrates how the RMSE varies for different algorithms at different SNRs. As depicted in the figure, the RMSE curves for CBF, MUSIC, and the proposed method all decrease as the SNR increases. Furthermore, when the SNR is very high, the RMSE for all three algorithms is significantly small. Notably, the RMSE curve for the proposed method remains consistently lower than those of the other algorithms, implying that it has superior estimation performance.

5. Discussion

Despite notable advances in array signal processing, the increasing complexity and diversity of signal environments pose challenges to the conventional narrowband signal-based DOA estimation algorithm. The method focuses on recovering underwater DOA estimation signals under pulse interference and presents a novel underwater target azimuth estimation algorithm based on SBL, which overcomes the limitations of traditional narrowband signal-based DOA estimation algorithms in complex signal environments. The simulation results demonstrate that the algorithm accurately recovers the source signal in both single and multi-snapshot scenarios. The algorithm's performance was evaluated using the RMSE, which demonstrated that the algorithm outperforms other algorithms in signal recovery, effectively addressing the problem of low estimation accuracy in pulse environments. The algorithm can be applied in underwater target tracking and localization. Future work will explore azimuth estimation and signal recovery under wideband signals.

Author Contributions: This research article was a collaborative effort that involved the contributions of several authors. L.D. and H.L. contributed equally to the work and are listed as the first and second authors, respectively. Z.L. served as the corresponding author and provided guidance throughout the study. H.L. was responsible for designing and conducting the experiments, analyzing the data, and writing the manuscript. Specifically, H.L. developed the research question and hypotheses, designed the study protocol, recruited participants, collected and managed the data, conducted statistical analyses, and interpreted the findings. H.L. also wrote the initial draft of the manuscript and revised it based on feedback from the other authors. L.D. and L.W. assisted with the experimental design, performed data analysis, and helped to revise the manuscript. Specifically, L.D. contributed to the development of the study protocol, assisted with data collection and management, conducted statistical analyses, and L.W. helped to interpret the findings. L.D. and L.W. also provided critical feedback on the manuscript and helped to revise it. Z.L. provided overall guidance throughout the study and served as the corresponding author. Specifically, Z.L. helped to conceive the study, provided input on the research question and hypotheses, supervised the experimental design and data collection, and provided guidance on the statistical analyses and interpretation of the findings. Z.L. and X.L. played key roles in revising and finalizing the manuscript. Each author has made significant contributions to the research and preparation of the manuscript. The contributions of each author reflect their individual expertise and skills and demonstrate their commitment to advancing scientific knowledge in their respective fields. In conclusion, this research article was a collaborative effort that involved the contributions of several authors. H.L. and L.D. contributed equally to the work, respectively. Z.L. served as the corresponding author and provided guidance throughout the study. Together, the authors developed the research question and hypotheses, designed the study protocol, collected and analyzed the data, and interpreted the findings. The contributions of each author were critical to the success of the study and demonstrate their commitment to advancing scientific knowledge in their respective fields. All authors have read and agreed to the published version of the manuscript.

Funding: This research was supported by several funding sources, including the Shandong Province "Double-Hundred" Talent Plan (WST2020002), Key R&D programs (2022YFC2808003; 2023YFE0201900) and the Open Project of the State Key Laboratory of Sound Field Acoustic Information (No. SKLA202203

Data Availability Statement: Not applicable.

Conflicts of Interest: The authors declare no conflict of interest.

References

1. Zhang, X.; Chen, H.; Qiu, X. *Array Signal Processing and MATLAB Implementation*; Publishing House of Electronics Industry: Beijing, China, 2014.
2. Lee, H.; Wengrovitz, M. Resolution threshold of beamspace MUSIC for two closely spaced emitters. *IEEE Trans. Acoust. Speech Signal Process.* **1990**, *38*, 1545–1559. [CrossRef]
3. Kawachi, T.; Malkin, R.; Takahashi, H.; Kikura, H. Prototype of an Echo-PIV Method for Use in Underwater Nuclear Decommissioning Inspections. *J. Flow Control. Meas. Vis.* **2019**, *07*, 28–43. [CrossRef]
4. Shostak, S.; Paul, V.; Starodubtsev, A.; Roman, N.A. Method for Assessing the Distance to the Underwater Object of the Curvature of the Wave Front in the Conditions of Exposure to Correlated Noise. *J. Sib. Fed. Univ. Eng. Technol.* **2019**, *12*, 138–145. [CrossRef]
5. Li, H.; Guo, X.; Ma, L.; Song, G. Estimating the parameters of the seabed using the spatial characteristics of ocean ambient noise. *MATEC Web Conf.* **2019**, *283*, 08004. [CrossRef]
6. Zhou, C.; Pang, Y.; Qian, L.; Chen, X.; Xu, Q.; Zhao, C.; Zhang, H.; Tu, Z.; Huang, J.; Gu, H.; et al. Demodulation of a Hydroacoustic Sensor Array of Fiber Interferometers Based on Ultra-Weak Fiber Bragg Grating Reflectors Using a Self-Referencing Signal. *J. Light. Technol.* **2018**, *37*, 2568–2576. [CrossRef]
7. Jianghui, L.; Paul, R.; White, J.; Bull, M.; Timothy, G.; Leighton, B.R.; John, W.D. Passive Acoustic Monitoring of Seabed Gas Seeps—Application of Beamforming Techniques. *Acoust. Soc. Am.* **2020**, *40*, 070008.
8. James, P.; Verdon, S.A.; Horne, A.C.; Anna, L.; Stork, A.; Baird, F.; Kendall, J.-M. Microseismic Monitoring Using a Fiber-optic Distributed Acoustic Sensor Array. *Geophysics* **2020**, *85*, KS89–KS99.
9. Matthew, E.; Schinault, S.M.; Penna, H.; Garcia, A.; Purnima, R. Investigation and Design of a Towable Hydrophone Array for General Ocean Sensing. In Proceedings of the OCEANS 2019-Marseille, Marseille, France, 17–20 June 2019; pp. 1–5.
10. Xie, Z.; Fan, C.; Zhu, J.; Huang, X. Robust beamforming for wideband array based on spectrum subspaces. *IET Radar Sonar Navig.* **2020**, *14*, 1319–1327. [CrossRef]

11. Zou, M.S.; Liu, S.X.; Jiang, L.W.; Huang, H. A mixed analytical-numerical method for the acoustic radiation of a spherical double shell in the ocean-acoustic environment. *Ocean. Eng.* **2020**, *199*, 107040. [CrossRef]
12. Yang, T.C. Deconvolution of decomposed conventional beamforming. *J. Acoust. Soc. Am.* **2020**, *148*, EL195–EL201. [CrossRef]
13. Xie, L.; Sun, C.; Tian, J.W. Deconvolved frequency-difference beamforming for a linear array. *J. Acoust. Soc. Am.* **2020**, *148*, EL440–EL446. [CrossRef]
14. Byun, G.; Hunter Akins, F.; Gemba, K.L.; Song, H.C.; Kuperman, W.A. Multiple constraint matched field processing tolerant to array tilt mismatch. *J. Acoust. Soc. Am.* **2020**, *147*, 1231–1238. [CrossRef]
15. Zhu, J.; Fan, C.; Song, Y.; Huang, X.; Zhang, B.; Ma, Y. Coordination of Complementary Sets for Low Doppler-Induced Sidelobes. *Remote Sens.* **2022**, *14*, 1549. [CrossRef]
16. Zhang, X.; Yang, P.; Sun, H. Frequency-domain multireceiver synthetic aperture sonar imagery with Chebyshev polynomials. *Electron. Lett.* **2022**, *58*, 995–998. [CrossRef]
17. Yi, W.; Fu, L.; García-Fernández, F.; Xu, L.; Kong, L. Particle filtering based track-before-detect method for passive array sonar systems. *Signal Process.* **2019**, *165*, 303–314. [CrossRef]
18. Huang, X.; Liao, B. One-Bit MUSIC. *IEEE Signal Process. Lett.* **2019**, *26*, 961–965. [CrossRef]
19. Cheng, X.; Wang, Y. Noise Suppression for Direction of Arrival Estimation in Co-located MIMO Sonar. *Sensors* **2019**, *19*, 1325. [CrossRef]
20. Li, T.; Han, P.; Zhou, G.; Yao, X. A Joint Processing-MUSIC Algorithm in Multipath Environment Based on Non-uniform Line Array. In Proceedings of the 2020 13th International Congress on Image and Signal Processing, BioMedical Engineering and Informatics (CISP-BMEI), Chengdu, China, 17–19 October 2020; pp. 495–500.
21. Zhu, B.; Han, G.; Cong, W. Method for Estimating Target Orientation in Single Snapshot and Coherent Echo Space. In Proceedings of the 2020 IEEE International Conference on Signal Processing, Communications and Computing (ICSPCC), Online, 21–24 August 2020; pp. 1–4.
22. Ahmed, N.; Wang, H.; Raja, M.A.Z.; Ali, W.; Zaman, F.; Khan, W.U.; He, Y. Performance Analysis of Efficient Computing Techniques for Direction of Arrival Estimation of Underwater Multi Targets. *IEEE Access* **2021**, *9*, 33284–33298. [CrossRef]
23. Nauman, A.; Huigang, W.; Rizwan, A.; Ali, A.S.; Rahisham, A.R.; Muhammad, K.; Shahzad, A.; Kwan, Y.L. High Resolution DOA Estimation of Acoustic Plane Waves: An Innovative Com-parison Among Cuckoo Search Heuristics and Subspace Based Algorithms. *PLoS ONE* **2022**, *17*, e0268786.
24. Yanyan, A.; Zhigang, S.; Fengmao, Y.; Bo, Z.; Yongjiao, W. High Resolution Near-field Localization Method Based on A Special Combination Linear Array. In Proceedings of the 2021 OES China Ocean Acoustics (COA), Harbin, China, 14–17 July 2021; pp. 782–786.
25. Wang, X.; Bai, H.; Zhang, Q.; Tian, Y. Off-grid DOA estimation based on sparse Bayesian learning under array mutual coupling. *J. Vib. Shock.* **2022**, *41*, 303–312.
26. Wan, Z.; Xing, C.; Jiang, S.; Yu, R. Azimuth estimation of hydroacoustic targets based on root-finding sparse Bayes. In Proceedings of the 2021~2022 Academic Conference of the Hydroacoustics Branch of the Chinese Society of Acoustics, Qingdao, China, 15 August 2022; pp. 8–11, Hydroacoustics Branch of Chinese Society of Acoustics, Shandong Society of Acoustics, Academic Committee of Ship Instrumentation of China Shipbuilding Engineering Society.
27. Bai, H. *DOA Estimation of Underwater Sensor Based on Sparse Bayesian Learning*; Qingdao University of Technology: Qingdao, China, 2022.
28. He, W.; Liang, L.; Gong, X. Sparse Bayesian DOA estimation under coherent source conditions. *Telecommun. Technol.* **2021**, *61*, 993–998.
29. Guo, Y.; Tianm, J.; Hu, G. DOA estimation of underdetermined broadband signals in unknown noise fields based on sparse Bayesian learning. *Lab. Res. Explor.* **2021**, *40*, 5–10.
30. Shen, X.; Zhao, J. Block sparse Bayesian learning DOA estimation based on subspace fitting. *Appl. Sci. Technol.* **2020**, *47*, 42–46.
31. Yun, Y.; Yang, C.; Qing, L.; Song, X. Horizontal Wavenumber Estimation Technique Based on Compressive Sensing in Shallow Water. In Proceedings of the 2019 IEEE International Conference on Signal, Information and Data Processing (ICSIDP), Chongqing, China, 11–13 December 2019; pp. 1–4.
32. Ming, M.; Rui, Z.; Haoyang, G. Synchronous Prestack Inversion for Automatic Extracting the Correlation of Elastic Parameters Using Block Sparse Bayesian Learning. In Proceedings of the SEG International Exposition and Annual Meeting, San Antonio, TX, USA, 16 September 2019; OnePetro: Richardson, TX, USA, 2019.
33. Zhu, J.; Song, Y.; Jiang, N.; Zhuang, X.; Fan, C.; Huang, X. Enhanced Doppler Resolution and Sidelobe Suppression Performance for Golay Complementary Waveforms. *Remote Sens.* **2023**, *15*, 2452. [CrossRef]
34. Jimenez-Martinez, M. Fatigue of offshore structures: A review of statistical fatigue damage assessment for stochastic loadings. *Int. J. Fatigue* **2020**, *132*, 105327. [CrossRef]
35. Zhang, X.; Yang, P. An Improved Imaging Algorithm for Multi-Receiver SAS System with Wide-Bandwidth Signal. *Remote. Sens.* **2021**, *13*, 5008. [CrossRef]
36. Tzikas, D.G.; Likas, A.C.; Galatsanos, N.P. The variational approximation for Bayesian inference. *IEEE Signal Process. Mag.* **2008**, *25*, 131–146. [CrossRef]
37. Neapolitan, R.E. *Learning Bayesian Networks*; Pearson Prentice Hall: Upper Saddle River, NJ, USA, 2004; Volume 38.
38. Bishop, C.M.; Nasser, M.N. *Pattern Recognition and Machine Learning*; Springer: New York, NY, USA, 2006; Volume 4.

39. Zhang, X.; Wu, H.; Sun, H.; Ying, W. Multireceiver SAS imagery based on monostatic conversion. *IEEE J. Sel. Top. Appl. Earth Obs. Remote Sens.* **2021**, *14*, 10835–10853. [CrossRef]

40. Cotter, S.F.; Rao, B.D.; Kjersti, E. Sparse solutions to linear inverse problems with multiple measurement vectors. *IEEE Trans. Signal Process.* **2005**, *53*, 2477–2488. [CrossRef]

41. Grosswald, E. The student t-distribution of any degree of freedom is infinitely divisible. *Probab. Theory Relat. Fields* **1976**, *36*, 103–109. [CrossRef]

42. Zhang, X.; Yang, P.; Huang, P.; Sun, H.; Ying, W. Wide-bandwidth signal-based multireceiver SAS imagery using extended chirp scaling algorithm. *IET Radar Sonar Navig.* **2022**, *16*, 531–541. [CrossRef]

43. Tipping, M.; Lawrence, N. Variational inference for Student-*t* models: Robust Bayesian interpolation and generalised component analysis. *Neurocomputing* **2005**, *69*, 123–141. [CrossRef]

44. Zhang, X.; Yang, P.; Feng, X.; Sun, H. Efficient imaging method for multireceiver SAS. *IET Radar Sonar Navig.* **2022**, *16*, 1470–1483. [CrossRef]

45. Gerstoft, P.; Mecklenbräuker, C.F. Wideband sparse Bayesian learning for DOA estimation from multiple snapshots. In Proceedings of the 2016 IEEE Sensor Array and Multichannel Signal Processing Workshop, Rio de Janeiro, Brazil, 10–13 July 2016; pp. 1–5.

46. Jiadong, S.; Zulin, W.; Qin, H. A robust algorithm for joint sparse recovery in presence of impulsive noise. *IEEE Signal Process. Lett.* **2015**, *22*, 1166–1170. [CrossRef]

47. Zhu, H.; Leung, H.; He, Z. A variational Bayesian approach to robust sensor fusion based on Student-t distribution. *Inf. Sci.* **2013**, *221*, 201–214. [CrossRef]

48. Zhang, X.; Yang, P.; Zhou, M. Multireceiver SAS imagery with generalized PCA. *IEEE Geosci. Remote Sens. Lett.* **2023**, 1. [CrossRef]

Article

Joint Detection and Reconstruction of Weak Spectral Lines under Non-Gaussian Impulsive Noise with Deep Learning

Zhen Li [1,2,3], Junyuan Guo [1,2,3] and Xiaohan Wang [1,2,3,*]

1. Acoustic Science and Technology Laboratory, Harbin Engineering University, Harbin 150001, China; greenhandli@hrbeu.edu.cn (Z.L.); guojunyuan@hrbeu.edu.cn (J.G.)
2. Key Laboratory of Marine Information Acquisition and Security (Harbin Engineering University), Ministry of Industry and Information Technology, Harbin 150001, China
3. College of Underwater Acoustic Engineering, Harbin Engineering University, Harbin 150001, China
* Correspondence: wangxiaohan@hrbeu.edu.cn

Abstract: Non-Gaussian impulsive noise in marine environments strongly influences the detection of weak spectral lines. However, existing detection algorithms based on the Gaussian noise model are futile under non-Gaussian impulsive noise. Therefore, a deep-learning method called AINP+LR-DRNet is proposed for joint detection and the reconstruction of weak spectral lines. First, non-Gaussian impulsive noise suppression was performed by an impulsive noise preprocessor (AINP). Second, a special detection and reconstruction network (DRNet) was proposed. An end-to-end training application learns to detect and reconstruct weak spectral lines by adding into an adaptive weighted loss function based on dual classification. Finally, a spectral line-detection algorithm based on DRNet (LR-DRNet) was proposed to improve the detection performance. The simulation indicated that the proposed AINP+LR-DRNet can detect and reconstruct weak spectral line features under non-Gaussian impulsive noise, even for a mixed signal-to-noise ratio as low as −26 dB. The performance of the proposed method was validated using experimental data. The proposed AINP+LR-DRNet detects and reconstructs spectral lines under strong background noise and interference with better reliability than other algorithms.

Keywords: non-Gaussian impulsive noises; detection and reconstruction of weak spectral lines; deep learning

Citation: Li, Z.; Guo, J.; Wang, X. Joint Detection and Reconstruction of Weak Spectral Lines under Non-Gaussian Impulsive Noise with Deep Learning. *Remote Sens.* **2023**, *15*, 3268. https://doi.org/10.3390/rs15133268

Academic Editor: Andrzej Stateczny

Received: 15 March 2023
Revised: 20 June 2023
Accepted: 22 June 2023
Published: 25 June 2023

1. Introduction

The single-frequency detection of underwater radiation noise with abundant single-frequency components is crucial for detecting quiet targets [1]. The time-frequency analysis is projected on the time and frequency planes to form a three-dimensional stereogram (lofargram). It presents the abundant features of underwater radiation noise [2]. Therefore, the lofargram is regularly employed to analyze its features for passive sonar signals. However, for low signal-to-noise ratios (SNRs), frequency fluctuations caused by a moving target [3] and a high amount of background noise may weaken spectral-line detection.

The detection of weak spectral lines using a lofargram has long been an attractive research topic. Image-processing methods, neural networks, and statistical models are applied to detect weak spectral lines in a lofargram. Image-processing and neural-network methods obtain spectral-line traces from complex image semantic features; however, their performance is usually unsatisfactory for low SNRs [4–7]. Some deep-learning methods [8–11] achieve good line-spectrum estimation, but the SNR requirement is relatively high. To overcome this limitation, deepLofargram was proposed to recover invisible and irregularly fluctuating frequency lines at low SNRs [12]. Furthermore, in lofargrams, when the weak spectral lines are far beyond the perceptual range of human vision, this is referred to as low SNR. A statistical model such as the hidden Markov model (HMM) can track the optimal spectral-line trajectory from multi-frame power-spectrum data [13,14]. Most of the aforementioned studies were applied to marine ambient noise following a Gaussian distribution. In particular, marine ambient noise presents strong impulsive characteristics owing to the superposition of seawater thermal noise, hydrodynamic noise, under-ice noise, biological noise, and other noises [15]. Existing underwater-acoustic-signal-processing methods may be invalidated under such non-Gaussian impulsive noise. To overcome this problem, several studies [16–18] have performed statistical analyses and models of non-Gaussian marine ambient noise. It was found that the generation and propagation of underwater impulsive noise are in accordance with the "heavy tail" statistical characteristics of the symmetric α-stable (SαS) distribution [19]. Furthermore, SNR and mixed signal-to-noise ratio (MSNR) have been used to characterize the energies of Gaussian noise and non-Gaussian impulsive noise [19]. Various preprocessors have been proposed to suppress non-Gaussian impulsive noise, which can be described by the SαS distribution, including the standard median filter (SMF) [20–22] and the memoryless analog nonlinear preprocessor (MANP) [23]. Nevertheless, weak spectral-line detection is unreliable at low MSNRs.

In recent years, with the introduction and development of deep convolutional structures such as UNet [24], SegNet [25], and LinkNet [26], image-semantic-segmentation technology based on deep learning has developed rapidly. In deep-learning semantic segmentation, the semantic features in an image are captured by finding semantic correlations between pixel points from global or local contextual information. In passive sonar-signal processing, weak spectral lines have time-frequency correlations, making them relatively continuous in a lofargram, even though they cannot be observed. Therefore, we argue that when combined with a preprocessor and a deep convolution structure, a lofargram would be able to handle the detection and reconstruction of weak spectral lines under non-Gaussian impulsive noise. Moreover, by "reconstruction," we mean that potential spectral-line features are recovered to output a lofargram with significant spectral lines.

In this study, we propose a novel method, called AINP+LR-DRNet, which is suitable for the detection and reconstruction of weak spectral lines under non-Gaussian impulsive noise. The spectral-line detection-and-reconstruction problem is redefined as a binary classification problem. First, an impulsive noise preprocessor (AINP) was applied to suppress the non-Gaussian impulsive noise. Second, a specially constructed DRNet was built to detect and reconstruct weak spectral lines. Third, a dual classification adaptive weighted loss was applied to obtain the optimal DRNet during training iterations. Fourth, the detection performance was further improved by the proposed LR-DRNet algorithm. Finally, we validated the ability of the proposed method to detect and reconstruct weak spectral lines under non-Gaussian-impulse noise using simulated and measured data sets.

2. Proposed Framework and Training

Considering deep learning (DL) techniques, we formulate the spectral-line detection-and-reconstruction problem in a lofargram as a binary classification problem. Thus, binary hypothesis testing can be performed, which is defined as follows:

$$
\begin{aligned}
H_1 &: \sum_{i,j}^{i=T,j=F} [s(t_i, f_j) + u(t_i, f_j)] \\
H_0 &: \sum_{i,j}^{i=T,j=F} u(t_i, f_j)
\end{aligned}
\tag{1}
$$

where H_1 and H_0 indicate the presence of spectral-line pixels and noise pixels in a lofargram, respectively. $\sum_{i,j}^{i=T,j=F} s(t_i, f_j)$ describes the set of spectral-line pixels, and $\sum_{i,j}^{i=T,j=F} u(t_i, f_j)$ describes the set of noise pixels.

Thus, the spectral-line detection-and-reconstruction framework are proposed to solve Equation (1). As shown in Figure 1, the proposed framework, with the sampling, detection, and reconstruction algorithm, is illustrated. In the sampling stage, the passive SONAR system collects the acoustic signals and noises. The received data are preprocessed by AINP to construct the dataset. Subsequently, a specially designed LR-DRNet is pre-trained to obtain the optimal model parameters in offline training by adding into an adaptive weighted loss function based on dual classification. The well-trained LR-DRNet is utilized to fine-tune the parameters to detect and reconstruct the measured unlabeled samples in online detection and reconstruction. More details are described below:

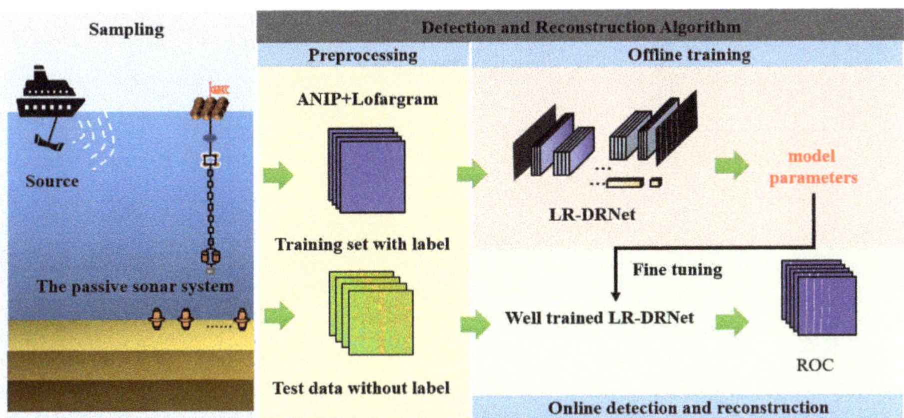

Figure 1. Proposed spectral-line detection-and-reconstruction framework.

2.1. Detection-and-Reconstruction Algorithm

2.1.1. Data Preprocessing

The heavy impulsive noise causes broadband interference in a lofargram. Therefore, appropriate preprocessing is required. In this study, the AINP method [24] is used to nonlinearly suppress the abnormal amplitude in the input signal $s(t)$, which is more prominent than the amplitude threshold $\theta(t)$. The influence function for the AINP is as follows:

$$e(t) = s(t) \begin{cases} 1, & |s(t)| \leq \theta(t) \\ \left(\frac{\theta(t)}{|s(t)|}\right)^2, & |s(t)| > \theta(t) \end{cases}, \tag{2}$$

where $\theta(t)$ can be obtained from Equation (3)

$$\theta(t) = (1 + 2\theta_0)Q_2(t). \tag{3}$$

In Equation (3), $Q_2(t)$ represents the second quartile of the absolute value of the input signal $|g(t)|$, and θ_0 is a coefficient, which is set to 1.5, as in [23].

2.1.2. DRNet Structure

The proposed DRNet is derived from LinkNet [26], including the shared encoder, detection decoder, and reconstruction decoder. One part of the shared encoder, illustrated in Figure 2a, is stacked with a series of residual convolution structures to extract spectral-line features. For spectral-line detection, the detection decoder is added to output result of spectral-line detection (for H_1 and H_0, respectively). As illustrated in Figure 2b, plugging into the squeeze-and-excitation (SE) blocks, the reconstruction decoder can perform channel enhancement by obtaining the importance of each channel through "squeeze" and "excitation" operations [27]. The reconstruction decoder outputs spectral line-reconstruction results through FinalConv, as shown in Figure 2c. Moreover, a reconstruction decoder is enabled when the detection decoder announces that H_1 holds during the online detection-and-reconstruction stage. The complete network structure is depicted in Figure 3. As shown in Figure 3, the proposed DRNet involves multi-task learning (MTL). For MTL, the model relies on the relative weighting between each task's loss, and manually adjusting these weights is difficult and time-consuming. Hence, inspired by [28], the adaptive weighted dual loss function is considered by modeling reconstruction-and-detection-task uncertainty. The details are described in Section 2.1.3.

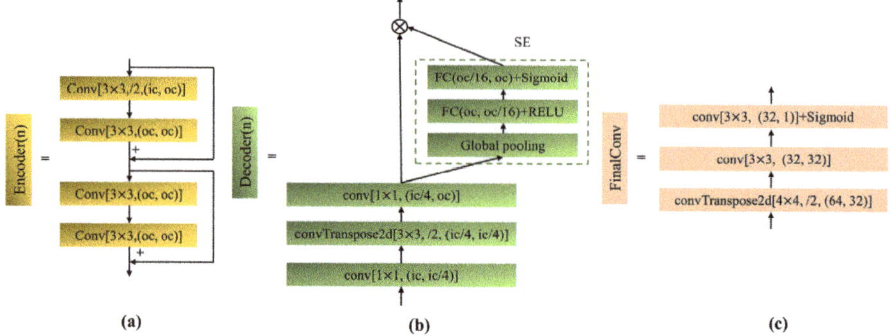

Figure 2. Structural diagram of each part in DRNet. (**a**) Structure of convolutional modules in Encoder (n). (**b**) Structure of the decoding layer. (**c**) Structure of the FinalConv layer.

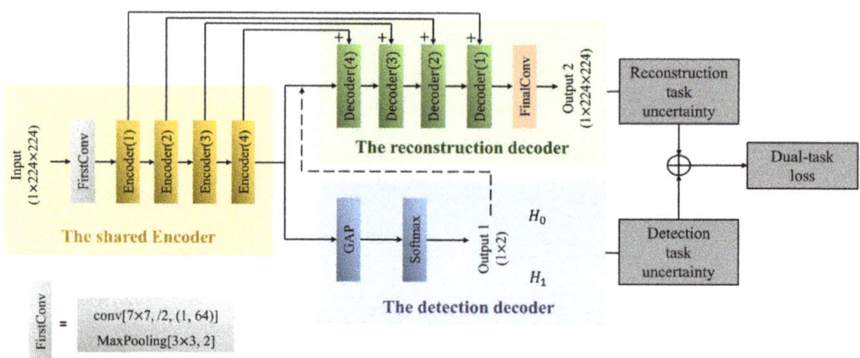

Figure 3. Architecture of DRNet.

2.1.3. Adaptive Weighted Loss Function Based on Dual Classification

For MTL, the loss function is weighted for each task loss. Thereafter, the MTL loss can be expressed as:

$$L = \sum_{i=1}^{I} w_i L_i, \tag{4}$$

where w_i and L_i denote the weight and loss of the i-th task, respectively. The I indicates the number of tasks. In this study, there are spectral-line detection and reconstruction tasks with different loss scales. A basic approach to overcoming the large loss difference between the detection and reconstruction tasks involves the model adaptively adjusting the weights w_i according to uncertainty of each task.

As the spectral-line detection-and-reconstruction problem is treated as a binary classification task, an adaptive weighted loss function based on a dual classification is used to train the model. Following the derivation in [28], when $f^W(x)$ is a sufficient statistic, the following multi-mask likelihood is expressed as

$$p(y_d, y_r | f^W(x)) = p(y_d | f^W(x)) p(y_r | f^W(x)), \tag{5}$$

where $f^W(x)$ represents the detection-and-reconstruction-prediction results of model with parameters W on input x. y_d and y_r are the ground-truth labels for detection and reconstruction tasks.

Under random noise, the log-likelihood of the detection and reconstruction task is output through a Softmax function, which can be expressed as

$$\log p(y_d|f^W(x)) = \log(\text{Softmax}(\frac{1}{\sigma_d^2}f_d^W(x))) = \frac{1}{\sigma_d^2}f_d^W(x) - \log\sum_{c_1}\exp(\frac{1}{\sigma_d^2}f_{c_1}^W(x)), \quad (6)$$

$$\log p(y_r|f^W(x)) = \log(\text{Softmax}(\frac{1}{\sigma_r^2}f_r^W(x))) = \frac{1}{\sigma_r^2}f_r^W(x) - \log\sum_{c_2}\exp(\frac{1}{\sigma_r^2}f_{c_2}^W(x)), \quad (7)$$

where c_1 and c_2 denote the categories of the detection and reconstruction tasks, respectively. The σ_d^2 and σ_r^2 denote the observation-noise parameters of the model for the detection and reconstruction tasks, respectively.

When the Softmax likelihood is modeled for the detection and reconstruction, the joint loss $L(W, \sigma_d, \sigma_r)$ is

$$
\begin{aligned}
L(W,\sigma_d,\sigma_r) &= -\log p(y_d, y_r|f^W(x)) \\
&= -\log[\text{Softmax}(\tfrac{1}{\sigma_d^2}f_d^W(x)) \cdot \text{Softmax}(\tfrac{1}{\sigma_r^2}f_r^W(x))] \\
&= \tfrac{1}{\sigma_d^2}f_d^W(x) - \log\sum_{c_1}\exp(\tfrac{1}{\sigma_d^2}f_{c_1}^W(x)) + \tfrac{1}{\sigma_r^2}f_r^W(x) - \log\sum_{c_2}\exp(\tfrac{1}{\sigma_r^2}f_{c_2}^W(x)) \\
&= \tfrac{1}{\sigma_d^2}[f_d^W(x) - \log\sum_{c_1}\exp(\tfrac{1}{\sigma_d^2}f_{c_1}^W(x))] + \tfrac{1}{\sigma_r^2}[f_r^W(x) - \log\sum_{c_2}\exp(\tfrac{1}{\sigma_r^2}f_{c_2}^W(x))] \\
&\quad + \log\sum_{c_1}\exp(\tfrac{1}{\sigma_d^2}f_{c_1}^W(x))/(\sum_{c_1}\exp(f_{c_1}^W(x)))^{\frac{1}{\sigma_d^2}} + \log\sum_{c_2}\exp(\tfrac{1}{\sigma_r^2}f_{c_2}^W(x))/(\sum_{c_2}\exp(f_{c_2}^W(x)))^{\frac{1}{\sigma_r^2}} \\
&\approx \tfrac{1}{\sigma_d^2}[-\log(\text{Softmax}(y_d, f_d^W(x)))] + \tfrac{1}{\sigma_r^2}[-\log(\text{Softmax}(y_r, f_r^W(x)))] + \log\sigma_d\sigma_r \\
&= \tfrac{1}{\sigma_d^2}L_d(W) + \tfrac{1}{\sigma_r^2}L_r(W) + \log\sigma_d\sigma_r
\end{aligned}
\quad (8)
$$

where $f_d^W(x)$ and $f_r^W(x)$ represent the outputs of detection and reconstruction in $f_{c_1}^W(x)$ and $f_{c_2}^W(x)$, respectively. Equation (8) can be applied for approximation, as follows:

$$\sum_{c_1}\exp(\frac{1}{\sigma_d^2}f_{c_1}^W(x))/\sum_{c_1}\exp(f_{c_1}^W(x))^{\frac{1}{\sigma_d^2}} \approx \sigma_d, \quad \sum_{c_2}\exp(\frac{1}{\sigma_r^2}f_{c_2}^W(x))/\sum_{c_2}\exp(f_{c_2}^W(x))^{\frac{1}{\sigma_r^2}} \approx \sigma_r, \quad (9)$$

where Equation (9) becomes equal when $\sigma_d, \sigma_r \to 1$.

Referring to the suggestion in [28], we set $s_d = \log\sigma_d^2$, $s_e = \log\sigma_e^2$. Accordingly, $L(W, \sigma_d, \sigma_r)$ can be rewritten as

$$L(W,\sigma_d,\sigma_r) = 2\exp(-s_d)L_d(W) + 2\exp(-s_r)L_r(W) + s_d + s_e. \quad (10)$$

For the detection and reconstruction loss functions $L_d(W)$ and $L_r(W)$, we adopt the two-class cross-entropy and the class-balanced cross-entropy loss functions in [12], as follows:

$$L_d(W) = -\sum_{i=1}^{N_1} h\log p + (1-h)\log(1-p), \quad (11)$$

where N_1 indicates the number of samples in the batch size. Moreover, $h \in \{0,1\}$ represents the H_0 and H_1 hypotheses, and p indicates the probability of the output sample class when using a Softmax function.

$$L_r(W) = \sum_{i=1}^{N_2}\lambda[\sum_{f,t\in G_+}\log p_{f,t} + (1-\lambda)\sum_{f,t\in G_-}\log(1-p_{f,t})], \quad (12)$$

where $\lambda = |G_-|/|G|$ and $1 - \lambda = |G_+|/|G|$. The $|G_+|$ and $|G_-|$ represent the spectral-line and noise ground-truth label sets, respectively. The $p_{f,t}$ indicates the predicted value of the H_1 samples at the (f, t) position by a sigmoid function.

According to Equations (10)–(12), the joint-loss form of the multi-task can be obtained. Simultaneously, two weight parameters, σ_d and σ_r, are adaptively adjusted during the training process. Thus, the purpose of adaptive loss weighting is achieved.

2.1.4. DRNet-Based Spectral-Line-Detection Algorithm

Inspired by the application of the CNN-based spectrum sensing algorithm [29] in narrowband spectrum sensing, which provides a path for detecting spectral lines in a lofargram, a LR-DRNet algorithm is proposed by considering only a single receiver hydrophone. In the proposed algorithm, we use DRNet for offline training and adopt a threshold-based mechanism for online detection.

Offline Training

In offline training, the dataset of the lofargram is constructed under H_0 and H_1 after applying AINP and labeled as follows:

$$(O_l, Z) = \left\{ (l^{(1)}, z^{(1)}), (l^{(2)}, z^{(2)}), ..., (l^{(M)}, z^{(M)}) \right\},$$ (13)

where O_l denotes the set of lofargrams l, and Z is its label. The $(l^{(m)}, z^{(m)})$ represents the m-th sample in the training set.

For the test statistic, the proposed LR-DRNet can extract weak spectral-line features in a lofargram. The output node of the detection decoder was set to 2 by converting spectral line detection into an image binary classification. After a series of convolutional layers, pooling layers, and activation functions, the probability that the lofargram belongs to H_0 or H_1 can be obtained. For the detection task, Equation (1) can be rewritten as follows:

$$\begin{aligned} H_1 &: P(z^{(m)} = 1 \,|\, l^{(m)}; \vartheta) = h_{\vartheta|H_1}(l^{(m)}) \\ H_0 &: P(z^{(m)} = 0 \,|\, l^{(m)}; \vartheta) = h_{\vartheta|H_0}(l^{(m)})' \end{aligned}$$ (14)

where $h_\vartheta(\cdot)$ represents a nonlinear expression of the model with parameters ϑ. After a Softmax function, the network's output layer has:

$$h_{\vartheta|H_1}(l^{(m)}) + h_{\vartheta|H_0}(l^{(m)}) = 1.$$ (15)

Next, Equation (11), as a training-error loss function, can be rewritten as:

$$J_{LR-DRNet}(\vartheta) = -\frac{1}{K} \sum_{k=1}^{K} \left\{ z^{(m)} \log h_{\vartheta|H_1}(l^{(m)}) + (1 - z^{(m)}) \log h_{\vartheta|H_0}(l^{(m)}) \right\}.$$ (16)

Training LR-DRNet minimizes the error loss in Equation (16) and maximizes the posterior probability of the parameter set ϑ. The optimal parameter set $\hat{\vartheta}$ can be obtained as follows:

$$\hat{\vartheta} = \mathrm{argmax} P(Z|L; \vartheta).$$ (17)

Based on the loss function in Equation (10), the backpropagation algorithm is employed to gradually update the parameters of LR-DRNet. Hence, the well-trained LR-DRNet can be illustrated as follows:

$$h_{\hat{\vartheta}|H_i}(l^{(m)}) = \begin{cases} h_{\hat{\vartheta}|H_1}(l^{(m)}) \\ h_{\hat{\vartheta}|H_0}(l^{(m)}) \end{cases}.$$ (18)

Considering the Bayesian and Neyman–Pearson (NP) criterion, and assuming that $P(H_0) = P(H_1)$, the test statistics under the proposed LR-DRNet can be acquired as:

$$\Lambda_{LR-DRNet} = \frac{h_{\hat{\vartheta}|H_1}(l^{(m)})}{h_{\hat{\vartheta}|H_0}(l^{(m)})} \gtrless \eta,$$ (19)

where η denotes the detection threshold. The presence or absence of spectral lines in the lofargram can be adjudicated by comparing the test statistic and detection threshold.

Next, the detection threshold should be determined. First, M, noise-sample data sets composed of H_0 lofargrams after applying AINP, are constructed. M lofargrams under H_0 after applying AINP are costructed.

$$O_n = \left\{ n^{(1)}, n^{(2)}, ..., n^{(M)} \right\}.$$ (20)

where O_n denotes the set of lofargrams n under H_0.

The probability of detection (P_D) and the false-alarm probability (P_f) are defined as follows:

$$P_D = P[\Lambda_{LR-DRNet}|H_1 > \eta],$$ (21)

$$P_f = P[\Lambda_{LR-DRNet}|H_0 > \eta].$$ (22)

Subsequently, the data set O_n is fed as samples into the pre-trained DRNet and the test statistics of all lofargrams under the H_0 hypothesis are obtained.

$$\Lambda_{LR-DRNet}|H_0 = \frac{h_{\hat{\vartheta}|H_1}(n^{(m)})}{h_{\hat{\vartheta}|H_0}(n^{(m)})}.$$ (23)

By arranging these values in descending order to form a sequence $T(m)$, the detection threshold of the artificially set false-alarm-probability value P_f can be acquired [29].

$$\Lambda_{LR-DRNet}|H_i(\tilde{z}) = \frac{h_{\hat{\gamma}|H_1}(l^{(m)})}{h_{\hat{\gamma}|H_0}(l^{(m)})} \overset{H_1}{\underset{H_0}{\gtrless}} \eta.$$ (24)

where $\lfloor \cdot \rfloor$ is the nearest smaller integer. The $\Lambda_{LR-DRNet}|H_0(\lfloor m \rfloor)$ denotes the m-th sample value of $T(m)$ in descending order.

Online Detection

According to Equation (24), a detection threshold η is set. The unlabeled lofargrams, denoted as $\tilde{z}$, are input into the well-trained LR-DRNet. Subsequently, online detection, based on LR-DRNet, is performed, that is,

$$\Lambda_{LR-DRNet}|H_i(\tilde{z}) = \frac{h_{\hat{\gamma}|H_1}(l^{(m)})}{h_{\hat{\gamma}|H_0}(l^{(m)})} \overset{H_1}{\underset{H_0}{\gtrless}} \eta.$$ (25)

When the test statistic is obtained, we can rapidly decide whether there are spectral lines in a lofargram by comparing it to the preset threshold.

2.2. Training Process

Training was optimized for the loss function in Equation (10) using the mini-batch gradient of the Adam optimizer [30], and by setting s_d and s_e to log2. The batch size was 128. Xavier weight initialization was performed [31]. As expressed in [12], the network's loss function is ineffective at converging at low SNRs. Hence, we first pre-trained the model with a learning rate of 10^{-4} for lofargrams with MSNR ranging from -19 dB to -22 dB. The model was then retrained with a learning rate of 10^{-5} for lofargrams with MSNR ranging from -23 dB to -26 dB. The learning rate was not fixed and was adjusted according to the cosine annealing warm restart [32] and gradual warmup [33]. Here, the gradual warmup was up to the 10th epoch, the initial restart epoch was set to 15, and the restart factor was set to 2. To prevent network overfitting and the problem of insufficient data, data augmentation was performed during training using methods such as horizontal and vertical flipping of images, random cropping, and grayscale maps. Both the above training procedures were terminated after approximately 300 epochs.

3. Simulation Analysis

This section first introduces the synthesis of the datasets and the network performance evaluation metrics. Subsequently, we illustrate the effectiveness of the proposed method by analyzing the effect of the network structure on performance. Finally, the performances of some existing methods are compared and analyzed through simulations.

3.1. Datasets

Non-Gaussian impulsive noise can be described by an α-stable distribution, whose characteristic function can be expressed as in [34].

$$\varphi(t) = \exp\left\{ jbt - |\gamma t|^{\alpha}[1 + j\beta\mathrm{sgn}(t)\omega(t,\alpha)] \right\}, \tag{26}$$

$$\omega(t,\alpha) = \begin{cases} -\tan(\frac{\pi\alpha}{2}), & \alpha \neq 1 \\ (\frac{2}{\pi})\log|t|, & \alpha = 1 \end{cases}, \tag{27}$$

where $0 < \alpha \leq 2$, $-1 \leq \beta \leq 1$, $\gamma > 0$, and $-\infty < b < \infty$. The characteristic exponent α determines the impulse intensity of the distribution; the higher the value of α, the lower the intensity. The position parameter b determines the center of the distribution. The dispersion coefficient γ measures the sample's degree of deviation by taking values relative to the mean, which is similar to the variance in a Gaussian distribution. The symmetry parameter β is used to describe the skewness of the distribution. When $\beta = 0$, the distribution is named the SαS distribution.

In this case of $0 < \alpha \leq 2$, only the first order is presented in α-stable distributed noise. Therefore, the SNR defined under traditional Gaussian noise is inapplicable. The mixed signal-to-noise ratio (MSNR) is defined as follows:

$$MSNR = 10\log_{10}(\frac{v_s^2}{\gamma}), \tag{28}$$

where v_s^2 denotes the signal variance.

The α-stable distribution degenerates into a Gaussian distribution when $\alpha = 2$. A conventional SNR measure of the relationship between the signal and noise power can be obtained as follows:

$$SNR = 10\log_{10}(\frac{a^2}{v_n^2}), \tag{29}$$

where a and v_n^2 denote the signal amplitude and the noise variance, respectively.

A low-frequency spectral line, radiated by underwater and surface vehicles, under Gaussian/non-Gaussian impulsive noise, is discussed in this study. Owing to the motion of vehicles (variable speed or steering), the spectral lines fluctuate even at low frequencies.

The fluctuating spectral lines can be simulated using a series of sinusoidal signals. The fluctuating spectral lines observed during the k-th time interval are described as:

$$s(t_k) = \sum_{i=1}^{I} a_i(t_k) \sin(2\pi f_i(t_k) + \varphi_k) + n(t_k), k = 0, 1, ..., T - 1, \tag{30}$$

where I represents the number of spectral lines. The $f_i(t_k)$ represents the frequency that subsequently varies t_k, meaning that the spectral line has unpredictable fluctuations. The $\varphi_k \in [0, 2\pi]$ is the initial phase, and $n(t_k)$ represents the sampling point of α-stable distribution noise in t_k-th.

The underwater acoustic channel contains Gaussian and non-Gaussian impulse noises. Before establishing the dataset, the modeling and statistical analysis of the measured marine environmental noise were performed. We first modeled three typical marine ambient noises with normal and α-stable distributions. Figure 4 shows that the α-stable distribution is approximate to the marine ambient noise, particularly at high pulse intensities, which conforms with the results reported in [14]. Subsequently, the characteristic function method [35] was applied to estimate the α-stable distribution parameters and statistically acquire the parameter-distribution regularities. Figure 5 presents the statistical conclusions for the four parameters estimated by the α-stable distribution. The α is distributed between 1.7 and 2.0, indicating that the analyzed marine-ambient-noise data contain weak pulse characteristics. The β is distributed at approximately 0, indicating that the SαS distribution can model the noise. The γ values are relatively low, ranging from 0 to 0.01. The data amplitudes are relatively concentrated, which is consistent with the weak pulse characteristics. The δ is distributed at approximately 0, indicating that the measured noise data are concentrated around the zero value. Therefore, the simulation dataset was synthesized according to the distribution regularities of the parameters above and Equation (30).

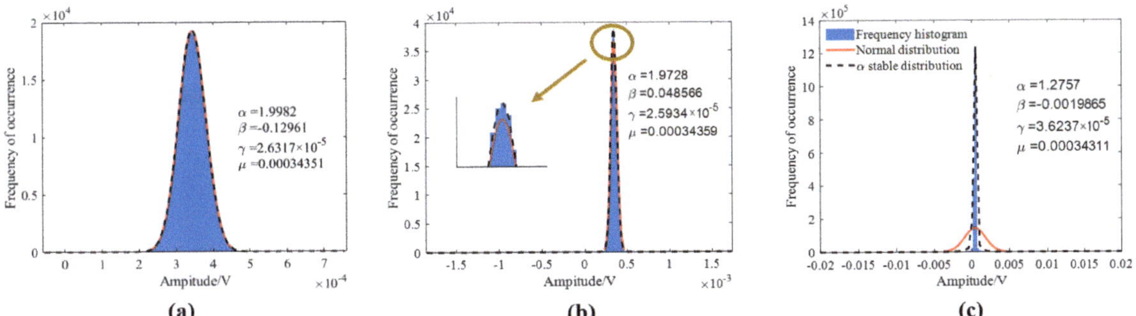

Figure 4. Comparison of the modeling of the normal distribution and α-stable distribution under various disturbances. (**a**) In a quiet environment; (**b**) in a ship-interference environment; (**c**) under airgun interference.

In the simulation of the SαS distribution noise, α was randomly selected in the range of [1.3, 2], β is set to 0, and γ, δ were set to 1 and 0, respectively. The fluctuating spectral lines within 100 Hz and MSNR in the range of $[-26, -19]$ dB were considered. The sampling rate f_s was 1000 Hz. Our synthetic dataset contained one to five fluctuating spectral lines, and multiple spectral lines had harmonic relations. The SαS distribution noise was added to the time-domain amplitude of sinusoids in the form of Equations (28) and (29) with the MSNR and SNR. Figure 6 presents the H_1 lofargrams of multiple sinusoidal signals of different MSNRs and H_0 lofargrams. The presence of spectral lines in lofargrams is not perceived through the visual senses below -22 dB. For MSNR in the range of -22 dB to -26 dB, we repeated the Monte Carlo simulation 1200 times to simulate the scenario under various parameters, splitting the dataset into 85% for training and 15% for testing.

Therefore, our training datasets comprised 9600 H_1 lofargrams and 6800 H_0 lofargrams, while the test set had 1440 H_1 lofargrams and 1200 H_0 lofargrams.

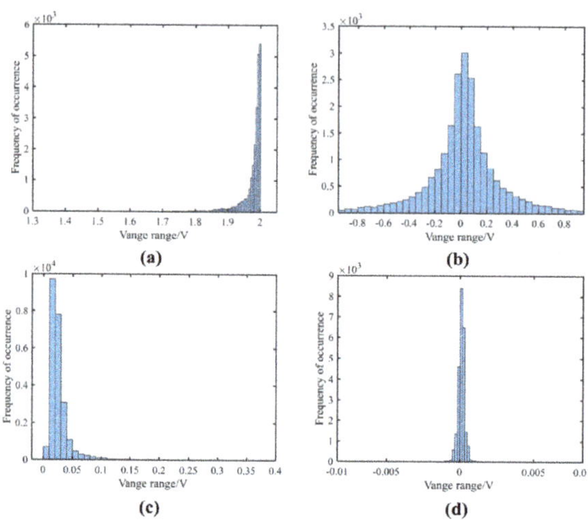

Figure 5. Estimation results of α-stable distribution parameters. (**a**) α-value distribution statistics; (**b**) β-value distribution statistics; (**c**) γ-value distribution statistics; (**d**) δ-value distribution statistics.

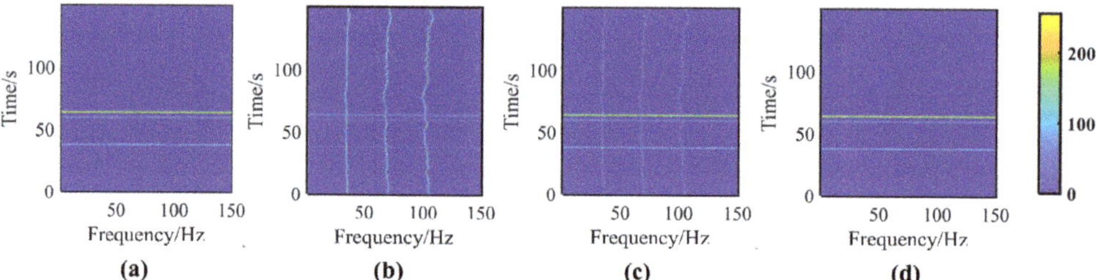

Figure 6. Lofargrams: (**a**) only the SαS distributed noise lofargram under H_0; (**b**) lofargram of the signal at MSNR = -5 dB under H_1; (**c**) lofargram of the same signal at MSNR = -15 dB under H_1; (**d**) lofargram of the same signal at MSNR = -22 dB under H_1.

3.2. Evaluation Metrics

The following assessment metrics were utilized to analyze the detection and reconstruction performance.

First, the receiver operating characteristic (ROC) curve was used to evaluate the detection performance. Using Equations (22)–(24), we set various P_f to obtain thresholds in the offline training stage. A serial set of P_f and P_D representing the points of the curve could be obtained, and these points together formed the ROCs.

Second, to evaluate the quality of the reconstruction lofargrams, the mIoU [36] and line-location accuracy (LLA) [37] were employed.

$$\text{mIOU} = \frac{1}{k+1}\sum_{i=0}^{k}\frac{TP}{FN + FP + TP}. \tag{31}$$

where TP, FN, FP, and TN denote the true positives, false negatives, false positives, and true negatives, respectively.

$$\text{LLA} = \frac{1}{\max(|B_1|, |B_2|)}\sum_{(m,n)\in G_1}\frac{1}{1 + \lambda\min_{(i,j)\in G_2}(\|[m, n] - [i, j]\|^2)}. \tag{32}$$

where $|B_1|$ and $|B_2|$ denote the accumulation of non-zero elements in the predicted lofargram map G_1 and actual lofargram map G_2, respectively. The $\| [m, n] - [i, j] \|^2$ indicates the Euclidean distance between the detected spectral lines and the actual spectral lines. We set $\lambda = 1$, as in [37].

3.3. Performance Analysis and Discussion

3.3.1. Necessity of AINP

Figures 7 and 8 compare the performances of the AINP under different intensity levels of impulse noise. As shown in Figure 7, heavy SαS noise creates broadband interference in lofargrams. At MSNR = -22 dB, the interference gradually increased as the value decreased. After the preprocessing with the AINP method, the broadband interference in the lofargrams was largely suppressed. However, the spectral-line pixels of the lofargram were still mixed with the low-amplitude impulse-noise pixels and were not visually distinguishable. The following LR-DRNet further processed lofargrams containing a significant amount of low-amplitude impulse-noise pixels. To further indicate the necessity of the AINP in the proposed method, Figure 8 presents a comparison of the performances of LR-DRNet and AINP+LR-DRNet. As shown in Figure 8, when $\alpha = 1.9$, for the cases of -22 and -23 dB, LR-DRNet and AINP+LR-DRNet exhibited comparable performances. As the MSNRs were further reduced to -25 and -26 dB, the performance of AINP+LR-DRNet was better than that of the LR-DRNet. This implies that LR-DRNet has the ability to adapt to weak impulse noise. However, when α decreased to 1.6, the performance of LR-DRNet degraded dynamically. Thus, it can be concluded that the AINP can effectively suppress the broadband interference caused by heavy SαS noise in lofargrams and is necessary for our method to detect and reconstruct weak spectral lines under non-Gaussian impulsive noise.

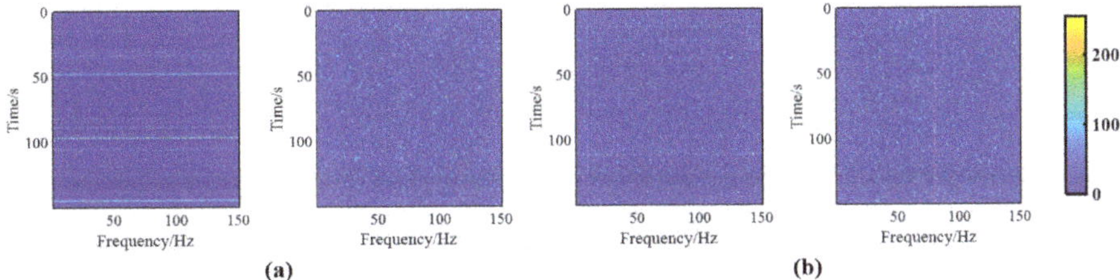

Figure 7. Comparison of original and AINP outcomes with different values of α at MSNR = -22 dB. (**a**) $\alpha = 1.6$; (**b**) $\alpha = 1.9$.

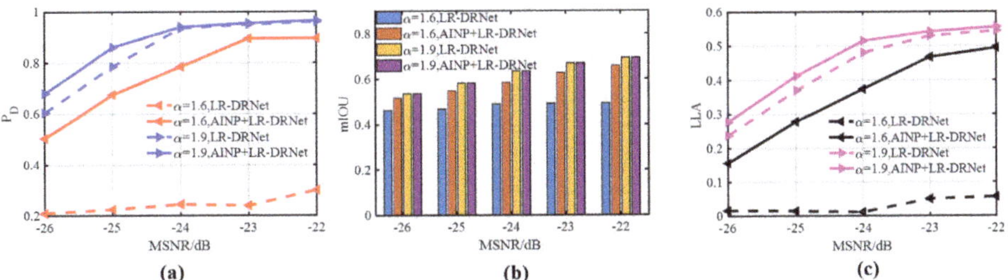

Figure 8. Performances of the LR-DRNet and the AINP+LR-DRNet under $\alpha = 1.6$ and $\alpha = 1.9$. (**a**) ROC; (**b**) mIOU; (**c**) LLA.

3.3.2. Network-Structure Analysis

Specific tasks may require suitable network structures. The simulation analyzed the appropriate network structure for spectral-line detection and reconstruction. The two network structures chosen for this analysis were LR-DRNet18 and LR-DRNet34. As shown in Figure 9, the impact of LR-DRNet depth on performance varies across all MSNRs. Compared with LR-DRNet18, the deeper LR-DRNet exhibited comparable performances in the cases of -22 and -23 dB, and exhibited better performances with -24, -25, and -26 dB, respectively. This shows that increasing the network depth improves network performance. Therefore, the coding layer of the LR-DRNet was set to 34 layers.

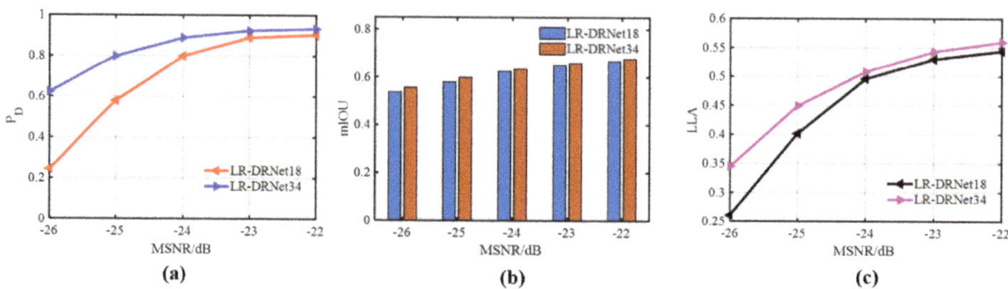

Figure 9. Performances of the AINP+LR-DRNet with 18 and 34 depths in different MSNRs. (**a**) ROC; (**b**) mIOU; (**c**) LLA.

Theoretically, the relevance of each feature channel can be automatically determined by the SE structure through learning. This learning of the SE structure determines the significance of each feature channel, which consequently strengthens the desirable features. Therefore, the SE structure needed to be analyzed to determine the performance of the proposed LR-DRNet. As shown in Figure 10, compared with LR-DRNet without SE, LR-DRNet with SE had a significant improvement in detection performance, especially at -25 and -26 dB, along with a slight improvement in reconstruction performance. The parameters of the SE structure participated in the end-to-end network parameter optimization process and optimized the encoder and decoder. Thus, we conclude that the SE structure can significantly enhance detection and reconstruction.

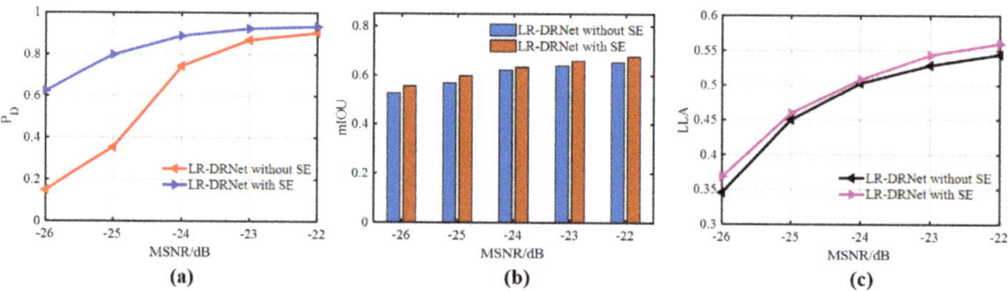

Figure 10. Performances of the AINP+LR-DRNet with and without SE in different MSNRs. (**a**) ROC; (**b**) mIOU; (**c**) LLA.

3.3.3. Detection and Reconstruction Performance Evaluation

With AINP used as a preprocessor, the outcomes of the proposed AINP+LR-DRNet were compared under various α values and MSNRs. As shown in Figure 11, the performance gradually decreased with decreases in alpha and MSNR, especially at $\alpha = 1.3$ and 1.5. The performances were comparable at high MSNR and α values, presenting more advantages at lower MSNR and α values. At an MSNR of -24 dB and an α of 1.7, the proposed AINP+LR-DRNet still had a P_D of approximately 78%, a mIOU of 0.59, and a LLA of 0.42. In particular, the stronger impulsive noise intensity and lower MSNR affected the feature-extraction ability of the network, encumbering the detection and reconstruction. Nevertheless, the proposed AINP+LR-DRNet is adaptable to low MSNR and strong impulse-noise intensity.

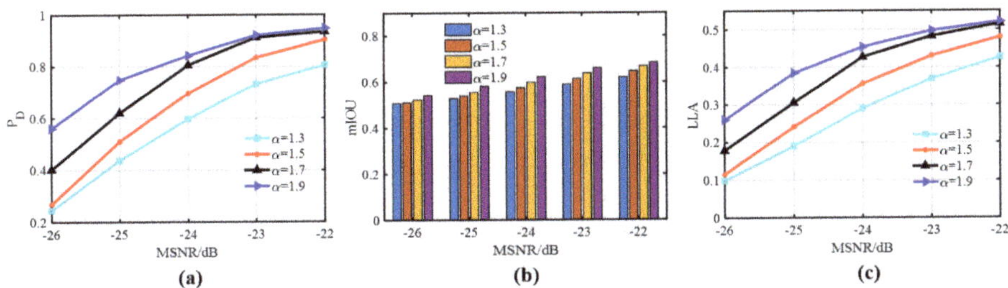

Figure 11. Detection and reconstruction with the SαS distribution noise, with α values of 1.3, 1.5, 1.7, and 1.9. (**a**) ROC; (**b**) mIOU; (**c**) LLA.

To verify the feasibility of the proposed LR-DRNet under Gaussian noise, its performance under Gaussian noise was compared with those of other methods. The LR-DRNet34 under a single detection task (LR-DNet34), HMM [14], UNet [24], SegNet [25], ResNet18 [38], ResNet34 [38], and LR-DRNet34 under a single reconstruction task (RNet34) were introduced for performance comparison. To ensure that this comparison was fair, ResNet used the same spectral-line-detection algorithm as LR-DRNet.

Figure 12 compares the detection performances of the proposed LR-DRNet with that of several deep learning methods under Gaussian noise. The proposed LR-DRNet achieved a higher detection rate, particularly at SNR values of -24 dB to -26 dB. Figure 13 presents the differences in the reconstruction performances of the five methods. The reconstruction performance of the proposed LR-DRNet was slightly better than that of RNet34, and better than that of the HMM and other deep-learning methods. In terms of reconstruction, as shown in Figure 14, the proposed LR-DRNet reconstructed weak spectral lines more accurately than the other methods at -25 and -26 dB, while exhibiting comparable performances at -22 and -23 dB. This was consistent with the analysis shown in Figure 13. The excellent detection and reconstruction performance of the proposed LR-DRNet and the superiority of MTL over single-task learning (STL) are illustrated. Thus, the feasibility of the proposed LR-DRNet method under Gaussian noise is illustrated.

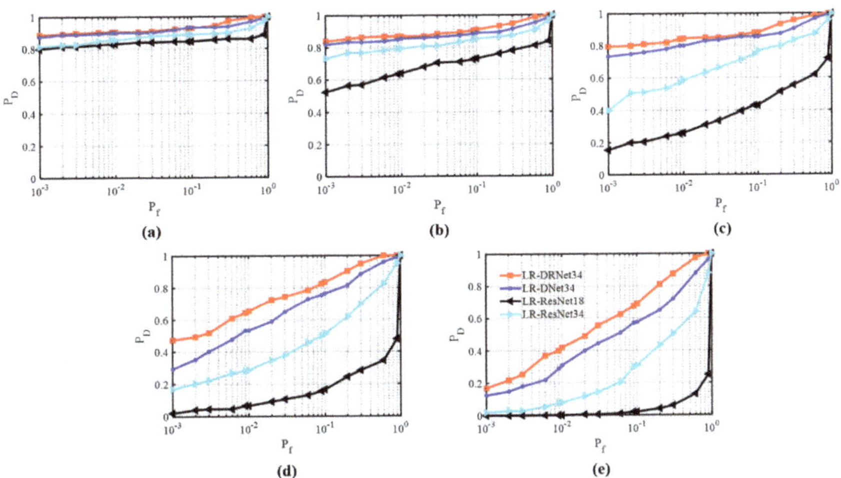

Figure 12. Comparison of ROCs of different methods under Gaussian noise in different SNRs. (**a**) SNR = -22 dB; (**b**) SNR = -23 dB; (**c**) SNR = -24 dB; (**d**) SNR = -25 dB; (**e**) SNR = -26 dB.

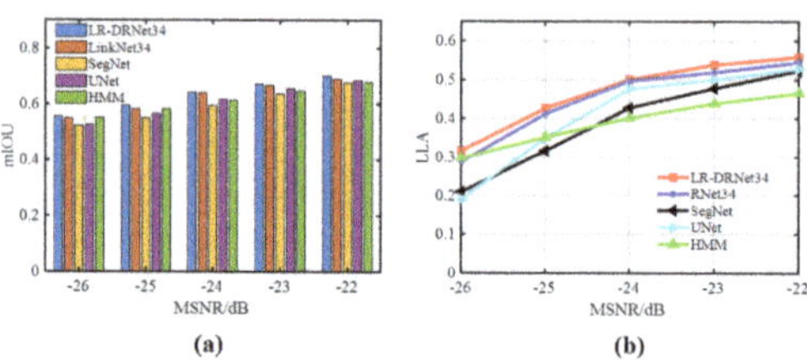

Figure 13. Comparison of the reconstruction performance of LR-DRNet, RNet34, SegNet, UNet, and HMM under different SNRs. (**a**) mIOU; (**b**) LLA.

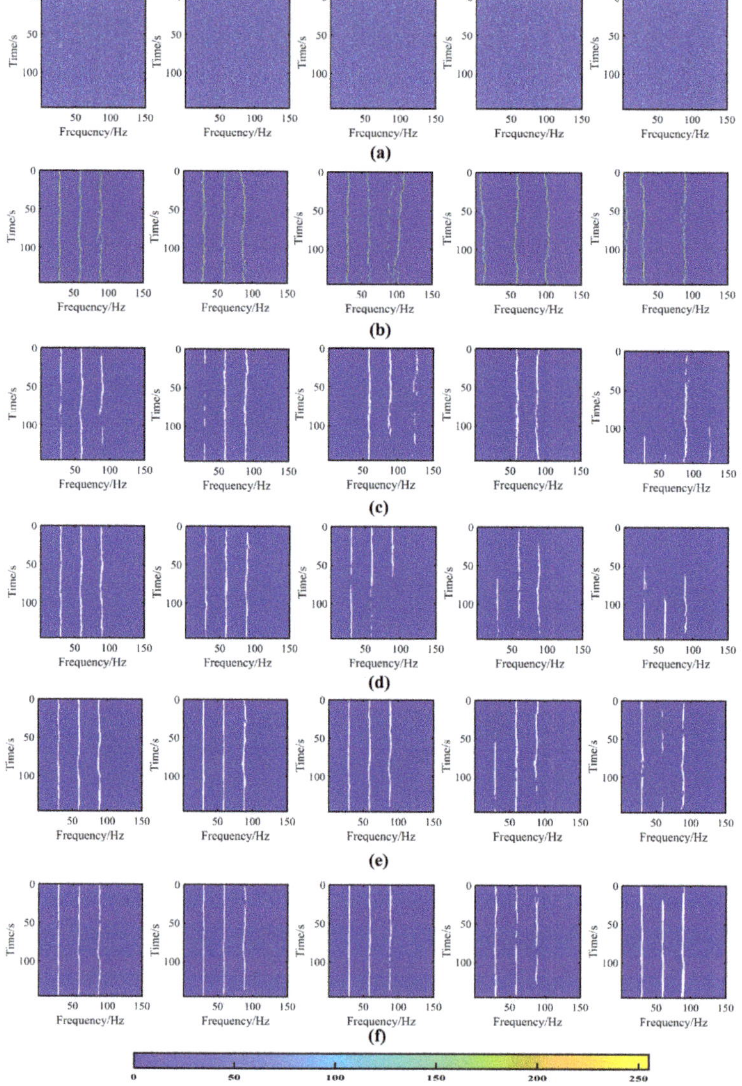

Figure 14. Reconstruction of five methods in different SNRs. The original lofargrams with SNR in the range of [−26, −22] dB are shown in (**a**). The same samples were reconstructed by HMM, SegNet, UNet, RNet34, and LR-DRNet34 in different SNRs, as shown in (**b**–**f**).

3.3.4. Comparison with Existing Methods

The detection and reconstruction performances of the proposed AINP+LR-DRNet were compared with those of other methods. Deep classification networks, such as ResNet34 [38] and DNet34, and a detector based on a Gaussian function (GF) [39] were introduced for the detection. Semantic segmentation structures, such as UNet [24], Seg-Net [25], RNet34, and HMM [14] were introduced for the reconstruction. In GF, the scale parameter c was set to 2.0, and the impulse intensity α and the dispersion coefficient γ were considered in plotting the ROC curve. The number of search times of the spectral line was set to four in the HMM. For a fair comparison, AINP and the algorithm in Section 2.1.4 were used for all the comparison algorithms, except GF.

First, we compared the detection performances of various methods with that of the proposed AINP+LR-DRNet. Figure 15 presents the ROCs of the four methods for MSNR from −22 dB to −26 dB. The AINP+LR-ResNet34, AINP+LR-DNet34, and the proposed AINP+LR-DRNet exhibited discrepancies, particularly at low MSNR values. Furthermore, the GF and the proposed AINP+LR-DRNet at the same P_f were compared. The GF detector filtered out impulse noise with large amplitudes via a nonlinear transformation, which suggests it had the worst performance. The superiority of the proposed AINP+LR-DRNet in detection is attributable to its specially designed network, which matches the spectral-line-detection algorithm, which is highly capable of feature extraction. The structures of other advanced networks and the disadvantages of the features of the traditional detection algorithm at a low MSNR may hinder detection.

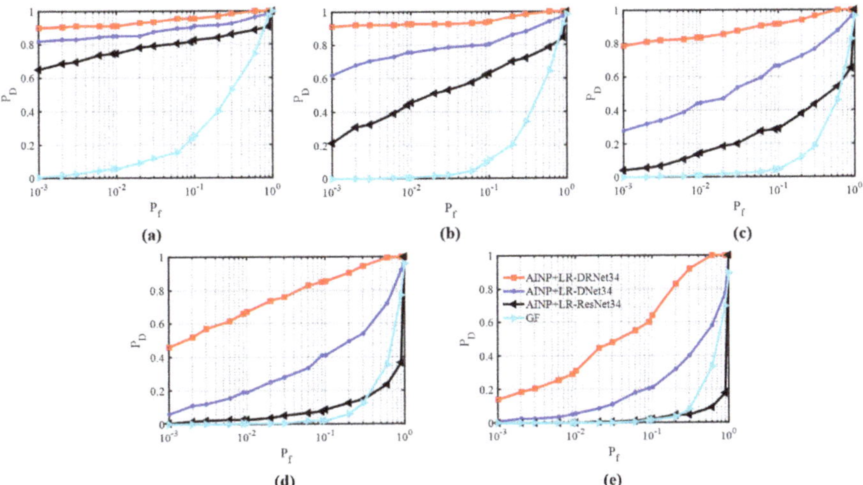

Figure 15. Comparison of ROCs of four methods under the SαS distribution noise for different MSNRs. (**a**) MSNR = −22 dB, (**b**) MSNR = −23 dB, (**c**) MSNR = −24 dB, (**d**) MSNR = −25 dB, and (**e**) MSNR = −26 dB.

Subsequently, we compared the different reconstruction methods. The SegNet, UNet, LinkNet34, and the proposed AINP+LR-DRNet are encoding and decoding networks, which segment features with different scales and complex boundaries by extracting the features of the encoding layer and reconstructing the decoding layer. As indicated in Tables 1 and 2, the proposed AINP+LR-DRNet outperformed the other methods by a considerable margin. Specifically, the mIOU and LLA of the AINP+HMM among the five MSNRs ranged from 0.4841 to 0.5566 and 0.2336 to 0.3985, respectively. The mIOU and LLA of AINP+SegNet and AINP+UNet in the five MSNRs were approximately 0.4917 to 0.6719 and 0.0246 to 0.5757, respectively. Accordingly, AINP+RNet34 was superior to the previous three methods, ranging from 0.5305 to 0.6881 and 0.2118 to 0.5859, respectively;

however, the proposed AINP+LR-DRNet achieved impressive performances, ranging from 0.5387 to 0.6932 and from 0.2777 to 0.5950, respectively. Figure 16 presents the lofargram reconstruction of the five methods. The lofargrams reconstructed by the AINP+HMM appeared as false spectral-line pixels after −23 dB, and the line profile became cluttered. The AINP+UNet and AINP+SegNet still worked at −22 dB, but the spectral line broke at varying degrees after −23 dB, and their integrity was reduced. The proposed AINP+LR-DRNet had a prominent spectral-line profile and a higher integrity at −22, −23, and −24 dB, respectively. At −25 and −26 dB, the spectral line could not be reconstructed in some positions because of the excessive background noise. The unique design of the network structure is more suited to reconstruction than those of other segmentation structures.

Table 1. mIOU values of different methods for different MSNRs.

Methods	MSNR/dB				
	−22	**−23**	**−24**	**−25**	**−26**
AINP+HMM	0.5566	0.5383	0.5236	0.5047	0.4841
AINP+SegNet	0.6584	0.5909	0.5301	0.5027	0.4917
AINP+UNet	0.6719	0.6205	0.5660	0.5205	0.4991
AINP+RNet34	0.6881	0.6655	0.6300	0.5833	0.5305
AINP+LR-DRNet34	0.6932	0.6688	0.6316	0.5905	0.5387

Table 2. LLA values of different methods for different MSNRs.

Methods	MSNR/dB				
	−22	**−23**	**−24**	**−25**	**−26**
AINP+HMM	0.3985	0.3599	0.3277	0.2840	0.2336
AINP+SegNet	0.5527	0.4182	0.1916	0.0734	0.0246
AINP+UNet	0.5757	0.5169	0.3406	0.1416	0.0499
AINP+RNet34	0.5859	0.5774	0.5221	0.4328	0.2118
AINP+LR-DRNet34	0.5950	0.5783	0.5373	0.4424	0.2777

Finally, we trained AINP+DNet34, AINP+RNet34, and the proposed AINP+LR-DRNet model separately to examine the validity of the MSL. As displayed in Figures 15 and 16, the proposed AINP+LR-DRNet improved the detection and reconstruction performances after utilizing an adaptive weighted loss function based on dual classification. Owing to the multi-task loss function, detection and reconstruction tasks complement each other by sharing valuable information.

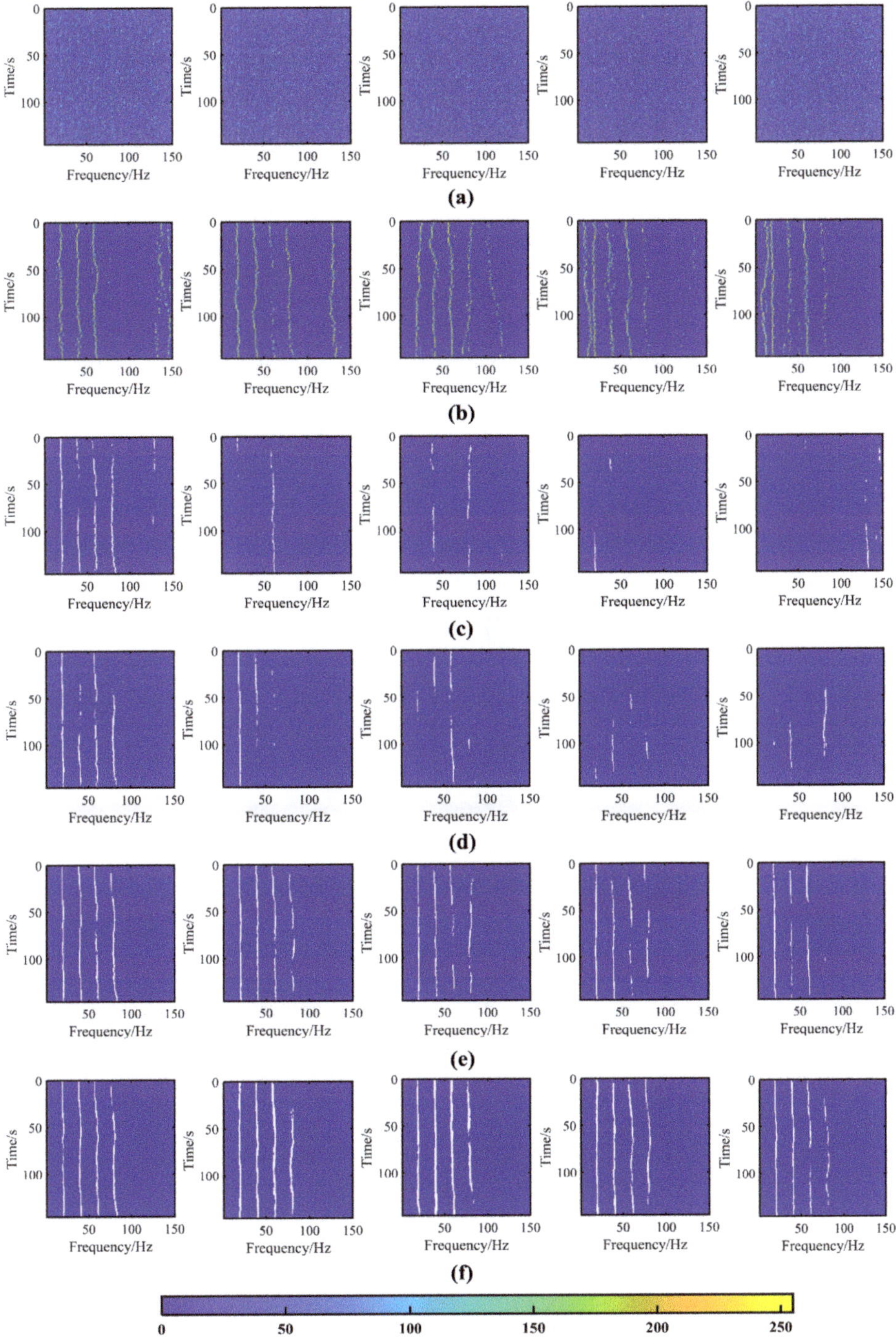

Figure 16. Reconstruction results of different methods in different MSNRs. The original lofargrams as shown in (**a**). Reconstructed by AINP+HMM, AINP+SegNet, AINP+UNet, AINP+RNet34, and the proposed AINP+LR-DRNet with MSNR in the range of [−26, −22] dB, as shown in (**b**–**f**).

4. Experimental Data Analysis

The detection and reconstruction of the proposed AINP+LR-DRNet in Gaussian/non-Gaussian impulsive noise were verified by the aforementioned simulation analysis. At this point, the weights of the pre-trained model in the simulation were fine-tuned using experimental data. The ability of the proposed AINP+LR-DRNet to detect and reconstruct single and multiple weak spectral lines was analyzed by employing two different experimental datasets, and the performances were compared with those of other methods.

4.1. Reconstruction of Weak Single Spectral Line from Strong Background Noise

The data for single-spectral-line detection and reconstruction were received from an experiment conducted in the South China Sea in July 2021. A vertical line array (VLA) composed of 32 hydrophones with an interval of 2 m was employed at a depth of 275–337 m. The sampling rate of the acoustic collector was 10 kHz. During the experiment, the sound source transmitted a single-frequency signal of 71 Hz and was towed 1.5 to 11 km away from the receiving array at a depth of approximately 20 m. Figure 17 displays the hydrophone arrays used in our experiment. We intercepted 2k signal and noise samples from VLA-1 to VLA-32, which formed the measured sample set.

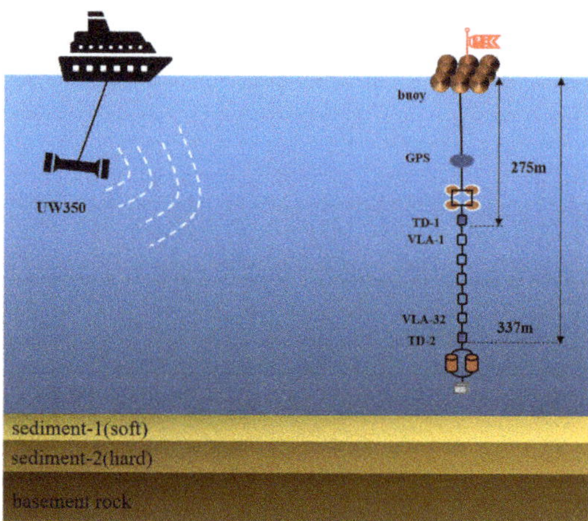

Figure 17. Schematic of ship movement and VLA deployment.

In Figure 18a, two unreconstructed lofargrams are displayed in the experimental data, with a relatively weak spectral line. The AINP+HMM, AINP+RNet34, and the proposed AINP+LR-DRNet effectively reconstructed the regions with obvious spectral lines, as shown to the left of Figure 18b–d. As the MSNR was low, the HMM reconstructed some false spectral-line pixels, and AINP+RNet34 reconstructed a few spectral-line pixels. The weak spectral line was reconstructed using the proposed AINP+LR-DRNet, despite the strong background noise in the two cases. Meanwhile, the experimental data show that MTL outperformed STL. Consequently, the proposed AINP+LR-DRNet is suitable for extracting weak single spectral lines from noise-dominated lofargrams.

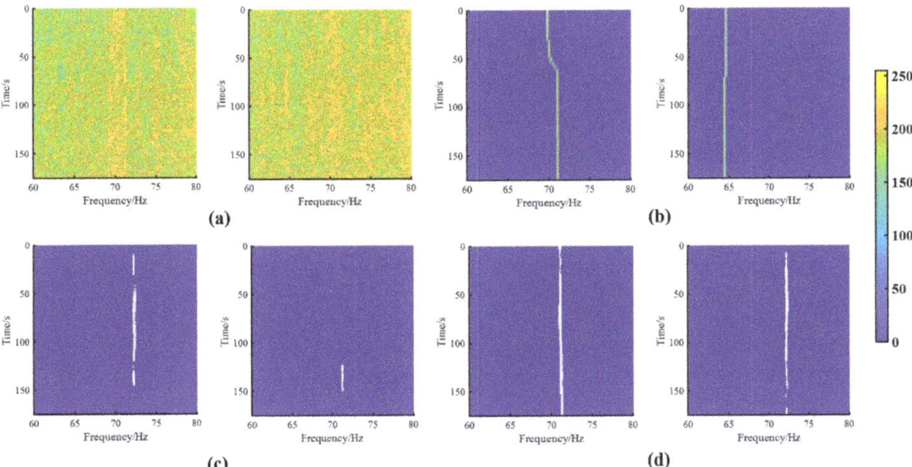

Figure 18. Lofargram reconstruction results of experimental data-1 by AINP+HMM, AINP+RNet34, and the proposed AINP+LR-DRNet, shown in (**b–d**), respectively. (**a**) Original lofargrams, which were not reconstructed.

4.2. Weak Multiple-Spectral-Line Reconstruction against Strong Interference Background

Another experiment conducted in the South China Sea in September 2021 was used to detect and reconstruct multiple spectral lines. An ocean-bottom seismmometer (OBS) was deployed every 5 km, with a line length of more than 100 km. The entire seabed was initially relatively flat, and it gradually became inclined near the destination. The sampling rate of the OBS was 100 Hz. As shown in Figure 19, the ship sailed along a straight line at a certain speed for the deployment and recovery of the OBS. Therefore, the OBS can collect ship-noise samples at low SNRs, as well as marine-ambient-noise samples. The test data set was formed with 5000 more signal and noise samples from OBS-1 to OBS-25.

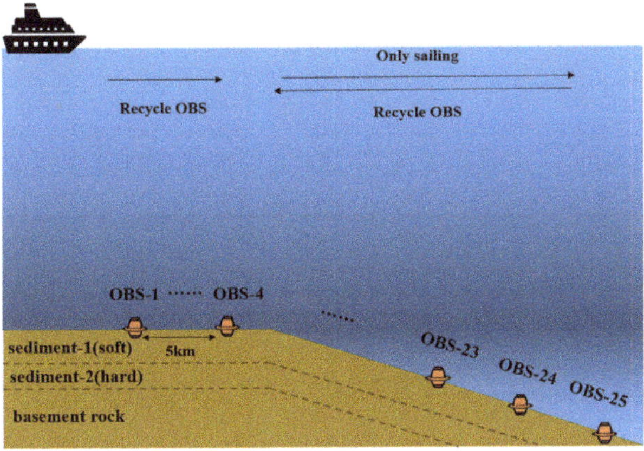

Figure 19. Schematic of OBS deployment and recovery in the experiment.

As shown in Figure 20a, the spectral lines of the ship were affected by the strong interference. In one of the cases on the left, the spectral lines at 25 Hz and 33 Hz were blurred on the original lofargram. Furthermore, in another case, the original lofargram did not present a spectral line at 16, 25, or 33 Hz. Figure 20b indicates that AINP+HMM can only reconstruct spectral lines with higher SNR, but becomes ineffective under strong interference. In addition, more spectral-line pixels were reconstructed using AINP+RNet34, as shown in Figure 20c. The proposed AINP+LR-DRNet is more appropriate for spectral line reconstruction than HMM and RNet34, thereby highlighting the spectral lines and suppressing noise. Hence, the proposed AINP+LR-DRNet is applicable for multiple weak ship spectral line reconstruction under intense interference.

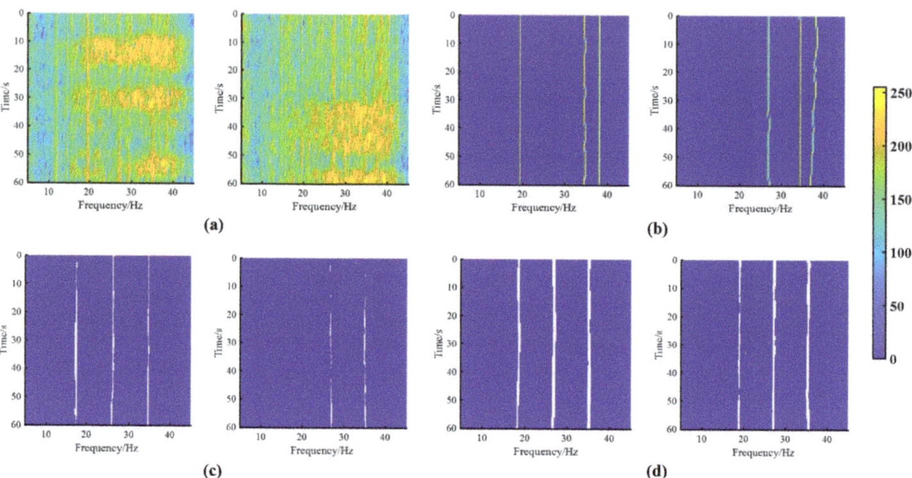

Figure 20. Lofargram reconstruction results of experimental data-2 by AINP+HMM, AINP+RNet34, and the proposed AINP+LR-DRNet in (**b–d**), respectively. (**a**) Two different original lofargrams.

4.3. Detection Performances with Two Real-World Data

Finally, the detection performances of the GF, AINP+ResNet18, and AINP+DNet34 were compared to evaluate the proposed AINP+LR-DRNet. The P_f was certain for a fixed test set. For a fair comparison, the GF was compared with the detection rate under the false alarm rate obtained by the proposed AINP+LR-DRNet.

As summarized in Table 3, GF displayed the lowest values at low SNR. Compared with GF, the P_D and P_f of AINP+LR-DNet34 were higher. The proposed AINP+LR-DRNet exhibited the highest P_D for the two measured datasets, reaching 94.73% and 94.49%, respectively. Values of P_f of 2.21% and 5.93% were also obtained, which was the best performance of all the methods. This analysis indicates that the proposed AINP+LR-DRNet has the advantage of detection at low SNR under MTL.

Table 3. Detection performances on practical data.

Methods	An Experiment in July 2021		An Experiment in September 2021	
	P_f	P_D	P_f	P_D
GF	2.21%	62.03%	5.93%	22.79%
AINP+LR-DNet34	11.0%	89.47%	14.83%	76.47%
AINP+LR-DRNet34	2.21%	94.73%	5.93%	94.79%

5. Conclusions

In this study, the joint detection and reconstruction of weak spectral lines under non-Gaussian impulsive noise using DL was investigated. First, with DL, the detection and reconstruction of spectral lines were formulated as a binary classification problem. Subsequently, a framework for weak-line-spectrum detection and reconstruction based on AINP and DRNet was developed. Under the developed framework, a LR-DRNet detection algorithm was designed, and the lofargrams after the AINP were used as the input of the LR-DRNet. In particular, LR-DRNet was trained by the dual classification adaptive loss to output high detection results and lofargrams with significant spectral lines. Finally, simulated data and real data from the South China Sea were used to verify the performance of AINP+LR-DRNet. The results show that the proposed AINP+LR-DRNet can effectively detect and reconstruct weak spectral lines under non-Gaussian impulsive noise.

In the future, various underwater acoustic signals and marine ambient noises following other distributions will be examined. Furthermore, weak-spectral-line detection based on unsupervised learning will be considered to alleviate the lack of underwater acoustic data and labeling requirements.

Author Contributions: Conceptualization, Z.L.; methodology, Z.L.; investigation, Z.L., X.W. and J.G.; data curation, X.W. and J.G.; writing—original draft preparation, Z.L.; writing—review and editing, X.W. and J.G.; visualization, Z.L.; simulations, Z.L.; supervision, X.W. and J.G. All authors have read and agreed to the published version of the manuscript.

Funding: We thank the staff of the South China Sea experiment in 2021 for their assistance in providing us with valuable data. This work was supported by the National Natural Science Foundation of China under grant no. 12174078.

Data Availability Statement: The data presented in this paper are available upon reasonable request to the corresponding author.

Conflicts of Interest: The authors declare no conflict of interest.

References

1. Li, Q.; Li, M.; Yang, X. The detection of single frequency component of underwater radiated noise of target: Theoretical analysis. *Acta Acust.* **2008**, *33*, 193–196.
2. Cohen, L. Time-frequency distributions—A review. *Proc. IEEE* **1989**, *77*, 941–981. [CrossRef]
3. Yu, G.; Yang, T.C.; Piao, S. Estimating the delay-Doppler of target echo in a high clutter underwater environment using wideband linear chirp signals: Evaluation of performance with experimental data. *J. Acoust. Soc. Am.* **2017**, *142*, 2047–2057. [CrossRef] [PubMed]
4. Abel, J.S.; Lee, H.J.; Lowell, A.P. An image processing approach to frequency tracking (application to sonar data). In Proceedings of the ICASSP-92: 1992 IEEE International Conference on Acoustics, Speech, and Signal Processing, San Francisco, CA, USA, 23–26 March 1992; IEEE: Piscataway, NJ, USA, 1992; Volume 2, pp. 561–564.
5. Gillespie, D. Detection and classification of right whale calls using an 'edge' detector operating on a smoothed spectrogram. *Can. Acoust.* **2004**, *32*, 39–47.
6. Khotanzad, A.; Lu, J.H.; Srinath, M.D. Target detection using a neural network based passive sonar system. In Proceedings of the International Joint Conference on Neural Networks, Washington, DC, USA, 18–22 June 1989; Volume 1, pp. 335–440.
7. Leeming, N. Artificial neural nets to detect lines in noise. In Proceedings of the International Conference on Acoustic Sensing and Imaging, London, UK, 29–30 March 1993; IET: London, UK, 1993; pp. 147–152.
8. Izacard, G.; Bernstein, B.; Fernandez-Granda, C. A learning-based framework for line-spectra super-resolution. In Proceedings of the ICASSP 2019—2019 IEEE International Conference on Acoustics, Speech and Signal Processing (ICASSP), Brighton, UK, 12–17 May 2019; IEEE: Piscataway, NJ, USA, 2019; pp. 3632–3636.
9. Izacard, G.; Mohan, S.; Fernandez-Granda, C. Data-driven estimation of sinusoid frequencies. In *Advances in Neural Information Processing Systems*; MIT Press: Cambridge, MA, USA, 2019; Volume 32.
10. Jiang, Y.; Li, H.; Rangaswamy, M. Deep learning denoising based line spectral estimation. *IEEE Signal Process. Lett.* **2019**, *26*, 1573–1577. [CrossRef]
11. Jiang, Y.; Zhang, T.; Zhang, W. Model-Based Neural Network and Its Application to Line Spectral Estimation. *arXiv* **2022**, arXiv:2202.06485.
12. Han, Y.; Li, Y.; Liu, Q. DeepLofargram: A deep learning based fluctuating dim frequency line detection and recovery. *J. Acoust. Soc. Am.* **2020**, *148*, 2182–2194. [CrossRef]
13. Paris, S.; Jauffret, C. Frequency line tracking using hmm-based schemes [passive sonar]. *IEEE Trans. Aerosp. Electron. Syst.* **2003**, *39*, 439–449. [CrossRef]
14. Luo, X.; Shen, Z. A sensing and tracking algorithm for multiple frequency line components in underwater acoustic signals. *Sensors* **2019**, *19*, 4866. [CrossRef]
15. Nikias, C.L.; Shao, M. *Signal Processing with Alpha-Stable Distributions and Applications*; Wiley-Interscience: Hoboken, NJ, USA, 1995.
16. Webster, R.J. A random number generator for ocean noise statistics. *IEEE J. Ocean. Eng.* **1994**, *19*, 134–137. [CrossRef]

17. Traverso, F.; Vernazza, G.; Trucco, A. Simulation of non-white and non-Gaussian underwater ambient noise. In Proceedings of the 2012 Oceans-Yeosu, Yeosu, Republic of Korea, 21–24 May 2012; IEEE: Piscataway, NJ, USA, 2012; pp. 1–10.
18. Song, G.; Guo, X.; Li, H. The α stable distribution model in ocean ambient noise. *Chin. J. Acoust.* **2021**, *40*, 63–79.
19. Wang, J.; Li, J.; Yan, S. A novel underwater acoustic signal denoising algorithm for Gaussian/non-Gaussian impulsive noise. *IEEE Trans. Veh. Technol.* **2020**, *70*, 429–445. [CrossRef]
20. Vijaykumar, V.R.; Mari, G.S.; Ebenezer, D. Fast switching based median–mean filter for high density salt and pepper noise removal. *AEU Int. J. Electron. Commun.* **2014**, *68*, 1145–1155. [CrossRef]
21. Sheela, C.J.J.; Suganthi, G. An efficient denoising of impulse noise from MRI using adaptive switching modified decision based unsymmetric trimmed median filter. *Biomed. Signal Process. Control* **2020**, *55*, 101657. [CrossRef]
22. Chanu, P.R.; Singh, K.M. A two-stage switching vector median filter based on quaternion for removing impulse noise in color images. *Multimed. Tools Appl.* **2019**, *78*, 15375–15401. [CrossRef]
23. Barazideh, R.; Sun, W.; Natarajan, B.; Nikitin, A.V.; Wang, Z. Impulsive noise mitigation in underwater acoustic communication systems: Experimental studies. In Proceedings of the 2019 IEEE 9th Annual Computing and Communication Workshop and Conference (CCWC), Las Vegas, NV, USA, 7–9 January 2019; IEEE: Piscataway, NJ, USA, 2019; pp. 880–885.
24. Ronneberger, O.; Fischer, P.; Brox, T. U-net: Convolutional networks for biomedical image segmentation. In *Medical Image Computing and Computer-Assisted Intervention–MICCAI 2015, Proceedings of the 18th International Conference, Munich, Germany, 5–9 October 2015*; Part III 18; Springer International Publishing: Berlin/Heidelberg, Germany, 2015; pp. 234–241.
25. Badrinarayanan, V.; Kendall, A.; Cipolla, R. Segnet: A deep convolutional encoder-decoder architecture for image segmentation. *IEEE Trans. Pattern Anal. Mach. Intell.* **2017**, *39*, 2481–2495. [CrossRef] [PubMed]
26. Chaurasia, A.; Culurciello, E. Linknet: Exploiting encoder representations for efficient semantic segmentation. In Proceedings of the 2017 IEEE Visual Communications and Image Processing (VCIP), St. Petersburg, FL, USA, 10–13 December 2017; IEEE: Piscataway, NJ, USA, 2017; pp. 1–4.
27. Hu, J.; Shen, L.; Sun, G. Squeeze-and-excitation networks. In Proceedings of the IEEE Conference on Computer Vision and Pattern Recognition, Salt Lake City, UT, USA, 18–23 June 2018; pp. 7132–7141.
28. Kendall, A.; Gal, Y.; Cipolla, R. Multi-task learning using uncertainty to weigh losses for scene geometry and semantics. In Proceedings of the IEEE Conference on Computer Vision and Pattern Recognition, Salt Lake City, UT, USA, 18–23 June 2018; pp. 7482–7491.
29. Liu, C.; Wang, J.; Liu, X. Deep CM-CNN for spectrum sensing in cognitive radio. *IEEE J. Sel. Areas Commun.* **2019**, *37*, 2306–2321. [CrossRef]
30. Kingma, D.P.; Ba, J. Adam: A method for stochastic optimization. *arXiv* **2014**, arXiv:1412.6980.
31. Glorot, X.; Bengio, Y. Understanding the difficulty of training deep feedforward neural networks. In Proceedings of the Thirteenth International Conference on Artificial Intelligence and Statistics, Sardinia, Italy, 13–15 May 2010; JMLR Workshop and Conference Proceedings. pp. 249–256.
32. Goyal, P.; Dollár, P.; Girshick, R. Accurate, large minibatch sgd: Training imagenet in 1 hour. *arXiv* **2017**, arXiv:1706.02677.
33. Loshchilov, I.; Hutter, F. Sgdr: Stochastic gradient descent with warm restarts. *arXiv* **2016**, arXiv:1608.03983.
34. Samorodnitsky, G.; Taqqu, M.S.; Linde, R.W. Stable non-gaussian random processes: Stochastic models with infinite variance. *Bull. Lond. Math. Soc.* **1996**, *28*, 554–555.
35. Koutrouvelis, I.A. An iterative procedure for the estimation of the parameters of stable laws: An iterative procedure for the estimation. *Commun. Stat-Simul. Comput.* **1981**, *10*, 17–28. [CrossRef]
36. Garcia-Garcia, A.; Orts-Escolano, S.; Oprea, S. A review on deep learning techniques applied to semantic segmentation. *arXiv* **2017**, arXiv:1704.06857.
37. Lampert, T.A.; O'Keefe, S.E.M. A survey of spectrogram track detection algorithms. *Appl. Acoust.* **2010**, *71*, 87–100. [CrossRef]
38. He, K.; Zhang, X.; Ren, S. Deep residual learning for image recognition. In Proceedings of the IEEE Conference on Computer Vision and Pattern Recognition, Las Vegas, NV, USA, 27–30 June 2016; pp. 770–778.
39. Luo, J.; Wang, S.; Zhang, E. Signal detection based on a decreasing exponential function in alpha-stable distributed noise. *KSII Trans. Internet Inf. Syst. TIIS* **2018**, *12*, 269–286.

Article

Neural-Network-Based Equalization and Detection for Underwater Acoustic Orthogonal Frequency Division Multiplexing Communications: A Low-Complexity Approach

Mingzhang Zhou [1,2], Junfeng Wang [2,3,*], Xiao Feng [4], Haixin Sun [1,2], Jie Qi [5] and Rongbin Lin [1]

[1] School of Informatics, Xiamen University, Xiamen 361005, China; mzzhou@xmu.edu.cn (M.Z.); hxsun@xmu.edu.cn (H.S.); rblin@xmu.edu.cn (R.L.)

[2] Key Laboratory of Southeast Coast Marine Information Intelligent Perception and Application, Ministry of Natural Resources, Zhangzhou 363000, China

[3] School of Integrated Circuit Science and Engineering, Tianjin University of Technology, Tianjin 300384, China

[4] Key Laboratory of Broadband Wireless Communication and Sensor Network Technology, Nanjing University of Posts and Telecommunications, Nanjing 210023, China; fengxiao88702@163.com

[5] School of Electornic Science and Engineering (National Model Microelectronics College), Xiamen University, Xiamen 361005, China; qijie@xmu.edu.cn

* Correspondence: great_seal@163.com

Abstract: The performance of the underwater acoustic (UWA) orthogonal frequency division multiplexing (OFDM) system is often restrained by time-varying channels with large delays. The existing frequency domain equalizers do not work well because of the high complexity and difficulty of finding the real-time signal-to-noise ratio. To solve these problems, we propose a low-complexity neural network (NN)-based scheme for joint equalization and detection. A simple NN structure is built to yield the detected symbols with the joint input of the segmented channel response and received symbol. The coherence bandwidth is investigated to find the optimal hyperparameters. By being completely trained offline with real channels, the proposed detector is applied independently in both simulations and sea trials. The results show that the proposed detector outperforms the ZF and MMSE equalizers and extreme learning machine (ELM)-based detectors in both the strongly reflected channels of the pool and time-variant channels of the shallow sea. The complexity of the proposed network is lower than the MMSE and ELM-based receiver.

Keywords: underwater acoustic communication; subcarrier multiplexing; neural networks; coherence bandwidth; equalizers; detectors

Citation: Zhou, M.; Wang, J.; Feng, X.; Sun, H.; Qi, J.; Lin, R. Neural-Network-Based Equalization and Detection for Underwater Acoustic Orthogonal Frequency Division Multiplexing Communications: A Low-Complexity Approach. *Remote Sens.* **2023**, *15*, 3796. https://doi.org/10.3390/rs15153796

Academic Editor: Gabriel Vasile

Received: 27 June 2023
Revised: 25 July 2023
Accepted: 27 July 2023
Published: 30 July 2023

1. Introduction

With the increasing requirements for an Internet of Things in the oceans, efficient data processing and transmission become critical for ensuring the instantaneity for the underwater environment monitoring [1,2] and emergency rescue [3]. Orthogonal frequency division multiplexing (OFDM) has been a viable method in bandwidth-constrained underwater acoustic communications [4,5], as a result of its high spectral efficiency and ability to resist frequency selective fading. Nonetheless, the selective channels decided by variant parameters, such as distribution of sound speed [6], bottom reflection coefficient and surface waves [7], limit the performance improvement of the underwater acoustic (UWA) OFDM system [8,9].

To better detect the OFDM signals from UWA channels, variable equalizers have been applied, e.g., linear equalizers including zero-forcing (ZF) and minimum mean square error (MMSE) equalizers [10], and decision feedback equalizer (DFE) such as the Turbo equalizer [11,12]. The linear equalizers with simple structures are widely used in terrestrial communication links [13,14], whose performances rely on accurate channel estimations. Altough the DFEs show satisfactory performance without a channel estimator, it is at the expense of the requirement for higher computational complexity and extra channel coding [15,16]. In practice, real-time underwater acoustic communications (UAC) do not allow for a large number of online iterations [17–19], which still require linear equalizers. However, the noise amplification problem occurs when applying the ZF equalizer. Although MMSE equalizer overcomes this problem by considering the signal-to-noise ratio (SNR), it is hard to estimate the statistical values of the noise in real underwater environments including non-Gaussian and colored noise [20,21]. To further optimize the equalizer, deep learning (DL) and neural network (NN) have been developed [22–26]. With enough samples, a DL-based receiver can statistically learn to detect the symbols from the channel and other interference.

H. Ye et al. proposed a DL-based OFDM receiver [22], which used three fully connected layers to deal with a 64-subcarrier OFDM symbol with a block-type pilot. The bit error rate (BER) of the NN-based receiver was lower than the least square (LS) and MMSE estimation and detector in the simulation. To further obtain higher detecting accuracy, researchers have tried to substitute the whole communication system for the end-to-end networks [22,24,25,27]. The traditional digital modulation and subcarrier mapping has been replaced by the autoencoder (AE) [24]. A blind receiver without the pilot has been built with a convolutional neural network (CNN), which showed better performance compared with the traditional baselines. Similarly, an AE has been designed [25] to provide a modulation scheme for the multicarrier system. This work fed the decoder with both the channel state information (CSI) and received symbol, constructing a data-driven model for symbol detection. The simulations showed significant BER performance in additive white Gaussian noise (AWGN) channels. In B. Lin's work [28], a super-resolution channel reconstruction network was combined with AE for the marine communication system, proving its effectiveness in slow fading channels. H. Zhao et al. [29], J. Liu et al. [30], and Y. Zhang et al. [31] proposed different network structures for the UWA OFDM receiver, and trained them with the WATERMARK dataset. These studies focused on designing specific network structures to improve the performance of the OFDM receiever. Nevertheless, the theoretical explanation of the networks remains limited. There is no quantitative analysis for the hyperparameters.

Although the above NN-based communication systems show good performances in simulations, it is difficult to practically implement them, particularly in underwater acoustic channels because of the heavy computations and complex structures. For instance, Refs. [29–31] did not conduct sea trials. Another option is to build simple networks for the module optimization. M. Turhan et al. proposed an NN-based generalized frequency division multiplex with index modulation (GFDM-IM) detector to detect the symbols after a coarse detector [32]. With perfect CSI in the receiver, the simulation results showed lower BERs of this network than the ZF detector. T. Wang et al. have built a CNN for index modulated OFDM (IM-OFDM) detection, whose performance approximates the maximum likelihood (ML) detector [33]. A further option for NN is the extreme learning machine (ELM)-based receiver. This kind of receiver integrates the channel estimator and equalizer with a single layer NN, which is trained online for each time [34,35]. In L. Yang's work, with enough block pilots for training, a long frame with a large quantity of OFDM symbols was simulated [35], showing better performance than MMSE equalizer and NN-based detector proposed by H. Ye et al. [22]. Since the UWA channels were time-variant, the ELM detector was unable to show good performance, because the transmitted frames had to be kept short to reduce the influence of the time-variant channel. H. Zhao et al. [36] proposed a transfer strategy for the DNN-based OFDM receiver and tested it with the WATERMARK dataset and real experimental data. This study focused on the network retraining, and did not discuss the design of the applied NN structure. Y. Zhang et al. [37] focused on solving the channel sample augmentation problem for the NN-based channel estimator. Both [36,37] proposed an innovative strategy to solve the application problem of DNN-based receievers.

Despite the good simulation results produced by the above structures, problems still exist when the system is implemented in UWA channels.

- Firstly, the real dataset is difficult to obtain because the UAC links are usually one-way with no feedback. Hence, the uncertain time-variant channel states do not allow the system to obtain the samples in a short period.
- Secondly, it is not realistic for the NNs to be retrained in a high rate link because the computation loading is still heavy for real-time applications. Consequently, the data-driven works mentioned above barely discuss the performance of the system with real experiments.

In this paper, an attempt at the design and derivation of an NN-based receiver is made for the UWA OFDM system. A simple NN is proposed to integrate the equalization and symbol detection, containing only one fully-connected layer. Firstly, taking both channel frequency response and received symbol as input, the network learns a robust structure to output the symbol directly. To minimize the complexity of the network, the channel and received symbol are divided into blocks of the same size, matched with small-size networks. For attribution to the simple structure, the hyperparameters (mainly the hidden layer size and input dimension) are inferred according to the delay and coherence bandwidth of the channel. Thus, the channel-driven networks are constructed. After being trained with mixed channels and noise samples, the networks show robustness in both simulations and sea trials, performing better than the ZF and MMSE equalizers as well as the ELM-based detector in [35]. The contributions of this paper are listed as follows.

- We propose a low-complexity NN-based symbol detector for the UWA OFDM system. The network takes the segmented channel response and symbol block as input and integrates the equalization and detection processes. The small input dimension also reduces the requirement for the hidden neurons. The proposed detector shows lower computational complexity than the MMSE and ELM-based detectors.
- The NN-based detector is trained offline with a channel dataset containing simulated and real channels. Then the detector can be applied completely independently online with fixed hyperparameters, improving the efficiency of the online receiver. Under the same LS channel estimator, the trained network outperforms the ZF and MMSE equalizers, and the whole receiver is more reliable than the online ELM-based detector in both frequency selective channels in the pool and time-variant shallow sea channels.
- To obtain the optimal network structure, the block size of an OFDM symbol is associated with the coherence bandwidth. By testing each network with the input sizes in the range of less than the coherence bandwidth, the optimal hyperparameters can be found. The simulations verify the above configurations.

The remainder of this paper is organized as follows. Section 2 describes the UWA OFDM system. Section 3 discusses the UWA channels and the traditional detectors for symbols suffering from them. The network structure and training strategy are described in Section 4, while the result discussions of simulations and sea trials are included in Sections 5 and 6. Section 7 concludes our work.

2. Preliminary

An UWA OFDM system with frequency domain equalization is shown in Figure 1. The bit stream **b** to be transmitted is modulated to symbols with digital modulation. After the inverse fast Fourier transform (IFFT) is performed, the signal is up-converted to the carrier frequency. Then it is transmitted through the channel and suffers from noise. In the receiver, the signal is represented as

$$r(t) = h(t, \tau) \otimes s(t) + z(t), \tag{1}$$

where $s(t)$, $h(t, \tau)$ and $z(t)$ are the transmitted signal, channel impulsive response and additive noise. τ is the channel delay. $\otimes$ denotes convolution. After being down-converted to the baseband and performing the fast Fourier transform (FFT), the received symbol in the frequency domain can be written as

$$\mathbf{R} = \mathbf{HS} + \mathbf{Z}, \tag{2}$$

where $\mathbf{S}$, $\mathbf{H}$, and $\mathbf{Z}$ are the transmitted symbol, channel transfer function, and additive noise in frequency domain. Usually, the UWA channels and noise are different from those in terrestrial communications. With more powerful recognizable paths and impulsive noise, the OFDM system does not show good performance in shallow water. To reduce the influence of channels on symbols, before detection, the channel should be estimated and used to equalize the symbol with specific algorithms such as ZF and MMSE.

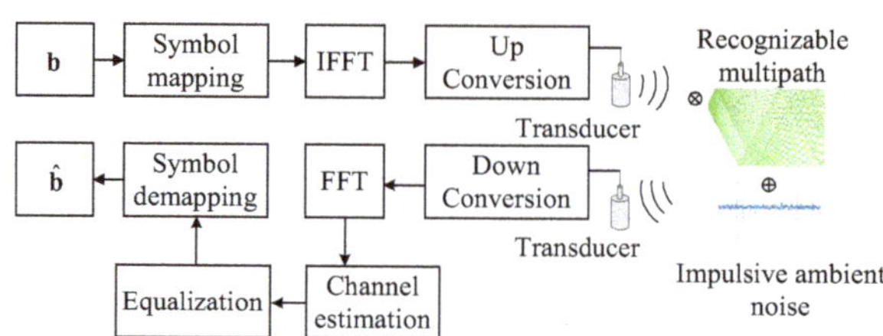

Figure 1. A classic UWA OFDM system. The performance of the channel estimator and equalizer are challenged by the UWA channels with large delays and impulsive noise.

3. UWA Receiver Structure

3.1. Signals Suffering from UWA Channel with Large Delays

The UWA channel differs from the terrestrial electromagnetic channel [38]. The ray theory reveals that the delay of an underwater channel is decided by the path length and sound speed [7], and in addition, according to [39], the motion of the transmitter/receiver pair, the scattering of the moving sea surface and the refraction due to sound speed variations. For a received symbol, the time-varying UWA channel impulse response (CIR) can be written as

$$h(t, \tau) = \sum_{i=0}^{N_p-1} c_i(t)\delta(\tau - \tau_i(t)), \tag{3}$$

where N_p is the number of paths, and $\tau_i(t) \approx \tau_i - a_i t$ is the time-varying delay of the i-th path, and a_i is the Doppler factor. $c_i(t)$ is the channel coefficient of each path varying with time. In the receiver, after resampling, FFT, and low-pass filtering [4], the channel function in frequency domain is written as

$$H(t, f) = \sum_{i=0}^{N_p-1} c_i(t)e^{j2\pi f\tau_i(t)}. \tag{4}$$

As a result of the slow sound speed and low reflection loss, $\tau_i(t)$ is large in long distances. These recognizable paths result in a small coherence bandwidth for the OFDM symbol. Figure 2 shows a CIR and corresponding transfer function caught in the water tank of Xiamen University. It can be seen from Figure 2b that the frequency selectivity is severe because of the long delay of the recognizable paths in Figure 2a.

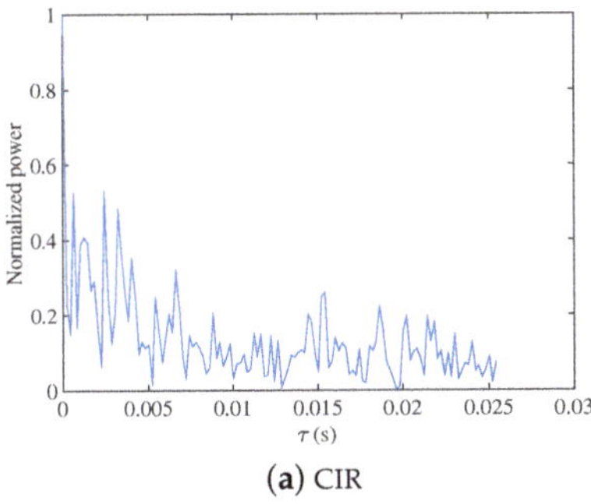
(a) CIR

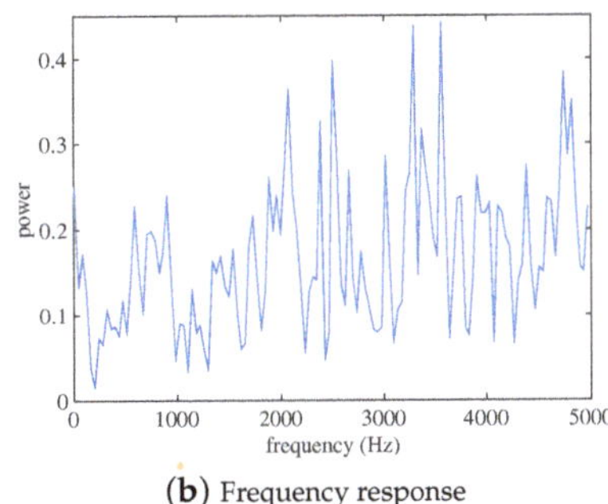
(b) Frequency response

Figure 2. Time and frequency response of a typical UWA channel: (**a**) is the CIR of the channel, which contains many strongly reflected paths with large delays. Its frequency response in (**b**) shows non-negligible frequency selectivity.

3.2. Signal Detection in UWA Environments

For the received signal described in Equation (1), using the minimum squared Euclidian distance is considered to detect the above symbol, the optimization problem can be expressed as

$$\hat{S}(n) = \underset{S^m(n), m \in [1, M]}{\arg\min} \ \|R(n) - H(n)S^m(n)\|^2, \tag{5}$$

where $\|\cdot\|$ represents 2-norm. $H(n)$ is the channel frequency response in any one OFDM symbol. M is the modulation order, and $S^m(n)$ is the m-th referred symbol. Further, the squared Euclidian distance can be written as

$$E_{Sk}^2 = \begin{cases} \|(H(n) - 1)S^m(n) + Z(n)\|^2 & k = m \\ \left\|H(n)S^k(n) - S^m(n) + Z(n)\right\|^2 & k \neq m \end{cases}. \tag{6}$$

Considering a system utilizing frequency domain equalization (FDE) for better performance, for the signal in Equation (6), a multiplier $G(n)$ is included in the detector, yielding

$$E_{Sk}^2 = \begin{cases} \|(H(n)G(n) - 1)S^m(n) + Z(n)G(n)\|^2 & k = m \\ \left\|H(n)G(n)S^k(n) - S^m(n) + Z(n)G(n)\right\|^2 & k \neq m \end{cases}. \tag{7}$$

Considering $G(n)$ as the entry of multiplier matrix **G**, for ZF equalizer,

$$\mathbf{G} = \left(\hat{\mathbf{H}}^{H}\hat{\mathbf{H}}\right)^{-1}\hat{\mathbf{H}}^{H}, \tag{8}$$

where $\hat{\mathbf{H}}$ is the estimated channel matrix. Multiplying **R** in Equation (2) with **G**, and expending the equation, the equalized symbol can be expressed as

$$\begin{bmatrix} \hat{s}_1 \\ \hat{s}_2 \\ \vdots \\ \hat{s}_N \end{bmatrix} = \begin{bmatrix} 1 & 0 & \cdots & 0 \\ 0 & 1 & \cdots & 0 \\ \vdots & \vdots & \ddots & \vdots \\ 0 & 0 & \cdots & 1 \end{bmatrix} \begin{bmatrix} s_1 \\ s_2 \\ \vdots \\ s_N \end{bmatrix} + \begin{bmatrix} g_{11} & g_{12} & \cdots & g_{1N} \\ g_{21} & g_{22} & \cdots & g_{2N} \\ \vdots & \vdots & \ddots & \vdots \\ g_{N1} & g_{N2} & \cdots & g_{NN} \end{bmatrix} \begin{bmatrix} z_1 \\ z_2 \\ \vdots \\ z_N \end{bmatrix}, \tag{9}$$

where g_{nn} is the element of **G**. It can be seen in Equation (10) that the equalized symbol includes amplified noise which will influence the detection. In the noiseless channel, the second term is zero.

For the MMSE equalizer, there is

$$\mathbf{G} = \hat{\mathbf{H}}^{\mathrm{H}} \left(\hat{\mathbf{H}}\hat{\mathbf{H}}^{\mathrm{H}} + \frac{\sigma^2}{P}\mathbf{I} \right)^{-1}, \tag{10}$$

where σ^2 and P are powers of noise and signal, respectively, and $\mathbf{I}$ is the identity matrix. With Equation (11), the MMSE-equalized symbol can be expressed as

$$\begin{bmatrix} \hat{s}_1 \\ \hat{s}_2 \\ \vdots \\ \hat{s}_N \end{bmatrix} = \begin{bmatrix} \frac{1}{|H_{11}|+\frac{\sigma_n^2}{\sigma_s^2}} & 0 & \cdots & 0 \\ 0 & \frac{1}{|H_{22}|+\frac{\sigma_n^2}{\sigma_s^2}} & \cdots & 0 \\ \vdots & \vdots & \ddots & \vdots \\ 0 & 0 & \cdots & \frac{1}{|H_{NN}|+\frac{\sigma_n^2}{\sigma_s^2}} \end{bmatrix} \left(\begin{bmatrix} s_1 \\ s_2 \\ \vdots \\ s_N \end{bmatrix} + \begin{bmatrix} g_{11} & g_{12} & \cdots & g_{1N} \\ g_{21} & g_{22} & \cdots & g_{2N} \\ \vdots & \vdots & \ddots & \vdots \\ g_{N1} & g_{N2} & \cdots & g_{NN} \end{bmatrix} \begin{bmatrix} n_1 \\ n_2 \\ \vdots \\ n_N \end{bmatrix} \right). \tag{11}$$

The first term of Equation (11) includes a factor matrix that only contains positive coefficients. As Figure 3 shows, for a single received point in the decision regions, the elements of this factor matrix linearly scale the received point to draw it closer to the reference point, which does not change its quadrant. Consequently, for the low-level constellations that can decide the symbols according to the quadrants they lie, such as BPSK and QPSK, the MMSE does not perform better than the ZF equalizer. In addition, the MMSE equalizer requires a priori SNR, which is difficult to obtain in time-varying UWA channels.

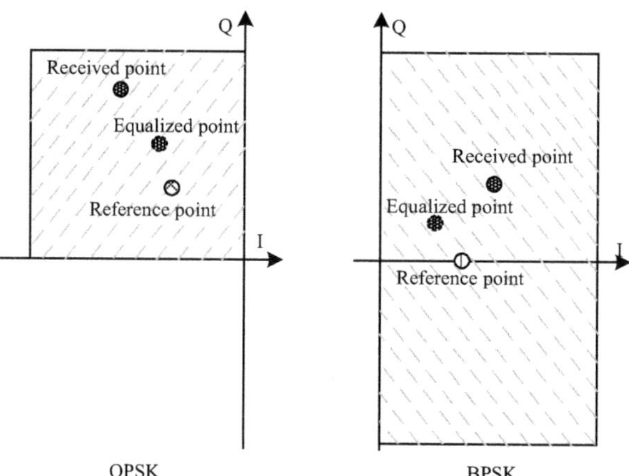

Figure 3. A decision region of the QPSK constellation.

When the frequency selectivity becomes strong, with imperfectly estimated channels, both ZF and MMSE equalizers cannot recover the symbols effectively [40]. However, it is still necessary to develop a more effective frequency domain equalizer in underwater acoustic channels because of the attractive low complexity. Variable NN structures provide new solutions to such interference elimination problems. The NN now has been proved to learn one or more nonlinear processes well with a proper structure [33]. With an intelligent interference simulating model [41], it is possible to train an equalization network offline, which can be applied independently online without extra a priori environment information.

4. NN-Based Joint Equalization and Detection

Unlike the frequency domain equalizers mentioned above, this paper combines equalization and symbol detection and implements a joint detector with the NN. The detected symbol can be written as

$$S_d(n) = \underset{D[\hat{H}(n),R(n)]}{\arg\min}\; L\{D[\hat{H}(n),R(n)],S(n)\}, \tag{12}$$

where $L\{\cdot\}$ is the loss function and $D[\cdot]$ represents the process of the proposed network.

It has been proved that a simple network structure is enough to well solve the receiver problems [33,42]. Inspired by this, the proposed joint detector utilizes a single-layered network. As shown in Figure 4, after channel estimation, $R(n)$ and $H(n)$ are sent to the network for detection. Moreover, a block-input strategy is proposed to further decrease the complexity of the network. N_b is the number of blocks and N_L denotes the number of neurons. The structure configurations of the network are described as follows.

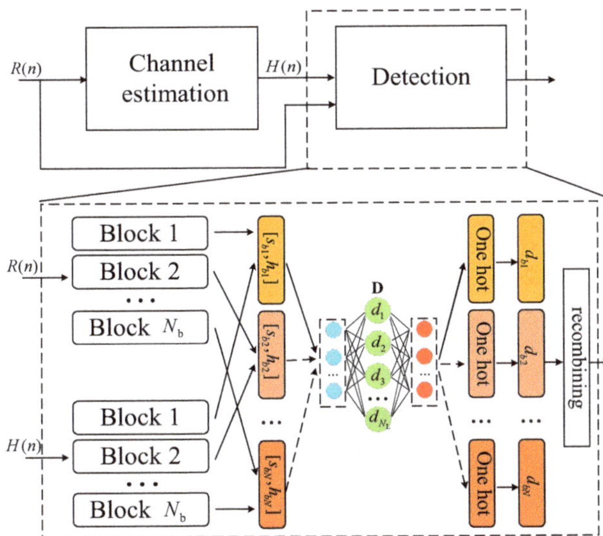

Figure 4. Block diagram of the proposed scheme on joint equalization and detection based on the NN. The detection process integrates the equalization and symbol demapping, which is completed by a low-complexity network, taking the segmented subcarriers as input.

Input: Before input to the network, the received OFDM symbol $R(n)$ with N subcarriers is firstly divided into N_b blocks. Each block contains N_c subcarriers. The same process is conducted with the corresponding estimated channel function $\hat{H}(n)$. Thus, a small reusable network can be designed for each combination of the data and channel block. To further determine the optimal hyperparameters, a proper N_c should be defined to balance the computation and accuracy of the network, i.e., two rules are proposed:

(1) N_c should optimize the performance of the joint detector with estimated channel.
(2)   The required number of neurons is positively correlated to the input dimension, and N_c should be as small as possible to minimize the computation of the network.

Based on the analysis above, the narrow coherence bandwidth W_c of the channel is considered, which always limits the performance of the underwater OFDM system. Because the subcarriers in each coherence frequency band suffer from relatively flat fading, an opportunity is found to find the optimal N_c [43]. By setting N_c in the range of coherence bandwidth, the joint detection network is able to deal with the symbol blocks separately in a flat fading channel. Figure 5a shows the frequency response of a simple channel with one path. A_1 to D_3 are values of the frequency points. According to [43], for a channel that follows the homogeneous assumption, the coherence bandwidth is inversely proportional to the maximum channel delay T_d, which can be approximately represented as

$$W_c \approx \frac{1}{T_d}. \tag{13}$$

For the channel in Figure 5a, W_c is easily observed as the frequency range between B_2 and D_3, written as

$$W_c = f_{D_3} - f_{B_2}. \tag{14}$$

In practice, the UWA channel contains more paths as shown in Figure 5b, whose frequency response is the sum of more than one paths with time-varying coefficients distributions [39]. With the same estimated W_c as Figure 5a, different situations are listed in Table 1 by taking different sections as blocks. It should be noted that the starting point of the first block should always be A_1, which is also the first subcarrier of an OFDM symbol. When the width of the block $W_b = W_c$, block $[A_1, C_2]$ suffers from selective fading because the section includes inflection points of the frequency response. The same result occurs when $W_b = W_c/2$ because $[B_1, B_3]$ includes an inflection point. When $W_b = W_c/3$, no sections include an inflection point or experience flat fading. This consistency of flat fading is more conducive to the symbol detection [44].

According to the discussion above, there is a W_b in the range $(0, W_c]$, which decides the optimal N_c, yielding

$$W_b = N_c \Delta f, \tag{15}$$

where Δf is the frequency interval of two contiguous subcarriers. The NN with input length N_c deals with blocks that all suffer from flat fading. Therefore, although the accurate W_c is difficult to find, its estimate can be an upper bound for finding the optimal input dimension for the network.

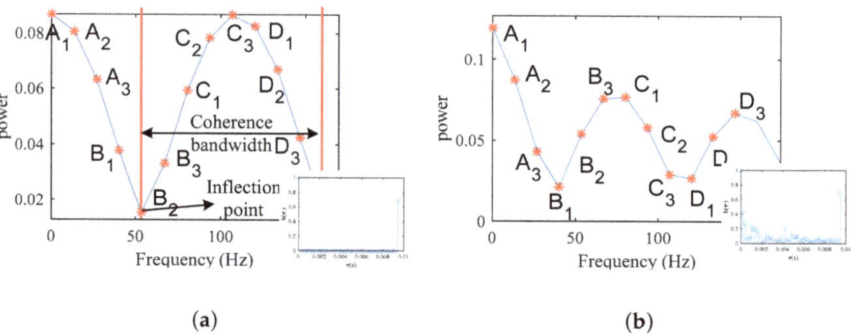

Figure 5. Time and frequency response of a simple and a complex channel: (**a**) is the response of the channel, which contains the main path and one reflected path. (**b**) is the response of a complex channel, which includes many other paths besides the paths in (**a**).

Consequently, the input block size N_c can be quickly found by going through the coherence bandwidth of the estimated channel. Algorithm 1 shows the steps to find N_c. N_{W_c} represents the number of subcarriers contained in the range of the empirical coherence bandwidth. The threshold C_{Th} is used to filter the paths with low power which does not affect the performance of the system. $\mathbf{I}_{Th}$ contains the indexes of the recognizable paths. $\mathrm{BER}(\cdot)$ represents the process to calculate the BER. Because there is only one optimal N_c

for each W_c that varies slowly in a short duration in fixed locations, steps 7 to 13 can be conducted independently with offline training.

Table 1. Fading situations of different sections.

Block	Type of Fading	Width of a Block
$[A_1, C_2]$	Selective fading	W_c
$[A_1, B_1], [B_2, C_2], [C_3, D_3]$	Flat fading	$2W_c/3$
$[A_1, A_3], [C_1, C_3], [D_1, D_3]$	Flat fading	$W_c/2$
$[B_1, B_3]$	Selective fading	$W_c/2$
$[A_1, A_2], [A_3, B_1]$	Flat fading	$W_c/3$
$[B_2, B_3], [C_1, C_2]$	Flat fading	$W_c/3$
$[C_3, D_1], [D_2, D_3]$	Flat fading	$W_c/3$

Algorithm 1 Finding the optimal input block length according to the estimated channel

---

Require: The estimated channel response $H(n)$;
Ensure: Power threshold C_{Th}, Bandwidth of an OFDM symbol W, Training symbol matrix $\{\mathbf{S}_t,$ Channel matrix $[\mathbf{H}_t], \mathbf{b}_t\}, N, P_b = 0.5$;
1: $h(n) = \text{IFFT}[H(n)]$;
2: $\mathbf{h}_{Th}, \mathbf{I}_{Th} = \text{find}(h(n) > C_{Th})$;
3: $T_d = \max(\mathbf{I}_{Th})$;
4: $W_c = 1/T_d$;
5: $N_{w_c} = NW_c/W$;
6: $N_c = N_{w_c}$;
7: While $N_c \geq 1$
8: $\mathbf{b}_{tn} = D[\mathbf{S}_t]$ with N_c as block length;
9: $P_{bn} = \text{BER}(\mathbf{b}_{tn}, \mathbf{b}_n)$;
10: if $P_{bn} < P_b$
11: $P_b = P_{bn}$;
12: end if
13: $N_c = N_c - 1$;
14: **return** N_c;

After N_c is decided, the input matrix is the combination of the symbol block and channel block. To input these complex symbols and channels to the real-value NN, the real and imaginary parts of the symbol and channel blocks are extracted and rearranged, which can be written as

$$
\mathbf{X} = \begin{bmatrix}
\text{Re}\{R_b^1(1)\} & \text{Im}\{R_b^1(1)\}... & \text{Re}\{R_b^1(N_c)\} & \text{Im}\{R_b^1(N_c)\} \\
& \vdots & & \vdots \\
\text{Re}\{R_b^B(1)\} & \text{Im}\{R_b^B(1)\}... & \text{Re}\{R_b^B(N_c)\} & \text{Im}\{R_b^B(N_c)\} \\
\text{Re}\{H_b^1(1)\} & \text{Im}\{H_b^1(1)\} ... & \text{Re}\{H_b^1(N_c)\} & \text{Im}\{H_b^1(N_c)\} \\
& \vdots & & \vdots \\
\text{Re}\{H_b^B(1)\} & \text{Im}\{H_b^B(1)\} ... & \text{Re}\{H_b^B(N_c)\} & \text{Im}\{H_b^B(N_c)\}
\end{bmatrix} ,
$$

$$
= \begin{bmatrix} \mathbf{R}_b^{ri}(i) & \hat{\mathbf{H}}_b^{ri}(i) \end{bmatrix} \tag{16}
$$

where $R_b^k(n)$ and $H_b^k(n)$ are the n-th received symbol block and estimated channel block of k-th training batch. B is the batch size. Since $H(n)$ is input directly, the network does not need to learn the changing channel characteristics [25]. Instead, a generative analytical process can be learned to be adaptive to any kind of channel.

Network configurations: As Figure 4 shows, the single-layered network proposed in this paper contains a fully connected layer with N_L neurons. To detect a symbol block, the input firstly multiples a weight vector $\mathbf{W}_L$, then adds bias vectors $\mathbf{B}_L$, yielding

$$
\mathbf{Y}_m = g(\mathbf{X}^H \mathbf{W}_L + \mathbf{B}_L), \tag{17}
$$

where $g(\cdot)$ is the activation function. To transform $\mathbf{Y}_{\mathrm{m}}$ to the output with required length, another linear map is built:

$$\mathbf{Y} = \mathbf{Y}_{\mathrm{m}}{}^{\mathrm{H}}\mathbf{W}_{\mathrm{out}} + \mathbf{B}_{\mathrm{out}}, \tag{18}$$

where $\mathbf{W}_{\mathrm{out}}$ and $\mathbf{B}_{\mathrm{out}}$ are weights and bias of the output layer. $\mathbf{Y}$ contains N_{c} symbols, which are further input with the one-hot format reference symbols to calculate the cross entropy as the loss function, yielding

$$L_D = -\sum_{i}^{B}\sum_{j}^{M} S_{ij}^{onehot}(n) \log\left\{ \mathrm{softmax}\left[Y_{ij}^{\mathrm{s}}(n) \right] \right\}, \tag{19}$$

where $S_{ij}^{onehot}(n)$ and $Y_{ij}^{\mathrm{s}}(n)$ are the one-hot reference symbol and the corresponding output data. M is the modulation order; $\mathrm{softmax}[\cdot]$ is a function to map $Y_{ij}^{\mathrm{s}}(n)$ to the range $(0,1)$ [28]. The use of cross entropy can make the network converge quickly.

Training strategy: For each N_{c}, there are two tasks for the network. One is to find the optimal N_{L} and another is to train the parameter matrix $\mathbf{D}$. Both tasks can be finished offline with one training. Figure 6 shows the training strategy of the network. The samples are constructed with three parts: random data symbols, different types of channels and noise. The channel samples contains Rayleigh distributed ones and underwater acoustic channels collected in the pool, artificial lake, and Wuyuanwan Bay, Xiamen, while the noise includes Gaussian distributed noise and impulsive noise generated by the GAN in [41].

The network for the given input dimension is trained with the range $[N_{\mathrm{bottom}}, N_{\mathrm{up}}]$ processed with a step to find the optimal number of neurons N_{Lopt}. N_{bottom} and N_{up} are lower and upper bound of the possible N_{Lopt}, respectively. Meanwhile, the optimal weight matrix $\mathbf{D}_{\mathrm{opt}}$ trained with N_{Lopt} is memorized. Then both $\mathbf{D}_{\mathrm{opt}}$ and N_{L} are delivered to the online network to detect the real received symbols.

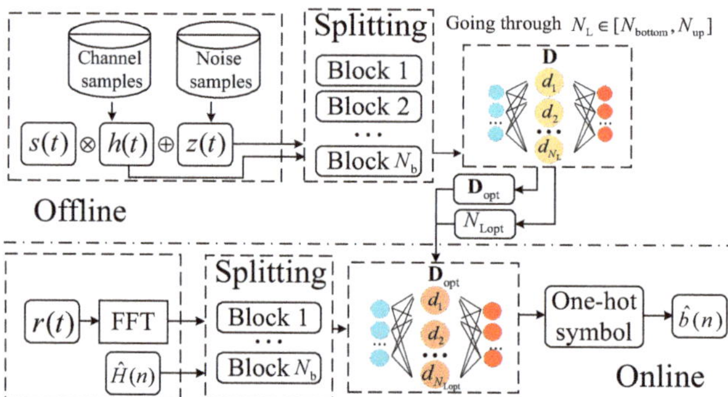

Figure 6. Block diagram of training strategy. All networks are only trained once offline and tested with fixed hyperparameters online in the simulations and sea trials.

Online applications: Unlike the NN-based detector, which takes the only the received symbol as the input, the estimated channel is included in the input in the proposed detector. This configuration offers more information for the network to detect symbols, determining a low-complexity semi-analytical detection network, which is generalized to different channels. Therefore, the final goal of this study was to train the networks which could be independently applied online without retraining. The simulations and experiments are described and discussed below to demonstrate the performance of the proposed structure.

5. Complexity Analysis

To analyze the time-complexity of the proposed NN-based detector, it should be firstly noted that in the following simulations and sea trials, the networks were all trained once in advance and were not retrained online. All parameters of the networks was fixed when conducting simulations and experiments. Thus the large batch size of the dataset was not included in the computation. Hence, only N_L and N were considered. The computations of both the LS estimator and minimum distance detector were included for the calculation of complexity of other equalizers. Consider the following LS channel estimator

$$\hat{h}(t) = \frac{r(t)}{s(t)}. \tag{20}$$

The number of multiply-accumulate operations (MACC) of the above process is the number of subcarriers N. For the minimum distance detector, each symbol experiences M times of complex operations, which contains $3MN + 5N$ MACC.

As a similar implementation of the NN-based equalizer, the ELM-based detector in [34] was considered for performance comparisons. According to [34], the number of MACC of the ELM-based detector was derived and listed in Table 2. Ignoring the addition with constant terms, the time complexities of different detectors are compared in Table 2. It can be seen that the computation of the proposed NN-based detectors is less than that of MMSE for 1 order of magnitude. The complexity of the ELM-based detector is the highest, which is 2 orders of magnitude larger than the proposed NN-based detector. Although the ZF equalizer shows the lowest complexity, it has been proved to be a suboptimal algorithm in noisy channels, which could be substituted by more advanced methods.

Table 2. Time complexity of different detectors.

Detector	Number of MACC	Time Complexity
ELM	$N(\frac{N_L^3}{3} + 2N_L{}^2 + \frac{N_L{}^2}{2} + 8N_L + \frac{5N_L}{6} + 6 + 3M)$	$O(N^4)$
ZF	$7N + 3MN$	$O(N)$
MMSE	$3(N^3 + 3N^2 + 2N + MN)$	$O(N^3)$
Proposed NN	$(2 + M)NN_L + 3N$	$O(N^3)O(NN_L)$

6. Numerical Simulations

The configurations of the simulated system are listed in Table 3. An OFDM system with the bandwidth 5000 Hz was built. BPSK and QPSK were chosen as the digital modulations. The number of subcarriers, which was 384, should be divisible by N_c, whereas several possible N_cs were chosen to build the networks.

Table 3. Parameters of the simulated OFDM system.

Item		Value
Bandwidth	B	5000 Hz
Modulation	$-$	QPSK/BPSK
Number of subcarriers	N	384
Block size	N_c	1/2/3/4/6/8/12/16
Proportion of training/testing set	$-$	3/1

Before comparing the performance, the network was firstly trained with mixed samples. As mentioned in Section 4, besides Rayleigh channels generated with MATLAB, the real channels collected in the pool, artificial lake, and Wuyuanwan Bay, Xiamen, were taken as samples. All channel samples were mixed randomly in the proportion 1:1:1:1. For Rayleigh channel samples, the maximum CFO was set to 100 Hz. Moreover, Figure 7a–c shows the real scenarios to collect channels. The average depths of the pool, artificial lake, and testing sea area were 1 m, 5 m, and 8 m, while the depths of the transmitter and the receiver in the three areas were 0.5 m, 0.8 m, and 1.5 m. The only factor which affected the pool channel was the hard wall and bottom made of tiles. In addition, the average wind power in the artificial lake and testing sea area was <level 3. The outdoor tests were all performed in sunny days.

Different distances were covered to obtain different maximum delays shown in Figure 7d–f. Furthermore, the power threshold C_{Th} was set to 0.01 to filter the paths with low power.

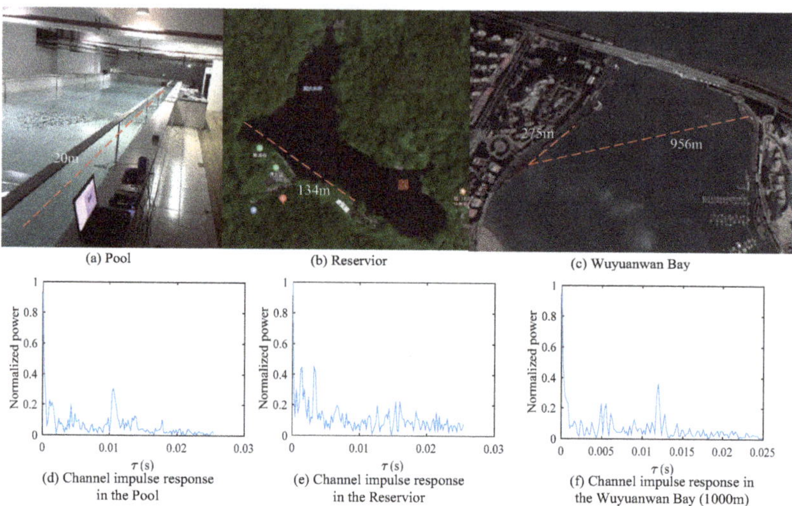

Figure 7. Scenes and their channel impulsive response: (**a**,**b**) are the scenes of pool and artificial lake in Xiamen University, while (**c**) is the testing area in Wuyuanwan Bay. (**d**–**f**) are three of the CIRs caught in these spots, which show large delays.

Taking mixture of the AWGN and impulsive noise generated with the GAN in [41] as noise samples, the dataset was finally constructed. Table 4 shows the training parameters for the network. The networks were trained with dynamic SNR and mean square errors (MSE) of the estimated channels. To accelerate convergence, ReLU was taken as the activation function, along with the Adam optimizer. The whole training process was conducted in Python with TensorFlow.

Table 4. Parameters for network training.

Item	Value
Input size	$2N_c$
Number of neurons	N_L
output size	MN_c
Activation function	ReLU
Optimizer	Adam
SNR for training (dB)	[20, 30]
Predetermined channel estimation MSE	[0, 0.03]
Epoch	300
learning rate	0.001
Platform	Python with TensorFlow

Figure 8 shows the BERs of the given N_c changing with N_L. The SNR was 25 dB and the assumed MSE of the channel estimated was 0.004. The step size for N_L was 4. It can be seen that large input dimensions, such as $N_c = 12$ and $N_c = 16$, required more neurons to reach the best performance. In addition, N_c for QPSK was larger than that of BPSK. To be clear, the estimated N_{Lopt}s from Figure 8 are listed in Table 5. For each N_c, QPSK needed at least 20 more neurons than BPSK.

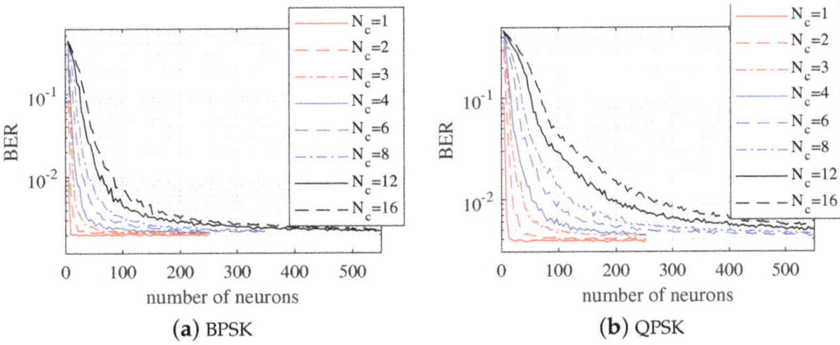

(a) BPSK   (b) QPSK

Figure 8. BER of different N_c varying with N_L.

Table 5. N_{Lopt} for different N_cs.

N_c	1	2	3	4	6	8	12	16
N_{Lopt} for BPSK	28	64	120	181	220	280	400	488
N_{Lopt} for QPSK	52	80	148	200	240	300	420	516

It should be noted that the optimal N_c of both modulations in Figure 8 is 1. This is because the training set contained multiple CIRs, including Rayleigh channels, with different delays. In this situation, the coherence bandwidth of the channel samples was limited to a small value. To further demonstrate the influence of the coherence bandwidth, the trained networks were further used to detect the symbol from the channels with specific maximum delays. The pool channels in Figure 7a were cut off with lengths of 48, 64, and 77 points, the corresponding N_{w_c}s of which were 8, 6, and 5. Figure 9a–f show the BERs with different N_c in BPSK-OFDM and QPSK-OFDM systems. The BER curves of $SNR = [15, 25]$ dB are enlarged in Figure 9b,d,f. The MSE of the channels estimated was set to 0.01. It can be seen from the figures that the optimal N_c varies with modulation and N_{w_c}. According to Algorithm 1 , the optimal N_cs are listed in Table 6. It can be found that in all situations, the networks with $N_c > N_{w_c}$ showed poor performance. These results prove the analysis in Section 4, and further demonstrate the feasibility of Algorithm 1.

Table 6. The optimal N_c for different systems.

	BPSK	QPSK
$N_{w_c} = 8$	2	2
$N_{w_c} = 6$	1	3
$N_{w_c} = 5$	4	3

In addition, the networks with optimal N_cs were compared with the ZF and MMSE equalizers. The pool channel with $N_{w_c} = 8$ was used. Figure 10 shows the BERs of different equalizers. The proposed network showed lower BERs than the ZF and MMSE equalizers both under perfect channel estimation (MSE = 0) and MSE = 0.01 of the estimated channel. In particular, when SNR = 20 dB and MSE = 0, the BER of QPSK detection network was 18.66% lower than that of the ZF and MMSE equalizers, while the BER of BPSK detection network was 14.23% lower than that of the ZF and MMSE equalizers. When SNR = 20 dB and MSE = 0.01, the BER of QPSK detection network was 29.26% lower than that of the ZF and MMSE equalizers and the BER of BPSK detection network was 22.16% lower than that of the ZF and MMSE equalizers. In addition, it can be seen from Figure 10 that the BER curve of MMSE equalizer for each estimation error is almost the same as the ZF equalizer, which confirms the discussion in Section 2.

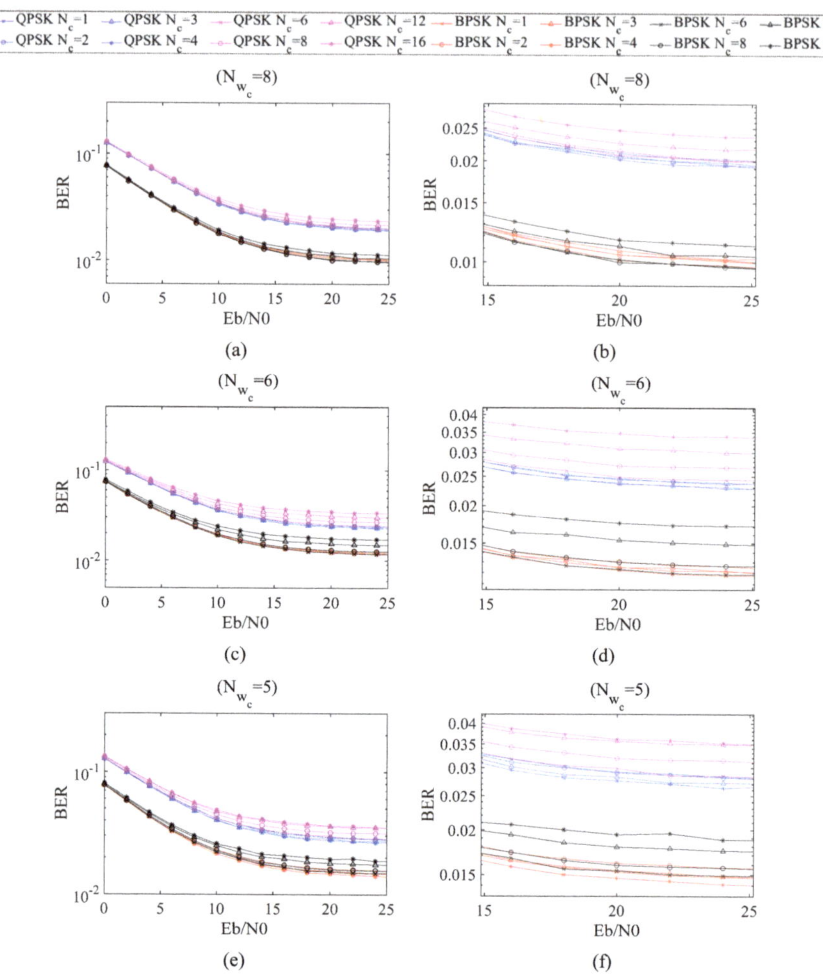

Figure 9. BER of different N_c in channels with specific maximum delays. (**a**,**c**,**e**) are complete BER curves of different N_{w_c}s. The BER in the range of [15, 20] dB are enlarged in (**b**,**d**,**f**), respectively.

Further, the above equalizers were tested with the LS estimation. The ELM-based detector in [35] was also compared. All receivers apply the minimum Euclidean distance method to detect symbols. Figure 11 shows the BER for these detectors. It can be seen that the ELM-based detector showed the worst performance in pool channels. The NN-based receiver still showed low BERs. When SNR = 20 dB, the BERs of NN were lower than ZF and MMSE equalizers for 25.92% and 30.99% under BPSK and QPSK modulations.

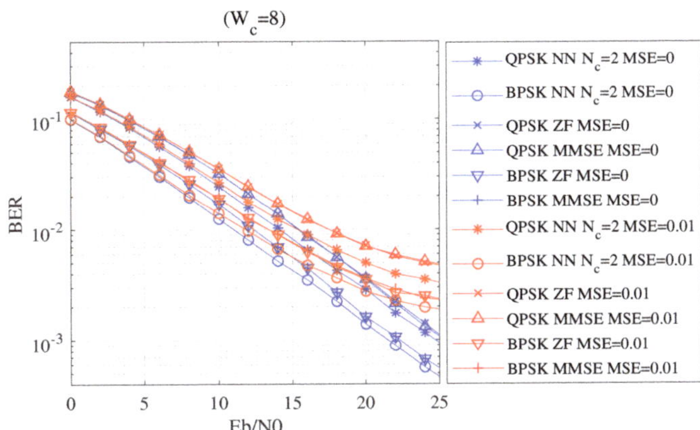

Figure 10. BER of different detectors with given MSE of channel estimations.

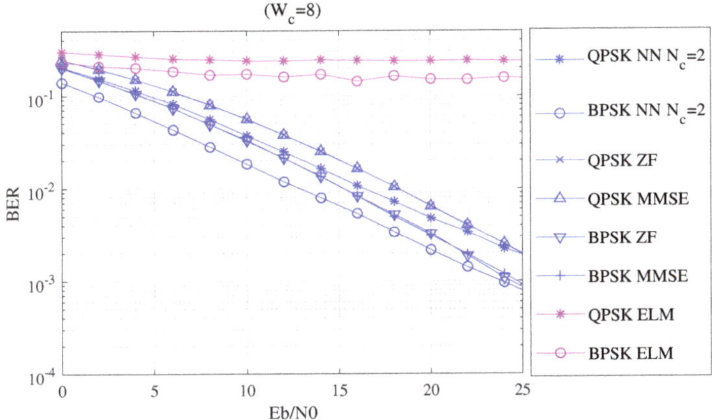

Figure 11. BERs of different detectors with the LS channel estimator. The BER curves of the ZF equalizer are almost overlapped by that of the MMSE equalizers, while the NN shows the lowest BER both with BPSK and QPSK. The ELM-based equalizer shows highest BER.

The above results have proved the efficiency of the proposed NN-based detector. Furthermore, the underwater trials introduced in Section 6 show the practicability of the proposed detector.

7. Underwater Trials

The above systems were first tested in the pool in Figure 7. The signals were transmitted with the carrier frequency $F_c = 12$ kHz and the sampling frequency F_s was 100 kHz. The parameters shown in Table 3 were taken to generate baseband signals. The block-type pilots were used for estimation of the time-invariant channels, and the length of cyclic prefixes was 1/3 of the symbol length. Table 7 lists the BERs of different receivers. It should be noted that not all the NN-based detectors were retrained, which was the same as the trained networks in the last section. The optimal N_cs for the NN was found to be 2. This was because the hard wall of the pool caused strong reflections that resulted in a small coherence bandwidth. By adjusting the transmitting power, two groups of signals were tested with SNR = 30 dB and 5 dB. The ELM-based receiver still showed the worst performance. It could also be found that the proposed NN showed low BERs in all conditions, while the

ZF and MMSE equalizers showed relatively high BERs. Compared with the simulations, the gap of the performance between the proposed NN and other equalizers became smaller. This was because the delays of the real channels were much longer than the simulated ones, which caused intersymbol interference (ISI) besides intra-symbol interference. Because the NN and traditional equalizers are only designed to eliminate the interference in each symbol, the influence of ISI could not be well equalized.

Table 7. BER of systems tested in the pool.

	SNR = 30 dB		SNR = 5 dB	
	BPSK	**QPSK**	**BPSK**	**QPSK**
Proposed NN	0.00206	0.0073	0.0193	0.0276
ZF	0.00209	0.0074	0.0205	0.0295
MMSE	0.00208	0.0074	0.0206	0.0293
ELM	0.1385	0.4023	0.1431	0.4051

Further, the receivers were tested in real sea. The spot chosen was Xiamen Bay near the location in Figure 7c and the communication distance was 1 km. In addition to the block-type pilots, the comb-type pilots were applied to make more accurate estimations, for the channels were time-variant in the shallow sea. In addition, the LS estimator was used for both pilot types. The SNR was controlled intentionally to 4.5 dB for comparison of the performances in hostile environments, leading to the received signals in Figure 12b. The impulse interference can be observed, which severely affects the performance of the receiver.

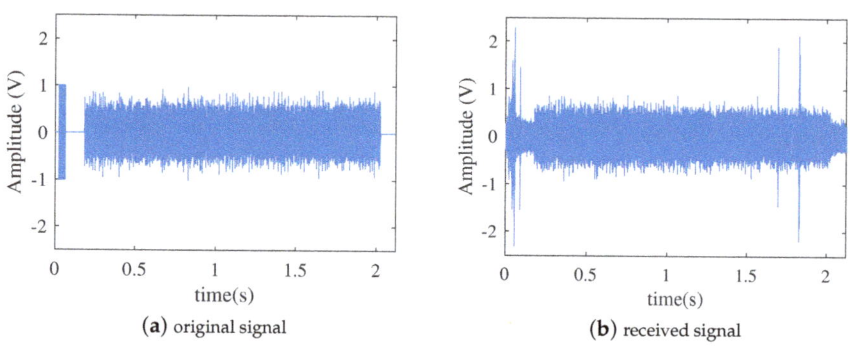

(**a**) original signal

(**b**) received signal

Figure 12. The original and received signals in time domain. Compared with (**a**), the signal in (**b**) suffers from impulses as well as fading.

Table 8 lists the BERs for block-type pilots, $N_c = 16$, and for comb-type pilots, $N_c = 4$. It can be seen that the received symbols with block-type pilots can hardly be detected with all detectors, although the proposed NN performs best. With comb-type pilots, the results are better. Since the ELM-based detector can only use the block-type pilots, the BERs with comb-type ones are ignored. Nevertheless, the BERs of the NN-based detector are the lowest among all equalizers. These results show the generation of the proposed NN-based detector in different UWA environments.

Table 8. BER of systems tested in the shallow sea.

	Block-Type Pilot		Comb-Type Pilot	
	BPSK	**QPSK**	**BPSK**	**QPSK**
Proposed NN	0.1602	0.2902	0.0970	0.1196
ZF	0.1867	0.3009	0.0988	0.1250
MMSE	0.1865	0.3010	0.0988	0.1249
ELM	0.4465	0.4905	-	-

It can be seen from Table 8 that compared with the ZF and MMSE equalizers, the proposed NN-based detector shows less difference than Figure 11. This is because the signals transmitted in sea trials experience more complex interferences caused by the time-variant sound speed field and noise distributions. In this situation, the signals suffer from more distortions than simulations, such as inter-carrier interference (ICI), which should be reduced by other algorithms.

To show the generation of the NN-based detector, the networks were tested with the data transmitted in Xiamen Bay in June 2018. The signals were transmitted at 500 m with level 3 sea conditions. A pair of NI-6341 data acquisition cards were connected with 30 kHz transducers as the transmitter and the receiver, shown in Figure 12b. The data were modulated by QPSK and the numbers of subcarriers were set as 128 and 512. A comb-type pilot was applied for the LS channel estimation. The NN-based detector was applied to replace the original MMSE equalizer and Euclidean distance detector. The SNR at the receiver was estimated as 32 dB. Table 9 shows the BERs of different detectors. Both detectors could detect the symbols well and the NN-based detector outperformed the original detector with both 128 and 512 subcarriers.

Table 9. BERs of different detectors with data in June 2018.

Number of Subcarriers	Proposed NN	MMSE
128	0.006185	0.006673
512	0.007568	0.007894

8. Conclusions

In this paper, a low-complexity NN-based detector has been proposed to be implemented in an OFDM system. The segmented channel responses and received symbols are input to the single-layered network, which directly outputs the detected symbols. By associating the network with the assist of coherence bandwidth of the estimated channel, an algorithm is built to find optimal hyperparameters. The networks are all trained offline, which are applied for both simulations and sea trials with fixed parameters. The quantitative simulations have compared the ZF, MMSE equalizers, and the ELM-based detector with the proposed NN-based detector, and the results show that the proposed detector reaches the lowest BER in the tested UWA channels. The same results can be found in the sea trials. With the best performance among the tested equalizers and detectors, the proposed detector has lower computational complexity than MMSE and ELM-based detectors. The proposed NN-based detector needs accurate estimated channels for better performance, which leads future research to focus on the construction of the optimization of the channel estimator.

Author Contributions: Conceptualization, M.Z. and J.W.; methodology, M.Z. and H.S.; software, M.Z.; validation, M.Z., X.F. and R.L.; formal analysis, M.Z.; investigation, M.Z.; resources, M.Z.; data curation, H.S.; writing—original draft preparation, M.Z.; writing—review and editing, J.W. and J.Q.; visualization, M.Z.; supervision, H.S.; project administration, H.S.; funding acquisition, H.S. All authors have read and agreed to the published version of the manuscript.

Funding: This research was funded by the Key Program of Marine Economy Development Special Foundation of Department of Natural Resources of Guangdong Province (GDNRC [2023]24), the Key Laboratory of Southeast Coast Marine Information Intelligent Perception and Application, MNR, NO. 220202, National Natural Science Foundation of China (NSFC) 62271426 and the Natural Resources Science and Technology Innovation Project of Fujian KY-080000-04-2022-025.

Data Availability Statement: Not applicable.

Conflicts of Interest: The authors declare no conflict of interest.

References

1.   Zhang, X.; Yang, P.; Zhou, M. Multireceiver SAS Imagery with Generalized PCA. *IEEE Geosci. Remote Sens. Lett.* **2023**, *20*, 1502205. [CrossRef]
2.   Zhang, X.; Wu, H.; Sun, H.; Ying, W. Multireceiver SAS Imagery Based on Monostatic Conversion. *IEEE J. Sel. Top. Appl. Earth Obs. Remote Sens.* **2021**, *14*, 10835–10853. [CrossRef]
3.   He, Y.; Han, G.; Tang, Z.; Martínez-García, M.; Peng, Y. State Prediction-Based Data Collection Algorithm in Underwater Acoustic Sensor Networks. *IEEE Trans. Wirel. Commun.* **2022**, *21*, 2830–2842. [CrossRef]
4.   Zhou, S.; Wang, Z. *OFDM for Underwater Acoustic Communications*; John Wiley & Sons.: Hoboken, NJ, USA, 2014.
5.   Li, J.; Bai, Y.; Zhang, Y.; Qu, F.; Wei, Y.; Wang, J. Cross power spectral density based beamforming for underwater acoustic communications. *Ocean. Eng.* **2020**, *216*, 107786. [CrossRef]
6.   Gul, S.; Zaidi, S.S.H.; Khan, R.; Wala, A.B. Underwater acoustic channel modeling using BELLHOP ray tracing method. In Proceedings of the 2017 14th International Bhurban Conference on Applied Sciences and Technology (IBCAST), Islamabad, Pakistan, 10–14 January 2017; pp. 665–670.
7.   Shehwar, D.E.; Gul, S.; Zafar, M.U.; Shaukat, U.; Syed, A.H.; Zaidi, S.S.H. Acoustic Wave Analysis In Deep Sea And Shallow Water Using BELLHOP Tool. In Proceedings of the 2021 OES China Ocean Acoustics (COA), Harbin, China, 14 July 2021; pp. 331–334.
8.   Stojanovic, M. OFDM for underwater acoustic communications: Adaptive synchronization and sparse channel estimation. In Proceedings of the 2008 IEEE International Conference on Acoustics, Speech and Signal Processing, Las Vegas, NV, USA, 31 March–4 April 2008; pp. 5288–5291.
9.   Ur Rehman Junejo, N.; Esmaiel, H.; Zhou, M.; Sun, H.; Qi, J.; Wang, J. Sparse Channel Estimation of Underwater TDS-OFDM System Using Look-Ahead Backtracking Orthogonal Matching Pursuit. *IEEE Access* **2018**, *6*, 74389–74399. [CrossRef]
10. Ma, X.; Zhao, C.; Qiao, G. The Underwater Acoustic OFDM Channel Equalizer Basing On Least Mean Square Adaptive Algorithm. In Proceedings of the 2008 IEEE International Symposium on Knowledge Acquisition and Modeling Workshop, Wuhan, China, 21–22 December 2008; pp. 1052–1055.
11. Zhao, S.; Yan, S.; Xi, J. Adaptive Turbo Equalization for Differential OFDM Systems in Underwater Acoustic Communications. *IEEE Trans. Veh. Technol.* **2020**, *69*, 13937–13941. [CrossRef]
12. Yang, G.; Wang, L.; Qiao, P.; Liang, J.; Chen, T. Joint Multiple Turbo Equalization for Harsh Time-Varying Underwater Acoustic Channels. *IEEE Access* **2021**, *9*, 82364–82372. [CrossRef]
13. Song, J.; Huang, S.; Wang, J.; Zhang, C.; Wang, J. The Noise Transfer Analysis in Frequency Domain Zero-Forcing Equalization. *IEEE Trans. Commun.* **2013**, *61*, 1–12. [CrossRef]
14. Kang, S.W.; Imn, S.B.; Choi, H.J. Frequency Domain MMSE Equalization with Moving FFT for MBOK DS-UWB System. In Proceedings of the 2006 International Conference on Software in Telecommunications and Computer Networks, Split, Croatia, 29 September–1 October 2006; pp. 286–290.
15. Nelson, J.; Singer, A.; Koetter, R. Linear turbo equalization for parallel ISI channels. *IEEE Trans. Commun.* **2003**, *51*, 860–864. [CrossRef]
16. Nakamura, Y.; Ueda, J.; Okamoto, Y.; Osawa, H.; Muraoka, H. Nonbinary LDPC Coding System With Symbol-By-Symbol Turbo Equalizer for Shingled Magnetic Recording. *IEEE Trans. Magn.* **2013**, *49*, 3791–3794. [CrossRef]
17. Zhang, Y.; Xie, L.; Chen, H.; Cui, J.H. On the use of sliding LT code in underwater acoustic real-time data transfer with high propagation latency. In Proceedings of the 2014 Oceans-St. John's, St. John's, NL, Canada, 14–19 September 2014; pp. 1–5.
18. Demirors, E.; Sklivanitis, G.; Santagati, G.E.; Melodia, T.; Batalama, S.N. A High-Rate Software-Defined Underwater Acoustic Modem With Real-Time Adaptation Capabilities. *IEEE Access* **2018**, *6*, 18602–18615. [CrossRef]
19. Albarakati, H.; Ammar, R.; Elfouly, R.; Rajasekaran, S. Real-Time Decision Making for Underwater Big Data Applications Using the Apriori Algorithm. In Proceedings of the 2019 IEEE Symposium on Computers and Communications (ISCC), Barcelona, Spain, 29 June–3 July 2019; pp. 1–7.
20. Abraham, D.A., Underwater Acoustic Signal and Noise Modeling. In *Underwater Acoustic Signal Processing: Modeling, Detection, and Estimation*; Springer International Publishing: Cham, Switzerland, 2019; pp. 349–456.
21. Xie, Z.; Xu, Z.; Han, S.; Zhu, J.; Huang, X. Modulus Constrained Minimax Radar Code Design Against Target Interpulse Fluctuation. *IEEE Trans. Veh. Technol.* **2023**, 1–6. [CrossRef]
22. Ye, H.; Li, G.Y.; Juang, B.H. Power of deep learning for channel estimation and signal detection in OFDM systems. *IEEE Wirel. Commun. Lett.* **2017**, *7*, 114–117. [CrossRef]

23. Zhao, H.; Yang, C.; Xu, Y.; Ji, F.; Wen, M.; Chen, Y. Model-Driven Based Deep Unfolding Equalizer for Underwater Acoustic OFDM Communications. *IEEE Trans. Veh. Technol.* **2023**, *72*, 6056–6067. [CrossRef]
24. Ye, H.; Li, G.Y.; Juang, B.H.F. Deep learning based End-to-End wireless communication systems without pilots. *IEEE Trans. Cogn. Commun. Netw.* **2021**, *7*, 702–714. [CrossRef]
25. Van Luong, T.; Ko, Y.; Matthaiou, M.; Vien, N.A.; Le, M.T.; Ngo, V.D. Deep learning-aided multicarrier systems. *IEEE Trans. Wirel. Commun.* **2020**, *20*, 2109–2119. [CrossRef]
26. Wang, Z.; Xu, Z.; He, J.; Delingette, H.; Fan, J. Long Short-Term Memory Neural Equalizer. *IEEE Trans. Signal Power Integr.* **2023**, *2*, 13–22. [CrossRef]
27. Gao, X.; Jin, S.; Wen, C.K.; Li, G.Y. ComNet: Combination of Deep Learning and Expert Knowledge in OFDM Receivers. *IEEE Commun. Lett.* **2018**, *22*, 2627–2630. [CrossRef]
28. Lin, B.; Wang, X.; Yuan, W.; Wu, N. A novel OFDM autoencoder featuring CNN-based channel estimation for internet of vessels. *IEEE Internet Things J.* **2020**, *7*, 7601–7611. [CrossRef]
29. Zhao, H.; Ji, F.; Wen, M.; Yu, H.; Guan, Q. Multi-task learning based underwater acoustic OFDM communications. In Proceedings of the 2021 IEEE International Conference on Signal Processing, Communications and Computing (ICSPCC), Xi'an, China, 17–19 August 2021; IEEE: Piscataway, NJ, USA, 2021; pp. 1–5.
30. Liu, J.; Ji, F.; Zhao, H.; Li, J.; Wen, M. CNN-based underwater acoustic OFDM communications over doubly-selective channels. In Proceedings of the 2021 IEEE 94th Vehicular Technology Conference (VTC2021-Fall), Virtual, 27 September–28 October 2021; IEEE: Piscataway, NJ, USA, 2021; pp. 1–6.
31. Zhang, Y.; Chang, J.; Liu, Y.; Xing, L.; Shen, X. Deep learning and expert knowledge based underwater acoustic OFDM receiver. *Phys. Commun.* **2023**, *58*, 102041. [CrossRef]
32. Turhan, M.; Öztürk, E.; Çırpan, H.A. Deep convolutional learning-aided detector for generalized frequency division multiplexing with index modulation. In Proceedings of the 2019 IEEE 30th Annual International Symposium on Personal, Indoor and Mobile Radio Communications (PIMRC), Istanbul, Turkey, 8–11 September 2012; IEEE: Piscataway, NJ, USA, 2019; pp. 1–6.
33. Wang, T.; Yang, F.; Song, J.; Han, Z. Deep convolutional neural network-based detector for index modulation. *IEEE Wirel. Commun. Lett.* **2020**, *9*, 1705–1709. [CrossRef]
34. Liu, J.; Mei, K.; Zhang, X.; Ma, D.; Wei, J. Online extreme learning machine-based channel estimation and equalization for OFDM systems. *IEEE Commun. Lett.* **2019**, *23*, 1276–1279. [CrossRef]
35. Yang, L.; Zhao, Q.; Jing, Y. Channel equalization and detection with ELM-based regressors for OFDM systems. *IEEE Commun. Lett.* **2019**, *24*, 86–89. [CrossRef]
36. Zhang, Y.; Wang, H.; Li, C.; Chen, D.; Meriaudeau, F. Meta-learning-aided orthogonal frequency division multiplexing for underwater acoustic communications. *J. Acoust. Soc. Am.* **2021**, *149*, 4596–4606. [CrossRef]
37. Zhang, Y.; Wang, H.; Li, C.; Meriaudeau, F. Data augmentation aided complex-valued network for channel estimation in underwater acoustic orthogonal frequency division multiplexing system. *J. Acoust. Soc. Am.* **2022**, *151*, 4150–4164. [CrossRef]
38. Han, J.; Zhang, L.; Leus, G. Partial FFT Demodulation for MIMO-OFDM over Time-Varying Underwater Acoustic Channels. *IEEE Signal Process. Lett.* **2016**, *23*, 282–286. [CrossRef]
39. Li, B.; Zhou, S.; Stojanovic, M.; Freitag, L.; Willett, P. Multicarrier Communication Over Underwater Acoustic Channels With Nonuniform Doppler Shifts. *IEEE J. Ocean. Eng. J. Devoted Appl. Electr. Electron. Eng. Ocean. Environ.* **2008**, *33*, 198–209.
40. Kuchi, K. Limiting Behavior of ZF/MMSE Linear Equalizers in Wideband Channels with Frequency Selective Fading. *IEEE Commun. Lett.* **2012**, *16*, 929–932. [CrossRef]
41. Zhou, M.; Wang, J.; Feng, X.; Sun, H.; Li, J.; Kuai, X. On Generative-Adversarial-Network-Based Underwater Acoustic Noise Modeling. *IEEE Trans. Veh. Technol.* **2021**, *70*, 9555–9559. [CrossRef]
42. Zou, J.; Qu, J.; Guo, Y.; Liu, G.; Shu, F. Index Modulation Based on Four-dimensional Spherical Code and its DNN-based Receiver Design. *IEEE Trans. Veh. Technol.* **2021**, *70*, 13401–13405. [CrossRef]
43. Zhao, J.; Ran, R.; Oh, C.H.; Seo, J. Analysis of the Effect of Coherence Bandwidth on Leakage Suppression Methods for OFDM Channel Estimation. *J. Inf. Commun. Converg. Eng.* **2014**, *12*, 221–227.
44. Kochanska, I.; Schmidt, J.H. Estimation of Coherence Bandwidth for Underwater Acoustic Communication Channel. In Proceedings of the 2018 Joint Conference-Acoustics, Ustka, Poland, 11–14 September 2018; pp. 1–5.

remote sensing

Article

An Image Quality Improvement Method in Side-Scan Sonar Based on Deconvolution

Jia Liu [1,*], Yan Pang [2], Lengleng Yan [3] and Hanhao Zhu [3]

[1] Ocean Acoustic Technology Laboratory, Institute of Acoustics, Chinese Academy of Sciences, Beijing 100190, China

[2] Qingdao Branch of Institute of Acoustics, Chinese Academy of Sciences, Qingdao 266114, China; pangyan@mail.ioa.ac.cn

[3] Marine Science and Technology College, Zhejiang Ocean University, Zhoushan 316022, China; yanlengleng@zjou.edu.cn (L.Y.); zhuhanhao@zjou.edu.cn (H.Z.)

* Correspondence: liujia@mail.ioa.ac.cn

Abstract: Side-scan sonar (SSS) is an important underwater imaging method that has high resolutions and is convenient to use. However, due to the restriction of conventional pulse compression technology, the side-scan sonar beam sidelobe in the range direction is relatively high, which affects the definition and contrast of images. When working in a shallow-water environment, image quality is especially influenced by strong bottom reverberation or other targets on the seabed. To solve this problem, a method for image-quality improvement based on deconvolution is proposed herein. In this method, to increase the range resolution and lower the sidelobe, a deconvolution algorithm is employed to improve the conventional pulse compression. In our simulation, the tolerance of the algorithm to different signal-to-noise ratios (SNRs) and the resolution ability of multi-target conditions were analyzed. Furthermore, the proposed method was applied to actual underwater data. The experimental results showed that the quality of underwater acoustic imaging could be effectively improved. The ratios of improvement for the SNR and contrast ratio (CR) were 32 and 12.5%, respectively. The target segmentation results based on this method are also shown. The accuracy of segmentation was effectively improved.

Keywords: side-scan sonar; deconvolution; image quality; contrast ratio; object segmentation

Citation: Liu, J.; Pang, Y.; Yan, L.; Zhu, H. An Image Quality Improvement Method in Side-Scan Sonar Based on Deconvolution. *Remote Sens.* **2023**, *15*, 4908. https://doi.org/10.3390/rs15204908

Academic Editor: Andrzej Stateczny

Received: 7 September 2023
Revised: 29 September 2023
Accepted: 1 October 2023
Published: 11 October 2023

1. Introduction

Side-scan sonar (SSS) is a method using a device that scans and generates sonar images of the seabed, which emits an acoustic pulse signal toward the seabed perpendicular to the path of the sensor and records the intensity of the acoustic pulses reflected from the seabed [1–3]. These pings, recorded by the SSS, are constantly updated in real time and stitched together into sonar waterfall images to provide a clear view of the seabed landform and objects such as small targets, pipelines, and shipwrecks. When the SSS works, acoustic pulse signals of different pulse widths and frequencies are transmitted by the transducers on both sides of the tow body in the form of a fan-beam, and the acoustic signals propagate in the form of spherical waves [4]. The transmitted signals are usually a Continuous Wave (CW) signal and a Linear Frequency Modulated (LFM) signal. The time that passes, from the moment of pulse emission to the moment the intensity sample is obtained, is translated into a measure of distance (half the time that passes multiplied by the speed of sound in water) that can be used to approximate the actual distance of the waterfall images in the lateral direction [5].

A typical SSS system consists of a data display and recording unit, data transmission and towing cable, and underwater towfish, including an electronics compartment, transducers, and a well-sealed shell. Due to its convenient operation, wide coverage, and clear imaging, SSS is widely used in underwater search and rescue, underwater hazard removal, and so on. In practical applications, the higher imaging resolutions and clearer sonar images of SSS are conducive to achieving ideal results. At the same time, horizontal beam resolution and range resolution are important factors affecting imaging resolution [6–8]. The horizontal beam resolution (the angular resolution) is mainly affected by the array form of the horizontal beam open angle. The horizontal beam resolution will increase as the horizontal beam opening angle decreases, which is generally less than $0.5°$. In addition, to obtain a high resolution in the horizontal direction, synthetic aperture sonar [9] has been developed. At the same time, direction of arrival (DOA) estimation algorithms such as Minimum Variance Distortionless Response (MVDR) [10,11] and Multiple Signal Classification (MUSIC) [12,13] have also been developed. However, for common SSS, the increase in horizontal beam resolution is limited by the operating frequency and array length.

Range resolution is mainly affected by the bandwidth of the sonar transmission signal. In general, the range resolution of SSS is consistent with the variation trend of the transmitted signal bandwidth, and the range resolution of the SSS will increase with the increase in the transmitted signal bandwidth. But, in previous studies, range resolution has not received much attention because it is generally thought to have little impact on imaging performance. However, with the wide use of SSS, especially in the development of autonomous detection algorithms in side-scan sonar image processing, the image quality of side-scan sonar has become a vital foundational factor that affects the performance of artificial intelligence algorithms in detecting and recognizing targets in SSS images [14–17]. This improvement in image quality has a positive effect on improving the performance of autonomous detection. In the practical application of SSS, the sidelobe plays an important role in the image quality of SSS. Namely, the contour boundary of the targets and seafloor background are blurred in high sidelobe conditions. Normally, the bandwidth of the signal transmitted by SSS is fixed, so the range resolution can be improved by decreasing the sidelobe intensity and further increasing the main-to-side lobe ratio (MSLR). Especially when there is strong interference in the seabed, the sidelobe of interference will suppress the weak target signals nearby, which is worse in the case of very shallow water. This is because in very shallow water, the seabed is complex, and there are more targets in the seabed caused by human activities.

Methods that improve range resolution by suppressing sidelobe interference in the range direction include weighted processing [18–20], signal coding [21], etc. Weighted processing includes the weighted transmitted signal and weighted received signal, but at the expense of the width of the main lobe. Signal coding is mainly used to transmit pseudorandom code signals, which can improve range resolution while suppressing the sidelobe, but at the cost of increasing system complexity. When using SSS or other sonar imaging systems for underwater exploration, the receiving waveform and transmitting template will greatly differ due to the uneven frequency response of the transducer and insufficient power when transmitting long pulses. To obtain a higher range resolution, a common signal processing method is to match the data collected by SSS, namely pulse compression technology [22–24]. However, conventional pulse compression techniques have limited range resolution and an increased sidelobe; in the case of strong interference, the influence of the sidelobe is more obvious and even affects the accuracy of autonomous detection. In recent years, deconvolution has been mainly used to improve angular resolution, being widely used in medical imaging, disturbance suppression, and other areas to improve image quality [25–27]. In addition, it has been applied in the field of underwater acoustic signal processing, such as in three-dimensional imaging of synthetic aperture sonar [28–30], location or separation of mixed sources of linear array [31–33], power spectral estimation [34], linear array beamforming [35,36], imaging of forward-look sonar, or MIMO sonar [37,38]. However, there is limited research on the application of the deconvolution algorithm to improve the image quality of SSS.

To solve the abovementioned problems, we propose an improved pulse compression technology based on deconvolution. The main contributions are as follows:

(1) A high-quality imaging method suitable for a sonar system is proposed. To overcome the limitations of conventional pulse compression techniques, in which the distance resolution and "main lobe to side lobe ratio" are hard to improve, a deconvolution-based method is adopted to improve the imaging quality of side-scan sonar. There are two main benefits. Firstly, this method improves distance resolution under certain bandwidth and array length conditions. Secondly, it improves the "main lobe to side lobe ratio", thereby improving imaging clarity.

(2) An improved deconvolution method is proposed based on the frequency response function of the sonar system. Based on the characteristics of the operating frequency response curve of actual sonar systems, an optimal reception response function suitable for sonar systems is proposed, which improves the practical adaptability of the method.

(3) In order to verify the effectiveness of the proposed method, we used simulation and sea trial datasets. Firstly, the performance of this method under different signal-to-noise ratios and strong interference conditions was analyzed through numerical simulation. Then, by processing the sea trial data, the impact of this method on the imaging quality of large-scale imaging and small underwater targets, as well as on the imaging and autonomous segmentation of small underwater targets, was analyzed. Autonomous segmentation is an important component of autonomous detection. Good image quality helps to achieve high-precision object segmentation, which has positive significance for autonomous detection in downstream applications.

The remainder of this paper is organized as follows: Section 2 introduces the proposed imaging method, including the introduction of the range resolution of the SSS, proposed pulse compression technology based on deconvolution, and calculation of deconvolution. Section 3 presents the numerical simulation and sea trial datasets, where the analysis and demonstration of this method are provided. Section 4 presents our final conclusions.

2. Methods

2.1. Range Resolution in Side-Scan Sonar Imaging

The image quality of SSS is mainly determined by the angular and range resolution. In the case of a certain angular resolution, the range resolution is the main parameter that determines the imaging resolution of the SSS system.

In consideration of the process of signal transition, as shown in Figure 1, the transmitted signal is a frequency modulation signal, which can be expressed as

$$s(t) = u(t)\exp(\mathrm{j}(\theta(t))) = \mathrm{rect}(\frac{t}{T})[\exp(\mathrm{j}2\pi(f_0 t + \frac{1}{2}kt^2))], \tag{1}$$

where f_0 is the center frequency, k is the frequency modulation coefficient, and T is the period.

$$\mathrm{rect}(\frac{t}{T}) = \begin{cases} 1, & |t| \leq \dfrac{T}{2} \\ 0, & \mathrm{others} \end{cases}, \tag{2}$$

In the target detection model shown in Figure 1, the received signal can be denoted as

$$x(t) = s(t) * Q(t), \tag{3}$$

where $Q(t)$ is the objective function, which demonstrates the reflection characteristics of the targets.

As the pulse length of the frequency modulation signal is long, the distance resolution is not good for imaging. In a traditional sonar system, the distance resolution of the received signal can be reached by pulse compression, which is expressed as follows:

$$y(t) = x(t) * h(t), \tag{4}$$

where $h(t)$ is the response of the best receiver and can be calculated by

$$h(t) = \text{rect}(\frac{t}{T})[\exp(j2\pi(f_0 t + \frac{1}{2}kt^2))].$$ (5)

Then, the pulse compression output can be obtained as

$$y(t) = \sqrt{kT^2} \frac{\sin \frac{2\pi(\xi + kt)T}{2}}{\frac{2\pi(\xi + kt)T}{2}} \exp(-j\pi kt^2)\exp(\frac{j\pi}{4}),$$ (6)

The above equation clearly shows that the envelope shape of the time domain signal $y(t)$ after pulse compression is a sine function.

The width of the main lobe is about $1/B$, and the width of the main lobe is the main factor affecting the imaging resolution; that is, the range resolution d is

$$d = \frac{1}{2} \cdot c \cdot \frac{1}{B}.$$ (7)

At the same time, the value of the main side ratio is about -14 dB. In the case of strong interference, the sidelobe of a strong target will seriously affect the image quality and influence target detection properties.

When the sonar works, the transmitter for the sonar needs to go through the drive circuit and transducer electroacoustic conversion, and the receiving signal also needs to go through electroacoustic conversion and filtering processes. Due to the frequency band response of the transducer and the influence of the transmitter and receiver circuits, the received signal waveform will be distorted, which leads to a risk of the main lobe widening and the sidelobe increasing. Thus, in the actual world, the image quality may be worse.

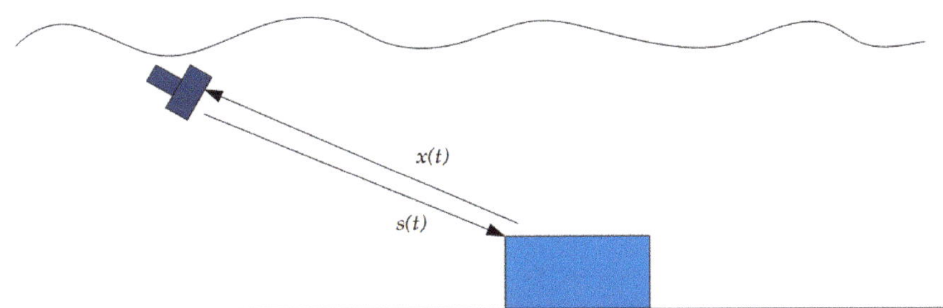

Figure 1. The geometry of target detection by sonar.

2.2. High-Resolution Pulse Compression Technology Based on Deconvolution

As previously mentioned, the pulse compression process can be written as

$$y(t) = x(t) * h(t) = x(t) * s(\tau - t) = \int_{-\infty}^{\infty} x(u) * s(\tau - t + u)du = R_{xs}(t - \tau),$$ (8)

where $R_{xs}(t-\tau)$ is the correlation function.

When the signal and noise are not correlated, and assuming that the echo of interest is a simple delay model, the output of the matching filter can be written as an auto-correlation function of the transmitted signal $s(t)$.

$$y(t) = aR_s(t - \tau) = R_s(t) * a\delta(t - \tau).$$ (9)

When M targets are present, the output of pulse compression is

$$y(t) = \sum_{k=1}^{M} aR_s(t - \tau_k) = R_s(t) * \sum_{k=1}^{M} a\delta(t - \tau_k),$$ (10)

where τ_k is the echo delay of the kth target.

Let the objective function be

$$Q(t) = \sum_{k=1}^{M} a\delta(t - \tau_k),\tag{11}$$

Thus, the received signal of the imaging sonar can be rewritten as

$$y(t) = R_s(t) * Q(t),\tag{12}$$

where $Q(t)$ is an ideal pulse-compressed output with the characteristics of a unitary impulse response function. It can reflect the acoustic reflection structure of the target well. For the imaging sonar, $Q(t)$ can reflect the shape of the target well. So, the method aims to solve $Q(t)$.

2.3. Calculation of Deconvolution

A deconvolution method is introduced to solve $Q(t)$. The Richardson–Lucy (R–L) algorithm [39–41], which is an optimal iterative algorithm based on Bayesian theory, is applied to solve the deconvolution problem. It is widely used in image restoration and has stable performance under low SNRs.

Firstly, denote $L(\cdot)$ as the convergence distance between the true value and the estimated

$$L(p(x), q(x)) = \int_{-\infty}^{\infty} p(x) \log \frac{p(x)}{q(x)} dx,\tag{13}$$

where $p(x), q(x)$ is a non-negative probability density function, $L(i)$ is a monotonically decreasing and non-negative function, and $L(p(x), q(x)) = 0$ if and only if $p(x) = q(x)$.

Then, we can obtain the following solution formula:

$$\lim_{n \to \infty} Q^n(t) = \mathrm{argmin} L(y(t), R_s(t) * Q(t)),\tag{14}$$

$$Q^{n+1}(t) = Q^n(t)[R_s(t) \cdot \frac{y(t)}{R_s(t) \cdot Q^K(t)}],\tag{15}$$

where n represents the number of iterations.

As the iteration number increases, $Q^n(t)$ can converge to a unique solution by minimizing the Csiszar discrimination.

Figure 2 represents the workflow of the proposed method. Firstly, the pulse compression is adapted for the receiving signal. Secondly, we need to estimate the Point Spread Function (PSF), which is necessary for the R–L algorithm. To estimate the PSF, the model is made to follow the transmit signal, and then modified by the response of the system. Thirdly, the calculation of deconvolution is performed by the R–L algorithm. Finally, the high-resolution data and image from the side-scan sonar are acquired.

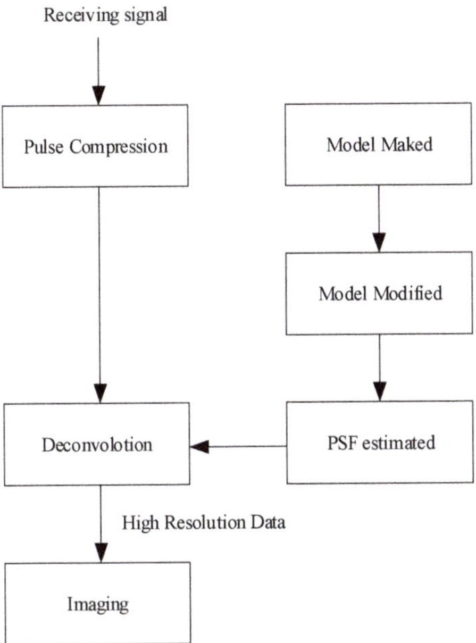

Figure 2. The workflow diagram of the proposed method.

3. Results

3.1. Numerical Simulation

The performance under different SNRs was calculated by simulation. The simulation conditions of this section were set as follows: the transmitted signal was a LFM signal, the center frequency and signal bandwidth were 500 and 50 kHz, respectively, and the pulse width length of the signal was 3.2 ms.

3.1.1. The Influence of Different SNRs

Firstly, we discuss the performance of pulse compressions with the deconvolution technique under different noise conditions, and the results of the traditional pulse compression technique are compared with the proposed method, as shown in Figure 3. The statistical results of the main flap width and main side ratio for both methods at different signal-to-noise ratios are shown in Table 1.

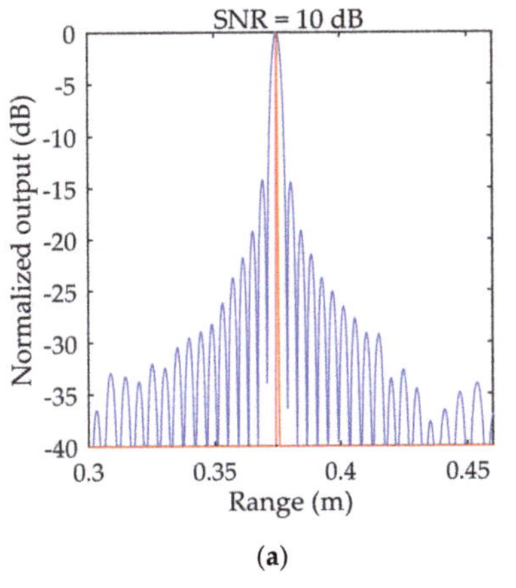

(a)

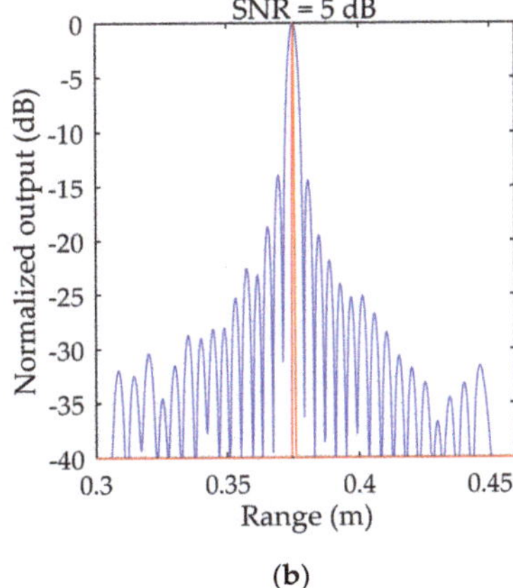

(b)

Figure 3. *Cont.*

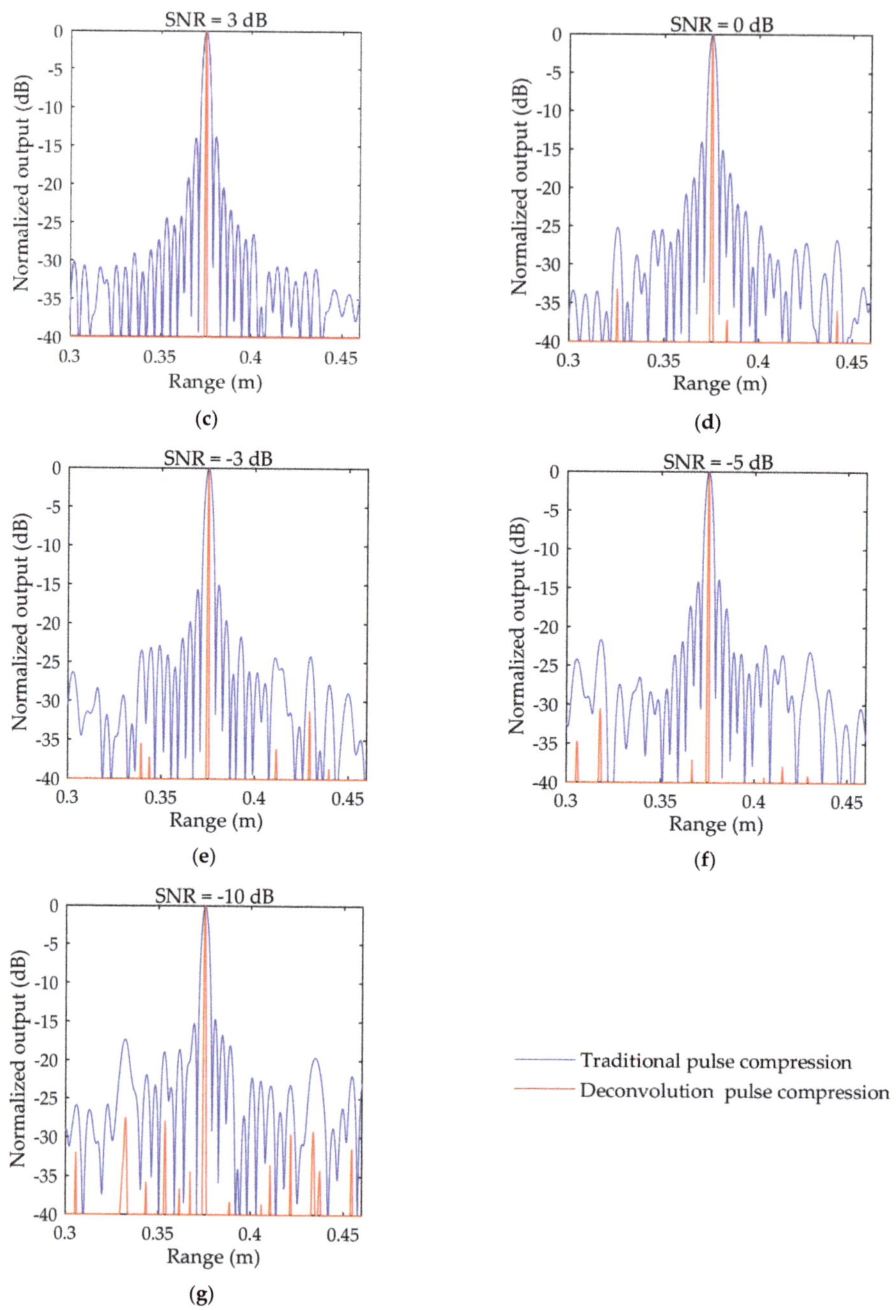

Figure 3. The simulation results of pulse compression and deconvolutional pulse compression under different SNRs: (**a**) simulation results under 10 dB; (**b**) simulation results under 5 dB; (**c**) simulation results under 3 dB; (**d**) simulation results under 0 dB; (**e**) simulation results under −3 dB; (**f**) simulation results under −5 dB; and (**g**) simulation results under −10 dB.

Table 1. Comparison of the main lobe width and MSLR between the two methods.

SNR	Method	Main Lobe Width	MSLR
10 dB	Traditional pulse pressure	3 mm	−14 dB
	Deconvolutional pulse pressure	0.4 mm	−40 dB
5 dB	Traditional pulse pressure	3 mm	−14 dB
	Deconvolutional pulse pressure	0.4 mm	−40 dB
3 dB	Traditional pulse pressure	3 mm	−14 dB
	Deconvolutional pulse pressure	0.4 mm	−40 dB
0 dB	Traditional pulse pressure	3 mm	−14 dB
	Deconvolutional pulse pressure	0.4 mm	−36 dB
−3 dB	Traditional pulse pressure	3 mm	−14 dB
	Deconvolutional pulse pressure	0.5 mm	−30 dB
−5 dB	Traditional pulse pressure	3 mm	−14 dB
	Deconvolutional pulse pressure	0.5 mm	−28 dB
−10 dB	Traditional pulse pressure	3 mm	−14 dB
	Deconvolutional pulse pressure	0.5 mm	−24 dB

Compared with the processing results in Figure 3 and Table 1, the main lobe width of the deconvolution method is similar to the impact function, and it shows better range resolution and can truly reflect the objective function. The MSLR of the deconvolution method is lower and can reach more than 40 dB, which can effectively reduce the interference of strong targets on the SSS sonogram. Moreover, the algorithm has a good tolerance-to-noise ratio and maintains good performance at a SNR of −10 dB.

3.1.2. Considering the Influence of Sonar Parameters

In the actual sonar system, the signal transmission and reception process includes transmission drive, transmission electroacoustic conversion, receiving electroacoustic conversion, receiving circuit reception, etc., as shown in Figure 4.

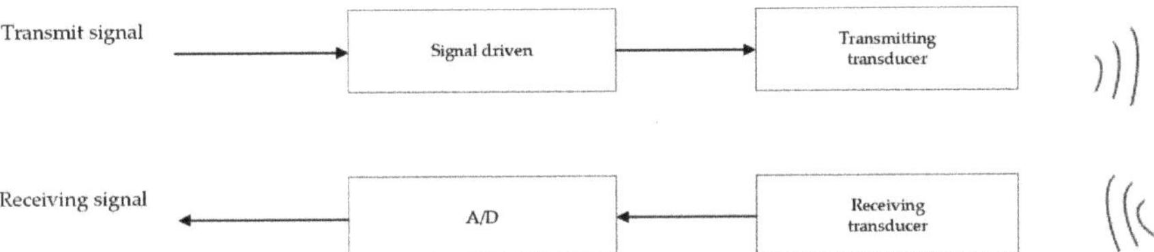

Figure 4. The signal generation process of the SSS system.

Among these, the frequency band response of the transducer is generally not very flat. The transmission drive also leads to a distortion in the transmitted signal due to the influence of charge and discharge, and the comprehensive sonar system response leads to a mismatch in the conventional pulse compression signal model, increasing the main lobe width of pulse compression and decreasing the MSLR. This decrease in the MSLR has an adverse impact on image quality. To calculate and analyze the effect of the algorithm on practical engineering applications through simulations, the band response of the sonar system to the signal was comprehensively considered in the simulation experiment, and the simulation values are shown in Figure 5. The output results of the deconvolution method and comparison method under these conditions are also shown in Figure 5.

Through comparative analysis, we see that under the simulation conditions, due to the frequency band response caused by the sonar system, the sidelobe of conventional pulse compression technology is irregular, and the performance is significantly reduced compared with the ideal pulse compression results; however, the deconvolution technology overcomes the influence of model mismatch due to its multiple iterations of optimization and maintains a good main lobe width and MSLR.

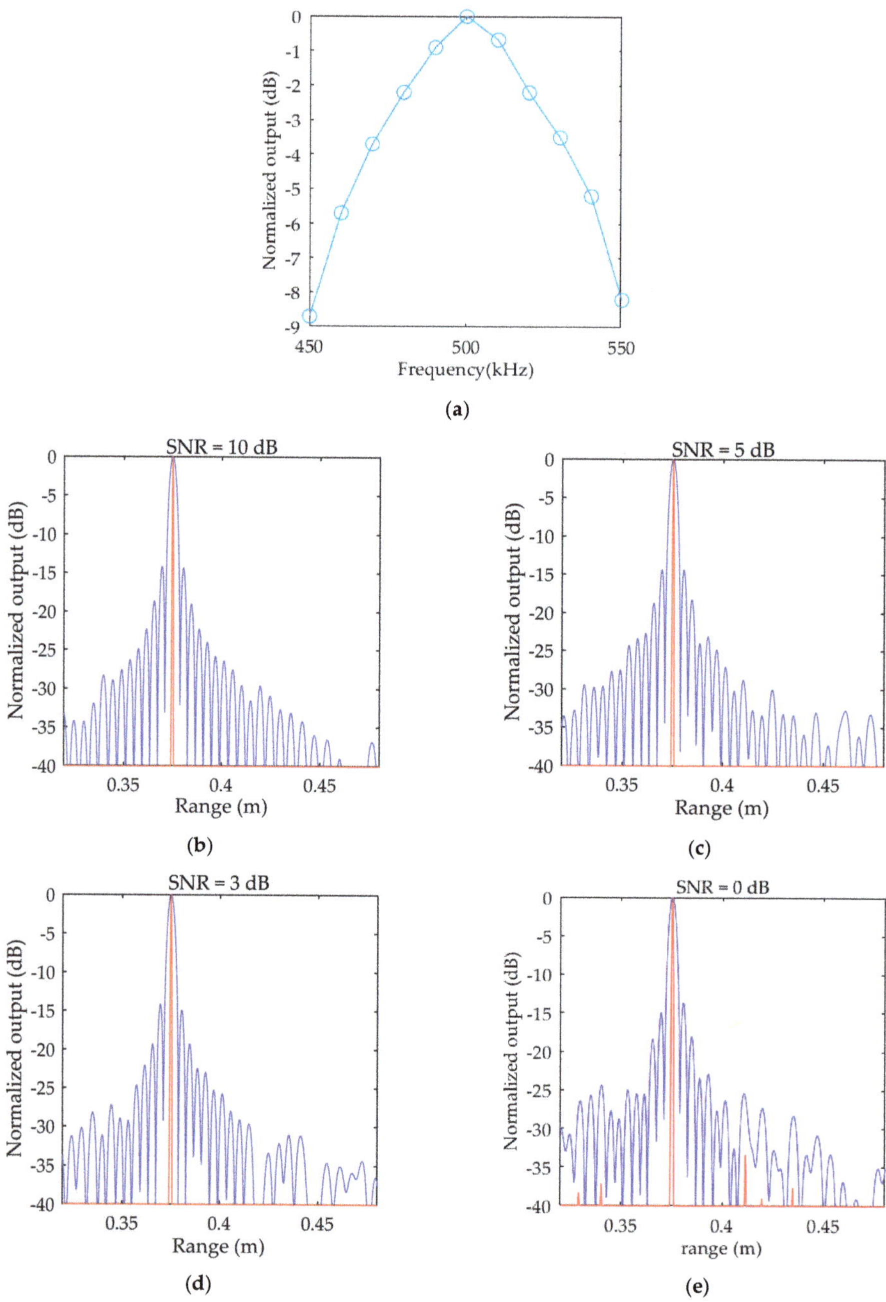

Figure 5. *Cont.*

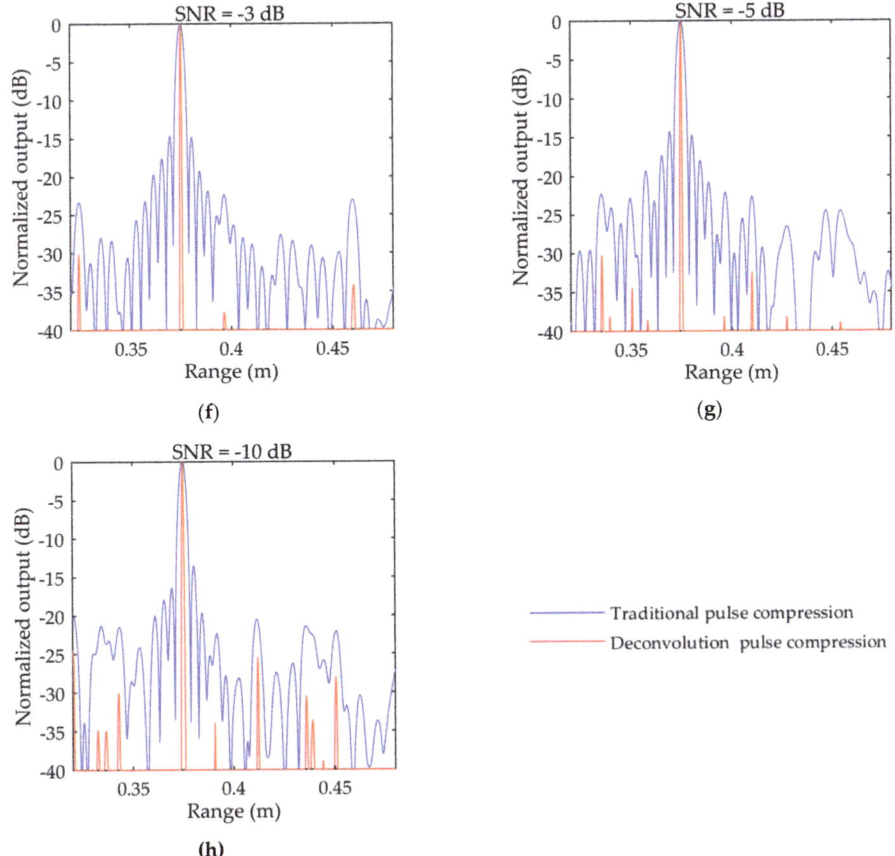

Figure 5. Comparison of deconvolution results and conventional pulse pressure under the influence of sonar parameters: (**a**) frequency response of the sonar; (**b**) simulation results under 10 dB; (**c**) simulation results under 5 dB; (**d**) simulation results under 3 dB; (**e**) simulation results under 0 dB; (**f**) simulation results under −3 dB; (**g**) simulation results under −5 dB; and (**h**) simulation results under −10 dB.

3.1.3. The Impact on Strong Interference

In practice, strong interference may lead to poor target resolution. To analyze the impact of strong interference, a simulation was performed. In the simulation, there was a target and an interference with the same intensity. The results are depicted in Figure 6.

Due to the good range resolution ability of the deconvolutional pulse compression algorithm, the distinction between the targets and interference was clearer than that of conventional pulse compression technology, and this method worked well under different SNRs.

3.1.4. The Impact on Imaging

To analyze the effect of the proposed algorithm on actual imaging results, side-scan sonar imaging results were simulated, and a comparison graph of the imaging results of the two methods is shown in Figure 7. Three targets are included in the simulation graph, two of which are on the same time slice (Ping).

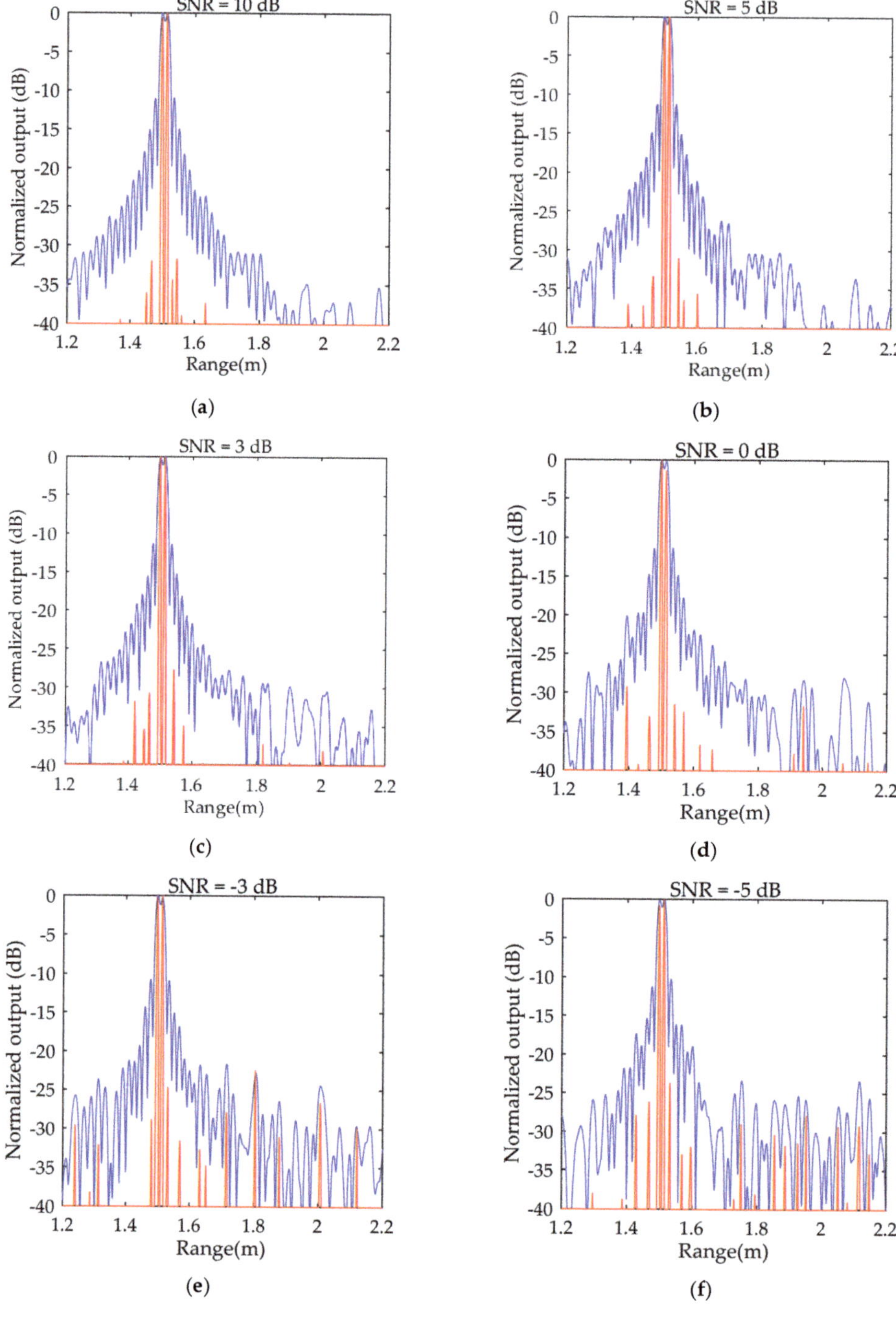

Figure 6. *Cont.*

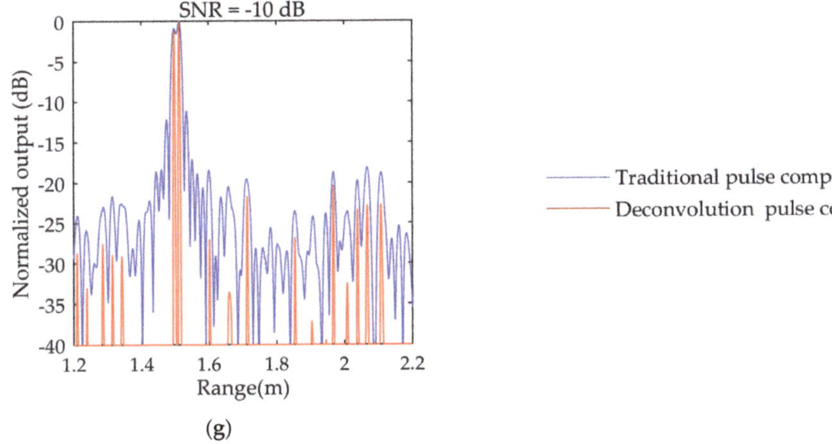

(**g**)

Figure 6. Comparison of deconvolution results with conventional pulse compression under the influence of strong interference: (**a**) simulation results under 10 dB; (**b**) simulation results under 5 dB; (**c**) simulation results under 3 dB; (**d**) simulation results under 0 dB; (**e**) simulation results under −3 dB; (**f**) simulation results under −5 dB; and (**g**) simulation results under −10 dB.

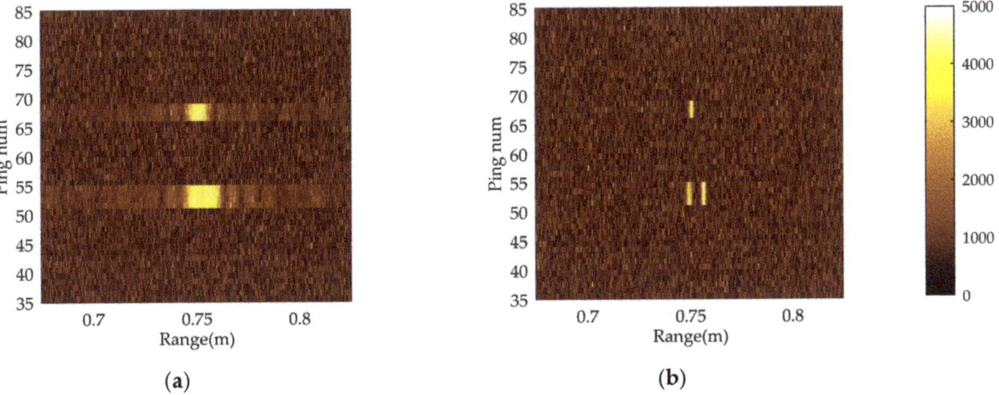

(**a**)　　　　　　(**b**)

Figure 7. The simulated images with targets were obtained by (**a**) the traditional method and (**b**) the proposed method.

In comparison to Figure 7a, Figure 7b has a clearer target contour because of the improved high-resolution algorithm, and the targets at different distances at the same time can be clearly resolved. Figure 7a, with its conventional processing method, cannot realize multi-target resolution due to the lower distance resolution and stronger sidelobe interference.

3.2. Sea Experiment

For the image of SSS, there are two main types of applications. One is the large-scale imaging of the seabed, which completes the exploration of seabed topography, including the imaging detection of shipwrecks, seamounts, and trenches; the second is to search and explore small underwater targets, such as pipelines and mines. To demonstrate the performance of the algorithm, seabed and small target imaging were performed separately.

These data were obtained through the sea trials. The water depth in the trial was below 20 m.

3.2.1. The Influence of the Algorithm on Seabed Imaging Evaluation Indicators

The imaging quality of our method was analyzed in the following via the imaging effect of seabed wrecks. There are several commonly used image metrics to evaluate the clarity and prominence of objects in the image background in the field of image processing, including the SNR, contrast ratio (CR) [42], and contrast–shadow ratio (CSR) [43]. CSR is a method used to evaluate the contrast between the target and shadow. The evaluation results given are similar to those of visual evaluations, and are often used for quantitative evaluation of image quality control, performance comparison, and defect detection. The higher the value of CSR, the better the image quality. Moreover, the sharpness of the image is improved with an increase in CR. This results in a clearer outline of the target in the image and more details displayed.

The specific calculation is as follows:

$$SNR = (I_M - \overline{I_B})/\overline{I_B},\tag{16}$$

$$CR = (\overline{I_H} - \overline{I_B})/\overline{I_B},\tag{17}$$

$$CSR = (\overline{I_H} - \overline{I_S})/\overline{I_S},\tag{18}$$

where I_M is the maximum intensity value of the highlighted areas; $\overline{I_H}$ $\overline{I_B}$, and $\overline{I_S}$ are the average intensity of the highlighted areas, seabed background areas, and seabed shadow areas, respectively.

To analyze the ability of the algorithm to maintain image sharpness, the algorithm's mean square error (MSE), peak signal-to-noise ratio (PSNR), and structural similarity (SSIM) performance were analyzed.

For two $m \times n$ images I and K, the mean square error (MSE) was defined as

$$MSE = \frac{1}{mn} \sum_{i=0}^{m-1} \sum_{j=0}^{n-1} [I(i,j) - K(i,j)]^2,\tag{19}$$

To locate the peak signal-to-noise ratio (PSNR), we used

$$PSNR = 10 \cdot \log_{10}\left(\frac{MAX_I^2}{MSE}\right) = 20 \cdot \log_{10}\left(\frac{MAX_I}{\sqrt{MSE}}\right),\tag{20}$$

where MAX_I is the maximum value that represents the color of the image; if each sample point is represented by 8 bits, then this value is 255. Thus, the smaller the MSE, the larger the PSNR; the larger the PSNR, the better the quality of the image is represented.

The structural similarity (SSIM) was calculated by

$$SSIM(x,y) = [l(x,y)]^{\alpha}[c(x,y)]^{\beta}[s(x,y)]^{\gamma},\tag{21}$$

where $\alpha > 0$, $\beta > 0$, and $\gamma > 0$.

$$l(x,y) = \frac{2\mu_x\mu_y + c1}{\mu_x^2 + \mu_y^2 + c1},\tag{22}$$

$$c(x,y) = \frac{2\sigma_{xy} + c2}{\sigma_x^2 + \sigma_y^2 + c2},\tag{23}$$

$$s(x,y) = \frac{\sigma_{xy} + c3}{\sigma_x\sigma_y + c3}\tag{24}$$

where I(x,y) is the luminance comparison, c(x,y) is the contrast comparison, and s(x,y) is the structure comparison. μ_x and μ_y represent the mean of x and y, respectively, and σ_x and σ_y represent the standard deviation of x and y, respectively. σ_{xy} represents the covariance between x and y. The denominator is constant to avoid the systematic error caused by zero. $c1$, $c2$, and $c3$ are constants to avoid the systematic error caused by the denominator being zero.

Comparative Analysis of Imaging Quality

As shown in Figure 8, the SSS image processed by pulse compression technology had a fuzzy boundary between the target and seabed background, and the target was almost integrated with the background. However, in the SSS image processed by the algorithm proposed in this paper, the sidelobe had less interference in the image, due to a reduction in sidelobe intensity; the range resolution in the lateral direction of the image was improved so that the boundary between the contour edge of the target and seabed background was more obvious; and the shape of the target in SSS image was closer to that of the actual target.

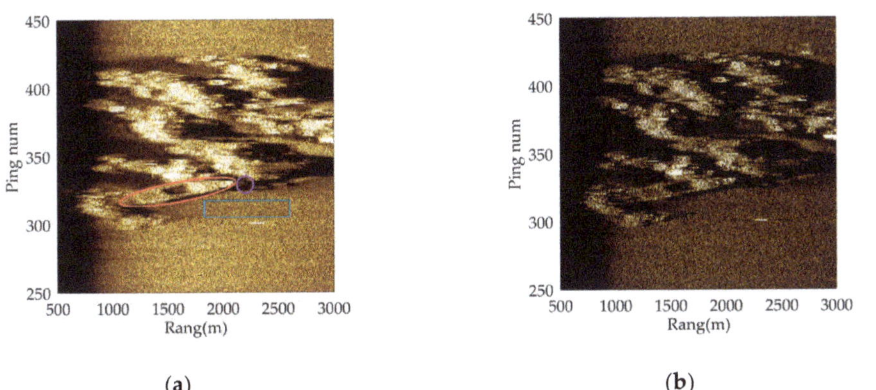

(a)   (b)

Figure 8. SSS image processing results: (**a**) SSS images processed by conventional pulse compression, where the red oval represents the target area, the purple circle represents the shadow area, and the blue rectangle represents the botttom area, respectively; (**b**) SSS images processed by the proposed method.

The area in the red oval in Figure 8 was taken as the area inside the target, the area in the blue box was taken as the seabed area around the target, and the area in the purple circle beside the red box was taken as the shadow of the target. The three indicators of the image obtained by the two methods were calculated, and the results are shown in Table 2.

Table 2. Performance metrics of the images of SNR, CR and CSR.

Method	SNR	CR	CSR
Conventional Pulse Compression	18.7	1.6	4.8
Proposed Method	24.7	1.88	25.8
Ratio of Improvement	32%	12.5%	437%

Table 2 shows that the SNR, CR, and CSR of the proposed algorithm were 24.7, 1.88, and 25.8, respectively, which were improved compared with the results of the conventional pulse compression method. The ratios of improvement for SNR, CR, and CSR were 32, 12.5, and 437%, respectively.

The proposed method was compared with the conventional image mean square sharpening method, and a comparison graph is shown in Figure 9, where Figure 9a shows the results of conventional image sharpening and Figure 9b shows the results of the proposed method.

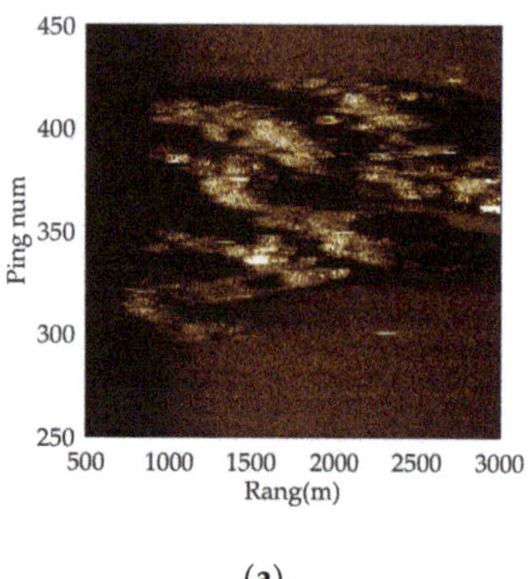 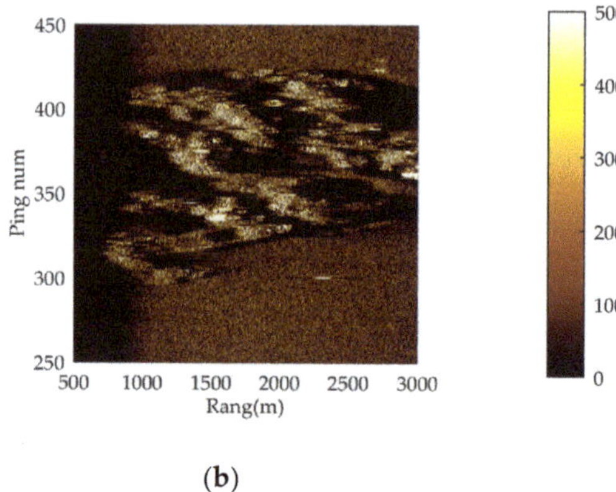

(**a**) (**b**)

Figure 9. The SSS image processing results: (**a**) SSS images processed by the sharpening method and (**b**) SSS images processed by the proposed method.

The calculated values were analyzed and are shown in Table 3.

Table 3. Performance metrics of the images of MSE, PSNR and SSIM.

Method	MSE	PSNR	SSIM
Common Sharpening Method	3.34	12.83	0.59
Proposed Method	1.52	16.30	0.79

This comparison shows that the improved method proposed in this paper has a larger PSNR and SSIM and smaller MSE for the image compared with the conventional method, indicating that our proposed method is superior. This is because the deconvolution method improves the resolution in the original echo domain during processing. Therefore, it achieves a better signal-to-noise ratio and improved structural features than the original image and introduces smaller errors.

3.2.2. The Influence of the Algorithm on Target Imaging

In recent years, with the continuous progress of unmanned technology, autonomous detection technology based on SSS has developed higher requirements. Detection technology usually includes segmentation, detection, identification, and other steps. The image quality of SSS is the foundation of autonomous detection and identification algorithms.

In this section, real small targets on the seabed were imaged using the deconvolutional pulse compression method and traditional pulse compression method. The difference in image quality was assessed by independent segmentation of each acoustic image. The segmentation algorithm adopted the region growth method [44–46], which is characterized by ensuring the continuity of the target.

Figures 10 and 11 are the processing results of the two sets of data (Data I and Data II), including the processing acoustic image of conventional pulse compression technology, the processing acoustic image of deconvolutional pulse compression algorithm, and the corresponding autonomous segmentation results. Figure 10 shows the results of Data I with a single target and image segmentation, and Figure 11 shows the imaging and segmenting results of Data II with a small target with strong interference nearby.

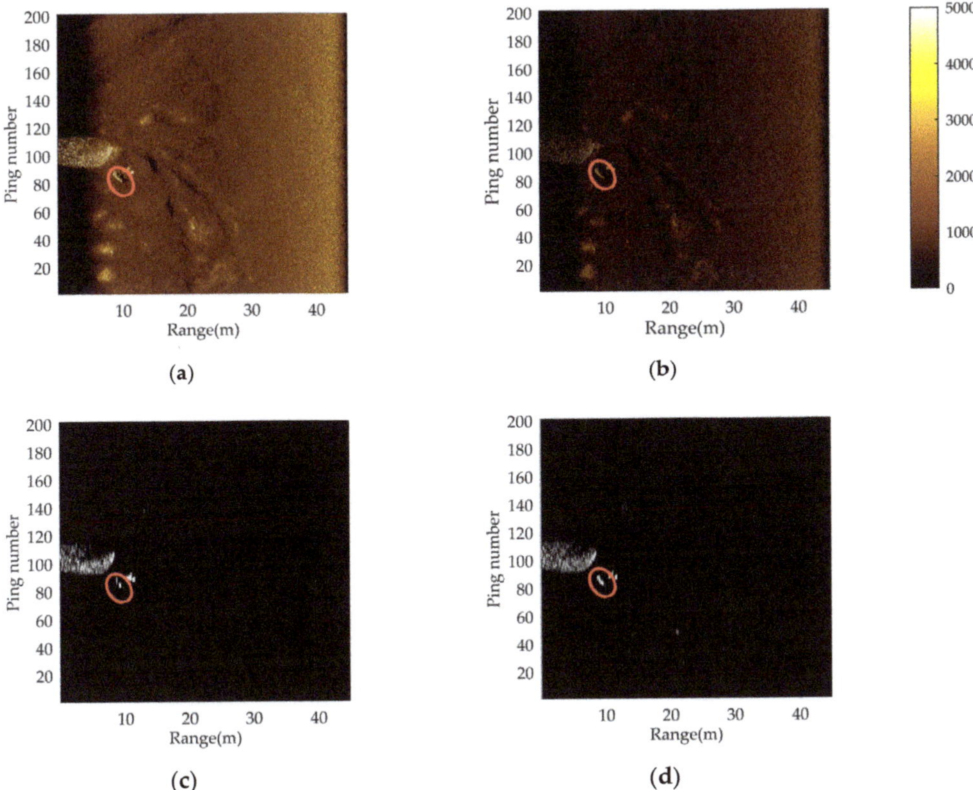

Figure 10. A comparison of processing results between the conventional pulse compression technique and proposed method: (**a**) imaging results of conventional pulse compression technology; (**b**) imaging results of the proposed method; (**c**) image segmentation results after imaging by conventional pulse compression technology; and (**d**) image segmentation results after imaging by the proposed method. The red ovals represent the small target area.

By comparing the processing results shown in Figure 10, we see that deconvolutional pulse compression has better resolution, and the shape of the target can be clearly described using this method. The shape of the target after autonomous segmentation is more accurate and closer to the real target, which is beneficial to the extraction of accurate target information. The follow-up detection process can effectively avoid missing judgment and misjudgment in the process of autonomous detection and processing.

Through the comparison in Figure 11, we see that in the original pulse compression processing results, due to the influence of strong interference, the target shape is difficult to clearly identify and the outline of the target is not obvious. The jamming target next to it is more similar to the target to be detected in the SSS image than the real target. Furthermore, the scale of the target after autonomous segmentation was severely distorted and was unable to maintain the target's true form, which interferes with subsequent detection. Contrary to the results of pulse compression processing, the image obtained by deconvolutional pulse compression had smaller sidelobe interference of the target and was more clearly visible. Meanwhile, the contour was more distinct than what was obtained by conventional pulse compression processing, which is a better representation of the original shape of the target and is more conducive to subsequent segmenting. Compared with the segmentation results of pulse-compressed images, images using our proposed method were closer to the real target after segmentation.

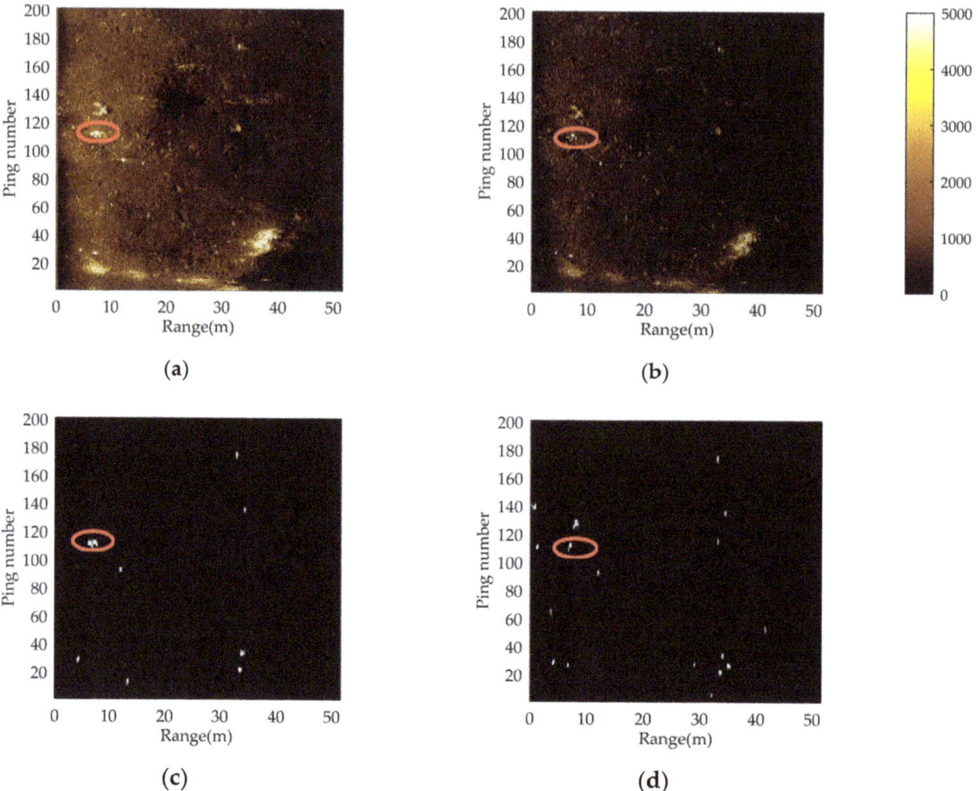

Figure 11. A comparison of processing results under strong interference conditions between the conventional pulse compression technique and proposed method: (**a**) imaging results of conventional pulse compression technology; (**b**) imaging results of the proposed method; (**c**) image segmentation results after imaging by conventional pulse compression technology; and (**d**) image segmentation results after imaging by the proposed method. The red ovals represent the small target area.

4. Conclusions

The performance of the range resolution in high-frequency image sonar systems greatly affects image quality. As one of the important indicators for evaluating image quality, range resolution is limited by sonar system parameters. The pulse compression technology commonly used in conventional sonar signal processing is limited by the system bandwidth in terms of range resolution, and the main-to-side lobe ratio is generally high. When there is strong interference or a large quantity of debris on the seabed, it can affect the quality of the observation image. In this paper, a pulse compression technique based on deconvolution was proposed to achieve ideal objective functions. This method could be used to overcome the limitations of inherent sonar system parameters and improve range resolution without affecting system complexity.

In our simulation results, the proposed algorithm improved the main lobe width of pulse compression, and the main-to-side lobe ratio also significantly increased. In the processing results of data from actual sea trials, the proposed method effectively improved image quality. In the results of the data processing, the ratios of improvement for SNR, CR, and CSR were 32, 12.5, and 437%, respectively. Combined with object segmentation methods, the object segmentation accuracy was effectively improved, and the segmented target largely retained the morphological information of the actual target. This shows that our proposed method has great potential in autonomous detection. In addition, some high-precision ocean surveys, exploration, and seabed sedimentological surveys may also be potential applications of this method.

Author Contributions: J.L. provided the ideas, wrote the manuscript, and created the figures, Y.P. wrote the manuscript and created the figures, H.Z. revised the paper and guided the research, and L.Y. revised the paper. All authors have read and agreed to the published version of the manuscript.

Funding: This research was funded by the Youth Innovation Promotion Association (No. 2020023, National Natural Science Foundation of China (Grant No.: 12374425)), the Stable Supporting Fund of National Key Laboratory of Underwater Acoustic Technology (JCKYS2023604SSJS016), and the Science and Technology program of Zhoushan City (2023C41025).

Data Availability Statement: The data presented in this study are available on request from the corresponding author. The data are not publicly available due to privacy or ethical restrictions.

Acknowledgments: We would like to thank engineers Xiaohui Yu and Yao Niu of Haiying Enterprise Group Co., Ltd. (Wuxi, China) for their support in data acquisition.

Conflicts of Interest: The authors declare no conflict of interest.

References

1. Nguyen, V.D.; Luu, N.M.; Nguyen, Q.K.; Nguyen, T.-D. Estimation of the Acoustic Transducer Beam Aperture by Using the Geometric Backscattering Model for Side-Scan Sonar Systems. *Sensors* **2023**, *23*, 2190. [CrossRef]
2. Meng, X.; Xu, W.; Shen, B.; Guo, X. A High–Efficiency Side–Scan Sonar Simulator for High–Speed Seabed Mapping. *Sensors* **2023**, *23*, 3083. [CrossRef]
3. Wang, Z.; Zhang, S.; Gross, L.; Zhang, C.; Wang, B. Fused Adaptive Receptive Field Mechanism and Dynamic Multiscale Dilated Convolution for Side-Scan Sonar Image Segmentation. *IEEE Trans. Geosci. Remote Sens.* **2022**, *60*, 5116817. [CrossRef]
4. Zhu, H.H.; Cui, Z.Q.; Liu, J.; Liu, X.; Wang, J.H. A Method for Inverting Shallow Sea Acoustic Parameters Based on the Backward Feedback Neural Network Model. *J. Mar. Sci. Eng.* **2023**, *11*, 1340. [CrossRef]
5. Borrelli, M.; Smith, T.L.; Mague, S.T. Vessel-Based, Shallow Water Mapping with a Phase-Measuring Sidescan Sonar. *Estuaries Coast.* **2022**, *45*, 961–979. [CrossRef]
6. Zheng, G.; Zhang, H.; Li, Y.; Zhao, J. A Universal Automatic Bottom Tracking Method of Side Scan Sonar Data Based on Semantic Segmentation. *Remote Sens.* **2021**, *13*, 1945. [CrossRef]
7. Cheng, Z.; Huo, G.; Li, H. A Multi-Domain Collaborative Transfer Learning Method with Multi-Scale Repeated Attention Mechanism for Underwater Side-Scan Sonar Image Classification. *Remote Sens.* **2022**, *14*, 355. [CrossRef]
8. Shang, X.; Zhao, J.; Zhang, H. Automatic Overlapping Area Determination and Segmentation for Multiple Side Scan Sonar Images Mosaic. *IEEE J. Sel. Top. Appl. Earth Obs. Remote Sens.* **2021**, *14*, 2886–2900. [CrossRef]
9. Brown, D.C.; Gerg, I.D.; Blanford, T.E. Interpolation Kernels for Synthetic Aperture Sonar Along-Track Motion Estimation. *IEEE J. Ocean. Eng.* **2020**, *45*, 1497–1505. [CrossRef]
10. Tuladhar, S.R.; Buck, J.R. Unit Circle Rectification of the Minimum Variance Distortionless Response Beamformer. *IEEE J. Ocean. Eng.* **2020**, *45*, 500–510. [CrossRef]
11. Wang, Q.L.; Zhu, H.H.; Chai, Z.G.; Cui, Z.Q.; Wang, Y.F. Analysis of VLF Wave Field Components and Characteristics Based on Finite Element Time-Domain Method. *J. Sensors.* **2023**, *2023*, 7702342. [CrossRef]
12. Elbir, A.M. DeepMUSIC: Multiple Signal Classification via Deep Learning. *IEEE Sens. Lett.* **2020**, *4*, 7001004. [CrossRef]
13. Yang, X.B.; Wang, K.; Zhou, P.Y.; Xu, L.; Liu, J.L.; Sun, P.P.; Su, Z.Q. Ameliorated-multiple signal classification (Am-MUSIC) for damage imaging using a sparse sensor network. *Mech. Syst. Signal Process.* **2022**, *163*, 108154. [CrossRef]
14. Yin, J.W.; Guo, K.; Han, X.; Yu, G. Fractional Fourier transform based underwater multi-targets direction-of-arrival estimation using wideband linear chirps. *Appl. Acoust.* **2020**, *169*, 107477. [CrossRef]
15. Thanh Le, H.; Phung, S.L.; Chapple, P.B.; Bouzerdoum, A.; Ritz, C.H.; Tran, L.C. Deep Gabor Neural Network for Automatic Detection of Mine-Like Objects in Sonar Imagery. *IEEE Access.* **2020**, *8*, 94126–94139. [CrossRef]
16. Zhu, H.H.; Xue, Y.Y.; Ren, Q.Y.; Liu, X. Inversion of shallow seabed structure and geoacoustic parameters with waveguide characteristic impedance based on Bayesian approach. *Front. Mar. Sci.* **2023**, *10*, 1104570. [CrossRef]
17. Yu, Y.C.; Zhao, J.H.; Gong, Q.H.; Huang, C.; Zheng, G.; Ma, J. Real-time underwater maritime object detection in side-scan sonar images based on transformer-YOLOv5. *Remote Sens.* **2021**, *13*, 3555. [CrossRef]
18. Połap, D.; Wawrzyniak, N.; Włodarczyk-Sielicka, M. Side-scan sonar analysis using roi analysis and deep neural networks. *IEEE Trans. Geosci. Remote Sens.* **2022**, *60*, 4206108. [CrossRef]
19. Sun, Y.; Liu, Q.; Cai, J.; Long, T. A Novel Weighted Mismatched Filter for Reducing. *IEEE Trans. Aerosp. Electron. Syst.* **2019**, *55*, 1450–1460. [CrossRef]
20. Xia, D.; Zhang, L.; Wu, T.; Hu, W. An interference suppression algorithm for cognitive bistatic airborne radars. *J. Syst. Eng. Electron.* **2022**, *33*, 585–593. [CrossRef]

21. Lønmo, T.I.B.; Austeng, A.; Hansen, R.E. Improving swath sonar water column imagery and bathymetry with adaptive beamforming. *IEEE J. Ocean. Eng.* **2020**, *45*, 1552–1563. [CrossRef]
22. Guan, C.Y.; Zhou, Z.M.; Zeng, X.W. A phase-coded sequence design method for active sonar. *Sensors* **2020**, *20*, 4659. [CrossRef] [PubMed]
23. Zhang, X.; Ying, W.; Dai, X. High-resolution imaging for the multireceiver SAS. *J. Eng. Technol.* **2019**, *19*, 6057–6062. [CrossRef]
24. Zeng, X.; Liu, J.; Gao, B. Three-dimensional Imaging Sonar Signal Processing System Based on Blade Server. *J. Phys. Conf. Ser.* **2019**, *1213*, 042061. [CrossRef]
25. Sun, D.; Ma, C.; Mei, J.; Shi, W. Improving the resolution of underwater acoustic image measurement by deconvolution. *Appl. Acoust.* **2020**, *165*, 107292. [CrossRef]
26. Bai, C.; Liu, C.; Jia, H. Compressed blind deconvolution and denoising for complementary beam subtraction light-sheet fluorescence microscopy. *IEEE. Trans. Biomed. Eng.* **2019**, *66*, 2979–2989. [CrossRef]
27. Xue, Y.Y.; Zhu, H.H.; Wang, X.H. Bayesian geoacoustic parameters inversion for multi-layer seabed in shallow sea using underwater acoustic field. *Front. Mar. Sci.* **2023**, *10*, 1058542. [CrossRef]
28. Wang, P.; Chi, C.; Ji, Y.Q.; Huang, Y.; Huang, H.N. Two-dimensional deconvoled beamforming for the high-resolution underwater three-dimensional acoustical imaging. *Acta Acust.* **2019**, *44*, 613–625.
29. Zhu, J.H.; Song, Y.P.; Jiang, N.; Xie, Z.; Fan, C.Y.; Huang, X.T. Enhanced Doppler Resolution and Sidelobe Suppression Performance for Golay Complementary Waveforms. *Remote Sens.* **2023**, *15*, 2452. [CrossRef]
30. Zhang, X.B.; Yang, P.X.; Sun, H.X. An omega-k algorithm for multireceiver synthetic aperture sonar. *Electron. Lett.* **2023**, *59*, e12859. [CrossRef]
31. Mei, J.D.; Shi, W.P.; Ma, C.; Sun, D.J. Near-field beamforming acoustic image measurement based on decovolution. *Acta Acust.* **2020**, *45*, 15–28.
32. Teng, T.T.; Liu, H.M.; Sun, D.S.; Xi, J.C.; Qu, G.Y.; Yang, S. Active sonar de-convolution matched filtering method. In Proceedings of the 2018 IEEE International Conference on Signal Processing, Communications and Computing (ICSPCC), Qingdao, China, 14–16 September 2018.
33. Shang, Z.G.; Qu, X.H.; Qiao, G.; Hao, C.P. Localizing mixed far- and near-field sources using beamforming deconvolution techniques. *Acta Acust.* **2023**, *48*, 447–458.
34. Guo, W.; Piao, S.C.; Yang, T.C.; Guo, J.Y.; Iqbal, K. High-resolution power spectral estimation method using deconvolution. *IEEE J. Ocean. Eng.* **2020**, *45*, 489–499. [CrossRef]
35. Yang, T.C. Deconvolved conventional beamforming for a horization line array. *IEEE J. Ocean. Eng.* **2018**, *43*, 160–172. [CrossRef]
36. Sun, D.J.; Ma, C.; Yang, T.C.; Mei, J.D.; Shi, W.P. Improving the performance of a vector sensor line array by deconvolution. *IEEE J. Ocean. Eng.* **2020**, *45*, 1063–1077. [CrossRef]
37. Liu, X.H.; Fan, J.H.; Sun, C.; Yang, Y.X.; Zhuo, J. High-resolution and low-sidelobe forward-look sonar imaging using deconvolution. *Appl. Acoust.* **2021**, *178*, 107986. [CrossRef]
38. Liu, X.H.; Shi, R.W.; Sun, C.; Yang, Y.X.; Zhuo, J. Using deconvolution to suppress range sidelobes for MIMO sonar imaging. *Appl. Acoust.* **2022**, *186*, 108491. [CrossRef]
39. Richardson, W.H. Bayesian-based iterative method of image restoration. *J. Opt. Soc. Am.* **1972**, *62*, 55–59. [CrossRef]
40. Lucy, L.B. An iterative technique for the rectification of observed distributions. *Astron. J.* **1974**, *79*, 745–754. [CrossRef]
41. Fish, D.A.; Brinicomde, A.M.; Pike, E.R. Blind deconvolution by means of the Richards-Lucy algorithm. *J. Opt. Soc. Am.* **1995**, *12*, 58–65. [CrossRef]
42. Yang, L.; Yang, Y.; Hasna, M.O. Coverage, probability of SNR gain, and DOR analysis of RIS-aided communication systems. *IEEE Wireless Commun. Lett.* **2020**, *9*, 1268–1272. [CrossRef]
43. Wang, X.Y.; Wang, L.Y.; Li, G.L.; Xie, X. A robust and fast method for sidescan sonar image segmentation based on region growing. *Sensors* **2021**, *21*, 6960. [CrossRef] [PubMed]
44. Li, J.; Jiang, P.; Zhu, H. A local region-based level set method with Markov random field for side-scan sonar image multi-level segmentation. *IEEE Sens. J.* **2020**, *21*, 510–519. [CrossRef]
45. Wang, H.; Gao, N.; Xiao, Y. Image feature extraction based on improved FCN for UUV side-scan sonar. *Mar. Geophys. Res.* **2020**, *41*, 18. [CrossRef]
46. Wang, Z.; Guo, J.; Huang, W. Side-scan sonar image segmentation based on multi-channel fusion convolution neural networks. *IEEE Sens. J.* **2022**, *22*, 5911–5928. [CrossRef]

Article

On the 2D Beampattern Optimization of Sparse Group-Constrained Robust Capon Beamforming with Conformal Arrays

Yan Dai [1,2], Chao Sun [1,2,*] and Xionghou Liu [1,2]

[1] School of Marine Science and Technology, Northwestern Polytechnical University, Xi'an 710072, China; dy1036987828@mail.nwpu.edu.cn (Y.D.); xhliu@nwpu.edu.cn (X.L.)

[2] Shaanxi Key Laboratory of Underwater Information Technology, Xi'an 710072, China

* Correspondence: csun@nwpu.edu.cn

Abstract: To overcome the problems of the high sidelobe levels and low computational efficiency of traditional Capon-based beamformers in optimizing the two-dimensional (elevation–azimuth) beampatterns of conformal arrays, in this paper, we propose a robust Capon beamforming method with sparse group constraints that is solved using the alternating-direction method of multipliers (ADMM). A robustness constraint based on worst-case performance optimization (WCPO) is imposed on the standard Capon beamformer (SCB) and then the sparse group constraints are applied to reduce the sidelobe level. The constraints are two sparsity constraints: the group one and the individual one. The former was developed to exploit the sparsity between groups based on the fact that the sidelobe can be divided into several different groups according to spatial regions in two-dimensional beampatterns, rather than different individual points in one-dimensional (azimuth-only) beampatterns. The latter is considered to emphasize the sparsity within groups. To solve the optimization problem, we introduce the ADMM to obtain the closed-form solution iteratively, which requires less computational complexity than the existing methods, such as second-order cone programming (SOCP). Numerical examples show that the proposed method can achieve flexible sidelobe-level control, and it is still effective in the case of steering vector mismatch.

Keywords: robust Capon beamforming (RCB); sparse group constraints; alternating-direction method of multipliers (ADMM); two-dimensional beampatterns

Citation: Dai, Y.; Sun, C.; Liu X.H. On the 2D Beampattern Optimization of Sparse Group-Constrained Robust Capon Beamforming with Conformal Arrays. *Remote Sens.* **2024**, *16*, 421. https://doi.org/10.3390/rs16020421

Academic Editors: Jiahua Zhu, Xiaotao Huang, Jianguo Liu, Xinbo Li, Gerardo Di Martino, Shengchun Piao, Junyuan Guo and Wei Guo

Received: 20 December 2023
Revised: 12 January 2024
Accepted: 15 January 2024
Published: 21 January 2024

1. Introduction

In the past decades, conformal arrays have been widely used in sonars [1–3] and radars [4,5] because they improve the dynamic characters of vehicles and offer three-dimensional observation. The core function of conformal arrays is beamforming, which performs weighted summation on the received data of arrays to suppress noise and interference [6], improving the postprocessing performance in applications such as detection [7]. The beampattern can evaluate the performance of the spatial filtering of the beamforming, which is worthy of proper design. Unlike the azimuth-only beampatterns of linear and circular arrays, the beamformers of conformal arrays can be steered at an arbitrary spatial angle without the direction ambiguity found in elevation–azimuth beampatterns, which has attracted much attention in relevant fields.However, the implementation of the beamforming of conformal arrays is more difficult than for other arrays due to the complexity of their array geometry and computation. Up to now, beamforming algorithms have been mainly applied to linear and circular arrays, so the study of the beamforming of conformal arrays is urgently needed.

Among the many beamfomers, theoretically, the Capon beamformer [8] can adaptively suppress interference and minimize the output noise power while maintaining the distortionless response of the desired signal. However, a Capon beamformer often suffers from severe performance degradation in practical applications, which is mainly caused by various types of steering vector (SV) mismatch. This SV mismatch not only causes signal self-cancellation in the mainlobe, but it also leads to an intolerable increase in the sidelobe level (SLL). Several approaches have been developed to improve robustness against SV mismatch. The diagonal loading technique (DL) [9] is adopted to alleviate the imperfect information in the covariance matrix of training data. To determine the diagonal loading factor reasonably, a series of algorithms based on uncertainty sets of SVs have been proposed, such as worst-case performance optimization (WCPO) [10,11] and robust Capon beamforming (RCB) [12]. On this basis, several approaches to synthesizing beampatterns with robust sidelobe control [13,14] have been developed.

Even if the problem of SV mismatch has been resolved, the inherent SLL of the beampatterns of conformal arrays is still too high to meet practical requirements, being restricted by geometry [15], aperture [16], and other factors [17]. It is necessary to artificially impose constraints to further optimize the SLL of the beampattern. Sidelobe control algorithms based on adaptive array theory [18,19] can be applied to arbitrary geometry arrays by adding virtual interference in the sidelobe region to reduce the SLL of the beampattern. The drawbacks of these algorithms are that the convergence is highly reliant on the iteration gain and that there is no clear criterion to determine the mainlobe region in each iteration.To reduce the computational complexity and improve flexibility, accurate array response control algorithms [20–22] have been developed; these are able to control multipoint responses simultaneously using closed-form solutions. For the design of two-dimensional (2D) beampatterns of conformal arrays, increased constraints in the sidelobe region reduce the degree of freedom of the beamformer, and a higher amount of computation is consumed. Consequently, finding a principle to reduce the SLL with as few constraints as possible has become an important issue. One method is to use the concept of the sparsity constraint [23], which refers to the requirement that the vector being sought or optimized must have as few non-zero entries as possible. Recently, this has been used for sidelobe suppression [24–26], and it was shown to be flexible and effective for the adjustment of the SLL and increased robustness. The sidelobes in azimuth-only beampatterns can be discretized into different single directions, and l_1 regularization [27] is usually employed to achieve individual sparsity. The sidelobes in 2D beampatterns can be further divided into block-regions composed of local directions, which suggests that group sparsity [28] with $l_{2,1}$ regularization is more appropriate for describing the features of the sidelobes in 2D beampatterns than individual sparsity. Up until now, group sparsity has not been utilized in beampattern optimization.

Besides the above problems, computational complexity is another important issue in the study of 2D beampattern optimization with conformal arrays. The number of spatial angles scanned using the elevation–azimuth beampattern of conformal arrays increases exponentially compared with that scanned using the azimuth-only beampattern of linear or circular arrays; hence, the design of 2D beampatterns entails a high computational cost. The algorithms in Refs. [10,12,13,24–26] can be easily transformed into the form of either second-order cone programming (SOCP) or semidefinite programming (SDP), followed by the use of software packages such as CVX [29]. The interior-point method used in CVX suffers from a high computational burden, so it is not applicable in the scenario of 2D beampattern optimization. To alleviate the complexity arising from additional computation, recently, the alternating-direction method of multipliers (ADMM) [30] has attracted much attention and has been applied to RCB [31] and beampattern synthesis [32–36], as well as other areas of signal processing [37,38]. The ADMM decouples the global problem into several more local subproblems that are easier to solve, and it obtains the solution of the global problem by coordinating the solutions to the subproblems. Benefiting from fast processing and good convergence, it is suitable for solving large-scale beamforming optimization problems. However, ADMM has not been exploited to solve the optimization problem of a 2D beampattern with sparsity constraints.

This paper is dedicated to the 2D beampattern optimization problem of a Capon beamformer with conformal arrays. We developed a robust Capon beamformer utilizing sparse group constraints that can reduce the SLL flexibly and achieve a robustness of interference suppression in the case of the SV mismatch. We first explore the properties of the 2D beampattern, introducing the group sparsity constraints [39] into the optimization problem, which forms the sparse group constraints together with the individual sparsity constraint. Based on the RCB, the sparse group-constrained robust Capon beamformer (SG-RCB) is then proposed. In order to reduce the computational complexity of the SG-RCB, the ADMM is employed to determine its weight vector. The optimization problem of the SG-RCB is divided into two independent subproblems with the help of a generalized sidelobe canceler (GSC) [40]; one is the WCPO, used to obtain the SV of the desired signal, and the other is the sparse group least absolute shrinkage and selection operator (SGLASSO) [41], used to solve adaptive weight in the GSC. Combining the solutions of the two subproblems, the weight vector is derived in closed form. We show that the proposed beamformer can dramatically reduce the SLL of the 2D beampattern without requiring heavy computational cost, which is significant in practical applications.

The rest of the paper is organized as follows. In Section 2, we review the signal model and concepts on the Capon beamforming and beampatterns. In Section 3, the RCB with sparse group constraints is proposed. In Section 4, the ADMM is introduced to solve the optimization problem of the SG-RCB. In Section 5, we demonstrate the improvement of the proposed method on the 2D beampattern of a conformal array. In Section 6, conclusions are drawn.

Notation 1. *Let us denote matrices and vectors as bold upper-case and lower-case letters, respectively. In particular,* **1** *denotes an array of all ones and* I *stands for the identity matrix.* $j \triangleq \sqrt{-1}$. $\mathbb{R}$ *and* $\mathbb{C}$ *denote the sets of real and complex numbers, respectively.* $\Re(\cdot)$ *and* $\Im(\cdot)$ *are the real part and imaginary part of the argument, respectively. The superscripts* $(\cdot)^T$ *and* $(\cdot)^H$ *denote the transpose operator and the conjugate-transpose operator, respectively.* $\|\cdot\|_p$ *denotes the* l_p *norm of the input vector* $(p = 0, 1, 2)$. $\mathcal{P}(\cdot)$ *is the principle component of the input matrix.*

2. Problem Formulation

2.1. Signal Model

Consider an array with an arbitrary configuration composed of identical M omnidirectional hydrophones. The m hydrophone's position in the three-dimensional Cartesian coordinate system is represented as

$$\mathbf{p}_m = (p_x, p_y, p_z)^T = [r_m \sin\theta_m \cos\phi_m, r_m \sin\theta_m \sin\phi_m, r_m \cos\theta_m]^T \tag{1}$$

where

r_m: the length of the mth hydrophone's radius vector.

(θ_m, ϕ_m): the elevation and azimuth angles of the m hydrophone, respectively.

Suppose that a source in the farfield is in the direction of (θ, ϕ) where $\theta \in [0, \pi]$ and $\phi \in [0, \pi]$. The SV of the plane wave of this source is defined as

$$a(k, \theta, \phi) = \left[e^{-jk^T \mathbf{p}_1}, e^{-jk^T \mathbf{p}_2}, \cdots, e^{-jk^T \mathbf{p}_M} \right]^T \tag{2}$$

where $k = \dfrac{2\pi f}{c} \cdot [\sin\theta \cos\phi, \sin\theta \sin\phi, \cos\theta]^T$ represents the wave number of the plane wave, where f is the frequency and c is the speed of sound.

As shown in Figure 1, suppose there are D far-field narrowband uncorrelated signals (plane waves) impinging on the array, one of which is the desired signal while the other $D-1$ are interferences. The signals received by the array can be written as [6]

$$\mathbf{y}(t) = \mathbf{a}_0 s_0(t) + \sum_{d=1}^{D-1} \mathbf{a}_d s_d(t) + \mathbf{n}(t) \tag{3}$$

where

t: the arbitrary sampling time.

$\mathbf{y}(t) \in \mathbb{C}^{M \times 1}$: the received data sampled by the array.

$\mathbf{a}_0$: the actual SV of the desired signal.

$\mathbf{a}_d$: the SV of the dth interference.

$s_0(t)$: the wavefront of the desired signal.

$s_d(t)$: the wavefront of the dth interference.

$\mathbf{n}(t)$: the zero-mean Gaussian white noise, representing additive noise in the environment received by the array.

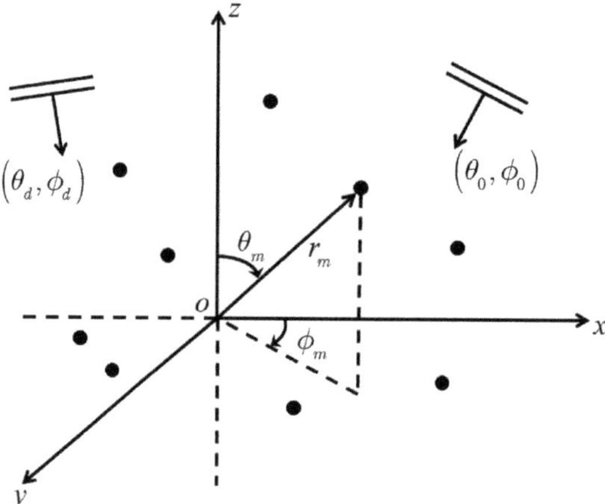

Figure 1. Illustration of the array with arbitrary geometry, in which (θ_0, ϕ_0) is the incidence angle of the desired signal and (θ_d, ϕ_d) is the incidence angle of the dth interference.

Assuming that the desired signal, interferences, and noise are uncorrelated with each other, the noise on each hydrophone is also uncorrelated. The covariance matrix of $\mathbf{y}(t)$ is given as

$$\begin{aligned}
\mathbf{R}_{\mathbf{y}} &= \mathbb{E}\left[\mathbf{y}(t)\mathbf{y}^H(t)\right] \\
&= \mathbf{R}_{\mathrm{s}} + \mathbf{R}_{\mathrm{int+n}} \\
&= \sigma_{\mathrm{s}}^2 \mathbf{a}_0 \mathbf{a}_0^H + \left(\sum_{d=1}^{D-1} \sigma_d^2 \mathbf{a}_d \mathbf{a}_d^H + \sigma_{\mathrm{n}}^2 \mathbf{I}\right)
\end{aligned} \tag{4}$$

where

$\mathbb{E}[\cdot]$: the statistical expectation.

$\mathbf{R}_{\mathrm{s}} = \sigma_{\mathrm{s}}^2 \mathbf{a}_0 \mathbf{a}_0^H$: the covariance matrix of the desired signal.

$\mathbf{R}_{\mathrm{int+n}} = \sum_{d=1}^{D-1} \sigma_d^2 \mathbf{a}_d \mathbf{a}_d^H + \sigma_{\mathrm{n}}^2 \mathbf{I}$: the covariance interference-plus-noise matrix.

$\sigma_{\mathrm{n}}^2 \mathbf{I}$: the covariance matrix of the Gaussian white noise.

σ_{s}^2: the power of the desired signal.

σ_d^2: the power of the dth interference.

σ_{n}^2: the power of noise.

The signal-to-noise ratio (SNR) and interference-to-noise ratio (INR) of the dth interference are defined as [6], respectively, σ_s^2/σ_n^2 and σ_d^2/σ_n^2. In practice, the covariance matrix of $\mathbf{y}(t)$ is estimated with a finite number of samples. The sample covariance matrix of $\mathbf{y}(t)$ is written as

$$\hat{R} = \frac{1}{L}\sum_{t=1}^{L}\mathbf{y}(t)\mathbf{y}^H(t) \tag{5}$$

where L denotes the sample size.

2.2. The Two-Dimensional Beampattern

Beamforming is a technique of weighted summation of the received signals to obtain the beam output of the array [6]. The output of the beamformer is

$$p = \mathbf{w}^H\mathbf{y}(t) \tag{6}$$

where $\mathbf{w} \in \mathbb{C}^{M\times 1}$ denotes the weight vector of the beamformer.

The beampattern is an important measurement to evaluate the performance of the beamformer, which describes the response of the beamformer to a signal impinging on the array in the direction of (θ, ϕ). The response is defined as

$$b(\theta,\phi) = \mathbf{w}^H\mathbf{a}(\theta,\phi) \tag{7}$$

For symmetric configuration arrays such as uniform linear arrays (ULA), the beamformer can be only steered in the azimuth. The resulting beampattern is mathematically represented as a one-dimensional vector $\mathbf{b}\left(\theta,\phi_{1:N_\phi}\right) = \left[b(\theta,\phi_1), b(\theta,\phi_2),\cdots,b\left(\theta,\phi_{N_\phi}\right)\right]$, where N_ϕ is the number of scanning points in the azimuth (as shown in Figure 2a).

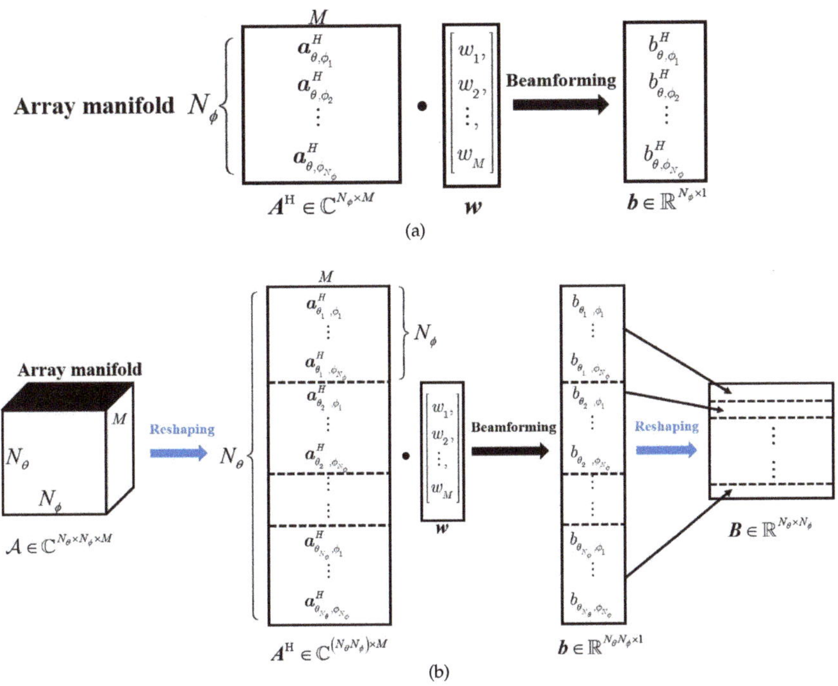

Figure 2. Implementation of (**a**) one-dimensional (azimuth-only) beampatterns; (**b**) 2D beampatterns. The arrows in blue represent the reshaping operation and the arrows in black represent the beamforming operation.

As a result, it can not be employed in applications where full beam steering in three-dimensional space is needed. Arrays similar to that shown in Figure 1 offer three-dimensional steering of the beamformer, which is defined as the elevation–azimuth steering in the coordinate system. The resulting beampattern is mathematically represented as a two-dimensional matrix (as shown in Figure 2b).

$$B(\theta, \phi) = \left[b^T\left(\theta_1, \phi_{1:N_\phi}\right), b^T\left(\theta_2, \phi_{1:N_\phi}\right), \cdots, b^T\left(\theta_{N_\theta}, \phi_{1:N_\phi}\right) \right]^T \tag{8}$$

where N_θ is the number of scanning points in the elevation. It can be seen from Figure 2 that a beampattern is derived by solving a weight vector w.

2.3. Capon-Based Beamforming

The standard Capon beamformer (SCB) minimizes the output power of the interference plus-noise under the distortionless constraint of the desired signal. Then, the w of the SCB is obtained by solving the following optimization problem:

$$\min_{w} \; w^H R_{\text{int}+\text{n}} w, \quad \text{s.t.} \; w^H a_0 = 1 \tag{9}$$

In practice, the $R_{\text{int}+\text{n}}$ is often unavailable and replaced with the sample covariance matrix $\hat{R}$. Substituting $\hat{R}$ into (9), the solution to (9) is given by

$$w_{\text{SCB}} = \frac{\hat{R}^{-1} a_0}{a_0^H \hat{R}^{-1} a_0} \tag{10}$$

The SCB should achieve the optimal performance if the covariance matrix and steering vector are accurately known. In practical scenarios, the estimated matrix $\hat{R}$ carries imprecise knowledge of the real one, leading to an increase in the SLL of the beampattern and affecting its suppression ability. There exists a mismatch between the presumed steering vector and the actual one; the beampattern of the SCB rejects the desired signal as interference and suffers robustness degradation.

When there is a mismatch between the presumed SV and the actual one (denoted as $\tilde{a}$), an improved Capon beamformer against the SV mismatch can be designed by

$$\begin{aligned} \min_{w} \quad & w^H \hat{R} w \\ \text{s.t.} \quad & a_0^H w - 1 \geq \varepsilon_0 \|w\|_2 \end{aligned} \tag{11}$$

where ε_0 specifies the uncertainty level of the norm of difference between the actual SV and the presumed one. The optimization problem (11) is well known as WCPO [10], which has been proven to be a diagonal loading method [9]. The weight vector of the WCPO improves the robustness of the SCB to a certain extent, but the 2D beampattern's SLL of the WCPO is still high.

It can be seen from Figure 2 that the implementation of the 2D beampattern is more complex and computationally intensive than that of the one-dimensional beampattern in the past. Based on (11), the objective of beamforming in this paper is to find a weight vector w through a computational efficient algorithm, such that it can reduce the SLL of the 2D beampattern on the basis of maintaining robustness.

3. The Proposed Robust Capon Beamformer with Sparse Group Constraints

3.1. Individual Sparsity Constraint

The sidelobe control is an important task of beampattern optimization. For the desired beampattern, we hope that the SLL is as low as possible to suppress noise and interference. Now, we consider the fact that the mainlobe region is much smaller than the sidelobe region in the spatial domain, and the average beam response of the mainlobe is much higher than that of the sidelobe. Therefore, the beam response of the 2D beampattern has the property of sparsity distribution, that is, the values of the responses in the mainlobe are far greater than zero, and the rest is equal or approximate to zero. The sparsity constraint is introduced to reduce the SLL based on (11), and the optimization problem (11) is rewritten as

$$
\min_{w} \quad w^H \hat{R} w + \lambda_1 \left\| A_{\mathrm{SL}}^H w \right\|_1
$$

$$
\text{s.t.} \quad a_0^H w - 1 \geq \varepsilon_0 \| w \|_2
$$

(12)

where

λ_1: the individual sparse parameter that is usually determined empirically.

A_{SL}: the array manifold matrix of the sidelobe region.

Mathematically, l_1 norm regularization is adopted to describe the sparsity constraint. The smaller the l_1 norm of the beam response in the sidelobe region, the lower the SLL of the 2D beampattern. The objective function in (12) is a remarkable model called the least absolute shrinkage and selection operator (LASSO) [42], and (12) is called the robust Capon beamforming with a sparse constraint (S-RCB). The constraint in (12) assumes that the elements in the constrained vector are independent of each other. Such constraint emphasizes controlling each beam response individually in sidelobe region, which meets the condition of the design of the one-dimensional (or azimuth-only) beampattern. This sparsity can be further subdivided into individual sparsity.

3.2. Group Sparsity Constraint

Compared with one-dimensional beampatterns, the property of the sidelobe region of 2D beampatterns is changed. Figure 3 shows the sketch of a 2D beampattern, and we will interpret this property in combination with it.

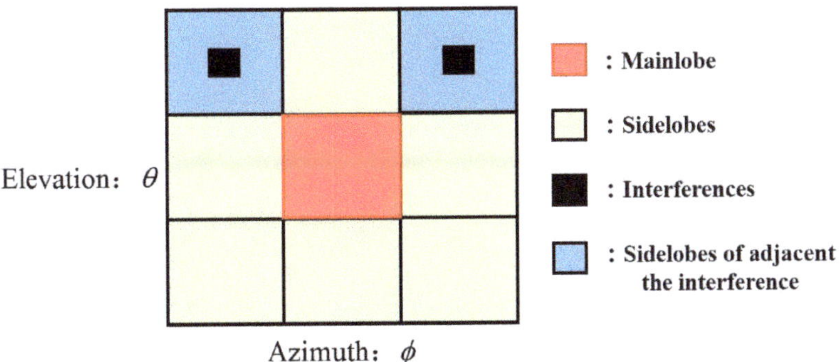

Figure 3. Sketch of the top view of a 2D beampattern.

In Figure 3, we assume that the ordinate is the elevation and the abscissa is the azimuth, both of which are discretized into several scanning points (the number of scanning points is N_θ and N_ϕ, respectively). Any set of coordinates on the two-dimensional grid determines one direction of the spatial angles. A complete 2D beampattern is composed of the beam responses in all scanned directions.

Interference can be properly suppressed without errors according to the nature of Capon beamforming. If the errors are considered, the directions of the interference estimated from the sample covariance matrix are disturbed, which are too "small" to identify on the whole grid plane. In order to improve the tolerance of interference suppression, the interference directions are expanded into small areas centered on the presumed interference directions (black regions in Figure 3). The sidelobe region can be further divided into normal sidelobe regions (green regions in Figure 3) and regions adjacent to the interferences (blue regions in Figure 3) due to the presence of interference.

The sidelobe regions of a 2D beampattern can be briefly divided into the three types mentioned above. The beam responses within each region have approximately the same design requirements, while regions are independent due to their different locations. We hope that the beam responses in the interference regions are lowest in the beampattern. At the same time, the beam responses of the regions adjacent to the interferences should be lower than the normal regions to improve the robustness of interference suppression and the display effect of the beampattern.

Combining the above with the introduction of Figure 3, it can be concluded that the sidelobe of the 2D beampattern is expected to have "regional sparsity" in addition to individual sparsity, and the object subject to sparsity constraints is expanded from a single azimuth to a region composed of multiple spatial angles. Mathematically, this "regional sparsity" is defined as group sparsity [28], which acts like the individual sparsity at the group level. Replacing the individual sparsity in (12) with the group sparsity, the optimization problem (12) is rewritten as

$$\min_{w} \quad w^H \hat{R} w + \lambda_2 \sum_{q=1}^{Q} \sqrt{n_q} \cdot \left\| A_q^H w \right\|_2 \tag{13}$$

$$\text{s.t.} \quad a^H w - 1 \geq \varepsilon_0 \|w\|_2$$

where

λ_2: the parameter of group sparsity that is usually determined empirically.

A_q: the qth array manifold matrix of A_{SL}, i.e., $A_{\text{SL}} = \left[A_1, A_2, \cdots, A_q, \cdots, A_Q \right]^T$, where Q is the number of sidelobe regions.

n_q: the regional sparsity parameter of the qth sidelobe region, and we define $n_Q = \left[n_1, n_2, \cdots, n_q, \cdots, n_Q \right]^T$.

The $l_{2,1}$ norm is introduced in (13), which is a rotational invariant of the l_1 norm [43] and defined as

$$\|x\|_{2,1} = \sum_{q=1}^{Q} \|x_q\|_2 \tag{14}$$

where $x = \left[x_1, x_2, \cdots, x_q, \cdots, x_Q \right]^T$. The $l_{2,1}$ norm in (13) represents the l_1 norm of beam responses of the Q sidelobe regions. Therefore, the sparse constraint in (13) is applied to sidelobe regions rather than directions of spatial angles. The objective function in (13) is called the group LASSO (G-LASSO) model [44], based on the LASSO. When $n_q = 1$ and $Q = N_{\text{SL}} = N_\theta \cdot N_\phi$, (13) is equivalent to (12). Equation (13) is called the robust Capon beamforming with a sparse group constraint (G-RCB).

3.3. Robust Capon Beamforming with Sparse Group Constraints (SG-RCB)

Combining (12) and (13), the optimization problem of the 2D beampattern with both sparsity constraints in this paper is finally given by

$$\min_{w} \quad w^H \hat{R} w + \lambda_1 \left\| A_{\text{SL}}^H w \right\|_1 + \lambda_2 \sum_{q=1}^{Q} \sqrt{n_q} \cdot \left\| A_q^H w \right\|_2 \tag{15}$$

$$\text{s.t.} \quad a^H w - 1 \geq \varepsilon_0 \|w\|_2$$

The objective function in (15) is called the sparse group LASSO (SG-LASSO) model [41], which takes advantage of the sparsity at both the group and individual level within groups in the 2D beampattern. Now, we finally refer to the method of 2D beampattern optimization as the robust Capon beamforming with the sparse group constraints (SG-RCB), which can be seen as a generalization of the S-RCB($\lambda_2 = 0$) and G-RCB($\lambda_1 = 0$). Optimization problems like (12), (13), and (15) are usually solved by convex optimization toolboxes utilizing the interior-point method (IPM). In this paper, the ADMM framework is adopted to obtain iterative solutions with lower computational complexity that are suitable for large-element array processing.

4. The Solution of the SG-RCB via ADMM

4.1. The Generalized Sidelobe Canceler

To simplify the process of solving the optimization problem (15), first we introduce the GSC. The weight vector to be solved in (15) is written as

$$w = \tilde{a}/M - Ug \tag{16}$$

where $\tilde{a}$ represents the actual steering vector and is also the unadaptive weight in the GSC; g is called the adaptive weight of the GSC; $U \in \mathbb{C}^{M \times (M-1)}$ represents the block matrix, which is a semi-unitary matrix orthogonal to $\tilde{a}$, i.e., $U^H U = I$ and $U^H \tilde{a} = 0$. Here, U is selected as the principal component of $P_{\tilde{a}}^{\perp}$: $U = \mathcal{P}\left(P_{\tilde{a}}^{\perp}\right)$, where $P_{\tilde{a}}^{\perp} = I - \tilde{a}\left(\tilde{a}^H \tilde{a}\right)^{-1}\tilde{a}^H$. The structure of the GSC is shown in Figure 4.

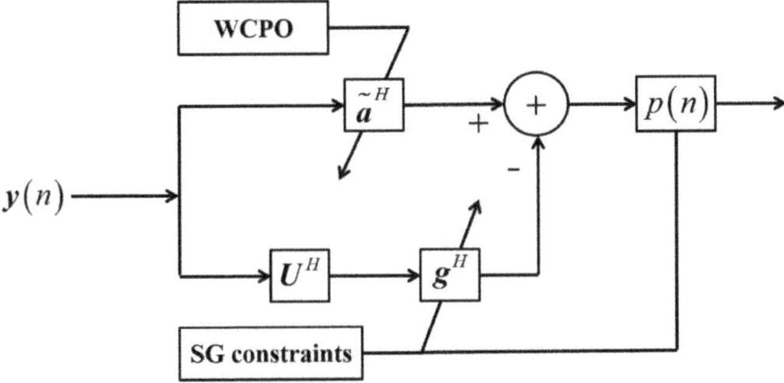

Figure 4. Structure of the SG-RCB based on GSC.

Substituting (16) into (15), (15) is rewritten as

$$\min_{\tilde{a},g} \ \frac{1}{2}\left\| \sqrt{2}\hat{R}^{1/2}\tilde{a}/M - \sqrt{2}\hat{R}^{1/2}Ug \right\|_2^2$$
$$+ \lambda_1 \left\| A_{SL}^H \tilde{a}/M - A_{SL}^H Ug \right\|_1 + \lambda_2 \sum_{q=1}^{Q} \sqrt{n_q} \cdot \left\| A_q^H \tilde{a}/M - A_q^H Ug \right\|_2 \tag{17}$$
$$\text{s.t.} \ \ a^H \tilde{a}/M - a^H Ug - 1 \geq \varepsilon_0 \|\tilde{a}/M - Ug\|_2$$

Combining (16) and (17) with Figure 4, the following corollaries can be drawn:

Corollary 1. *The value of $\tilde{a}$ essentially depends on the source, propagation environment, and physical feature of the array. Assuming that they remain unchanged in the problem, once $\tilde{a}$ is determined, it remains unchanged in the whole optimization problem, and so does U. Therefore, the solving process of $\tilde{a}$ is independent to that of g and the sparsity constraints, and $\tilde{a}$ can be solved by transforming WCPO into RCB [12].*

Corollary 2. *The GSC appropriately adjusts g according to the received data and constraints to meet the design requirements. Combined with Figure 4 and Corollary 1, it can be seen that the sparsity constraint can only affect the value of g.*

Corollary 3. *The bottom branch of the GSC is to suppress noise and interference in the received data according to the nature of the Capon beamformer. Combining Corollary 1, Corollary 2, and Figure 4, it can be seen that the influence of the robustness constraint on the value of g is fixed once the values of $\tilde{a}$ and U are known; that is to say, the robustness constraint only affects the value of g through the quadratic term of the objective function in (15) or (17).*

Corollary 4. *Based on Corollaries 1–3, the unadaptive weight and the adaptive weight are orthogonal and can be solved independently because $\tilde{a}^H U g = 0$.*

Although $\tilde{a}$ and g are coupled in (17), due to the separable structure of the GSC, the optimization problem (17) can be divided into two independent subproblems, one for solving $\tilde{a}$ and the other for solving g, which can be efficiently executed through the ADMM framework.

4.2. Subproblem One: Solve $\tilde{a}$ Using the ADMM-RCB

Ignoring the sparse constraints, (17) is simplified to the WCPO of (11). The WCPO is equivalently written as

$$
\begin{aligned}
&\min_{\tilde{a}} \quad \tilde{a}^H \hat{R}^{-1} \tilde{a} \\
&\text{s.t.} \quad \|\tilde{a} - a\|_2^2 \leq \varepsilon_0^2
\end{aligned}
\tag{18}
$$

Let us define the steering vector and the sample covariance matrix in real domain:

$$
\overline{a} = \left[\Re\left(a^T\right), \Im\left(a^T\right) \right]^T
\tag{19}
$$

$$
\overline{\tilde{a}} = \left[\Re\left(\tilde{a}^T\right), \Im\left(\tilde{a}^T\right) \right]^T
\tag{20}
$$

$$
\overline{\hat{R}^{-1}} = \begin{bmatrix} \Re\left(\hat{R}^{-1}\right) & -\Im\left(\hat{R}^{-1}\right) \\ \Im\left(\hat{R}^{-1}\right) & \Re\left(\hat{R}^{-1}\right) \end{bmatrix}
\tag{21}
$$

where $\overline{a}, \overline{\tilde{a}} \in \mathbb{R}^{2M \times 1}$ and $\overline{\hat{R}^{-1}} \in \mathbb{R}^{2M \times 2M}$.

Let $\delta = \overline{\tilde{a}} - \overline{a}$. Constructing the auxiliary variable z and substituting (19)–(21) into (18), (18) is rewritten as

$$
\begin{aligned}
&\min_{\delta} \quad \delta^H \overline{\hat{R}^{-1}} \delta + \delta^H \overline{\hat{R}^{-1}} \overline{a} + \overline{a}^H \overline{\hat{R}^{-1}} \delta + \overline{a}^H \overline{\hat{R}^{-1}} \overline{a} \\
&\text{s.t.} \quad \delta = z \\
&\qquad \|z\|_2^2 \leq \varepsilon_0^2
\end{aligned}
\tag{22}
$$

The augmented Lagrangian function (ALM) corresponding to (22) is

$$
\mathcal{L}_\rho(\delta, z, u) = \delta^H \overline{\hat{R}^{-1}} \delta + \delta^H \overline{\hat{R}^{-1}} \overline{a} + \overline{a}^H \overline{\hat{R}^{-1}} \delta + u^H(\delta - z) + \frac{\rho}{2} \|\delta - z\|_2^2
\tag{23}
$$

where

u: Lagrangian multiplier.

ρ: penalty factor.

The ADMM fixes the remaining variables to remain unchanged when updating a variable in one iteration, iterating the unknown variables in the objective function alternately until all variables converge. One iteration is as follows:

Step 1: Updating δ.

$$
\delta^{(k+1)} = \underset{\delta}{\arg\min} \, \mathcal{L}_\rho\left(\delta, z^{(k)}, u^{(k)}\right)
\tag{24}
$$

Step 2: Updating z.

$$z^{(k+1)} = \underset{z}{\operatorname{argmin}} \mathcal{L}_\rho\left(\delta^{(k+1)}, z, u^{(k)}\right) \tag{25}$$

$$\text{s.t.} \quad \|z\|_2^2 \leq \varepsilon_0^2$$

Step 3: Updating u.

$$u^{(k+1)} = \underset{u}{\operatorname{argmin}} \, \mathcal{L}_\rho\left(\delta^{(k+1)}, z^{(k+1)}, u\right) \tag{26}$$

where k denotes iterations. The specific process of each step is described below.

Step 1: Updating δ.

In iteration $k+1$, let the partial derivative of $\mathcal{L}_\rho(\delta, z^{(k)}, u^{(k)})$ with respect to δ equal to zero. Then, δ in iteration $k+1$ is obtained:

$$\delta^{(k+1)} = \left(\overline{\hat{R}^{-1}} + \frac{\rho}{2}I\right)^{-1}\left(\frac{\rho}{2}z^{(k)} - u^{(k)} - \overline{\hat{R}^{-1}\overline{a}}\right) \tag{27}$$

Step 2: Updating z.

In the $(k+1)$th iteration, we ignore the terms unrelated to z and write the ALM corresponding to (25):

$$\mathcal{L}_z = \left(u^{(k)}\right)^H\left(\delta^{(k+1)} - z^{(k)}\right) + \frac{\rho}{2}\left\|\delta^{(k+1)} - z^{(k)}\right\|_2^2 + \lambda\left[\left(z^{(k)}\right)^H z^{(k)} - \varepsilon_0^2\right] \tag{28}$$

Let $\dfrac{\partial \mathcal{L}_z}{\partial z} = 0$ The updated z in iteration $k+1$ is then derived by

$$z_i^{(k+1)} = \frac{u_i^{(k)} + \rho/2}{\lambda + \rho/2} \tag{29}$$

where z_i is the ith element of $z^{(k+1)}$.

If $\left\|z^{(k+1)}\right\|_2^2 < \varepsilon_0^2$, the inequality constraint in (25) is not activated, and the obtained $z^{(k+1)}$ is a solution that satisfies (25). On the contrary, the value of $\left\|z^{(k+1)}\right\|_2^2$ is obtained on the boundary of the inequality constraint, i.e., $\left\|z^{(k+1)}\right\|_2^2 = \varepsilon_0^2$, and substituting it into (28) to obtain λ, the updated z_i in z in the $(k+1)$th iteration can be represented by

$$z_i^{(k+1)} = \varepsilon_0 \cdot \left(u_i^{(k)} + \frac{\rho}{2}\delta_i^{(k+1)}\right) \Big/ \left\|u_i^{(k)} + \frac{\rho}{2}\delta_i^{(k+1)}\right\|_2 \tag{30}$$

Step 3: Updating u.

Let $\dfrac{\partial \mathcal{L}_\rho\left(\delta^{(k+1)}, z^{(k+1)}, u\right)}{\partial u} = 0$. The updated u in the $(k+1)$th iteration is then derived as

$$u^{(k+1)} = u^{(k)} + \rho(\delta^{(k+1)} - z^{(k+1)}) \tag{31}$$

In the ADMM of subproblem one, steps 1–3 are alternately cycled until the following two termination conditions are met simultaneously:

$$e^{primal} = \left\|\delta^{(k+1)} - z^{(k+1)}\right\|_2 \leq \zeta^{primal}$$

$$e^{dual} = \left\|z^{(k+1)} - z^{(k)}\right\|_2 \leq \zeta^{dual} \tag{32}$$

where $\zeta^{primal} > 0$ and $\zeta^{dual} > 0$ are the tolerances of the feasibility conditions, respectively, and "primary" and "dual" refer to the primal feasibility and the dual feasibility, respectively. Their values are determined jointly by the absolute tolerance and the relative tolerance [30]. The complex solution of subproblem one is finally obtained according to the relationship shown in (19) and (20):

$$\tilde{a} = a + \delta(1:M) + j \cdot \delta(M+1:2M) \tag{33}$$

U is also determined with $\tilde{a}$ derived. The algorithm of solving subproblem one is called the robust Capon beamforming based on the alternating-direction method of multipliers, abbreviated as the ADMM-RCB and summarized in Algorithm 1.

Algorithm 1: ADMM-RCB

Input: the sample covariance matrix $\hat{R}$, the presumed SV of the desired signal a, and the uncertainty level ε_0;
Output: the actual SV of the desired signal $\tilde{a}$;

1: Perform the eigenvalue-decomposition of $\hat{R}$ and obtain $\overline{\hat{R}^{-1}}$, $\overline{a}$, and $\overline{\overline{a}}$ defined by (19)–(21);
2: Let $\delta = \overline{\overline{a}} - \overline{a}$ and initialize δ, z and u;
3: **while** $e^{primal} > \zeta^{primal}$ or $e^{dual} > \zeta^{dual}$ **do**
4:      Update $\delta^{(k+1)}$ by (27);
5:      Update $z_i^{(k+1)}$ in $z^{(k+1)}$ by (30);
6:      Update $u^{(k+1)}$ by (31);
7:      $k \leftarrow k+1$;
8: **end while**

4.3. Subproblem Two: Solve g Using the ADMM-SGLASSO

When $\tilde{a}$ and U have been solved, (17) can be simplified as the standard SGLASSO:

$$\min_{g} \quad \frac{1}{2}\left\|\sqrt{2}\hat{R}^{1/2}\tilde{a}/M - \sqrt{2}\hat{R}^{1/2}Ug\right\|_2^2$$
$$+ \lambda_1\left\|A_{\mathrm{SL}}^H\tilde{a}/M - A_{\mathrm{SL}}^H Ug\right\|_1 + \lambda_2\sum_{q=1}^{Q}\sqrt{n_q}\cdot\left\|A_q^H\tilde{a}/M - A_q^H Ug\right\|_2 \tag{34}$$

We now define the following real variables:

$$\overline{g} = \left[\Re\left(g^T\right),\Im\left(g^T\right)\right]^T \tag{35}$$

$$\overline{U} = \begin{bmatrix} \Re(U) & -\Im(U) \\ \Im(U) & \Re(U) \end{bmatrix} \tag{36}$$

$$\overline{A_{\mathrm{SL}}} = \begin{bmatrix} \Re(A_{\mathrm{SL}}) & -\Im(A_{\mathrm{SL}}) \\ \Im(A_{\mathrm{SL}}) & \Re(A_{\mathrm{SL}}) \end{bmatrix} \tag{37}$$

where $\overline{g} \in \mathbb{R}^{(2M-2)\times 1}$, $\overline{U} \in \mathbb{R}^{2M\times(2M-2)}$, and $\overline{A_{\mathrm{SL}}} \in \mathbb{R}^{2M\times 2N_{SL}}$. By substituting (20), (21), and (35)–(37) into (34), it can be rewritten as

$$\min_{g,r} \quad \frac{1}{2}\left\|\sqrt{2}\overline{\hat{R}^{1/2}}\overline{a}/M - \sqrt{2}\overline{\hat{R}^{1/2}}\,\overline{U}\overline{g}\right\|_2^2 + \lambda_1\left\|\overline{A_{\mathrm{SL}}}^H\overline{a}/M - \overline{A_{\mathrm{SL}}}^H\overline{U}\overline{g}\right\|_1 + \lambda_2\sum_{q=1}^{Q}\sqrt{n_q}\cdot\|r_q\|_2 \tag{38}$$

$$\text{s.t.} \quad \overline{A_{\mathrm{SL}}}^H\overline{a}/M - \overline{A_{\mathrm{SL}}}^H\overline{U}\overline{g} = r$$

where $\widehat{\boldsymbol{R}}^{1/2}$ is derived by substituting $\hat{\boldsymbol{R}}^{1/2}$ into (21), $\boldsymbol{r} = \left[\boldsymbol{r}_1^T, \boldsymbol{r}_2^T, \cdots, \boldsymbol{r}_q^T, \cdots, \boldsymbol{r}_Q^T\right]^T$ represents the auxiliary variable, and $\boldsymbol{r}_q$ is the qth group of auxiliary variables, which corresponds to the beam responses of each spatial angle in the qth sidelobe region. The ALM corresponding to (38) is written as follows:

$$
\begin{aligned}
\mathcal{L}_\rho(\overline{\boldsymbol{g}}, \boldsymbol{r}, \boldsymbol{u}_0) =& \frac{1}{2}\left\|\sqrt{2}\overline{\widehat{\boldsymbol{R}}^{1/2}}\overline{\boldsymbol{a}}/M - \sqrt{2}\overline{\widehat{\boldsymbol{R}}^{1/2}}\overline{\boldsymbol{U}}\overline{\boldsymbol{g}}\right\|_2^2 + \lambda_1\left\|\overline{\boldsymbol{A}_{\mathrm{SL}}}^H\overline{\boldsymbol{a}}/M - \overline{\boldsymbol{A}_{\mathrm{SL}}}^H\overline{\boldsymbol{U}}\overline{\boldsymbol{g}}\right\|_1 + \lambda_2\sum_{q=1}^{Q}\sqrt{n_q}\cdot\|\boldsymbol{r}_q\|_2 \\
& + \boldsymbol{u}_0^H\left(\overline{\boldsymbol{A}_{\mathrm{SL}}}^H\overline{\boldsymbol{a}}/M - \overline{\boldsymbol{A}_{\mathrm{SL}}}^H\overline{\boldsymbol{U}}\overline{\boldsymbol{g}} - \boldsymbol{r}\right) + \frac{\rho}{2}\left\|\overline{\boldsymbol{A}_{\mathrm{SL}}}^H\overline{\boldsymbol{a}}/M - \overline{\boldsymbol{A}_{\mathrm{SL}}}^H\overline{\boldsymbol{U}}\overline{\boldsymbol{g}} - \boldsymbol{r}\right\|_2^2
\end{aligned}
\tag{39}
$$

where $\boldsymbol{u}_0$ is the Lagrange multiplier. One iteration of the subproblem two is as follows:

Step 1: Updating $\overline{\boldsymbol{g}}$.

$$
\overline{\boldsymbol{g}}^{(k+1)} = \underset{\overline{\boldsymbol{g}}}{\arg\min}\ \mathcal{L}_\rho\left(\overline{\boldsymbol{g}}, \boldsymbol{r}^{(k)}, \boldsymbol{u}_0^{(k)}\right)
\tag{40}
$$

Step 2: Updating $\boldsymbol{r}$.

$$
\boldsymbol{r}^{(k+1)} = \underset{\boldsymbol{r}}{\arg\min}\ \mathcal{L}_\rho\left(\overline{\boldsymbol{g}}^{(k+1)}, \boldsymbol{r}, \boldsymbol{u}_0^{(k)}\right)
\tag{41}
$$

Step 3: Updating $\boldsymbol{u}_0$.

$$
\boldsymbol{u}_0^{(k+1)} = \underset{\boldsymbol{u}_0}{\arg\min}\ \mathcal{L}_\rho\left(\overline{\boldsymbol{g}}^{(k+1)}, \boldsymbol{r}^{(k+1)}, \boldsymbol{u}_0\right)
\tag{42}
$$

The specific process of each step is described below:

Step 1: Updating $\overline{\boldsymbol{g}}$.

Constructing the auxiliary variable $\boldsymbol{z}$, in the $(k+1)$th iteration, (40) can be equivalently expressed as

$$
\overline{\boldsymbol{g}}^{(k+1)} = \underset{\overline{\boldsymbol{g}}}{\arg\min}\ \mathcal{L}_\rho\left(\overline{\boldsymbol{g}}, \boldsymbol{r}^{(k)}, \boldsymbol{u}_0^{(k)}\right), \quad \text{s.t.}\ \ \overline{\boldsymbol{A}_{\mathrm{SL}}}^H\overline{\boldsymbol{a}}/M - \overline{\boldsymbol{A}_{\mathrm{SL}}}^H\overline{\boldsymbol{U}}\overline{\boldsymbol{g}} = \boldsymbol{z}
\tag{43}
$$

which can also be iteratively solved by the ADMM. The ALM corresponding to (43) is

$$
\begin{aligned}
\mathcal{L}_\rho\left(\overline{\boldsymbol{g}}, \boldsymbol{r}^{(k)}, \boldsymbol{u}_0^{(k)}, \boldsymbol{z}, \boldsymbol{u}_1\right) =& \frac{1}{2}\left\|\sqrt{2}\overline{\widehat{\boldsymbol{R}}^{1/2}}\overline{\boldsymbol{a}}/M - \sqrt{2}\overline{\widehat{\boldsymbol{R}}^{1/2}}\overline{\boldsymbol{U}}\overline{\boldsymbol{g}}\right\|_2^2 + \lambda_1\|\boldsymbol{z}\|_1 \\
& + \left[\boldsymbol{u}_0^{(k)} + \boldsymbol{u}_1\right]^H\left(\overline{\boldsymbol{A}_{\mathrm{SL}}}^H\overline{\boldsymbol{a}}/M - \overline{\boldsymbol{A}_{\mathrm{SL}}}^H\overline{\boldsymbol{U}}\overline{\boldsymbol{g}}\right) - \left[\boldsymbol{u}_0^{(k)}\right]^H\boldsymbol{r}^{(k)} - \boldsymbol{u}_1^H\boldsymbol{z} \\
& + \frac{\rho}{2}\left[\left\|\overline{\boldsymbol{A}_{\mathrm{SL}}}^H\overline{\boldsymbol{a}}/M - \overline{\boldsymbol{A}_{\mathrm{SL}}}^H\overline{\boldsymbol{U}}\overline{\boldsymbol{g}} - \boldsymbol{r}^{(k)}\right\|_2^2 + \left\|\overline{\boldsymbol{A}_{\mathrm{SL}}}^H\overline{\boldsymbol{a}}/M - \overline{\boldsymbol{A}_{\mathrm{SL}}}^H\overline{\boldsymbol{U}}\overline{\boldsymbol{g}} - \boldsymbol{z}\right\|_2^2\right]
\end{aligned}
\tag{44}
$$

where $\boldsymbol{u}_1$ is the Lagrange multiplier. For step 1, $\boldsymbol{r}^{(k)}$ and $\boldsymbol{u}_0^{(k)}$ are regarded as constants, and $\overline{\boldsymbol{g}}, \boldsymbol{z}$, and $\boldsymbol{u}_1$ are the variables that need to be iteratively solved. This is similar to subproblem two; step 1 in subproblem two can be solved by the ADMM as follows:

Substep 1.1: Updating $\overline{\boldsymbol{g}}$.

$$
\overline{\boldsymbol{g}}^{(k'+1)} = \underset{\overline{\boldsymbol{g}}}{\arg\min}\ \mathcal{L}_\rho\left(\overline{\boldsymbol{g}}, \boldsymbol{r}^{(k)}, \boldsymbol{u}_0^{(k)}, \boldsymbol{z}^{(k')}, \boldsymbol{u}_1^{(k')}\right)
\tag{45}
$$

Substep 1.2: Updating $\boldsymbol{z}$.

$$
\boldsymbol{z}^{(k'+1)} = \underset{\boldsymbol{z}}{\arg\min}\ \mathcal{L}_\rho\left(\overline{\boldsymbol{g}}^{(k'+1)}, \boldsymbol{r}^{(k)}, \boldsymbol{u}_0^{(k)}, \boldsymbol{z}, \boldsymbol{u}_1^{(k')}\right)
\tag{46}
$$

Substep 1.3: Updating u_1.

$$u_1^{(k'+1)} = \underset{u_1}{\arg\min} \, \mathcal{L}_\rho\left(\overline{g}^{(k'+1)}, r^{(k)}, u_0^{(k)}, z^{(k'+1)}, u_1\right) \tag{47}$$

where k' denotes the iterations in step 1. The following describes the specific process:

Substep 1.1: Updating $\overline{g}$.

In the $(k'+1)$th iteration, by taking the partial derivative of the ALM in (45) with respect to $\overline{g}$ and then making it zero, the expression of $\overline{g}^{(k'+1)}$ is derived as

$$\overline{g}^{(k'+1)} = Q^{-1} b^{(k')} \tag{48}$$

where

$$Q = \left[2\overline{U}^H\left(\overline{R} + \rho\overline{A_{\mathrm{SL}}}\,\overline{A_{\mathrm{SL}}}^H\right)\overline{U}\right] \tag{49}$$

$$b^{(k')} = 2\overline{U}^H\overline{R}\overline{a}/M + \rho\overline{U}^H\overline{A_{\mathrm{SL}}}\left[2\overline{A_{\mathrm{SL}}}^H\overline{a}/M + \left(u_0^{(k)} + u_1^{(k')}\right)\middle/\rho - r^{(k)} - z^{(k')}\right] \tag{50}$$

where $\overline{R}$ is in the real form of $\hat{R}$.

Substep 1.2: Updating z.

In the $(k'+1)$th iteration, ignoring terms unrelated to z, (46) is written as

$$\begin{aligned}
z^{(k'+1)} &= \underset{z}{\arg\min} \, \mathcal{L}_\rho\left(\overline{g}^{(k'+1)}, r^{(k)}, u_0^{(k)}, z, u_1^{(k')}\right) \\
&= \underset{z}{\arg\min}\left[\frac{2\lambda_1}{\rho}\|z\|_1 + \left\|z - \left(u_1^{(k')}\middle/\rho + \overline{A_{\mathrm{SL}}}^H\overline{a}/M - \overline{A_{\mathrm{SL}}}^H\overline{U}\overline{g}^{(k'+1)}\right)\right\|_2^2\right]
\end{aligned} \tag{51}$$

Let $t^{(k')} = u_1^{(k')}\middle/\rho + \overline{A_{\mathrm{SL}}}^H\overline{a}/M - \overline{A_{\mathrm{SL}}}^H\overline{U}\overline{g}^{(k'+1)}$. The last row of (51) is the proximal mapping of $t^{(k')}$. For a given $t^{(k')}$, the elements in $z^{(k'+1)}$ can be expressed by soft thresholding as

$$z_i^{(k'+1)} = S_{\lambda_1/\rho}\left[t_i^{(k')}\right] = \begin{cases} t_i^{(k')} - \lambda_1/\rho & t_i^{(k')} > \lambda_1/\rho \\ 0 & \left|t_i^{(k')}\right| \le \lambda_1/\rho \\ t_i^{(k')} + \lambda_1/\rho & t_i^{(k')} < -\lambda_1/\rho \end{cases} \tag{52}$$

where $S_{\lambda_1/\rho}\left[t_i^{(k')}\right]$ represents the soft thresholding operator, the diagram of which is shown in Figure 5a. It can be seen from Figure 5a that the operator performs a "zero" on some elements of the argument, thereby satisfying the sparsity constraint.

Substep 1.3: Updating u_1.

Let the partial derivative of the ALM regarding u_1 in (47) be zero. Then, u_1 in the $(k+1)$th iteration is derived as

$$u_1^{(k'+1)} = u_1^{(k')} + \rho\left(\overline{A_{\mathrm{SL}}}^H\overline{a}/M - \overline{A_{\mathrm{SL}}}^H\overline{U}\overline{g}^{(k'+1)} - z^{(k'+1)}\right) \tag{53}$$

Substeps 1.1 to 1.3 are alternately cycled until both of the following iteration termination conditions are met:

$$e^{primal} = \left\| \overline{A_{SL}}^H \overline{\tilde{a}}/M - \overline{A_{SL}}^H \overline{U}\overline{g}^{(k'+1)} - z^{(k'+1)} \right\|_2 \leq \eta^{primal}$$

$$e^{dual} = \left\| z^{(k'+1)} - z^{(k')} \right\|_2 \leq \eta^{dual} \tag{54}$$

where $\eta^{primal} > 0$ and $\eta^{dual} > 0$ are the tolerances of the feasibility conditions, respectively. $\overline{g}$ is yielded as the output of the $(k+1)$th iteration in subproblem two.

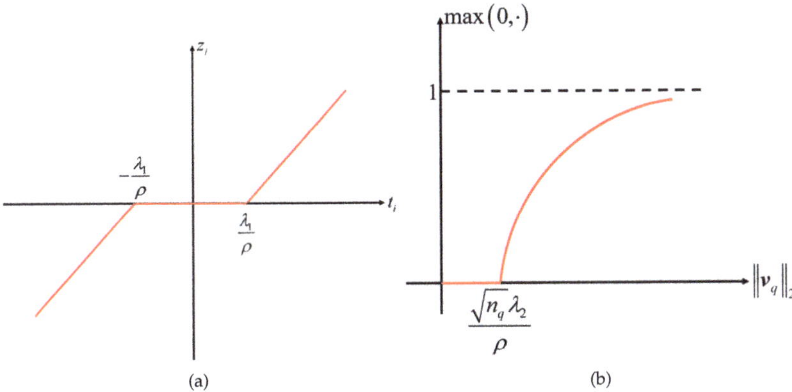

Figure 5. Diagrams of two soft thresholding operators: (**a**) the soft thresholding operator $S_{\lambda_1/\rho}\left[t_i^{(k')}\right]$; (**b**) the block soft thresholding operator $\mathcal{S}_{\sqrt{n_q}\lambda_2/\rho}\left[v_q^{(k)}\right]$].

Step 2: Updating r.

According to (38), the auxiliary variable r is divided into Q groups for the group sparsity, and each r_q has its own regional sparsity parameter, so it needs to be calculated separately. Ignoring the terms unrelated to r, in the $(k+1)$th iteration, the optimization problem on r_q in (41) can be expressed as

$$
\begin{aligned}
r_q^{(k+1)} &= \underset{r_q}{\arg\min}\, \mathcal{L}_\rho\left(\overline{g}^{(k+1)}, r_q, u_{0,q}^{(k)}\right) \\
&= \underset{r_q}{\arg\min}\left[\frac{2\sqrt{n_q}\lambda_2}{\rho}\|r_q\|_2 + \left\|r_q - \left(u_{0,q}^{(k)}/\rho + \overline{A_{SL,q}}^H\overline{\tilde{a}}/M - \overline{A_{SL,q}}^H\overline{U}\overline{g}^{(k+1)}\right)\right\|_2^2\right]
\end{aligned}
\tag{55}
$$

where $u_{0,q}^{(k)}$ represents the part of $u_0^{(k)}$ corresponding to $r_q^{(k+1)}$. Similar to (51), let $v_q^{(k)} = u_{0,q}^{(k)}/\rho + \overline{A_{SL,q}}^H\overline{\tilde{a}}/M - \overline{A_{SL,q}}^H\overline{U}\overline{g}^{(k+1)}$. Then, the elements in $r_q^{(k+1)}$ can be expressed by block soft thresholding as

$$
\begin{aligned}
r_q^{(k+1)} &= \mathcal{S}_{\sqrt{n_q}\lambda_2/\rho}\left[v_q^{(k)}\right] \\
&= \begin{cases} 0 & \left\|v_q^{(k)}\right\|_2 = 0 \\ \max\left(0, 1 - \left(\sqrt{n_q}\lambda_2/\rho\right)\Big/\left\|v_q^{(k)}\right\|_2\right)\cdot v_q^{(k)} & \text{otherwise} \end{cases}
\end{aligned}
\tag{56}
$$

where $\mathcal{S}_{\sqrt{n_q}\lambda_2/\rho}\left[v_q^{(k)}\right]$ is the block soft thresholding operator. $\max(0, \cdot)$ is the function that indicates the maximum value after the input and zero are compared. The diagram of $\mathcal{S}_{\sqrt{n_q}\lambda_2/\rho}\left(v_q^{(k)}\right)$ is shown in Figure 5b. Here, it can be seen that the value less than $\sqrt{n_q}\lambda_2/\rho$

in the augment is set to zero, and the remaining value is reduced. $r^{(k+1)}$ is obtained by performing the operator in (56) on each $r_q^{(k+1)}$ corresponding to the qth region and combining them together.

Step 3: Updating u_0.

Let the partial derivative of the ALM with respect to u_0 in (42) be zero. Then, $u_0^{(k+1)}$ is expressed as

$$u_0^{(k+1)} = u_0^{(k)} + \rho\left(\overline{A_{SL}}^H\overline{\tilde{a}}/M - \overline{A_{SL}}^H\overline{U}\overline{g}^{(k+1)} - r^{(k+1)}\right) \tag{57}$$

In subproblem two, steps 1–3 are performed alternately until the following two iteration termination conditions are met simultaneously:

$$e^{primal} = \left\|\overline{A_{SL}}^H\overline{\tilde{a}}/M - \overline{A_{SL}}^H\overline{U}\overline{g}^{(k+1)} - r^{(k+1)}\right\|_2 \leq \epsilon^{primal}$$
$$e^{dual} = \left\|r^{(k+1)} - r^{(k)}\right\|_2 \leq \epsilon^{dual} \tag{58}$$

where $\epsilon^{primal} > 0$ and $\epsilon^{dual} > 0$ are the tolerances of feasibility conditions, respectively. The solution of subproblem two is finally obtained according to the vector relationship shown in (35):

$$g = \overline{g}(1:M-1) + j \cdot \overline{g}(M:2M-2) \tag{59}$$

The complete method for solving subproblem two is called the sparse group LASSO based on the alternating-direction method of multipliers, abbreviated as the ADMM-SGLASSO and summarized in Algorithm 2. Substituting (33) and (59) into (16) yields the weight vector of the SG-RCB beamformer. The flowchart of the SG-RCB is shown in Figure 6.

Algorithm 2 : ADMM-SGLASSO

Input: the sample covariance matrix $\hat{R}$, the actual SV of the desired signal $\tilde{a}$, the block
       matrix U, the array manifold matrix of the sidelebe region A_{SL} and $\lambda_1, \lambda_2, n_q$

Output: the adaptive weight g;

1: Obtain $\overline{\tilde{a}}, \hat{R}^{1/2}, \overline{g}, \overline{U}$ and $\overline{A_{SL}}$ defined by (20), (21), and (35)–(37);
2: Initialize $\overline{g}, r = \left[r_1^T, r_2^T, \cdots, r_q^T, \cdots, r_Q^T\right]^T, u_0$ and Q;
3: **while** $e^{primal} > \epsilon^{primal}$ or $e^{dual} > \epsilon^{dual}$ **do**
4:        Update $\overline{g}^{(k+1)}$ by ADMM for $\overline{g}$:
5:        Initialize z and u_1;
6:   **while** $e^{primal} > \eta^{primal}$ or $e^{dual} > \eta^{dual}$ **do**
7:           Calculate $b^{(k')}$ by (50);
8:           Update $\overline{g}^{(k'+1)}$ by (48);
9:           Update $z_i^{(k'+1)}$ in $z^{(k'+1)}$ by (52);
10:         Update $u_1^{(k'+1)}$ by (53);
11:         $k' \leftarrow k' + 1$;
12: **end while**
13:        Update $r_q^{(k+1)}$ in $r^{(k+1)}$ by (56);
14:        Update u_0^{k+1} by (57);
15:        $k \leftarrow k + 1$;
16: **end while**

4.4. Computational Complexity Analysis

We describe the computational complexity of the proposed SG-RCB as measured by the number of multiplication operations. The SG-RCB consists of two algorithms, the

ADMM-RCB and the ADMM-SGLASSO. As Algorithms 1 and 2 show, each algorithm is divided into two stages: preprocessing and iteration. For simplicity, the complexity of one iteration is discussed.

In the preprocessing stage of the ADMM-RCB, $\hat{R}^{-1}$ and $\hat{R}^{-1}\overline{a}$ are calculated in advance, resulting in costs of $O(M^3)$ and $O(M^2)$, respectively. In one complete iteration, the computational costs of steps 4–6 in the ADMM-RCB are $O(M^2)$, $O(M)$, and $O(1)$, respectively.

In the preprocessing stage of the ADMM-SGLASSO, Q^{-1}, $\overline{U}^H\overline{R}\overline{a}$, $\overline{A_{SL}}^H\overline{U}$, and $\overline{A_{SL}}^H\overline{a}$ are calculated and fixed in irritations, which cost $O(M^3 + M^2 N_{SL}) + O(M^2 + MN_{SL})$ multiplications in total. At the iteration stage, the computational costs of steps 7–10 in the ADMM-SGLASSO are $O(MN_{SL})$, $O(M^2)$, $O(MN_{SL})$, and $O(1)$, respectively.

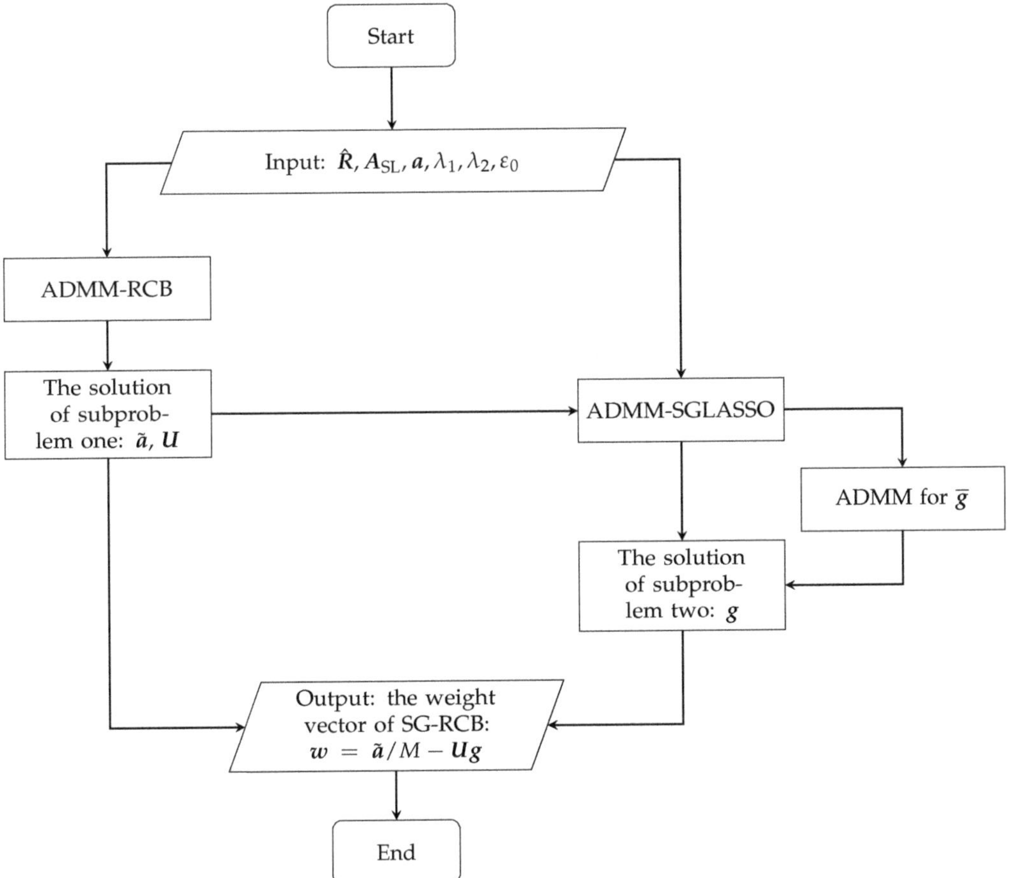

Figure 6. Flowchart of the SG-RCB.

The computational costs of steps 13 and 14 in the ADMM-SGLASSO are $O(MN_{SL}) + O(N_{SL})$ and $O(1)$, respectively. Therefore, the dominant order of the per-iteration computational complexity of the proposed SG-RCB is $O(M^2 + MN_{SL})$. The computation in the preprocessing only needs to be calculated once, which has little impact on the overall complexity of the SG-RCB, although its complexity increases rapidly with the increase in the dimension of variables.

Now, let us compare the SG-RCB with other ADMM-based beamforming methods, for instance, the methods proposed in Refs. [35,36]. In the preprocessing stage, the dominant order of the computational costs of the method in Ref. [35] is $O(M^3 + MN_{SL}^2)$, while the computational complexity in this stage is not discussed in [36]. In practice, the number of hydrophones is lower than the number of scanning directions in the sidelobes; thus, the computational complexity of the SG-RCB in the preprocessing stage is lower than that of the method in Ref. [35]. In the iteration stage, both methods proposed in Refs. [35,36] have same dominant cost of $O\left(M^2 + MN_{SL}\right)$ in per iteration, which is also equal to that of the SG-RCB.

Next, we discuss the complexity of the SOCP for comparison. In the preprocessing stage, the computational cost of the SOCP is of the same order as that of the SG-RCB, i.e., $O\left(M^3 + M^2N_{SL}\right) + O\left(M^2\right)$. In the iteration stage, however, the SOCP adopts the IPM, which costs $O\left(M^2N_{SL}\right)$ at each iteration. The overall computational complexity of the SG-RCB is lower than that of the SOCP solved by the IPM with the variable dimension becoming larger; thus, the proposed SG-RCB has a significant advantage in computational complexity in the case of large-element conformal arrays.

5. Simulation Results

5.1. Parameter Setting

A half-cylindrical conformal array consisting of 40 hydrophones is analyzed in this section. The geometry and parameters of the conformal array are shown in Figure 7. Table 1 presents the relevant parameters shown in Figure 7d.

We consider the simulation being implemented in an underwater free field where the sound velocity is constant at 1500 m/s. The point sources are single-frequency with a frequency of 10 kHz. The receiving array is located in the far field relative to the source, and the transmission loss is not considered.

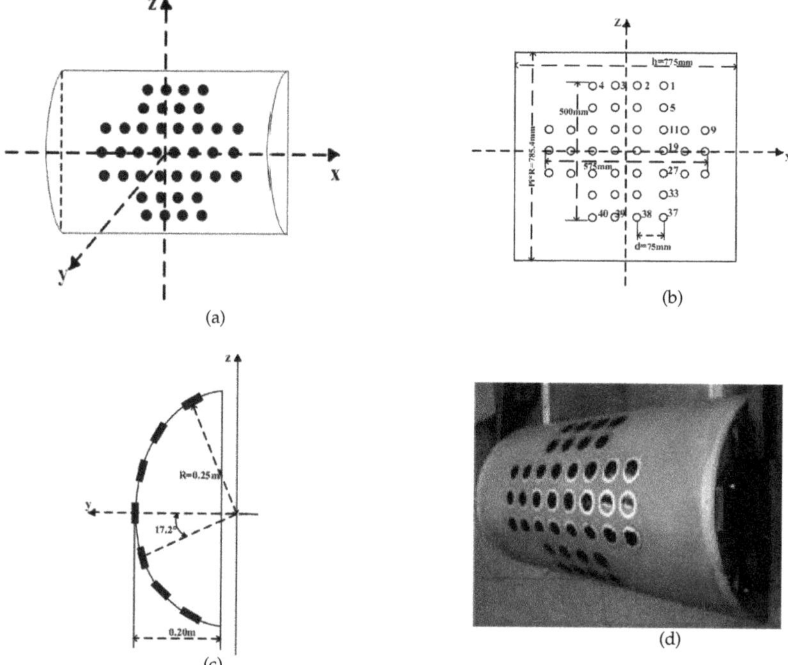

Figure 7. Diagram of the half-cylindrical conformal array (HCCA) configuration: (**a**) three-dimensional diagram (solid black dots represent hydrophones); (**b**) front view (the top left circle represents hydrophone number 1, the bottom right circle represents hydrophone number 40, and the labels of hydrophones increase from top left corner to bottom right corner); (**c**) side view; (**d**) the real object.

Table 1. The physical parameters of the HCCA in Figure 7.

Length of the Bus Bar	Radius	Number of Hydrophones	Spacing between Hydrophones	Operating Frequency
$h = 0.775$ m	$r = 0.25$ m	40	$d = 0.075$ m	6–10 kHz

Assuming that the look direction is $(45°, 45°)$, the signal-to-noise ratio (SNR) of the desired signal is 10 dB. There are two interferences in the direction of $(-30°, 60°)$ and $(-10°, 30°)$, and their interference-to-noise ratios (INRs) are 35 dB. The grouping of the sidelobe is shown in Figure 8, in which the number of regions of sidelobes and interferences are indicated, respectively. The scanning intervals of elevation and azimuth are $1°$, respectively, and the interval of the interference regions in the elevation and azimuth are both $10°$. Our codes were written in MATLAB R2023a. All our numerical experiments were conducted on a laptop computer with an AMD CPU (3.20 GHz) and 32 GB RAM running Windows 11.

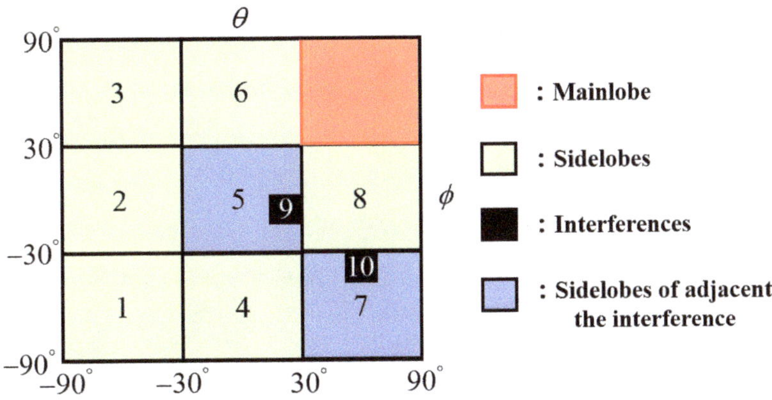

Figure 8. Grouping of sidelobe regions. The numbers indicate the sequence numbers of the different sidelobe regions.

5.2. Cpu Times

Some existing methods were selected to compare the ADMM to in order to show its advantage in terms of computational efficiency. Here, the ADMM-RCB in subproblem one is compared with the RCB with the SOCP solved by CVX [29], which is abbreviated as the SOCP-RCB, and the ε_0 in (18) is set to 1. For the SG-RCB, the SOCP is selected for comparison, which is abbreviated as the SOCP-SG-RCB. The λ_1 and λ_2 in (12), (13), and (15) are set to 0.01 and 0.1 [41,45], respectively. In the ADMM-based methods, the values of the absolute tolerance and the relative tolerance are unified and set to 10^{-4} and 10^{-2}, respectively [30].

Figure 9 shows the computing time (CPU time) of the different studied methods. We compare the ADMM-RCB with the SOCP-RCB in Figure 9a. It is easily observed that the CPU time of the ADMM is much shorter than that of the SOCP. The average time of the SOCP-RCB is 0.3293 s, while that of the ADMM-RCB is 0.009 s.

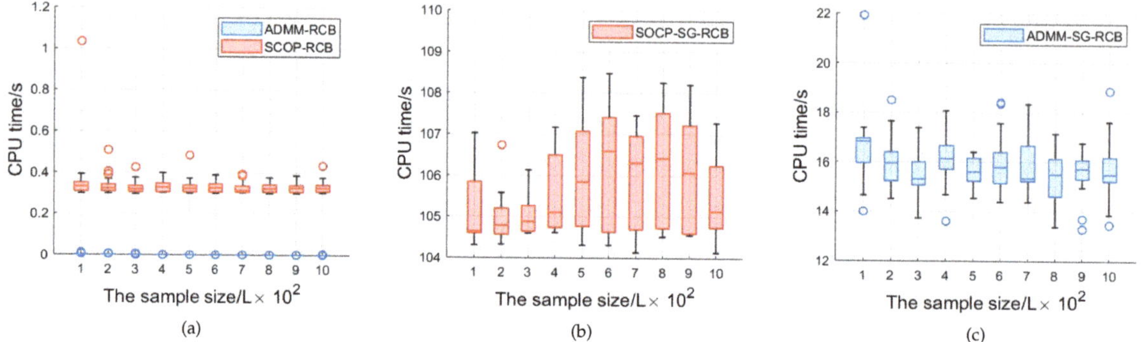

Figure 9. Boxcharts of the computing times of different methods: (**a**) ADMM–RCB and SOCP–RCB; (**b**) SOCP–SG–RCB; (**c**) ADMM–SG–RCB.

Next, we compare the computing time of different methods to solve the SG-RCB. Figure 9b,c show the CPU time of the SOCP, as well as of the ADMM for solving the SG-RCB. The SOCP-SG-RCB involves a long running time, and its average time is 105.6675 s. The SOCP belongs to interior-point method; it is not applicable to the 2D beampatterns because of its high computational complexity. Comparing Figure 9c with Figure 9b, it can be found that the CPU time of the ADMM-SG-RCB is much less than that of the SOCP-SG-RCB (15.8016 s).

5.3. Beampatterns of Different Methods

In this subsection, we further explore the 2D beampatterns of different beamformers. All used parameters are the same as those in Section 5.2. First, we present the beampatterns of some fundamental beamformers in Figure 10, in which the beam responses of two interferences are labeled. It is observed that the CBF beampattern can not suppress two interferences, while the SCB beampatterns distort and their SLL is even higher than the response in the actual signal direction. In accordance with Figure 9a, the ADMM-RCB beampattern is superior to that of the SOCP-RCB, even though their beampatterns are close.

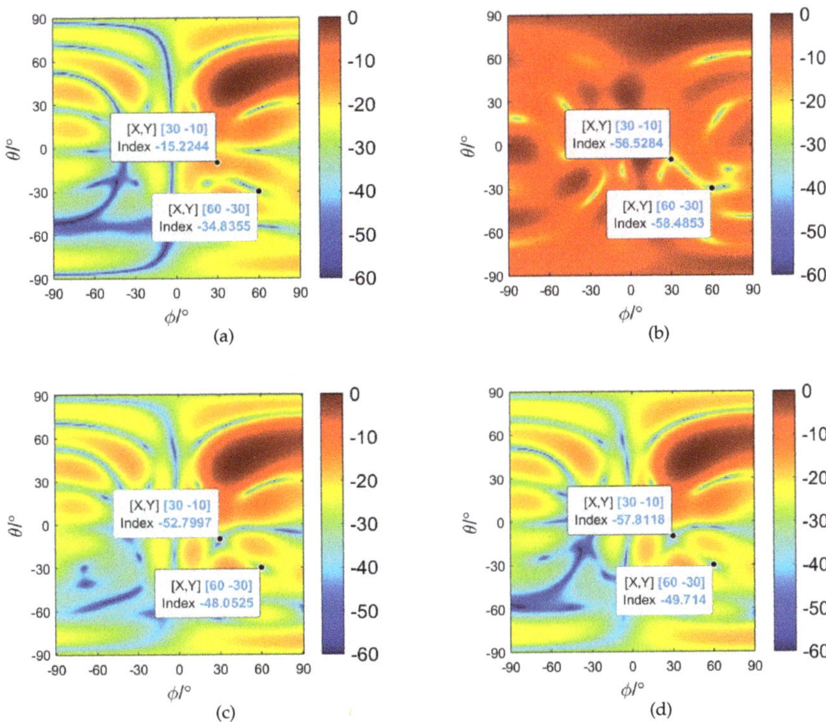

Figure 10. Beampatterns of some fundamental beamformers: (**a**) CBF; (**b**) SCB; (**c**) SOCP–RCB; (**d**) ADMM–RCB.

Then, the beampattern of the SG-RCB is analyzed, while the S-RCB and G-RCB are run for comparison. Figure 11 shows the beampatterns of the three beamformers, in which the first row is the three-dimensional view and the second is the top view. It can be seen from Figure 11d that the S-RCB makes the beam responses of sidelobe regions as low as possible at the cost of increasing the mainlobe width. In Figure 11e, the mainlobe width of the G-RCB is narrower than that of the S-RCB, but its SLL is higher. The constraint of the G-RCB is imposed on regions of the sidelobe and interferences separately, and the beam responses of the interferences of the G-RCB are lower than those of the sidelobes. The SG-RCB's beampattern in Figure 11c,f can be seen as a combination of the above two beamformers, which ensures a low SLL and a narrow mainlobe width while keeping the beamformer from losing its interference suppression capability.

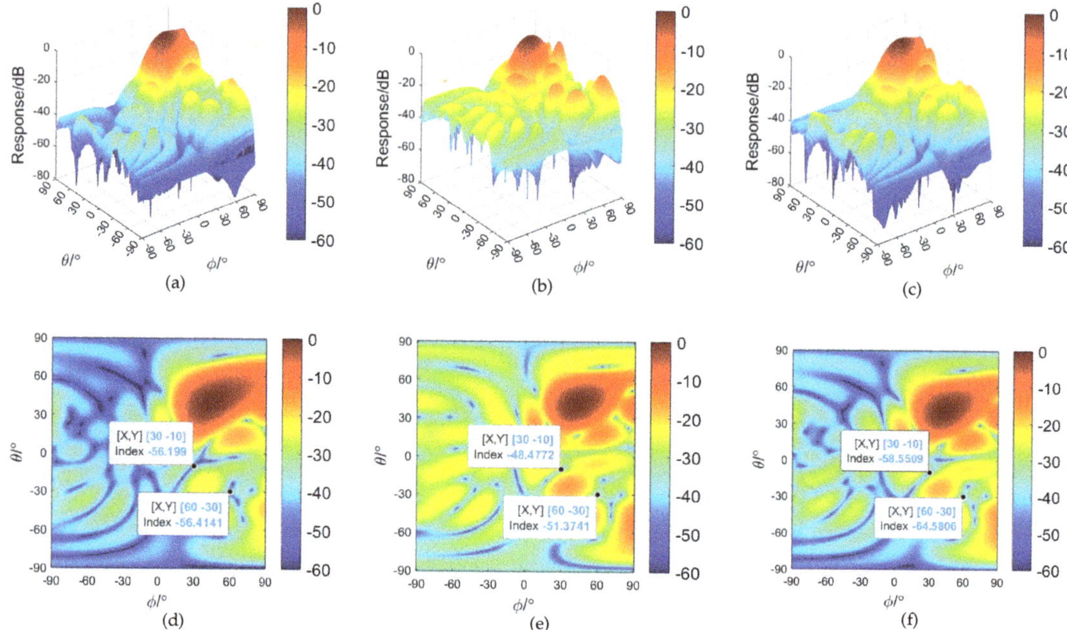

Figure 11. Beampatterns among the S–RCB, G–RCB and SG–RCB: (**a**) 2D beampattern of the S–RCB ($\lambda_1 = 0.01$ and $\lambda_2 = 0$); (**b**) 2D beampattern of the G–RCB ($\lambda_1 = 0$ and $\lambda_2 = 0.1$); (**c**) 2D beampattern of the SG–RCB ($\lambda_1 = 0.01$ and $\lambda_2 = 0.1$); (**d**) top view of the S–RCB result; (**e**) top view of the G–RCB result; (**f**) top view of the SG–RCB result.

In Table 2, we compare the specific performance of the beampattern obtained by different beamformers in terms of the width of the mainlobe (taking a -3 dB beam width as an example ($\mathrm{BW}_{-3\mathrm{dB}}$)), the SLL, and the beam responses of the interferences. It is shown that the beampattern optimized by the G-RCB has the narrowest mainlobe width instead of that of the CBF among the beamformers in Table 2 with respect to the conformal array. The similarity of the three sparsity-constrained beamformers is that the SLL is significantly reduced, in which the SLL of the S-RCB is more than 4 dB lower than that of the conventional beamforming (CBF). The SG-RCB combines the advantages of the S-RCB and G-RCB, which can achieve the optimal interference suppression while maintaining the mainlobe width almost unchanged. It is also implied that the SG-RCB can make a flexible tradeoff between the mainlobe width and the SLL by adjusting the values of λ_1 and λ_2 according to the actual situation.

5.4. SINR$_{out}$ versus SNR and the Sample Size

Now, we compare SINR$_{out}$ of the SG-RCB with other methods, which is calculated by

$$\mathrm{SINR}_{\mathrm{out}} = \frac{\boldsymbol{w}^H \boldsymbol{R}_s \boldsymbol{w}}{\boldsymbol{w}^H \boldsymbol{R}_{\mathrm{int+n}} \boldsymbol{w}} = \frac{\sigma_s^2 \left| \boldsymbol{w}^H \boldsymbol{a}_0 \right|^2}{\boldsymbol{w}^H \boldsymbol{R}_{\mathrm{int+n}} \boldsymbol{w}} \tag{60}$$

Particularly, the optimal SINR$_{out}$ is calculated by the SCB with the theoretical covariance matrix $\boldsymbol{R}_y$ for performance evaluation. It is observed from Figure 12a that the SCB suffers a poor performance. The SINR$_{out}$ of other methods is similar, with a low SNR(≤ 10 dB). With the increase in the SNR, an inaccurate estimation of the sample covariance matrix is amplified so that the SINR$_{out}$ of these methods slightly decreases. The SINR$_{out}$ of the S-RCB, G-RCB, and SG-RCB is higher than that of the RCB; furthermore, the SINR$_{out}$ of the SG-RCB increases by approximately 3 dB compared to the RCB. Figure 12b shows that three sparsity-class methods do not only quickly converge to their corresponding optimal values, but their $SINR_{out}$ is also higher than that of the RCB in the case of a low number of samples. Figure 12 indicates that the RCB with sparsity constraints is basically effective in optimizing 2D beampatterns, although a certain $SINR_{out}$ is lost when SNR is high.

Table 2. Performance comparison of beampatterns of different beamformers. The red indicates the optimal value in this column.

Beamformers	$BW_{-3dB}/°$ (Elevation, Azimuth)	SLL/dB	Beam Response of Interference One/dB	Beam Response of Interference Two/dB
CBF	(22.17, 35.45)	−11.7779	−15.2244	−34.8355
SOCP-RCB	(22.49, 34.11)	−11.1561	−52.7997	−48.0525
ADMM-RCB	(22.44, 34.22)	−11.7801	−57.8118	−49.714
S-RCB ($\lambda_1 = 0.01, \lambda_2 = 0$)	(23.48, 31.55)	−15.9865	−56.199	−56.4141
G-RCB ($\lambda_1 = 0, \lambda_2 = 0.1$)	(20.76, 28.63)	−13.7339	−48.4772	−51.3741
SG-RCB ($\lambda_1 = 0.01, \lambda_2 = 0.1$)	(21.92, 29.93)	−14.171	−58.5509	−64.5806

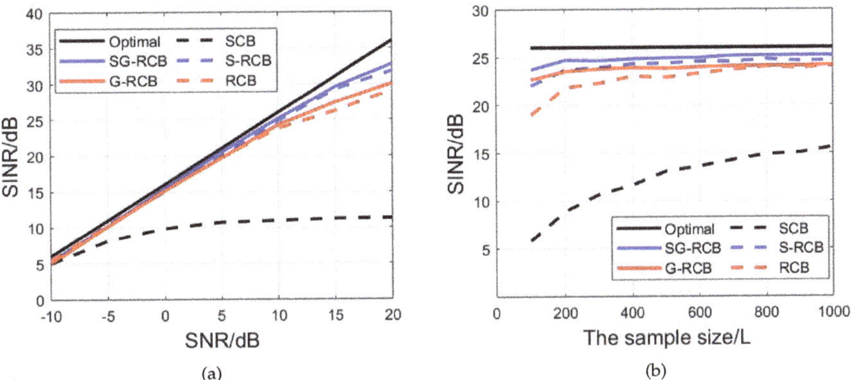

Figure 12. Performance comparison of different methods: (**a**) $SINR_{out}$ versus SNR, L = 800; (**b**) $SINR_{out}$ versus the sample size. SNR = 10 dB.

5.5. Beampatterns with Different Regional Sparsity Parameters

In Sections 5.2–5.4, we uniformly set the regional sparsity parameters n_Q to $\mathbf{1}_{10\times1}$, i.e., $n_Q = [n_1, n_2, \cdots, n_{10}]^T = \underbrace{[1, 1, \cdots, 1]}_{10\times1}^T$. In this subsection, the influence of the regional sparsity parameters on the beampattern is verified.

According to the grouping in Figure 8, one value of n_Q in this subsection is set as follows:

$$n_{Q,1} = [1, 1, 1, 1, 1, 1, 1, 1, 10, 10]^T \tag{61}$$

where the two number 10s represent the parameters of the interference regions (black regions) in Figure 8. Substituting (61) into (15), the beampattern of the SG-RCB based on (61) is shown in Figure 13, in which the regions of the sidelobes adjacent to interferences and the interferences are marked out by the red dashed line and the red solid line, respectively.

It can been seen from Figure 13b that the nulls in the interference regions are widened so that the interferences can be correctly suppressed.

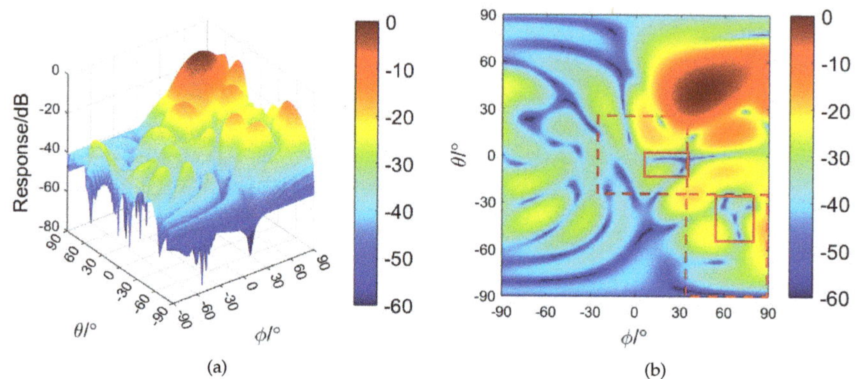

Figure 13. Influence of $n_{Q,1}$ on the beampattern of the SG–RCB: (**a**) two–dimensional view; (**b**) top view.

The other value of n_Q is set blow.

$$n_{Q,2} = [1,1,1,1,5,1,5,1,10,10]^T \tag{62}$$

where the two number 5s represent the parameters of the sidelobes adjacent to the interferences (blue regions in Figure 8). The beampatterns of the SG-RCB using $n_{Q,2}$ are shown in Figure 14. It is observed that the beam responses in the fifth and seventh sidelobe regions marked by the red dashed line decrease further, but the cost is that the mainlobe is obviously widened. As a summary, the beam responses in the sidelobe regions can be adjusted by the regional sparsity parameters. The greater the parameters, the lower the responses, but their values need to balance the relationship between the mainlobe width and the SLL.

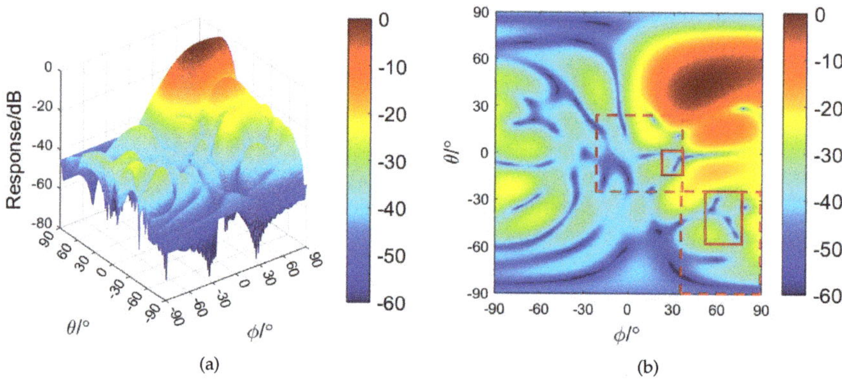

Figure 14. Influence of $n_{Q,2}$ on the beampattern of the SG–RCB: (**a**) two–dimensional view; (**b**) top view.

5.6. Interference Suppression in the Presence of the SV Mismatch

The array manifold of the conformal array is susceptible to distortion caused by various factors, especially the scattering of baffles. The robustness of the SG-RCB is tested in this subsection. The SVs of the desired signal and interferences in this subsection are obtained by the finite-element software COMSOL [46] to guarantee the model in practice [47], which converts the physical model into the finite-element mesh model and then carries out the numerical calculation.

First, we consider $n_Q = \mathbf{1}_{10\times 1}$. Substituting such SVs into the optimization problem of the SG-RCB, the resulting beampatterns are shown in Figure 15. The mainlobes of the beampatterns are not distorted under the robustness constraint, but each beamformer has poor performance in interference suppression due to the SV mismatch.

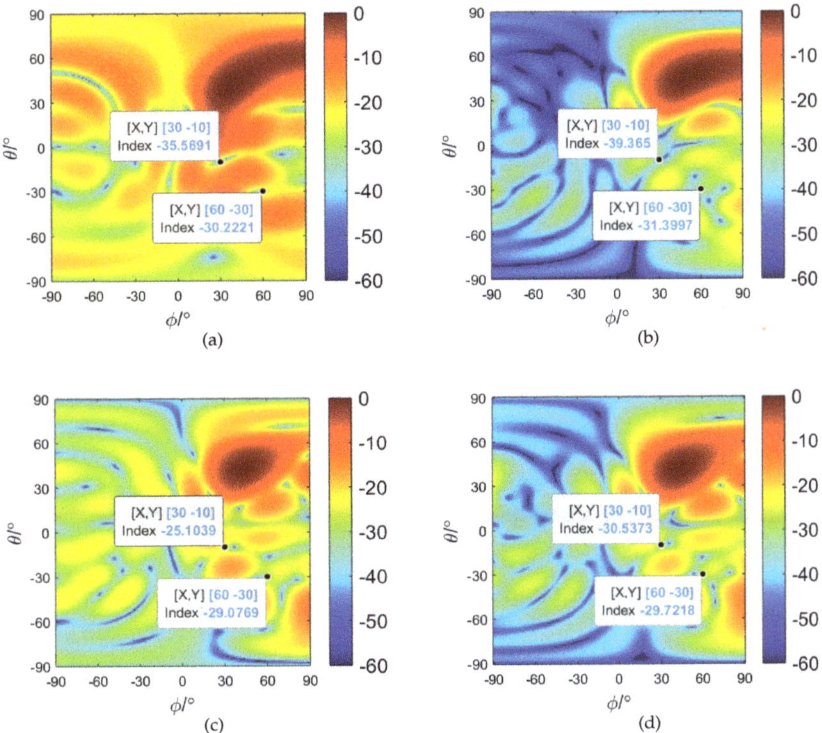

Figure 15. Beampatterns in the case of the SV mismatch: (**a**) ADMM–RCB; (**b**) S–RCB; (**c**) G–RCB; (**d**) SG–RCB.

Then, we investigate the effect of $n_{Q,1}$ and $n_{Q,2}$ on interference suppression. The beampattern using $n_{Q,1}$ is shown in Figure 16. It is observed that the beam responses of the two interferences are reduced by about 10 dB compared with the results of Figure 15d.

Furthermore, by replacing $n_{Q,1}$ with $n_{Q,2}$ in the SG-RCB, the resulting beampattern is obtained, as shown in Figure 17. It can be seen that the beam responses of the interferences and the sidelobes decrease further at the cost of the mainlobe widening. It can be seen from Figures 16 and 17 that setting the regional sparsity parameters properly can improve the interference suppression without changing the overall performance of the 2D beampattern when the SV mismatch exists.

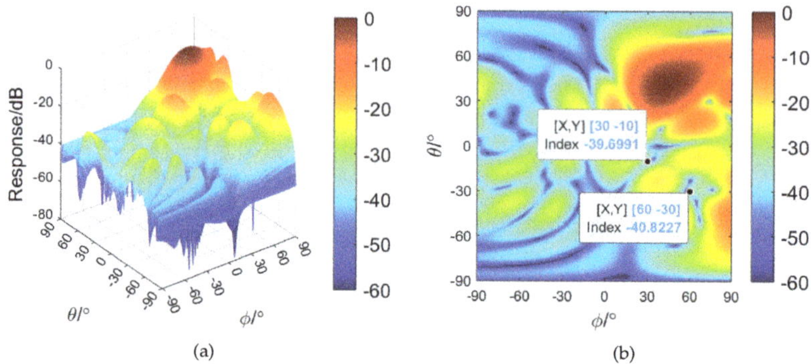

Figure 16. Performance of interference suppression of the SG–RCB using $n_{Q,1}$: (**a**) two–dimensional view; (**b**) top view.

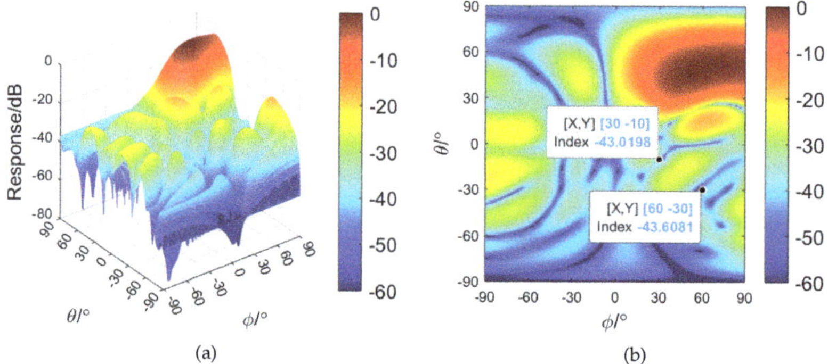

Figure 17. Performance of interference suppression of the SG–RCB using $n_{Q,2}$: (**a**) two–dimensional view; (**b**) top view.

6. Conclusions

In this paper, we developed the SG-RCB, which utilizes sparse group constraints based on the RCB to reduce the SLL of the 2D beampattern for conformal arrays. By introducing the GSC framework, the original optimization problem was divided into two subproblems. The first is the RCB problem and the second is the SGLASSO problem. To handle these problems, the ADMM was employed to solve them in closed form.

The main advantages of the proposed method for 2D beampattern optimization are as follows: The SLL of the 2D beampattern is greatly reduced by the sparse group constraints. The ADMM is applied to solve the optimization problem, which greatly improves the computational efficiency compared with other existing methods. The interference suppression of the proposed method in the presence of the SV mismatch can be recovered by adjusting the regional sparsity parameters.

The SG-RCB has broad application prospects in various underwater scenarios. For example, the SG-RCB can be applied to the arrays of autonomous underwater vehicles (AUV) or unmanned underwater vehicles (UUV) to realize the detection and direction of arrival (D.O.A) estimation of the target. In addition, the SG-RCB has potential applications in the real-time processing of received signals due to its significant computational efficiency.

Author Contributions: Conceptualization, Y.D.; methodology, Y.D. and C.S.; validation, Y.D.; formal analysis, Y.D.; investigation, Y.D.; resources, C.S.; data curation, Y.D.; writing—original draft preparation, Y.D.; writing—review and editing, C.S. and X.L.; supervision, C.S.; project administration, C.S.; funding acquisition, X.L. All authors have read and agreed to the published version of the manuscript.

Funding: This work was supported by the National Natural Science Foundation of China (U2341203, 12274346).

Institutional Review Board Statement: Not applicable.

Informed Consent Statement: Not applicable.

Data Availability Statement: The data presented in this study are available on request from the corresponding author.

Conflicts of Interest: The authors declare no conflicts of interest.

References

1. Frank, T.H.; Kesner, J.W.; Gruen, H.M. Conformal array beam patterns and directivity indices. *J. Acoust. Soc. Am. Mar.* **1978**, *63*, 841–847. [CrossRef]
2. Traweek, C.M. Optimal Spatial Filtering for Design of a Conformal Velocity Sonar Array. Ph.D. Thesis, Pennsylvania State University, University Park, PA, USA, 2003.
3. Yang, Y.; Wang, Y.; Ma, Y.; Sun, C. *Experimental Study on Robust Supergain Beamforming for Conformal Vector Arrays*; OCEANS: San Diego, CA, USA, 2013; pp. 1–5.
4. Josefsson, L.; Persson, P. *Conformal Array Antenna Theory and Design*; John Wiley & Sons Ltd.: Hoboken, NJ, USA, 2006.
5. Zhang, H.; Guo, D.; Cao, X. An Airborne Conformal Array Beampattern Optimization Algorithm Based on Convex Optimization. In Proceedings of the 2022 IEEE 5th Advanced Information Management, Communicates, Electronic and Automation Control Conference (IMCEC), Chongqing, China, 16–18 December 2022.
6. Van Trees, H.L. *Optimum Array Processing: Part IV of Detection, Estimation, and Modulation Theory*; John Wiley & Sons Ltd.: New York, NY, USA, 2002.
7. Zhu, J.; Song, Y.; Jiang, N.; Xie, Z.; Fan, C.; Huang, X. Enhanced Doppler Resolution and Sidelobe Suppression Performance for Golay Complementary Waveforms. *Remote Sens.* **2023**, *15*, 2452. [CrossRef]
8. Capon, J. High-resolution frequency-wavenumber spectrum analysis. *Proc. IEEE* **1969**, *57*, 1408–1418. [CrossRef]
9. Carlson, B.D. Covariance matrix estimation errors and diagonal loading in adaptive arrays. *IEEE Trans. Aerosp. Electron. Syst.* **1988**, *24*, 397–401. [CrossRef]
10. Vorobyov, S.A.; Gershman, A.B.; Luo, Z.-Q. Robust adaptive beamforming using worst-case performance optimization: A solution to the signal mismatch problem. *IEEE Trans. Signal Process.* **2003**, *51*, 313–324. [CrossRef]
11. Huang, Y.; Fu, H.; Vorobyov, S.A.; Luo, Z.Q. Robust Adaptive Beamforming via Worst-Case SINR Maximization With Nonconvex Uncertainty Sets. *IEEE Trans. Signal Process.* **2023**, *71*, 218–232. [CrossRef]
12. Li, J.; Stoica, P.; Wang, Z. On robust Capon beamforming and diagonal loading. *IEEE Trans. Signal Process.* **2003**, *51*, 1702–1715.
13. Zhang, X.; He, Z.; Zhang, X.; Xie, J. Robust Sidelobe Control via Complex-Coefficient Weight Vector Orthogonal Decomposition. *IEEE Trans. Antennas Propag.* **2019**, *67*, 5411–5425. [CrossRef]
14. Li, H.; Ran, L.; He, C.; Ding, Z.; Chen, S. Adaptive Beamforming with Sidelobe Level Control for Multiband Sparse Linear Array. *Remote Sens.* **2023**, *15*, 4929. [CrossRef]
15. Wu, R.; Bao, Z.; Ma, Y. Control of peak sidelobe level in adaptive arrays. *IEEE Trans. Antennas Propag.* **1996**, *44*, 1341–1347.
16. Yan, S.F. *Optimal Array Signal Processing: Beamforming Design Theory and Methods*; Science Press: Beijing, China, 2018.
17. Liu, L.; Liang, X.; Li, Y.; Liu, Y.; Bu, X.; Wang, M. A Spatial-Temporal Joint Radar-Communication Waveform Design Method with Low Sidelobe Level of Beampattern. *Remote Sens.* **2023**, *15*, 1167. [CrossRef]
18. Olen, C.A.; Compton, R.T. A numerical pattern synthesis algorithm for arrays. *IEEE Trans. Antennas Propag.* **1990**, *38*, 1666–1676. [CrossRef]
19. Liu, Y.; Wang, C.; Gong, J.; Tan, M.; Chen, G. Robust Suppression of Deceptive Jamming with VHF-FDA-MIMO Radar under Multipath Effects. *Remote Sens.* **2022**, *14*, 942. [CrossRef]
20. Zhang, X.; He, Z.; Liao, B.; Zhang, X.; Peng, W. Pattern Synthesis With Multipoint Accurate Array Response Control. *IEEE Trans. Antennas Propag.* **2017**, *65*, 4075–4088. [CrossRef]
21. Zhang, X.; He, Z.; Xia, X.-G.; Liao, B.; Zhang, X.; Yang, Y. OPARC: Optimal and Precise Array Response Control Algorithm—Part II: Multi-Points and Applications. *IEEE Trans. Signal Process.* **2019**, *67*, 668–683. [CrossRef]
22. Peng, W.; Zhang, X.; He, Z.; Xie, J.; Han, C. Beampattern synthesis for large-scale antenna array via accurate array response control. *Digit. Signal Process.* **2021**, *117*, 103152. [CrossRef]
23. Rao, B.D. Signal processing with the sparseness constraint. In Proceedings of the 1998 IEEE International Conference on Acoustics, Speech and Signal Processing, ICASSP'98, Seattle, WA, USA, 12–15 May 1998; Volume 3, pp. 1861–1864.
24. Zhang, Y.; Zhao, H.; Ng, B.P.; Lie, J.P.; Wan, Q. Robust Beamforming Technique with Sidelobe Suppression Using Sparse Constraint on Beampattern. *ACES J.* **2010**, *25*, 947–955.
25. Liu, Y.; Wan, Q. A Robust Beamformer Based on Weighted Sparse Constraint. *Prog. Electromagn. Res. Lett.* **2010**, *16*, 53–60. [CrossRef]
26. Wang, Y.; Zhu, S.; Lan, L.; Li, X.; Liu, Z.; Wu, Z. Range-Ambiguous Clutter Suppression via FDA MIMO Planar Array Radar with Compressed Sensing. *Remote Sens.* **2022**, *14*, 1926. [CrossRef]

27. C.; ès, E.J.; Wakin, M.B.; Boyd, S.P. Enhancing Sparsity by Reweighted l_1 Minimization. *J. Fourier Anal Appl.* **2008**, *14*, 877–905. [CrossRef]

28. Bakin, S. Adaptive Regression and Model Selection in Data Mining Problems. Ph.D. Thesis, School of Mathematical Sciences, Australian National University, Canberra, Australia, 1999.

29. Grant, M.C.; Boyd, S.P. *The CVX Users' Guide, Release 2.2*; CVX Research, Inc.: Austin, TX, USA, 2020. Available online: http://cvxr.com/cvx (accessed on 14 January 2024).

30. Boyd, S.; Parikh, N.; Chu, E.; Peleato, B.; Eckstein, J. Distributed Optimization and Statistical Learning via the Alternating Direction Method of Multipliers. *Now Found. Trends Mach. Learn.* **2011**, *3*, 1122.

31. Fan, W.; Liang, J.; Yu, G.; So, H.C.; Li, J. Robust Capon Beamforming via ADMM. In Proceedings of the ICASSP 2019—2019 IEEE International Conference on Acoustics, Speech and Signal Processing (ICASSP), Brighton, UK, 12–17 May 2019; pp. 4345–4349.

32. Liang, J.; Fan, X.; Fan, W.; Zhou, D.; Li, J. Phase-Only Pattern Synthesis for Linear Antenna Arrays. *IEEE Antennas Wirel. Propag. Lett.* **2017**, *16*, 3232–3235. [CrossRef]

33. Yang, J.; Lin, J.; Shi, Q.; Li, Q. An ADMM-Based Approach to Robust Array Pattern Synthesis. *IEEE Signal Process. Lett.* **2019**, *26*, 898–902. [CrossRef]

34. Tian, X.; Chen, H.; He, M.M.; Wang, W.Q. Fast Beampattern Synthesis Algorithm for Flexible Conformal Array. *IEEE Signal Process. Lett.* **2022**, *29*, 2417–2421. [CrossRef]

35. Wang, W.; Yan, S.; Mao, L.; Guo, X. Robust Minimum Variance Beamforming With Sidelobe-Level Control Using the Alternating Direction Method of Multipliers. *IEEE Trans. Aerosp. Electron. Syst.* **2021**, *57*, 3506–3519. [CrossRef]

36. Zhang, W.; Lin, J.; Wu, X.; Pan, Y. A distributed approach to robust minimum variance distortionless response beamforming in large-scale arrays. *IET Commun.* **2023**, *17*, 950–959. [CrossRef]

37. Chen, C.; Liu, T.; Liu, Y.; Yang, B.; Su, Y. Learning-Based Clutter Mitigation with Subspace Projection and Sparse Representation in Holographic Subsurface Radar Imaging. *Remote Sens.* **2022**, *14*, 682. [CrossRef]

38. Zhao, Y.; Liu, Q.; Tian, H.; Ling, B.W.-K.; Zhang, Z. DeepRED Based Sparse SAR Imaging. *Remote Sens.* **2024**, *16*, 212. [CrossRef]

39. Wu, T.T.; Lange, K.L. Coordinate descent algorithms for lasso penalized regression. *Ann. Appl. Stat.* **2008**, *2*, 224244. [CrossRef]

40. Griffiths, L.; Jim, C. An alternative approach to linearly constrained adaptive beamforming. *IEEE Trans. Antennas Propag.* **1982**, *30*, 27–34. [CrossRef]

41. Simon, N.; Friedman, J.; Hastie, T.; Tibshirani, R. A Sparse-Group Lasso. *J. Comput. Graph. Stat.* **2013**, *22*, 231–245. [CrossRef]

42. Tibshirani, R. Regression Shrinkage and Selection Via the Lasso. *J. R. Stat. Soc. Ser. (Methodol.)* **1996**, *58*, 267–288. [CrossRef]

43. Ding, C.; Zhou, D.; He, X.; Zha, H. R1-PCA: Rotational invariant L1-norm principal component analysis for robust subspace factorization. In Proceedings of the 23rd international conference on Machine learning (ICML'06), Pittsburgh, PA, USA, 21–26 June 2006; Association for Computing Machinery: New York, NY, USA, 2006; pp. 281–288.

44. Yuan, M.; Lin, Y. Model Selection and Estimation in Regression with Grouped Variables. *J. R. Stat. Soc. Ser. Stat. Methodol.* **2006**, *68*, 49–67. [CrossRef]

45. Liu, J.; Ji, S.; Ye, J. SLEP: Sparse Learning with Efficient Projections. Arizona State University. 2009. Available online: http://yelabs.net/software/SLEP (accessed on 14 January 2024).

46. COMSOL Multiphysics® v.5.2. Acoustics Module Users' Guide. COMSOL AB, Stockholm, Sweden. 2021. Available online: https://cn.comsol.com/models/acoustics-module (accessed on 14 January 2024).

47. Yang, B.; Sun, C.; Chen, Y.L. Conformal Array Beampattern Optimization Method and Experimental Research Based on Sound Field Forecast. *Torpedo Technol.* **2006**, *14*, 18–20.

remote sensing

Article

Detecting Weak Underwater Targets Using Block Updating of Sparse and Structured Channel Impulse Responses

Chaoran Yang [1,2,3], Qing Ling [1,2,3], Xueli Sheng [1,2,3,*], Mengfei Mu [1,2,3] and Andreas Jakobsson [4]

1   National Key Laboratory of Underwater Acoustic Technology, Harbin Engineering University, Harbin 150001, China
2   Key Laboratory for Polar Acoustics and Application of Ministry of Education, Harbin Engineering University, Ministry of Education, Harbin 150001, China
3   College of Underwater Acoustic Engineering, Harbin Engineering University, Harbin 150001, China; yangchaoran_heu@163.com (C.Y.); 18846130480@hrbeu.edu.cn (M.M.)
4   Center for Mathematical Statistics, Lund University, 22100 Lund, Sweden; andreas.jakobsson@matstat.lu.se
*   Correspondence: shengxueli@hrbeu.edu.cn

Abstract: In this paper, we considered the real-time modeling of an underwater channel impulse response (CIR), exploiting the inherent structure and sparsity of such channels. Building on the recent development in the modeling of acoustic channels using a Kronecker structure, we approximated the CIR using a structured and sparse model, allowing for a computationally efficient sparse block-updating algorithm, which can track the time-varying CIR even in low signal-to-noise ratio (SNR) scenarios. The algorithm employs a conjugate gradient formulation, which enables a gradual refinement if the SNR is sufficiently high to allow for this. This was performed by gradually relaxing the assumed Kronecker structure, as well as the sparsity assumptions, if possible. The estimated CIR was further used to form a residual signal containing (primarily) information of the time-varying signal responses, thereby allowing for the detection of weak target signals. The proposed method was evaluated using both simulated and measured underwater signals, clearly illustrating the better performance of the proposed method.

Keywords: time-varying impulse response; drift compensation; structured channel estimate; underwater sonar; weak target detection

Citation: Yang, C.; Ling, Q.; Sheng, X.; Mu, M.; Jakobsson, A. Detecting Weak Underwater Targets Using Block Updating of Sparse and Structured Channel Impulse Responses. *Remote Sens.* **2024**, *16*, 476. https://doi.org/10.3390/rs16030476

Academic Editor: Andrzej Stateczny

Received: 14 December 2023
Revised: 16 January 2024
Accepted: 18 January 2024
Published: 26 January 2024

1. Introduction

Numerous underwater applications, ranging from monitoring the marine environment, for instance to detect pollution, to underwater communication, depend on accurate and reliable estimates of the underwater channel impulse response (CIR), detailing the time- and location-dependent multipath wave propagation typical of such an environment [1–6]. The channel is notably affected by numerous factors, ranging from the depth and salinity of the water to sea structures, thermoclines, sea mammals, and ships, as well as experiences strong noise and interference signals and often also varies due to ship or sonar motions. An accurate estimation of the CIR is critical to allow for the detection of the energy of weak echo signals, such as the backscattered signal of underwater targets, which will appear as a corresponding variation in the resulting CIR estimate [7,8]. Due to the importance of the CIR estimate, notable efforts have been made to construct reliable estimation technologies for underwater CIRs, with recent efforts mainly focusing on exploiting the typical sparse structure of the CIRs, such as in the compressed sensing formulation in [9], where an extended orthogonal matching pursuit method was proposed. Other common alternatives include adaptive estimation methods, such as the one proposed by Tian et al. in [10], which combined a least-mean-squares (LMS) formulation with the use of an adaptive complex-valued penalty term. In [11–13], the authors imposed sparsity by making use of a step-size selection, which varied with the magnitude of the CIR coefficients.

As an alternative, recursive least-squares (RLS) formulations may be used, typically having a notably faster convergence, although at the price of increased computational complexity. Given the time-varying nature of underwater CIRs, several sparse RLS formulations have been developed, striving to retain the fast convergence while imposing the sparse structure of the CIRs [14–16]. An interesting alternative formulation is that of the sparse conjugate gradient (SCG) algorithm proposed in [17], which employs an affine scaling transform (AST) to enforce sparsity and combines the advantages of the low complexity of the LMS-based methods and the fast convergence of the RLS-based methods. The resulting estimator has been found to offer better performance compared to several sparse RLS formulations, such as l_0-RLS and l_1-RLS [18,19], as well as l_1-RRLS [19]. Notably, most of the derived methods strive to update the estimated CIR on a sample-to-sample basis, i.e., as each new sample becomes available, imposing the gradual forgetting of the previously observed measurements. This is typically not the case for active sonar measurement, where a batch of data is collected, resulting from each of the transmitted sonar pulses, necessitating a block (pulse) updating scheme, wherein earlier block measurements are gradually forgotten instead. Similar situations occur also in other fields, and several block-updating versions of CIR estimators have been investigated in the literature (see, e.g., [20–22]).

It is worth noting that drift causes a slight Doppler shift in the resulting signal, but also a time-varying time delay between the transmitter and receiver due to the varying distance of the propagation path. In the studied measurements, the latter (typically dominant) form of time delay shifting was our primary concern, as it affects the CIR's sparsity over multiple transmissions. The Doppler effect of the drift here causes a slight mismatch between the transmitted and matched signals, somewhat increasing the resulting line widths. In order to exploit the sparsity of the CIR, the time delay drift has to be taken into account and compensated.

In this work, we examined this form of block-based updating scheme, while also incorporating recent developments in acoustic channel estimation, wherein structural information is imposed on the impulse response. In [23], the authors showed that an acoustic impulse response can be well modeled using a summed Kronecker structure, thereby significantly reducing the number of parameters required for the detail ofthe channel. These efforts have since been extended to incorporating such a structure in a variety of estimators [24,25], including both an RLS-based [26] and an LMS-based estimator [27]. To further this development, this work presents a joint estimation framework for time-varying channels, which is robust against the low SNR scenarios typically encountered in underwater measurements, as well as a detection algorithm for weak underwater targets, for active sonar systems. Figure 1 illustrates the flow diagram of the proposed combined methods.

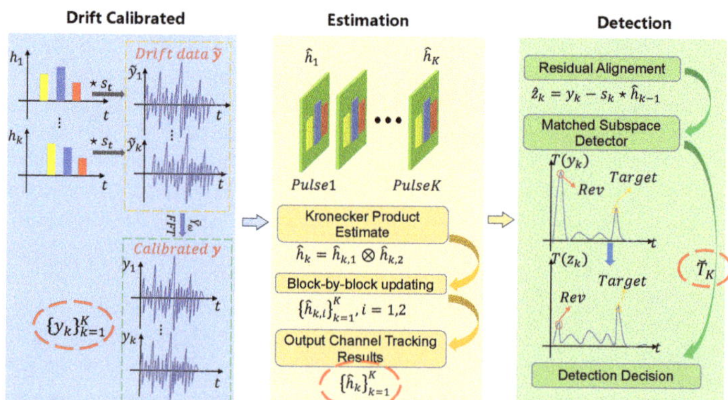

Figure 1. Block diagram of the processing chain for the proposed joint estimation and detection framework.

The main contributions are summarized as follows:

(1) *Drift compensation:* At the initial stage, we compensated the measured signal for the typically present drift. This was achieved by estimating the pulse-to-pulse time delay of the measured signal using a Fourier-based technique [28]; this allows the measured signal to be shifted accordingly before inverse transformation, thereby compensating for the drift.

(2) *Block updating the CIR estimate:* Using the drift-compensated signals, we proceeded to formulate the proposed sparse and structural CIR estimator. The resulting Kronecker-based block SCG (KBSCG) estimator is able to model the CIR even for low signal-to-noise ratios (SNRs) well, here defined as $\text{SNR} = 10 \log \frac{P_s}{P_e}$, where P_s and P_e denote the power of the signal and the noise, respectively. Although, given the inevitable model mismatch with the true CIR, which generally does not exhibit an exact sparse Kronecker structure, the method will be unable to continue to improve with the increasing SNR. Furthermore, in cases where the channel is changing rapidly, we found that the block-based improved proportionate NLMS (IPNLMS) [29] will react faster to these changes than the KBSCG estimator, although not being able to achieve the same level of performance after convergence. Therefore, in order to ascertainthe better performance of the method, first, we compared the model fit with that of the IPNLMS, to see if the latter approach managed to adapt faster. If it did, we proceeded to also compare it to the proposed block-based SCG (BSCG), which does not impose a Kronecker structure on the CIR. In the high SNR case, the BSCG estimate will be able to offer improved modeling, and the proposed estimator, therefore, proceeds to use the BSCG update instead. For an even higher SNR, the assumption of sparsity may also be relaxed without suffering from numerous spurious estimates in the CIR; in such a case, the updating may instead employ the block-based RLS (BRLS) to allow further refinement. We term the resulting combined estimator the block combined estimator (BCE). As we were mainly interested in the low SNR case, the later steps were less often applicable, but are included here to also allow for such cases.

(3) *Detecting weak targets' echo:* We illustrate the effectiveness of the CIR estimate by using it to form the residual between the observed data and the reconstructed data using the CIR estimate from the preceding pulse. The use of the preceding CIR estimates ensures that any channel variations, such as those resulting from a moving target, will remain in the resulting residual. We illustrate this by implementing a matched subspace detector to determine the presence of the target when using different forms of CIR estimates, using both simulated and measured underwater data, showing that the proposed CIR estimator offers a better performance.

The remainder of this paper is organized as follows: In the following section, we detail the problem formulation and derive the proposed block-updating CIR estimate. Then, in Section 2.2, we proceed to introduce the matched subspace detector. Sections 3 and 3.4 illustrate the performance of the proposed CIR estimator and the resulting detector using both simulated and measured underwater data. Finally, Section 4 gives our conclusions.

2. Materials and Methods

2.1. Estimating the Time-Varying CIR

In practice, underwater sensor equipment is always impacted by water flow, resulting in variations in the signal propagation. The effect is illustrated in Figure 2, showing the CIR estimated using the least squares (LS), for each pulse, in a real sea experiment. As can be seen, the CIR exhibits a sparse structure, but one that varies slowly in between pulses due to the drift.

This time delay effect also differs from the Doppler effect that occurs, as the latter primarily involves the effect of the relative velocity between the source and the observer, causing a corresponding signal distortion. As recursive algorithms rely on the assumption of a constant or slowly varying system, one has to compensate for such time delay perturbations since these otherwise introduce abrupt changes in the system dynamics, which can lead to inaccurate estimates and even unstable behavior in the recursive estimation process.

In order to model the time-varying drift, we considered the mth tap of the CIR for the kth pulse, $\mathbf{h}_k(m)$, modeled as

$$\mathbf{h}_k(m) = \mathbf{h}_{k-1}(m - \lfloor f_s \tau_k \rfloor) + \boldsymbol{\Delta}_k \tag{1}$$

where $\tau_k \in \mathbb{R}$ denotes the time delay of the CIR for pulse k, with $\lfloor \cdot \rfloor$ denoting the rounding down (floor) operation, and f_s the sampling frequency, where $\boldsymbol{\Delta}_k \in \mathbb{N}(0, \sigma^2)$ denotes the channel perturbation as compared to the previous CIR (taking the drift and amplitude variation into account). As a result, the L-dimensional measured signal resulting from the kth pulse, $\tilde{\mathbf{y}}_k$, may be modeled as

$$\tilde{\mathbf{y}}_k = \mathbf{s}_k \star \mathbf{h}_k + \mathbf{n}_k \tag{2}$$

where $\mathbf{s}_k$ is the N-dimensional transmitted waveform, $\mathbf{h}_k$ the M-dimensional CIR, and $\mathbf{n}_k$ an additive noise component, for the kth pulse, with $\star$ denoting the convolution. In order to determine the time delay drift, we employed the classical Fourier-based time delay estimator presented in [28], which allows for a computationally efficient and reliable estimate of the (possibly non-integer) shift between $\tilde{\mathbf{y}}_k$ and $\tilde{\mathbf{y}}_{k-1}$. By shifting the Fourier-transformed representation of the measured signal with the estimated shift, $\hat{\tau}_k$, the drift-compensated measured signal, $\mathbf{y}_k$, may then be formed using an inverse Fourier transform. We then proceeded to form a block-by-block time updating of the CIR using $\mathbf{y}_k$, thereby allowing us to exploit the sparse structure of the drift-compensated CIR, which is illustrated in Figure 3 for the CIR shown in Figure 2.

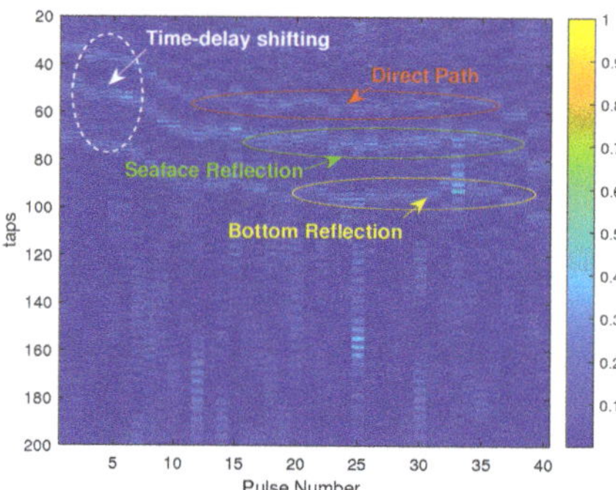

Figure 2. Pulse-to-pulse LS estimate of the CIR from high SNR sea data. The figure shows both the underlying sparse nature of the underwater CIR and the channel drift in between pulses. The numerous spurious components visible in the CIR estimates are due to the non-sparse nature of the LS estimate.

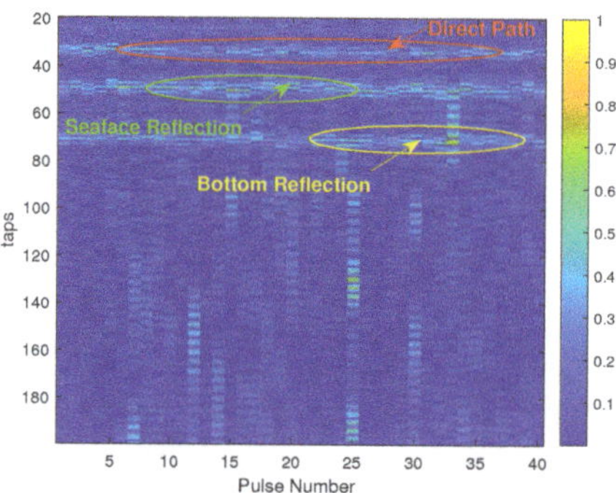

Figure 3. Pulse-to-pulse LS estimate of the drift-compensated CIR from high SNR sea data.

Next, we extended the SCG algorithm presented in [17] to a block-by-block formulation incorporating a summed Kronecker structure on the CIR. The SCG is formulated similarly to the RLS, but instead uses a conjugate gradient formulation to minimize the resulting weighted quadratic cost function [17,30]. Extending this formulation to a block-by-block update, the block-SCG (BSCG) is formed as

$$\hat{\mathbf{h}}_k = \arg\min_{\mathbf{h}_k} J_k(\mathbf{W}_k\mathbf{h}_k)$$

$$= \arg\min_{\mathbf{h}_k} \frac{1}{2}\mathbf{h}_k^T\mathbf{W}_k\mathbf{R}_k\mathbf{W}_k\mathbf{h}_k - \mathbf{h}_k^T\mathbf{W}_k\mathbf{r}_k \tag{3}$$

where

$$\mathbf{R}_k = \lambda\mathbf{R}_{k-1} + \mathbf{S}_k\mathbf{S}_k^T \tag{4}$$

$$\mathbf{r}_k = \lambda\mathbf{r}_{k-1} + \mathbf{S}_k\mathbf{y}_k \tag{5}$$

with $0 < \lambda \leq 1$ denoting the blockwise forgetting factor, and

$$\mathbf{S}_k = \begin{bmatrix} \mathbf{s}_k(1) & \cdots & \mathbf{s}_k(N) & 0 & 0 \\ \vdots & \ddots & \ddots & \ddots & \vdots \\ 0 & \cdots & \mathbf{s}_k(1) & \cdots & \mathbf{s}_k(N) \end{bmatrix} \tag{6}$$

where $\mathbf{s}_k(\ell)$ denotes the ℓth index in the kth transmitted pulse. The weighting matrix, $\mathbf{W}_k$, in (3) is formed as the $M \times M$ diagonal matrix with the vector $\mathbf{w}_k$ along the diagonal, where

$$\mathbf{w}_k(\ell) = \frac{1}{\xi}\left(\left|\hat{\mathbf{h}}_{k-1}(\ell)\right| + c\right)^{1-\xi} \tag{7}$$

for $\ell = 1,\ldots,M$, with $c > 0$ denoting a small regularization constant introduced for stability purposes and $0 < \xi \leq 1$ a factor used to control the sparsity of the solution (with a smaller ξ promoting stronger sparsity). Forming the gradient of $J_k(\mathbf{W}_k\mathbf{h}_k)$ with respect to $\mathbf{h}_k$ as

$$\mathbf{g}_k = \nabla_{\mathbf{h}}J_k(\mathbf{W}_k\mathbf{h}) = \mathbf{W}_k(\mathbf{R}_k\mathbf{h}_{k-1} - \mathbf{r}_k) \tag{8}$$

the blockwise update $\hat{\mathbf{h}}_k$ may be formed as

$$\hat{\mathbf{h}}_k = \hat{\mathbf{h}}_{k-1} + \alpha_k\mathbf{W}_k\mathbf{p}_k \tag{9}$$

where

$$\alpha_k = \frac{\mathbf{p}_k^T\mathbf{g}_k}{\mathbf{p}_k^T\mathbf{W}_k\mathbf{R}_k\mathbf{W}_k\mathbf{p}_k + \delta} \tag{10}$$

$$\mathbf{p}_k = \beta_k\mathbf{p}_{k-1} - \mathbf{g}_k \tag{11}$$

$$\beta_k = \frac{\mathbf{p}_{k-1}^T\mathbf{W}_k\mathbf{R}_k\mathbf{W}_k\mathbf{g}_k}{\mathbf{p}_{k-1}^T\mathbf{W}_k\mathbf{R}_k\mathbf{W}_k\mathbf{p}_{k-1} + \delta} \tag{12}$$

where $\delta > 0$ is a small regularization constant preventing division by zero. Further details on the step size α_k and the scaling factor β_k to sustain the Markov conjugacy can be found in [30]. The method was further extended to incorporate the additionally assumed summed Kronecker product structure, reminiscent of the development in [23], such that we proceeded to model

$$\mathbf{h}_k = \sum_{p=1}^{P} \mathbf{h}_{k,2,p} \otimes \mathbf{h}_{k,1,p} \tag{13}$$

where $\otimes$ denotes the Kronecker product, with $\mathbf{h}_{k,2,p}$ and $\mathbf{h}_{k,1,p}$, for $p = 1, 2, \ldots, P$, denoting two shorter impulse responses of lengths M_2 and M_1, respectively. Relaxing the formulation in [23], we here assumed that $M_1 M_2 \leq M$, with $M_1 \geq M_2 \geq P$. Following [31], (13) is, thus, rewritten as

$$\mathbf{h}_k = \sum_{p=1}^{P} \mathbf{H}_{k,2,p} \mathbf{h}_{k,1,p} = \sum_{p=1}^{P} \mathbf{H}_{k,1,p} \mathbf{h}_{k,2,p} \tag{14}$$

where $\mathbf{H}_{k,2,p} = \mathbf{h}_{k,2,p} \otimes \mathbf{I}_{M_1}$ and $\mathbf{H}_{k,1,p} = \mathbf{I}_{M_2} \otimes \mathbf{h}_{k,1,p}$ are matrices of sizes $M_1 M_2 \times M_1$ and $M_2 \times M_1 M_2$, respectively. This reformulation allowed us to separate the contributions from the two sets of CIRs in (3), such that

$$J_k(\mathbf{W}_k \mathbf{h}_k) = J_{k,1} + J_{k,2} \tag{15}$$

We proceeded to perform the update by keeping one of the sets of CIRs fixed at a time, minimizing the other, such that, when keeping $\mathbf{h}_{k,2,p}$ fixed, we only needed to minimize

$$J_{k,1} = \frac{1}{2} \underline{\mathbf{h}}_{k,1}^T \mathbf{W}_{k,1} \mathbf{R}_{k,1} \mathbf{W}_{k,1} \underline{\mathbf{h}}_{k,1} - \underline{\mathbf{h}}_{k,1}^T \mathbf{W}_{k,1} \mathbf{r}_{k,1} \tag{16}$$

where

$$\underline{\mathbf{h}}_{k,1} = \begin{bmatrix} \mathbf{h}_{k,1,1}^T & \cdots & \mathbf{h}_{k,1,P}^T \end{bmatrix}^T \tag{17}$$

$$\mathbf{R}_{k,1} = \lambda \mathbf{R}_{k-1,1} + \underline{\mathbf{S}}_{k,2} \underline{\mathbf{S}}_{k,2}^T \tag{18}$$

$$\mathbf{r}_{k,1} = \lambda \mathbf{r}_{k-1,1} + \underline{\mathbf{S}}_{k,2} \mathbf{y}_k \tag{19}$$

$$\underline{\mathbf{S}}_{k,2} = \begin{bmatrix} \mathbf{S}_{k,2,1}^T & \cdots & \mathbf{S}_{k,2,P}^T \end{bmatrix}^T \tag{20}$$

$$\mathbf{S}_{k,2,p} = \mathbf{S}_k^T \mathbf{H}_{k,2,p} \tag{21}$$

Here, the weighting matrix for the first set of CIRs, $\mathbf{W}_{k,1} = \mathrm{diag}(\mathbf{w}_{k,1})$, is formed using only this set, such that

$$\mathbf{w}_{k,1}(\ell) = \frac{1}{\xi_1} \left(\left| \hat{\mathbf{h}}_{k-1,1}(\ell) \right| + c_1 \right)^{1-\xi_1} \tag{22}$$

This led to the updating of the first set of CIR estimates:

$$\underline{\mathbf{h}}_{k,1} = \underline{\mathbf{h}}_{k-1,1} + \mathbf{W}_{k,1} \mathbf{p}_{k,1} \alpha_{k,1} \tag{23}$$

where

$$\alpha_{k,1} = \frac{\mathbf{p}_{k,1}^T \mathbf{g}_{k,1}}{\mathbf{p}_{k,1}^T \mathbf{W}_{k,1} \mathbf{R}_{k,1} \mathbf{W}_{k,1} \mathbf{p}_{k,1} + \delta} \tag{24}$$

$$\mathbf{p}_{k,1} = \beta_{k,1} \mathbf{p}_{k-1,1} - \mathbf{g}_{k,1} \tag{25}$$

$$\beta_{k,1} = \frac{\mathbf{p}_{k-1,1}^T \mathbf{W}_{k,1} \mathbf{R}_{k,1} \mathbf{W}_{k,1} \mathbf{g}_{k,1}}{\mathbf{p}_{k-1,1}^T \mathbf{W}_{k,1} \mathbf{R}_{k,1} \mathbf{W}_{k,1} \mathbf{p}_{k-1,1} + \delta} \tag{26}$$

$$\mathbf{g}_{k,1} = \mathbf{W}_{k,1} (\mathbf{R}_{k,1} \underline{\mathbf{h}}_{k-1,1} - \mathbf{r}_{k,1}) \tag{27}$$

Similarly, fixing the first set of CIRs, the second set may be updated as

$$\underline{\mathbf{h}}_{k,2} = \underline{\mathbf{h}}_{k-1,2} + \mathbf{W}_{k,2} \mathbf{p}_{k,2} \alpha_{k,2} \tag{28}$$

with similar definitions as for the first set. The Kronecker-based BSCG (KBSCG) estimator was, thus, formed by updating both sets of CIR estimates separately. As both (23) and (28) depend on the other set of CIRs, the estimates of (24)–(27) and the corresponding equations for the second set were alternatively computed using a bilinear optimization strategy [32], until practical convergence, prior to forming the updating in (23) and (28).

As shown in the following, the proposed KBSCG allowed for an accurate representation of the CIR, especially in low SNR cases, but did not converge as fast as the IPNLMS method [29], which has been found to allow for the tracking of rapid CIR changes, although then with only limited accuracy. To also allow for rapid changes in the CIR, the proposed combined estimator, therefore, compares the fitting of the KBSCG estimate with that of the IPNLMS estimate, using the reconstruction error

$$\eta = \frac{\left\| \mathbf{S}_k^T \hat{\mathbf{h}}_k - \mathbf{y}_k \right\|^2}{\left\| \mathbf{y}_k \right\|^2} \tag{29}$$

If the IPNLMS is deemed to offer an improved fit, which will be the case if the CIR has changed rapidly, this estimate will then be used instead. Furthermore, it is worth noting that the BSCG strives to exploit the sparse structure of the CIR, whereas the KBSCG will also impose a structured form of the CIR, which will further reduce the number of parameters that need to be estimated. One may, therefore, expect that these estimators will achieve good performance in low SNR conditions, as is also shown to be the case in the following, but will have performance limitations at higher SNRs, as the assumptions will not necessarily match the true CIR, thereby imposing constraints on the performance. This is illustrated in the numerical section, where it is shown that the sparse and structured KBSCG and the BSCG estimates will offer better performance in the low SNR cases, but will not be able to achieve such high-quality estimates as alternative formulations as the SNR increases. In this work, we were primarily interested in these low SNR cases, as these are the ones typically occurring in underwater measurements. However, in the interest of generality, one may in the higher SNR cases also improve on the found estimates. This may be accomplished by examining if the BSCG estimate, initiated using the KBSCG estimate, offers an improved fit of the observed data, which will be the case if the SNR is sufficiently high. If so, the combined estimator then uses the BSCG estimate instead.

Next, the BSCG estimate was used as an initialization of the BRLS update; if this estimate offers a lower reconstruction error, the resulting block combined estimator (BCE) instead uses this update as the current estimate. This gradual performance improvement for the discussed estimators is illustrated in Figure 4, where it may be seen that the BCE will havethe best estimate, for each SNR.

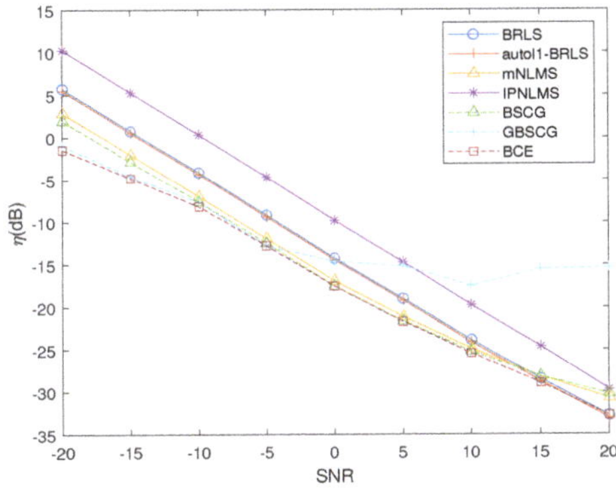

Figure 4. Performance of the discussed CIR estimators for the simulated data, as a function of the SNR.

The proposed method is summarized in Algorithm 1. As discussed above, the algorithm depends on a number of user-defined parameters. In Section 3.2, we discuss how these parameters may be suitably selected.

Algorithm 1 The BCE algorithm (our Matlab implementation will be provided on the authors' web pages upon publication).

1: Input: $\lambda, \delta, c_1, c_2, \xi_1, \xi_2, \mathbf{s}(k), \tilde{\mathbf{y}}(k), M_1, M_2, P$
2: Output: $\hat{\mathbf{h}}_k$.
3: Estimate τ_k, and form $\mathbf{y}_k$.
4: Form $\mathbf{S}_{k,1}, \mathbf{S}_{k,2}, \mathbf{W}_{k,1}, \mathbf{W}_{k,2}$, etc.
5: **while** Still converging **do**
6:   Update (23) and (28), keeping $\mathbf{h}_{k,2,p}$ fixed.
7:   Keep $\mathbf{h}_{k,1,p}$ fixed, and update the corresponding equations.
8: **end while**
9: Update $\underline{\mathbf{h}}_{k,1}$ and $\underline{\mathbf{h}}_{k,2}$ using (23) and (28).
10: Compute η^{KBSCG} using (29).
11: Form the IPN update using [29], and compute the η^{IPN}
12: **if** $\eta^{IPN} < \eta^{KBSCG}$ **then**
13:   Form the BSCG update using (9), and compute η^{BSCG}.
14:   **if** $\eta^{BSCG} < \eta^{IPN}$ **then**
15:     Form the BRLS update, and compute η^{BRLS}.
16:     **if** $\eta^{BRLS} < \eta^{BSCG}$ **then**
17:       $\hat{\mathbf{h}}_k = \hat{\mathbf{h}}_k^{BRLS}$.
18:     **else**
19:       $\hat{\mathbf{h}}_k = \hat{\mathbf{h}}_k^{BSCG}$.
20:     **end if**
21:   **else**
22:     $\hat{\mathbf{h}}_k = \hat{\mathbf{h}}_k^{IPN}$.
23:   **end if**
24: **else**
25:   $\hat{\mathbf{h}}_k = \hat{\mathbf{h}}_k^{KBSCG}$.
26: **end if**

2.2. Detecting Weak Underwater Targets

In order to illustrate the better performance of the introduced CIR estimator, we proceeded to examine how the found estimates may be used to detect a weak moving target. Traditionally, such a detection may be formed by applying a matched filter (MF) to the measured signal. However, as this signal will be severely corrupted and have a blurring effect due to the reverberation, such a detector will perform poorly if one does not compensate for the effect of the CIR. In order to do so, we proceeded to remove the influence of the channel, forming the residual:

$$\hat{\mathbf{z}}_k = \mathbf{y}_k - \hat{\mathbf{y}}_{k|k-1} = \mathbf{y}_k - \mathbf{s}_k \star \hat{\mathbf{h}}_{k-1} \tag{30}$$

where $\hat{\mathbf{h}}_{k-1}$ denotes the BCE estimate of the CIR at time $k-1$. The reason for using the CIR estimate from pulse $k-1$ when forming the kth residual is to allow the detector to determine the relative change in the CIR between pulses $k-1$ and k, thereby enabling the detection of weak moving targets. During much of the time, the resulting residual will not contain any of the sought targets; although, the resulting residual will still contain a notable structure due to noise and weak reflectors not captured by the CIR estimate. Using a low-order autoregressive (AR) model structure to detail the resulting residual, one may, reminiscent of the procedure in [33], form a pre-whitened version of the residual as $A(z)\hat{z}_k(\ell) = \hat{n}_k(\ell)$, where $\hat{z}_k(\ell)$ denotes the ℓth sample of $\hat{\mathbf{z}}_k$, $\hat{n}_k(\ell)$ is assumed to be well modeled as a circularly symmetric complex white Gaussian noise with variance σ_n^2, and the pre-whitening filter, $A(z)$, may be formed as (Using the measured sea data, we determined that a reasonable model for $A(z)$, for this measurement, is $A(z) = 1 + 0.061z^{-2} + 0.034z^{-4}$. This polynomial will depend on the assumed underwater conditions and should be determined for each setup using a small amount of secondary data, wherein no target is deemed present).

$$A(z) = 1 + \sum_{\ell=1}^{d} a_\ell z^{-\ell} \tag{31}$$

As any target will cause a response that is a scaled and delayed version of the transmitted pulse, $\mathbf{s}_k$, the target signal is known to lie in a (one-dimensional) subspace, $\mathbf{H}$, spanned by this signal. This allows the resulting detection problem to be formulated as

$$H_0: \quad \hat{\mathbf{n}}_k = \mathbf{w}_k$$
$$H_1: \quad \hat{\mathbf{n}}_k = \mathbf{H}\theta + \mathbf{w}_k$$

where $\mathbf{w}_k \in \mathcal{N}(\mathbf{0}, \mathbf{R}_n)$ and θ denotes the corresponding scaling of $\mathbf{H}$, with $\mathbf{R}_n$ denoting the (true) residual covariance matrix, here modeled as a white process, $\mathbf{R}_n = \sigma_n^2 \mathbf{I}$. Let $\hat{\mathbf{n}}_k(\tau_t)$ denote the part of the residual signal $\hat{\mathbf{n}}_k$ corresponding to a delay of τ_t, i.e.,

$$\hat{\mathbf{n}}_k(\tau_t) = \left[\hat{\mathbf{n}}_k\right]_{\ell}, \quad \ell = \xi_t, \dots, \xi_t + N \tag{32}$$

where $\xi_t = \lfloor f_s \tau_t \rfloor$, $[\cdot]_\ell$ is the ℓth index in the vector, and N is the length of the transmitted signal, $\mathbf{s}_k$.

This allows an (approximative) generalized likelihood ratio test (GLRT), assuming a target reflection at delay τ_t, to be formed as (see, e.g., [34])

$$\psi(\hat{\mathbf{n}}_k(\tau_t)) = (N-1)\frac{\hat{\mathbf{n}}_k(\tau_t)^T \Pi_{\mathbf{H}} \hat{\mathbf{n}}_k(\tau_t)}{\hat{\mathbf{n}}_k(\tau_t)^T \Pi_{\mathbf{H}}^\perp \hat{\mathbf{n}}_k(\tau_t)} \tag{33}$$

where $\Pi_{\mathbf{H}}$ denotes the projection onto $\mathbf{H}$, formed as

$$\Pi_{\mathbf{H}} = \mathbf{H}\left(\mathbf{H}^T\mathbf{H}\right)^{-1}\mathbf{H}^T \tag{34}$$

with $\Pi_{\mathbf{H}}^\perp = \mathbf{I} - \Pi_{\mathbf{H}}$ denoting the projection onto the space orthogonal to $\mathbf{H}$. Using (33), the target is, therefore, deemed present if and only if $\psi(\hat{\mathbf{n}}_k(\tau_t)) > \gamma_\psi$, otherwise not, where γ_ψ is a predetermined threshold value reflecting the acceptable probability of a false alarm (p_f), where, under the assumptions made, $p_f = Q_{F_{1,N-1}}(\gamma_\psi)$, with $Q_{F_{r,\ell}}(\gamma_\psi)$ denoting the complementary cumulative distribution function for an F-distribution with r numerator degrees of freedom and ℓ denominator degrees of freedom [34].

As the delay of the target reflection, τ_t, is unknown, we proceeded to evaluate $\hat{\mathbf{n}}_k(\tau_t)$ over the shifted version of the received measurement, $\mathbf{y}_k$. An illustration of the resulting detection variable with a target present at $\tau_t = 0.02$ s is shown in Figure 5, where a target with a signal-to-reverberation ratio (SRR) of 3 dB was added, with the SRR being defined as

$$\mathrm{SRR} = 10\log\frac{P_s}{P_r} \tag{35}$$

where P_s and P_r denote the power of the target signal and the reverberation, respectively. As can be seen in Figure 5, the resulting detection variable is quite noisy due to the imperfections in canceling the minor components in the CIR; to reduce the influence of such unmodeled reflections, we formed a P-step sliding median measure of the formed detection variable, notably reducing the influence of these reflections.

The resulting detection variable, T_k, for the kth pulse is then formed as this measure, i.e.,

$$T_k = \mathrm{median}_P\left[\psi(\hat{\mathbf{n}}_k(\tau))\right] \tag{36}$$

where $\mathrm{median}_P[\cdot]$ denotes the P-step sliding median filter. In this work, we used $P = 100$.

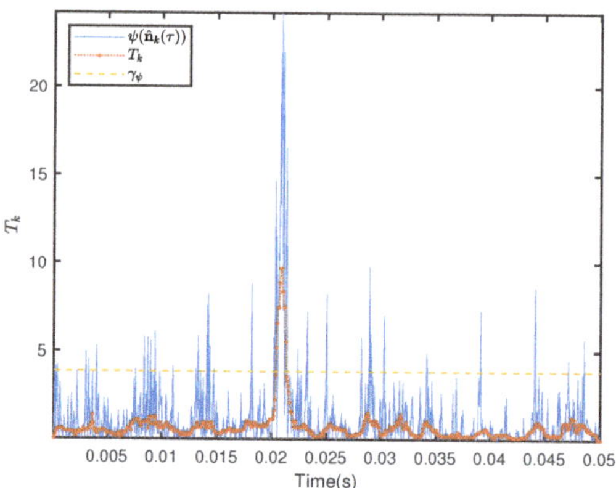

Figure 5. Illustration of the detection variable, $\psi(\hat{\mathbf{n}}_k(\tau))$, as compared to the median filtered detection variable, T_k, and the threshold γ_ψ for $p_f = 5\%$. The target is present here at $\tau_t = 0.02$ s, for SNR $= -10$ dB and SRR = 3 dB.

3. Results and Discussion

In this section, we examine the performance of both the proposed CIR estimator and the corresponding detection algorithm using simulated sea data.

3.1. Drift Compensation

We initially examined the drift compensation, which is illustrated in Figures 2 and 3 for the real sea data (see below for more details on these measurements). Here, the CIR was estimated for each pulse using the LS; as can be seen in Figure 2, the CIR experienced a notable drift due to the motion of the transmitter and the receiver. By using the proposed drift compensation of the measured data, this drift can be substantially reduced, as illustrated in Figure 3. In order to evaluate this performance gain using simulated data, we proceeded to simulate a channel mimicking the observed sea data channel, where the transmitted signal is a wideband chirp signal, such that

$$s(t) = exp\left[j\left(2\pi f_0 t + \pi k t^2\right)\right], \quad 0 \le t \le T_p \tag{37}$$

where $f_0 = 3$ kHz and $k = 20$ kHz/s represent the starting frequency and the chirp rate, respectively, using a pulse width of $T_p = 200$ ms, as was also used in the real experiment. We simulated a channel with $M = 800$ taps, based on the dominant components in Figure 2 (only showing the initial 200 taps), with uniformly distributed amplitudes with the variance equal to the envelope of the LS estimated CIR channel, each CIR shifted using a sequentially increasing delay to mimic the motion of the underwater acoustic channel resulting from the movements of the transmitting and receiving platforms, and using a maximum drift of no more than 5 ms. Each shift also included a random component to model the fluctuating nature of the amplitude, simulated using a zero-mean unit variance uniformly distributed random variable. Table 1 summarizes the reconstruction error, η, for the resulting CIR estimate, $\hat{\mathbf{h}}_k$, as compared to the CIR used to generate the simulations, $\mathbf{h}_k$, for SNR = 0 dB. The values given are the average η for all simulations. Here, the additive noise was simulated as a white additive Gaussian noise. As is clear from the table, the resulting CIR estimates suffered a notable loss of performance if the drift was not compensated for, as such errors then accumulated over time, degrading the performance further and further for each consecutive pulse. As shown in the table, the drift compensation was able to adequately compensate for the time-varying propagation delay, thereby allowing for more-accurate CIR estimates.

Table 1. The η of the CIR estimate for different algorithms, with or without drift compensation, for SNR = 0 dB.

Algorithm	η (dB)	
	SNR = 0 dB	
	Without	With
LS	3.74	−4.47
BRLS [20]	0.91	−9.49
ℓ_1−BRLS [19]	0.89	−9.63
IPNLMS [29]	2.39	−6.69
mNLMS [30]	0.58	−10.84
BSCG	0.39	−11.17
KBSCG	0.81	−9.59
BCE	2.38	**−11.59**

Next, we examined how fast the CIR estimates were able to adapt to a change in the true CIR. Figure 6 shows the η for the estimated CIR when the channel was abruptly changed at pulse 40, with the amplitudes of the channels redrawn. As can be seen in the figure, the sparse and the structured estimates were able to both adapt faster to the changing CIR and to yield a more-accurate estimate than the BRLS. Here, the SNR = 0 dB.

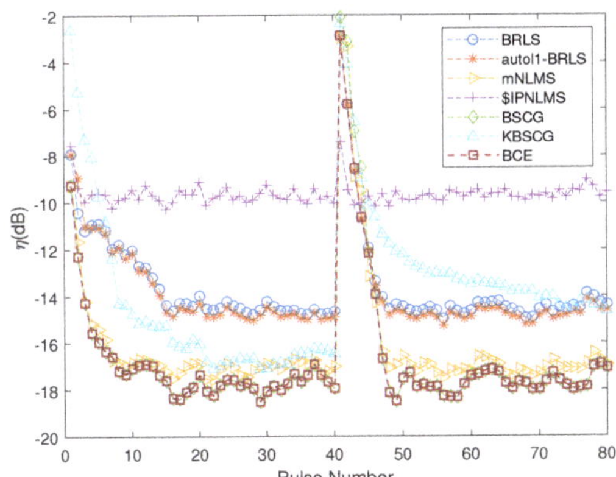

Figure 6. Performance comparison for the discussed algorithms when the CIR changes abruptly at pulse 40, for SNR = 0 dB.

3.2. Selecting Suitable Parameters

As has been noted, the proposed BCE algorithm, as well as the BSCG and KBSCG variant, will depend on several user-specified parameters, which will all affect the resulting performance of the estimator. For all estimators, the choice of λ in (4) and (5) will affect the overall speed and variability of the algorithm, similar to all forms of exponentially forgetting algorithms. Figure 7 examines how the performance is affected by the forgetting factor for the simulated sea measurements. Here, as the channel was varying fairly slowly (after compensating for the drift), it can be seen that a blockwise forgetting factor around $0.7 \leq \lambda \leq 0.9$ is preferable. Thus, we selected $\lambda = 0.7$ to be able to track the real sonar data well.

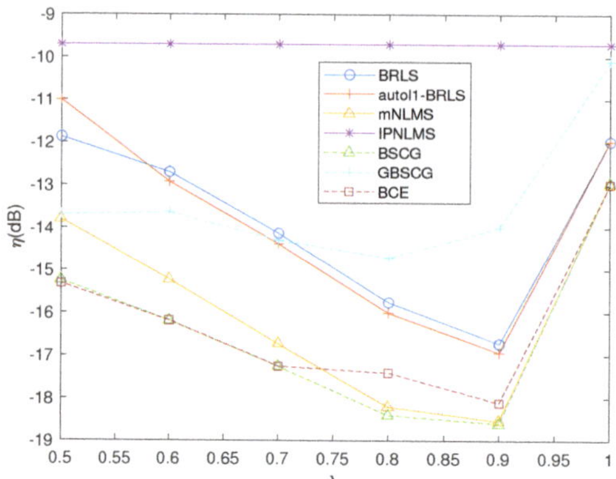

Figure 7. Effect of forgetting factor on proposed algorithms in time-varying CIR case, for SNR = 0 dB.

Next, we initially examined the BSCG estimator, where the resulting update depends on the sparsity parameter ζ and constant c, as shown in (7). The constants c and δ were only included for stability purposes and will only mildly affect the resulting estimates. Here, we selected these as $c = \delta = 10^{-4}$. The ζ parameter promotes a sparser solution and should, in our experience, be selected in the range $0.6 < \zeta \leq 0.8$. In this work, we used $\zeta = 0.7$. Figure 8 illustrates how η evolves for an increasing number of pulses for different values of ζ, for SNR = 0 dB, supporting this recommendation. As may be seen in the figure, the BSCG fit was, for this SNR, preferable with respect to the BRLS, for all settings of ζ (see also Figure 8). The KBSCG estimator will be affected similarly by the choice of ζ_1 and ζ_2, but also requires the selection of the model orders M_1, M_2, and P in (13).

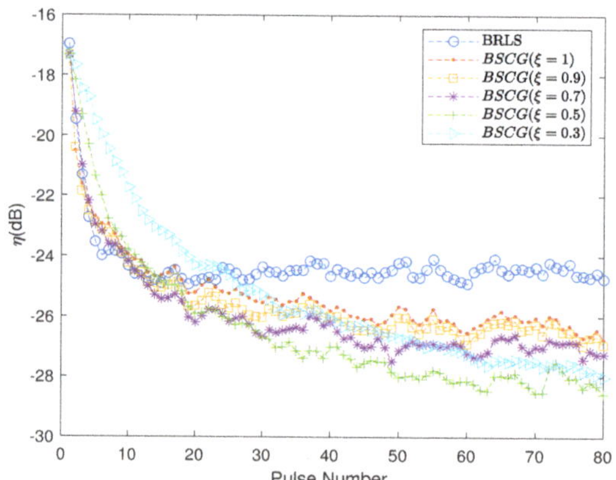

Figure 8. Effect of ζ on the BSCG algorithm, for SNR = 0 dB.

These parameters will affect the total number of unknown parameters to be estimated, Ψ, where $\Psi = (M_1 + M_2)P$. Clearly, it is preferable to reduce the overall number of parameters in order to simplify the estimation procedure and, therefore, also speed up the adaptation to changes in the CIR. In order to determine a suitable model order, we formed the Bayesian Information Criteria (BIC) selection rule [35]:

$$\text{BIC}(\Psi) = \Psi ln(L) + L \log(\eta(\Psi)) \tag{38}$$

where $\eta(\Psi)$ denotes the error for the model using the Ψ parameters (which is uniquely defined given the constraint that $M_1 \geq M_2 \geq P$) and L is the number of measured samples per pulse, in our case $L = 4000$. Figure 9 illustrates the resulting model order selection rule, suggesting that $\Psi = 280$, which implies that $M_1 = 30$, $M_2 = 26$, and $P = 5$ are suitable model ordersto minimize the fitting error while keeping the model order low. It may be noted from the figure that, due to the various parameter combinations possible to form Ψ, the resulting BIC curve will be non-smooth; although, this may, as can be seen, still be used to determine suitable model orders.

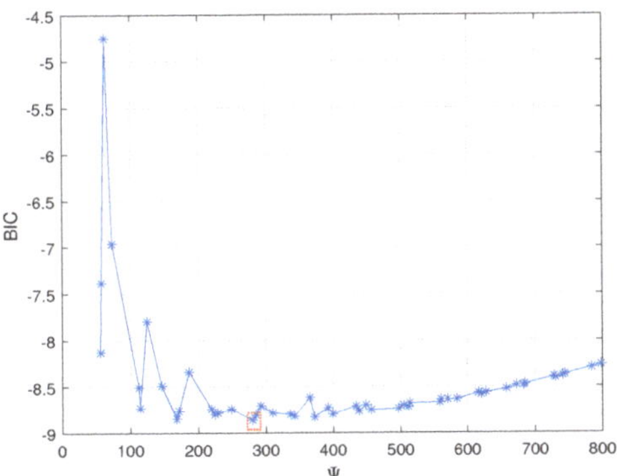

Figure 9. The BIC curve for SNR = 0 dB, illustrating the preferable choice of the model orders.

3.3. Matched-Filter-Based Detection

Next, we examined how the proposed CIR estimates can be used to detect weak moving targets. In doing so, we added a simulated target signal to the measured signal, where the target signal is modeled as the reflection of a moving target between the transmitter and receiver, gradually moving away from the receiver such that the relative reflection is shifted 3 ms per consecutive pulse. In order to mimic the local scattering of the target, the primary target response was modeled here as a delay of 33 ms with unit amplitude. The local scattering of the target was modeled using (normally distributed) randomly generated weak reflections following the main reflection. Figure 10 illustrates a typical example of the (noise-free) target impulse response; the measured target response signal was modeled as the convolution of this response by the transmitted signal, being scaled to yield the examined SRR. The target response was added here at the eight pulse to allow the algorithm to converge prior to forming the detection variable at this time. In these simulations, we used an SNR of -15 dB and an SRR of -3 dB.

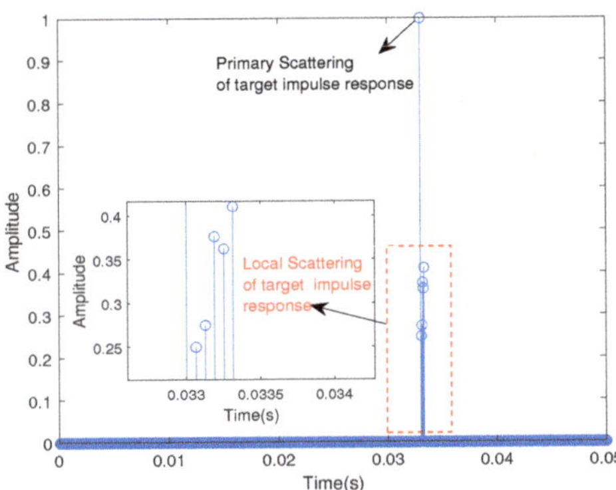

Figure 10. An example of the simulated noise-free target impulse response.

Figure 11 shows the resulting receiver operating characteristic (ROC) curve formed using MC = 1000 Monte Carlo simulations for the detection variable T_k, defined in (36), when employing the matched subspace detector to the residual $\hat{\mathbf{n}}_k(\tau_t)$ in (32), formed using the discussed CIR estimators. As a comparison, the figure also shows the performance of this detector when applied instead directly to the measured signal, $\mathbf{y}_k$, without any CIR compensation. This detector is denoted here as the uncompensated matched filter (UMF). From the figure, it is clear that the CIR compensation notably improved the detector performance, with the more-accurate CIR estimates gradually improving the detection performance.

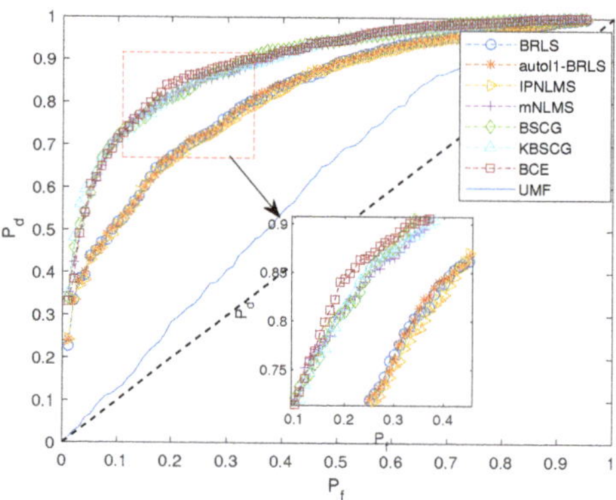

Figure 11. Estimated ROC for the resulting detectors, using a simulated CIR, when observing a single pulse containing the target moving with velocity 1.5 m/s, for SNR = −15 dB and SRR = −3 dB.

The poor performance of the UMF was a result of the reverberation, causing the measured signal to be formed and overlapped by multiple shifted versions of the transmitted signal; when forming the detection variable, one cannot, as a result, distinguish the reflections from the target from those of the reverberation, which caused the poor performance. The proposed method instead uses the residual signal after the CIR compensation, reducing the influence of the reverberation, to form the detection variable, thereby allowing for the more-robust decision. Given that the reverberation is relatively stationary as compared to the reflections of the moving reflector, one may, in this way, exploit the changes in the sound field to detect the moving target.

3.4. Experimental Results

We finally examined a real sea measurement in a shallow sea in May 2022, in Laoshan, Qingdao, China, having a depth of 10 m, where a single transmitter placed at a depth of 4 m was transmitting a linear frequency-modulated pulse sequence covering 3 to 7 kHz, using a sampling rate of $f_s = 16$ kHz. The pulse repeat time (PRT) used was 3.3 s, with a pulse width of 0.2 s. The transmitted signal was measured using the same sampling rate by a receiver positioned 5 km away from the transmitter, at a depth of 2 m. The main system parameters are listed in Table 2.

Table 2. Underwater experiment parameters.

Parameter	Value
Starting frequency	3 kHz
Bandwidth	4 kHz
Sampling frequency	16 kHz
Pulse width	200 ms
PRT	3.3 s
Tx depth	4 m
Rx depth	2 m
Sea depth	10 m
Horizontal range of Tx and Rx	5 km

Both the transmitter and the receiver were almost static, being fixed to an underwater chain experiencing only a mild drift; although, as shown in Figure 2, the resulting channel was still drifting notably. As can be seen in Figure 2, as well as in the drift-compensated CIR estimate shown in Figure 3, the CIR had a few dominant reflections, corresponding to the direct path, the bottom reflection, as well as the surface reflection. Figure 12 shows the resulting BCE estimate, using the noted settings, for the measurement data. As can be seen in the figure, the estimator was able to estimate the sparse structure of the CIR well, without suffering from the notable spurious estimates present in the LS estimate.

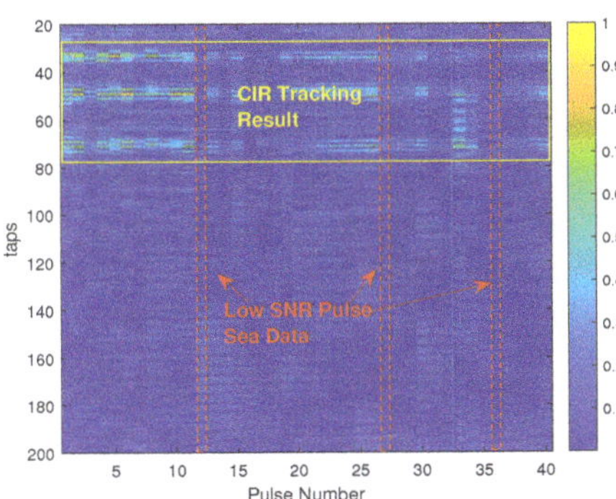

Figure 12. Pulse-to-pulse BCE estimate of the drift-compensated CIR from high SNR sea data.

Figure 13 illustrates how well the discussed methods were able to fit the observed data as the SNR was reduced; here, we normalized the power of the measured signal and added an additive white Gaussian noise corresponding to the noted SNR, computing the reconstruction error using the difference between the sea data (without additive noise) and the reconstructed signal. As can be seen in Figure 13, paralleling the simulation results in Figure 4, the proposed estimator, as expected, offered superior performance in the low SNR case, where the use of the structured sparsity was beneficial.

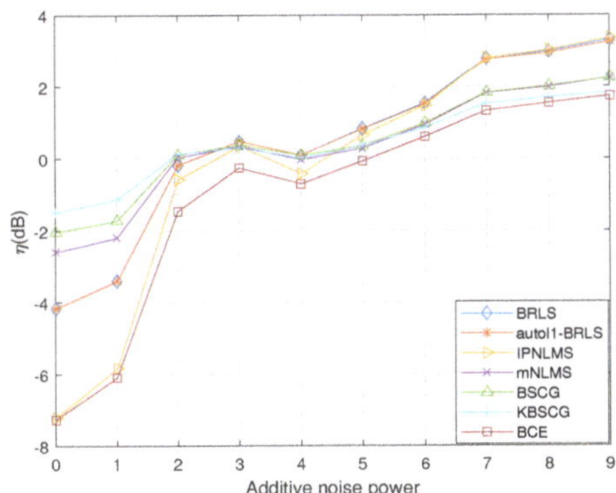

Figure 13. Performance comparison of the CIR estimates for the measured sea data, for varying noise levels.

Next, we used the measured sea data illustrated in Figure 2, but now included the reflection of a target (modeled as above), moving away from the transmitter with a velocity 1.5 m/s, with an SRR of -2.5 dB. It is worth noting that, as the residual in (30) was formed using the CIR estimate from pulse $k - 1$, the residual will contain a contribution both from the new location of the target, as well as the lack of a contribution at its earlier location, thereby increasing the power of the target signal in (33). Figure 14 shows the resulting detection variable T_k, defined in (36), for the 8th, 9th, and 10th pulses, when using the BCE estimator. As can be seen in the figure, although the CIR estimate was able to describe parts of the sea channel, the residual $\hat{n}_k$ still retained some reflections, partly due to the varying SNR of the sea measurements between pulses, but also due to the unexplained parts of the sea channel, causing large values of T_k for the initial part of $\hat{n}_k$ for all three pulses. The reflecting target can be seen in Figure 14 at a delay of about 0.033 s, corresponding to sample 528, for the eight pulse, and then moving 48 samples (3 ms) in each of the following pulses due to the target's velocity (it should be noted that the approach will also work for the detection of reflectors moving with non-constant velocity, although at some loss, as the shifted reflection will align less ideally in such a case).

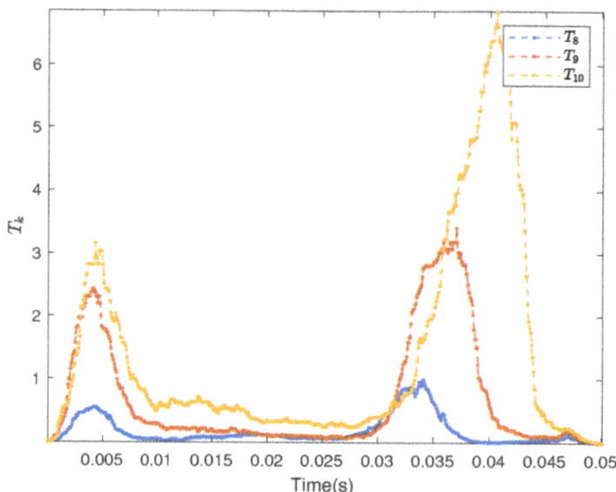

Figure 14. The detection variable T_k, defined in (36), for three consecutive pulses for a simulated target moving with velocity 1.5 m/s, with SNR = -12.3 dB, and SRR = -2.5 dB.

As the unexplained part of the channel caused significant reflections, the detector may well fail to detect the presence of weaker targets. In order to improve the detection, we proceeded to exploit the motion of the target, assuming that it will exhibit a reasonably constant velocity during a short interval. We, therefore, formed the cross-correlation between consecutive pulses, $\rho_{T_k,T_{k-1}}$, and then, determined the common shift between pulses, reflecting the target's constant velocity, as

$$k_{\tau_K} = \max \sum_{k=1}^{K} \rho_{T_k,T_{k-1}} \tag{39}$$

for K pulses. By then circularly shifting the kth detection variable, T_k, with k_{τ_K}, forming $\tilde{T}_k$, all the shifted detection variables $\tilde{T}_k$ may be summed coherently, creating a new detection variable $\breve{T}_K$, which is then used to determine the presence or the absence of a target in the K measurements. The shifted $\tilde{T}_k$ for the 8th, 9th, and 10th pulses, together with the resulting $\breve{T}_K$ are shown in Figure 15, illustrating how the resulting detection variable can efficiently exploit the response of the target in the pulses, reducing the influence of the still unexplained sea channel.

Figure 16 shows the resulting detection performance of the detection variable $\breve{T}_K$, using $K = 4$ pulses with a simulated target with SRR = -2.5 dB. In each of the MC = 1000 Monte Carlo simulations used to form the ROC, an additive Gaussian noise was added to the real measurement, yielding an SNR of -12.3 dB. For this low SRR and SNR, the UMF failed to allow for a viable detection and was, therefore, in the interest of clarity, omitted from the figure. As can be seen in Figure 16, the combined detection variable, $\breve{T}_K$, was able to accurately detect the weak target using the structured CIR estimates, again showing the excellent performance of the proposed detector.

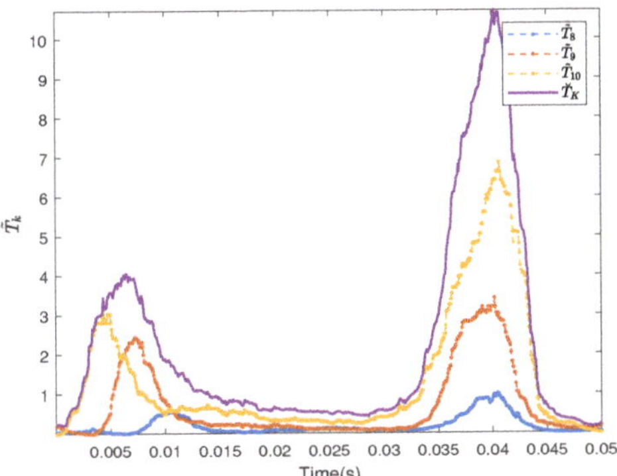

Figure 15. The shifted detection variables $\tilde{T}_k$ for the three pulses shown in Figure 14, together with the combined detection variable $\tilde{T}_K$.

Consistent with the simulation results shown in Figure 11, one may note that the structured CIR estimators were also able to track the channel fluctuations for the measured sea data sufficiently fast to allow for an improved detection, as compared to the non-structured estimators.

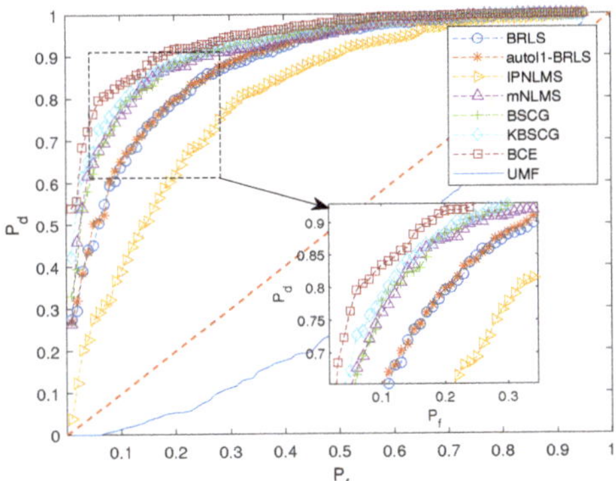

Figure 16. Estimated ROC for the resulting detectors, using the discussed CIR estimates with the real sea data, when observing $K = 4$ pulses containing a target moving with velocity 1.5 m/s, with SNR = -12.3 dB and SRR = -2.5 dB.

4. Conclusions

In this paper, we introduced a sparse and structured block-updating channel impulse response (CIR) estimator. By exploiting a sparse and structured approximation of the CIR, we formulated a block-updated conjugate gradient formulation that allowed the CIR estimator to provide accurate performance even in noisy environments, whereas the estimator also allowed for a gradual relaxation of these constraints for higher SNR cases, enabling the estimator to retain the better performance of the estimators without posing such restrictions. We also included a discussion of how the required user parameters should be selected and how these affect the performance of the method, as well as introduced a matched subspace detector formed on the resulting channel residual. The proposed CIR estimate was evaluated by comparing it to several recent CIR estimators, for both simulated and measured sea data, illustrating both the better performance of the CIR estimate and the resulting improved detection performance. In the future, we will aim to incorporate an adaptive selection of the hyperparameters used, as well as examine how the target motion affects the detectability of a target.

Author Contributions: C.Y. contributed mainly to the work and is listed as the first author. X.S. served as the corresponding author and provided guidance throughout the study. C.Y. and A.J. were responsible for designing and conducting the experiments, analyzing the data, and writing the manuscript. Specifically, Q.L., X.S. and A.J. developed the research question and hypotheses, designed the study protocol, recruited the participants, collected and managed the data, conducted the statistical analyses, and interpreted the findings. C.Y. also wrote the initial draft of the manuscript and revised it based on feedback from the other authors. M.M. assisted with the experimental design and performed the data analysis. All authors have read and agreed to the published version of the manuscript.

Funding: This research was funded by the National Natural Science Foundation of China (Grant No. U20A20329) and China Scholarship Council under Grant No. 202106680039.

Data Availability Statement: The data presented in this study are available in article.

Conflicts of Interest: The authors declare no conflicts of interest.

References

1. Jiang, W.; Zheng, S.; Zhou, Y.; Tong, F.; Kastner, R. Exploiting time varying sparsity for underwater acoustic communication via dynamic compressed sensing. *J. Acoust. Soc. Am.* **2018**, *143*, 3997–4007. [CrossRef] [PubMed]
2. Yu, G.; Zhao, D.; Piao, S. Target detection method using multipath information in an underwater waveguide environment. *IET Radar Sonar Navig.* **2020**, *14*, 226–232. [CrossRef]
3. Junejo, N.U.R.; Sattar, M.; Adnan, S.; Sun, H.; Adam, A.B.; Hassan, A.; Esmaiel, H. A Survey on Physical Layer Techniques and Challenges in Underwater Communication Systems. *J. Mar. Sci. Eng.* **2023**, *11*, 885. [CrossRef]
4. Jing, S.; Hall, J.; Zheng, Y.R.; Xiao, C. Signal detection for underwater IoT devices with long and sparse channels. *IEEE Internet Things J.* **2020**, *7*, 6664–6675. [CrossRef]
5. Lin, J.; Wang, G.; Zheng, Z.; Ye, R.; He, R.; Ai, B. Modeling and channel estimation for piezo-acoustic backscatter assisted underwater acoustic communications. *China Commun.* **2022**, *19*, 297–307. [CrossRef]
6. Yuan, X.; Guo, L.; Luo, C.; Zhou, X.; Yu, C. A survey of target detection and recognition methods in underwater turbid areas. *Appl. Sci.* **2022**, *12*, 4898. [CrossRef]
7. Li, W.; Zhou, S.; Willett, P.; Zhang, Q. Preamble detection for underwater acoustic communications based on sparse channel identification. *IEEE J. Ocean. Eng.* **2017**, *44*, 256–268. [CrossRef]
8. Tian, Y.; Han, X.; Vorobyov, S.A.; Yin, J.; Liu, Q.; Qiao, G. Wideband signal detection in multipath environment affected by impulsive noise. *J. Acoust. Soc. Am.* **2022**, *152*, 445–455. [CrossRef]
9. Zhang, Y.; Venkatesan, R.; Dobre, O.A.; Li, C. Efficient Estimation and Prediction for Sparse Time-Varying Underwater Acoustic Channels. *IEEE J. Ocean. Eng.* **2020**, *45*, 1112–1125. [CrossRef]
10. Tian, Y.; Han, X.; Yin, J.; Li, Y. Adaption penalized complex LMS for sparse under-ice acoustic channel estimations. *IEEE Access* **2018**, *6*, 63214–63222. [CrossRef]
11. Duttweiler, D.L. Proportionate normalized least-mean-squares adaptation in echo cancelers. *IEEE Trans. Speech Audio Proc.* **2000**, *8*, 508–518. [CrossRef]
12. Benesty, J.; Gay, S.L. An improved PNLMS algorithm. In Proceedings of the IEEE International Conference on Acoustics, Speech, and Signal Processing, Orlando, FL, USA, 13–17 May 2002; Volume 2, pp. 1881–1884.
13. Hoshuyama, O.; Goubran, R.A.; Sugiyama, A. A generalized proportionate variable step-size algorithm for fast changing acoustic environments. In Proceedings of the IEEE International Conference on Acoustics, Speech, and Signal Processing, Montreal, QC, Canada, 17–21 May 2004; Volume 4, pp. 161–164.
14. Qi, C.; Wang, X.; Wu, L. Underwater acoustic channel estimation based on sparse recovery algorithms. *IET Signal Process.* **2011**, *5*, 739–747. [CrossRef]
15. Zakharov, Y.V.; Li, J. Sliding-window homotopy adaptive filter for estimation of sparse UWA channels. In Proceedings of the IEEE Sensor Array and Multichannel Signal Processing Workshop, Rio de Janerio, Brazil, 10–13 July 2016; pp. 1–4.
16. Qiao, G.; Gan, S.; Liu, S.; Ma, L.; Sun, Z. Digital self-interference cancellation for asynchronous in-band full-duplex underwater acoustic communication. *Sensors* **2018**, *18*, 1700. [CrossRef] [PubMed]
17. Lee, C.; Rao, B.D.; Garudadri, H. A sparse conjugate gradient adaptive filter. *IEEE Signal Process. Lett.* **2020**, *27*, 1000–1004. [CrossRef]
18. Eksioglu, E.M.; Tanc, A.K. RLS algorithm with convex regularization. *IEEE Signal Process. Lett.* **2011**, *18*, 470–473. [CrossRef]
19. Eksioglu, E.M. Sparsity regularised recursive least squares adaptive filtering. *IET Signal Process.* **2011**, *5*, 480–487. [CrossRef]
20. Montazeri, M.; Duhamel, P. A set of algorithms linking NLMS and block RLS algorithms. *IEEE Trans. Signal Process.* **1995**, *43*, 444–453. [CrossRef]
21. Wang, Z.; Zhou, S.; Preisig, J.C.; Pattipati, K.R.; Willett, P. Clustered Adaptation for Estimation of Time-Varying Underwater Acoustic Channels. *IEEE Trans. Signal Process.* **2012**, *60*, 3079–3091. [CrossRef]

22. Qiao, G.; Song, Q.; Ma, L.; Liu, S.; Sun, Z.; Gan, S. Sparse Bayesian learning for channel estimation in time-varying underwater acoustic OFDM communication. *IEEE Access* **2018**, *6*, 56675–56684. [CrossRef]
23. Paleologu, C.; Benesty, J.; Ciochină, S. Linear system identification based on a Kronecker product decomposition. *IEEE/ACM Trans. Audio Speech Lang. Proc.* **2018**, *26*, 1793–1808. [CrossRef]
24. Elisei-Iliescu, C.; Paleologu, C.; Benesty, J.; Stanciu, C.; Anghel, C.; Ciochină, S. A multichannel recursive least-squares algorithm based on a Kronecker product decomposition. In Proceedings of the 43rd International Conference on Telecommunications and Signal Processing, Milan, Italy, 7–9 July 2020; pp. 14–18.
25. Wang, X.; Huang, G.; Benesty, J.; Chen, J.; Cohen, I. Time difference of arrival estimation based on a Kronecker product decomposition. *IEEE Signal Process. Lett.* **2020**, *28*, 51–55. [CrossRef]
26. Elisei-Iliescu, C.; Paleologu, C.; Benesty, J.; Stanciu, C.; Anghel, C.; Ciochină, S. Recursive least-squares algorithms for the identification of low-rank systems. *IEEE/ACM Trans. Audio Speech Lang. Process.* **2019**, *27*, 903–918. [CrossRef]
27. Bhattacharjee, S.S.; George, N.V. Nearest Kronecker product decomposition based normalized least mean square algorithm. In Proceedings of the IEEE International Conference on Acoustics, Speech and Signal Processing, Barcelona, Spain, 4–8 May 2020; pp. 476–480.
28. Marple, S.L. Estimating group delay and phase delay via discrete-time "analytic" cross-correlation. *IEEE Trans. Signal Process.* **1999**, *47*, 2604–2607. [CrossRef]
29. Dong, Y.; Zhao, H. A new proportionate normalized least mean square algorithm for high measurement noise. In Proceedings of the 2015 IEEE International Conference on Signal Processing, Communications and Computing, Ningbo, China, 19–22 September 2015; IEEE: Piscataway, NJ, USA, 2015; pp. 1–5.
30. Variddhisaï, T.; Mandic, D.P. On an RLS-like LMS adaptive filter. In Proceedings of the 2017 22nd International Conference on Digital Signal Processing, London, UK, 23–25 August 2017; IEEE: Piscataway, NJ, USA, 2017; pp. 1–5.
31. Harville, D.A. *Matrix Algebra from a Statistician's Perspective*; Taylor & Francis: Abingdon-on-Thames, UK, 1998.
32. L. Van, F.C. The ubiquitous Kronecker product. *J. Comput. Appl. Math.* **2000**, *123*, 85–100. [CrossRef]
33. Jakobsson, A.; Mossberg, M.; Rowe, M.; Smith, J.A.S. Exploiting Temperature Dependency in the Detection of NQR Signals. *IEEE Trans. Signal Process.* **2006**, *54*, 1610–1616. [CrossRef]
34. Kay, S.M. *Fundamentals of Statistical Signal Processing, Volume II: Detection Theory*; Prentice-Hall: Englewood Cliffs, NJ, USA, 1998.
35. Stoica, P.; Moses, R.L. *Spectral Analysis of Signals*; Pearson Prentice Hall: Upper Saddle River, NJ, USA, 2005.

remote sensing

Communication

Source Depth Discrimination Using Intensity Striations in the Frequency–Depth Plane in Shallow Water with a Thermocline

Xiaobin Li [1,2] and Chao Sun [1,2,*]

1 School of Marine Science and Technology, Northwestern Polytechnical University, Xi'an 710072, China; xblee@mail.nwpu.edu.cn
2 Shaanxi Key Laboratory of Underwater Information Technology, Xi'an 710072, China
* Correspondence: csun@nwpu.edu.cn

Abstract: A source depth discrimination method based on intensity striations in the frequency–depth plane with a vertical linear array in a shallow water environment is proposed and studied theoretically and experimentally. To quantify the orientation of the interference patterns, a generalized waveguide variant (GWV) η is introduced. Due to the different dominance of the mode groups, the GWV distribution in the surface source is sharply peaked, indicating the presence of striations in the interferogram and the slope associated with the source–array range, while the distribution of the submerged source is more diffuse, and its interferogram is chaotic. The existence or lack of a distinct peak is used to separate the surface and submerged source classes. The method does not demand prior knowledge of the sound speed profile or the relative movement between the source and the array. In addition, it is the presence of the striations, not the value of η, that is exploited to separate the surface and submerged source classes, which means the source–array range can be unknown. The proposed method is validated using experimental data on the towing ship in SWellEx–96 and numerical modeling. The method's performance under noise situations and for different source–array ranges is also investigated.

Keywords: source depth discrimination; modal interference in frequency–depth plane; generalized waveguide variant; shallow water with a thermocline

Citation: Li, X.; Sun, C. Source Depth Discrimination Using Intensity Striations in the Frequency–Depth Plane in Shallow Water with a Thermocline. *Remote Sens.* **2024**, *16*, 639. https://doi.org/10.3390/rs16040639

Academic Editors: Gerardo Di Martino, Jiahua Zhu, Xinbo Li, Shengchun Piao, Junyuan Guo, Wei Guo, Xiaotao Huang and Jianguo Liu

Received: 21 December 2023
Revised: 5 February 2024
Accepted: 7 February 2024
Published: 8 February 2024

1. Introduction

Source depth discrimination in shallow water has significant research value, aiming to distinguish surface sources from submerged ones rather than calculating depth. The distinction between these two classes of sources is based on their respective mode spectrum excitation patterns. It exploits the difference in energies of low-order normal modes (also known as trapped modes [1], TMs) and high-order normal modes (non–trapped modes, NTMs), since the surface source cannot excite TMs due to their evanescent mode amplitudes near the surface [2]. In contrast, a submerged source can excite both TMs and NTMs.

Publications have explicitly used the numerical representation of the energy difference for source depth discrimination. A horizontal line array (HLA) at the endfire was utilized to build the mode subspace projections and estimate the energy ratio between these two groups of modes for discrimination [1], requiring the inputs of an approximate sound speed profile, water depth, and bottom type. Mode filtering was also used to build the trapped energy ratio with an HLA close to the endfire [3]. This demands prior knowledge of the mode characteristics, which cannot be precisely obtained, due to the uncertainty in the acoustic model or the environmental mismatch.

The application of the waveguide invariant β [4] in depth discrimination implicitly exploits the aforementioned difference, which suggests whether the source is near the surface or submerged, depending on its value. The invariance quantifies the orientation of the intensity striations caused by modal interferences in the frequency–range $(f - r)$

plane. It is found that, for a surface source, the distribution of the β peaks are at different values when the receiver is above or below the thermocline [5]. According to the reciprocity principle, the distribution will peak, depending on the depth of the source for a fixed near-surface receiver. β can also be extracted by using the warping transform in the frequency domain from the interference pattern without knowing the source range [6]. However, β requires relative movement between the source and receiver, since the modal interferences occur in the range domain.

Note that there are modal interferences in the frequency–depth ($f - z$) plane that behave comparably to those in the $f - r$ plane, determined by the dominant modes associated with the source depth. This paper discusses the discrimination between surface and submerged sourcesin shallow water, which are above or below the thermocline, respectively. The group of NTMs, which dominate the surface source, produces an observable striation pattern in the acoustic intensity, while the interferogram of the submerged source is chaotic. The proposed method utilizes a vertical line array (VLA) to "capture" these interference patterns, which are described by a generalized waveguide variant (GWV) distribution. The existence or lack of a distinct peak in the distribution represents the presence or absence of the striation, which is further used to discriminate surface and submerged sources. The method is valid for a higher frequency source, as its NTMs have similar characteristics compared to those from low-frequency sources. The accumulation in the depth domain allows for some robustness against noise. In addition, the value of the GWV is related to the source range; however, the discrimination does not require this value, which means that the source range can be unknown.

The paper is organized as follows. In Section 2, the GWV η is derived in the $f - z$ plane with interferometric signal processing. The method based on η for source depth discrimination and the requirements of method based on the source frequency are presented. In Section 3, the proposed method is performed on the data from the towing ship in the SWellEx–96 experiment. The striation pattern and η of the intensity distribution in the $f - z$ plane generated by the surface source are verified. In Section 4, the numerical modeling results for the same experimental situation are presented. Complementing the simulations of the submerged source, which does not meet the implementation requirement in the experiment, the proposed discrimination method is validated. Furthermore, the performance under noise conditions and for different source ranges is investigated. A summary is presented in Section 5.

2. Discrimination Using Intensity Striations in the $f - z$ Plane

2.1. Generalized Waveguide Variant Describing the Intensity Striations

For a point source at depth z_s, the acoustic intensity at depth z and range r can be expressed as [7]:

$$I(r, z, f, z_s) = \sum_{m=1}^{M} \sum_{n=1}^{M} B_m(r, z_s)\phi_m(z)B_n^*(r, z_s)\phi_n^*(z)e^{i(k_{rm}-k_{rn})r}, \tag{1}$$

where

$$B_m(r, z_s) = \sqrt{2\pi / k_{rm}r}\phi_m(z_s), \tag{2}$$

M is the total number of normal modes excited by the source, and $\phi_m(z)$ and k_{rm} are the depth function and the horizontal wavenumber of the mth mode, respectively. $(\cdot)^*$ is the conjugate operator.

The intensity maximum in the frequency–depth ($f - z$) plane is determined by the following condition

$$dI = \frac{\partial I}{\partial z}dz + \frac{\partial I}{\partial f}df = 0. \tag{3}$$

The slope of the striations κ is

$$\kappa = \frac{df}{dz} = -\frac{\partial I/\partial z}{\partial I/\partial f}$$

$$= -\frac{2 \sum\limits_{m=1}^{M} \sum\limits_{n=1}^{M} B_m B_n^* \phi_n^*(z) e^{i(k_{rm}-k_{rn})r} \partial \phi_m/\partial z}{r \sum\limits_{m=1}^{M} \sum\limits_{n=1}^{M} B_m \phi_m B_n^* \phi_n^* e^{i(k_{rm}-k_{rn})r} \left(S_{g,m} - S_{g,n}\right)}, \tag{4}$$

where $S_{g,m}$ is the group slowness of the mth mode.

A generalized waveguide variant (GWV) η of the $f - z$ plane is defined as

$$\eta(z, f | z_s) = \kappa * \frac{r}{2} = -\frac{\sum\limits_{m=1}^{M} \sum\limits_{n=1}^{M} B_m B_n^* \phi_n^* e^{i(k_{rm}-k_{rn})r} \partial \phi_m/\partial z}{\sum\limits_{m=1}^{M} \sum\limits_{n=1}^{M} B_m \phi_m B_n^* \phi_n^* e^{i(k_{rm}-k_{rn})r} \left(S_{g,m} - S_{g,n}\right)}. \tag{5}$$

Equation (5) can be rewritten as a sum of weighted components η_{mn} with coefficients α_{mn} representing their contributions to the variant (similar to the derivation of β in [8]), given by

$$\eta(z, f | z_s) = \sum\limits_{m=1}^{M} \sum\limits_{n=1}^{M} \alpha_{mn} \eta_{mn}, \tag{6}$$

where

$$\alpha_{mn} = \frac{B_m \phi_m B_n^* \phi_n^* e^{i(k_{rm}-k_{rn})r} \left(S_{g,m} - S_{g,n}\right)}{\sum\limits_{m=1}^{M} \sum\limits_{n=1}^{M} B_m \phi_m B_n^* \phi_n^* e^{i(k_{rm}-k_{rn})r} \left(S_{g,m} - S_{g,n}\right)}, \tag{7}$$

$$\eta_{mn}(f, z) = \frac{-\partial \phi_m/\partial z}{\phi_m \left(S_{g,m} - S_{g,n}\right)}. \tag{8}$$

Under the WKB [7] approximation, the mode function $\phi_m(z)$ can be expressed as

$$\phi_m(f, z) = \sin[k_{zm}(z)z], \tag{9}$$

where $k_{zm}(f, z) = \sqrt{[2\pi f/c(z)]^2 - k_{rm}^2}$ is the vertical wavenumber of the mth mode. Consider the situation when the VLA is deployed below the thermocline, which means the receivers are below all the turning points of trapped modes, making $c(z)$ and k_{zm} both constants. Therefore,

$$\eta_{mn}(f, z) = -\frac{k_{zm} \cot(k_{zm} z)}{S_{g,m} - S_{g,n}}. \tag{10}$$

Since many η_{mn} values, as pairs of modes (m, n), contribute to the GWV, a distribution of η denoted by E_η better quantifies this complex striation pattern, similar to E_β proposed in [9–11].

As mentioned above, the GWV is converted from the slope of the interference striations, which can be calculated by using two-dimensional Fast Fourier Transform [12] (2D–FFT) on the intensity distribution $I(f, z)$. The corresponding algorithm follows [9,10] and will be briefly reviewed below.

The 2D–FFT of $I(f, z)$ with depth aperture D and bandwidth B is defined by

$$I(x, y) = \left| \int_{f_m - B/2}^{f_m + B/2} \int_{z_m - D/2}^{z_m + D/2} I(f, z) e^{-i2\pi(xz+yf)} \, dz \, df \right|, \tag{11}$$

where f_m and z_m are the mean values of axis f and z, and x (in m^{-1}) and y (in s) are the FFT variables conjugate to the depth and frequency, respectively. We replace the slope in Equation (5) with its expression in the Fourier domain and obtain

$$\eta = -\frac{r}{2}\frac{x}{y},\tag{12}$$

and the GWV can be represented in another set of variables related to the polar coordinate system by choosing

$$K = \sqrt{x^2 + y^2}.\tag{13}$$

The GWV distribution E_η is given by summing up K in the (η, K) plane, which is the result of the polar coordinate transform. The presence of a clear peak indicates the existence of striations and the corresponding slope at η.

2.2. Discrimination Based on the GWV

In shallow water with a thermocline, when the source is located above the thermocline, only the first several NTMs exist, since the TMs are poorly excited, and the higher order modes attenuate rapidly during the propagation. These NTMs can be regarded as a group of modes due to their similar k_zs, which will further behave similarly in $\cot(k_z z)$ and η_{mn}. As long as the sample locations are not at the depths where the η_{mn} approaches 0, the enhancement of these η_{mn} provides a proper value of η (a peak of the GWV distribution) and corresponding striations in the $f - z$ plane.

However, for the submerged source, which excites both TMs and NTMs, the k_zs of the modes vary by an order of magnitude, resulting in the change in the function period. Therefore, the sum of η_{mn} containing $\cot(k_z z)$ with different periods cannot make η a certain value, which means that the interferogram will be chaotic, and no striations will exist.

As a crucial parameter, shown in Equation (10), for the proposed method, k_{zm} determines η_{mn} and η for the fixed receiver depths. In general, k_z increases with the source frequency, but its rate decreases. For a higher-frequency surface source, the k_zs of NTMs vary slowly during the frequency band, resulting in similar periods of $\cot(k_z z)$, which further ensure the enhancement of those η_{mn}. In the case of discriminating the lower-frequency surface source, the fact that the k_zs of NTMs in the processing band vary greatly, and the periods of $\cot(k_z z)$ change rapidly, means the summation of η_{mn} is like that of the submerged source, and it fails to distinguish between these two source classes.

As shown in Equations (5) and (12), η is related to the source–array range r, which implies that the slope of the striation is scaled up/down by the ratio of the range (if the range is estimated before or later, which is outside the scope of this paper). However, the scale change does not affect the striation's existence. It is the fact that the distinct peak of η_{mn} exists and not the value itself that is an important clue to whether the target is on the surface or submerged. These above observations are verified using experimental and simulated data in the following sections.

3. Experimental Data Analysis

The SWellEx–96 experiment [13] was conducted near San Diego, CA, in May of 1996. The SSP can be approximately regarded as a typical downward refracting profile with a thermocline. The VLA was deployed from a depth of 94.125 m to a depth of 212.25 m and contained 21 elements that were evenly spaced (ignoring the small vertical tilt). The range in the depth was nearly the same as in the situation discussed in Section 2, since it was below the thermocline. The SSP and the array configuration are presented in Figure 1.

One second of data were analyzed, which was 73 min after the start of event S5, where the towing ship (R/V Sproul) was 2.323 km from the VLA. The analyzed data involved one signal with the band $\Delta f = 150$ Hz (600–750 Hz) radiated by the towing ship (at a depth of 2.9 m [14]), which was regarded as a surface source. The reasons for not choosing the two experiment sources were as follows: (i) a broadband source was required, since it had the premise of an interference structure, while the shallow source transmitted nine frequencies between 109 Hz and 385 Hz; (ii) although the deep source stopped projecting CW tones and started projecting FM chirps (200–400 Hz) at the beginning, midway point, and end of the track, its frequency was not applicable (too low) in this scenario.

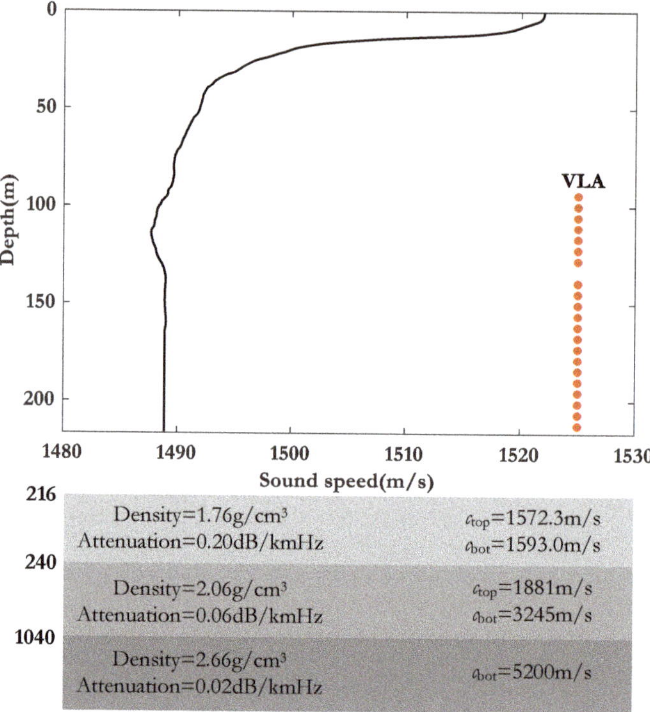

Figure 1. The SSP of SWellEx–96 and the arrangement of VLA.

The results of the experimental data processing are shown in Figure 2. The intensity distribution $I(f,z)$ Equation (1), 2D–FFT of $I(f,z)$ Equation (12), 2D–FFT in the polar coordinates, and the GWV distribution E_η are shown for the assessing procedure.

One can observe the intensity striations in Figure 2a (to show the striations more clearly, we show the image with a larger bandwidth (550–900 Hz), which is symmetrical at about $f = 750$ Hz, since the sample frequency of the data is 1500 Hz). It is worth mentioning that the striations were not caused by frequency shifting, although there were several tones (such as 605 Hz and 677 Hz) projected by the towing ship, since the speed of the ship was 2.5 m/s (5 knots), and the length of the data was 1s.

Figure 2b shows the result of the 2D–FFT of the region enclosed by the white dashed lines in Figure 2a and exhibits a vertical line resulting from the background noise.

We removed this vertical line and performed the polar coordinate transform (the transformations later were all conducted after vertical line removal) in Figure 2c. Figure 2d shows the GWV distribution E_η of the data we analyzed, and the peak $\eta_{ex} = 68.4$.

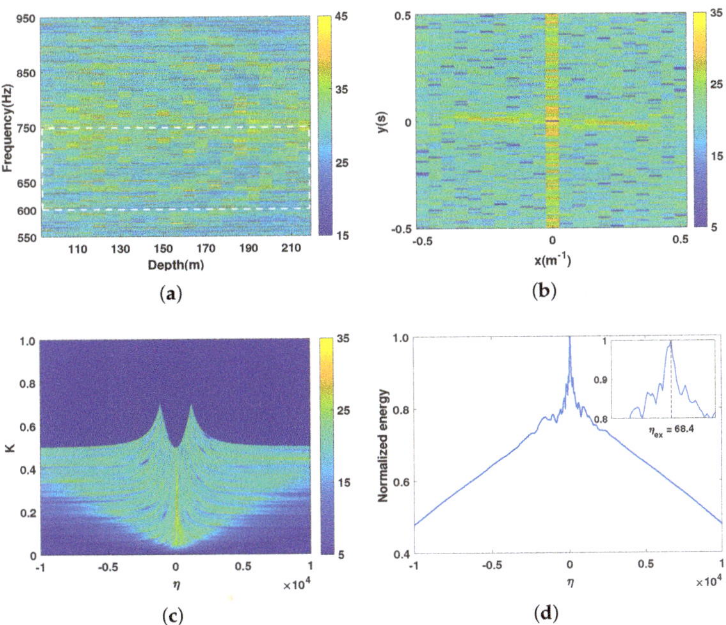

Figure 2. The results of the experimental data: (**a**) $I(f,z)$; (**b**) 2D−FFT of $I(f,z)$; (**c**) 2D−FFT in the polar coordinate; (**d**) the GWV distribution E_η and zoom of peak.

4. Numerical Modeling Results

For numerical simulations, we used the acoustic environment in SWellEx–96, considering the same VLA as that deployed during the experiment. The source frequency band was the same as noted above, $\Delta f = 150$ Hz (600–750 Hz). The horizontal distance between the source and the VLA was 2.323 km. The KRAKEN [15] was used to calculate the pressure field.

4.1. The Mode Functions, Normalized Amplitudes of Modes, and η_{mn}

Figure 3a displays the mode depth functions for the central frequency $f_c = 675$ Hz, with black lines marking the depths of 3 m and 54 m.

Figure 3b,c show the normalized amplitudes of the normal modes excited by a surface source and a submerged source, respectively, with marked TMs and NTMs. As can be seen, the two source classes differ in the dominance of excited modes, providing the basis for depth discrimination.

Figure 3d,e show $\eta_{43,44}$ (typical dominant interference modes of the surface source) as a function of the frequency and the water depth, and $\eta_{43,44}$ versus depth for $f_c = 675$ Hz, with red circles representing the VLA receivers, respectively. One can note that the η_{mn} of 600–750 Hz show periodicity and share a similar period, since there is $\cot(k_{zm}z)$ in η_{mn}, and k_{zm} varies slowly during the processing bandwidth, which shows the potential to make their combination E_η have a sharp peak. The VLA receivers are mostly not located near the zero point of $\eta_{43,44}$ versus depth for $f_c = 675$ Hz, letting η avoid being 0. The high $\eta_{43,44}$ (the dazzling line in $f = 746$ Hz) in Figure 3d is due to the tiny difference (0.0274 m/s) in group speeds of the 43rd and 44th modes, which happen to be the first two NTMs.

Figure 3f,g show $\eta_{9,10}$ with a larger period, which exhibited an abnormal situation at depths of 115–130m, caused by $c(z)$, leading to unusual k_{zm} and $\cot(k_{zm}z)$, and $\eta_{9,10}$ versus depth for $f_c = 675$ Hz with red circles representing the VLA receivers, respectively. More importantly, for the submerged source that excited both TMs and NTMs, the period of $\cot(k_{z9}z)$ (seen in the Figure 3g) was nearly four times that of $\cot(k_{z43}z)$ or several times that of the other $\cot(k_{zm}z)$, which means that each type of interference mode contributes its own peak, resulting in multiple sidelobes in the distribution E_η .

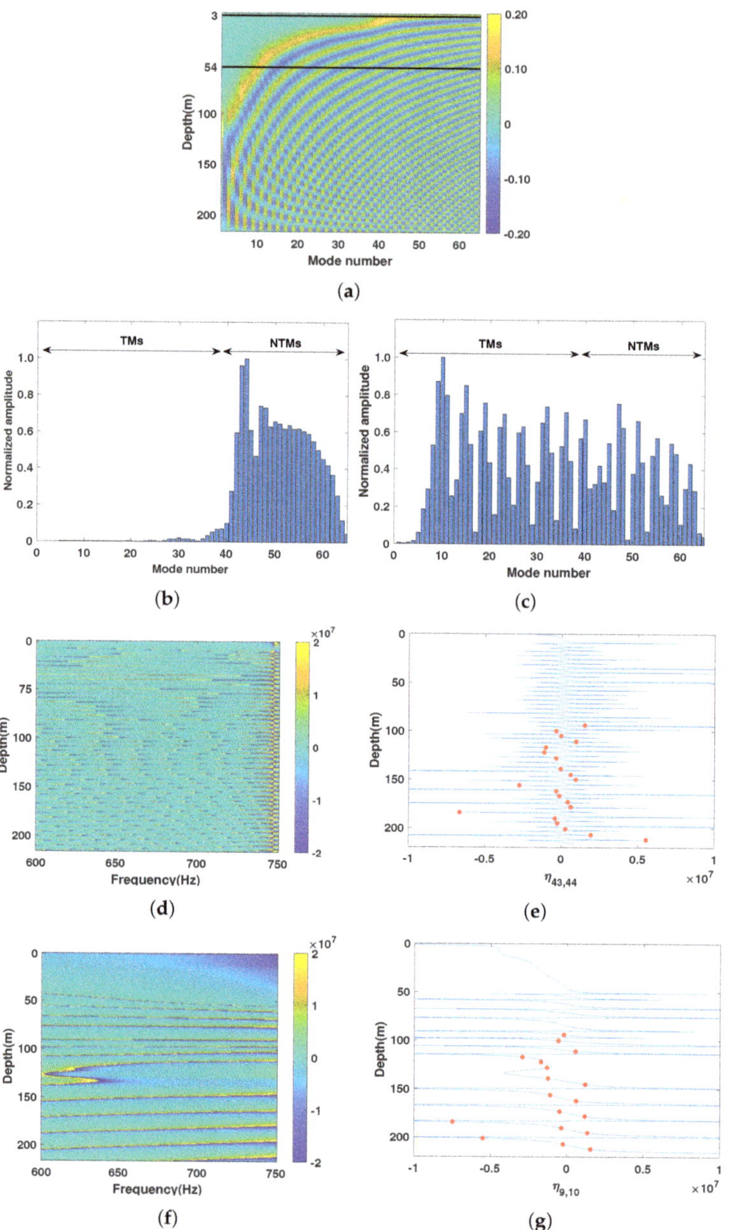

Figure 3. Mode depth functions for f_c=675Hz (**a**); normalized amplitude of the normal modes excited by the surface/submerged source (**b,c**); $\eta_{43,44}$ and $\eta_{9,10}$ as a function of the frequency and the water depth (**d,f**); $\eta_{43,44}$ and $\eta_{9,10}$ versus depth for f_c=675Hz with red dots representing the values of $\eta_{m,n}$ at the depths of the VLA receivers, respectively (**e,g**).

Table 1 presents the total number of modes, the dominant interference modes, and the corresponding period of $\cot(k_{zm}z)$ for different surface source frequencies. In underwater acoustics, in a general sense, a frequency above 500 Hz can be referred to as a high frequency (mid-high frequency). In the scenario discussed here, it needs to be discussed in combination with the specific SSP, such as shown in Table 1. Under the same bandwidth, the period of the dominant interference at different frequencies changes with a smaller period (600–750 Hz, $5.26/4.42 * 100\% \approx 119\%$) is called higher-frequency, and the opposite is called lower-frequency (150-300 Hz, and 300-450 Hz). There is no absolutely clear boundary between higher and lower frequencies here. For the surface source with lower frequencies (for example, 150–450 Hz), the period of $\cot(k_{zm}z)$ in the dominant interference modes decreases quickly ($15.47/9.52 * 100\% \approx 163\%$, $9.52/6.69 * 100\% \approx 142\%$). This difference between these periods makes the superposition of the NTMs of the surface source behave like those of the TMs and NTMs of the submerged source, and the proposed method fails.

Table 1. Total number of modes, the dominant interference modes, and the corresponding period of $\cot(k_{zm}z)$ for different surface source frequencies.

Surface Source Frequency	150 Hz	300 Hz	450 Hz	600 Hz	750 Hz
Total number of modes	18	32	46	61	73
Dominant interference modes (m,n)	(14,15)	(22,23)	(31,32)	(38,39)	(46,47)
Corresponding period of $\cot(k_{zm}z)$	15.47 m	9.52 m	6.69 m	5.26 m	4.42 m

4.2. Performance Study under the SSP from the Experiment

Two cases are considered here: (I) one case of a surface source at a depth of 3 m corresponding to the towing ship; (II) the other of a submerged source at a depth of 54 m.

The interferogram in Figure 4a, 2D–FFT of $I(f,z)$ in Figure 4b, 2D–FFT in the polar coordinates in Figure 4c, and the GWV distribution E_η in Figure 4d correspond to case I. Compared to the intensity striations in Figure 2a, those interference structures are more obvious, due to the stationary spectra used in the simulation. Being free from background noise, the vertical line (as in Figure 2b, caused by the noise) disappears. The highest energy of E_η denotes the presence of striation, and the number of $\eta_{\text{simu}} = 68.4$, which is the same as the experimental data result ($\eta_{\text{ex}} = 68.4$). The agreement between the simulation of the surface source and the experimental data analysis proves that there are intensity striations in the $f - z$ plane and verifies the effectiveness of the simulation.

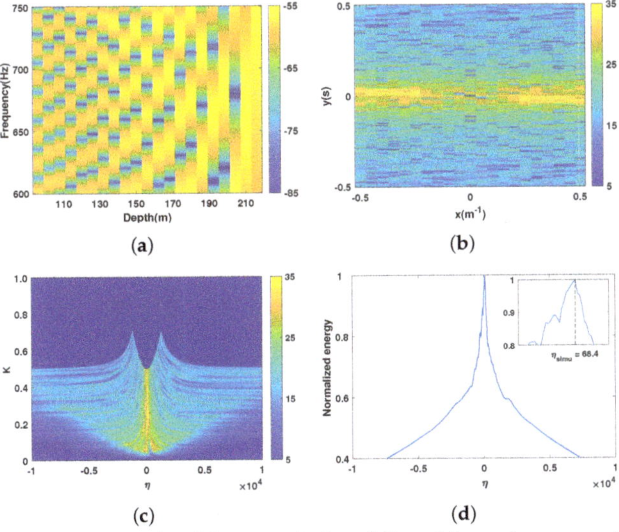

Figure 4. The results of the numerical modeling of the surface source (3 m): (**a**) $I(f,z)$; (**b**) 2D−FFT of $I(f,z)$; (**c**) 2D−FFT in the polar coordinate; (**d**) the GWV distribution E_η and zoom of peak.

The intensity distribution $I(f,z)$, 2D–FFT of it, 2D–FFT in the polar coordinates, and the GWV distribution E_η are shown in Figure 5a–d for case II. The interferogram is chaotic. The intensity striation can hardly be found in the picture, let alone its slope. The distribution E_η in Figure 5d has many peaks with similar values (differences in the second decimal point), which implies that there are not striations associated with the GWV.

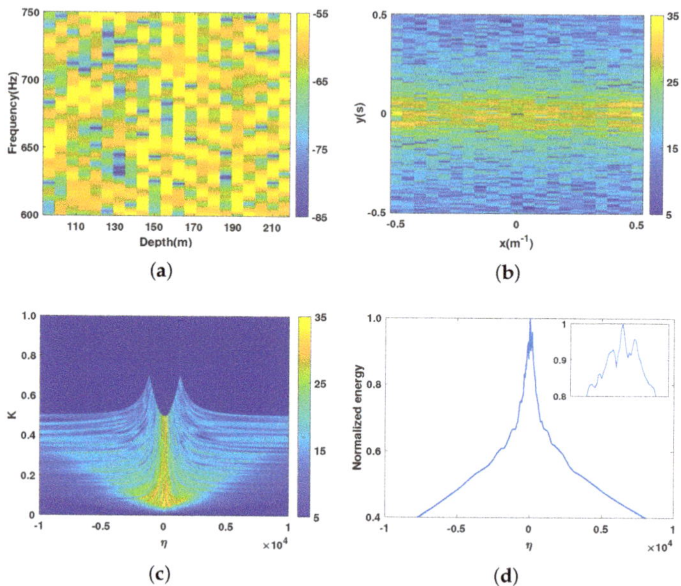

Figure 5. The results of the numerical modeling of the submerged source (54 m): (**a**) $I(f,z)$; (**b**) 2D−FFT of $I(f,z)$; (**c**) 2D−FFT in the polar coordinate; (**d**) the GWV distribution E_η and zoom of peak.

4.3. The Effect of the Noise and Source Range on the Performance

The performance of the proposed method under noise conditions and for different source ranges is described in this section.

We define the signal–noise ratio (SNR) as

$$\text{SNR} = 10 \log \left(\frac{\sum_1^L s_l^2}{L} \Big/ \sigma^2 \right), \tag{14}$$

where l and L represent the array element index and total number of elements, respectively, s_l^2 is the signal power on the lth element, and σ^2 is the noise power.

The results of the processing for two SNRs ($\text{SNR}_1 = -3$ dB and $\text{SNR}_2 = -10$ dB) are shown in Figures 6 and 7, respectively. The results of each polar transform are omitted here.

For the case of $\text{SNR}_1 = -3$dB, there are still observable striations in the intensity distribution of the surface source (Figure 6a), and the peak $\eta_1 = 68.5$ is similar to the $\eta_{\text{ex}} = 68.4$ in the absence of noise. The interferogram of the submerged source (Figure 6d) is still chaotic, and the peak in E_η (Figure 6f) cannot be identified.

For the case of $\text{SNR}_2 = -10$ dB, under such noise conditions, the interferograms (Figure 7a,d) are chaotic, and no peak can be identified in either Figure 7c nor Figure 7f. The method works poorly at a low SNR, since its sample aperture is restricted below the thermocline and limited by the water depth, unlike β sampling in the r domain for an extendable distance, which enhances the SNR. In addition, it is neither practical nor economical to densely deploy receivers in the z domain.

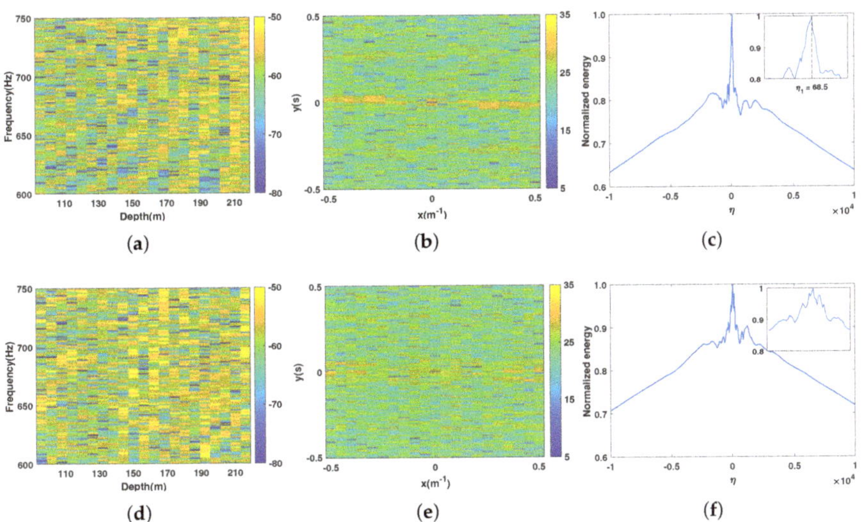

Figure 6. The results of the numerical modeling of sources (3 m and 54 m) for $\mathrm{SNR}_1 = -3$ dB: $I(f,z)$ (**a**,**d**); 2D$-$FFT of $I(f,z)$ (**b**,**e**); E_η and zoom of peak (**c**,**f**).

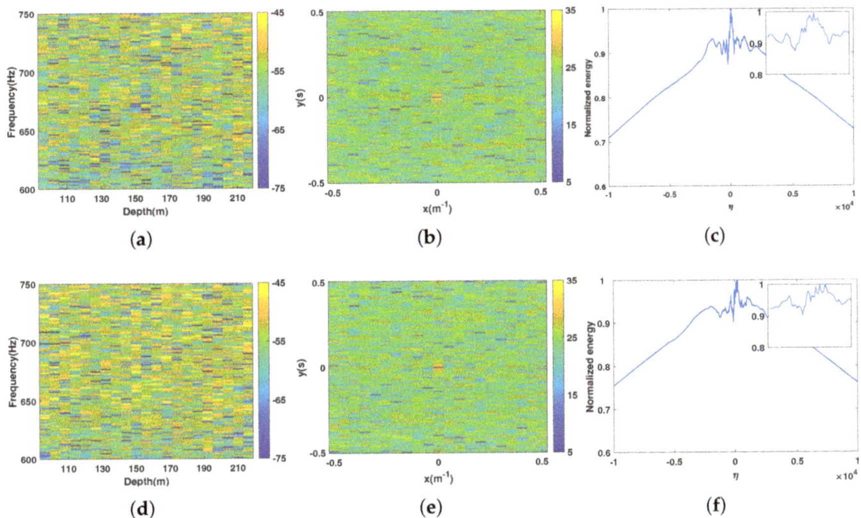

Figure 7. The results of the numerical modeling of sources (3 m and 54 m) for $\mathrm{SNR}_2 = -10$ dB: $I(f,z)$ (**a**,**d**); 2D$-$FFT of $I(f,z)$ (**b**,**e**); E_η and zoom of peak (**c**,**f**).

The performance for a different source range $r_1 = 3$ km is studied here. Figure 8 shows the results and also omits the polar transform. The interference patterns still exist (Figure 8a), and the peak $\eta_{3\mathrm{km}} = 87.9$ is proportional to η_{simu}, with the ratio between two source ranges ($\eta_{3\mathrm{km}}/\eta_{\mathrm{simu}} = 87.9/68.4 \approx 1.29 \approx r_{3\mathrm{km}}/r = 3/2.323$). The interferogram of the submerged source (Figure 8d) is chaotic, and the peak in E_η (Figure 8f) can not be identified.

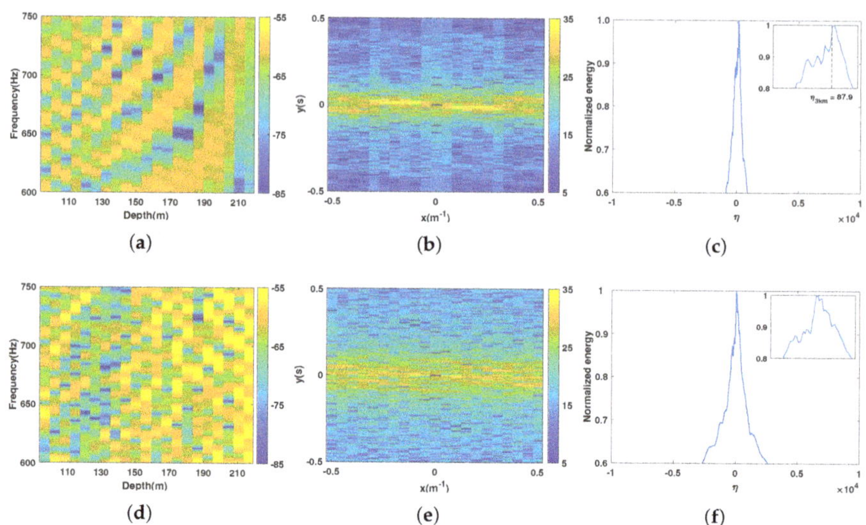

Figure 8. The results of the numerical modeling of sources (3 m and 54 m) for r_{3km}: $I(f,z)$ (**a,d**); 2D−FFT of $I(f,z)$ (**b,e**); E_η and zoom of peak (**c,f**).

5. Conclusions

A method for source depth discrimination is presented for VLA based on the presence of intensity striations extracted from the frequency–depth plane. The orientation of this intensity interference pattern is characterized as a generalized waveguide variant called η, which was derived in this paper, dominated by different types of normal modes excited by a surface/submerged source. Analytical expressions illustrate that for the higher-frequency surface source, the source interferogram shows the intensity striation patterns clearly and the distribution of η peaks associated with the source range. However, for the submerged source, the interferogram is chaotic, and the distribution of η does not show the peak (there are many high sidelobes).

This method was verified with experimental data and simulated data with reasonable success. For the surface source, there is a good agreement between the experimental intensity striation patterns and those predicted by the theory, as well as the peaks of η in each situation. The successful discrimination with a low noise background and different source ranges further indicates the potential of the method on real data. It should be pointed out that, although η is related to the source–array range r, it is the presence of the striations, not the value of its slope, that we use to determine the depth class of the source.

Author Contributions: Conceptualization, X.L.; methodology, X.L. and C.S.; validation, X.L.; formal analysis, X.L.; investigation, X.L.; resources, C.S.; data curation, X.L.; writing—original draft preparation, X.L.; writing—review and editing, C.S.; supervision, C.S. All authors have read and agreed to the published version of the manuscript.

Funding: This research received no external funding.

Data Availability Statement: The data presented in this study are available on request from the corresponding author.

Acknowledgments: The authors would like to thank the SWellEx-96 team for making the experiment data publicly available.

Conflicts of Interest: The authors declare no conflicts of interest.

References

1. Premus, V.E.; Helfrick, M.N. Use of mode subspace projections for depth discrimination with a horizontal line array: Theory and experimental results. *J. Acoust. Soc. Am.* **2013**, *133*, 4019–4031. [CrossRef] [PubMed]
2. Zhang, R.H. Smooth-averaged sound field in underwater sound channel. *Acta Acust.* **1979**, *4*, 102–108. Available online: https://en.cnki.com.cn/Article_en/CJFDTOTAL-XIBA197902002.htm (accessed on 4 July 2023).
3. Conan, E.; Bonnel, J.; Chonavel, T.; Nicolas, B. Source depth discrimination with a vertical line array. *J. Acoust. Soc. Am.* **2016**, *140*, EL434–EL440. [CrossRef] [PubMed]
4. Chuprov, S.; Brekhovskikh, L. Interference structure of a sound field in a layered ocean. *Ocean. Acoust. Curr. State* **1982**, 71–91.
5. Turgut, A.; Fialkowski, L.T. Depth discrimination using waveguide invariance. *J. Acoust. Soc. Am.* **2012**, *132*, 2054–2054. [CrossRef]
6. Liu, Z.T.; Guo, L.H.; Yan, C. Source depth discrimination in negative thermocline using waveguide invariant. *Acta Acust.* **2019**, *44*, 925–933. [CrossRef]
7. Jensen, F.B.; Kuperman, W.A.; Porter, M.B.; Schmidt, H. *Computational Ocean Acoustics*; Springer: New York, NY, USA, 2011.
8. Song, W.H.; Hu, T.; Guo, S.M.; Li, M. Time–varying characteristics of the waveguide invariant under internal wave condition in the shallow water area. *Acta Phys. Sin.* **2014**, *63*, 194303-1–194303-9. [CrossRef]
9. Rouseff, D.; Spindel, R.C. Modeling the Waveguide Invariant as a Distribution. *AIP Conf. Proc.* **2002**, *621*, 137–150. [CrossRef]
10. Emmetière, R.; Bonnel, J.; Géhant, M.; Cristol, X.; Chonavel, T. Understanding deep-water striation patterns and predicting the waveguide invariant as a distribution depending on range and depth. *J. Acoust. Soc. Am.* **2018**, *143*, 3444–3454. [CrossRef] [PubMed]
11. Pereselkov, S.A.; Kuz'kin, V.M. Interferometric processing of hydroacoustic signals for the purpose of source localization. *J. Acoust. Soc. Am.* **2022**, *151*, 666–676. [CrossRef] [PubMed]
12. Rouseff, D. Effect of shallow water internal waves on ocean acoustic striation patterns. *Waves Random Media* **2001**, *11*, 377–393. [CrossRef]
13. Marine Physical Lab. The SWellEx-96 Experiment. Available online: http://swellex96.ucsd.edu/index.htm (accessed on 4 July 2023).
14. University of California. R/V Robert Gordon Sproul Specifications. Available online: https://scripps.ucsd.edu/ships/sproul/specifications (accessed on 4 July 2023).
15. Porter, M.B. *The KRAKEN Normal Mode Program*; Technical Report; Naval Research Lab: Washington, DC, USA, 1992.

remote sensing

MDPI

Communication

Underwater Acoustic Nonlinear Blind Ship Noise Separation Using Recurrent Attention Neural Networks

Ruiping Song [1,2], Xiao Feng [3], Junfeng Wang [2,4], Haixin Sun [1,2,*], Mingzhang Zhou [1,2] and Hamada Esmaiel [5,6]

[1]   School of Informatics, Xiamen University, Xiamen 361005, China; songruiping@stu.xmu.edu.cn (R.S.)
[2]   Key Laboratory of Southeast Coast Marine Information Intelligent Perception and Application, Ministry of Natural Resources, Zhangzhou 363000, China
[3]   School of Communication and Information Engineering, Nanjing University of Posts and Telecommunications, Nanjing 210003, China
[4]   School of Integrated Circuit Science and Engineering, Tianjin University of Technology, Tianjin 300384, China
[5]   Department of Electronics and Communication Engineering, College of Engineering, A'Sharqiyah University, P.O. Box 42, Ibra 400, Oman
[6]   Department of Electrical Engineering, Faculty of Engineering, Aswan University, Aswan 81542, Egypt
*   Correspondence: hxsun@xmu.edu.cn

Abstract: Ship-radiated noise is the main basis for ship detection in underwater acoustic environments. Due to the increasing human activity in the ocean, the captured ship noise is usually mixed with or covered by other signals or noise. On the other hand, due to the softening effect of bubbles in the water generated by ships, ship noise undergoes non-negligible nonlinear distortion. To mitigate the nonlinear distortion and separate the target ship noise, blind source separation (BSS) becomes a promising solution. However, underwater acoustic nonlinear models are seldom used in research for nonlinear BSS. This paper is based on the hypothesis that the recovery and separation accuracy can be improved by considering this nonlinear effect in the underwater environment. The purpose of this research is to explore and discover a method with the above advantages. In this paper, a model is used in underwater BSS to describe the nonlinear impact of the softening effect of bubbles on ship noise. To separate the target ship-radiated noise from the nonlinear mixtures, an end-to-end network combining an attention mechanism and bidirectional long short-term memory (Bi-LSTM) recurrent neural network is proposed. Ship noise from the database ShipsEar and line spectrum signals are used in the simulation. The simulation results show that, compared with several recent neural networks used for linear and nonlinear BSS, the proposed scheme has an advantage in terms of the mean square error, correlation coefficient and signal-to-distortion ratio.

Keywords: nonlinear blind source separation; ship-radiated noise; underwater acoustic nonlinear propagation; attention mechanism; recurrent neural networks

Citation: Song, R.; Feng, X.; Wang, J.; Sun, H.; Zhou, M.; Esmaiel, H. Underwater Acoustic Nonlinear Blind Ship Noise Separation Using Recurrent Attention Neural Networks. *Remote Sens.* **2024**, *16*, 653. https://doi.org/10.3390/rs16040653

Academic Editor: Andrzej Stateczny

Received: 19 December 2023
Revised: 6 February 2024
Accepted: 6 February 2024
Published: 9 February 2024

1. Introduction

Acoustic signals are the main carriers of information and the best means of communication in underwater environments. Acoustic signal processing is the most popular means for the detection of human underwater activities. However, in an underwater acoustic environment, the target signal undergoes non-negligible distortion and is usually mixed with heavy noise or interference, which makes it difficult to detect [1–4]. As a result, signal recovery is crucial in many underwater applications, such as communication, detection and localization [5–10]. For active target detection, the design of a detection waveform is crucial and, to a great extent, determines the detection accuracy [11–13]. Similarly, waveform recovery plays a crucial role in passive scenes, and reliable detection cannot be achieved without the precise separation of the target signal from the received mixture. The application of multiple receivers can greatly improve the quality of the receiving signals and can even be used in imaging and array signal processing in synthetic aperture sonar (SAS) [14,15]. In this situation, blind source separation (BSS) based on multiple receivers becomes one of the candidates to solve this problem. BSS is effective in recovering the original signals from the mixture, and it was first introduced to solve the Cocktail Party Problem [16]. Nowadays, BSS is widely used and performs well with the assumption of a linear mixing procedure [17–21], in which nonlinear components are neglected. However, in fact, the nonlinear effect is non-negligible in underwater acoustic channels, such as nonlinear distortion caused by hydrodynamics and the adiabatic relation between pressure and density [22], nonlinearity in devices like power amplifiers [23–26], the nonlinear interaction of collimated plane waves [27], the thermal current and nonlinear fluids such as relaxing fluids, bubbly liquids and fluids in saturated porous solids [28–31]. Linear methods cannot separate the signals with nonlinear components to a high degree of accuracy.

Many scholars have published methods and validations in the field of underwater acoustic BSS. For conventional algorithms, such as in [32], the researchers build an improved non-negative matrix factorization (NMF)-based BSS algorithm on a fast independent component analysis (FastICA) machine learning backbone to obtain a better signal-to-noise reduction and separation accuracy. Furthermore, a low-complexity method based on Probabilistic Stone's Blind Source Separation (PS-BSS) is proposed in [33] to be used in multi-input multi-output (MIMO) orthogonal frequency division modulation (OFDM) in the Internet of Underwater Things (IoUT). Artificial neural networks are also frequently used in similar situations. The method of time–frequency domain source separation is utilized in [34], by using deep bidirectional long short-term memory (Bi-LSTM) recurrent neural networks (RNN) to estimate the ideal amplitude mask target. In addition, [35] uses a Bi-LSTM approach to explore the features of a time–frequency (T-F) mask and applies it for signal separation. The detection and recognition of underwater creatures can also adopt a BSS approach. The researchers in [36] apply ICA to separate the snapping shrimp sound from mixed underwater sound for passive acoustic monitoring (PAM). Moreover, the ICA based method is also utilized in [37] to separate spiny lobster noise from mixed underwater acoustic sound in a PAM application. However, the studies above fail to take the nonlinear characteristics of underwater acoustic channels into consideration.

To better recover the nonlinear component in blind mixtures, many nonlinear BSS methods have been invented. The authors in [38] extend the standard NMF and propose a BSS/BMI approach so as to jointly handle LQ mixtures and arbitrary source intraclass variability. Moreover, the work in [39] theoretically validates that a cascade of linear principal component analysis (PCA) and ICA can solve a nonlinear BSS problem when mixtures are generated via nonlinear mappings with sufficient dimensions. By using information theoretic learning methods, scholars have explored the use of the Epanechnikov kernel in kernel density estimators (KDE) applied to equalization and nonlinear blind source separation problems [40]. Furthermore, the useful signals from the complex nonlinear mixtures are separated by applying a three-layer deep recurrent neural network to achieve single-channel BSS in [41]. Additionally, nonlinearity can be fairly approximated using a Taylor series and an end-to-end RNN that learns a nonlinear BSS system [42]. Even so, the performance can still be further improved for nonlinear BSS.

As a powerful candidate in artificial neural network methods, the Transformer is utilized in many fields [43]. In particular, the Transformer is widely used in BSS for mixing signals. For example, [44] proposes a three-way architecture that incorporates a pre-trained dual-path recurrent neural network and Transformer. A Transformer network-based plane-wave domain masking approach is utilized to retrieve the reverberant ambisonic signals from a multichannel recording in [45]. The researchers in [46] propose a deep stripe feature learning method for music source separation with a Transformer-based architecture. Similarly, the work in [47] designs a reasonable densely connected U-Net combining multi-head attention and a dual-path Transformer to capture the long-term characteristics in music signal mixtures and separate sources. A slot-centric generative model for blind source separation in the audio domain is built by using a Transformer architecture-based encoder network in [48]. Ref. [49] extends the Transformer module and exploits the use of several efficient self-attention mechanisms to reduce the memory requirements significantly in speech separation. However, most of these Transformer-based BSS studies do not explicitly explore the nonlinearity widely existing in real situations like underwater acoustic channels, which is discussed in this paper.

Ship recognition is vital in underwater acoustic signal processing and it is based on its radiated noise, which is mainly caused by propeller blades. When the propellers generate noise, the blades generate bubbles due to cavitation, which also causes erosion [50,51]. The softening effect by bubbles has a nonlinear impact on acoustic signals. This type of nonlinearity is investigated in this paper. To better separate and recover the original ship-radiated noise before distortion and mixing, an end-to-end nonlinear BSS network based on an attention mechanism is proposed in this paper. Due to the fact that the Transformer has a shortcoming in capturing local self-dependency and performs well in learning long-term or global dependencies, while convolutional neural networks (CNN) and RNN behave in the opposite way [52–56], an end-to-end network is utilized combining an RNN and multi-head self-attention, i.e., recurrent attention neural networks (RANN). The recurrent attention mechanism is used in image aesthetics, target detection, flow forecasting and time series forecasting [57–61], but it has not been used in nonlinear BSS yet. In order to simulate real ship noise as much as possible, the ShipsEar database is used and two classes of ship-radiated noise are selected to act as original ship noise [62]. Based on this noise and nonlinear model, a dataset is generated and used in the neural network training, validation and testing. The simulation results indicate that the proposed network performs better in terms of separation accuracy than networks purely based on RNNs [42], the classical Transformer [43] or a recently published end-to-end BSS U-net [47]. The advantages of the proposed scheme are its lower mean square error (MSE), higher correlation coefficient and higher signal-to-distortion ratio (SDR).

The rest of this paper is organized as follows. Section 2 describes the nonlinear model of the underwater acoustic channel and nonlinear BSS, as well as the proposed recurrent attention neural networks. Section 3 displays the simulation configuration, while the results and discussion are given in Section 4. The paper is concluded in Section 5.

2. Materials and Methods

To model the nonlinear effect in an underwater channel, the nonlinear model based on the softening effect of bubbles in water is utilized, which is derived in [30,63]. The post-nonlinear (PNL) model is used as a generic framework in nonlinear BSS. To achieve the recovery and separation of ship-radiated noise, a RANN combining an RNN and attention mechanism is designed and proposed in this paper, and it is first utilized in nonlinear BSS. The design of the network structure is also introduced in this section.

2.1. Nonlinear Underwater Acoustic Channel Model

It has been proven in [30,63] that bubbles in water have a softening effect on sound pressure, and a model of varying sound pressure was derived. The same model is therefore used in this paper. To simplify the discussion, only one-dimensional space is considered. Usually, it is assumed that bubbles have the same radius and are uniformly distributed in seawater. The model of the nonlinear distortion caused by the softening effect is derived from the wave equation and Rayleigh–Plesset equation:

$$\frac{\partial^2 p(x,t)}{\partial x^2} - \frac{1}{c_{0l}^2}\frac{\partial^2 p(x,t)}{\partial t^2} = -\rho_{0l} N_g \frac{\partial^2 v(x,t)}{\partial t^2} \tag{1}$$

$$\frac{\partial^2 v(x,t)}{\partial t^2} + \delta\omega_{0g}\frac{\partial v(x,t)}{\partial t} + \omega_{0g}^2 v(x,t) + \eta p(x,t)$$
$$= av(x,t)^2 + b\left(2v(x,t)\frac{\partial^2 v(x,t)}{\partial t^2} + \left(\frac{\partial v(x,t)}{\partial t}\right)^2\right) \tag{2}$$

In the equations above, $p(x, t)$ is the sound pressure, which varies with the coordinates and time, similar to the volume variation of bubbles $v(x, t) = V(x, t) - v_{0g}$, with the present volume $V(x, t)$ and the initial one v_{0g}, $v_{0g} = 4\pi R_{0g}^3/3$, with the initial radius R_{0g} of bubbles. c_{0l} and ρ_{0l} are, respectively, the sound speed in sea water and the density of the medium. N_g is the number of bubbles per unit volume. δ denotes the viscous damping coefficient in sea water and ω_{0g} represents the resonance angular frequency of bubbles. $a = (\gamma_g + 1)\omega_{0g}^2/(2v_{0g})$, $b = 1/(6v_{0g})$ and $\eta = 4\pi R_{0g}/\rho_{0l}$ are nonlinear coefficients defined to be convenient, in which γ_g is the specific heat ratio. Using i and j as space and time indices, the first-order and second-order partial derivatives relative to time and coordinates can be converted into a discrete format and expressed as

$$\frac{\partial^2 p_i}{\partial x^2} = \frac{p_{i+1} - 2p_i + p_{i-1}}{h^2} \tag{3}$$

in which p_i represents $p(x = i, t)$. Similarly, p_j represents $p(x, t = j)$ and the in time domain

$$\frac{\partial^2 p_j}{\partial t^2} = \frac{p_{j+1} - 2p_j + p_{j-1}}{\tau^2}$$

$$\frac{\partial^2 v_j}{\partial t^2} = \frac{v_{j+1} - 2v_j + v_{j-1}}{\tau^2}$$

$$\frac{\partial p_j}{\partial t} = \frac{p_j - p_{j-1}}{\tau} \tag{4}$$

$$\frac{\partial v_j}{\partial t} = \frac{v_j - v_{j-1}}{\tau}$$

in which h and τ are the space and time interval, respectively.

Finally, sound pressure in any coordinates and at any time can be derived as

$$p_{i,j+1} = \left[\tau^2 p_{i+1,j} + \tau^2 p_{i-1,j} + 2\left(\frac{h^2}{c_{0l}^2} - \tau^2\right)p_{i,j} - \frac{h^2}{c_{0l}^2}p_{i,j-1}\right.$$
$$\left. - 2\rho_{0l}N_g h^2 v_{i,j} + \rho_{0l}N_g h^2 v_{i,j-1} + \rho_{0l}N_g h^2 v_{i,j+1}\right] / (h^2/c_{0l}^2) \tag{5}$$

where the volume of bubbles at the same point is

$$v_{i,j+1} = \left[(1 - \delta\omega_{0g}\tau - bv_{i,j-1})v_{i,j-1} + \eta\tau^2 p_{i,j} + (-a\tau^2 + 3b)v_{i,j}^2\right.$$
$$\left. + (-2 + \delta\omega_{0g}\tau + \omega_{0g}^2\tau^2)v_{i,j}\right] / (2bv_{i,j} - 1) \tag{6}$$

2.2. Nonlinear Mixing Model in BSS

In the BSS problem, the generic and instantaneous nonlinear mixing model can be defined as

$$X = \varphi(S) \tag{7}$$

where $X = [x_1(n)x_2(n)\dots x_M(n)]^T$ and $S = [s_1(n)s_2(n)\dots s_N(n)]^T$ are M mixtures and N sources. $\varphi(.)$ is a generic nonlinear mixing function. The purpose of nonlinear BSS is to find a nonlinear unmixing function $\psi(.) = \varphi^{-1}(.)$, which is the inverse of the mixing function, to recover the sources as precisely as possible by applying $Y = \psi(X)$.

Many types of nonlinear mixing models are used in BSS, such as the linear quadratic, bi-linear and post-nonlinear (PNL) models [64,65]. The PNL model is utilized in this paper and its structure is shown in Figure 1. In the mixing system, the sources are multiplied by an M-by-N mixing matrix to be linearly mixed based on transmitting attenuation before undergoing nonlinear distortion by softening effect. In the separating part, the observed signals are first nonlinearly recovered, which is the inverse of nonlinear distortion, and then linearly unmixed by an N-by-M matrix.

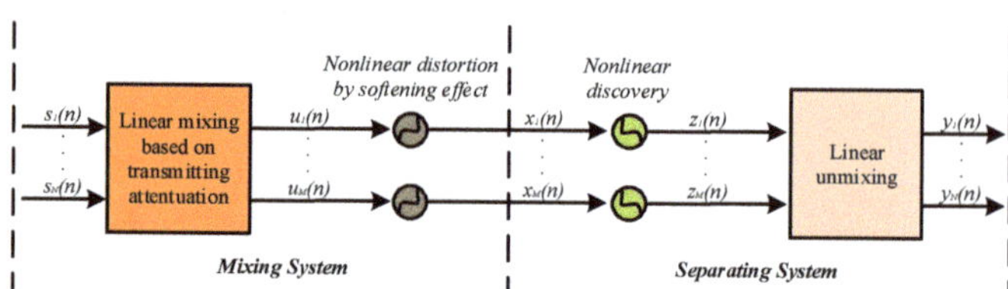

Figure 1. The sketch of the PNL model used in this paper.

2.3. Recurrent Attention Neural Networks

Due to the advantages of RNNs in capturing local information and the Transformer's excellent performance in acquiring global dependencies, a recurrent attention neural network (RANN) is designed in this paper, which is first utilized in nonlinear BSS. To achieve the recovery and separation of ship-radiated noise, the design of the network structure is as shown in Figure 2 and introduced as follows.

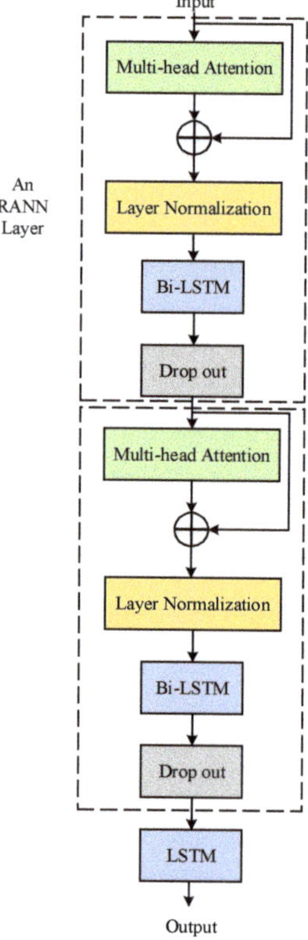

Figure 2. The model of the proposed recurrent attention neural network, containing two layers of recurrent attention neural network and an LSTM layer.

The multi-head attention mechanism shows the best performance in processing one-dimensional temporal signals and it has been applied in many fields. A multi-head attention layer is first used to process the input signals, which is followed by a residual connection and layer normalization. However, the attention mechanism performs weakly in acquiring the local information. Bi-LSTM has been proven to have excellent performance in nonlinear BSS [42] and it is successful in extracting the local information but weak in acquiring global information; thus, it could work well to complement the attention mechanism. As a result, a Bi-LSTM layer is then utilized in the network, which is followed by a drop-out layer to avoid overfitting. Then, the layer of the RANN is completed.

The recurrent attention process is repeated once to improve the performance, as it will be worse without the repetition. However, more repetitions are not necessary because this may heavily increase the number of parameters used but only slightly improve the performance, with more details shown in Section 4. An LSTM layer is finally used to obtain the separated results.

The whole procedure of signal processing is described as follows. The mixture signal is denoted as $X \in R^{M \times L}$, where M is the number of mixtures and L is the length of the signal in the time domain. Then, X is segmented into parts with length Ns and, for convenience in training, the size of batch bs is set. Thus, X is converted to $D \in R^{bn \times bs \times M \times Ns}$ with a permutation operation and bn is the number of batches. The total bn batches are divided into training, validation and testing sets in a proportion of 7:1:2 and used as inputs of the network for various purposes. In the simulation below, L is set to 500 and Ns is set to 40 to simplify the calculation. There are 1000 samples of mixture data used in total and they are divided into various datasets with the proportion of 7:1:2.

Take the training progress as an example. The training batch $D_1 \in R^{bs \times M \times Ns}$ is extended to h channels as features, $D_2 \in R^{bs \times h \times Ns}$, before the attention mechanism, where h is the number of hidden elements in the attention and RNN layers. The shape of the feature tensors will not change until the LSTM layer, whose output is $D_n \in R^{bs \times N \times Ns}$, and N is the number of expected separated signals. Lastly, the outputs of the network are concatenated and recovered to $Y \in R^{N \times L}$. The progress is similar in the validation and testing stages.

3. Simulation Configuration

3.1. Original Signals, Distortion and Mixtures

To simplify the discussion, the smallest number of signals is selected, i.e., $m = n = 2$ in Figure 1. To better validate the separation accuracy of the proposed network, the real recorded ship-radiated noise from the database ShipsEar [62] is used. Entries with index 6 and 22 are used as original signals, where the types of ships are passengers and ocean liners, respectively.

In order to maintain generality, various propagation distances are applied to the original signals, with 10 km for source $s1$ and 12 km for $s2$. Due to the fact that the propagation distance will influence the nonlinear distortion, as shown in Equations (5) and (6), nonlinear distorting functions will be different for various linear mixing signals $u(n)$ in Figure 1. For linear mixing matrix A, the coefficients are randomly selected and A must be fully ranked. When A equals $[0.45, 0.13; 0.21, 0.43]$ and white Gaussian noise with 15 dB is added, the spectra of the original and mixed ship noise signals are as shown in Figure 3. It can be concluded from Figure 3 that the characteristics of the original ship noise are severely hidden in the mixture spectra. Only frequencies below 2 kHz are displayed as the ship-radiated noise is mainly distributed in this low-frequency region. This is used as the first dataset.

Due to the fact that ship-radiated noise consists of strong line spectra and weak continuous ones, and the former is the main basis of ship recognition [66,67], signals with line spectra are also generated as originals of the second dataset. Usually, a line spectrum consists of a fundamental frequency and resonance frequencies, with a strong relationship with the propeller shaft frequency and blade frequency. In this set of data, the fundamental frequencies are selected as 100 Hz and 120 Hz for both sources, and the amplitude attenuates as the frequency increases. Parameters used for the environment are the same as in the set of real ship noise discussed above. The spectra of the original and mixed signals are shown in Figure 4. A similar conclusion is obtained that the line spectra characteristics are heavily distorted in the mixtures.

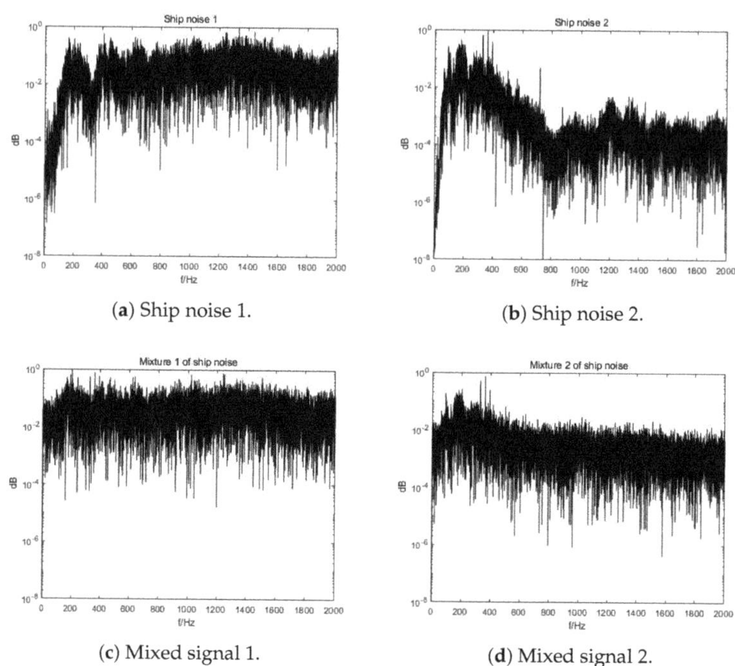

(**a**) Ship noise 1.

(**b**) Ship noise 2.

(**c**) Mixed signal 1.

(**d**) Mixed signal 2.

Figure 3. Spectra of original and mixed ship noise, with characteristics severely hidden in the mixture spectra.

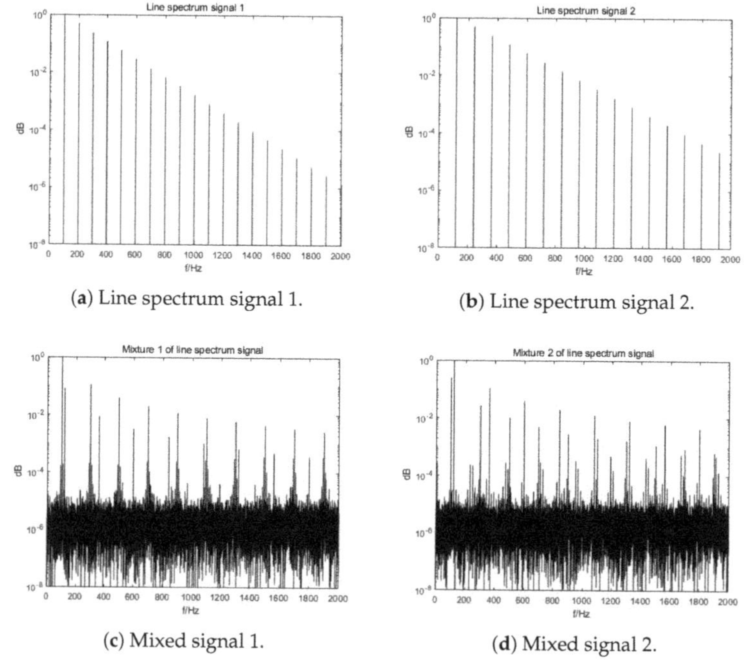

(**a**) Line spectrum signal 1.

(**b**) Line spectrum signal 2.

(**c**) Mixed signal 1.

(**d**) Mixed signal 2.

Figure 4. Spectra of original and mixed line spectrum signals, with line spectra characteristics heavily distorted in mixtures.

3.2. Configuration and Referred Networks

In the proposed network, the output size of each recurrent unit is 256, as well as the number of hidden elements in the attention mechanism. The number of heads in attention is 8 and a 50% dropping rate is used in each drop-out layer. The learning rate is set as 10^{-4} at the beginning and multiplies by 0.8 after every 10 epochs, with a total of 100 epochs used in the training process. The Adam optimizer is also used to reduce the impact of the learning rate [68]. The mean square error (MSE) is used as the loss function and is displayed in Equation (8). The simulations are conducted on a Windows deep-learning server with 2 Intel(R) E5−2603 1.7 GHz CPUs (Santa Clara, CA, USA), 8 Samsung DDR4 16 GB RAM (Suwon, Republic of Korea) and 4 Nvidia GTX 1080Ti GPUs (Santa Clara, CA, USA).

The referred networks should be representative of different types. The candidate in [42], combining Bi-LSTM, LSTM and a drop-out layer, shows good performance in end-to-end nonlinear blind source separation, denoted as RNN in the description hereafter. The classical Transformer network [43] is another good candidate to test the performance and is denoted as Transformer. A recently published end-to-end U-net combining a CNN and multi-head attention, but used in linear BSS for music separation, is also selected as a referred network [47] and denoted as U-net. The total numbers of parameters used in the different networks are calculated by the Python package fvcore and shown in Table 1.

Table 1. Number of parameters used in different networks.

Network	Parameters
RNN	0.3 M
Transformer	24 K
U-net	3.4 M
RANN (1 layer)	1.1 M
Proposed RANN (2 layers)	2.1 M
RANN (3 layers)	3.2 M
RANN (4 layers)	4.2 M

The metrics of the separation accuracy consist of the MSE, the correlation coefficient ρ used in [69] and the signal-to-distortion ratio (SDR). The lower the MSE or the greater the ρ and SDR, the better the performance. The metrics are calculated as follows and s and y represent the original signals and separated ones with length L:

$$MSE = \frac{1}{L} \sum_{i=1}^{L} (s_i - y_i)^2 \tag{8}$$

$$\rho = \frac{\sum_{i=1}^{L} (s_i \cdot y_i)}{\sum_{i=1}^{L} (|s_i| \cdot |y_i|)} \tag{9}$$

$$SDR = 10 \log_{10} \frac{\sum_{i=1}^{L} (s_i - y_i)^2}{\sum_{i=1}^{L} s_i^2} \tag{10}$$

4. Results and Discussion

4.1. Metrics of Results

The metrics of the separation results are shown in Tables 2 and 3. U-net performs the worst because it is designed for linear BSS and the nonlinear channel is not taken into account. The proposed network performs better than the RNN and Transformer because the RNN is successful in capturing local dependencies and the Transformer performs well in acquiring global ones. The RNN performs much better than the Transformer and slightly worse than the proposed network because the local dependencies are much stronger than the global relation in the considered underwater acoustic channels, as expressed in Equations (5) and (6). For the RANN with different layers of recurrent attention, the network with two layers is the best candidate because it performs much better than those with one layer, especially in line spectrum signal separation, and fewer parameters are used than in those with more layers.

Table 2. Separation results of different networks for ship-radiated noise.

Network	MSE	ρ	SDR (dB)
RNN	0.007	0.817	4.495
Transformer	0.037	0.001	−2.947
U-net	0.053	0.006	−4.453
RANN (1 layer)	0.007	0.830	4.492
Proposed RANN (2 layers)	0.006	0.840	4.651
RANN (3 layers)	0.006	0.842	4.674
RANN (4 layers)	0.006	0.843	4.684

Table 3. Separation results of different networks for line spectrum signals.

Network	MSE	ρ	SDR (dB)
RNN	0.035	0.929	8.621
Transformer	0.311	0.015	−0.842
U-net	0.504	−0.002	−2.939
RANN (1 layer)	0.020	0.967	11.003
Proposed RANN (2 layers)	0.014	0.981	12.699
RANN (3 layers)	0.013	0.987	12.850
RANN (4 layers)	0.012	0.992	12.922

4.2. Separated Waveforms

The waveforms of the signals separated by different networks are displayed. By randomly selecting time segments, the original and separated ship-radiated noise from various networks is as shown in Figures 5 and 6. In Figure 5, it can be seen that the waveform separated by the proposed network (in pink line) is slightly nearer the original waveform (in red line) than in those by the RNN (in blue line) and it performs much better than any other candidates. Conclusions can be made that, for ship noise, the proposed network separates them in with slightly higher accuracy than the RNN and is better than all others. For the line spectrum signals in Figure 6, the waveform separated by the proposed network (in pink line) shows the best performance with the least distortion from the original waveform (in red line) and it performs much better than any other method, including the RNN. In conclusion, the proposed network apparently works better than other candidates.

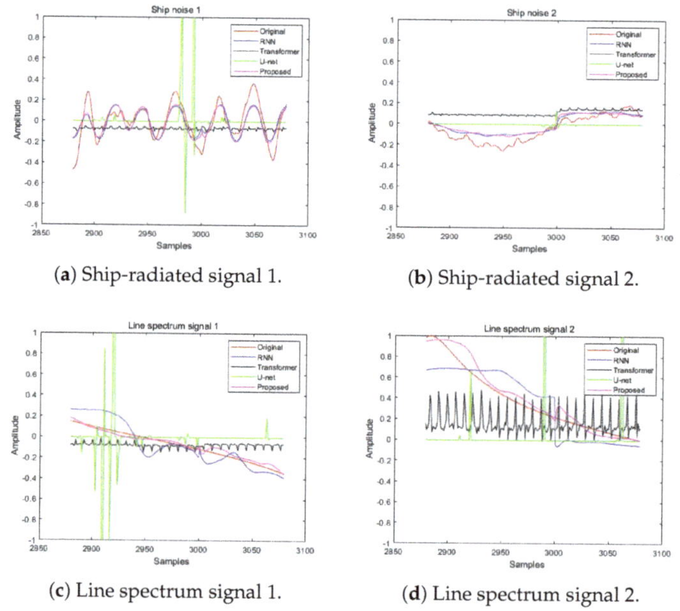

(**a**) Ship-radiated signal 1.

(**b**) Ship-radiated signal 2.

(**c**) Line spectrum signal 1.

(**d**) Line spectrum signal 2.

Figure 5. Fragment 1 of original and separated waveform, with the proposed network performing the best.

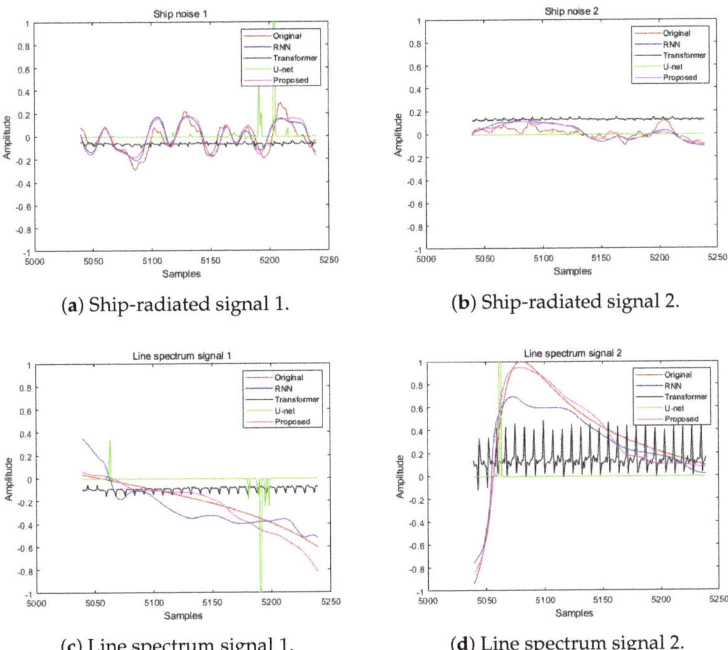

(a) Ship-radiated signal 1.

(b) Ship-radiated signal 2.

(c) Line spectrum signal 1.

(d) Line spectrum signal 2.

Figure 6. Fragment 2 of original and separated waveform, with the proposed network performing the best.

5. Conclusions

In this paper, the softening effect of bubbles in water is considered in underwater nonlinear blind ship-radiated noise separation, and an end-to-end recurrent attention neural network is proposed combining the advantages of the RNN and Transformer. According to the simulation results, high accuracy of separation is obtained by the proposal in terms of the MSE, correlation coefficient and SDR, compared with other networks including the RNN, Transformer and U-net, which are effective in other BSS scenes. It is found that the proposed scheme performs better than other candidates obviously in line spectrum signal separation and has a slight advantage over the RNN in separating the real ship noise. In the future, networks with greater compatibility with the underwater acoustic channel can be explored to obtain higher separation accuracy.

Author Contributions: Conceptualization, R.S. and X.F.; methodology, R.S., J.W. and H.E.; software, R.S. and M.Z.; validation, R.S. and H.E.; formal analysis, R.S., X.F. and H.E.; investigation, R.S. and X.F.; resources, R.S., M.Z. and H.S.; data curation, R.S. and H.S.; writing—original draft preparation, R.S.; writing—review and editing, J.W., M.Z. and H.E.; visualization, R.S., J.W. and M.Z.; supervision, H.S.; project administration, H.S.; funding acquisition, H.S. All authors have read and agreed to the published version of the manuscript.

Funding: This research was funded by the Key Program of Marine Economy Development Special Foundation of Department of Natural Resources of Guangdong Province (GDNRC [2023]24) and the National Natural Science Foundation of China (NSFC) 62271426.

Data Availability Statement: Data are contained within the article.

Conflicts of Interest: The authors declare no conflicts of interest.

Abbreviations

The following abbreviations are used in this paper:

SAS	Synthetic Aperture Sonar
BSS	Blind Source Separation
NMF	Non-Negative Matrix Factorization
FastICA	Fast Independent Component Analysis
MIMO	Multi-Input Multi-Output
OFDM	Orthogonal Frequency Division Modulation
IoUT	Internet of Underwater Things
CNN	Convolutional Neural Network
RNN	Recurrent Neural Network
RANN	Recurrent Attention Neural Network
Bi-LSTM	Bidirectional Long Short-Term Memory
PAM	Passive Acoustic Monitoring
BMI	Blind Mixture Identification
PCA	Principal Component Analysis
PNL	Post-Nonlinear
MSE	Mean Square Error
SDR	Signal-to-Distortion Ratio

References

1. Yin, F.; Li, C.; Wang, H.; Nie, L.; Zhang, Y.; Liu, C.; Yang, F. Weak Underwater Acoustic Target Detection and Enhancement with BM-SEED Algorithm. *J. Mar. Sci. Eng.* **2023**, *11*, 357. [CrossRef]
2. Yin, F.; Li, C.; Wang, H.; Zhou, S.; Nie, L.; Zhang, Y.; Yin, H. A Robust Denoised Algorithm Based on Hessian-Sparse Deconvolution for Passive Underwater Acoustic Detection. *J. Mar. Sci. Eng.* **2023**, *11*, 2028. [CrossRef]
3. Chu, H.; Li, C.; Wang, H.; Wang, J.; Tai, Y.; Zhang, Y.; Yang, F.; Benezeth, Y. A deep-learning based high-gain method for underwater acoustic signal detection in intensity fluctuation environments. *Appl. Acoust.* **2023**, *211*, 109513. [CrossRef]
4. Zhou, M.; Wang, J.; Feng, X.; Sun, H.; Qi, J.; Lin, R. Neural-Network-Based Equalization and Detection for Underwater Acoustic Orthogonal Frequency Division Multiplexing Communications: A Low-Complexity Approach. *Remote Sens.* **2023**, *15*, 3796. [CrossRef]
5. Yonglin, Z.; Chao, L.; Haibin, W.; Jun, W.; Fan, Y.; Fabrice, M. Deep learning aided OFDM receiver for underwater acoustic communications. *Appl. Acoust.* **2022**, *187*, 108515. [CrossRef]
6. Wang, Z.; Li, Y.; Wang, C.; Ouyang, D.; Huang, Y. A-OMP: An Adaptive OMP Algorithm for Underwater Acoustic OFDM Channel Estimation. *IEEE Wirel. Commun. Lett.* **2021**, *10*, 1761–1765. [CrossRef]
7. Atanackovic, L.; Lampe, L.; Diamant, R. Deep-Learning Based Ship-Radiated Noise Suppression for Underwater Acoustic OFDM Systems. In Proceedings of the Global Oceans 2020: Singapore—U.S. Gulf Coast, Biloxi, MS, USA, 5–30 October 2020; pp. 1–8. [CrossRef]
8. Li, P.; Zhang, X.G.; Zhang, X.Y.; Liu, Y. Vertical array signal recovery method based on normalized virtual time reversal mirror. *J. Phys. Conf. Ser.* **2023**, *2486*, 012070. [CrossRef]
9. Yuqing, Y.; Peng, X.; Nikos, D. Underwater localization with binary measurements: From compressed sensing to deep unfolding. *Digit. Signal Process.* **2023**, *133*, 103867. [CrossRef]
10. Zonglong, B. Sparse Bayesian learning for sparse signal recovery using l1/2-norm. *Appl. Acoust.* **2023**, *207*, 109340. [CrossRef]
11. Zhu, J.H.; Fan, C.Y.; Song, Y.P.; Huang, X.T.; Zhang, B.B.; Ma, Y.X. Coordination of Complementary Sets for Low Doppler-Induced Sidelobes. *Remote Sens.* **2022**, *14*, 1549. [CrossRef]
12. Zhu, J.H.; Song, Y.P.; Jiang, N.; Xie, Z.; Fan, C.Y.; Huang, X.T. Enhanced Doppler Resolution and Sidelobe Suppression Performance for Golay Complementary Waveforms. *Remote Sens.* **2023**, *15*, 2452. [CrossRef]
13. Xie, Z.; Xu, Z.; Han, S.; Zhu, J.; Huang, X. Modulus Constrained Minimax Radar Code Design Against Target Interpulse Fluctuation. *IEEE Trans. Veh. Technol.* **2023**, *72*, 13671–13676. [CrossRef]
14. Zhang, X.B.; Wu, H.R.; Sun, H.X.; Ying, W.W. Multireceiver SAS Imagery Based on Monostatic Conversion. *IEEE J. Sel. Top. Appl. Earth Obs. Remote Sens.* **2021**, *14*, 10835–10853. [CrossRef]
15. Zhang, X.B.; Yang, P.X.; Zhou, M.Z. Multireceiver SAS Imagery With Generalized PCA. *IEEE Geosci. Remote Sens. Lett.* **2023**, *20*, 1502205. [CrossRef]
16. Cherry, E.C. Some experiments on the recognition of speech, with one and with two ears. *J. Acoust. Soc. Am.* **1953**, *25*, 975–979. [CrossRef]
17. Zhang, W.; Tait, A.; Huang, C.; Ferreira de Lima, T.; Bilodeau, S.; Blow, E.C.; Jha, A.; Shastri, B.J.; Prucnal, P. Broadband physical layer cognitive radio with an integrated photonic processor for blind source separation. *Nat. Commun.* **2023**, *14*, 1107. [CrossRef] [PubMed]

18. Kumari, R.; Mustafi, A. The spatial frequency domain designated watermarking framework uses linear blind source separation for intelligent visual signal processing. *Front. Neurorobot.* **2022**, *16*, 1054481. [CrossRef] [PubMed]
19. Erdogan, A.T. An Information Maximization Based Blind Source Separation Approach for Dependent and Independent Sources. In Proceedings of the ICASSP 2022—2022 IEEE International Conference on Acoustics, Speech and Signal Processing (ICASSP), Singapore, 23–27 May 2022; pp. 4378–4382. [CrossRef]
20. Boccuto, A.; Gerace, I.; Giorgetti, V.; Valenti, G. A Blind Source Separation Technique for Document Restoration Based on Image Discrete Derivative. In *International Conference on Computational Science and Its Applications*; Springer: Berlin/Heidelberg, Germany, 2022; pp. 445–461.
21. Martin, G.; Hooper, A.; Wright, T.J.; Selvakumaran, S. Blind Source Separation for MT-InSAR Analysis with Structural Health Monitoring Applications. *IEEE J. Sel. Top. Appl. Earth Obs. Remote Sens.* **2022**, *15*, 7605–7618. [CrossRef]
22. Yao, H.; Wang, H.; Zhang, Z.; Xu, Y.; Juergen, K. A stochastic nonlinear differential propagation model for underwater acoustic propagation: Theory and solution. *Chaos Solitons Fractals* **2021**, *150*, 111105.
23. Naman, H.A.; Abdelkareem, A.E. Variable direction-based self-interference full-duplex channel model for underwater acoustic communication systems. *Int. J. Commun. Syst.* **2022**, *35*, e5096. [CrossRef]
24. Shen, L.; Henson, B.; Zakharov, Y.; Mitchell, P. Digital Self-Interference Cancellation for Full-Duplex Underwater Acoustic Systems. *IEEE Trans. Circuits Syst. II Express Briefs* **2020**, *67*, 192–196. [CrossRef]
25. Yang, S.; Deane, G.B.; Preisig, J.C.; Sevüktekin, N.C.; Choi, J.W.; Singer, A.C. On the Reusability of Postexperimental Field Data for Underwater Acoustic Communications R&D. *IEEE J. Ocean. Eng.* **2019**, *44*, 912–931. [CrossRef]
26. Ma, X.; Raza, W.; Wu, Z.; Bilal, M.; Zhou, Z.; Ali, A. A Nonlinear Distortion Removal Based on Deep Neural Network for Underwater Acoustic OFDM Communication with the Mitigation of Peak to Average Power Ratio. *Appl. Sci.* **2020**, *10*, 4986. [CrossRef]
27. Campo-Valera, M.; Rodríguez-Rodríguez, I.; Rodríguez, J.V.; Herrera-Fernández, L.J. Proof of Concept of the Use of the Parametric Effect in Two Media with Application to Underwater Acoustic Communications. *Electronics* **2023**, *12*, 3459. [CrossRef]
28. Yao, H.; Wang, H.; Xu, Y.; Juergen, K. A recurrent plot based stochastic nonlinear ray propagation model for underwater signal propagation. *New J. Phys.* **2020**, *22*, 063025.
29. Cheng, Y.; Shi, J.; Deng, A. Effective Nonlinearity Parameter and Acoustic Propagation Oscillation Behavior in Medium of Water Containing Distributed Bubbles. In Proceedings of the 2021 OES China Ocean Acoustics (COA), Harbin, China, 14–17 July 2021; IEEE: Piscataway, NJ, USA, 2021; pp. 345–350.
30. Yu, J.; Yang, D.; Shi, J. Influence of softening effect of bubble water on cavity resonance. In Proceedings of the 2021 OES China Ocean Acoustics (COA), Harbin, China, 14–17 July 2021; IEEE: Piscataway, NJ, USA, 2021; pp. 351–355.
31. Li, J.; Yang, D.; Chen, G. Study on the acoustic scattering characteristics of the parametric array in the wake field of underwater cylindrical structures. In Proceedings of the 2021 OES China Ocean Acoustics (COA), Harbin, China, 14–17 July 2021; IEEE: Piscataway, NJ, USA, 2021; pp. 340–344.
32. Li, D.; Wu, M.; Yu, L.; Han, J.; Zhang, H. Single-channel blind source separation of underwater acoustic signals using improved NMF and FastICA. *Front. Mar. Sci.* **2023**, *9*, 1097003. [CrossRef]
33. Khosravy, M.; Gupta, N.; Dey, N.; Crespo, R.G. Underwater IoT network by blind MIMO OFDM transceiver based on probabilistic Stone's blind source separation. *ACM Trans. Sens. Netw. (TOSN)* **2022**, *18*, 1–27. [CrossRef]
34. Zhang, W.; Li, X.; Zhou, A.; Ren, K.; Song, J. Underwater acoustic source separation with deep Bi-LSTM networks. In Proceedings of the 2021 4th International Conference on Information Communication and Signal Processing (ICICSP), Shanghai, China, 24–26 September 2021; pp. 254–258. [CrossRef]
35. Chen, J.; Liu, C.; Xie, J.; An, J.; Huang, N. Time–Frequency Mask-Aware Bidirectional LSTM: A Deep Learning Approach for Underwater Acoustic Signal Separation. *Sensors* **2022**, *22*, 5598. [CrossRef]
36. Hadi, F.I.M.A.; Ramli, D.A.; Azhar, A.S. Passive Acoustic Monitoring (PAM) of Snapping Shrimp Sound Based on Blind Source Separation (BSS) Technique. In *Proceedings of the 11th International Conference on Robotics, Vision, Signal Processing and Power Applications: Enhancing Research and Innovation through the Fourth Industrial Revolution*; Springer: Berlin/Heidelberg, Germany, 2022; pp. 605–611.
37. Hadi, F.I.M.; Ramli, D.A.; Hassan, N. Spiny Lobster Sound Identification Based on Blind Source Separation (BSS) for Passive Acoustic Monitoring (PAM). *Procedia Comput. Sci.* **2021**, *192*, 4493–4502. [CrossRef]
38. Deville, Y.; Faury, G.; Achard, V.; Briottet, X. An NMF-based method for jointly handling mixture nonlinearity and intraclass variability in hyperspectral blind source separation. *Digit. Signal Process.* **2023**, *133*, 103838. [CrossRef]
39. Isomura, T.; Toyoizumi, T. On the achievability of blind source separation for high-dimensional nonlinear source mixtures. *Neural Comput.* **2021**, *33*, 1433–1468. [CrossRef]
40. Moraes, C.P.; Fantinato, D.G.; Neves, A. Epanechnikov kernel for PDF estimation applied to equalization and blind source separation. *Signal Process.* **2021**, *189*, 108251. [CrossRef]
41. He, J.; Chen, W.; Song, Y. Single channel blind source separation under deep recurrent neural network. *Wirel. Pers. Commun.* **2020**, *115*, 1277–1289. [CrossRef]
42. Zamani, H.; Razavikia, S.; Otroshi-Shahreza, H.; Amini, A. Separation of Nonlinearly Mixed Sources Using End-to-End Deep Neural Networks. *IEEE Signal Process. Lett.* **2020**, *27*, 101–105. [CrossRef]

43. Vaswani, A.; Shazeer, N.; Parmar, N.; Uszkoreit, J.; Jones, L.; Gomez, A.N.; Kaiser, L.u.; Polosukhin, I. Attention is All you Need. In *Advances in Neural Information Processing Systems*; Guyon, I., Luxburg, U.V., Bengio, S., Wallach, H., Fergus, R., Vishwanathan, S., Garnett, R., Eds.; Curran Associates, Inc.: Red Hook, NY, USA, 2017; Volume 30, pp. 1–11.

44. Ansari, S.; Alnajjar, K.A.; Khater, T.; Mahmoud, S.; Hussain, A. A Robust Hybrid Neural Network Architecture for Blind Source Separation of Speech Signals Exploiting Deep Learning. *IEEE Access* **2023**, *11*, 100414–100437. [CrossRef]

45. Herzog, A.; Chetupalli, S.R.; Habets, E.A.P. AmbiSep: Ambisonic-to-Ambisonic Reverberant Speech Separation Using Transformer Networks. In Proceedings of the 2022 International Workshop on Acoustic Signal Enhancement (IWAENC), Bamberg, Germany, 5–8 September 2022; pp. 1–5. [CrossRef]

46. Qian, J.; Liu, X.; Yu, Y.; Li, W. Stripe-Transformer: Deep stripe feature learning for music source separation. *EURASIP J. Audio Speech Music. Process.* **2023**, *2023*, 2. [CrossRef]

47. Wang, J.; Liu, H.; Ying, H.; Qiu, C.; Li, J.; Anwar, M.S. Attention-based neural network for end-to-end music separation. *CAAI Trans. Intell. Technol.* **2023**, *8*, 355–363. [CrossRef]

48. Reddy, P.; Wisdom, S.; Greff, K.; Hershey, J.R.; Kipf, T. AudioSlots: A slot-centric generative model for audio separation. *arXiv* **2023**, arXiv:2305.05591.

49. Subakan, C.; Ravanelli, M.; Cornell, S.; Grondin, F.; Bronzi, M. Exploring Self-Attention Mechanisms for Speech Separation. *IEEE/ACM Trans. Audio Speech Lang. Process.* **2023**, *31*, 2169–2180. [CrossRef]

50. Melissaris, T.; Schenke, S.; Van Terwisga, T.J. Cavitation erosion risk assessment for a marine propeller behind a Ro–Ro container vessel. *Phys. Fluids* **2023**, *35*, 013342. [CrossRef]

51. Abbasia, A.A.; Viviania, M.; Bertetta, D.; Delucchia, M.; Ricottic, R.; Tania, G. Experimental Analysis of Cavitation Erosion on Blade Root of Controlable Pitch Propeller. In Proceedings of the 20th International Conference on Ship & Maritime Research, Genoa, La Spazia, Italy, 15–17 June 2022; pp. 254–262.

52. Wang, Y.; Zhang, H.; Huang, W. Fast ship radiated noise recognition using three-dimensional mel-spectrograms with an additive attention based transformer. *Front. Mar. Sci.* **2023**, 1–15. [CrossRef]

53. Pu, X.; Yi, P.; Chen, K.; Ma, Z.; Zhao, D.; Ren, Y. EEGDnet: Fusing non-local and local self-similarity for EEG signal denoising with transformer. *Comput. Biol. Med.* **2022**, *151*, 106248:1–106248:9. [CrossRef] [PubMed]

54. Woo, B.J.; Kim, H.Y.; Kim, J.; Kim, N.S. Speech separation based on dptnet with sparse attention. In Proceedings of the 2021 7th IEEE International Conference on Network Intelligence and Digital Content (IC-NIDC), Beijing, China, 17–19 November 2021; IEEE: Piscataway, NJ, USA, 2021; pp. 339–343.

55. Liu, Y.; Xu, X.; Tu, W.; Yang, Y.; Xiao, L. Improving Acoustic Echo Cancellation by Mixing Speech Local and Global Features with Transformer. In Proceedings of the ICASSP 2023—2023 IEEE International Conference on Acoustics, Speech and Signal Processing (ICASSP), Rhodes Island, Greece, 4–10 June 2023; IEEE: Piscataway, NJ, USA, 2023; pp. 1–5.

56. Yin, J.; Liu, A.; Li, C.; Qian, R.; Chen, X. A GAN Guided Parallel CNN and Transformer Network for EEG Denoising. *IEEE J. Biomed. Health Inform.* **2023**, 1–12, *Early Access*. [CrossRef] [PubMed]

57. Ji, W.; Yan, G.; Li, J.; Piao, Y.; Yao, S.; Zhang, M.; Cheng, L.; Lu, H. DMRA: Depth-induced multi-scale recurrent attention network for RGB-D saliency detection. *IEEE Trans. Image Process.* **2022**, *31*, 2321–2336. [CrossRef] [PubMed]

58. Reza, S.; Ferreira, M.C.; Machado, J.J.M.; Tavares, J.M.R. A multi-head attention-based transformer model for traffic flow forecasting with a comparative analysis to recurrent neural networks. *Expert Syst. Appl.* **2022**, *202*, 117275. [CrossRef]

59. Geng, X.; He, X.; Xu, L.; Yu, J. Graph correlated attention recurrent neural network for multivariate time series forecasting. *Inf. Sci.* **2022**, *606*, 126–142. [CrossRef]

60. De Maissin, A.; Vallée, R.; Flamant, M.; Fondain-Bossiere, M.; Le Berre, C.; Coutrot, A.; Normand, N.; Mouchère, H.; Coudol, S.; Trang, C.; et al. Multi-expert annotation of Crohn's disease images of the small bowel for automatic detection using a convolutional recurrent attention neural network. *Endosc. Int. Open* **2021**, *9*, E1136–E1144. [CrossRef] [PubMed]

61. Zhang, X.; Gao, X.; Lu, W.; He, L.; Li, J. Beyond vision: A multimodal recurrent attention convolutional neural network for unified image aesthetic prediction tasks. *IEEE Trans. Multimed.* **2020**, *23*, 611–623. [CrossRef]

62. Santos-Domínguez, D.; Torres-Guijarro, S.; Cardenal-López, A.; Pena-Gimenez, A. ShipsEar: An underwater vessel noise database. *Appl. Acoust.* **2016**, *113*, 64–69. [CrossRef]

63. Yu, J.; Yang, D.; Shi, J.; Zhang, J.; Fu, X. Nonlinear sound field under bubble softening effect. *J. Harbin Eng. Univ.* **2023**, *44*, 1433–1444.

64. Deville, Y.; Duarte, L.T.; Hosseini, S. *Nonlinear Blind Source Separation and Blind Mixture Identification: Methods for Bilinear, Linear-Quadratic and Polynomial Mixtures*; Springer: Berlin/Heidelberg, Germany, 2021.

65. Moraes, C.P.; Saldanha, J.; Neves, A.; Fantinato, D.G.; Attux, R.; Duarte, L.T. An SOS-Based Algorithm for Source Separation in Nonlinear Mixtures. In Proceedings of the 2021 IEEE Statistical Signal Processing Workshop (SSP), Rio de Janeiro, Brazil, 11–14 July 2021; IEEE: Piscataway, NJ, USA, 2021; pp. 156–160.

66. Wang, H.; Guo, L. Research of Modulation Feature Extraction from Ship-Radiated Noise. *Proc. J. Phys. Conf. Ser.* **2020**, *1631*, 012130. [CrossRef]

67. Peng, C.; Yang, L.; Jiang, X.; Song, Y. Design of a ship radiated noise model and its application to feature extraction based on winger's higher-order spectrum. In Proceedings of the 2019 IEEE 4th Advanced Information Technology, Electronic and Automation Control Conference (IAEAC), Chengdu, China, 20–22 December 2019; IEEE: Piscataway, NJ, USA, 2019; Volume 1, pp. 582–587. [CrossRef]

68. Kingma, D.P.; Ba, J. Adam: A method for stochastic optimization. *arXiv* **2014**, arXiv:1412.6980.
69. Cao, Y.; Zhang, H.; Qin, Y.; Zhu, H.; Cao, J.; Ma, N. Joint Denoising Blind Source Separation Algorithmfor Anti-jamming. In Proceedings of the 2021 4th International Conference on Information Communication and Signal Processing (ICICSP), Shanghai, China, 24–26 September 2021; IEEE: Piscataway, NJ, USA, 2021; pp. 27–35. [CrossRef]

remote sensing

Article

Transfer-Learning-Based Human Activity Recognition Using Antenna Array

Kun Ye [1,2,3], Sheng Wu [1,2,3], Yongbin Cai [1,2,3], Lang Zhou [1,3,4], Lijun Xiao [1,2,3], Xuebo Zhang [3,5,*], Zheng Zheng [1,2,3] and Jiaqing Lin [1,2,3]

[1] Shenzhen Research Institute, Xiamen University, Shenzhen 518000, China; yekun@stu.xmu.edu.cn (K.Y.); wusheng@stu.xmu.edu.cn (S.W.); caiyongbin@stu.xmu.edu.cn (Y.C.); zhoulang@stu.xmu.edu.cn (L.Z.); 23320191153346@stu.xmu.edu.cn (L.X.); 23320211154270@stu.xmu.edu.cn (Z.Z.); jquite@xmu.edu.cn (J.L.)
[2] School of Informatics, Xiamen University, Xiamen 361005, China
[3] Key Laboratory of Southeast Coast Marine Information Intelligent Perception and Application, Ministry of Natural Resources, Xiamen 363000, China
[4] School of Electronic Science and Engineering (National Model Microelectronics College), Xiamen University, Xiamen 361005, China
[5] Whale Wave Technology Inc., Kunming 650200, China
* Correspondence: xby_zhang@nwnu.edu.cn

Abstract: Due to its low cost and privacy protection, Channel-State-Information (CSI)-based activity detection has gained interest recently. However, to achieve high accuracy, which is challenging in practice, a significant number of training samples are required. To address the issues of the small sample size and cross-scenario in neural network training, this paper proposes a WiFi human activity-recognition system based on transfer learning using an antenna array: Wi-AR. First, the Intel5300 network card collects CSI signal measurements through an antenna array and processes them with a low-pass filter to reduce noise. Then, a threshold-based sliding window method is applied to extract the signal of independent activities, which is further transformed into time–frequency diagrams. Finally, the produced diagrams are used as input to a pretrained ResNet18 to recognize human activities. The proposed Wi-AR was evaluated using a dataset collected in three different room layouts. The testing results showed that the suggested Wi-AR recognizes human activities with a consistent accuracy of about 94%, outperforming the other conventional convolutional neural network approach.

Keywords: activity recognition; WiFi sensing; transfer learning; CSI; ResNet18

Citation: Ye, K.; Wu, S.; Cai, Y.; Zhou, L.; Xiao, L.; Zhang, X.; Zheng, Z.; Lin, J. Transfer-Learning-Based Human Activity Recognition Using Antenna Array. *Remote Sens.* **2024**, *16*, 845. https://doi.org/10.3390/rs16050845

Academic Editors: Prasad S. Thenkabail, Gerardo Di Martino, Jiahua Zhu, Xinbo Li, Shengchun Piao, Junyuan Guo, Wei Guo, Xiaotao Huang and Jianguo Liu

Received: 15 December 2023
Revised: 15 February 2024
Accepted: 21 February 2024
Published: 28 February 2024

1. Introduction

A variety of scenarios have drawn intense attention to human activity recognition (HAR), including health monitoring [1], smart homes [2], and fall detection [3]. In general, traditional HAR systems are based on wearable devices [4–6] or cameras [7,8]. However, in camera-based systems, users run the risk of compromising their privacy.

Due to the low cost and low equipment requirement of Received Signal Strength Indication (RSSI) and Channel State Information (CSI), the RSSI and CSI from commercial WiFi antenna array devices have become widely used in activity recognition. For example, PAWS [9] and WiFinger [10] are RSSI-based [11] methods, whose recognition accuracy is relatively low due to the limited perception performance of the RSSI. WiFall [12], CARM [13], and TensorBeat [14] are CSI-based methods that have higher accuracy and data resolution than the RSSI-based methods. The CSI-based works have been used widely in WiFi sensing, such as gestures [15], gait [16], and breath rate [17], which is also considered in the activity-recognition model proposed in this paper.

Due to the ubiquity of WiFi signals, CSI-based activity recognition utilizes only the wireless communication function and does not require any physical sensors, which provides a great improvement in the security and protection performance of privacy. CSI-based activity-recognition schemes are often composed of four steps, i.e., data processing, activity classification, feature extraction, and activity detection [12,18]. The corresponding works use primitive signal features, which carry much information caused by human activity and the environment changing with different layouts. At the same time, the CSI extracted by the antenna array is affected by environmental changes, which results in different impacts on the wireless link by human behavior in different scenarios, shown in Figure 1, where two people are performing walking and squatting in two different rooms. It is called the cross-scene problem. The cross-scene problem refers to the ability of the model to generalize across different environments or scenarios and to handle the transition from one scenario to another, which also means that the system needs to be adaptable to recognize activities in multiple scenarios. The multipath channel caused by a specific activity varies with changing environment deployment in different scenes [19,20]. Current works have applied machine learning to solve this problem [21–24]. A hybrid image dataset (ADORESet) has been proposed, which combines real and synthetic images to improve object recognition in robotics, bridging the gap between real and simulation environments [25]. The possibility of connecting object visual recognition with physical attributes such as weight and center of gravity has been explored to improve object manipulation performance via deep neural networks [26]. However, much of this work relies on many training samples to improve accuracy, which is unrealistic when collecting data in reality. The experimental environments are all single scenes, which cannot verify the generalization ability of the models. Therefore, more flexible methods need to be developed for CSI-based human activity recognition with small samples and across scenes.

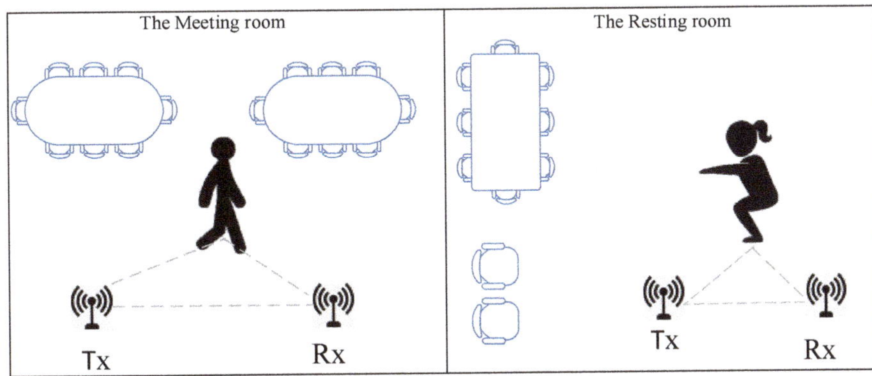

Figure 1. Cross-scene activity recognition.

Addressing the problem of recognizing human activity in cross-scenario and small sample environments, this paper proposes a transfer-learning-based activity-recognition system using an antenna array: Wi-AR. The proposed structure uses the pretrained network to reduce the system's computational complexity instead of training it from scratch [27], which avoids the problem of overfitting. In the method we propose, the original CSI data collected through the antenna array are first processed by a low-pass filter for noise reduction. The purpose of the threshold-based sliding window technique is to determine the beginning and conclusion of activity in a protracted signal. We can then extract the valid segment of activity from the CSI data. Time–frequency diagrams are then created by applying the short-time Fourier transform (STFT) on the four segmented datasets. Finally, the time–frequency diagrams are fed into the pretrained ResNet18 for identification and classification. Based on the simulation results, it is possible to reach 94.2% precision with the proposed Wi-AR system, which is superior to other convolutional neural network (CNN) models. This paper contributes the following:

(1) This paper proposes a low-cost, non-intrusive human activity-recognition system called Wi-AR, which uses antenna arrays to detect WiFi signals without the need for any devices.

(2) An activity feature extraction algorithm is proposed to perform the feature segmentation of different activities to detect start and end moments in noisy environments. Using a threshold-based sliding window approach, activity periods can be extracted from CSI data more efficiently.

(3) The transfer learning strategy employs the fine-tuning of the CNN for training small samples in a changing environment, which improves the accuracy and robustness of activity recognition and avoids overfitting during the training process.

This paper's remaining sections are arranged as follows: Section 2 presents the proposed scheme and the preliminaries. The data collection and preprocessing are investigated in Section 3. In Section 4, the activity-recognition model is proposed. Section 5 shows the experimental validation results, and the last section concludes the paper.

2. Proposed Scheme and Preliminaries

2.1. Proposed Scheme

Wi-AR uses WiFi devices to recognize human movement without the need for a device. As shown in Figure 2, CSI is collected and then processed to reduce the noise. There are four steps in Wi-AR, among which two steps are essential, i.e., activity segmentation and activity recognition. More specifically, the whole Wi-AR system framework can be concluded as follows:

CSI Collection: Wi-AR system collects CSI by Intel 5300 network interface card and WiFi devices.

Data Preprocessing: In terms of amplitude information, the Butterworth filter is chosen to reduce the noise.

Activity Segmentation: Utilizing the processed CSI series, Wi-AR divides the whole CSI series into four segments, which represent four different activities. To determine the beginning and conclusion of each action, Wi-AR uses a sliding window technique based on thresholds.

Activity Recognition: The time–frequency diagram generated by STFT is used for classification. For activity recognition, Wi-AR uses a deep convolutional neural network that has been trained using ResNet18.

2.2. Preliminary

(1) *Channel State Information:* The proposed Wi-AR exploits commercial WiFi devices to obtain CSI. In 802.11n, each multiple-input multiple-output (MIMO) link comprising multiple subcarriers uses orthogonal frequency division multiplexing (OFDM) technology. Each link has a unique channel frequency response caused by the CSI. Due to the 802.11n protocol, the WiFi network has 56 OFDM subcarriers in a 20 MHz band. Using the development tools of the Intel5300 network interface card [28], we can obtain the CSI from 30 subcarriers of the antenna array. Let the number of transmitter and reception antennae be denoted by N_t and N_r. If X_i denotes the transmit signal vector of each packet i and Y_i denotes the receive signal vector of each packet i, the received signal of the network interface card can be represented as:

$$Y_i = H_i X_i + N_i, \quad i \in [1, N] \tag{1}$$

where N_i is the white Gaussian noise vector, H_i is the CSI channel matrix, and N is the total number of received packets. Consequently, for every communication link, the total 30 subcarriers can be obtained, $H = [H_1, H_2 \ldots H_{30}]$ is an expression for the CSI channel matrix, and the total $N_T \times N_R \times 30$ CSI values are finally obtained. The CSI value for each subcarrier, including amplitude and phase information, is denoted by H_i in Equation (1) and can be expressed as $H_i = \|H_i\| e^{j(2\pi f_i + \theta_i)}$, where the magnitude frequency and phase of the i-th subcarrier are represented by $\|H_i\|$, f_i, and θ_i.

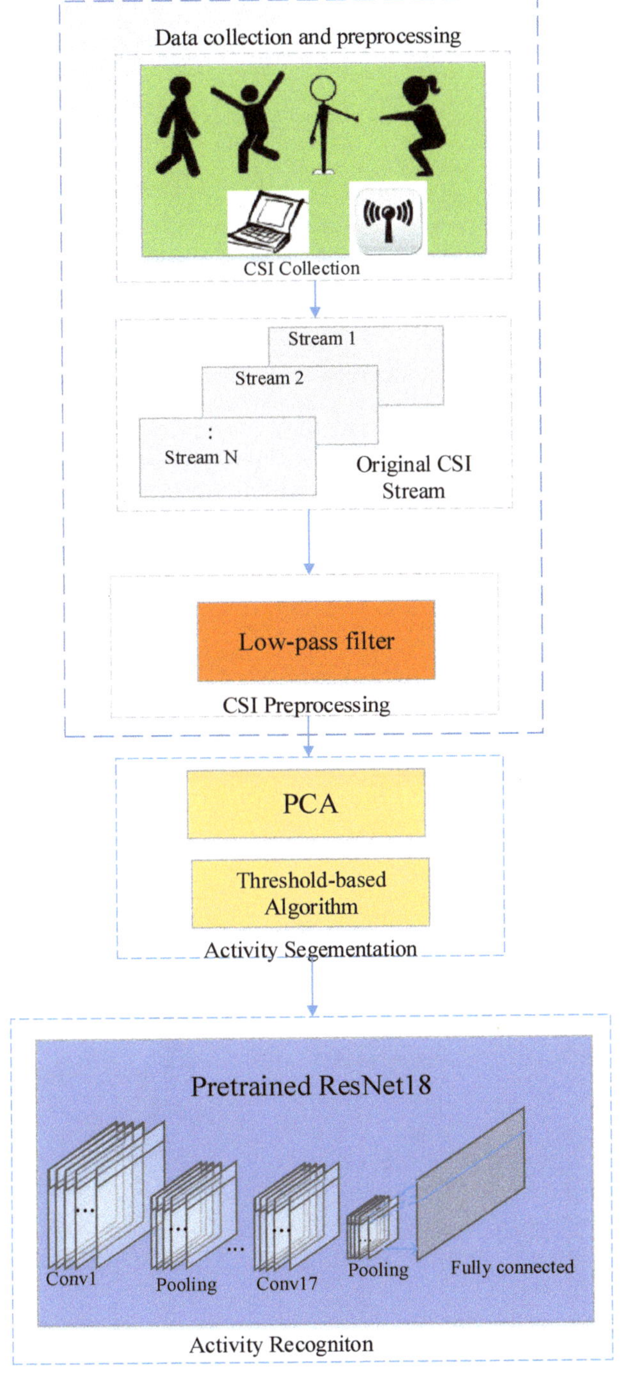

Figure 2. System overview.

(2) *Reflection of Activity on CSI:* In a WiFi system, the CSI is sensitive to environmental changes, and thus, it can be used to describe the channel frequency response (CFR). Assume $H(f,t)$ characterizes the CFR. It can be described as follows:

$$H(f,t) = \sum_{k=1}^{K} \alpha_k(t)e^{-j2\pi f \tau_k(t)},$$ (2)

where $e^{-j2\pi f \tau_k(t)}$ is the phase difference given by carrier frequency shift, $\alpha_k(t)$ is the channel attenuation, and K is the total of the multipath numbers. Human activity can impact the WiFi signal on many pathways between the WiFi transmitter and the WiFi receiver. Let λ and $d(t)$ represent the signal length and the change in the reflection path length, respectively, to better explain the link between human activity and the variations in the WiFi signal. Given that $f_D = -\frac{1}{\lambda} \cdot \frac{d}{dt}d(t)$ is the frequency shift, we obtain:

$$H(f,t) = e^{-2\pi\Delta ft}\left(H_s(f) + \sum_{k\in p_d} \alpha_k(t)e^{j2\pi \int_{-\infty}^{t} fD_k(u)du}\right),$$ (3)

where P_d denotes the dynamic pathways, $H_s(f)$ represents the total CFR of the static paths, and Δf is the carrier frequency offset (CFO). Preprocessing can be used to filter out the high-frequency components in static response, as the CFR power fluctuates mostly due to human activity. Because of multipath effects, the value of CFR fluctuates with dynamic components, which may be used to detect human activity. Based on the Friis free space propagation equation [29], as illustrated in Figure 3, the power of receiver is defined as:

$$P_d = \frac{p_t G_t G_r \lambda^2}{(4\pi)^2(d+4h+\Delta)^2},$$ (4)

where d is the distance between the transceiver pair and λ is the signal wavelength. The WiFi transmitter and reception powers are denoted by the variables P_t and P_r, respectively. The transmitter and receiver gains are G_t and G_r, while the vertical distance is indicated by h. The reflection path's length is Δ. The receiving power varies with the distance between the transceiver pair, as shown by Equation (4). As a result, the shift in CSI may be used to identify human activities.

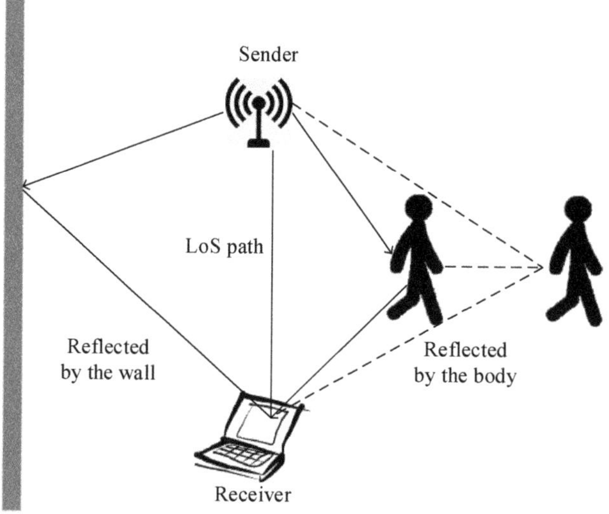

Figure 3. WiFi signal reflection scenario.

3. Data Collection and Preprocessing

3.1. CSI Collection and Denoising

The proposed Wi-AR collects the CSI with N_t transmitting antennae, N_r receiving antennae, and 30 CSI subcarriers reported by Intel5300 network interface card, and it can obtain $30 \times N_t \times N_r$ CSI streams for all communication links. First, a Butterworth low-pass filter is used to reduce noise and remove the high frequency. In our experiments, we designed the Butterworth filter as a low-pass filter, in which case the noise is usually considered to be a high-frequency signal, while the human activity signal that we wish to preserve is a low-frequency signal. Specifically, the frequency response of the Butterworth filter is smooth without abrupt jumps, which gives it an advantage in processing human activity signals. Its main advantage is that it has the flattest frequency response within the pass bandwidth, which means that it processes all frequencies consistently within this range. The original CSI stream and the denoised CSI stream are depicted in Figures 4a and 4b, respectively. It can be observed that CSI generated by human activity is covered by noise. Since the Butterworth filter can maximize the passband flatness of the filter and reduce the high-frequency noise, it is exploited before activity segmentation. The expression for the Butterworth low-pass filter is

$$|\mathbb{H}(\omega)|^2 = \frac{1}{1 + (\frac{\omega}{\omega_c})^{2n}} = \frac{1}{1 + \epsilon^2(\frac{\omega}{\omega_p})^{2n}}, \tag{5}$$

where the filter order is denoted by n, the cut-off frequency is ω_c, the passband edge frequency is ω_p, and the value of $|\mathbb{H}(\omega)|^2$ at the passband edge is $1 + \epsilon^2$.

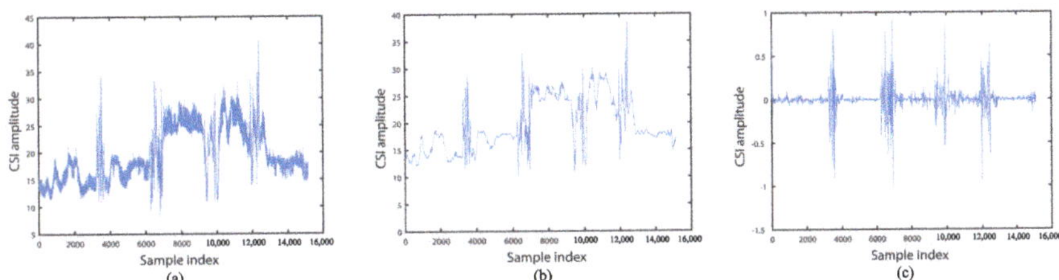

Figure 4. CSI signal preprocessing. (**a**) The original CSI signal. (**b**) The low-pass filtering signal. (**c**) The first-order difference signal.

While the high-frequency noise can be successfully reduced by the signal following the Butterworth low-pass filter, it cannot reflect the characteristic change in the signal from the waveform. To fully reflect the amplitude information of the low-pass filtering signal, it is necessary to process the first-order difference of the signal. Figure 4b,c represents the filtered CSI signal and the first-order difference CSI signal, respectively. It can be seen that the signal after the first-order difference can reflect the signal characteristics of the four behaviors, providing favorable conditions for the activity segmentation. The definition of the first-order difference can be expressed as $y(m) = x(m) - x(m-1)$, where $x(m)$ represents the CSI value corresponding to the m-th sample index.

3.2. Activity Segmentation Based on Domain Adaptation

The purpose of activity segmentation is to truncate the start and end moments of human activity from a long signal to extract the complete signal containing the whole behavior. In this paper, to improve the robustness of the segmentation algorithm according to different room layouts, a threshold-based sliding window method is adopted, shown in Algorithm 1. First, principal component analysis (PCA) is used to extract features, and then several components of PCA are chosen to calculate variance. The first PCA component is not used as it contains very little useful information [13]. Second, the moving variance of the total of the aforementioned primary components is computed using a sliding window. PCA can transform high-dimensional data into low-dimensional data while retaining as much variation information as possible from the original data. This allows us to reduce computational complexity while still retaining the main features of the data. At the same time, PCA can effectively remove noise by retaining the main components of the data. Moving variance can adapt to changes in the data in real time, and due to the adjustability of the window size, this makes it very flexible when dealing with different datasets. The variance of PCA components is shown in Figure 5a. Moving variance depicts the difference of packets reflected by the activity. The variance of PCA is defined as follows:

$$\sigma_i^2 = \frac{\sum_1^m \left(x_{i+j-1} - \bar{x}\right)^2)}{m}, \quad (i = 1, 2, \ldots, n - m) \tag{6}$$

where $\bar{x}$ is the mean value of samples and m represents step size. Then, the median of the variance is calculated by the sliding window, which also reflects the changing trend. The threshold is calculated according to the data in the priority stationary environment. In general, One-tenth of the maximum value of the data at static time is used as the standard. Large threshold standards are chosen to avoid the effects of static mutations, and compensation needs to be made before the beginning and end of the behavior. For this purpose, the compensation number is set to half of the sampling rate, i.e., 100 sampling points. Therefore, the real start and the end are $sta - 100$ and $fin + 100$. In this way, we finally obtain the result of activity segmentation in the original CSI amplitude, shown in Figure 5b. The figure illustrates that the green dotted line and the red dotted line, respectively, represent the beginning and the end of the activity, and the segmented signal contains complete activities. To be suitable for different scenes, different thresholds are selected. The activity extraction results in a time–frequency domain are shown in Figure 5c. The relevant content of the time–frequency domain diagram will be introduced in Section 3.3.

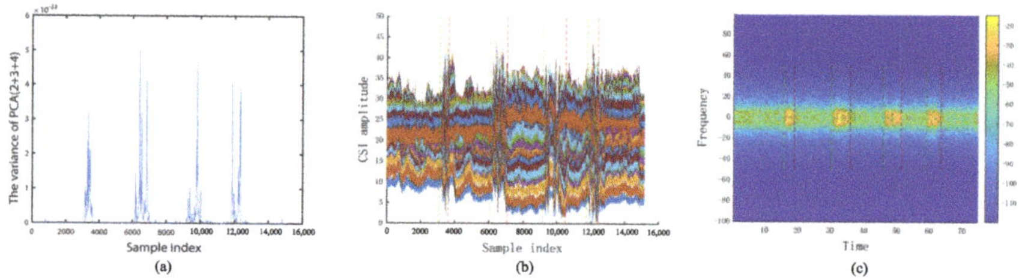

Figure 5. Activity segmentation. (**a**) The variance of principal component sum. (**b**) Activity segmentation on original CSI. (**c**) Time–frequency feature segmentation diagram.

Algorithm 1 Activity segmentation algorithm.

---

input: The amplitude $\alpha(f, t)$;
   The length of variance window and stride w_1, s_1;
   The length of the median window w_2, s_2;
   The Minimum interval between two actions;
   The variation between the maximum and
   minimum values of the stationary environment;
output: The start and finish time index sta, fin;

1: $[,score] = pca(\alpha)$
2: $pca(a) = score(:, 2) + score(:, 3) + score(:, 4)$
3: $n = 1$;
4: **for** $i = 1 : w_1 : length(pcadata) - w_1$ **do**
5: $pcavar(n) = var(pcadata(i : i - 1 + w1))$;
6: $n = n + 1$;
7:**end**
8:$sta = [], fin = []$;

Algorithm 1 *Cont.*

$9{:}n = 1;$
$10{:}\textbf{for } ii = 1 : s_2 : length(pcavar) - s_2 \textbf{ do}$
$11{:} \quad index(n) = median(pcavar(ii : ii + s_2));$
$12{:} \quad temp = 1 + (ii - 1) \times w_1;$
$13{:} \textbf{ if } [length(sta) > lenghth(fin)]\,and$
$\qquad [index(n) > threshold] \textbf{ then}$
$14{:} \quad sta = [sta, temp]$
$15{:} \quad \textbf{if } length(sta) > length(fin)\,and$
$\qquad\quad index(n) < threshold \textbf{ then}$
$16{:} \qquad \textbf{if } temp - sta(end) >= minint \textbf{ then}$
$17{:} \qquad\quad fin = [fin, temp];$
$18{:} \qquad\qquad n = n + 1$
$19{:} \qquad\quad \textbf{end}$
$20{:} \qquad \textbf{end}$
$21{:} \quad \textbf{end}$
$22{:}\textbf{end for}$
$23{:}fin = fin + 100;$
$24{:}sta = sta - 100;$

3.3. STFT Transform

Wi-AR converts the waveforms to time–frequency diagrams using the STFT to extract the signal's combined time–frequency properties. Considering the trade-off of the frequency-time resolution, Wi-AR sets the sliding window step size of 256 samples in this paper.

4. Activity Recognition Model

The proposed Wi-AR adopts the ResNet18 trained by ImageNet as a classification network combined with transfer learning. To avoid losing the source weights, the classifier is first trained using the initial parameters to train all the network's weights at a low learning rate. Then, the last fully connected layer is modified to suit the target dataset. More specifically, the learned features and weights in the pretraining process are transferred to the recognition network of human activity. After that, the time-frequency diagrams of the CSI are input into the pretrained ResNet18 to train the recognition model. In Figure 6, the layer with a complete connection number is finally substituted with the categories of human activity.

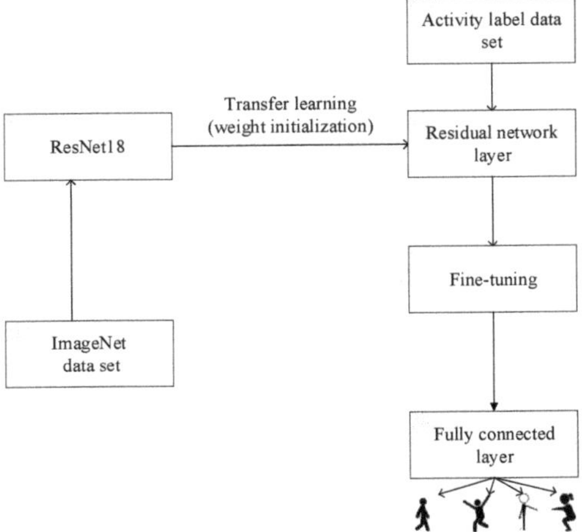

Figure 6. Model flow chart.

4.1. ResNet18

In deep learning, the complexity of traditional CNN increases with layers. It is suggested to use a deep residual network (ResNet) to address this situation [30]. Compared with traditional CNN, ResNet is easier to train and has a faster convergence since the whole model only needs to pay attention to the difference between input and output. Moreover, the increase in depth does not increase the amount of computation but increases the accuracy and the efficiency of network training. ResNet combines a deep convolutional neural network with a specially designed residual structure, which can achieve an intense network [31]. To obtain the deep features in the image, which can have a better characterization of the human activity, and considering the depth and calculation of the network, the ResNet18 is finally chosen as the activity classification model, which includes 17 convolutional layers and one full connection layer. The learning process is simplified because ResNet mainly learns residual rather than the complete output. The convolution operation computation is described as

$$X_j^L = f\left(\sum_{i \in M_j} X_i^L \times K_{i,j}^L + b_j^L \right), \tag{7}$$

where M_j is the input feature map, L denotes the number of layers in the neural network, and $K_{i,j}^L$ suggests the convolution kernel. The activation function is f, and the unique offset b is output for each layer of the feature graph.

To extract more features, the number of convolutional layers increases. However, with an increasing number of convolutional layers, there is a risk of gradient dispersion and gradient explosion. The residual unit in ResNet18 can effectively solve the above problem. The core idea is to divide network output into two parts: identity mapping and residual mapping. The definition of the residual unit is

$$X_{k+1} = f(F(X_k, W_k) + h(X_k)), \tag{8}$$

where X_k and $X_k + 1$ stand for the input and output of the k-th residual unit, respectively. $F(X_k, W_k)$ is the residual mapping that must be learned. The activation function is denoted by f and the convolution kernel by W. Figure 7a displays the CNN learning block with less stacked non-linear layers through a direct mapping $x \to y$ representing $F(x)$ and x as stacked non-linear layers and the identity function, respectively. Figure 7b shows the identity mapping through the residual function $F(x)$, where $y = F(x) + x$, as proposed in [32].

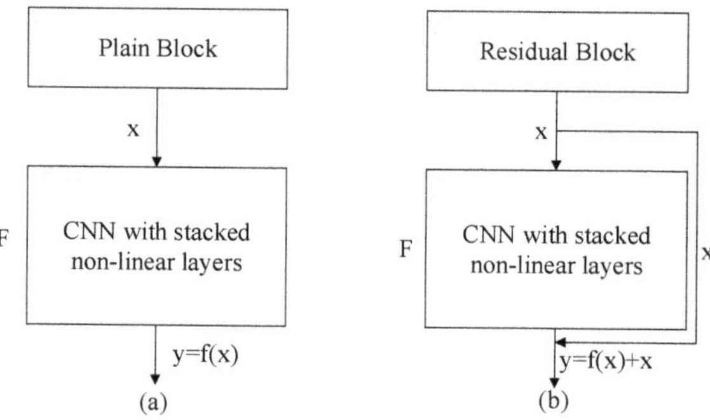

Figure 7. (**a**) Direct mapping in plain CNN (**b**) Identity mapping in ResNet.

In ResNet18, both the max-pooling and the average-pooling techniques are used. Reducing training parameters in the network is the aim of the max-pooling layer, which comes after the convolutional layer [33]. The last completely linked layer, which consists of four nodes symbolizing four distinct activities, is added to receive the output of the human activity. Figure 8 illustrates this process.

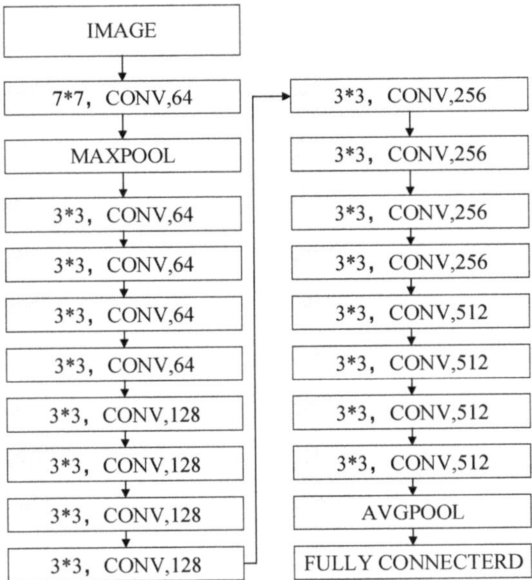

Figure 8. ResNet18 network structure.

4.2. Transfer Learning

One common deep learning technique that is frequently applied to tiny training samples is transfer learning. To accomplish the migration, model-based transfer learning is used [34], which looks for pretrained parameter weights that the network's bottom layer may share. After obtaining the pretrained model with the ImageNet dataset, we replace the random initialization model parameters with the new parameters except for the last fully connected layer. Since each CNN in the ImageNet dataset is trained using 1000 classes, and our activities include four classes, fine-tuning is used to make sure that the network weights will not change too quickly and will fit our data without altering the original model. Therefore, a small learning rate is set to avoid gradient vanishing.

5. Experimental Evaluation

To show the model's capacity for generalization, the performance of the suggested Wi-AR is first assessed on a dataset that the user has collected, and then it is tested on additional datasets. We also contrast our model's accuracy with that of other CNN models in this section.

5.1. Experiment Setup

In the experiment, we collected CSI from various rooms using a computer equipped with an Intel 5300 network interface card and a commercial WiFi router. The router is equipped with three antennae, and the reason for using three antennae is mainly related to the working principle of the MIMO (Multiple-Input Multiple-Output) system. In a MIMO system, multiple antennae can send and receive multiple streams of data at the same time, therefore increasing the transmission rate and reliability of the system. Specifically, three antennae can form three independent antenna links, each of which can receive an independent data stream. These three links form 90 subcarriers, i.e., for each timestamped data, they are composed of 90 subcarriers. In this way, by analyzing the CSI of each subcarrier, we can obtain more detailed and accurate information about the channel state and thus better understand and utilize the wireless channel. In addition, multiple antennae can provide more spatial diversity and spatial multiplexing, thus improving the capacity and anti-interference capability of the system. Spatial diversity improves the reception quality of the signal, while spatial multiplexing increases the system's data transmission rate. The router is equipped with three antennae, forming an antenna array. The room is furnished with tables and chairs, as shown in Figure 9. The data gathered in the meeting and rest areas serve as the training set for the experiment, while the data gathered in three separate rooms serves as the testing set. A total of 120 pieces of data were randomly chosen from the three scenarios, and 240 samples were chosen from each of the first two scenarios to represent the model's generalization. As a result, there are 480 samples in the training set and 120 samples in the testing set. Three antennae are installed on the WiFi transmitter's receiving side and two on the transmitter side. The sampling rate is 200 packets/s, and 30 subcarriers from each transceiver pair are obtained. At the same time, three volunteers with different body shapes are asked to perform four kinds of activities. Four behaviors contain two coarse-grained activities (jumping and walking) and two relatively fine-grained activities (squatting and leg lifting). The volunteers are asked to perform each activity one by one and keep them for five seconds. Each task has a total of 150 samples split into a test set and a training set. If the generated time–frequency diagrams are not normalized, the ranges of feature value distribution will vary differently. To avoid such a problem, Wi-AR first normalizes all time–frequency diagrams and further resizes them to suit the pretrained model. Lastly, the amount of human activity changes the final completely connected layer.

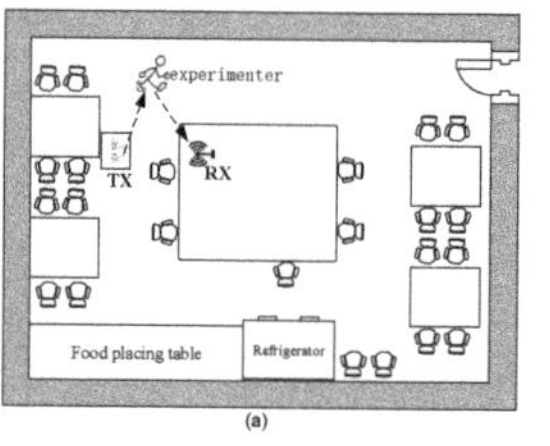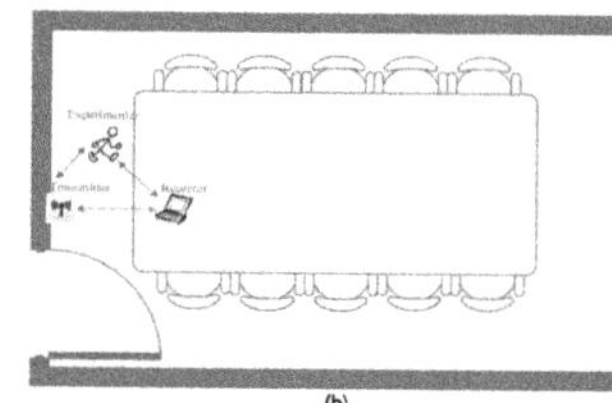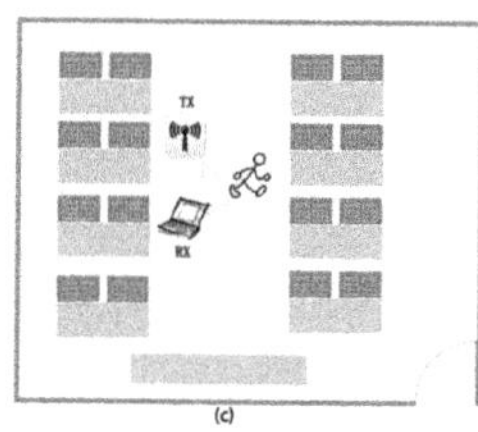

Figure 9. Different room layout for data collection. (**a**) The rest-room layout. (**b**) The meeting room layout. (**c**) The class-room layout.

Before the experiment, several initial parameters need to be defined, which are listed in Table 1. The batch size is set to 16, and the iteration value is 30. In this research, a reduced starting learning rate of 0.001 is chosen to prevent overfitting.

Table 1. Training parameters.

Parameter	Value
Image size	224 × 224
Epoch	30
Batch size	16
Initial learning rate	0.001
Lr-function	StepLR

5.2. Experimental Validation

The proposed Wi-AR produces time-frequency diagrams by STFT. Figure 10 displays the time–frequency diagrams for four different types of activities.

The scene's computers are equipped with an NVIDIA GeForce GTX 1660S GPU, an Intel 10500 CPU, and 32 GB of RAM. Then, use the cross-entropy loss function. If the predicted value is the same as the true value, it approaches 0. If the predicted value is different from the true value, the cross-entropy loss function will become very large. Training and testing curves are, respectively, recorded in Figures 11 and 12, which represent the value changes in accuracy and loss in the training and testing process. Both can converge after 20 iterations. In addition, a detailed evaluation of the classification result is conducted using three types of metrics. The metrics are defined as follows: (1) Precision is defined as $\dfrac{TP}{TP+FP}$, where the ratios of correctly marked activities (TP) and erroneously marked activities (FP) are expressed. (2) $\dfrac{TP}{TP+FN}$ is the definition of recall, where FN represents erroneously promoted negative samples. (3) The formula for the F1-score (F1) is $F1 = \dfrac{2 \times PR \times RE}{PR+RE}$. Table 2 tabulates the testing data and shows the high precision and F1-score. Figure 13a shows the confusion matrix whose rows represent the predicted activity and columns refer to the actual activity. We find that the accuracy of leg lifting and squatting can achieve 100% due to their obvious features, and the average accuracy is 94.2%. Walking and jumping have similar movements, so the accuracy of recognition is not as good as it is for the other two activities. The dispersion of various activity data is displayed using the ResNet18 model under T-SNE visualization in Figure 13b. It can be intuitively seen that the model can distinguish different activities well.

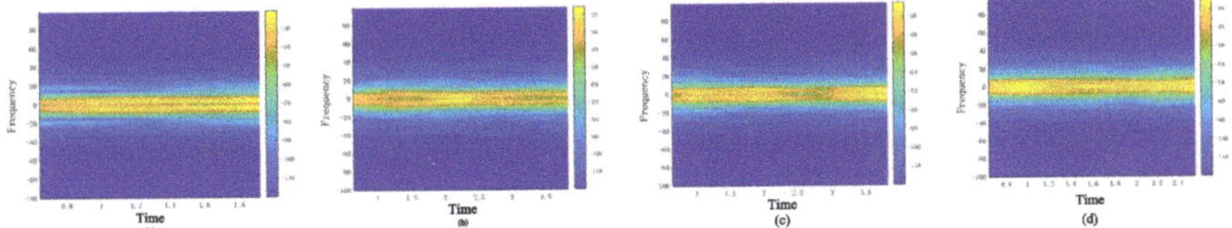

Figure 10. The time-frequency diagrams of four kinds of activities. (**a**) Jumping. (**b**) Walking. (**c**) Squatting. (**d**) Leg lifting.

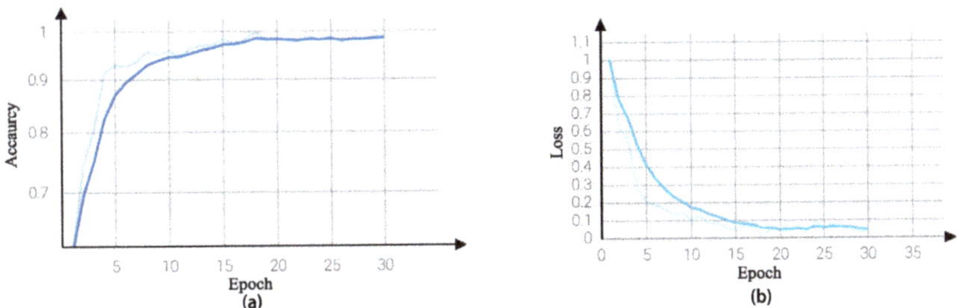

Figure 11. The training g accuracy and loss. (**a**) The accuracy of training. (**b**) The training accuracy and loss.

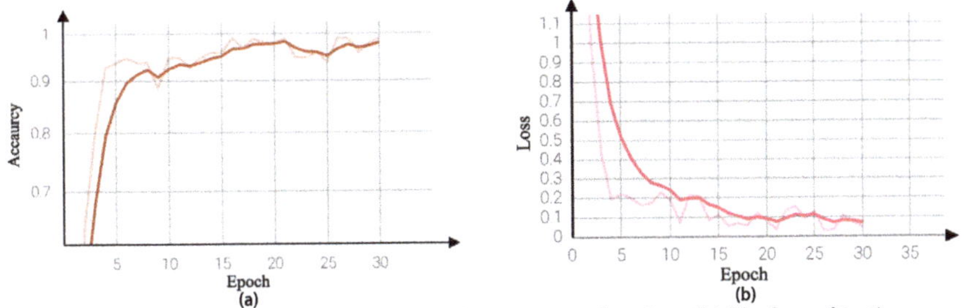

Figure 12. The testing accuracy and loss. (**a**) The accuracy of testing. (**b**) The loss of testing.

Table 2. Testing Results of the ResNet18.

Human Activity	Precision	Recall	F1-Score
Jumping	0.90	0.87	0.93
Leg Lifting	1.00	0.97	0.98
Squatting	1.00	1.00	0.97
Walking	0.83	0.93	0.92

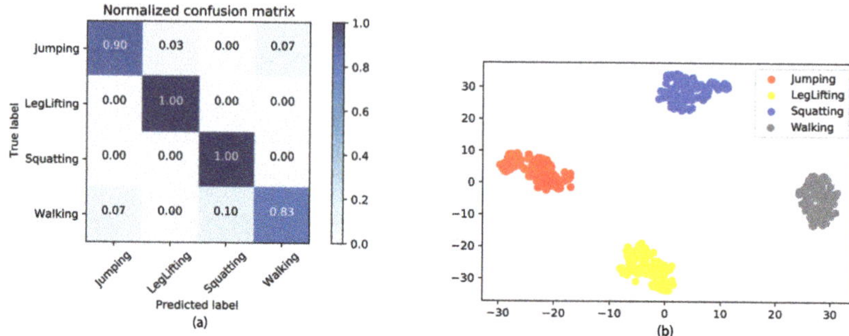

Figure 13. (**a**) The confusion matrix of human activity recognition. (**b**) T-SNE visualization after the ResNet18 model for four kinds of activities.

To verify the generalization ability of the recognition model, the proposed Wi-AR is also tested on the dataset collected by the team of Chunjing Xiao [35] for more evaluation. The differences are that their sampling rate is 50 packets/s, and the WiFi router has only one antenna. Ten different types of activities make up the dataset, five of which are fine-grained and five of which are coarse-grained. The experiment is conducted using two relatively fine-grained activities (hand swing and drawing O) and two coarse-grained activities (running and squatting). The result of testing is tabulated in Table 3, from which we can also see high accuracy, achieving more than 94% accuracy among each activity. And since the drawing circle has unique movement characteristics, the recognition accuracy achieves 100% for the given dataset. As a result, the validation results show that the recognition model has the capacity for generalization.

Table 3. Testing results of the ResNet18 on another dataset.

Human Activity	Precision	Recall	F1-Score
Drawing O	1.00	1.00	1.00
Hand Swing	0.96	0.87	0.91
Running	0.94	0.97	0.95
Squatting	0.94	1.00	0.97

To compare the accuracy of the pretrained networks, we choose different CNN models, such as Aexmet, VGG11, and ResNet34. Meanwhile, some classical machine learning algorithms, such as decision trees, random forests, SVMs, etc., are also used as comparative tests. The results of the CNN classifiers in terms of accuracy and time consumed are shown in Table 4 to show the performance of the different CNN networks in comparison with the model in this paper. The accuracy of classical machine learning algorithms is shown in Table 5. The experimental results show that the recognition accuracy of classical machine learning algorithms is generally low because the action features carried by CSI signals are significantly reduced after going through the wall, and ordinary machine learning algorithms cannot accurately classify them, and more complex deep networks are needed to extract their features. Among the CNN classifiers, the accuracy of ResNet18 is 0.8% higher than that of VGG19, but the time consumed is about 25% of that of

VGG19. In addition, ResNet18 shows higher accuracy and less time consumption than other ResNet models.

Table 4. The accuracy of each CNN classification.

CNN Models	Accuracy Rate	Time Consumption
AlexNet	85.00%	4 m 22 s
Vgg11	91.67%	7 m 33 s
Vgg13	92.50%	9 m 16 s
Vgg16	92.67%	10 m 45 s
Vgg19	93.33%	13 m 10 s
ResNet18	94.17%	3 m 29 s
ResNet34	90.00%	5 m 23 s
ResNet50	84.17%	6 m 43 s
ResNet101	86.67%	8 m 17 s
ResNet152	92.50%	10 m 23 s

Table 5. Accuracy of classical machine learning algorithms

Machine Learning Algorithms	Accuracy Rate
Naive Bayesian	44%
KNN	58%
Decision tree	65%
SVM	74%
Proposed	94%

Moreover, MF-ABLSTM [36] leverages attention-based long short-term memory neural network and time–frequency domain features for small CSI sample-based activity recognition, achieving 92% with 490 training and testing samples after 200 iterations. Due to the proposed method being able to train very deep neural networks, it avoids the problem of gradient vanishing and improves the model's expressive power and performance. It uses residual connections to preserve the original features, making the learning of the network smoother and more stable, further improving the accuracy and generalization ability of the model. During training, gradient vanishing and exploding problems can be avoided, accelerating network convergence. Therefore, Wi-AR achieves 94.2% with 600 samples after 30 iterations. The results of the comparison are, respectively, shown in Figure 14a,b, which demonstrate that our proposed Wi-AR scheme achieves higher accuracy with fewer iterations for small sample-based activity recognition.

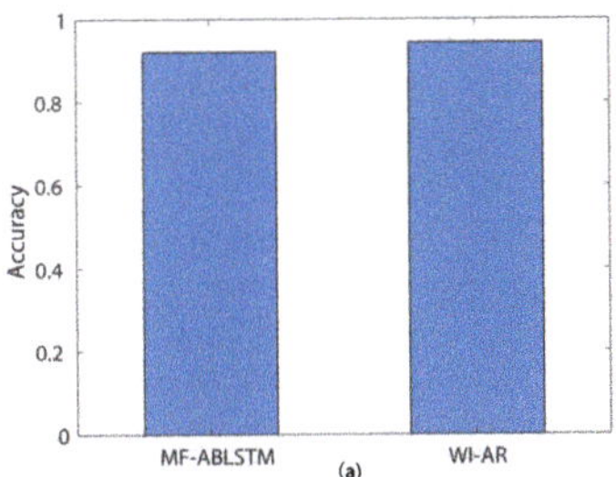

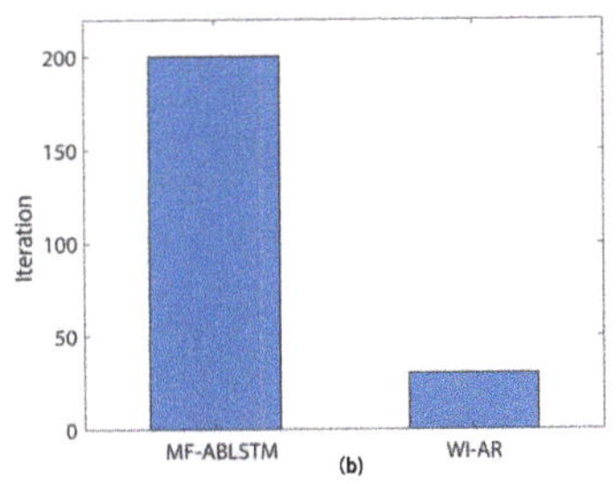

Figure 14. (**a**) The comparison of activity-recognition accuracy. (**b**) The comparison of training iterations.

6. Conclusions

In response to the small sample size and cross-scenario issues in activity recognition, this paper proposes the Wi-AR human activity-recognition system, which is based on channel state data and antenna arrays. Wi-AR collects CSI data from an array of antennae for a sequence of four kinds of activities, preprocesses the collected CSI signal, transforms it into time–frequency diagrams, and marks samples for supervised machine learning. The experimental results show that this method based on transfer learning can achieve 94% accuracy with a small number of samples. We can see that Wi-AR is relevant in single-person multi-scene environments. In future work, we will consider more realistic multi-user human activity-recognition scene recognition. Meanwhile, it is also a challenging problem to do effective differentiation for some similar actions. For the problem of difficult label annotation of sensing data, semi-supervised learning is also an effective solution to deal with this difficulty, which is also the focus of our future work.

Author Contributions: Conceptualization, K.Y., L.X. and X.Z.; methodology, K.Y., L.X. and X.Z.; software, K.Y.; validation, K.Y., Y.C. and Z.Z.; formal analysis, K.Y.; investigation, K.Y.; resources, K.Y.; data curation, Y.C.; writing—original draft preparation, K.Y. and L.X.; writing—review and editing, S.W. and L.Z.; visualization, K.Y.; supervision, S.W.; project administration, J.L.; funding acquisition, J.L. All authors have read and agreed to the published version of the manuscript.

Funding: This research was funded by the Key Program of Marine Economy Development Special Foundation of the Department of Natural Resources of Guangdong Province (GDNRC [2022]19) and the Natural Resources Science and Technology Innovation Project of Fujian KY-080000-04-2022-025.

Data Availability Statement: The data presented in this study are available on reasonable request from the corresponding author.

Conflicts of Interest: Author Xuebo Zhang was employed by the company Whale Wave Technology Inc. The remaining authors declare that the research was conducted in the absence of any commercial or financial relationships that could be construed as a potential conflict of interest.

References

1. Wang, X.; Yang, C.; Mao, S. PhaseBeat: Exploiting CSI phase data for vital sign monitoring with commodity WiFi devices. In Proceedings of the 2017 IEEE 37th International Conference on Distributed Computing Systems (ICDCS), Atlanta, GA, USA, 5–8 June 2017; pp. 1230–1239.
2. Zou, H.; Zhou, Y.; Yang, J.; Jiang, H.; Xie, L.; Spanos, C.J. WiFi-enabled device-free gesture recognition for smart home automation. In Proceedings of the 2018 IEEE 14th International Conference on Control and Automation (ICCA), Anchorage, AK, USA, 12–15 June 2018; pp. 476–481.
3. Palipana, S.; Rojas, D.; Agrawal, P.; Pesch, D. FallDeFi: Ubiquitous fall detection using commodity WiFi devices. *Proc. ACM Interact. Mob. Wearable Ubiquitous Technol.* **2018**, *1*, 155. [CrossRef]
4. Yatani, K.; Truong, K.N. Bodyscope: A wearable acoustic sensor for activity recognition. In Proceedings of the 2012 ACM Conference on Ubiquitous Computing, Pittsburgh, PA, USA, 5–8 September 2012; pp. 341–350.
5. Fortino, G.; Galzarano, S.; Gravina, R.; Li, W. A framework for collaborative computing and multi-sensor data fusion in body sensor networks. *Inf. Fusion* **2015**, *22*, 50–70. [CrossRef]
6. Ghasemzadeh, H.; Panuccio, P.; Trovato, S.; Fortino, G.; Jafari, R. Power-aware activity monitoring using distributed wearable sensors. *IEEE Trans. Hum.Mach. Syst.* **2014**, *44*, 537–544. [CrossRef]
7. Bodor, R.; Jackson, B.; Papanikolopoulos, N. Vision-based human tracking and activity recognition. In Proceedings of the 11th Mediterranean Conference on Control and Automation, Rhodes, Greece, 18–20 June 2003; Volume 1.
8. De Sanctis, M.; Cianca, E.; Di Domenico, S.; Provenziani, D.; Bianchi, G.; Ruggieri, M. Wibecam: Device free human activity recognition through WiFi beacon-enabled camera. In Proceedings of the 2nd Workshop on Workshop on Physical Analytics, Florence, Italy, 22 May 2015; pp. 7–12.
9. Gu, Y.; Ren, F.; Li, J. Paws: Passive human activity recognition based on WiFi ambient signals. *IEEE Internet Things J.* **2015**, *3*, 796–805. [CrossRef]
10. Tan, S.; Yang, J. WiFinger: Leveraging commodity WiFi for fine-grained finger gesture recognition. In Proceedings of the 17th ACM International Symposium on Mobile Ad Hoc Networking and Computing, Paderborn, Germany, 5–8 July 2016; pp. 201–210.
11. Yang, Z.; Zhou, Z.; Liu, Y. From RSSI to CSI: Indoor localization via channel response. *ACM Comput. Surv.* **2013**, *46*, 25. [CrossRef]
12. Chen, Z.; Zhang, L.; Jiang, C.; Cao, Z.; Cui, W. WiFi CSI based passive human activity recognition using attention based BLSTM. *IEEE Trans. Mob. Comput.* **2018**, *18*, 2714–2724. [CrossRef]

13. Wang, W.; Liu, A.X.; Shahzad, M.; Ling, K.; Lu, S. Understanding and modeling of WiFi signal based human activity recognition. In Proceedings of the 21st Annual International Conference on Mobile Computing and Networking, Paris, France, 7–11 September 2015; pp. 65–76.

14. Wang, X.; Yang, C.; Mao, S. TensorBeat: Tensor decomposition for monitoring multiperson breathing beats with commodity WiFi. *ACM Trans. Intell. Syst. Technol.* **2017**, *9*, 8. [CrossRef]

15. Feng, C.; Arshad, S.; Yu, R.; Liu, Y. Evaluation and improvement of activity detection systems with recurrent neural network. In Proceedings of the 2018 IEEE International Conference on Communications (ICC), Kansas City, MO, USA, 20–24 May 2018; pp. 1–6.

16. Abdelnasser, H.; Youssef, M.; Harras, K.A. WiGest: A ubiquitous WiFi-based gesture recognition system. In Proceedings of the 2015 IEEE Conference on Computer Communications (INFOCOM), Hong Kong, China, 26 April–1 May 2015; pp. 1472–1480.

17. Liu, X.; Cao, J.; Tang, S.; Wen, J. Wi-Sleep: Contactless sleep monitoring via WiFi signals. In Proceedings of the 2014 IEEE Real-Time Systems Symposium, Rome, Italy, 2–5 December 2014; pp. 346–355.

18. Duan, S.; Yu, T.; He, J. WiDriver: Driver activity recognition system based on WiFi CSI. *Int. J. Wirel. Inf. Netw.* **2018**, *25*, 146–156. [CrossRef]

19. Zhang, Y.; Wang, X.; Wang, Y.; Chen, H. Human Activity Recognition Across Scenes and Categories Based on CSI. *IEEE Trans. Mob. Comput.* **2020**, *21*, 2411–2420. [CrossRef]

20. Wang, J.; Chen, Y.; Hu, L.; Peng, X.; Philip, S.Y. Stratified transfer learning for cross-domain activity recognition. In Proceedings of the 2018 IEEE International Conference on Pervasive Computing and Communications (PerCom), Athens, Greece, 19–23 March 2018; pp. 1–10.

21. Zheng, V.W.; Hu, D.H.; Yang, Q. Cross-domain activity recognition. In Proceedings of the 11th International Conference on Ubiquitous Computing, Orlando, FL, USA, 30 September–3 October 2009; pp. 61–70.

22. Hu, D.H.; Zheng, V.W.; Yang, Q. Cross-domain activity recognition via transfer learning. *Pervasive Mob. Comput.* **2011**, *7*, 344–358. [CrossRef]

23. Wang, J.; Zhang, L.; Gao, Q.; Pan, M.; Wang, H. Device-free wireless sensing in complex scenarios using spatial structural information. *IEEE Trans. Wirel. Commun.* **2018**, *17*, 2432–2442. [CrossRef]

24. Ding, X.; Jiang, T.; Zhong, Y.; Wu, S.; Yang, J.; Xue, W. Improving WiFi-based Human Activity Recognition with Adaptive Initial State via One-shot Learning. In Proceedings of the 2021 IEEE Wireless Communications and Networking Conference (WCNC), Nanjing, China, 29 March–1 April 2021; pp. 1–6.

25. Bayraktar, E.; Yigit, C.B.; Boyraz, P. A hybrid image dataset toward bridging the gap between real and simulation environments for robotics: Annotated desktop objects real and synthetic images dataset: ADORESet. *Mach. Vis. Appl.* **2019**, *30*, 23–40. [CrossRef]

26. Bayraktar, E.; Yigit, C.B.; Boyraz, P. Object manipulation with a variable-stiffness robotic mechanism using deep neural networks for visual semantics and load estimation. *Neural Comput. Appl.* **2020**, *32*, 9029–9045. [CrossRef]

27. Liang, H.; Fu, W.; Yi, F. A survey of recent advances in transfer learning. In Proceedings of the 2019 IEEE 19th International Conference on Communication Technology (ICCT), Xi'an, China, 16–19 October 2019; pp. 1516–1523.

28. Halperin, D.; Hu, W.; Sheth, A.; Wetherall, D. Tool release: Gathering 802.11 n traces with channel state information. *ACM SIGCOMM Comput. Commun. Rev.* **2011**, *41*, 53–53. [CrossRef]

29. Rappaport, T.S. Wireless Communications–Principles and Practice, (The Book End). *Microw. J.* **2002**, *45*, 128–129.

30. He, K.; Zhang, X.; Ren, S.; Sun, J. Delving deep into rectifiers: Surpassing human-level performance on imagenet classification. In Proceedings of the IEEE International Conference on Computer Vision, Santiago, Chile, 7–13 December 2015; pp. 1026–1034.

31. Li, Y.; Ding, Z.; Zhang, C.; Wang, Y.; Chen, J. SAR ship detection based on resnet and transfer learning. In Proceedings of the IGARSS 2019—2019 IEEE International Geoscience and Remote Sensing Symposium, Yokohama, Japan, 28 July–2 August 2019; pp. 1188–1191.

32. He, K.; Zhang, X.; Ren, S.; Sun, J. Deep residual learning for image recognition. In Proceedings of the IEEE Conference on Computer Vision and Pattern Recognition, Las Vegas, NV, USA, 27–30 June 2016; pp. 770–778.

33. Song, X.; Chen, K.; Cao, Z. ResNet-based Image Classification of Railway Shelling Defect. In Proceedings of the 2020 39th Chinese Control Conference (CCC), Shenyang, China, 27–29 July 2020; pp. 6589–6593.

34. Zhu, Y.; Chen, Y.; Lu, Z.; Pan, S.J.; Xue, G.R.; Yu, Y.; Yang, Q. Heterogeneous transfer learning for image classification. In Proceedings of the Twenty-Fifth AAAI Conference on Artificial Intelligence, San Francisco, CA, USA, 7–11 August 2011.

35. Xiao, C.; Lei, Y.; Ma, Y.; Zhou, F.; Qin, Z. DeepSeg: Deep-Learning-Based Activity Segmentation Framework for Activity Recognition Using WiFi. *IEEE Internet Things J.* **2020**, *8*, 5669–5681. [CrossRef]

36. Tian, Y.; Li, S.; Chen, C.; Zhang, Q.; Zhuang, C.; Ding, X. Small CSI Samples-Based Activity Recognition: A Deep Learning Approach Using Multidimensional Features. *Secur. Commun. Netw.* **2021**, *2021*, 5632298. [CrossRef]

Article

Broadband Source Localization Using Asynchronous Distributed Hydrophones Based on Frequency Invariability of Acoustic Field in Shallow Water

Hui Li [1,2,*], Jun Huang [1,2], Zhezhen Xu [1,2], Kunde Yang [2,3] and Jixing Qin [4]

1   School of Marine Science and Technology, Northwestern Polytechnical University, Xi'an 710072, China; junhuang@mail.nwpu.edu.cn (J.H.); zzxu@mail.nwpu.edu.cn (Z.X.)
2   Key Laboratory of Ocean Acoustics and Sensing (Northwestern Polytechnical University), Ministry of Industry and Information Technology, Xi'an 710072, China; ykdzym@nwpu.edu.cn
3   Ocean Institute of Northwestern Polytechnical University, Taicang 215400, China
4   State Key Laboratory of Acoustics, Institute of Acoustics, Chinese Academy of Sciences, Beijing 100190, China; qjx@mail.ioa.ac.cn
*   Correspondence: lihui2018@nwpu.edu.cn

Abstract: This paper introduces a model-independent passive source localization method, employing asynchronous distributed hydrophones in shallow water. Based on the frequency invariability of the acoustic field, assuming the correct source range information, the warped spectra of received signals at distributed hydrophones exhibit identical shapes. Subsequently, a cost function is formulated to mutually align the warped spectra, with its maximum point indicating the source location. The proposed method can locate the source in two-dimensional horizontal space without requiring either angle- or time-synchronization information. Numerical simulations are conducted to demonstrate the performance of the proposed method.

Keywords: asynchronous distributed hydrophones; shallow water; passive source localization; frequency invariability of acoustic field

Citation: Li, H.; Huang, J.; Xu, Z.; Yang, K.; Qin, J. Broadband Source Localization Using Asynchronous Distributed Hydrophones Based on Frequency Invariability of Acoustic Field in Shallow Water. *Remote Sens.* **2024**, *16*, 982. https://doi.org/10.3390/rs16060982

Academic Editors: Gerardo Di Martino, Jiahua Zhu, Xinbo Li, Shengchun Piao, Junyuan Guo, Wei Guo, Xiaotao Huang and Jianguo Liu

Received: 2 December 2023
Revised: 27 February 2024
Accepted: 8 March 2024
Published: 11 March 2024

1. Introduction

In shallow water, passive source localization remains a topic of ongoing interest and concern for many researchers. One classical method for addressing this issue is matched field processing (MFP) [1], where the received acoustic field is matched with model-calculated replicas to determine the source location based on the best matching point. However, despite its effectiveness, the performance of MFP significantly degrades if the environment model used for replica calculation is mismatched. Additionally, this class of methods requires a large-aperture array comparable to the water depth to adequately sample the acoustic field, thereby limiting its practical application. These limitations of MFP have spurred the development of model-independent localization methods.

In shallow water, normal mode theory is commonly employed to explain low-frequency sound propagation, where the received field can be described as the summation of a series of normal modes [2]. The interference characteristics among these modes contain information about the ocean environment and the source location. As a result, the source can be located by exploiting the interference characteristics among the modes. The methods of source localization in shallow water could be broadly categorized into two groups: (1) waveguide invariant [3] or array invariant-based methods [4–6] and (2) warping transform-based methods [7–9]. The first category estimates the source range based on the relationship between the source range and the value of the waveguide/array invariant, with the waveguide invariant obtained as a priori environmental information. The second category, warping transform-based methods, utilize the invariability of the characteristic frequency of the normal modes and determines the source range by matching the measured and standard warped spectra. In this localization process, the standard warped spectrum must be obtained in advance by using a guide source at a known range. Both types of methods can estimate the source range without requiring sound propagation modeling. However, the majority of existing methods are designed for a single receiving platform and focus on locating the source in range space. Additionally, some a priori information (e.g., the value of waveguide invariant or a guide source) is required. However, a two-dimensional (2D) location (i.e., both x and y coordinates) is unquestionably more useful than range information in practical applications such as attack, communication, and monitoring. To achieve 2D localization, an additional array with a horizontal aperture can be used to determine the azimuthal angle of the source. Another effective approach is to observe the source using spatially distributed hydrophones. In this case, information about the mode interference characteristics is sufficient for 2D localization.

Passive localization using distributed sensors has been extensively studied. A traditional methodology for this task involves combining the target's direction of arrival (DOA) measured by the distributed sensors [10–16]. The cost function can be constructed using these measured bearings to derive the source position. Weighted least-squares (WLS) methods [10–12] and maximum likelihood methods [13] are commonly used to solve this problem. Another common approach for this task involves measuring the time differences of the sensors [17–22]; then, the source position can be solved by intersecting a set of hyperbolic curves [17]. However, most of the algorithms mentioned above are derived under free-field and plane-wave assumptions, which may lead to performance degradation in practical shallow water environments. Specifically, the measured angle of arrival (AOA) for a horizontal line array could deviate from the true counterpart due to boundary reflections, rendering AOA-based methods biased or invalid. For the time difference of arrival-based algorithms, the primary challenge is clock synchronization among the distributed sensors. Additionally, performance degradation can occur due to time-delay estimation errors and sound speed mismatches in practical applications. In shallow water, propagation time differences between individual modes can be utilized to locate broadband sources using distributed hydrophones [23,24]. Benefitting from the stability of the acoustic field, this kind of algorithm can locate sources without either angle or time synchronization information. However, this algorithm requires separating propagation modes and determining their arrival times, which discourages its practical application.

This paper introduces a passive model-independent source localization method based on several spatially distributed hydrophones in shallow water. The proposed method determines the 2D source location based on the invariability of the characteristic frequency of the normal modes, thus neither the angle nor time synchronization information is required. Compared to the method in Ref. [23], the proposed method introduces the warping transform of the autocorrelation function (ACF) and directly utilizes the warped spectra to construct the cost function, simplifying its implementation in practice and extending its applicability to non-impulse sources. Specifically, for an assumed source location, the method applies the warping transform to the ACF of the received signal of each hydrophone and calculates the cost function by mutually matching the warped spectra from different hydrophones. Due to the invariability of the characteristic frequency of the acoustic field, the warped spectra from different hydrophones will have the same shape if the correct source location is assumed, enabling the maximum point of the cost function to indicate the source location. The main innovations of the method in this paper are as follows: (1) The proposed method is able to achieve source localization by exploiting the acoustic field feature without either the angle- or time-synchronization information. (2) The proposed method works on the signal autocorrelation rather than on the raw signal. Numerical simulations are conducted in a shallow water environment to analyze the performance of the proposed method. The results can be summarized as the following two points. Firstly, the proposed method performs as expected when the received field is dominated by the reflected modes but may be seriously degraded when the refracted modes dominate the received field. Secondly, the source in the detection area (i.e., the area enclosed by the connection line of the hydrophones) can be located unambiguously when the hydrophones are deployed as a regular polygon.

The remainder of this paper is organized as follows: The basic theories are presented in Section 2. In Section 3, the principle and the localization procedures are described. Numerical simulations and corresponding performance analysis are presented in Section 4, followed by a short conclusion in Section 5.

2. Basic Theory

2.1. Application Scene and Normal Mode Theory

The application scene is illustrated in Figure 1, where M distributed hydrophones are deployed at a common depth to record the signal radiated from a remote broadband source. The source is located at $p_s = [x_s, y_s, z_s]^T$. The emitting signal of the source is denoted as $s(t)$ in the time domain and $S(f)$ in the frequency domain. One assumes that the location of the mth hydrophone is $p_m = [x_m, y_m, z_r]^T$, in which z_r (positive downwards) denotes the common receiver depth. The received signal of the mth hydrophone can be denoted as $x_m(t)$ in the time domain and $p_m(f)$ in the frequency domain. What we need to do is to estimate $[x_s, y_s]^T$ based on the observed acoustic field $x_m(t)$ [or $p_m(f)$], $m = 1, 2, \ldots, M$, without either the angle- or time-synchronization information.

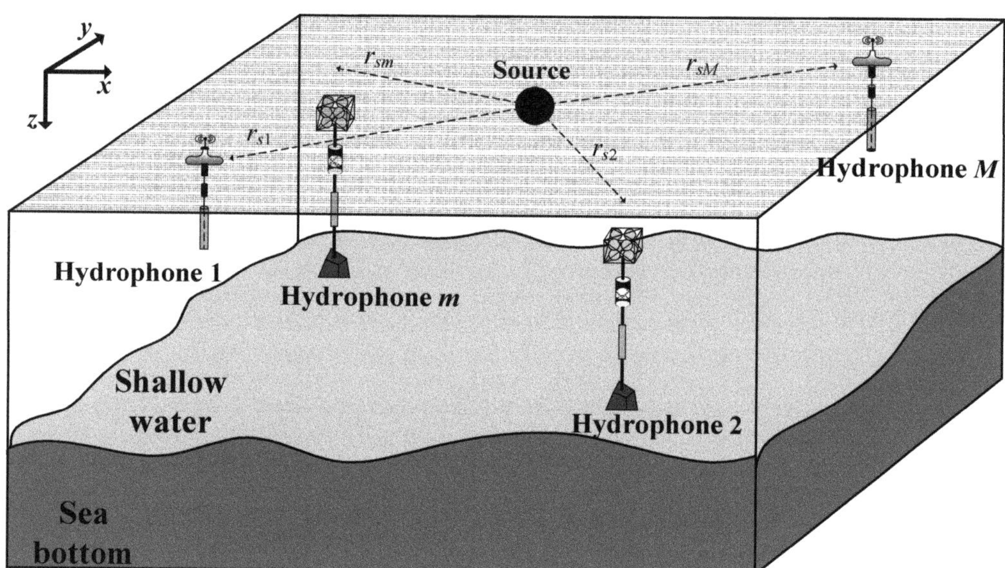

Figure 1. Application scene.

According to normal mode theory, the received field in shallow water can be expressed by a superposition of a series of normal modes. Under the specific boundary conditions, the eigenvalue and the eigenfunction of the normal mode can be determined by solving the wave equation. The received field $p_m(r_{sm}, z_m, f)$ on the mth hydrophone in the range-independent environment can be expressed as follows [2]:

$$
\begin{aligned}
p_m(r_{sm}, z_r, f) &= S(f) \sum_{n=1}^{N} A_{mn}(f) e^{jk_{rn}(f)r_{sm}} \\
&= S(f) \frac{je^{-j\pi/4}}{\rho(z_s)\sqrt{8\pi r_{sm}}} \sum_{n=1}^{N} \psi_n(z_s)\psi_n(z_r) \frac{e^{jk_{rn}(f)r_{sm}}}{\sqrt{k_{rn}(f)}},
\end{aligned}
\tag{1}
$$

where N denotes the number of the propagation modes, $k_{rn}(f)$ and $\psi_n(z)$ denote the eigenvalue (horizontal wavenumber) and eigenfunction (mode depth function) of the nth mode, respectively, and $A_{mn}(f)$ is the mode amplitude.

$$A_{mn}(f) = \frac{je^{-j\pi/4}}{\rho(z_s)\sqrt{8\pi r_{sm}}} \frac{\psi_n(z_s)\psi_n(z_r)}{\sqrt{k_{rn}(f)}}. \tag{2}$$

The medium density around the source is denoted as $\rho(z_s)$, and r_{sm} is the horizontal range between the source and the mth hydrophone.

$$r_{sm} = \sqrt{(x_s - x_m)^2 + (y_s - y_m)^2}. \tag{3}$$

2.2. Warping Transform of the Autocorrelation Function

On the basis of the normal mode representation [illustrated in Equation (1)], the ACF of the received field on the mth hydrophone can be shown as:

$$
\begin{aligned}
\xi_m(r_{sm}, z_r, t) &= \int_{-\infty}^{+\infty} p_m(r_{sm}, z_r, f) p_m^*(r_{sm}, z_r, f) df \\
&= \int_{-\infty}^{+\infty} |S(f)|^2 \sum_{n=1}^{N} \sum_{k=1}^{N} A_{mn}(f) A_{mn}^*(f) e^{j(k_{rn}(f)-k_{rk}(f))r_{sm}} e^{j2\pi ft} df
\end{aligned} \tag{4}
$$

Equation (4) contains the self-interference terms ($n = k$) and the cross-interference terms ($n \neq k$). The unilateral ACF (referred to as ACF below) is introduced herein to eliminate the useless self-interference terms [8], which is defined as:

$$
\begin{aligned}
\tilde{\xi}_m(r_{sm}, z_r, t) &= \int_{-\infty}^{+\infty} |S(f)|^2 \sum_{n=1}^{N} \sum_{k>n}^{N} A_{mn}(f) A_{mn}^*(f) e^{j(k_{rn}(f)-k_{rk}(f))r_{sm}} e^{j2\pi f(t-t_{m0})} df \\
&= \sum_{l=1}^{L} B_{ml}\left(\mu_l \frac{t}{\sqrt{t^2-t_{m0}^2}}\right) \frac{t_{m0}\sqrt{\mu_l}}{\sqrt{2}(t^2-t_{m0}^2)^{3/4}} e^{j2\pi\mu_l\sqrt{t^2-t_{m0}^2}},
\end{aligned} \tag{5}
$$

where

$$
\begin{aligned}
B_{ml}(f) &= |S(f)|^2 A_{mn}(f) A_{mk}^*(f) \\
\mu_l &= \sqrt{v_n^2 - v_k^2},
\end{aligned} \tag{6}
$$

In Equation (5), $t_{m0} = r_{sm}/c$ is the propagation time, c is the sound speed, v_n is the characteristic frequency of the nth mode, μ_l denotes the interference characteristic frequency between the nth and kth mode, $l = 0, 1, 2, \ldots, L$ $L = C_N^2$ denotes the number of the possible combinations, and the superscript * denotes the complex conjugation operation.

As seen in Equation (5), the time-dependent phase of the signal ACF is $2\pi\mu_l\sqrt{t^2 - t_{m0}^2}$. Given a known source location, the warping transform in the time domain can be applied to $\tilde{\xi}_m(r_{sm}, z_r, t)$, based on the resampling function.

$$h_m(t) = \sqrt{t^2 + t_{m0}^2}. \tag{7}$$

The output can be shown as:

$$(W_h\tilde{\xi}_m)(t, r_{sm}) = \sum_{l=1}^{L} \left|\frac{\partial h_m(t)}{\partial t}\right|^{1/2} B_{ml}\left(\mu_l\frac{\sqrt{t^2+t_{m0}^2}}{t}\right) \frac{t_{m0}\sqrt{\mu_l}}{\sqrt{2}t^{3/2}} e^{j2\pi\mu_l t}. \tag{8}$$

It is obviously shown in Equation (8) that the spectrum of the warped ACF (named as F_TW_T spectrum) supplies a stationary monochromatic output with the intrinsic frequency μ_l. Here, the interference characteristic frequency μ_l also bears the property of frequency invariability [8].

As shown in Equation (7), source range is involved in the warping transform. Actually, if the source range is correctly given, the obtained interference characteristic frequencies will only be determined by the environment and is independent of the source-receiver geometry. On the contrary, with the wrong source range information, the obtained interference characteristic frequencies will deviate from the standard values. Following the conclusions in Ref. [8], if the true source range is r_s, while one conducts the warping transform with an assumed range r_a, then the obtained F_TW_T spectrum will peak at the biased interference characteristic frequencies as follows:

$$\mu_{al} \approx \sqrt{\frac{r_s}{r_a}}\mu_l,\tag{9}$$

where μ_l is the standard interference characteristic frequencies, which only relates to the environment, and μ_{al} is the biased counterpart when using the incorrect source range in Equation (7).

3. Proposed Localization Method

In the case that the source is detected by multiple spatially distributed hydrophones simultaneously, its location can then be determined by mutually matching the F_TW_T spectra from different hydrophones. Specifically, assuming a source location $p_{sa} = [x_{sa}, y_{sa}]^T$ (depth coordinates of the source and sensors omitted for brevity), one can calculate its ranges to the distributed hydrophones and then apply the warping transform to the ACF of each hydrophone. If the assumed source location is correct, the characteristics of the obtained F_TW_T spectra obtained from different hydrophones should be consistent, as all hydrophones can extract standard interference characteristic frequencies determined solely by the environment. Otherwise, if the assumed source location is incorrect, the abovementioned consistency will be disrupted. In other words, the accuracy of an assumed source location can be assessed by evaluating the consistency of the F_TW_T spectra from different hydrophones. The cost function is constructed as:

$$F(p_{sa}) = \exp\left[\sum_{m=1}^{M}\sum_{n>m}^{M}\frac{\int_f FW_m(p_{sa},f)FW_n(p_{sa},f)df}{\sqrt{\int_f FW_m(p_{sa},f)df\int_f FW_n(p_{sa},f)df}}\right],\tag{10}$$

where p_{sa} is the assumed source location, $FW_m(p_{sa}, f)$ denotes the F_TW_T spectrum for the mth hydrophone that can be calculated by applying Fourier/wavelet transform to the corresponding warping output $(W_h\widetilde{\zeta}_m)(t, \|p_{sa} - p_m\|)$, and $\|\cdot\|$ is the Euclidean norm.

Overall, the proposed method can be summarized in the following seven steps:

(1) Deploy M hydrophones at a common depth to record the signal radiated by a broadband source. The location and received signal of the mth hydrophone are denoted as $p_m = [x_m, y_m]^T$ and $x_m(t)$, respectively;
(2) Calculate the unilateral ACF denoted as $\widetilde{\zeta}_m(t)$;
(3) Divide the area of interest into grid points, denoted as $[x_i, y_j]$, $i = 1, 2, \ldots, L_x, j = 1, 2, \ldots, L_y$, where L_x and L_y are the number of the grid on the x and y axes, respectively;
(4) Calculate the range between p_{sa} and p_m for each grid point $p_{sa} = [x_i, y_j]^T$

$$\begin{aligned}r_{am} &= \|p_{sa} - p_m\| \\ &= \sqrt{(x_i - x_m)^2 + (y_j - y_m)^2},\end{aligned}\tag{11}$$

and apply the warping transform to $\widetilde{\zeta}_m(t)$ based on r_{am} to obtain the warped ACF $(W_h\widetilde{\zeta}_m)(t, r_{am})$. The resampling function of the warping is given as:

$$h_{am}(t) = \sqrt{t^2 + \left(\frac{r_{am}}{c_a}\right)^2},\tag{12}$$

where c_a is the sound speed used in the warping transform;

(5) Apply the Fourier/wavelet transform to $(W_h\tilde{\xi}_m)(t, r_{am})$ to calculate the F_TW_T spectrum $FW_m\,(p_{sa}, f)$;

(6) Calculate the cost function Equation (10) based on the F_TW_T spectra obtained by all M hydrophones;

(7) Conduct steps (4)–(6) for each scanning point to obtain the localization ambiguity surface and determine the source location by the maximum point.

The diagram of the proposed method is shown in Figure 2.

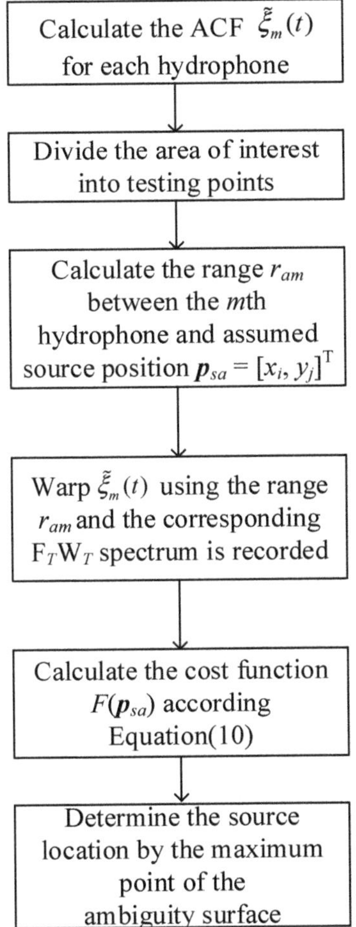

Figure 2. The diagram the proposed method.

4. Simulation Demonstration

4.1. Effectiveness Verification

This section conducts numerical simulations in shallow water to validate the proposed method. A Pekeris waveguide, illustrated in Figure 3a, is considered in the simulation. The water depth is $H = 100$ m with a constant sound speed $c = 1500$ m/s and a constant density $\rho_1 = 1.0$ g/cm^3. The seabed is modeled as a soft half space. The density, sound speed, and attenuation coefficient are $\rho_2 = 1.0$ g/cm^3, $c_2 = 1600$ m/s, $\alpha = 0.14$ dB/λ, respectively.

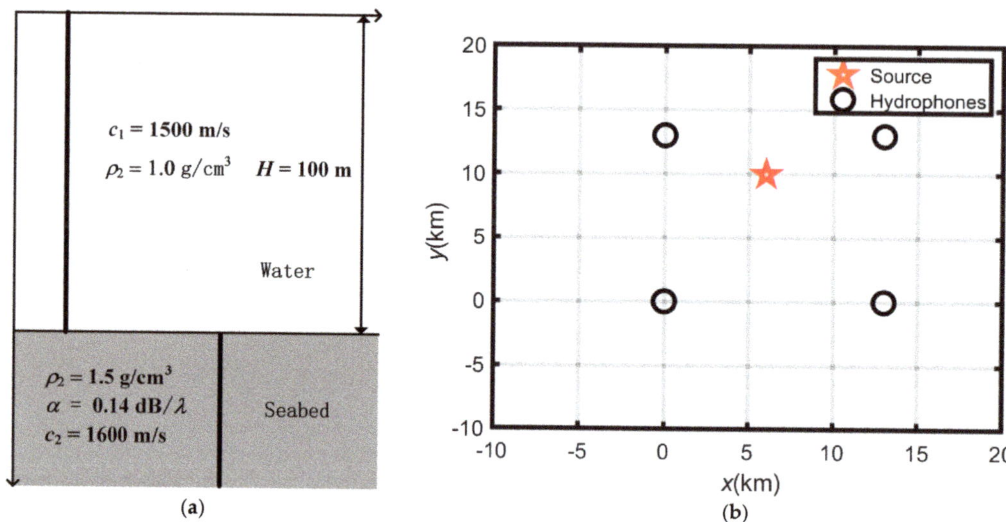

Figure 3. Simulation Scene: (**a**) a Pekeris waveguide and (**b**) the top view of the source and distributed hydrophones.

The top view of the simulation scene is illustrated in Figure 3b. The source is located at $p_s = [6, 10]^T$ km. The four hydrophones are at $p_1 = [0, 0]^T$ km, $p_2 = [13, 0]^T$ km, $p_3 = [0, 13]^T$ km, and $p_4 = [13, 13]^T$ km, respectively. Since the modal depth function varies with the receiving depth, the method requires all hydrophones to be deployed at the same depth. Both the source and the hydrophones are deployed at a depth of 20 m. The signal emitted by the source is modeled as a linear frequency modulation (LFM) signal. The duration and modulation band of the source signal are 3 s and [50, 150] Hz, respectively. The received field on the hydrophones is calculated by the KRAKEN normal mode code [25] with signal-to-noise (SNR) being set at 10 dB.

Figure 4 presents the normalized $F_T W_T$ spectra from the distributed hydrophones for the scenario described in Figure 3b. Figure 4a,b correspond to the cases where the source location is correctly and incorrectly assumed, respectively. As can be seen, if the source location is correctly assumed (i.e., Figure 4a), the $F_T W_T$ spectra for the different hydrophones shows the same shape. Otherwise, the consistency among the $F_T W_T$ spectra will be disrupted as shown in Figure 4b. The results in Figure 4 convincingly demonstrate the feasibility of the proposed method.

Divide the area of interest (i.e., -10 km $< x < 20$ km, -10 km $< y < 20$ km) into grids with a step of 0.4 km and calculate the ambiguity surface using the algorithm shown in Figure 2. The sound speed used in the warping transform is $c_a = 1500$ m/s. The obtained localization ambiguity surface is shown in Figure 5, wherein the cost function is normalized in decibels with a dynamic range of 10 dB. In Figure 5, the black asterisk donates the true source location, and the yellow circles represent the locations of the hydrophones. As can be seen, the ambiguity surface clearly peaks at the location of the source, demonstrating the effectiveness of the proposed method. In this simulation example, the localization result is $[6.17, 10]^T$ km with a localization error of 0.17 km. It is worth mentioning that the basic idea of the method is to mutually match the warped ACF obtained by distributed hydrophones, and the warping processing has nothing to do with the source depth. In other words, the method proposed in this paper can estimate the two-dimensional position of the sound source in the xoy plane when the source depth is unknown.

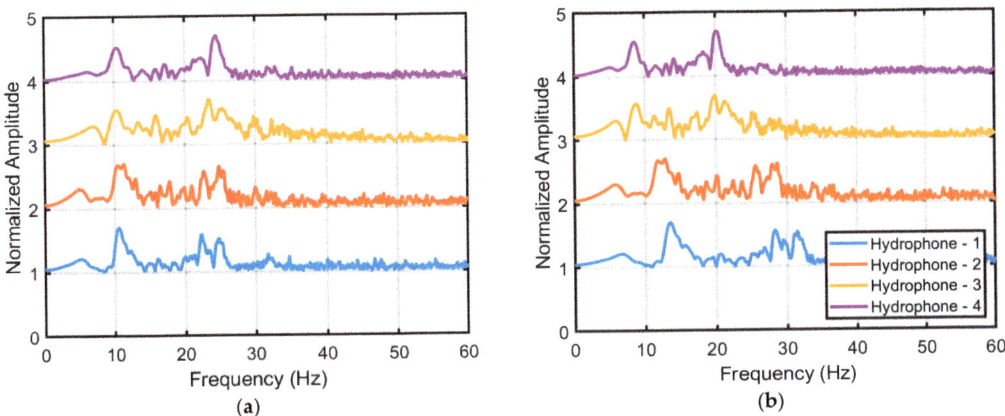

Figure 4. Normalized FTWT spectra from distributed hydrophones. (**a**) With the correct source location; (**b**) with an incorrect source location, i.e., $\boldsymbol{p}_{sa} = [5, 5]^{T}$ km.

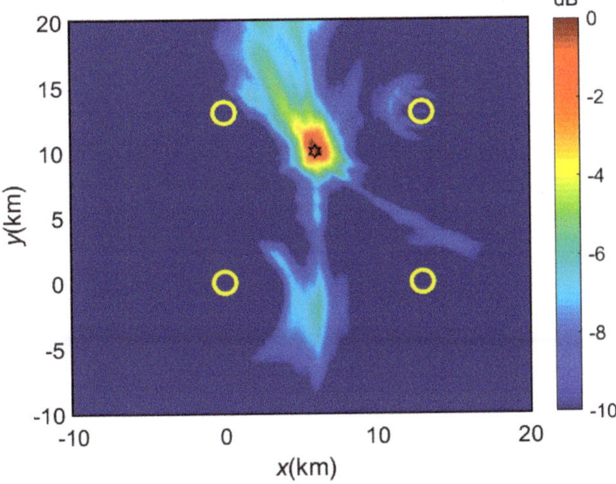

Figure 5. Localization ambiguity surface. The black asterisk and yellow circles indicate the true source location and the location of the hydrophones, respectively. The black asterisk and yellow circles indicate the true source location and the location of the hydrophones, respectively.

The result of the warping transform is closely related to the assumed sound speed, i.e., a larger (or smaller) c_a is equivalent to a smaller (or larger) assumed horizontal range r_{sam}, thus resulting in a larger (or smaller) interference characteristic frequency (as shown in Equation (9)). Nevertheless, despite the influence of sound-speed mismatch on the result of the $\mathrm{F}_T\mathrm{W}_T$ spectrum, the relative positions of the $\mathrm{F}_T\mathrm{W}_T$ spectrum peak among sensors are independent of c_a. Therefore, the proposed method still works as intended even if c_a is mismatched with the true sound speed. The simulation results shown in Figure 6 confirm this speculation, where higher and lower sound speeds are used, respectively.

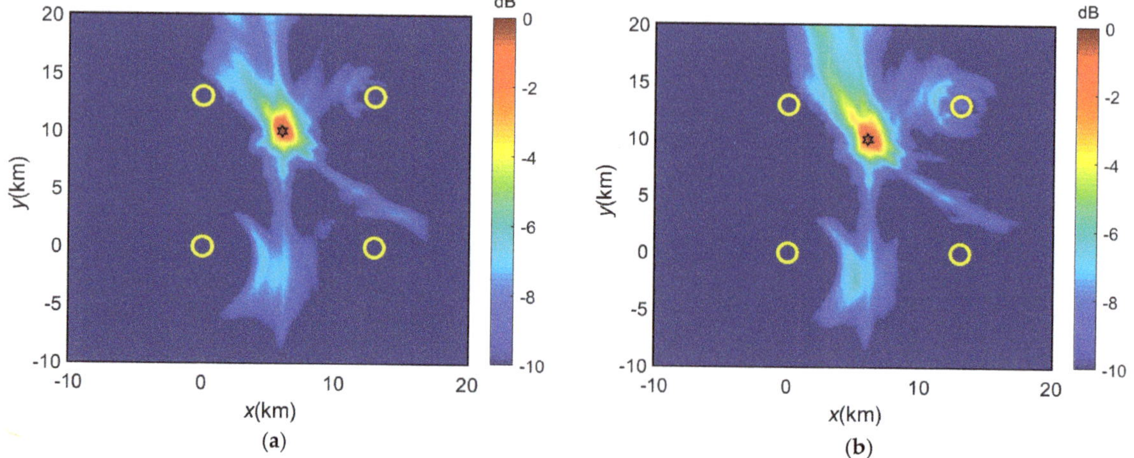

Figure 6. Localization ambiguity surfaces with mismatched sound speed: (**a**) lower sound speed, i.e., $c_a = 1480$ m/s; (**b**) higher sound speed, i.e., $c_a = 1520$ m/s. The black asterisk and yellow circles indicate the true source location and the location of the hydrophones, respectively.

4.2. Performance Analysis

- Hydrophone Distribution

The hydrophone distribution is one of the important factors that affects the performance of the method. In this section, the localization ambiguity under different distributions of hydrophones will be analyzed through numerical simulations. According to the theoretical analysis in Section 3, the cost function reaches its maximum when the search grid point is identical to the source position ($p_{sa} = p_s$). However, according to Equation (9), the cost function will also reach its maximum (equivalent to $p_{sa} = p_s$) if the following equation holds:

$$\frac{r_{s1}}{r_{sa1}} = \frac{r_{s2}}{r_{sa2}} = \cdots = \frac{r_{sM}}{r_{saM}}. \tag{13}$$

Therefore, if there exists more than one point that satisfies Equation (13) in the area of interest, the ambiguity surface will exhibit more than one peak, and the localization result will become ambiguous. For example, if only two hydrophones are used, the points satisfying Equation (13) form a circle characterized by a radius of $R = QL/|1 - Q^2|$, where Q is the distance ratio between the source and the two hydrophones, and L is the distance between these two hydrophones. Thus, the proposed method becomes invalid when only two hydrophones are used, as shown in Figure 7a. A fuzzy band emerges in the localization ambiguity surface and the true source location cannot be identified.

When three hydrophones are deployed in a linear distribution, localization ambiguity arises, similar to the problem of port and starboard ambiguity for the DOA estimation using a linear array. The corresponding simulation result is shown in Figure 7b. As seen, a fuzzy source symmetric to the true counterpart appears. Changing the distribution of the hydrophones to a non-colinear arrangement allows unambiguous source localization, as shown in Figure 7c. Therefore, in practical applications, colinear distribution of hydrophones should be avoided. Finally, it is noteworthy that the colinear distribution can locate the source unambiguously when and only when the source is also on the same line, as shown in Figure 7d. But the performance of the method in this case is somewhat unsatisfying.

Define the area enclosed by the hydrophones as the detection area for the nonlinear distribution. The localization ambiguity of the proposed method can be numerically analyzed based on Equation (13). Specifically, one can count the number of the points (except p_s) that satisfy Equation (13) in the detection area. The simulation results under different numbers of the hydrophones are shown in Figure 8, wherein the hydrophones are deployed through a regular M polygon. As seen with a M-polygon distribution, the proposed method can provide an unambiguous localization in the detection area.

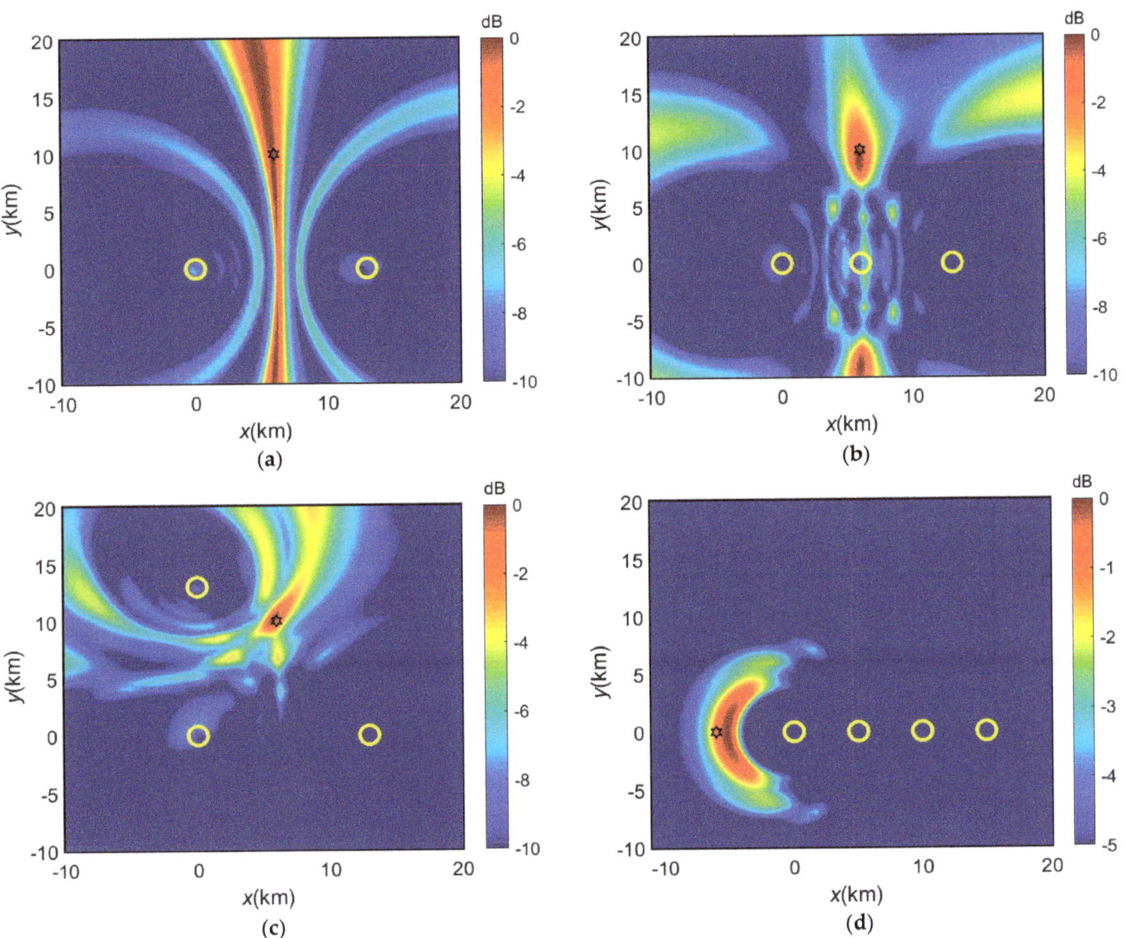

Figure 7. The localization results under different hydrophones distributions: (**a**) two hydrophones; (**b**) three hydrophones with a colinear distribution; (**c**) three hydrophones with a non-colinear distribution; (**d**) four hydrophones with a colinear distribution and the source is also at this line. The black asterisk and yellow circles indicate the true source location and the location of the hydrophones, respectively. The black asterisk and yellow circles indicate the true source location and the location of the hydrophones, respectively.

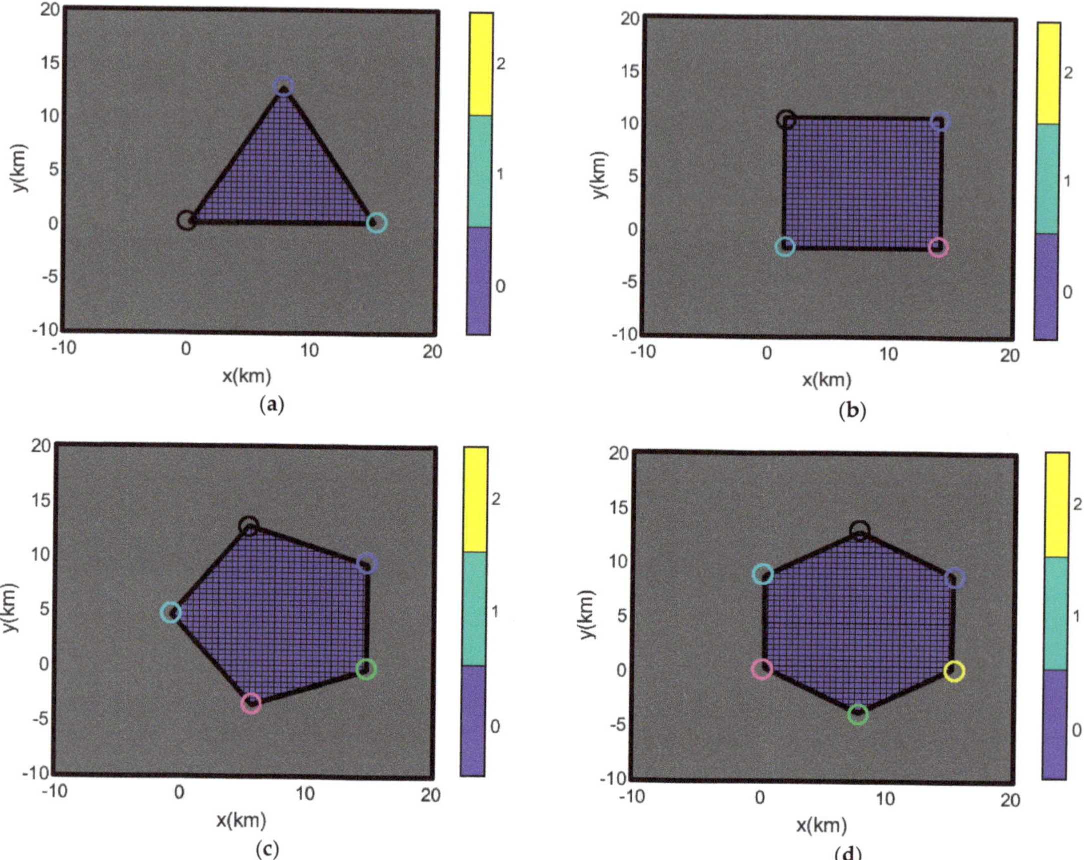

Figure 8. The theoretical localization ambiguity of the method in the detection area under different number of hydrophones. (**a**–**d**) correspond to the cases where the number of the hydrophones is 3–6. The color bar indicates the number of the fuzzy peaks (i.e., the number of the points (except p_s) that satisfy Equation (13)). The circles indicate the location of the hydrophones.

The above analysis leads to the conclusion that at least three sensors must be deployed to support the proposed method. In the colinear distribution scenario, a fuzzy peak symmetric to the true source will appear, except when the source is located on the line linking the hydrophones. When the hydrophones are deployed with a regular M polygon, there is no localization ambiguity in the detection area, i.e., the area enclosed by the hydrophones.

- Hydrophone Depth

The simulation environment in Section 4.1 is characterized by an isovelocity waveguide, where the reflected modes dominate the received field. However, in the non-isovelocity waveguide, the types of modes that dominate the received field depend on the depths of both the source and receiver. Different types of modes present different interference characteristics, which will affect the applicability of the proposed method. In this section, we take a classical thermocline waveguide as an example to analyze the effect of source and receiver depth on the performance of the proposed method.

The thermocline waveguide used in simulations is shown in Figure 9, which exhibits a classical downward-refracting sound speed profile (SSP) with a mixed layer depth down to 30 m. The sound speed reduces linearly from 1520 m/s to 1480 m/s as the depth changes from 30 m to 70 m. The sound speed, density, and the attenuation coefficient of the seabed are $\rho_2 = 1.0$ g/cm^3, $c_2 = 1600$ m/s, and $\alpha = 0.14$ dB/λ, respectively. Four different source–receiver configurations (a)–(d) are displayed in Figure 9. Assuming the source location is the same as the case in Figure 3b, the corresponding localization results are shown in Figure 10, where Figure 10a–d correspond to the four cases in Figure 9, one by one. The black asterisk and yellow circles indicate the true source location and the location of the hydrophones, respectively.

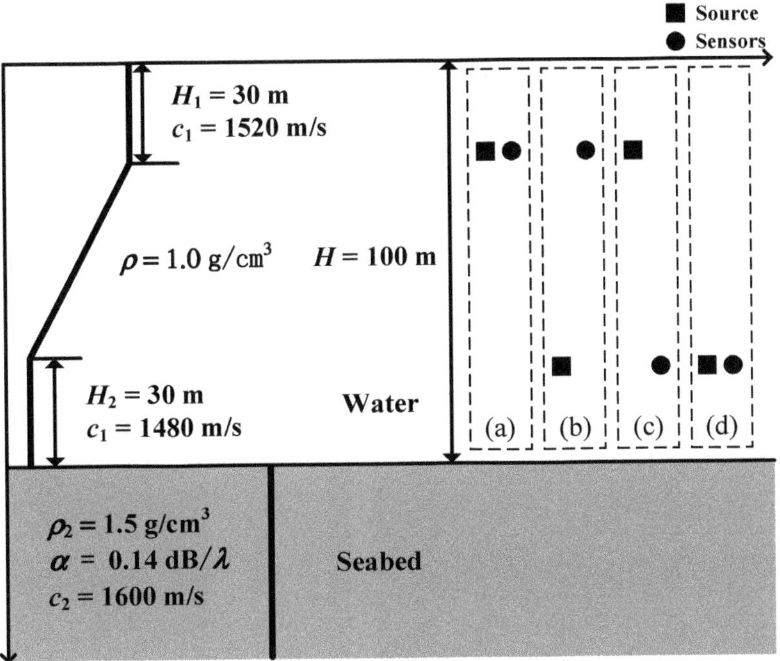

Figure 9. The thermocline environment and different source-sensors depths configuration: (**a**) $z_s = 20$ m, $z_r = 20$ m, (**b**) $z_s = 70$ m, $z_r = 20$ m, (**c**) $z_s = 20$ m, $z_r = 70$ m, and (**d**) $z_s = 70$ m, $z_r = 70$ m.

As shown in Figure 10, the proposed method works as expected in the following two cases: (1) both the source and hydrophones are above the thermocline layer; (2) the source is above (or below) the thermocline layer while the hydrophones are below (or above) the thermocline layer. The proposed method becomes invalid if both the source and the hydrophones are below the thermocline layer.

Normal mode theory could be used to account for these simulation results. The modes distribution at 125 Hz for the waveguide in Figure 9 is depicted in Figure 11, which involves (a) the modes depth function and (b) the relationship between phase and group slowness. As can be seen in Figure 11, when both the source and hydrophones are deployed above the thermocline layer, or the source is above (or below) the thermocline layer while the hydrophones are below (or above) this layer, the received field is predominantly influenced by the higher order modes (i.e., $n \geq 3$). When both the source and hydrophones are below the thermocline layer, the lower order (i.e., 1st and 2nd) modes dominate the field. The higher order modes correspond to the reflected modes while the lower order modes are the refracted modes. Following the conclusion in Ref. [26], the time warping function shown in Equation (7) is exclusively associated with the reflected modes, rendering it ineffective for the refracted modes. Therefore, the proposed method works well in the case that the reflected modes dominate the received field (i.e., the source and/or hydrophones are above the thermocline layer) and becomes invalid if the received field is dominated by the refracted modes (i.e., both the source and hydrophones are below the thermocline layer). As a result, it is recommended to deploy the hydrophones above the thermocline layer in the thermocline waveguide to improve the applicability of the proposed method.

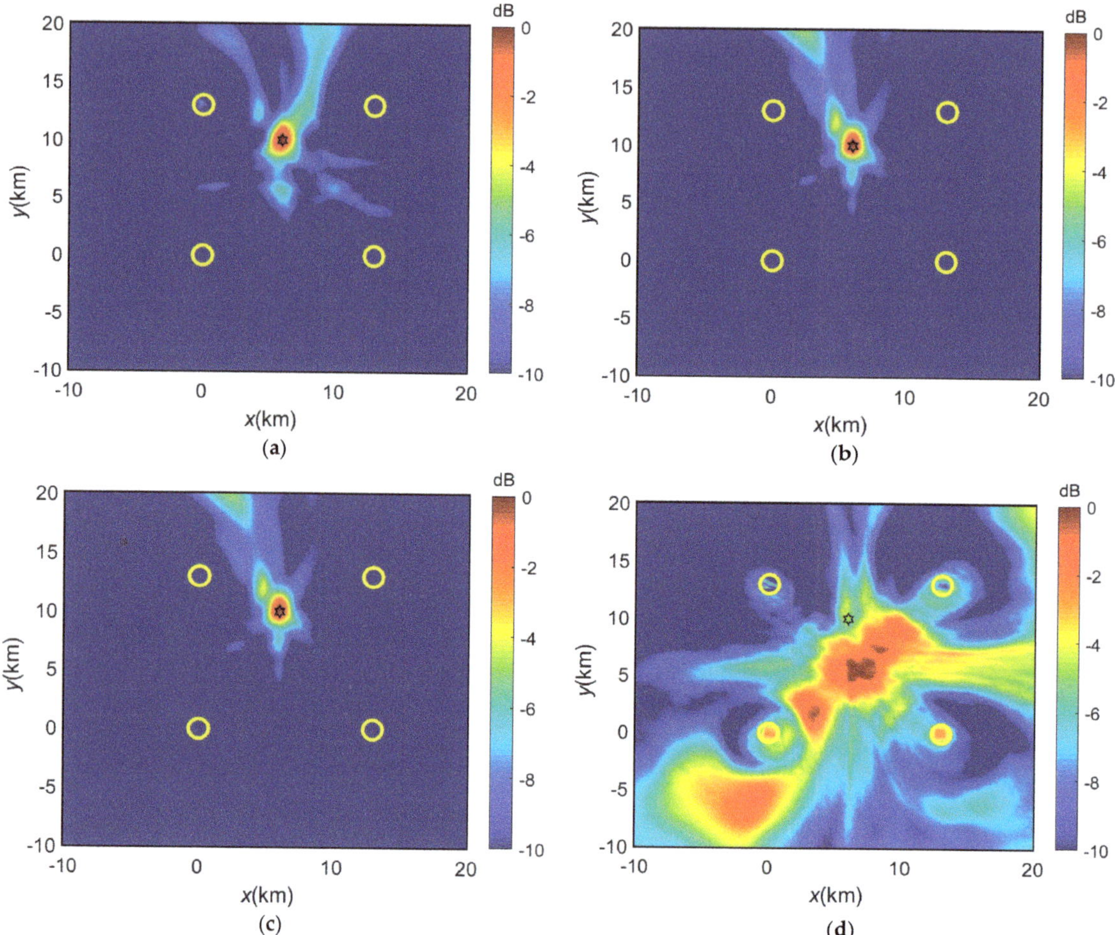

Figure 10. The localization results with different source-sensors depths configurations under the thermocline waveguide: (**a**) $z_s = 20$ m, $z_r = 20$ m, (**b**) $z_s = 70$ m, $z_r = 20$ m, (**c**) $z_s = 20$ m, $z_r = 70$ m, and (**d**) $z_s = 70$ m, $z_r = 70$ m. The black asterisk and yellow circles indicate the true source location and the location of the hydrophones, respectively. Due to the general principle of reciprocity, scene (**c**) has the same result as (**b**).

Similar simulations have been conducted to analyze the applicability of the proposed method under the positive/negative-gradient SSP waveguide. The results can be concluded as follows: (1) Modes generated under a negative-gradient SSP waveguide presents similar characteristics to those in the thermocline waveguide. Therefore, it is suggested to deploy the hydrophones near the sea surface to guarantee the effectiveness of the proposed method; (2) The depth function of the refracted mode generated under a positive-gradient SSP waveguide remains high amplitude near the sea surface while decaying exponentially near the seabed. Therefore, it is suggested to deploy the hydrophones near the seabed. All in all, in a non-isovelocity environment, hydrophones should be deployed at the depth with higher sound speed to guarantee the effectiveness of the proposed method.

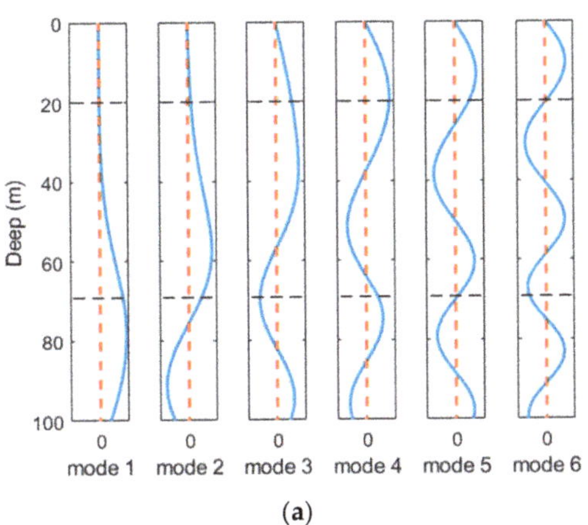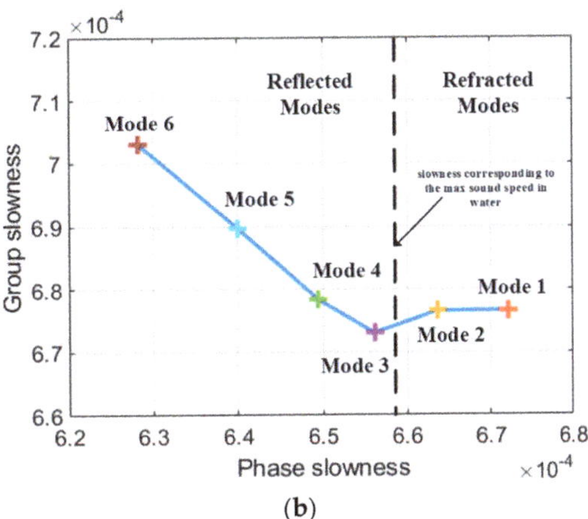

Figure 11. The (**a**) modes depth function and (**b**) modes group (phase) slowness in the waveguide shown in Figure 9.

5. Conclusions

In shallow water, the mode interference characteristic frequencies present an invariability property that is independent of the source–receiver geometry. Based on this phenomenon, a passive model-independent broadband source localization method utilizing distributed hydrophones is proposed in this paper. The basic idea of the proposed method is to mutually match the $F_T W_T$ spectra from different hydrophones and determine the source location by verifying the invariability of the mode interference characteristic frequencies. Specifically, for an assumed source location, the time warping transform is firstly applied to the signal ACF to extract the interference characteristic frequencies for each hydrophone and then, a cost function is calculated to verify the consistency of the $F_T W_T$ spectra from different hydrophones. The maximum point of the cost function indicates the location of the source.

Numerical simulations are conducted to demonstrate the performance of the method and the results can be concluded as follows: (1) the proposed method can locate the broadband source successfully in the classical Pekeris waveguide; (2) the method works as advertised in case that the reflected modes dominate the received field in the non-isovelocity environment (e.g., the thermocline waveguide). Thus, it is suggested to deploy the hydrophones at a depth with a higher sound speed to guarantee the effectiveness of the method; (3) the localization ambiguity can be avoided in the detection area (i.e., area enclosed by the hydrophones) when the hydrophones are distributed through a regular M polygon ($M \geq 3$).

The proposed method theoretically circumvents the environmental mismatch problem by not relying on prior environment information. In addition, benefitting from exploiting the interference characteristics of the acoustic field, the proposed methods require neither the angle- nor time-synchronization information in its localization procedure. However, the drawbacks of this method are also obvious. Since the modal depth function varies with the receiving depth, the method in this paper requires all hydrophones to be deployed at the same depth. Meanwhile, the hydrophones should not be colinearly distributed.

In our work getting under the way, it is expected to analyze the localization ambiguity in detail with an arbitrary hydrophone distribution. Moreover, due to the fact that the method becomes invalid when the acoustic field is dominated by the refracted modes, how to exploit the interference characteristics of these refracted modes to locate the source is still an ongoing work.

Author Contributions: H.L.: Conceptualization, supervision, writing—review, funding acquisition. J.H.: Writing—editing, data curation. Z.X.: methodology, software, investigation, writing—original draft. K.Y.: Resources, project administration. J.Q.: Validation, funding acquisition. All authors have read and agreed to the published version of the manuscript.

Funding: This research was funded by the State Key Laboratory of Acoustics, Chinese Academy of Sciences (Grant No. SKLA202307), the National Natural Science Foundation of China (Grant No. 61901383 and No. 52231013), the Fundamental Research Funds for the Central Universities (Grant No. 3102021HHZY030011), and the Youth Innovation Promotion Association, Chinese Academy of Sciences (No. 2021023).

Data Availability Statement: Data will be made available on request.

Conflicts of Interest: The authors declare no conflicts of interest.

References

1. Livingston, E. Effects of sound-speed mismatch in the lower water column on matched-field processing. *J. Acoust. Soc. Am.* **1992**, *91*, 2363–2364. [CrossRef]
2. Jensen, F.B.; Kuperman, W.A.; Poter, M.B.; Schmidt, H. *Computational Ocean Acoustics*, 2nd ed.; Springer: New York, NY, USA, 2011.
3. Cockrell, K.L.; Schmidt, H. Robust passive range estimation using the waveguide invariant. *J. Acoust. Soc. Am.* **2010**, *127*, 2780–2789. [CrossRef]
4. Song, H.C.; Cho, C. Array invariant-based source localization in shallow water using a sparse vertical array. *J. Acoust. Soc. Am.* **2017**, *141*, 183–188. [CrossRef] [PubMed]
5. Song, H.C.; Cho, C.; Byun, G.; Kim, J.S. Cascade of blind deconvolution and array invariant for robust source-range estimation. *J. Acoust. Soc. Am.* **2017**, *141*, 3270–3273. [CrossRef] [PubMed]
6. Song, H.C.; Cho, C. The relation between the waveguide invariant and array invariant. *J. Acoust. Soc. Am.* **2015**, *138*, 899–903. [CrossRef] [PubMed]
7. Qi, Y.-B.; Zhou, S.; Ren, Y.; Liu, J.-J.; Wang, D.-J.; Feng, X.-Q. Passive source range estimation with a single receiver in shallow water. *Acta Acust.* **2015**, *40*, 144–152.
8. Zhou, S.; Qi, Y.; Ren, Y. Frequency invariability of acoustic field and passive source range estimation in shallow water. *Sci. China-Phys. Mech. Astron.* **2014**, *57*, 225–232. [CrossRef]
9. Julien, B.; Aaron, T.M.; Susanna, B.B.; Katherine, K.; Michael, M.A. Range estimation of bowhead whale (Balaena mysticetus) calls in the Arctic using a single hydrophonea. *J. Acoust. Soc. Am.* **2014**, *136*, 145–155.
10. Doğançay, K. Bearings-only target localization using total least squares. *Signal Process.* **2005**, *85*, 1695–1710. [CrossRef]
11. Dogancay, K.; Ibal, G. Instrumental Variable Estimator for 3D Bearings-Only Emitter Localization. In Proceedings of the 2005 International Conference on Intelligent Sensors, Sensor Networks and Information Processing, Melbourne, Australia, 5–8 December 2005; pp. 63–68.
12. Hawkes, M.; Nehorai, A. Wideband source localization using a distributed acoustic vector-sensor array. *IEEE Trans. Signal Process.* **2003**, *51*, 1479–1491. [CrossRef]
13. Kaplan, L.M.; Le, Q.; Molnar, N. Maximum likelood methods for bearings-only target localization. In Proceedings of the 2001 IEEE International Conference on Acoustics, Speech, and Signal Processing (ICASSP), Salt Lake City, UT, USA, 7–11 May 2001; Volume 5, pp. 3001–3004.
14. Kaplan, L.M.; Le, Q. On exploiting propagation delays for passive target localization using bearings-only measurements. *J. Frankl. Inst.* **2005**, *342*, 193–211. [CrossRef]
15. Wang, Z.; Luo, J.; Zhang, X. A novel location-penalized maximum likelihood estimator for bearing-only target localization. *IEEE Trans. Signal Process.* **2012**, *60*, 6166–6181. [CrossRef]
16. Luo, J.; Shao, X.; Peng, D.; Zhang, X. A Novel Subspace Approach for Bearing-Only Target Localization. *IEEE Sens. J.* **2019**, *19*, 8174–8182. [CrossRef]
17. Chan, Y.T.; Ho, K.C. A simple and efficient estimator for hyperbolic location. *IEEE Trans. Signal Process.* **1994**, *42*, 1905–1915. [CrossRef]
18. Foy, W.H. Position-Location Solutions by Taylor-Series Estimation. *IEEE Trans. Aerosp. Electron. Syst.* **1976**, *AES-12*, 187–194. [CrossRef]
19. Ho, K.C. Bias Reduction for an Explicit Solution of Source Localization Using TDOA. *IEEE Trans. Signal Process.* **2012**, *60*, 2101–2114. [CrossRef]
20. Schau, H.; Robinson, A. Passive source localization employing intersecting spherical surfaces from time-of-arrival differences. *IEEE Trans. Acoust. Speech Signal Process.* **1987**, *35*, 1223–1225. [CrossRef]
21. Laurinolli, M.H.; Hay, A.E. Localisation of right whale sounds in the workshop bay of fundy dataset by spectrogram cross-correlation and hyperbolic fixing. *Can. Acoust.* **2004**, *32*, 132–136.
22. Desharnais, F.; Côté, M.; Calnan, C.J.; Ebbeson, G.R.; Thomson, D.J.; Collison, N.E.; Gillard, C.A. Right whale localisation using a downhill simplex inversion scheme. *Can. Acoust.* **2004**, *32*, 137–145.

23. Bonnel, J.; Touze, G.L.; Nicolas, B.; Mars, J.I.; Gervaise, C. Automatic and passive whale localization in shallow water using gunshots. In Proceedings of the OCEANS 2008, Quebec City, QC, Canada, 15–18 September 2008; pp. 1–6.
24. Gervaise, C.; Vallez, S.; Stephan, Y.; Simard, Y. Robust 2D localization of low-frequency calls in shallow waters using modal propagation modelling. *Can. Acoust.* **2008**, *36*, 153–159.
25. Porter, M.B. *The KRAKEN Normal Mode Program*; Technical Report; Naval Research Laboratory: Washington, DC, USA, 1992.
26. Qi, Y.; Zhou, S.; Zhang, R. Warping transform of the refractive normal mode in a shallow water waveguide. *Acta Phys. Sin.* **2016**, *65*, 134301. [CrossRef]

Technical Note

Gridless Underdetermined DOA Estimation for Mobile Agents with Limited Snapshots Based on Deep Convolutional Generative Adversarial Network

Yue Cui [1], Feiyu Yang [1], Mingzhang Zhou [2,3], Lianxiu Hao [4], Junfeng Wang [2,5,*], Haixin Sun [2,3], Aokun Kong [1] and Jiajie Qi [1]

[1] College of Computer and Information Engineering, Tianjin Normal University, Tianjin 300387, China; cuiyue@tjnu.edu.cn (Y.C.); 2111090052@stu.tjnu.edu.cn (F.Y.); 2211090049@stu.tjnu.edu.cn (A.K.); 2111090053@stu.tjnu.edu.cn (J.Q.)
[2] Key Laboratory of Southeast Coast Marine Information Intelligent Perception and Application, Ministry of Natural Resources, Zhangzhou 363000, China; mzzhou@xmu.edu.cn (M.Z.); hxsun@xmu.edu.cn (H.S.)
[3] School of Informatics, Xiamen University, Xiamen 361005, China
[4] School of Electrical and Information Engineering, Tianjin University, Tianjin 300072, China; tjhlx@163.com
[5] School of Integrated Circuit Science and Engineering, Tianjin University of Technology, Tianjin 300384, China
* Correspondence: jfwang@tjut.edu.cn

Abstract: Deep learning techniques have made certain breakthroughs in direction-of-arrival (DOA) estimation in recent years. However, most of the current deep-learning-based DOA estimation methods view the direction finding problem as a grid-based multi-label classification task and require multiple samplings with a uniform linear array (ULA), which leads to grid mismatch issues and difficulty in ensuring accurate DOA estimation with insufficient sampling and in underdetermined scenarios. In order to solve these challenges, we propose a new DOA estimation method based on a deep convolutional generative adversarial network (DCGAN) with a coprime array. By employing virtual interpolation, the difference co-array derived from the coprime array is extended to a virtual ULA with more degrees of freedom (DOFs). Then, combining with the Hermitian and Toeplitz prior knowledge, the covariance matrix is retrieved by the DCGAN. A backtracking method is employed to ensure that the reconstructed covariance matrix has a low-rank characteristic. We performed DOA estimation using the MUSIC algorithm. Simulation results demonstrate that the proposed method can not only distinguish more sources than the number of physical sensors but can also quickly and accurately solve DOA, especially with limited snapshots, which is suitable for fast estimation in mobile agent localization.

Keywords: direction-of-arrival (DOA) estimation; deep learning; matrix recovery

Citation: Cui, Y.; Yang, F.; Zhou, M.; Hao, L.; Wang, J.; Sun, H.; Kong, A.; Qi, J. Gridless Underdetermined DOA Estimation for Mobile Agents with Limited Snapshots Based on Deep Convolutional Generative Adversarial Network. *Remote Sens.* **2024**, *16*, 626. https://doi.org/10.3390/rs16040626

Academic Editor: Andrzej Stateczny

Received: 19 December 2023
Revised: 19 January 2024
Accepted: 22 January 2024
Published: 8 February 2024

1. Introduction

In the past few decades, direction-of-arrival (DOA) estimation has emerged a critical issue across various domains, including radar, sonar, mobile communication and localization. To perform DOA estimation in actual environments, researchers have conducted in-depth studies and developed two main types of methods: physical model-driven methods [1–5] and data-driven methods [6–8]. The DOA estimation method based on phase interferometry is proposed in [1] for real-time localization. This method can compute the DOA in real time with lightweight architecture and full-digital dedicated hardware. However, it has implications for phase ambiguity and phase error, and could only distinguish a low number of receivers, with no ability to accurately estimate more DOAs at the same time. The high-resolution DOA estimation methods, such as the multiple signal classification (MUSIC) algorithm in [2] and the estimation of signal parameters via rotational invariance techniques (ESPRIT) algorithm in [3], could estimate more DOAs of signals and achieve more accurate performance. Nevertheless, the number of DOAs they can distinguish is still limited by the number of physical sensors, and the computational complexity remains very high. To cope with a multipath environment, the forward/backward spatial smoothing (FBSS) algorithm is proposed in [4] to decorrelate the coherent signals, but its degrees of freedom (DOFs) are reduced, and the required SNR is slightly higher. In [5], the DOA estimation algorithm for coherent GPS signals not only employs Toeplitz decorrelation but also oblique projection to suppress noise at low SNR. The aforementioned physical model-driven methods usually require a number of snapshots and have high computational complexity and lengthy solution time. Moreover, those methods are based on a rigorous physical model; once in non-ideal conditions, such as a limited number of snapshots or model mismatch, their estimation performance would be degraded obviously. Among data-driven methods, deep neural network models have shown excellent performance and lower computational complexity. The literature [8] introduces a deep neural network (DNN) model, which exhibits robustness in the presence of defective arrays. In [9], the authors used a convolutional neural network (CNN) for DOA estimation in low SNR conditions. In [10], it is proved that the columns of the covariance matrix can be expressed as linear measurements of undersampling noise in the spatial spectrum, and a deep convolutional neural network (DCNN) was built using sparse priors. In response to the significant reduction in estimation accuracy of existing methods for a multipath environment, reference [11] proposes a phase enhancement model based on a CNN for coherent DOA estimation which improves DOA estimation accuracy by enhancing phase and reducing phase distortion. In addition, the authors evaluate the importance of the phase feature for DOA estimation accuracy and demonstrated that the amplitude feature is redundant for DOA estimation. In [12], residual neural networks (ResNet) were used to achieve DOA estimation in a single snapshot. In [13], deep learning was applied to DOA estimation in underwater acoustic arrays. In [14], the authors present a novel DOA estimation framework that utilizes a complex-valued deep learning technique. In [15], researchers used the upper triangular region data of the received signal covariance matrix for training, effectively reducing training complexity and accelerating training speed. In [16], an angle separation deep learning method is proposed to achieve near-real-time DOA estimation for coherent signal sources. Furthermore, the lightweight DNN DOA estimation method for array imperfection correction has lower computational complexity and faster running speed, making it suitable for real-time signal processing application [17]. In [18], deep residual learning was used to achieve wideband DOA estimation. In addition, the DOA estimation method based on unsupervised learning with sparse array employs ResNet, which can effectively cope with low SNR and few snapshots scenarios [19]. However, the aforementioned methods did not consider the underdetermined scenario.

With the continuous development of the Internet of Things (IoT) and Internet of Vehicles (IoV), the number of intelligent mobile agents is growing constantly. In the process of localization and communication, the number of estimated targets is often greater than the number of array sensors, which results in the frequent occurrence of underdetermined situations. Moreover, the mobile agents require fast calculation speed with limited snapshots, which places higher requirements on the running speed of the DOA estimation algorithm. However, most of the current deep-learning-based DOA estimation methods use CNN models [10,11,20], treating the direction finding problem as a multi-label classification task and requiring multiple samplings with a uniform linear array (ULA). The output of the network in these methods is the probability associated with each corresponding label. These methods not only suffer from grid mismatch problems but are also unable to distinguish all targets in underdetermined situations, which would decrease the estimation accuracy dramatically. In [20], the sparse array was adopted, and its covariance matrix was recovered from the first row using a CNN-based regression method. Then, the DOA was obtained with the Root-MUSIC algorithm from the recovered covariance matrix. This approach has the ability to cope with underdetermined situations but cannot guarantee the low-rank characteristic of the recovered covariance matrix, so its DOA estimation accuracy is constrained, especially with limited snapshots.

Therefore, in order to address the aforementioned challenges, a virtual ULA was constructed in this study by filling the virtual sensors into the difference co-array derived from the coprime array, which can obtain more DOFs and improve DOA estimation resolution. The deep convolutional generative adversarial network (DCGAN) was adopted to recover the data associated with the virtual sensors and rebuild the covariance matrix of the virtual ULA using the Hermitian and Toeplitz prior knowledge. In order to ensure the low-rank characteristic of the covariance matrix, the output data of the DCGAN were further processed using the low-rank matrix optimization algorithm. Finally, DOA estimation was performed using the MUSIC algorithm. The proposed method not only has the ability to cope with underdetermined scenarios but can also improve the accuracy and estimation speed with limited snapshots.

The remaining sections of this paper are organized as follows. Section 2 introduces the signal model. Section 3 elaborates on the structure and processing details of the proposed method. Section 4 describes the loss function used by the network and some important parameters. Section 5 provides experimental results. The last section summarizes the entire paper.

2. Signal Model

It is presumed that K far-field narrow-band source signals impinge onto an M-element array antenna ($K>M$), and the received signal at the array is given by:

$$X(t) = A(\theta)s(t) + \mathbf{n}(t), t = 1, 2, \ldots, T, \tag{1}$$

where θ, A, and T represent the source direction vector, array manifold matrix, and snapshot number, respectively. $s(t)$ and $\mathbf{n}(t)$ denote the spatial signal vector and additive Gaussian white noise vector at time t, respectively. The k-th column of the array manifold matrix A can be represented as $\mathbf{a}(\theta) = \left[e^{\frac{-j*2\pi u_1 d\sin(\theta)}{\lambda}}, e^{\frac{-j*2\pi u_2 d\sin(\theta)}{\lambda}}, \ldots, e^{\frac{-j*2\pi u_M d\sin(\theta)}{\lambda}} \right]^T$, where $u_i, (i = 1, 2, \ldots M)$ represents the i-th element position.

A coprime array is constructed with two sparse uniform linear sub-arrays with $I + J - 1$ sensors, the first sub-array being $[0, Id, 2Id, \ldots, (J-1)Id]$ and the second sub-array being $[0, Jd, 2Jd, \ldots, (I-1)Jd]$, where I and J are coprime integers. The two sub-arrays do not overlap except for position 0. The structure of the coprime array is depicted in Figure 1a. The covariance matrix of the received signal $X(t)$ with the coprime array can be expressed as

$$R_X = E\left[X(t)X^H(t) \right] = \sum_{k=1}^{K} p_k \mathbf{a}(\theta_k)\mathbf{a}^H(\theta_k) + \sigma^2 I, \tag{2}$$

where p_k denotes the power of the k-th source signal, and I denotes the identity matrix. Afterward, by vectorizing the covariance matrix R_X and taking the distinct elements, the equivalent virtual signal of the difference co-array can be obtained as

$$\mathbf{y}_d = A_d \mathbf{p} + \sigma^2 \mathbf{i}, \tag{3}$$

where $A_d = [\mathbf{a}^*(\theta_1) \otimes \mathbf{a}(\theta_1), \mathbf{a}^*(\theta_2) \otimes \mathbf{a}(\theta_2), \ldots, \mathbf{a}^*(\theta_K) \otimes \mathbf{a}(\theta_K)] \in \mathbb{C}^{[2I(J-1)+1] \times K}$, $\otimes$ denotes the Kronecker product, $\mathbf{p} = [p_1, p_2, \ldots p_K]^T$ and $\mathbf{i} = vec(I)$. The difference co-array contains a few missing elements that are called holes. The array structure of the difference co-array is shown in Figure 1b.

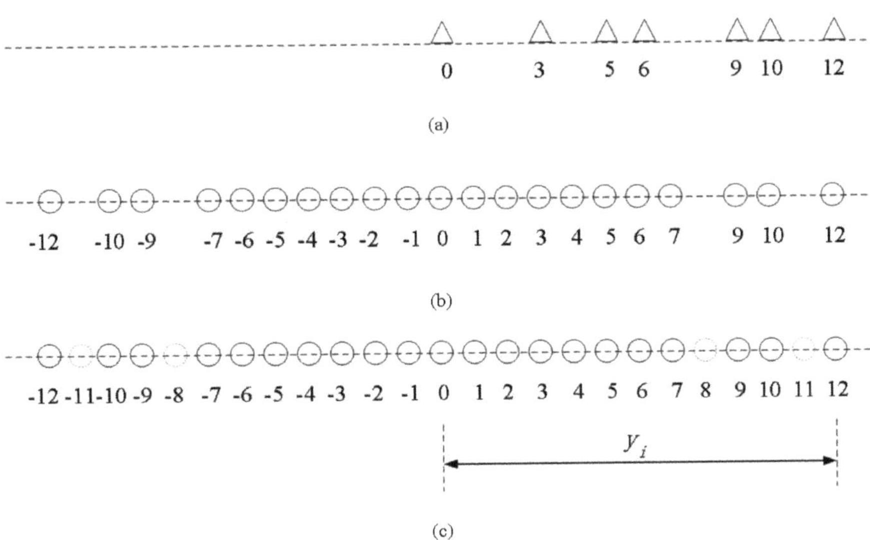

Figure 1. Array structures. (**a**) The coprime array for I = 3, J = 5. (**b**) The difference co-array derived from coprime array. (**c**) The virtual ULA when the number of sensors is 13.

So as to fully utilize the available information and increase DOFs, by filling the interpolated virtual sensors, the model can be extended further as a virtual ULA with $N = \max\{(I-1)J, (J-1)I\} + 1$ sensors, as shown in Figure 1c. The virtual ULA corresponds to a binary vector v of 0s and 1s, in which 0 represents the interpolated virtual element and 1 stands for the others. Correspondingly, the received signal $\mathbf{y}_d$ is extended to the N dimension vector $\mathbf{y}_i$, which has some zero elements corresponding to the virtual received signal of interpolated virtual sensors. As demonstrated in [21], the covariance matrix R_v of the received signal with the virtual ULA is equal to the Toeplitz matrix $\mathcal{T}(\mathbf{y}_i)$ with vector $\mathbf{y}_i$ as its first row, which can be represented as

$$\mathcal{T}(\mathbf{y}_i) = R_v. \tag{4}$$

In actual application, because the received signals of the interpolated virtual sensors in virtual ULA default to 0, some elements in covariance matrix R_v are also set to 0. Compared with the covariance matrix R of the actual ULA with N physical sensors, the covariance matrix of the virtual ULA and actual ULA has the following relationship

$$\mathcal{T}(\mathbf{y}_i) = R_v = R \odot L, \tag{5}$$

where $\odot$ denotes the Hadamard product, $L = v * v^T$ is a binary matrix to imply the zero and non-zero elements in R_v and R is the covariance matrix associated with the actual ULA with N elements. Our focus is to rebuild the covariance matrix R of the virtual ULA from $\mathcal{T}(\mathbf{y}_i)$ with some missing elements.

As a priori knowledge, a covariance matrix should be a Hermitian matrix with a Toeplitz structure and has a low-rank characteristic in theory. Therefore, in order to reconstruct the covariance matrix accurately and quickly, we adopted some measures to ensure that the recovered R_{res} has the above characteristics. Here, we took the average of the values in the conjugate symmetric part of the generated matrix so as to limit the changes in the non-missing part to the minimum range. Finally, the backtracking method further ensures the positive definiteness of the covariance matrix.

3. The Proposed Method

Depicted in Figure 2, the proposed model framework consists of three components: preprocessing, the DCGAN structure and post-processing.

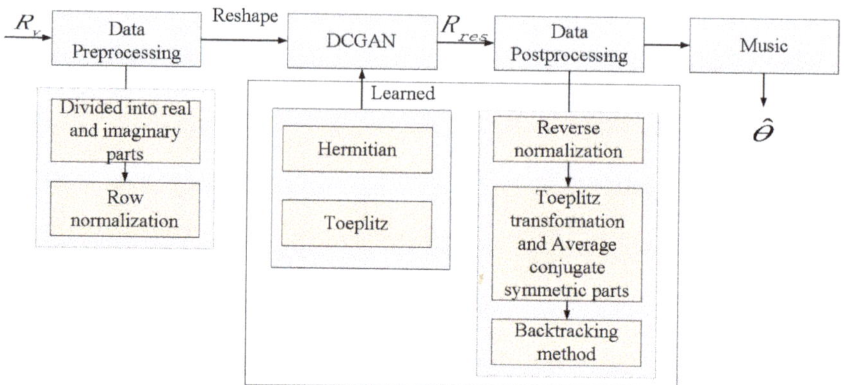

Figure 2. Framework of proposed model.

Firstly, we assume that the signal X is collected by T snapshots with a coprime array. The preprocessing part calculates the covariance matrix through the raw data, which is then normalized to the range of $[-1, 1]$. This is to reduce the range of values for different features to the same range in order to accelerate training speed and improve model stability. Subsequently, the covariance matrix R_v is transformed into a two-channel tensor. The DCGAN structure is responsible for reconstructing the covariance matrix with a virtual ULA from noise signals. The generator produces a result that is similar to the real covariance matrix. Finally, the post-processing part ensures the low-rank characteristic of the recovered covariance matrix and solves the DOA using the generated output.

3.1. Data Preprocessing

In order to adapt to the input requirements of the DCGAN, we used R_v and R' as two inputs for the DCGAN, both of which are real tensors. In the experiment, since the covariance matrix R is a theoretical value and unknown, its sampled value R' was used with N-elements ULA. The first dimension represents the real part matrix $R_v[1, :, :] = Real(R_v)$, and the second dimension represents the imaginary part matrix $R_v[2, :, :] = Imag(R_v)$. According to the structure of the DCGAN generator, the generator restricts the output data to the range of $[-1, 1]$. In order to speed up the training process, we performed row-wise normalization of the real and imaginary parts. It is also helpful to create different features with the same scale, which leads to easier optimization. Moreover, normalizing the input data can effectively prevent gradient explosion and mode collapse, which can better balance the generator and discriminator and improve the stability and robustness of the model.

3.2. DCGAN Structure

For the DCGAN, the proposed design is illustrated in Figure 3.

We approach the covariance matrix reconstruction as a restoration task aiming to compute the mapping correlation between R_v and R', so that the generated R_{res} is as close as possible to R'. The DCGAN consists of a generator with a transposed convolutional structure and a discriminator with a convolutional structure. The transposed convolutional structure in the generator allows for a more suitable upsampling method based on the dataset. Following each transposed convolutional layer in the generator, a ReLU activation function and a batch normalization layer are applied. The final layer of the generator network utilizes a Tanh activation function. A convolutional structure was adopted for the discriminator. Following each convolutional layer, a LeakyReLU activation function and a batch normalization layer are utilized. The final layer uses a Sigmoid activation function. The LeakyReLU activation function retains a small gradient for the negative part, facilitating higher quality recovery by the generator. Additionally, a dropout layer is incorporated into the discriminator to balance training.

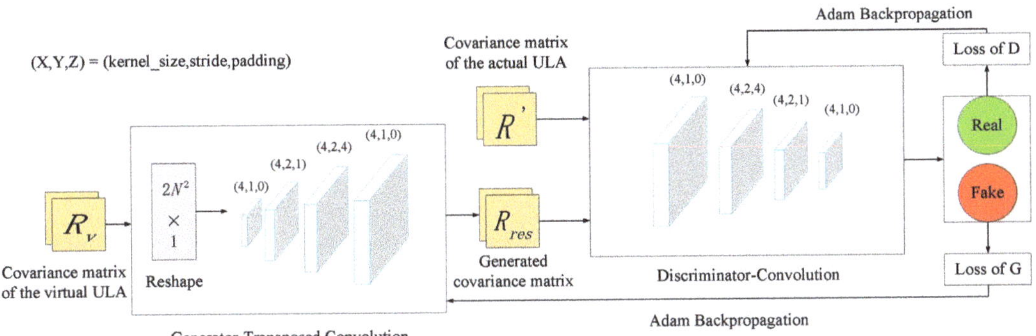

Figure 3. DCGAN structure.

The proposed model widely uses batch normalization layers due to their ability to prevent overfitting and accelerate the training and convergence process. However, it is important to note that batch normalization layers are not used in the input layer of the generator or the output layer of the discriminator, since this may cause sample oscillation and model instability. The DCGAN structure does not have pooling layers or fully connected layers because pooling operations may lose some important information, and the use of fully connected layers is prone to overfitting. The final output shape of the generator is $2 \times N \times N$.

3.3. Data Post-Processing

Finally, for the data post-processing part, it should be noted that the last layer of the generator uses the Tanh activation function. Therefore, we used the saved parameters to reverse-normalize the network output back to its original values. In addition, although the generated data roughly conform to the distribution of $R^{'}$, the data do not strictly satisfy the conjugate symmetry. Therefore, the average of the conjugate symmetric parts of the real part matrix was directly calculated. The diagonal data of the imaginary matrix were set to 0, the absolute values of the conjugate symmetric parts of other data were taken, the average was calculated and positive and negative signs were assigned. Strict adherence to this property was ensured. Furthermore, since the training strategy involves real and imaginary dual channels, the two-channel real value data for each recovered covariance matrix were combined into a complex-valued matrix for DOA estimation. Finally, in order to ensure positive definiteness of the complex-valued matrix, we utilized the low-rank matrix optimization algorithm to regularize this matrix.

4. Training Approach

4.1. Loss Function

For small-scale tasks, cross-entropy loss is sufficient for network training. However, during the experimental process, it was found that a single cross-entropy loss led to difficulty in limiting the recovery direction of the covariance matrix. Therefore, in this study, a combined approach of generator loss, discriminator loss, context loss, perceptual loss and nuclear norm loss was adopted for training. Both generator loss and discriminator loss are cross-entropy losses, and the input of the cross-entropy loss is a pair of outputs from the generator or discriminator and the corresponding size label. The labels of the real data have been smoothed using 0.9 to maintain balance at both ends. The perceptual loss is generated by the DCGAN itself and can be represented as

$$L_{perceptual} = \log(1 - \mathbf{D}(\mathbf{G}(\mathbf{R}_v))), \tag{6}$$

where $\mathbf{D}(\cdot)$ represents the discriminator, and $\mathbf{G}(\cdot)$ stands for the generator.

The context loss constrains the consistency of the non-missing parts of the covariance matrix and aims to minimize changes in non-missing parts during the recovery process. The L_2 norm is employed to calculate the loss, and the inputs of the context loss are $\mathbf{R}'$, $\mathbf{G}(\mathbf{R}_v)$, and $\mathbf{L}$, which can be represented as

$$L_{contextual} = \left\| \mathbf{L} \odot \mathbf{G}(\mathbf{R}_v) - \mathbf{L} \odot \mathbf{R}' \right\|_2. \tag{7}$$

The nuclear norm loss serves as a regularization constraint to reduce the rank of the restored covariance matrix. It can reduce the number of unknown values that need to be restored. Additionally, the nuclear norm loss helps control the complexity of the matrix to avoid overfitting. The nuclear norm loss can be represented as

$$\|\mathbf{R}_{res}\|_* = \sum_{i=1}^{N} \sigma_i(\mathbf{R}_{res}), \tag{8}$$

where $\mathbf{R}_{res}$ is the covariance matrix generated by the generator, N represents the number of rows and columns in the covariance matrix and $\sigma_i(\mathbf{R}_{res})$ represents the i-th singular value of $\mathbf{R}_{res}$.

Cross-entropy loss provides backpropagation gradients and other parameters. The entire restoration task's loss function can be represented as

$$L_{total} = L_{contextual} + \lambda_1 L_{perceptual} + \lambda_2 L_{norm}, \tag{9}$$

where λ_1 and λ_2 are hyperparameters used to adjust the importance of the two losses.

Therefore, our goal is to ensure the stability of the non-missing parts while guiding the generator to produce globally consistent results with the real covariance matrix. This can further improve the accuracy of subsequent DOA estimation.

4.2. DCGAN Training

To construct the dataset, we randomly selected two angles within the range of $[-60°, 60°]$. Each data point was then generated based on the signal model. The dataset has SNR values ranging from -5 dB to 10 dB, with a size of 1,000,000. The training set consists of 80% of the data, and the remaining 20% are used for validation. The model employs the Adam optimizer to update the weights, and hyperparameters λ_1 and λ_2 of the total loss of the recovery task were all set to 0.1.

The model initializes its weights from a normal distribution $N(0, 0.02^2)$. After initialization, the model immediately applies these weights. Unlike generative tasks, the generator's input $\mathbf{R}_v$ is reshaped into a vector of shape $(N \times N \times 2, 1)$ rather than random noise, which can utilize the prior characteristics of the covariance matrix.

5. Simulation Results

We conducted several experiments to demonstrate the performance of the proposed method. Based on individual experiment results and quantitative experimental results, we compared this approach with some other methods. In this study, all experiments were conducted on a desktop computer equipped with an Intel Core i7-12700F processor running at 3.5 GHz, with 16 GB of RAM and an NVIDIA GeForce RTX 4060Ti GPU (Galax, Hong Kong, China). The operating system used is Windows 10. The software environment uses

Python 3.6.5 as the programming language and uses the PyTorch framework for training and testing deep learning models.

5.1. Single Experiment Results

The proposed method was tested with a physical array consisting of seven sensors. We conducted the experiments with a fixed snapshot count of 256 and SNR at 10 dB. Two scenarios were considered: one with five signal sources (less than seven) and another with eight signal sources (greater than seven). In both scenarios, we employed the following comparison algorithms: the MUSIC algorithm, the sparse representation with l_p-norm algorithm (MAP) [22], the sparse-recovery-based method (SR-D) [23] and the CNN-based DOA estimation method (CNN-D) [20].

As shown in Figure 4, when the number of signal sources is five (less than seven), we assume that five uncorrelated signals originate from $[-43°, -29°, 10°, 32°, 54°]$. It is visible that all of the aforementioned methods can achieve good performance and provide accurate DOA estimation.

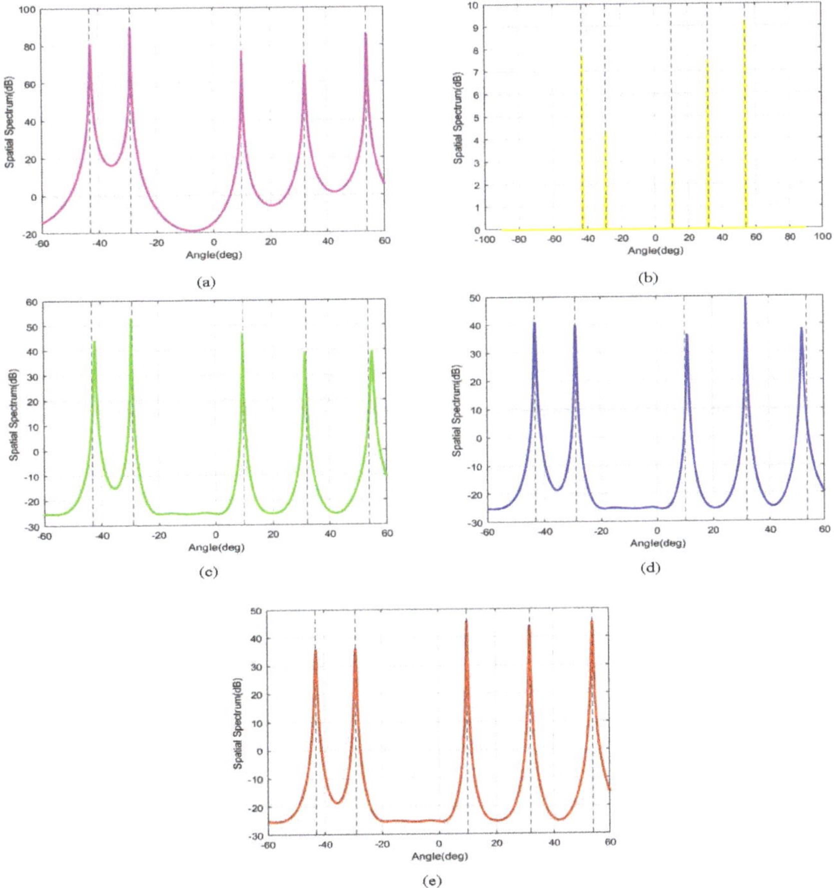

Figure 4. Spectrum of DOA estimation methods when the number of signal sources is five. (**a**) MUSIC. (**b**) MAP. (**c**) SR−D. (**d**) CNN−D. (**e**) Proposed method.

However, as depicted in Figure 5, when the number of signal sources increases to eight (greater than seven), these signal sources arrive from $[-43°, -29°, -16°, 0°, 10°, 21°, 32°, 54°]$. The spatial spectrum of the MUSIC algorithm becomes flattened, and some spectral peaks merge together. The MAP algorithm can only accurately estimate partial angles of arrival. Both of these two algorithms fail to distinguish more sources than the number of physical sensors. Although the CNN-D method and SR-D method can obtain eight spectral peaks, their peaks exhibit some bias, which leads to a decrease in the accuracy of these DOA estimation algorithms. In contrast, the proposed method still forms eight sharp peaks at the actual DOAs, which is more than the number of physical sensors (seven). It can achieve more accurate DOA estimation in underdetermined scenarios, which is because the proposed method extends the DOFs of the virtual ULA to 12 and reconstructs its covariance matrix accurately using the DCGAN with the prior knowledge.

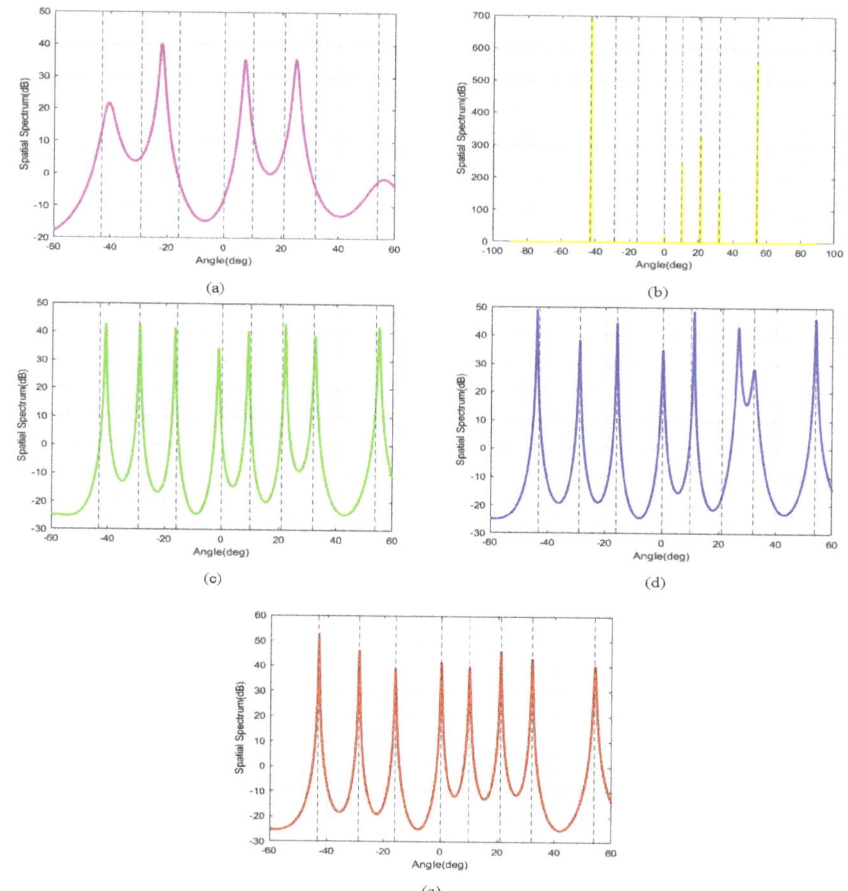

Figure 5. Spectrum of DOA estimation methods when the number of signal sources is eight. (**a**) MU-SIC. (**b**) MAP. (**c**) SR−D. (**d**) CNN−D. (**e**) Proposed method.

5.2. Quantitative Experimental Results

To evaluate the performance of the proposed DCGAN method, we compared it with two existing methods: CNN-D and SR-D. The evaluation is based on the root mean square error (RMSE) metric. We constructed a coprime array using the coprime pairs of 3 and 5, with the element positions being $\{0, 3, 5, 6, 9, 10, 12\}d$. Furthermore, experiments were conducted at SNR values of $[-5, 0, 5, 10]$ dB.

As presented in Figure 6, when the quantity of snapshots is held constant at 256, the performance of the proposed method improves consistently as the SNR increases and surpasses the other methods, especially in low SNR conditions.

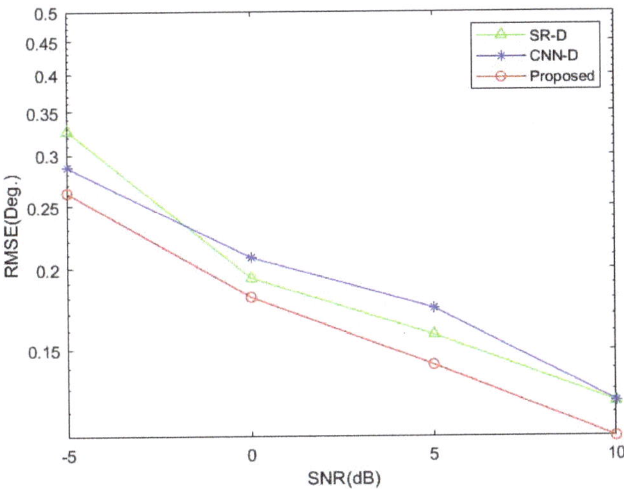

Figure 6. RMSE versus SNR.

Then, as illustrated in Figure 7, the SNR was fixed at 10 dB, and we compared the performance of the above methods with different numbers of snapshots. The proposed method does not experience a significant performance degradation as snapshots decrease and outperforms other methods. This is because the covariance matrix could be accurately rebuilt although with the limited snapshots, which preserves low-rank characteristics and more DOFs.

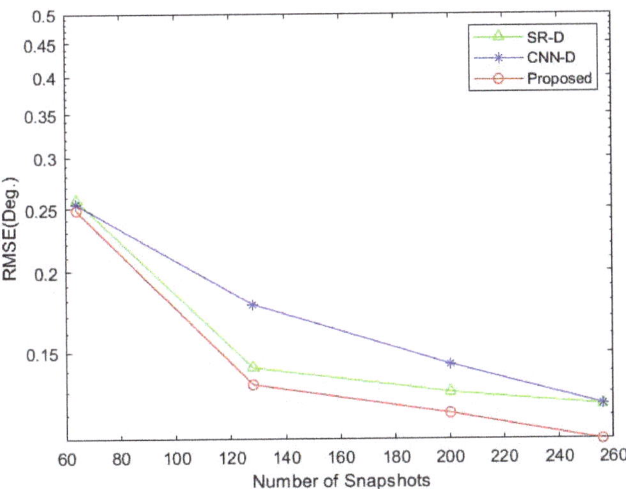

Figure 7. RMSE versus snapshots.

Next, as shown in Figure 8, we demonstrate the RMSE of these methods at different angle separation degrees. It is apparent that the proposed method exhibits robust performance at different resolutions, without significant fluctuations, and exhibits considerable robustness. The CNN-D method only uses the first-row elements to recover the covariance matrix and is incapable of guaranteeing the positive definiteness of the resulting covariance matrix.

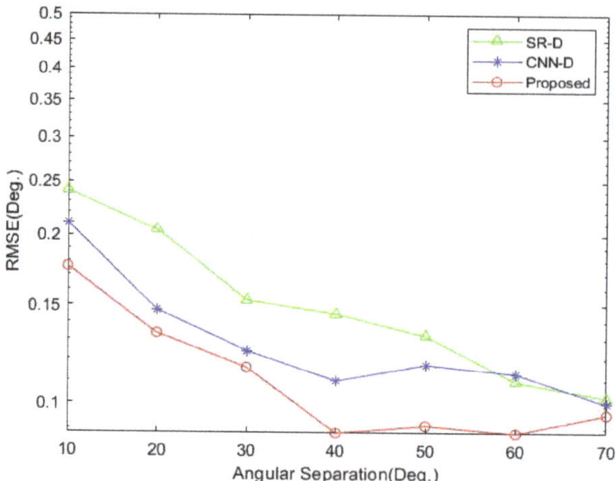

Figure 8. RMSE versus angular separation.

In the subsequent analysis, we investigated the influence of different snapshots and SNR levels on the performance of the proposed method. As shown in Figure 9, the performance of the proposed method improves with an increase in the number of snapshots. It can be observed from the figure that when the snapshot count is greater than 256, the performance of the proposed method stabilizes. Even with a relatively limited number of snapshots, the proposed method can achieve accurate DOA estimation without excessive performance loss. Furthermore, the performance of the proposed method continuously improves with an increase in SNR and stabilizes at a level of 10 dB.

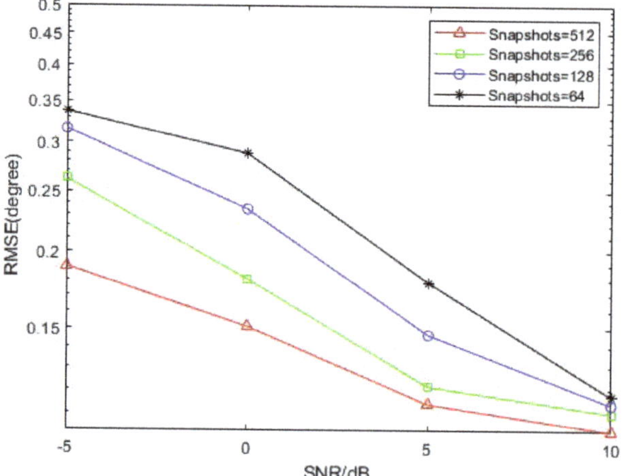

Figure 9. RMSE of the proposed method with different SNRs and snapshots.

Finally, we performed 10,000 Monte Carlo simulations and recorded the estimated total time results in Table 1. It should be noted that, to ensure the accuracy of the model, the deep learning methods mentioned above require a long training period, so well-trained models were used for testing. From the results, it can be seen that compared to the traditional physics-based model SR-D, the proposed method can achieve faster estimation time, with a decrease of about 30 times. Compared to the CNN-D method, especially at lower SNR from −5 dB to 5 dB, our estimation time is about 10–30 s faster, although, at

10 dB SNR, the proposed method also has a slight improvement. Combining Figure 7 and Table 1, it becomes evident that the proposed method is capable of achieving fast estimation even with limited snapshots. Therefore, it is suitable for fast DOA estimation in mobile agent localization scenarios.

Table 1. DOA estimation times.

Method	−5 dB	0 dB	5 dB	10 dB
SR-D	5890 s	5290 s	5720 s	5810 s
CNN-D	216.5672 s	209.0678 s	184.01 s	171.723 s
Proposed	192.8996 s	179.845 s	172.8436 s	171.667 s

6. Conclusions

This paper proposes a DOA estimation framework based on the DCGAN in underdetermined scenarios. Compared with most of the current DL-based methods, our proposed method transforms DOA estimation to a recovery task of a covariance matrix with more DOFs. Our method uses the DCGAN model and takes measures to preserve the Hermitian, Toeplitz and low-rank prior characteristics of the recovered covariance matrix. In underdetermined scenarios, the proposed method exhibits notable advantages in the fields of both accuracy and estimation speed, especially with limited snapshots. It is suitable for mobile agent localization.

Author Contributions: Conceptualization, Y.C. and F.Y.; methodology, Y.C. and F.Y. and A.K.; supervision, Y.C.; validation, Y.C., F.Y. and A.K.; formal analysis, J.W.; funding acquisition, Y.C., J.W. and H.S.; investigation, Y.C., F.Y., M.Z., J.W., H.S., A.K. and J.Q.; data curation, F.Y.; writing—original draft preparation, Y.C. and F.Y.; writing—review and editing, Y.C., F.Y., M.Z., L.H., J.W. and H.S. All authors have read and agreed to the published version of the manuscript.

Funding: This research is supported by National Natural Science Foundation of China (Grant No. 62201385), Technology Innovation Guidance Special Fund of Tianjin Science and Technology Plan Project (Grant No. 22YDTPJC00610), the Stable Supporting Fund of National Key Laboratory of Underwater Acoustic Technology (Grant No. JCKYS2023604SSJS008), the Key Laboratory of Southeast Coast Marine Information Intelligent Perception and Application, MNR, (Grant No. 220202) and the National Natural Science Foundation of China (Grant No. 62271426).

Data Availability Statement: Data are contained within the article.

Acknowledgments: We would like to express our gratitude to all members of the signal processing laboratory who have contributed to the experiment.

Conflicts of Interest: The authors declare no conflicts of interest.

References

1. Florio, A.; Avitabile, G.; Talarico, C.; Coviello, G. A Reconfigurable Full-Digital Architecture for Angle of Arrival Estimation. *IEEE Trans. Circuits Syst. Regul. Pap.* 2023, *in press.* [CrossRef]
2. Schmidt, R. Multiple emitter location and signal parameter estimation. *IEEE Trans. Antennas Propag.* **1986**, *34*, 276–280. [CrossRef]
3. Roy, R.; Kailath, T. ESPRIT-estimation of signal parameters via rotational invariance techniques. *IEEE Trans. Acoust. Speech Signal Process.* **1989**, *37*, 984–995. [CrossRef]
4. Ye, Z.; Xu, X. DOA Estimation by Exploiting the Symmetric Configuration of Uniform Linear Array. *IEEE Trans. Antennas Propag.* **2007**, *55*, 3716–3720. [CrossRef]
5. Cui, Y.; Liu, K.; Wang, J. Direction-of-arrival estimation for coherent GPS signals based on oblique projection. *Signal Process.* **2012**, *92*, 294–299. [CrossRef]
6. El Zooghby, A.H.; Christodoulou, C.G.; Georgiopoulos, M. Performance of radial-basis function networks for direction of arrival estimation with antenna arrays. *IEEE Trans. Antennas Propag.* **1997**, *45*, 1611–1617. [CrossRef]
7. Shieh, C.; Lin, C. Direction of arrival estimation based on phase differences using neural fuzzy network. *IEEE Trans. Antennas Propag.* **2000**, *48*, 1115–1124. [CrossRef]
8. Liu, Z.-M.; Zhang, C.; Yu, P.S. Direction-of-Arrival Estimation Based on Deep Neural Networks with Robustness to Array Imperfections. *IEEE Trans. Antennas Propag.* **2018**, *66*, 7315–7327. [CrossRef]

9.   Papageorgiou, G.K.; Sellathurai, M.; Eldar, Y.C. Deep Networks for Direction-of-Arrival Estimation in Low SNR. *IEEE Trans. Signal Process.* **2021**, *69*, 3714–3729. [CrossRef]

10.   Wu, L.; Liu, Z.-M.; Huang, Z.-T. Deep Convolution Network for Direction of Arrival Estimation with Sparse Prior. *IEEE Signal Process. Lett.* **2019**, *26*, 1688–1692. [CrossRef]

11.   Xiang, H.; Chen, B.; Yang, T.; Liu, D. Phase enhancement model based on supervised convolutional neural network for coherent DOA estimation. *Appl. Intell.* **2020**, *50*, 2411–2422. [CrossRef]

12.   Lima de Oliveira, M.L.; Bekooij, M.J.G. ResNet Applied for a Single-Snapshot DOA Estimation. In Proceedings of the 2022 IEEE Radar Conference (RadarConf22), New York, NY, USA, 21–25 March 2022; pp. 1–6.

13.   Peng, J.; Nie, W.; Li, T.; Xu, J. An end-to-end DOA estimation method based on deep learning for underwater acoustic array. In Proceedings of the OCEANS 2022, Hampton Roads, VA, USA, 17–20 October 2022; pp. 1–6.

14.   Cao, Y.; Lv, T.; Lin, Z.; Huang, P.; Lin, F. Complex ResNet Aided DoA Estimation for Near-Field MIMO Systems. *IEEE Trans. Veh. Technol.* **2020**, *69*, 11139–11151. [CrossRef]

15.   Zhao, F.; Hu, G.; Zhan, C.; Zhang, Y. DOA Estimation Method Based on Improved Deep Convolutional Neural Network. *Sensors* **2022**, *22*, 1305. [CrossRef]

16.   Xiang, H.; Chen, B.; Yang, M.; Xu, S. Angle Separation Learning for Coherent DOA Estimation with Deep Sparse Prior. *IEEE Commun. Lett.* **2021**, *25*, 465–469. [CrossRef]

17.   Fang, W.; Cao, Z.; Yu, D.; Wang, X.; Ma, Z.; Lan, B.; Xu, Z. A Lightweight Deep Learning-Based Algorithm for Array Imperfection Correction and DOA Estimation. *J. Commun. Inf. Netw.* **2022**, *7*, 296–308. [CrossRef]

18.   Yao, Y.; Lei, H.; He, W. Wideband DOA Estimation Based on Deep Residual Learning with Lyapunov Stability Analysis. *IEEE Geosci. Remote. Sens. Lett.* **2022**, *19*, 8014505. [CrossRef]

19.   Gao, S.; Ma, H.; Liu, H.; Yang, J.; Yang, Y. A Gridless DOA Estimation Method for Sparse Sensor Array. *Remote Sens.* **2023**, *15*, 5281. [CrossRef]

20.   Wu, X.; Yang, X.; Jia, X.; Tian, F. A Gridless DOA Estimation Method Based on Convolutional Neural Network with Toeplitz Prior. *IEEE Signal Process. Lett.* **2022**, *29*, 1247–1251. [CrossRef]

21.   Zhou, C.; Gu, Y.; Fan, X.; Shi, Z.; Mao, G.; Zhang, Y.D. Direction-of-Arrival Estimation for Coprime Array via Virtual Array Interpolation. *IEEE Trans. Signal Process.* **2018**, *66*, 5956–5971. [CrossRef]

22.   Liu, L.; Rao, Z. An Adaptive Lp Norm Minimization Algorithm for Direction of Arrival Estimation. *Remote Sens.* **2022**, *14*, 766. [CrossRef]

23.   Zhou, C.; Gu, Y.; Shi, Z.; Zhang, Y.D. Off-Grid Direction-of-Arrival Estimation Using Coprime Array Interpolation. *IEEE Signal Process. Lett.* **2018**, *25*, 1710–1714. [CrossRef]

MDPI

St. Alban-Anlage 66

4052 Basel

Switzerland

www.mdpi.com

Remote Sensing Editorial Office

E-mail: remotesensing@mdpi.com

www.mdpi.com/journal/remotesensing